D0518352

Biological Regulation and Development

Volume 1

Gene Expression

Biological Regulation and Development

Volume 1 GENE EXPRESSION
Edited by Robert F. Goldberger

Biological Regulation and Development

Volume 1

Gene Expression

Edited by
Robert F. Goldberger
National Cancer Institute
Bethesda, Maryland

PLENUM PRESS · NEW YORK AND LONDON

Library of Congress Cataloging in Publication Data

Main entry under title:

Gene expression.

(Biological regulation and development; v. 1)
Includes bibliographies and index.
1. Gene expression. 2. Genetic regulation. I. Goldberger, Robert F. II. Series.
QH450.G45 574.8'732 78-21893
ISBN 0-306-40098-7

First Printing – March 1979
Second Printing – March 1980

A Division of Plenum Publishing Corporation
227 West 17th Street, New York, N.Y. 10011

Printed in the United States of America

Contributors

Allan Campbell
Department of Biological Sciences
Stanford University
Stanford, California

Patricia H. Clarke
Department of Biochemistry
University College London
London, England

Riccardo Cortese
Institute of Biological Chemistry
Faculty of Medicine and Surgery
University of Naples
Naples, Italy
Present Address: MRC Laboratory of Molecular Biology
Cambridge, England

Robert F. Goldberger
Laboratory of Biochemistry
National Cancer Institute
National Institutes of Health
Bethesda, Maryland

Karl G. Lark
Department of Biology
University of Utah
Salt Lake City, Utah

O. Maaløe
Institute of Microbiology
University of Copenhagen
Copenhagen, Denmark

David Pribnow
Department of Molecular, Cellular, and Developmental Biology
University of Colorado
Boulder, Colorado

Michael A. Savageau
Department of Microbiology
The University of Michigan
Ann Arbor, Michigan

Henry M. Sobell
Department of Chemistry, and Department of Radiation Biology and Biophysics
University of Rochester School of Medicine and Dentistry
Rochester, New York

Dieter G. Söll
Department of Molecular
Biophysics and Biochemistry
Yale University
New Haven, Connecticut

Deborah A. Steege
Department of Biochemistry
Duke University Medical Center
Durham, North Carolina

Joan Argetsinger Steitz
Department of Molecular
Biophysics and Biochemistry
Yale University
New Haven, Connecticut

Peter H. von Hippel
Institute of Molecular Biology and
Department of Chemistry
University of Oregon
Eugene, Oregon

Preface

The motivation for us to produce a treatise on regulation was mainly our conviction that it would be fun, and at the same time productive, to approach the subject in a way that differs from that of other treatises. We had ourselves written reviews for various volumes over the years, most of them bringing together all possible facts relevant to a particular operon, virus, or biosynthetic system. And we were not convinced of the value of such reviews for anyone but the expert in the field reviewed. We thought it might be more interesting and more instructive—for both author and reader—to avoid reviewing topics that any one scientist might work on, but instead to review the various parts of what many different scientists work on. Cutting across the traditional boundaries that have separated the subjects in past volumes on regulation is not an easy thing to do—not because it is difficult to think of what interesting topics should replace the old ones, but because it is difficult to find authors who possess sufficient breadth of knowledge and who are willing to write about areas outside those pursued in their own laboratories. For example, no one scientist works on suppression *per se.* He may study the structure of suppressor tRNAs in *Escherichia coli,* he may study phenotypic suppression of various characters in drosophila, he may study polarity in gene expression, and so on. Anyone who takes on the task of reviewing suppression must be willing to weave together the various parts of the subject, picking up the threads from many different laboratories, and attempt to produce a fabric with a meaningful design. Finding people who are likely to succeed in such tasks was the most difficult part of our job, since the qualifications required are not the same as those by which we are accustomed to evaluating our colleagues. For example, a high degree of productivity as a research scientist is not nearly so important as is the ability to think deeply about scientific issues.

Having determined at least to make a try at accomplishing our goal, we were surprised and gratified to find that we had been sufficiently convincing to enlist the participation of a group of authors we considered truly outstanding for the enterprise. But after getting that far, we were a bit anxious about what the outcome would be; after all, no amount of editing can alter significantly the substance of what an author has to say. As the manuscripts appeared, we were relieved to find that the authors had taken to heart the charge we had given them. We noted, for example, the high frequency with which the same system was discussed from different points of view by different authors; the multiple instances in which not only a structure or phenomenon was discussed, but also its possible evolution; the willingness of authors to make interesting speculations; and so on. Thus, having gotten what we asked for from our authors, it is we who must take the blame from any reader who is dissatisfied. As for taking credit, we are afraid that it will belong to the authors themselves; although we took an active role in editing, it is only the authors' contributions that can make the basic philosophy of this treatise work.

Robert F. Goldberger
Paul Berg
Leroy E. Hood
Philip Leder
Kivie Moldave
Robert T. Schimke

Contents

1 Strategies of Genetic Regulation in Prokaryotes

ROBERT F. GOLDBERGER

2 Structure of Complex Operons

ALLAN CAMPBELL

3 Autogenous and Classical Regulation of Gene Expression: A General Theory and Experimental Evidence

MICHAEL A. SAVAGEAU

4 Regulation of Enzyme Synthesis in the Bacteria: A Comparative and Evolutionary Study

PATRICIA H. CLARKE

5 Importance of Symmetry and Conformational Flexibility in DNA Structure for Understanding Protein–DNA Interactions

HENRY M. SOBELL

6 Some Aspects of the Regulation of DNA Replication in *Escherichia coli*

KARL G. LARK

7 Genetic Control Signals in DNA

DAVID PRIBNOW

8 On the Molecular Bases of the Specificity of Interaction of Transcriptional Proteins with Genome DNA

PETER H. VON HIPPEL

9 Genetic Signals and Nucleotide Sequences in Messenger RNA

JOAN ARGETSINGER STEITZ

10 The Role of tRNA in Regulation

RICCARDO CORTESE

11 Suppression

DEBORAH A. STEEGE AND DIETER G. SÖLL

12 Regulation of the Protein-Synthesizing Machinery—Ribosomes, tRNA, Factors, and So On

O. MAALØE

1

Strategies of Genetic Regulation in Prokaryotes

ROBERT F. GOLDBERGER

1 *Introduction*

One of the most striking characteristics of living systems is that they function in an orderly manner despite their high degree of complexity. One workable definition of regulation, in fact, is the set of mechanisms that allows organisms to maintain this orderly functioning. It is important to realize, however, that regulation was not superimposed upon living systems; orderly processes are simply more successful than are disorderly ones, and therefore tend to be preserved through the evolutionary process by conferring advantages upon organisms that possess them. The thousands of chemical reactions occurring in cells are controlled by regulatory mechanisms that operate at many different levels. This introductory chapter focuses on those that operate at the level of gene expression and will introduce some of the strategies of genetic regulation that have evolved in prokaryotic organisms. Scanning the table of contents of this brief essay should suffice to tell the reader that a very general overview is in store for him. The renaissance in biological research that occurred in the last 25 years has been due mostly to the exciting studies concerning genetic regulation in prokaryotes. I have tried to abstract from those studies the most important basic principles they illustrate and to organize into a few generalizations the enormous body of data they have produced. I believe it is these principles and generalizations with which the reader will need to arm himself before proceeding further into this volume. It is to be hoped that the

ROBERT F. GOLDBERGER • Laboratory of Biochemistry, National Cancer Institute, National Institutes of Health, Bethesda, Maryland

necessarily simplistic view of regulation they provide will be preferable to the bewilderment that so often results from exhaustive reviews that include the details of many specific regulated systems.

2 *The Prokaryotic Chromosome and Its Genes*

Over the past two decades much has been learned about the mechanisms by which a protein is synthesized in accordance with the specifications of a so-called *structural gene.* First, RNA polymerase faithfully transcribes the sequence of deoxyribonucleotides of the gene into the ribonucleotide sequence of a messenger RNA (mRNA) molecule (see Chapters 7 and 8 of this volume). Next, ribosomes, transfer RNA, and amino acids participate, together with various other proteins and small molecules, in translating the nucleotide sequence of the mRNA into the amino acid sequence of a polypeptide chain (see Chapters 9, 10, 11, and 12 of this volume). The work of Anfinsen (1973) and his colleagues has demonstrated that once a polypeptide chain is formed it folds spontaneously, under physiological conditions, to achieve the three-dimensional conformation characteristic of the native protein (see Chapter 3 in Volume 2 of this treatise). Thus, in order to specify the complete and final structure of a native protein, a gene need do no more than specify the linear sequence of its amino acids. A few years ago it seemed clear that genetic information is stored (DNA) and transmitted (mRNA) in linear form and is expressed (protein) in three-dimensional form. More recent insights into the details of the structure and function of the genetic apparatus, however, have revealed a far more complex and interesting picture in which, for example, secondary and higher order structures of nucleic acids play an essential part in the functions of these macromolecules.

The prokaryotic chromosome is a single circular molecule of double-stranded DNA with a molecular weight of about two billion. It contains approximately 4000 genes. From the fact that this molecule is circular, one might make the naive guess that one strand of the DNA is the sense strand (the coding strand) and that transcription of this strand starts at one point on the circle and proceeds all the way around, in a manner similar to that for replication. Such a guess would be not only naive but also entirely incorrect. In fact, in certain parts of the chromosome the strand that runs clockwise is the sense strand and in other parts of the chromosome the strand that runs counterclockwise is the sense strand. Once this fact is appreciated, it becomes obvious that there must be signals in the DNA that direct the transcription apparatus where to begin transcribing. Although the direction of transcription is always 5′ to 3′, the correct strand must be selected for any given gene or group of genes. The fact that each gene or group of genes is transcribed independently is, of course, required if the cell is to be able to regulate the rate at which certain items of genetic information are utilized independently of the rate at which others are utilized. The mechanism by which the correct strand is selected involves the recognition of specific nucleotide sequences in the DNA, known as *promoters,* by the enzyme that carries out the transcription process, RNA polymerase. But it is not sufficient for the RNA polymerase to recognize where to begin transcription; if it is to be prevented from continuing its journey along the

DNA forever, it must also recognize where to stop. Signals in the form of specific nucleotide sequences fulfill this purpose, too. Such signals are known as *terminators.* Thus, the cell is equipped with the means for directing the polymerase to begin transcription of each gene in the right place (at the beginning of the gene and on the correct strand of DNA) and to stop transcription at the end of the gene or genes.

Acting as structural genes, which carry the specifications for the amino acid sequences of proteins, is not the only function of DNA in the cell. As mentioned earlier, there are also regions of DNA that serve as signals for the initiation of RNA synthesis—promoters—and for termination of RNA synthesis—terminators; there are genes for transfer RNA and for ribosomal RNA; and there are genes of another type—control genes—that regulate the frequency with which neighboring structural genes are transcribed into mRNA. Genes of this last type provide the cell with a means for regulating the intracellular concentrations of specific proteins. Many structural genes, perhaps the majority of them, are not regulated in this way. They are said to be *constitutive;* the proteins they specify are present at essentially constant concentrations under a wide variety of conditions. Although the intracellular concentration of any given constitutive gene product remains essentially constant, there are great differences in concentration among various constitutive gene products. Some proteins are always present in the cell at low concentration; some, always at high concentration. Such differences have generally been thought to reflect differences in the strengths of interaction of the various constitutive structural gene promoters with RNA polymerase (see Chapters 6 and 12 of this volume). However, it should be noted that the finding that the intracellular concentration of a given protein always remains constant does not in itself constitute evidence that the gene for that protein is expressed at a constant rate. (In fact, expression of a gene at a constant rate would give rise to fluctuations in the concentration of its mRNA, and hence of its protein, at different stages of the cell cycle.) There is one regulatory mechanism, *autogenous regulation,* that is particularly well suited for maintaining a constant concentration of specific mRNA in the cell (see Chapter 3 of this volume).

It is important to recognize that regulation of gene expression may be exerted both on the efficiency with which RNA polymerase initiates transcription of specific genes and on the efficiency with which RNA polymerase terminates transcription of specific genes. In this chapter I will focus primarily on regulation of transcriptional initiation.

3 Gene Clustering and the Operon Concept

Those structural genes for which the transcription frequency can be regulated have been the subject of intensive investigation for more than two decades. In a fairly high proportion of cases these genes have been found to be clustered on the bacterial chromosome with one or more other structural genes that specify functionally related proteins. A cluster of functionally related genes that is regulated and transcribed as a unit is known as an *operon* (Jacob and Monod, 1961). The proteins encoded in the genes of a given operon are ordinarily the enzymes that catalyze the several steps of a metabolic pathway. For example, all ten enzymes that catalyze the conversion of ATP and phosphoribosyl pyrophosphate to the amino acid, histidine, are encoded in the continguous genes of the histidine operon (Ames and Hartman, 1963); the same organization is found for the three enzymes that catalyze the degradation of galactose (Buttin, 1963), the six enzymes that catalyze the biosynthesis of tryptophan (Yanofsky, 1971), and the enzymes of many other metabolic pathways.

One of the important features of operons is that they are transcribed into polycistronic mRNA (Martin, 1963). The term *polycistronic* is used to indicate a single molecule of mRNA that carries the specifications for more than one polypeptide chain, each polypeptide chain being specified by one *cistron.* Because of the polycistronic nature of the mRNA from a given operon, the intracellular levels of all the enzymes encoded in the genes of that operon increase or decrease when the frequency of transcription of the operon increases or decreases, respectively.

4 *Regulatory Molecules and the Genes with Which They Interact*

4.1 *The Promoter*

The polycistronic nature of the mRNA transcribed from each operon is a reflection of the fact that each operon has only one primary promoter site at which RNA polymerase binds prior to initiating transcription (Jacob *et al.,* 1964; Epstein and Beckwith, 1968). Although other promoters have been identified within certain operons (Margolin and Bauerle, 1966; Atkins and Loper, 1970), these are usually secondary promoter sites for which the affinity of RNA polymerase is relatively low in comparison with the primary promoter at the beginning of the operon; therefore, they do not ordinarily function to a significant degree (see Chapter 2 of this volume).

4.2 *The Operator and Initiator Genes*

Since RNA polymerase binds at a single site (the promoter) for each operon, normally only one additional region, a regulatory region, located between the promoter and the structural gene(s), would be required to provide a mechanism by which the cell could regulate the frequency with which RNA polymerase molecules are permitted to transcribe the operon once they have become bound. This regulatory region, which serves to control the frequency of transcription for each operon, functions much as does a valve, allowing greater or smaller numbers of RNA polymerase molecules to traverse the operon per unit time, depending on whether it is more "open" or more "closed." Thus, the frequency with which an operon is transcribed depends not only on the affinity of its promoter for RNA polymerase, but also on the degree to which the regulatory region restricts the passage of RNA polymerase molecules from the promoter into the structural genes.

The regulatory regions are of two types. One of these, the *operator* gene, is a valve that by itself is "open," allowing RNA polymerase molecules through at a high frequency; it is regulated by being progressively closed. The other, the *initiator* gene (sometimes called a *positive activator* gene), is a valve that by itself is "closed," restricting the passage of RNA polymerase molecules to a low frequency; it is regulated by being progressively opened. In both cases, the mechanism for closing or opening the valve, altering the frequency with which RNA polymerase is able to transcribe the operon, involves a specific regulatory protein and a specific small molecule for each operon. Although I have described the regulatory region in terms of a valve that regulates the flow of RNA polymerase molecules from promoter to structural genes, the mechanism by which control is exerted at these regulatory sites is actually by influencing the efficiency of mRNA initiation at the promoter. Thus, operator and initiator genes may, at least partly, overlap the promoter so that the regulatory proteins that bind to them may influence the RNA polymerase directly. (See Chapter 7 of this volume for a full discussion of the mechanism of initiation of transcription.)

4.3 Regulatory Proteins and the Small Molecules That Affect Their Activities

Most regulatory proteins are bifunctional: They possess the capacity for interacting with specific sites in the genome, their interaction at such sites influencing transcription of the neighboring genes, *and* they possess the capacity for interacting with specific small molecules. Their interaction with such small molecules affects their affinity for specific sites in the genome and in this manner serves to communicate the metabolic needs of the cell to the chromosomal sites that regulate the cell's ability to respond to those needs. Thus, the regulatory protein serves as a Rosetta stone, translating metabolic needs (small molecule interaction) into changes in gene expression (DNA interaction).

The regulatory protein for the operator gene is known as a *repressor*. This protein binds specifically to the operator, closing it to the passage of RNA polymerase molecules, and thereby diminishing transcription of the operon. For the initiator, or positive activator gene, the regulatory protein is known as an *activator*. This protein binds specifically to the initiator gene, opening it to the passage of RNA polymerase molecules, and thereby increasing transcription of the operon. In both cases, the specificity of the interaction resides not only in the recognition of the DNA of the regulatory region by the regulatory proteins but also in the recognition of the small molecules that interact with and affect the properties of these proteins.

In the case of repressible operons, the regulatory protein has no activity by itself—it is called an *aporepressor*. It takes on the properties of a repressor only when it binds the appropriate small molecule, known as the *corepressor*. In general, the small molecule in a system of this type is the end product of a biosynthetic pathway or a molecule closely related to the end product. The pathway is, of course, the one catalyzed by the enzymes encoded in the regulated operon. In the case of inducible operons, the regulatory protein is, by itself, the active repressor. The small molecule, known as the *inducer*, binds to the repressor and thereby either renders it unable to bind to the operator gene *(negative control)* or imparts to it a new activity—namely, that of an activator with the ability to bind to the initiator gene *(positive control)*. In general, the small molecule in such systems is the substrate of a catabolic pathway or a molecule closely related to the substrate. Once again, the pathway is the one catalyzed by the enzymes encoded in the regulated operon.

5 Induction and Repression

In general, inducible systems are catabolic and provide the organism with the ability to adapt to changes in the availability of metabolic substrates in its environment. The small molecules that act as effectors for the regulatory proteins in such systems are ordinarily substrates for catabolic pathways. Repressible systems are generally biosynthetic and provide the organism with the ability to display frugality by utilizing metabolic end products that are present in its environment rather than synthesizing such products. The small molecules that act as effectors for the regulatory proteins in such systems are ordinarily the end products of biosynthetic pathways.

5.1 Induction

One of the evolutionary advantages of inducible systems is that they allow the organism to acquire quickly the ability to metabolize compounds not usually present in the environment while sparing the organism the waste of maintaining this ability all the time. For example, when *Escherichia coli* is exposed to the metabolic poison, D-serine, it quickly gains the capacity to destroy this compound by inducing the production of a specific degradative enzyme, but it maintains synthesis of the enzyme only as long as D-serine remains present in the environment (McFall, 1964). Another (more classical) example of an inducible system is the lactose operon of *E. coli* (Jacob and Monod, 1961). When glucose is available in the medium, the organism utilizes this sugar as its energy source, and the enzymes for uptake and catabolism of lactose are not made. When glucose is depleted, however, and lactose is present in the environment, lactose enters the cell, is converted in one step to β-allolactose, and this compound acts as inducer of the lactose operon (Jobe and Bourgeois, 1972). β-Allolactose binds to the specific repressor, removing it from the operator gene of the lactose operon, and thereby causing a greatly (several thousand-fold) increased frequency of transcription of the operon. The intracellular levels of the proteins involved in lactose degradation rise dramatically, allowing the organism to thrive in its new environment by utilizing lactose as a source of energy. Thus, as in the examples given earlier, inducible systems are ordinarily catabolic and function in an adaptive capacity. A model of a regulatory system of this type is presented in Fig. 1. It shows an operon that consists of a promoter, an operator, and two structural genes, and it also shows a regulatory gene that specifies the specific repressor for the operon. The repressor binds to the operator, preventing transcription of the operon by RNA polymerase. Repression is counteracted by the inducer, a small molecule that

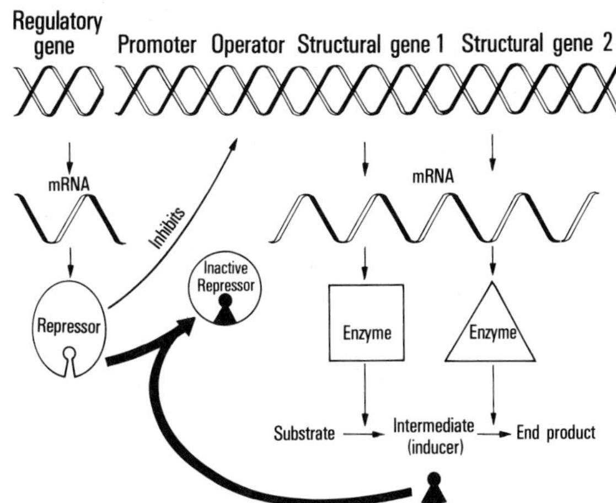

Figure 1. Schematic representation of a negatively controlled inducible system. This model presents, in simplified form, the hypothesis that was proposed by Jacob and Monod in 1961, supplemented with several additional features that have been discovered more recently. It shows a regulatory gene, a promoter, an operator, and two structural genes. (The smallest operon contains only one structural gene; the largest discovered so far contains nine.) The regulatory gene need not be located next to the others, but the promoter, operator, and structural genes function together as a unit known as an *operon*. Both structural genes are transcribed into a single molecule of mRNA, and this polycistronic mRNA is then translated into individual polypetide chains corresponding to the cistrons in the operon. The two structural genes shown here specify two enzymes that catalyze the two steps of a degradative pathway. The substrate (or a closely related molecule) acts as an inducer: It inactivates the repressor. The repressor is the protein specified by the regulatory gene for the operon, itself a structural gene. When the repressor is active (not combined with inducer) it binds to the operator gene and prevents transcription of the structural genes. When inducer is present, it interacts with the repressor, and the repressor–inducer complex is not able to bind to the operator. In this way, the inducer stops the repressor from acting.

interacts with the repressor and prevents it from binding to the operator. The inducer is the substrate (or closely related molecule) of the catabolic pathway catalyzed by the enzmyes encoded in the operon. Such a system is well suited for rapid production of specific enzymes when the substrate for these enzymes is available in the cell's environment. It should be noted, however, that the description of the lactose operon given herein is greatly simplified. In reality, the cell is rarely faced with the problem of producing enzymes for degrading a single energy source. More often, a variety of energy sources is available at the same time. Under such circumstances the cell does not use these energy sources indiscriminately. Instead, it is able to suppress the production of certain catabolic enzymes, even though the inducers may be present, when the availability of a "preferred" energy source renders their production both unnecessary and uneconomical. This complex mechanism is known as *catabolite repression* (see Section 7.1).

In the model shown in Fig. 1, the inducible system is regulated by negative control—that is, expression of the operon is inhibited by the regulatory protein. In other inducible systems the operon is regulated by positive control—that is, expression of the operon is facilitated by the regulatory protein. Figure 2 shows a model for an inducible system of this kind, based mainly on work of Englesberg and his colleagues on the arabinose operon of *E. coli* (see Englesberg and Wilcox, 1974). The arrangement of genes and the products they specify are similar to those for a negatively controlled system. The important difference is that when the repressor interacts with the inducer it does not simply become unable to bind to the operator; instead, it takes on a new activity, that of a positive activator. This positive activator is able to interact with a site within the regulatory region of the

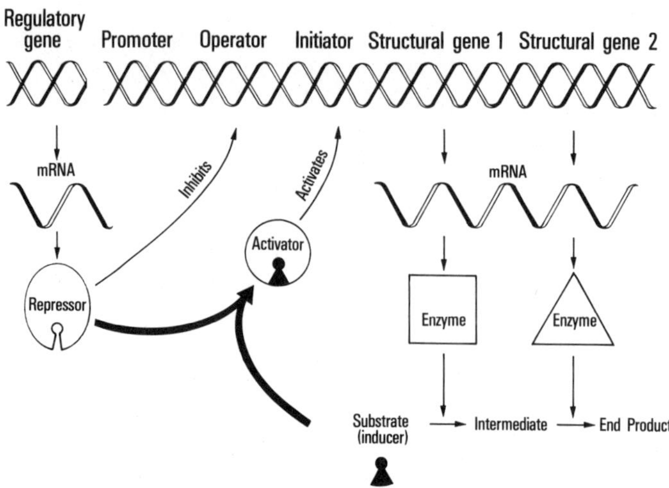

Figure 2. Schematic representation of a positively controlled inducible system. It shows a regulatory gene, a promoter, an operator, an initiator, and two structural genes. The regulatory gene need not be located next to the other genes, but the promoter, operator, initiator, and structural genes function together as an operon. Both structural genes of the operon are transcribed into a single molecule of mRNA, and this polycistronic mRNA is then translated into individual polypeptide chains corresponding to the cistrons in the operon. The two structural genes shown here specify two enzymes that catalyze the two steps of a degradative pathway. As in the situation with *negatively* controlled inducible systems, the regulatory gene specifies the structure of the repressor. Unlike the situation with negatively controlled systems, however, when the repressor interacts with the inducer, it does not simply become inactive; instead, it takes on a new activity, that of a *positive activator.* This positive activator interacts with the initiator, facilitating transcription of the structural genes.

operon known as the initiator gene, thereby facilitating transcription of the operon.

5.2 *Repression*

In contrast to inducible systems, repressible systems ordinarily involve biosynthetic pathways and spare the organism the energy it would expend in making the enzymes for synthesizing complex molecules when those molecules are available preformed in the environment. A model for a negatively controlled repressible system is shown in Fig. 3. In this model we have a biosynthetic pathway. It is the end product of this pathway, not the substrate, that plays a role in regulation. The regulatory gene specifies a protein, known as an aporepressor, that is not active in itself. It becomes an active repressor, capable of binding to the operator and preventing transcription of the operon, only when it is combined with the end product of the pathway, the corepressor.

The pathway for the synthesis of the amino acid, tryptophan, is among the systems regulated in this fashion (for review, see Yanofsky, 1976). The enzymes that catalyze the six steps of this pathway in *E. coli* are encoded in the genes of a single operon, controlled by one operator gene. The repressor, as is the rule for repressible systems, is composed of a protein, the aporepressor, and a small molecule, the corepressor. The corepressor in this case is the end product of the biosynthetic pathway, tryptophan. Thus, when the organism grows in the presence of exogenous tryptophan, the intracellular level of tryptophan is sufficiently high to maintain repression of the tryptophan operon; the enzymes for tryptophan biosynthesis are made at a relatively low (basal) rate, and the pathway functions at a low level. When the organism grows in the absence of exogenous tryptophan, however, the amount of this amino acid that is synthesized when the operon is expressed at the basal rate is not sufficient to saturate the aporepressor. Unable to form a sufficient amount of active repressor, the operon becomes *derepressed;* the enzymes for tryptophan biosynthesis are now produced at a higher rate, and the cell's capacity to synthesize tryptophan consequently increases. If tryptophan is added back to the derepressed culture or accumulates because of decreased consumption, the tryptophan operon becomes repressed once more. Thus, the rate at which the enzymes for tryptophan biosynthesis are manufactured is closely geared to both the demand for tryptophan and the availability of exogenous tryptophan. By this mechanism, the organism is able to conserve the energy that would be wasted in synthesizing the tryptophan enzymes when they are not needed. For a discussion of a second mechanism for regulating transcription of the tryptophan operon, one that involves the so-called *attenuator,* the reader is referred to Chapters 2, 7, and 10 of this volume and the references cited therein.

5.2.1 Multivalent and Cumulative Repression

In the model for a negatively controlled repressible system shown in Fig. 3 the biosynthetic pathway has only one end product, and that makes it rather simple

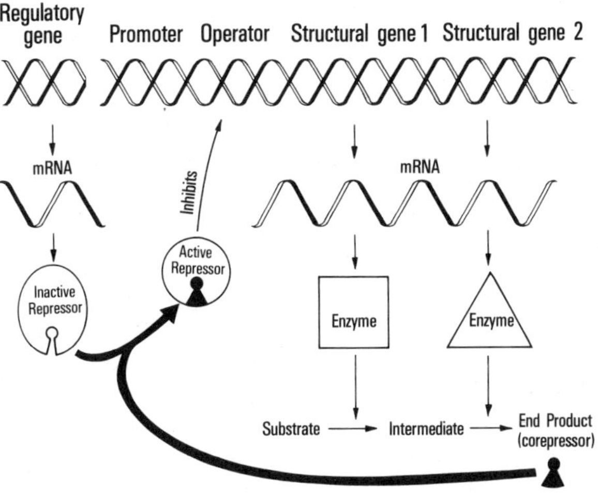

Figure 3. Schematic representation of a negatively controlled repressible system. Here we see the same arrangement of genes, forming products of the same kind, as that shown in Fig. 1. However, there are some important differences. In this case, the metabolic pathway catalyzed by the two enzymes encoded in the structural genes of the operon is a biosynthetic one. It is the product of the biosynthetic pathway, not the substrate, that plays a role in regulation. The regulatory gene specifies a protein that is not active by itself. It is called an *aporepressor.* It achieves the capacity to bind to the operator gene and prevent transcription of the structural genes only when it is combined with the end product of the biosynthetic pathway (or closely related molecule), the *corepressor.* The complex formea by interaction of aporepressor and corepressor has all the properties of the repressor described for negatively controlled systems

for the cell to exert control. But what of biosynthetic pathways that are branched, producing two or more products? If any one of the end products could act as corepressor by itself, accumulation of that product would suffice to shut off expression of the operon, and an incurable shortage of the other product(s) might ensue. In fact, at least two mechanisms have evolved for solving this problem. One is known as *multivalent* repression, an example of which is provided by the regulation of the enzymes for synthesis of isoleucine and valine (Freundlich *et al.*, 1962). The second mechanism is known as *cumulative* repression, an example of which is provided by regulation of the enzyme, carbamylphosphate synthetase (Piérard *et al.*, 1965). In multivalent repression, the aporepressor cannot be activated at all by any one of the end products of the biosynthetic pathway; it becomes an active repressor only upon binding all of the end products. In cumulative repression, the aporepressor becomes a partially active repressor upon binding any one of the end products of the biosynthetic pathway but becomes fully active only upon binding all of the end products. A schematic model for these two mechanisms is shown in Fig. 4.

5.2.2 *Natural Advantage of Repression*

The fact that repressible systems arose and were maintained through the evolutionary process has been explained by the hypothesis that organisms that

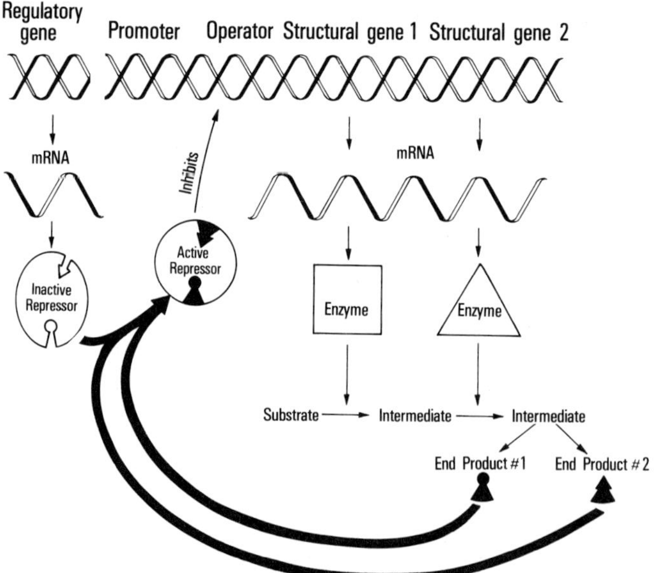

Figure 4. Schematic representation of multivalent and cumulative repression. The arrangement of genes and the products they specify is the same as that shown in Fig. 3, except that the biosynthetic pathway has two end products, both of which act as corepressors. The aporepressor has a site specific for binding each of the corepressors. In *multivalent repression,* the aporepressor achieves no repressor activity at all by interacting with either corepressor alone; it must form a complex with both of them to become active. In *cumulative repression,* the aporepressor becomes partially active as a repressor upon binding one or the other of the corepressors but becomes a fully active repressor only upon binding both of them.

possess regulated systems are at a selective advantage over organisms that do not. This hypothesis was tested experimentally by Zamenhof and Eichhorn (1967). They inoculated a culture with an equal number of cells of two different bacterial strains. These strains differed in only one respect, the ability to regulate tryptophan biosynthesis. One strain was mutated so that it was constitutive; it had lost the ability to repress the tryptophan operon. The other strain was normally repressible; its tryptophan operon became repressed whenever tryptophan was available exogenously. The mixed culture of the two strains was grown in liquid medium containing tryptophan. The cells of the first (constitutive) strain continued to make the enzymes for tryptophan biosynthesis at a high rate, even though their presence was superfluous since the amino acid was available in the culture medium; the second (normally regulated) strain produced these enzymes at the lowest possible (basal) rate. At the end of a few days the normally regulated strain was found to have overgrown the constitutive strain, taking over essentially full possession of the culture. Evidently, even the relatively small amount of energy saved by repressing a single operon under the appropriate conditions can be crucial. The difference in the growth rates of the two strains may have been extremely small, but still it was obviously sufficient to cost one strain extinction and bestow survival on the other.

5.3 *The Regulon*

Up to this point, I have relied heavily on genes clustered together in operons to illustrate various aspects of genetic regulation. It is important to recognize, however, that not all bacterial genes that are controlled as a unit are clustered. For example, the arginine system consists of an eight-step metabolic pathway that is catalyzed by enzymes encoded in genes scattered around the bacterial chromosome, yet the genes are controlled as a unit in a manner similar to that for genes clustered in a single operon. The mechanism by which the cell accomplishes this feat is by repeating the same operator sequence beside each of the arginine genes, so that transcription of all of them is responsive to the same regulatory protein. Under conditions in which the intracellular concentration of the arginine repressor is low, all of the arginine operators allow rapid transcription of the structural genes they control. Thus, the scattered arginine genes belong to a single unit of regulation, known as a *regulon* (Maas *et al.*, 1964). The fact that the example given here involves a repressible system rather than an inducible system is not relevant to the regulon concept. What is relevant is that the regulon provides a mechanism for the coordinated regulation of genes that are not clustered together. This mechanism may have especially important implications for control in higher organisms, in which coordinated regulation of disparate genes is recognized but in which operons have not yet been identified.

6 *Autogenous Regulation*

Autogenous regulation is a specific regulatory mechanism, common to a number of systems in both prokaryotic and eukaryotic organisms, that represents a variation on the theme of induction and repression developed earlier herein (Goldberger, 1974). In this mechanism, the product of a structural gene regulates expression of the operon in which that structural gene resides, as shown schematically in Fig. 5. Although the model depicted in Fig. 5 is based on a negatively regulated inducible system, similar models can be constructed for autogenous regulation of systems of any type—negative or positive, inducible or repressible. In many cases, the protein specified by the autogenous regulatory gene has a dual function, acting not only as a regulatory protein, but also as an enzyme, structural protein, and so on. In a few cases, this protein is the multimeric allosteric enzyme that catalyzes the first step of a metabolic pathway, gearing together the two most important mechanisms for controlling the biosynthesis of metabolites in bacterial cells, repression and feedback inhibition (see Chapter 1 of Volume 2 of this treatise). Autogenous regulation of particular systems provides the cell with a means for accomplishing a number of regulatory tasks that cannot be accomplished so efficiently, if at all, by classical regulation of gene expression. The unique properties of autogenously regulated systems result from the presence of a feedback loop within the regulatory circuit. These properties are discussed in detail in Chapter 3 of this volume.

7 *Integration of Regulatory Mechanisms*

In addition to regulatory mechanisms involving individual operons, there are a number of systems in prokaryotes for regulating the expression of several different operons at the same time. Whereas mechanisms of the first group provide the organism with a high degree of specificity in its response to the environment, those of the second group allow the organism to coordinate a set of responses.

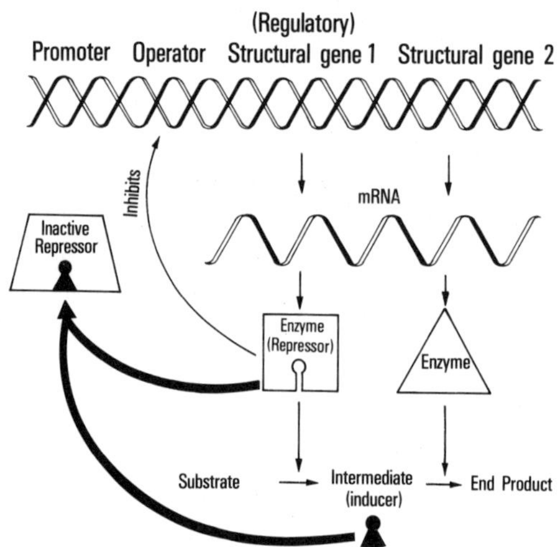

Figure 5. Autogenous regulation. In this schematic representation, the autogenously regulated system is negatively controlled and inducible. The genes and their products are very similar to those shown in Fig. 1. The important difference involves the regulatory gene; in this case it is located within the operon, forming a closed regulatory loop. In the model shown here, the product of the first structural gene of the operon plays a dual role, acting not only as the enzyme that catalyzes the first step of a catabolic pathway but also as the repressor for the operon in which it resides.

7.1 Catabolite Repression

The first coordinated set of responses to be recognized is known as *catabolite repression*. This mechanism involves positive regulation of a group of distinct but functionally related operons by a protein, the *catabolite repressor protein* (CRP), and a small molecule, cyclic AMP (Zubay *et al.*, 1970; Emmer *et al.*, 1970). The operons of this group, which we shall call the catabolic operons, contain the structural genes for the metabolic pathways by which various substrates may be degraded to yield energy to the organism. In each of the catablic operons there is a site located in or adjacent to the promoter at which the binding of CRP–cyclic AMP greatly increases the productive binding of RNA polymerase to the promoter (see Chapter 7 of this volume). When *E. coli* grows in the presence of glucose, it utilizes this sugar preferentially as its source of energy, and the enzymes encoded in the catabolic operons (the enzymes for degrading other potential energy sources) are repressed. The reason for this repression is that as long as *E. coli* utilizes glucose it does not maintain an intracellular concentration of cyclic AMP sufficiently high to saturate the CRP. By itself, CRP cannot act as a positive activator for the catabolic operons. When glucose is depleted from the medium, however, then the level of cyclic AMP rises, and CRP–cyclic AMP is now available to facilitate transcription of these operons. But CRP–cyclic AMP alone is still not sufficient. Induction of a specific catabolic operon also requires that the specific inducer (energy source) for that particular operon be present in the cell's environment. If, for example, lactose is present, it will enter the cell and will induce the lactose operon; if galactose is present, it will induce the galactose operon, and so on.* In other words, it takes CRP–cyclic AMP *and* a specific inducer to increase the frequency of transcription of a specific catabolic operon. Thus, catabolite repression provides the cell with the ability to utilize glucose preferentially, but to adapt quickly to the absence of glucose by allowing the cell to start manufacturing the enzymes for utilizing other energy-rich compounds when those compounds are available in the environment. Which set of enzymes is manufactured depends on which energy-rich compound is available in the environment to act as inducer. The mechanism is both adaptive and economical, since it saves the cell the waste of manufacturing all possible degradative enzymes when they are not needed.

But catabolite repression is even more versatile than that. It can also provide the cell with the means for discriminating among various energy-rich compounds when two or more different ones are present in the environment. When *E. coli* grows in the absence of glucose, it does not induce every catabolic operon just because the inducers happen to be present in its environment. That would still be a rather wasteful response to the absence of glucose. Instead, it has a clear program that determines which potential energy source in the environment will be allowed to induce its degradative enzymes in the cell (Alper and Ames, 1978). The priority order is set by the differential sensitivities of the catabolic operons to CRP–cyclic AMP. The most sensitive operon will be induced preferentially (if the inducer is present in the environment), whereas the least sensitive operon will be

*For the sake of simplicity, the substrates of the pathways of catabolic operons are referred to here as inducers. Actually, as pointed out in Section 4.3, it may be a molecule easily derived from the substrate, rather than the substrate itself, that acts as inducer.

induced only if none of the inducers of the more sensitive ones is present in the environment. Thus, catabolite repression is a complex mechanism that allows the organism to select its diet in gourmet fashion, even though the menu may be quite extensive.The reader is referred to the chapter by Robert Macnab in Volume 2 of this treatise for a discussion of chemotaxis, another process that is sensitive to catabolite repression. By virtue of this sensitivity, the bacterial cell is able not only to be selective in utilizing nutrients from its immediate environment but also to coordinate such selection with migration to a more favorable environment when one exists nearby.

7.2 Stringency

Two other nucleotides, guanosine tetraphosphate and guanosine pentaphosphate, play perhaps an even more central role in integrating the metabolic activities of bacterial cells than does cyclic AMP (Cashel and Gallant, 1969). The regulatory mechanism in which these nucleotides participate is known as *stringency* (for reviews, see Cashel and Gallant, 1974; Block and Hazeltine, 1974).

The most characteristic feature of stringency is that when the bacterial cell is faced with amino acid starvation, and its rate of growth is reduced or ceases, it stops manufacturing ribosomal RNA. This response is clearly economical, since the starving (nongrowing) cell has no need to increase its supply of ribosomes. In fact, it is not the lack of an amino acid *per se* that brings about this so-called *stringent response,* but an increase in the intracellular concentration of the corresponding tRNA in its nonaminoacylated form (Neidhardt, 1966). What is more, the stringent response involves a great many biological activities in addition to ribosomal RNA synthesis—alterations in transcription of tRNA, protein stability, coding fidelity, various transport systems, and others (see Cashel and Gallant, 1974). The stringent response depends on the activity of the *relA* gene. Mutation of this gene causes what is known as *relaxed* control, allowing the organism to accumulate stable RNA and to carry out a number of other functions normally restricted by the stringent response, under conditions of amino acid limitation. It appears that the stringent response is mediated by ppGpp and pppGpp and that these nucleotides are synthesized only in the presence of *stringent factor,* the protein specified by the *relA* gene (Block and Hazeltine, 1973). Through the mechanism of stringency, the bacterial cell couples its biosynthetic and metabolic activities with the growth conditions prevailing in its environment (see Chapters 10 and 12 of this volume).

8 Translational Control

For some time, molecular biologists have been interested in the question of whether specific enzyme synthesis in prokaryotic organisms is regulated not only by alterations in the rate of transcription of specific genes but also by alterations in the frequency with which specific mRNA molecules are translated. Among the early findings that led some workers to believe that translation is a regulated process was the discovery that the several genes of a single operon may direct the synthesis of different numbers of molecules of the proteins they specify. If ribosomes initiate translation of a polycistronic mRNA at one end and inevitably continue to the other end, one would expect that the numbers of molecules of all the polypeptide chains encoded in the RNA would be equal. Indeed, this appears to be true in some cases and under certain conditions. However, there are also well-documented cases in which the various cistrons of a polycistronic mRNA are utilized for the production of different numbers of molecules of the corresponding polypeptides. This *noncoordinacy,* as it is sometimes called, cannot be explained by the presence of internal promoters in the operon that give rise to shorter than normal molecules of mRNA because then the cell would always contain greater numbers of molecules of the proteins encoded in the distal portion of the operon; nor can it be explained by a certain fraction of the ribosomes becoming detached from the mRNA as they traverse it, because then the cell would always contain greater numbers of molecules of the proteins encoded in the proximal portion of the operon. The fact to be reckoned with is that in certain cases internal cistrons of an operon give rise to the greatest number of protein molecules. For example, in the bacteriophage R17, as in a whole group of RNA phages, the middle of the three cistrons is translated about three times more frequently than is its neighbor on one side, and about 20 times more frequently than is its neighbor on the other side (Lodish, 1968). This finding indicates, as had been suggested previously (Bretscher, 1968; Lodish and Robertson, 1969), that ribosomes may initiate translation of certain mRNA molecules not only at the 5′ end but also at internal sites, and that such internal sites may even be utilized preferentially. The work of Steitz and her colleagues (see Chapter 9 of this volume) has provided evidence that preferential translation of certain cistrons of RNA phages is due to the secondary structure of the RNA. How widely this mechanism may apply to translation of bacterial mRNA has not yet been determined.

In addition to the effects of RNA secondary structure, translation of specific mRNA molecules may also be modulated by specific proteins in a manner analogous to the modulation of transcription by specific regulatory proteins. There is even a well-documented example of an autogenously regulated system that operates at the level of translation: The protein specified by gene 32 of bacteriophage T4 "represses" the synthesis of additional copies of that very protein not by binding to a regulatory region in the DNA near its structural gene but by binding to the mRNA transcribed from its gene. In this way, gene 32 protein inhibits further translation of the message in which it is encoded (Russel *et al.,* 1976).

The extent to which translational control contributes to cellular regulation has not yet been fully elucidated. Initiation factors, interference factors, and other specific regulatory proteins that modulate translation of mRNA, as well as factors that affect the stability of mRNA, have all been explored. The data suggest a complex and interesting set of controls for the translation system.

9 *Conclusion*

The process by which transcription of a given gene results in a change in a specific cellular function is extremely complex. Regulation may occur at the very first step of the process, in which the interaction of RNA polymerase with the promoter may be influenced by the intracellular concentrations of various other proteins, such as CRP and other activators and repressors. The processes of repression and induction involve binding of specific regulatory proteins to specific regulatory regions in DNA. The effects of the regulatory proteins may be negative, restricting transcription, or positive, facilitating transcription. The regulatory protein may bind to only a single site on the chromosome, as in the case of the repressor or activator of a particular operon, or it may bind to a (small) number of specific sites on the chromosome. The former case allows the cell to react to its environment with a single response, whereas the latter case allows the cell to react to its environment with an organized set of responses.

Coordinated sets of responses to environmental needs have been recognized recently as an important aspect of cellular regulation. Catabolite repression, mediated through cyclic AMP, is one example of such coordination; the stringent response, mediated through ppGpp and pppGpp, is another. Small molecules such as cyclic AMP and ppGpp are among the so-called *alarmones* (Stephens *et al.*, 1975; Alper and Ames, 1978; Schindler and Sussman, 1977) that have far-reaching effects on prokaryotic cellular metabolism in a manner somewhat analogous to that of hormones in eukaryotic organisms. One of the generalizations that can be made from studies on genetic regulation is that alterations in gene expression are brought about by specific regulatory proteins. In a few systems (so-called *autonomous* systems), it may take no more than the regulatory protein to do the job required. In most systems, however, the activity of the regulatory protein must itself be regulated, and it is the various specific small molecules of the cell that carry out the latter regulatory task. In fact, it is primarily such small molecules that communicate the metabolic needs of the cell to the sites at which gene expression is controlled; therefore, any process that changes the intracellular concentration of either a regulatory protein or a small molecule of this kind is itself an important step at which regulation may be exerted.

Gene expression may be regulated not only by alterations in the frequency with which RNA polymerase successfully initiates transcription at specific genetic sites (as described in this chapter), but also by alterations in the efficiency with which the polymerase terminates transcription at specific genetic sites (as described in detail in Chapter 7 of this volume).

Translational control has been recognized in the RNA phages as well as in bacteria, and may be of even greater importance in the cells of higher organisms. Theoretically, gene expression could be modulated by alterations in the stability of mRNA, in the frequency with which specific portions of a polycistronic mRNA are translated, in the availability of the various species of tRNA or their synthetases, and in any of the many other catalytic events involved in the translation process. In eukaryotic cells, processing of RNA transcripts and transport of mRNA from nucleus to cytoplasm represent additional steps for potential regulation.

Folding of the newly synthesized polypeptide chain, leading to the formation of a native protein with a specific three-dimensional conformation and a specific catalytic or other function, is a step in protein synthesis that is very rapid; it does not appear to regulate the rate of synthesis of specific proteins. In many cases, the acquisition of quaternary structure by specific association of subunits is required before a protein can realize its functional potential. It is known that the processes of subunit aggregation and disaggregation may be subject to control mechanisms.

The activity of a protein may be modified by specific small molecules through noncovalent interactions, such as those involved in "feedback" inhibition, or through covalent changes, such as phosphorylation and adenylylation. Such modifications are extremely important in cellular regulation (see Chapter 1 of Volume 2 of this treatise).

There are two main lessons to be learned from studying regulation in prokaryotes. They require special mention here because the reader cannot learn them from the generalizations that characterize this brief review; they become apparent only by studying the details of a great many systems, including the many exceptions to the "rules" I have discussed, and by studying regulatory mechanisms that operate at many different levels, not just those that operate at the genetic level. These lessons are as follows: First, any given metabolic system is likely to be regulated by a hierarchical array of controls rather than by a single control mechanism. From this finding we conclude that the possibility of exerting control at multiple sites in a metabolic process is advantageous to the survival of the whole organism. It is not difficult to understand why survival may be facilitated by providing metabolic systems with the capacity to respond in several different ways (involving multiple molecular interactions) to an environmental stimulus, since such complex responsiveness gives the organism sophistication in dealing appropriately with its environment. Second, there is great diversity among the mechanisms by which the genetic potential of the cell is regulated. This diversity reflects the diverse needs of the organism. It is easily understandable that a regulatory device perfected through the evolutionary process for its ability to provide the organism with a constant high concentration of a certain protein may differ from a device selected for its ability to alter the concentration of a certain protein quickly in response to environmental emergencies. But even among those regulated systems that apparently play similar roles in the life of the cell there is a great deal of diversity at the molecular level. This fact reflects the ability of the evolutionary process to provide the cell with diverse solutions to apparently similar problems.

Because of the multiplicity of controls in each regulatory system and because of the diversity of control mechanisms among different regulatory systems, the field of biological regulation holds great excitement and great challenge for the scientist interested in vital functions. In fact, it is the regulatory devices of the cell and the coordination among them that provide the cell with the capacity to perform the balancing act that differentiates life from a conglomerate of chemical reactions.

References

Alper, M. D., and Ames, B. N., 1978, Transport of antibiotics and metabolite analogs by systems under cyclic AMP control: Positive selection of *Salmonella typhimurium cya* and *crp* mutants, *J. Bacteriol.* **133**:149.

Ames, B. N., and Hartman, P., 1963, The histidine operon, *Cold Spring Harbor Symp. Quant. Biol.* **28**:349.

Anfinsen, C. B., 1973, Principles that govern the folding of polypeptide chains, *Science* **181**:223.

Atkins, J. F., and Loper, J. C., 1970, Transcription initiation in the histidine operon of *Salmonella typhimurium, Proc. Natl. Acad. Sci. U.S.A.* **65**:925.

Block, R., and Hazeltine, W. A., 1973, Thermolability of the stringent factor in *rel* mutants of *Escherichia coli, J. Mol. Biol.* **77**:625.

Block, R., and Hazeltine, W. A., 1974, *In vitro* synthesis of ppGpp and pppGpp, in: *The Ribosome* (A. Tissiere, P. Lengyel, and M. Nomura, eds.), pp. 747–781, Cold Spring Harbor Lab., Cold Spring Habor, New York.

Bretscher, M. S., 1968, Direct translation of a circular messenger DNA, *Nature* **220**:1088.

Buttin, G., 1963, Méchanismes régulateurs dans la biosynthèse des enzymes du métabolisme du galactose chez *Escherichia coli,* K12 I. La biosynthèse induite de la galactokinase et l'induction simultanée de la séquence enzymatique, *J. Mol. Biol.* **7**:164.

Cashel, M., and Gallant, J., 1969, Two compounds implicated in the function of the RC gene of *E. coli, Nature* **221**:838.

Cashel, M., and Gallant, J., 1974, Cellular regulation of guanosine tetraphosphate and guanosine pentaphosphate, in: *The Ribosome* (A. Tissiere, P. Lengyel, and M. Nomura, eds.), pp. 733–745, Cold Spring Harbor Lab., Cold Spring Harbor, New York.

Emmer, M., de Crombrugghe, B., Pastan, I., and Perlman, R., 1970, Cyclic AMP receptor protein of *E. coli:* Its role in the synthesis of inducible enzymes, *Proc. Natl. Acad. Sci. U.S.A.* **66**:480.

Englesberg, E., and Wilcox, G., 1974, Regulation: Positive control, *Annu. Rev. Genet.* **8**:219.

Epstein, W., and Beckwith, J., 1968, Regulation of gene expression, *Annu. Rev. Biochem.* **37**:411.

Freundlich, M., Burns, R. O., and Umbarger, H. E., 1962, Control of isoleucine, valine, and leucine biosynthesis, I. Multi-valent repression, *Proc. Natl. Acad. Sci. U.S.A.* **48**:1804.

Goldberger, R. F., 1974, Autogenous regulation of gene expression, *Science* **183**:810.

Jacob, F., and Monod, J., 1961, Genetic regulatory mechanisms in the synthesis of proteins, *J. Mol. Biol.* **3**:318.

Jacob, F., Ullman, A., and Monod, J., 1964, Le promoteur, élément génétique necéssaire à l'expression d'un operon, *C.R. Acad. Sci.* **258**:3125.

Jobe, A., and Bourgeois, S., 1972, *Lac* repressor–operator interaction VI. The natural inducer of the *lac* operon, *J. Mol. Biol.* **75**:303.

Lodish, H. F., 1968, Bacteriophage f2 RNA: Control of translation and gene order, *Nature* **220**:345.

Lodish, H. F., and Robertson, H. D., 1969, Regulation of *in vitro* translation of bacteriophage f2 RNA, *Cold Spring Harbor Symp. Quant. Biol.* **34**:655.

Maas, W. K., Maas, R., Wiame, J. M., and Glansdorff, N., 1964, Studies on the mechanism of repression of arginine biosynthesis in *Escherichia coli,* I. Dominance of repressibility in zygotes, *J. Mol. Biol.* **8**:359.

Margolin, P., and Bauerle, R. H., 1966, Determinants for regulation and initiation of expression of tryptophan genes, *Cold Spring Harbor Symp. Quant. Biol.* **31**:311.

Martin, R. G., 1963, The one operon–one messenger theory of transcription, *Cold Spring Harbor Symp. Quant. Biol.* **28**:357.

McFall, E., 1964, Genetic structure of the D-serine deaminase system of *Escherichia coli, J. Mol. Biol.* **9**:746.

Neidhardt, F. C., 1966, Roles of amino acid activating enzymes in cellular physiology, *Bacteriol. Rev.* **30**:701.

Piérard, A., Glansdorff, N., Mergeay, M., and Wiame, J. M., 1965, Control of biosynthesis of carbamoyl phosphate in *Escherichia coli, J. Mol. Biol.* **14**:23.

Russel, M., Gold, L., Morrissett, H., and O'Farrell, P. Z., 1976, Translational, autogenous regulation of gene 32 expression during bacteriophage T4 infection, *J. Biol. Chem.* **251**:7263.

Schindler, J., and Sussman, M., 1977, Ammonia determines the choice of morphogenic pathways in *Dictyostelium discoidium, J. Mol. Biol.* **116**:161.

Stephens, J. C., Artz, S. W., and Ames, B. N., 1975, Guanosine 5′-diphosphate 3′-diphosphate (ppGpp): Positive effector for histidine operon transcription and general signal for amino-acid deficiency, *Proc. Natl. Acad. Sci. U.S.A.* **72**:4389.

Yanofsky, C., 1971, Tryptophan biosynthesis in *Escherichia coli.* Genetic determination of the proteins involved, *JAMA* **218**:1026.

Yanofsky, C., 1976, Regulation of transcription initiation and termination in the control of expression of the tryptophan operon of *E. coli,* in: *Molecular Mechanisms in the Control of Gene Expression* (D. P. Nierlich and W. J. Rutter, eds.), pp. 75–87, Academic Press, New York.

Zamenhof, S., and Eichhorn, H. H., 1967, Study of microbial evolution through loss of biosynthetic functions: Establishment of "defective" mutants, *Nature* **216**:456.

Zubay, G., Schwartz, D., and Beckwith, J., 1970, Mechanism of activation of catabolite-sensitive genes: A positive control system, *Proc. Natl. Acad. Sci. U.S.A.* **66**:104.

2

Structure of Complex Operons

ALLAN CAMPBELL

1 Evolution of the Operon Concept

In 1961 and 1965, Jacob and Monod unified and systematized many of the facts then known about genetic regulation in bacteria. Starting with two prototypic examples, the *lac* operon of *Escherichia coli* and the bacteriophage lambda (which is more complex, but has many features in common with the *lac* operon), these authors developed a general scheme for regulation that might have been extendable to all genes of all organisms. This scheme has provided a conceptual nucleus and a source of terminology for almost all subsequent discussions of regulation, even though further experimentation has revealed some facts that could not be directly assimilated into the original framework of this scheme without modification or expansion.

The *lac* operon consists of three closely linked genes *(lacZ, lacY, lacA)* that code for the three proteins that function in the uptake and metabolism of β-galactosides. These genes are transcribed onto one mRNA molecule, from a fixed initiation point (the promoter). Their transcription is coordinately regulated by the binding of a specific protein (the *lac* repressor) to a DNA segment (the operator) that lies close to the promoter. The *lac* repressor, when attached to the operator, sterically blocks the binding of RNA polymerase and thereby prevents initiation of transcription. Certain β-galactosides can bind to the *lac* repressor, causing a conformational change in this protein that reduces its affinity for the

ALLAN CAMPBELL • Department of Biological Sciences, Stanford University, Stanford, California

operator, thus releasing it from the DNA. Most of these facts, now thoroughly documented with direct physical evidence, were correctly inferred by Jacob and Monod from the *in vivo* results available in 1961. Their concept of operon regulation has since been expanded to encompass cases in which specific proteins exert positive as well as negative control over initiation, but it has not been profoundly modified in other respects.

Having described the basic units or atoms of regulation, the operons, Monod and Jacob (1961) proceeded to explore some of the ways in which these units could be used as building blocks for the intricately controlled sophisticated piece of machinery that constitutes a whole cell or a whole organism. These investigators displayed several ways in which individual operons might be linked together through regulatory interconnections, so that groups of operons might collectively exhibit characteristics that were thought to be of adaptive value in biological systems. For example, if one of the metabolic products of operon A were to inactivate the repressor of operon B, while a product of operon B could inactivate the repressor of operon C, then derepression of operon A could initiate the expression of these operons in a temporally ordered sequence. If two operons are so constituted that each one elaborates a product that represses the other, then under suitable kinetic conditions expression of the two will be mutually exclusive, thus constituting a developmental switch that can be flipped in either of two directions. Many of the circuits indicated either explicitly or implicitly in their discussion have since been found to exist in nature. Some are discussed in this chapter and in other chapters of this volume.

Conceptually, there are two possible ways to develop regulatory complexity starting from a simple operon such as *lac*. One of these, as just discussed, is to create regulatory connections between separate operons. The second is to complicate the structure of the operon itself by introducing control points within, rather than terminal to, the operon. It turns out that both tactics are used extensively by the bacteriophage λ (the author's favorite object of study) and jointly account for the ability of a single species of repressor protein to shut off permanently the sequential expression of genes that function in viral development.

Complexity of operon structure, the topic of this chapter, was not explicitly anticipated in 1961. Nor has the subject, to my knowledge, been previously discussed as such, although various examples are widely recognized (Levin, 1974). Operon complexity is perhaps best defined by the manner in which it deviates from the simple operon concept of Jacob and Monod. Consider a chromosomal segment with structural genes arranged in order:

$$ABCDEFGHIJ$$

From the classical Jacob–Monod scheme, we might expect that sufficiently intensive study of the expression of the genes in this segment would allow us to resolve the segment into one or more operons, each comprising a distinct subsegment and characterized by a unique order of transcription. We might, for example, group these genes into three operons:

$$\overleftarrow{ABCDE} \quad F \quad \overrightarrow{GHIJ}$$

where genes E through A are transcribed leftward, G through J are transcribed rightward, and F is expressed independently of the others. Such a classification is unambiguous only if a given gene is always transcribed in the same direction, on a message that always starts and terminates at the same fixed points. When these conditions fail, we consider the operon to be complex, as in the following examples: (1) if gene "*A*" can be transcribed under some circumstances from a promoter located between genes *E* and *F* as part of the "*ABCDE*" operon, but under other circumstances from a separate promoter located between genes *C* and *D* as part of an "*ABC*" operon (dual promoters); (2) if transcription of the *ABCDE* operon frequently terminates between genes *E* and *D* so that usually only *E* is transcribed (attenuation); (3) if leftward transcription of *ABCDE* and rightward transcription of *FGHIJ* arise from a common operator–promoter (divergent transcription); and (4) if each of the two gene blocks, *ABCDEF* (leftward) and *FGHIJ* (rightward), is transcribed as a unit from its own operator–promoter (overlapping transcription).

2 *Types of Complex Operons*

This chapter considers the extent of current information about some specific examples of complex operons and their possible significance. The particular examples used were selected partly on the basis of familiarity to the author and partly because they illustrate the methodology that may in the long run prove to be the most significant contribution of the study of prokaryotic genetics to biology I try to avoid discussing instances of mere noise in the transcription process, such as false starts *in vitro,* premature termination *in vivo,* and messenger degradation. However, to the extent that such processes occur preferentially at specific locations, they may be quite relevant. Also somewhat artibrarily excluded is information that seems more appropriately regarded as defining the internal structure of elements of the Jacob–Monod operon, such as information on the dissection of promoters into sites of polymerase entry, polymerase binding, and message initiation (Pribnow, 1975); the arrangement of multiple regulatory sites governing a single promoter, as in operons responsive both to repression and to cyclic AMP (Dickson *et al.,* 1975); and the translational overlap of genes with reading frames in different phases (Barrell *et al.,* 1976). The recently discovered phenomenon of messenger splicing (Berget *et al.,* 1977), known thus far only in eukaryotes, opens up additional possibilities for operon complexity. The subject is not covered in depth here because its genetic and functional correlates are still unknown.

If complexity in operon structure is of any real importance, each particular regulatory pattern it allows must be of adaptive value to the organism. I therefore devote considerable space to describing the functions of some of the operons treated in the overall biology of the organism, and to discussing what the adaptive values may be. In Section 3, I come to the question of how complexity may have developed during evolution.

2.1 *Internal Promoters and Multiple Promoters*

2.1.1 *The Internal Promoter of the trp Operon*

One of the first operons in which internal complexity came to light was the *trp* operon of *Salmonella typhimurium* (Margolin and Bauerle, 1966). This operon

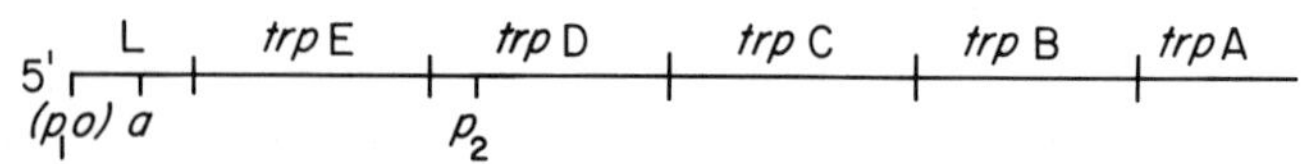

Figure 1. The tryptophan operon of *E. coli.* L, leader sequence (162 nucleotides preceding the first structural gene in the message). *trpEDCBA,* structural genes for enzymes of tryptophan biosynthesis. *trpD* and *trpC* specify bifunctional polypeptides that presumably evolved through gene fusion (see Section 3) (Crawford, 1975); *trpE* specifies anthranilate synthase (AS); *trpD* specifies phosphoribosyltransferase (PRT) and glutamine amidotransferase subunit of AS; *trpC* specifies indoleglycerolphosphate synthase (InGPS) and phosphoribosylanthranilate isomerase (PRAI); *trpB* specifies tryptophan synthetase (β subunit); and *trpA* specifies tryptophan synthetase (α subunit). *o,* operator; p_1, promoter controlled by operator; *a*, attenuator (site of premature termination located 136–141 nucleotides from the 5 ′ end of the message); p_2, weak constitutive promoter within *trpD.*

comprises five closely linked structural genes, coding for the polypeptides of the enzymes that catalyze the terminal steps in tryptophan biosynthesis (Fig. 1). Synthesis of all five polypeptides is at least 98% repressible by tryptophan. Repressibility by trytophan can be eliminated by a single operator mutation lying at one end of the gene cluster. Thus, all five genes appear, on first inspection, to lie in one simple operon.

One aspect of internal complexity became apparent during a study of deletion mutations that enter the operon from the operator end. According to the simple operon concept, a deletion that beheads an operon should have one of two consequences, depending on the nature of the DNA to which the remaining genes of the operon become connected. Either those genes should be turned off completely (because the promoter has been deleted) or else they should become fused to a nearby operon and transcribed from its promoter. In the case of the *trp* operon, however, all deletions that cut into the first two genes of the operon allow a low rate of expression of the last three genes. The products of the last three genes are formed at a rate that is about 1 to 2% of the rate in a derepressed wild type strain, whether or not tryptophan is present. Deletions that cut into the third gene completely prevent the formation of tryptophan synthetase, even though the two genes coding for that enzyme are both present and intact. It was therefore postulated that the *trp* operon, in addition to the strong repressible promoter (p_1) at the *trpE* end, contains a weak internal promoter (p_2) between *trpD* and *trpC,** expression of which is constitutive. The p_2 promoter functions both in wild type and in the deletion mutants just described. In the wild type it is less conspicuous because the small amount of transcription coming from p_2 is generally masked by that originating at p_1. However, when p_1 is maximally repressed, transcription from p_2 becomes significant, causing the last three genes of the operon to show a lower repression ratio than do the first two genes. Further work demonstrated that the *trp* operon of *E. coli,* like that of *S. typhimurium,* has an internal promoter for the last three genes of the operon. From the repression ratios observed in mutants with various internal deletions, Jackson and Yanofsky (1972) localized p_2 more precisely, showing that it lies at a site within the second gene of the operon

*For historical reasons, the *trp* genes of *Salmonella* have been assigned symbols that differ from those for the corresponding genes of *Escherichia coli.* For our purposes here, it is convenient to use a "universal" nomenclature (Crawford, 1975) in which the symbol used for the *E. coli* genes are also used for the analogous genes of other species (Fig. 1, cf. Fig. 9).

rather than between the second and the third genes. Weak constitutive promoters within bacterial operons seem to be rather common. For example, two such promoters have been detected in the histidine operon of *S. typhimurium* (Atkins and Loper, 1970).

Although it poses some interesting questions, the existence of the *trp* internal promoter did not seem to necessitate any profound revision of the idea that genomes could be decomposed into unitary operons. The internal promoter is very weak, accounting for only a small fraction of the total potential rate of operon expression. Furthermore, it is not known to be regulated in any manner. Its presence throughout Enterobacteriaceae (Largen and Belser, 1973), among species whose tryptophan operons have diverged substantially in other respects (Denney and Yanofsky, 1972; Crawford, 1975), suggests that it probably has some adaptive value. But the deviations from coordinacy that it generates are minor, and for most purposes the entire operon can still be treated as a single unit.

2.1.2 *Genes Transcribed from More Than One Independently Regulated Promoter*

Whereas the *trp* internal promoter is constitutive, there are also cases in which the same gene is transcribed from more than one promoter, each of which is regulated. Two examples are provided by bacteriophage lambda. In both instances, the genes are ones that function during the lysogenic cycle of viral development.

2.1.2a Lambda Repressor Synthesis. Bacteria that are lysogenic for coliphage lambda perpetuate the viral DNA as a physiologically inactive segment of the bacterial chromosome. Most of the viral genes are unexpressed because of a repressor protein (product of the λ*cI* gene) that binds to two specific regions of the viral DNA and prevents transcription from promoters within these regions (Ptashne, 1971). The *cI* gene of the prophage is, of course, not repressed. Other prophage genes that are transcribed in the lysogenic cell are *rex* (adjacent to *cI* and expressed coordinately with it) and *int.* The function of the *rex* gene in λ biology is poorly understood. We will return to the *int* gene in Section 2.1.2b. All the other genes of λ are turned off in the lysogenic cell. Not all of them are transcribed from promoters directly controlled by the *cI*-coded repressor. However, the two operons that are repressed contain genes whose expression is needed to activate the rest of the genome. Thus, inactivation of repressor (as, for instance, by raising the growth temperature of bacteria lysogenic for a mutant prophage that forms a thermolabile repressor) triggers the initiation of a complete cycle of viral development. As shown in Fig. 2, the two promoters controlled directly by repressor, p_L and p_R, lie close to the *cI* gene itself, and originate transcription away from the *cI* gene on both sides.

The *cI* gene thus makes a repressor that, directly or indirectly, controls the activity of all the other viral genes. What controls the *cI* gene itself? The significance of this question is more fully appreciated if we recognize that there are two stages in λ biology in which repressor synthesis is needed: first, in lysogenic bacteria, where repressor is continually produced at a constant rate; and second, in nonlysogenic bacteria that have recently been infected by phage. A fraction of such infected cells lyses and produces phage, but another fraction survives and becomes lysogenic. In cells of the latter fraction, repressor is acting not to *perpetuate* a steady state but rather to *establish* it by shutting off viral genes that were active up to that time. It turns out that repressor synthesis at these two stages (maintenance and establishment of lysogeny, respectively) depends on *cI* transcription originating at two distinct promoters, each of which responds to different controls. In an established lysogen, *cI* message is initiated from a promoter (p_{RM}) very close to the *cI* gene (Figs. 1 and 3). The rate of transcription is enhanced by the *cI* protein itself. This stimulation of *cI* transcription, as well as the repression of rightward transcription, is attributable to repressor bound at two of the three clustered repressor binding sites, o_R1 and o_R2 (Fig. 3) (Ptashne *et al.*, 1976). At high concentrations, repressor inhibits *cI* transcription by binding to the leftmost site, o_R3, which has a lower affinity for repressor than does o_R1. Thus, repressor both activates and represses its own synthesis.

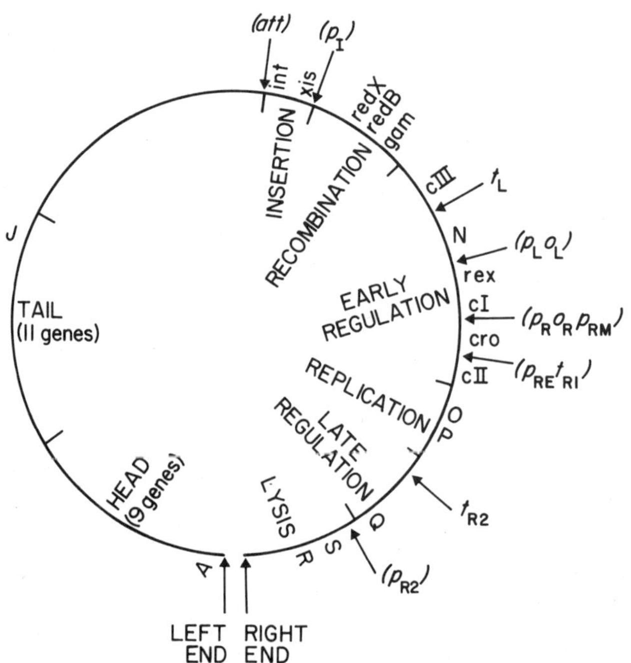

Figure 2. Genetic map of phage λ, drawn as an open circle. In the virion, the DNA is linear. In the infected cell, left and right ends are covalently joined. Inside the circle are indicated the functions of gene clusters in different segments of the genome. Outside are indicated certain genes. Only those relevant to the text are indicated individually. *int,* integrase; *xis,* excisionase; *redX,* exonuclease; *redB,* β protein (needed for recombination); *gam,* γ protein (inhibitor of host exonuclease V); *cII, cIII,* turn on of *cI; cI,* repressor; *N,* transcription extension; *rex,* exclusion of T4*rII; cro,* turn off of *cI* and early genes transcribed from p_L and p_R; *O,P,* replication; *Q,* turn on of late genes; *S,* destruction of cell membrane; *R,* endolysin (digestion of mucopolysaccharide). Arrows point to target sites in the DNA: *att,* insertion into bacterial chromosome; p_I, promoter for *int;* t_L, termination of leftward transcription (in the absence of gp*N*); p_L, promoter for leftward (anticlockwise) transcription of *N-int;* o_L, operator that binds repressor and controls p_L; p_R, promoter for rightward transcription of *cro-Q*; o_R, operator that controls p_R and p_{RM}; p_{RM}, promoter for transcription of *cI-rex* in lysogenic cells; p_{RE}, promoter for transcription of *cI-rex* during lysogenization; t_{R1}, t_{R2}, termination of rightward transcription in the absence of gp*N; p_{R2}*, promoter for transcription of *I-J*.

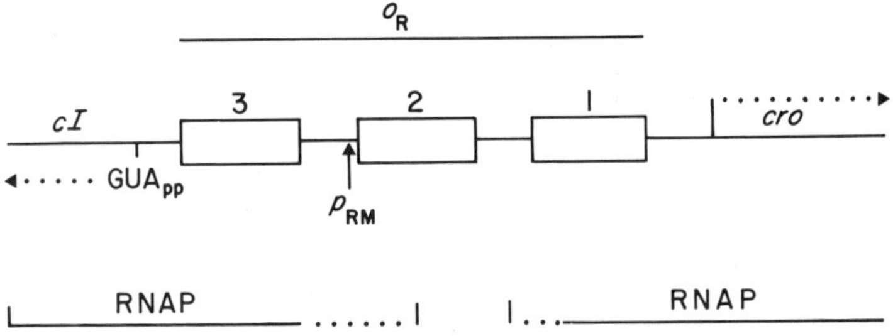

Figure 3. The λ rightward operator–promoter region $p_R o_R p_{RM}$ (modified from Ptashne *et al.*, 1976). The operator itself comprises three repressor-binding sites (1, 2, and 3), each 17 bases long, separated from each other by spacers five to seven bases long. RNAP, polymerase-binding site, as determined by polymerase protection of DNA from nuclease digestion (solid and dotted lines correspond to different nucleases). Within the left RNAP site lie one promoter mutation that has been sequenced (p_{RM}) and the origin of the *cI* message (GUA_{ppp}), 12 nucleotides to the left of o_R3.

Transcription from p_{RM}, while critical to the maintenance of lysogeny, probably plays no significant part in the decision of an infected cell to become lysogenic rather than to lyse. p_{RM} is a fairly weak promoter at best, the message initiated at this site lacks a strong ribosome-binding site preceding the *cI* gene (and is therefore translated inefficiently), and the activation of repressor synthesis by repressor poses a classical chicken–egg problem as to how the first *cI* transcript, or the first molecule of repressor, is to be formed.

This last problem is actually more quantitative than qualitative, because *cI* transcription, although enhanced by repressor, can take place at a low rate in the absence of repressor. At any rate, the phage avoids the dilemma by making the first molecules of repressor from a different *cI* message, a message that is initiated farther to the right on the phage chromosome and that responds to different controls. This message includes a strong ribosome-binding site proximal to the *cI* gene and is thus translated much more efficiently. Hence, it allows a transient rate of repressor synthesis that far exceeds the steady state value characteristic of p_{RM}. The high burst of repressor synthesis can direct the cell into the lysogenic state, even in the presence of gene products characteristic of the lytic cycle.

Repressor formation following infection strongly depends on the products of two genes (*cII* and *cIII*) that lie close to *cI*, on opposite sides. These genes belong to the two operons that are directly controlled by repressor. These operons also include the genes *N, O, P,* and *Q,* which play essential roles in lytic development.

Thus, following infection of a nonlysogenic cell, transcription initiated at the two early promoters, p_L and p_R, results in the formation of some proteins (gp*O*, gp*P*, gp*Q*)* that tend to push the cell toward lytic development and others (gp*cII*, gp*cIII*) that promote the buildup of repressor. If the latter are to win out, they must effect a higher concentration of repressor than that which is needed to maintain the repressed state in a lysogenic cell in which *O, P,* and *Q* are already shut off.

*Throughout our discussion the symbol "gp*X*" denotes "protein encoded by gene *X*." In some cases, as with the three listed here, chemical isolation of these products has not yet been achieved. They are probably proteins because of the occurrence of mutations suppressible by translational suppressors.

We know that the gp*cII*- and gp*cIII*-induced message starts farther to the right than p_{RM}: first, because a leftward message dependent on these gene products is hybridizable to DNA to the right of p_{RM} (Spiegelman *et al.,* 1972); and second, because mutations affecting the response of the DNA to these gene products lie to the right of gene *cro.* Mutations of three types have been characterized: (1) *cy* mutants have lost the ability to respond to gp*cII* and gp*cIII.* Their phenotype in single infection is like that of *cII* or *cIII* mutants. However, unlike *cII* or *cIII* mutants, they fail to complement *cI* mutants in mixed infection. (2) The *cin*-1 ("*cII*-independent") mutants make repressor even in the absence of gp*cII* (Wulff, 1976). (3) A mutation *cnc* reverses the effect of *cin*-1 (McDermit *et al.,* 1976). The disposition of these mutations with respect to one another is shown in Fig. 4.

The DNA site that is modified by *cy* and *cin* mutations has been called the promoter for repressor establishment (p_{RE}). It is not known whether gp*cII* and gp*cIII* initiate a message at this site (as the term "promoter" implies) or rather extend an 81-nucleotide message (*oop*) that otherwise terminates to the right of *cII* (Fig. 4) (Roberts, 1975; Honigman *et al.,* 1976). The antitermination mechanism is supported by the physical isolation of the extended message and also on the properties of some mutations (called *sar*) that map between genes *cII* and *O* (Honigman *et al.,* 1975). The exact location of these mutations with respect to p_o (Fig. 4) is unknown. The *sar* mutations were isolated as mutations that prevent plaque formation on a host that is forming gp*cro* constitutively. Whereas the molecular basis for that phenotype is unknown, the point of interest here is that a *sar* mutation, when coupled in cis to a *cy* mutation, reverses its phenotype; λ*cy*42 does not form repressor in response to gp*cII,* but λ*cy*42 *sar* does. The normal function of the p_{RE} region of Fig. 4 is unspecified in the antitermination model.

The general sequence of events during lysogenization is thus as follows: Transcripts initiated at p_L and p_R are translated to give proteins (gp*N,* gp*O,* gp*Q,* gp*cII,* and gp*cIII,* among others), some of which favor lysogeny and others of

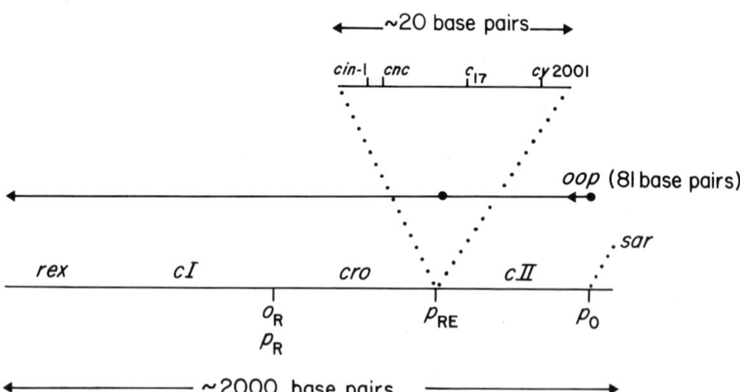

Figure 4. Fine structure of the p_{RE} promoter. *cin*-1, *cII*-independent production of *cI* message; *cnc,* suppressor of *cin*-1; c_{17}, mutation that creates a promoter for rightward transcription of *O* and *P*; *cy*2001, abolishes gp*cII*-promoted production of *cI* message; *oop,* 81-nucleotide message believed to serve as leader sequence for immunity message; p_o, initiation point for immunity message; *sar,* reverses effect of *cy* mutation. Length of the p_{RE} segment estimated, from recombination frequencies, to be 20 nucleotides. (Based on Wulff, 1976; Honigman *et al.,* 1975, 1976).

which favor lysis. If gp*cII* and gp*cIII* accumulate to a high enough concentration, they stimulate formation of *cI* message from p_{RE}. Translation of this message gives a high concentration of repressor. The repressor itself stops transcription from p_L and p_R, arresting further progress toward lysis and preventing production of more gp*cII* and gp*cIII*. It also initiates the maintenance mode of repressor synthesis by stimulating transcription from p_{RM}.

This circuitry may appear unduly complex. If the phage genome were so organized that lytic cycle genes were not transcribed in the first place, it would be unnecessary to build up a high concentration of repressor to shut them off. What the complexity of the actual process may accomplish is to allow the infected cell to make an informed decision as to whether to become lysogenic, rather than committing itself blindly (Dove, 1971; Echols, 1972). The fraction of infected cells that becomes lysogenic depends on the metabolic state of the cell, the ambient environment, and the population density of the virus in a manner that can be rationalized with general Darwinian principles. The sensitivity of viral development to these influences is perhaps monitored following infection while the phage temporizes by following a common pathway in all infected cells, before the master switch is finally flipped in one direction or the other.

The switch itself is complicated by additional features not yet mentioned here. Not only gp*cI*, gp*cII*, and gp*cIII*, but also gp*N*, gp*O*, gp*P*, and gp*cro* critically influence the decision. Some effects of gp*N* and gp*cro* are considered in Section 2.2.1.

At any rate, the details of phage biology are relevant here only insofar as they define the context in which to view the adaptive significance of the dual promoter arrangement. The important point is that genes *cI* and *rex* can be transcribed together from either of two promoters. Each promoter responds to specific controls and each plays an essential role in λ biology. There is the added feature that the *cI* gene is translated with very different efficiencies from the two transcripts on which it appears. In the lysogenic cell, transcription from p_{RM} occurs frequently, but each message is translated only a few times, assuring a fairly uniform distribution of repressor among cells, so that accidental loss of repression is relatively rare (Ptashne *et al.*, 1976). We may anticipate a similar strategy in biosynthetic operons for nutrilites required by the cell in very small amounts but required by every cell in adequate supply (Cleary *et al.*, 1972).

2.1.2b Lambda Integrase. There are two key events in lysogenization: establishment of repression and insertion of viral DNA into the host chromosome. Failure to establish repression results in cell death, usually accompanied by virus production. If repression succeeds but insertion fails, the viral genome generally remains intact in the infected cell, but it is unable to multiply and hence is diluted out in the growing population of cells. Insertion is effected by breaking viral and host DNA, each at a specific site, and then rejoining them as a single molecule. Insertion of lambda requires a protein, integrase, encoded by the viral gene *int*. The *int* gene lies in the leftward operon, adjacent to the insertion site *att* (Fig. 2). Transcription of *int* can occur either from the major leftward promoter p_L, or from a second promoter p_I that is slightly to the right of *int*. Knowledge about this second promoter is less complete than is that described earlier for p_{RM}. Nevertheless, current data fit a consistent simple picture that is readily rationalized with the biological role of integrase.

As with repressor, integrase is important at two different stages in the viral cycle: first, in the establishment of the lysogenic condition; and second, in the reversal of insertion (excision from the chromosome) that takes place when a lysogenic cell is derepressed. The latter process requires not only integrase but also the product of a second viral gene, *xis*. Excision should be regulated in such a way that it takes place rapidly following derepression, but that it seldom happens during growth of lysogenic bacteria. The difference in catalytic requirements for insertion and excision allows the phage to control not only the rate but also the direction of the process. Whereas an induced lysogen forms integrase and excisionase coordinately, an infected cell committed to lysogeny seems to use an independent pathway that produces integrase alone.

The major promoter for *int* transcription in both infected cells and induced lysogens is p_L. Transcription from p_L gives coordinated expression of *int* and *xis*. Several observations have demonstrated transcription from at least one other promoter: (1) A leftward promoter near the left end of the prophage is detectable because it causes a low rate of constitutive transcription originating in the prophage and extending into neighboring bacterial genes (best observed in unusual lysogens in which the prophage has become inserted within a bacterial operon rather than at its usual insertion site; see Fig. 5) (Shimada *et al.*, 1973). (2) Mutations *(int-c)* that were selected on the basis of an elevated rate of transcription of the adjacent bacterial gene, *trpB*, map within the *xis* gene and cause a high spontaneous rate of *int* transcription (Shimada and Campbell, 1974; Campbell *et al.*, 1977). (3) The kinetics of integrase production following infection differ from those of other proteins coded by genes transcribed from p_L. At early times, *int* is expressed coordinately with other genes of the leftward operon. At late times, the rate of synthesis of the other proteins decreases (because transcription from p_L is reduced by accumulation of gp*cro*), but integrase continues to be formed. This late integrase formation is not seen in cII^- or $cIII^-$ mutants, but does take place in a *cy* mutant in which gp*cII* and gp*cIII* cannot turn on *cI* transcription (Katzir *et al.*, 1976; Court *et al.*, 1977).

These observations are consistent with the hypothesis that the weak promoter activity observed in repressed lysogens represents the basal level of an inducible promoter located within *xis* that can be stimulated by gp*cII* and gp*cIII*. This hypothesis is not yet completely verified because the promoter observed in repressed lysogens, the locus of the *int-c* mutations, and the promoter affected by

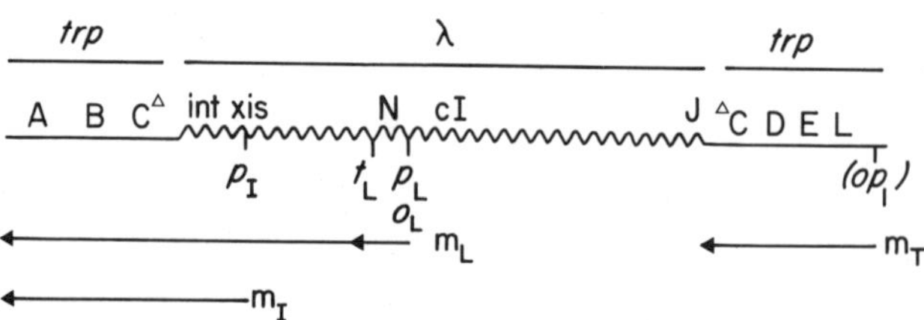

Figure 5. Transcription from λ prophage that has inserted into the *trpC* gene. Gene symbols as in Figs. 1 and 2. m_T, *trp* message (which would normally extend through *trpA*, but stops because it encounters a termination signal somewhere within the prophage); m_L, message from p_L, extended by gp*N* (observed only when repressor is inactivated or o_L has mutated to repressor insensitivity); m_I, message from the integrase promoter p_I (constitutively expressed at a low rate even in repressed cells).

gp*cII* and gp*cIII* have not yet been demonstrated to be identical. Some of the *int-c* mutations, in fact, do not alter the normal promoter but rather comprise insertions of the IS2 element, an element that introduces its own promoter within the *xis* gene. Others appear to be point mutations. These may lie in the normal promoter or in a regulatory site controlling it, but may instead represent new promoter sites (Campbell *et al.*, 1977).

There is also no direct evidence for a gp*cII* recognition site near *int.* Its existence is inferred from the fact that *cII* mutations influence integrase synthesis, whereas *cy* mutations do not. Reference to Fig. 4 shows that this observation is definitive only insofar as we may assume that neither the leftward transcript between *oop* and p_{RE} nor its translational products (if any) affect integrase.

The order of events during lysogenization then appears to be as follows: After infection, transcription is initiated leftward from p_L and rightward from p_R. The leftward message extends (following antitermination; Section 2.2.1) through genes *cIII, xis,* and *int,* and the rightward message extends through genes *cII, O, P,* and *Q.* As long as this transcription continues, integrase and excisionase are formed coordinately. Because excisionase appears to inhibit insertion, as well as to reverse it (Nash, 1975), little insertion takes place during this period. In those cells destined to become lysogenic, gp*cII* and gp*cIII* accumulate to high concentrations, thus inducing enough repressor to shut off p_L and p_R. The gp*cII* and gp*cIII* also act at p_I to induce rapid transcription of *int* (but not of the complete *xis* gene), thus specifically facilitating insertion. By using the same protein effectors to turn on both *cI* and *int,* the phage assures coordination between the two processes of repression and insertion.

The effect of gp*cII* and gp*cIII* on p_I, as on p_{RE}, may comprise extension of a preexisting message rather than initiation of a new one (Honigman *et al.*, 1977). In keeping with this view is the possibility that the point mutations in *xis* that produce high levels of integrase constitutively might remove a normal termination signal, obviating the need for gp*cII* and gp*cIII.* Thus, the *int* gene resembles *cI* in that it is transcribed from two different promoters at distinct times during viral development. Some of the results from *in vivo* experiments suggest that the message coming from p_I may differ from that coming from p_L in its efficiency of translation and perhaps even in the exact nature of its protein product, but direct evidence is not yet available (Court *et al.*, 1977).

Because integrase control is incompletely understood, some alternative hypotheses have been offered for the functional significance of p_I. One of these merits special attention because it concerns an important feature of the lysogenic condition and because I consider it highly unlikely on the basis of other experiments. According to this alternative hypothesis, the prophage in an established lysogen exists in a state of dynamic equilibrium between inserted and excised DNA, with the equilibrium far toward insertion. Constitutive integrase then serves to push the prophage back into the chromosome should it happen to excise. This idea is superficially plausible, especially as p_I was discovered by its low constitutive rate of expression in established lysogens. Just as the *cI* gene has both a promoter for repressor establishment and a promoter for repressor maintenance, one could imagine that p_L and p_I promote a change in state (excision) and perpetuation of the integrated state, respectively.

The reason for doubting this explanation is the evidence that prophage is not in a dynamic state, but rather is irreversibly inserted into the chromosome (Campbell, 1976). The most compelling argument comes from studies on bacterial strains with two equivalent sites for prophage insertion. From the rate of equilibration of prophage between an occupied and a vacant site, we can conclude that excision and reinsertion occur less than once per 100 cell generations, and probably much less. It is not known whether the low rate of integrase production in lysogens serves any purpose, or just represents an occasional accidental transcription of a promoter whose function is to make integrase following infection. However, it seems most unlikely that its purpose is to reinsert prophages that accidentally become excised, because in fact excision and reinsertion almost never occur.

2.1.3 P2 Integrase

P2 is a coliphage that is not detectably related to λ. Like λ, it lysogenizes by inserting its DNA into the host chromosome. The integrase gene of P2 has two properties of interest here: (1) As in λ, there is evidence that the P2 *int* gene can be transcribed from either of two promoters; and (2) the relationship of the two promoters to the overall biology of the phage is quite different from that in λ. As with λ, very little integrase activity is detectable (by the *in vivo* criteria to be discussed shortly) in an established P2 lysogen. The reason, however, seems to be quite different from that for λ: λ integrase is not formed abundantly in lysogens because p_L and p_I are controlled by repressor. Transcription from p_L is directly blocked by repressor, whereas transcription from p_I is blocked indirectly, because repressor prevents formation of *gpcII* and *gpcIII*. In P2, on the other hand, repression is less important than the inserted state of the prophage, apparently because insertion separates the integrase gene from its major promoter (Fig. 6). The evidence is as follows (Bertani, 1970): (1) When a P2 lysogen is derepressed by heating a *c ts* mutant (where the *c* gene makes a repressor analogous to the product of λ*cI*), the derepressed cells lyse, but very little active virus is formed. However, superinfection of the induced cell with an int^+ phage (but not an int^-

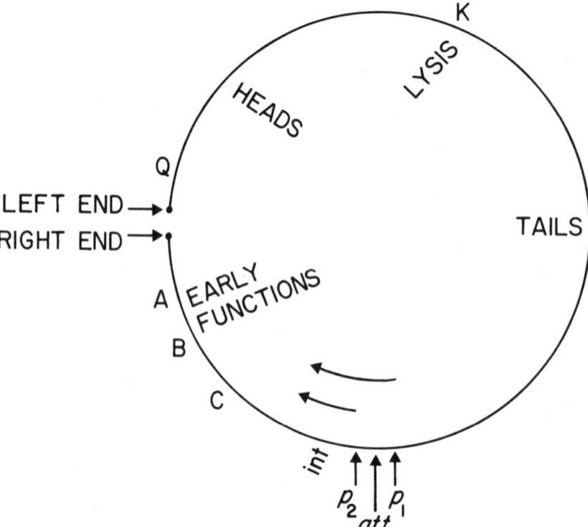

Figure 6. Abbreviated genetic map of phage P2, abstracted from Barrett *et al.* (1975), and including the location of the promoters for *int* transcription (p_1, p_2), as inferred from the studies of Bertani (1970, 1971).

one) increases the yield of phage of the prophage genotype. From the known properties of integrase, we may attribute the increase in phage production to integrase-catalyzed excision of viral DNA from the chromosome. This indicates that the *int* gene of the superinfecting phage is expressed more efficiently than is that of the prophage, although both are exposed to the same physiological stimuli. (2) Superinfection of a repressed lysogen with P2 int^+ results in substitution of the prophage by the superinfecting phage, at a frequency proportional to the multiplicity of superinfection. Thus, the *int* gene is expressed even when repressor is present. (3) Unlike λ, P2 does not form stable double lysogens in which the two prophages are adjacent to each other, in tandem. However, such double lysogens are found when one of the two prophages is int^- (Bertani, 1971). This makes sense because the middle *int* gene of the double lysogen should be connected to the promoter of the prophage carrying the other *int* gene. Whereas the first two lines of evidence might be misleading because of complications due to gene dosage or other quantitative problems, the third provides very strong support for a split operon model and seems to offer no simple alternative explanation.

Although *int* gene expression is reduced in an inserted prophage, it is not eliminated altogether. This is inferred from the fact that there is still some excision from an induced P2 int^+ lysogen, but not if the prophage is int^-. On these grounds a second, weaker promoter (p_2 in Fig. 6) is postulated. Neither of the two promoters (p_1 and p_2) is known to be regulated. The rate of integrase transcription is turned down in the lysogen because the gene becomes separated from its promoter in the insertion event itself.

In interpreting Bertani's result, it should be borne in mind that direct data on P2 *int* gene transcription are not available. Therefore, although it seems necessary to postulate a split operon, the location of the split could differ from that shown in Fig. 6. For example, the direction of transcription could be opposite to that shown in Fig. 6, with the integrase gene slightly overlapping the insertion site. The integrase formed from a prophage would then be abnormal at its C terminus and, by hypothesis, less active than is that produced from phage DNA. This would make the direction of transcription in P2 *int* the same as that of λ *int* (with respect to the attachment site and the repressor gene). The λ *int* gene lies very close to the insertion site and could overlap it, although *int* function is not strongly affected by insertion.*

2.1.4 Multiple Promoters in Bacteriophage T7

The genes of phage T7 can be divided into two categories on the basis of their modes of transcription: early genes, transcribed soon after infection by the host RNA polymerase; and late genes, whose transcription is accomplished by a phage-coded polymerase. The direction of transcription is the same for all genes of the phage. Characterization of the late messenger RNAs by molecular size and pattern of *in vitro* polypeptide synthesis shows that in several cases a single gene occurs on

*A "split gene" model for P2 integrase was considered by Bertani (1970), but seemed unlikely because the integrase activity of an inserted prophage, though low, is not zero. This argument is relevant only if the split gene produces a completely inactive product. As all known *int* mutations have been mapped on one side of the insertion site, the split must in any case lie near one terminus of the *int* gene, so formation of a partially active product seems quite plausible.

two or more distinct messenger species (Pachl and Young, 1976). Some of these species may arise by cleavage of longer transcripts. However, multiple promoters that generate overlapping messages are also observable *in vitro* (Golomb and Chamberlin, 1974), and some of the *in vitro* messages correspond with those made *in vivo*. Thus, a rather extensive use of multiple promoters seems likely for some of the T7 late genes. Insofar as these promoters all require a special polymerase, they can be considered to be regulated rather than constitutive. However, there is no evidence as yet for independent regulation of the different promoters.

2.1.5 *Multiple Promoters and Translational Control*

Multiple promoters clearly can afford the organism some advantages when the affected gene, such as *cI* or *int,* is expressed at more than one stage in the life cycle and in response to different stimuli. The *cI* case offers the additional feature of differential translatability on the two messages. For *int,* indirect evidence suggests a situation exactly opposite to that for *cI*. The extended message coming from p_L appears to supply integrase activity *in vivo* less efficiently than does the short message coming from p_I (Court *et al.,* 1977).

Whereas the integrase data may have alternative interpretations, they call attention to a possibility that we should expect to encounter in some multiple promoter systems. We should anticipate not only additional examples in which the initiation point of a short message separates a gene from its ribosome-binding site (as for *cI*), but also instances of the opposite type, in which a short message is translated more efficiently than is a long one. The simplest mechanism is one in which a transcript originates between a ribosome-binding site and an upstream nucleotide sequence that interferes with ribosome binding, such as the sequence observable *in vitro* in the RNA of phage MS2 (Fiers *et al.,* 1975).

2.2 *Antitermination and Antiattenuation*

In the preceding section I discussed some cases in which the same gene is transcribed from more than one promoter. Now we consider the opposite situation, in which two transcripts originating at a single promoter terminate at different sites. Our primary concern is not accidental premature termination, but termination points that are sufficiently specific to define discrete messenger species. Of greatest interest are those cases in which the position of termination is demonstrably under specific control. The existence of such control implies that those regulatory factors favoring larger messenger species must allow the RNA polymerase to continue past the DNA signals specifying termination at specific sites.

Two systems are discussed here: antitermination of the early genes of phage λ, and antiattenuation in the *trp* operon of *E. coli.* Whereas different names are applied in the two cases, many elements of mechanism are shared.

2.2.1 *Antitermination in Phage λ*

The discovery of antitermination in the early operons of phage λ is instructive as to genetic methodology. The critical initial observations were made by Luzzati

(1970) and represented an application of a general strategy for studying λ regulation that was used extensively by Thomas (1970).

Thomas' approach was to divide the genes of the λ lytic cycle into two categories: those that are directly shut off by repressor bound at p_L or p_R, and those that are indirectly controlled, requiring prior activity of some other genes. As discussed earlier, we expect that all these genes fall into one or the other category, because there is almost no phage-specific transcription in an established lysogen, but inactivation of repressor initiates a complete lytic cycle. Thomas noted that the critical situation that distinguishes direct from indirect control of a given gene is one in which a phage genome finds itself in a cell that contains not only repressor but also the products of all those phage genes whose expression might normally precede that of the gene under test. Experimentally, such a situation is created by infecting a lysogenic cell with a λ-related phage that differs from λ in two respects: First, it should be insensitive to λ repressor; and second, the gene under study should be mutant. The first criterion was met by using a phage (λ *imm*434) in which the *cI* gene and its flanking operator sites had been replaced, by genetic recombination, with the corresponding segment of a λ-related phage, 434, whose operator sites are insensitive to λ repressor. One can then ask whether or not a given gene in the prophage is turned on by gene products of the superinfecting phage.

When Luzzati (1970) applied this test to the *redX* gene of the leftward operon, the results clearly indicated direct control (Table I). This seemed incongruous with previous experiments that had shown equally clearly that formation of the *redX* product, exonuclease, required prior expression of gene *N* (see Table I, line 2). Since *N* is located between *redX* and the leftward promoter p_L (see Fig. 2), this implied that *N* is on the same transcript as *redX* and upstream from it. However, if the two genes are copied onto the same transcript, it is not obvious how the product of one can be needed for transcription of the other.

Further investigation of transcription (and consequent protein synthesis) of genes between *int* and *N* led to the following picture (Roberts, 1970; Szybalski *et al.*, 1970; Franklin, 1974; Adhya *et al.*, 1974): When *E. coli* RNA polymerase transcribes λ DNA either *in vitro* or in a fully infected cell, leftward transcription originating at p_L terminates at a site t_L that lies between genes *N* and *cIII*. Termination at t_L requires the bacterial protein rho. In an infected cell, once gp*N*

TABLE I. Induction of Prophage Exonuclease Gene by Superinfection

Strain[a]	Exonuclease increase after 40 min[b]	
	Repressed	Derepressed
W3350 (λ*cIts* N^+)	0	3.4
W3350 (λ*cIts* N^-)	0	0
W3350 (λ*cIts* N^-) superinfected with λ Δ*redX imm*434	0	1.8

Lysogenic bacteria were placed at 33°C (repressed) or 42°C (derepressed).

[a]W3350 is a standard $gal^-F^-Su^-$ derivative of *E. coli* K-12. Prophages carry the *cIts*857 mutation. N^-:*N am*7 *am*53. Cells in the third row were superinfected at time zero with λ*bio*11 *imm*434, a phage in which the exonuclease gene *redX* has been substituted with host DNA. Exonuclease is in arbitrary units above background activity in uninfected cells.

[b]Calculated from Luzzati (1970, Fig. 2) as arbitrary units above background. The background level of an uninduced lysogen is about 1.0 on this scale.

has been synthesized, it modifies the nature of transcription so that it becomes resistant to rho-mediated termination and hence continues beyond t_L.

Before Luzzati's experiment, the simplest explanation for the absence of exonuclease in N^- mutants was that gp*N* caused the initiation of transcription at some site between *N* and *redX*. However, if that were so, cells of a λ lysogen infected with λ*redX*$^-$*N*$^+$*imm*434 should produce exonuclease. They do not (Table I, line 3). Exonuclease synthesis requires the presence of an unrepressed p_L and an intact *redX* gene not only in the same cell but on the same DNA molecule. This formal genetic result is compatible with at least two biochemical mechanisms: (1) extension of a previous transcript, as implied here, or (2) direct activation of reinitiation near t_L by transcripts that originate at p_L and terminate at t_L. Rigorous demonstration of the former alternative requires isolation of the extended message to prove that its 5′ end comes from p_L. This goal has been frustrated by the extensive processing of leftward message that seems to follow *in vivo* transcription. However, analysis of transcripts produced in an RNase III$^-$ host indicates that, when processing is reduced, a single transcript going from p_L to *att* can be recovered intact (H. M. Lozeron and P. J. Anevski, quoted in Lozeron *et al.*, 1976).

Szybalski *et al.* (1970) pointed out that the genes from *int* through *N* are not properly termed an operon, which implies unitary control under all conditions. These authors proposed the name *scripton* for such a group of genes that are all transcribed from a common promoter, although controlled termination at specific points within the scripton generates different transcripts. Whereas the term scripton has not gained wide currency, it perhaps represents the first explicit recognition of structural complexity as an important feature of natural operons and their regulation.

Subsequent studies of λ*N* control have revealed some aspects of the mechanism, and have also extended our knowledge of the structure of λ early operons. Transcription extension by gp*N* occurs in both the leftward and the rightward operons. In an N^- mutant, rightward transcription stops at a site (t_{R1}) between genes *cro* and *cII*, though termination at t_{R1} is not completely effective; even in the absence of gp*N*, some transcripts continue through t_{R1} to a second termination site (t_{R2}) situated between genes *P* and *Q*. Mutations (*byp* and *nin*) that lie between *P* and *Q* apparently alter t_{R2} so that transcription can proceed even farther, into gene *Q* (Court and Sato, 1969; Butler and Echols, 1970). The leftward operon likewise seems to contain some stop signals downstream from t_L (L. Heffernan and A. Campbell, unpublished data).

Some insight into the mechanism of gp*N* action has come from studies in which bacterial genes are transcribed from p_L or p_R. For genes such as *gal* and *bio,* which lie close to the λ prophage, transcription from a λ promoter takes place naturally when lysogenic bacteria are derepressed (Buttin *et al.,* 1960; Yarmolinsky and Wiesmeyer, 1960). Such transcription can also be engineered through the construction of specialized transducing phages. Franklin (1974) has investigated a series of *trp* phages made from a λ-φ80 hybrid phage, in which the *trp* operon is connected to p_L at various fusion points (Fig. 7).

A critical experiment is to examine the effect of transcription from p_L through a mutation that is ordinarily polar on distal genes. Amber mutations in the *trpE* gene, for example, can reduce expression of downstream genes by a factor of 4 to 5. This polarity is seen when the *trp* operon is transcribed from its normal promoter (p_1 in Fig. 1). It is ascribable to rho-dependent termination since it can be suppressed by mutations in the structural gene for rho (Richardson *et al.*, 1975; Korn and Yanofsky, 1976). However, when *trp* DNA containing the same mutation is transcribed from p_L of λ in the presence of gp*N*, little, if any, polarity is observed (Table II). This observation, confirmed with other polar mutations and corroborated by similar results with the *gal* operon (Table II), indicates that the ability of gp*N* to extend transcription beyond t_L, t_{R1}, and t_{R2} does not derive from some special property of those termination sites, but rather that gp*N* renders transcription resistant to rho-dependent termination. However, the function of gp*N* is not simply to inhibit or inactivate rho. Rather, its antiterminating activity appears to be limited to transcription originating from certain promoters, such as p_L and p_R. The specificity of gp*N* for particular operons is indicated by the results of mixed infection with phage λ and the related phage, 21. Phage 21 is another in the series of natural phage isolates, such as phage 434, that will grow in the same cell as λ and can recombine genetically with λ, but that form repressors specific to the operator sites on their own DNA. Whereas the gp*N* made by phage 434 seems to be equivalent to that made by λ, the gp*N* made by phage 21 is different: In a cell

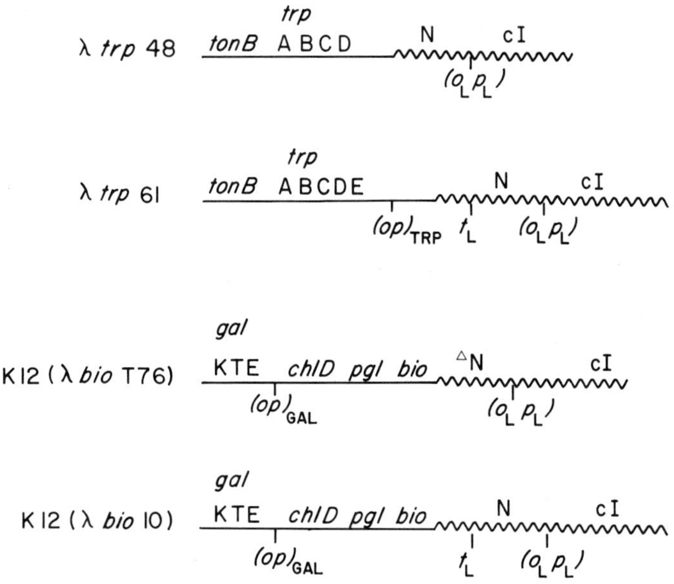

Figure 7. Diagrammatic structure of representative genetic fusions useful in studying promoter dependence of polarity relief by gp*N*. Two systems are shown that have in common the fact that a bacterial operon can be transcribed from p_L of λ with varying amounts of intervening DNA. Top two lines: λ*trp*48 and λ*trp*61 are two transducing phages that arose by abnormal excision of a λ-ϕ80 hybrid prophage inserted at the ϕ80 attachment site, near *trp*. Bottom two lines: Bacteria lysogenized at the λ site (near *gal*) by λ*bio* transducing phages, in which some of the viral DNA between *att* and p_L has been replaced by *bio* DNA. All strains have a complete *N* gene except K12 (λ*bio*T76), in which half of gene *N* has been deleted. In the bottom three lines, bacterial genes can be transcribed either from their normal promoter or from p_L; in λ*trp*48, the *trp* promoter is deleted, so that *trpABCD* can be transcribed only from p_L. *galK*, galactokinase; *galT*, galactose-1-phosphate uridyltransferase; *galE*, uridine diphosphogalactose-4-epimerase; *chlD*, resistance to chlorate; *pgl*, 6-phosphogluconolactonase; *tonB*, resistance to bacteriophage T1.

TABLE II. Relief from Polarity Caused by Upstream Nonsense Mutations in Two Bacterial Operons

Gene	Promoter	Polar mutation	λ*N*	Relative activity of *trpA* or *galK* product[a]
trpA[b]	p_{TRP}	None	−	(100)
		*trpD*9778	−	10
trpA[c]	p_{L}	None	+	(100)
			−	35
		*trpD*9778	+	81
			−	10
galK[d]	p_{GAL}	None	−	(100)
		*galE*95	−	7
galK[e]	p_{L}	None	+	(100)
		*galE*95	+	75

[a]Normalized to 100 as maximum value observed for the promoter in question. *trp* data from Franklin (1974), *gal* data from Adhya *et al.* (1974).
[b]Assayed in wild type or mutant bacteria starved for tryptophan.
[c]Assayed in bacteria infected with λ*trp*48 *cro*⁻ phage (Fig. 7), either wild type or mutant for *N* and *trpD*.
[d]Assayed in wild type or mutant bacteria lysogenic for λ*cIts*857, grown at 33°C, *gal* transcription induced with fucose.
[e]Same bacteria as in footnote d, grown at 42°C in the absence of fucose.

infected with both λN^- and 21N^+, the gp*N* of 21 does not affect the analogous sites of phage λ. Thus, a mixed infection between, say, λN^- and 21Q^- is unproductive. The gp*Q* of λ is not formed because rightward transcription terminates before reaching gene *Q*; the only *Q* gene transcribed is the mutant *Q* gene of phage 21 (Herskowitz and Signer, 1970*a*).

This experiment does not tell us whether the antiterminating activity of gp*N* is specific to the p_{R} and p_{L} promoters, or to some signal in the DNA between these sites and the first natural termination point. This question has been resolved to some extent through the study of deletion mutants and of transducing phages in which heterologous DNA is fused to p_{L}. For example, there are λ *trp* phages in which t_{L} and much of gene *N* itself have been deleted, so that genes of the *trp* operon are now fused directly to p_{L}, with no known natural terminator signals in between. In such phages, gp*N* remains able to overcome polarity of *trp* mutations, indicating that its specificity is directed either toward p_{L} itself or at least toward some site very early in the operon (N. Franklin, personal communication).* Corroborative evidence comes from the isolation of mutations within p_{R} that reduce the responsiveness of rightward transcription to gp*N* (Friedman *et al.*, 1976). Thus, gp*N* appears to recognize some base sequences near p_{L} and p_{R} and to alter the properties of RNA polymerase initiating transcription at those points in such a way that the polymerase no longer recognizes signals for rho-dependent

*The most definitive evidence comes from studies on a deletion mutation (*ninL*32) that leaves less than 200 nucleotides to the left of the *imm*434 substitution, yet still responds to gp*N*. The *trp*48 substitution (Fig. 7, Table II) eliminates t_{L}, but leaves the *N* gene intact (N. Franklin, personal communication). An equivalent conclusion can be drawn from the transcriptional readthrough of *gal* in bacteria lysogenic for the λ*bioT*76 prophage (Fig. 7), in which the left half of the λ*N* gene has been replaced with bacterial DNA (Adhya *et al.*, 1974).

termination. It is possible, for example, that gp*N* chemically modifies the polymerase, replaces one of its subunits, or precedes it down the DNA molecule. There is no reason to believe that the termination signals themselves are recognized by gp*N*, rather than just by rho. Rho itself is clearly implicated: Termination at t_L and t_{R1} *in vitro* requires rho (Roberts, 1970) and mutations in the host rho gene partially relieve the deleterious consequences of N^- mutations on λ growth (Korn and Yanofsky, 1976).

Whereas the λ-21 experiments show that gp*N* is operon specific, they do not prove that p_L and p_R are unique in responding to gp*N*. In fact, gp*N* influences the synthesis of the sigma subunit of RNA polymerase (Nakamura and Yura, 1976). Likewise, infection by λN^+ phage may overcome the effects of a rho-dependent attenuator in the *gal* operon (Petit-Koskas and Contesse, 1976). (See also Section 2.2.3.) Both observations might mean that certain bacterial promoters mimic p_L and p_R in their reaction to gp*N*. In both cases, alternative explanations are also plausible. The effect on sigma could also result from perturbation of the normal regulation of RNA polymerase; gp*N* may act by modifying RNA polymerase, and polymerase may influence the synthesis of its component polypeptides. Modification of even a small amount of polymerase could conceivably induce hyperproduction of sigma. The effect on *gal* has not been pinpointed to gp*N* itself. It might equally well be some indirect consequence of *N* action. As gp*N* is needed to turn on all the other genes of the infectious cycle, and as *gal* polarity is affected by metabolic changes, such as alterations in the cAMP level, this is a very serious concern. In any event, the effect of λ infection is at least somewhat operon specific, since effects such as those just described do not occur with the *lac* operon (Contesse *et al.,* 1973).

One subject on which divergent opinions are currently held is the relation between gp*N* and another λ regulatory protein, gp*cro,* which has several effects on λ development, one of which is to inhibit transcription from p_L and p_R. Thus, the effect of gp*cro* on expression of genes downstream from t_L and t_R is opposite to that of gp*N;* gp*N* turns these genes on, whereas gp*cro* progressively slows them down as it accumulates during the infectious cycle. Under conditions (such as infection of a *recA*$^-$ host) in which distal genes of the leftward operon are essential to viral development, N^- *nin* mutants do not form plaques, but N^- *cro*$^-$ *nin* phages do (Court and Campbell, 1972). Similarly, enhanced rightward transcription distal to t_{R1} and t_{R3} is the presumed basis for the ability of certain N^- *cro* phages to form plaques even without the *nin* mutation (Galland *et al.,* 1975).*

The point at issue is whether gp*cro* is a direct antagonist of gp*N* or whether it acts at some other level: Is the effect of gp*cro* solely on the rate of initiation at p_L and p_R or does gp*cro* also (or instead) change the probability that messages initiated at these promoters will terminate at rho-dependent sites, including t_L, t_{R1}, and t_{R2}? If the latter possibility is true, does gp*cro* favor termination only in the presence of gp*N,* or also in the absence of gp*N*? It seems unlikely that gp*N* and gp*cro* are simple antagonists of each other, because antitermination by gp*N* is observed even when a *cro*$^-$ phage is used for infection. Likewise, the small amount of residual transcription seen in an N^- phage is greater if the phage is *cro*$^-$ than if

*The failure of a simple λN^- to plate seems to be due mainly to a deficiency in gp*Q* production, which is alleviated by deletion of t_{R2} in the *nin* mutant. The degree of readthrough of $t_{R\,1}$, even in the absence of *N*, is sufficient to allow growth by λN^- *nin*.

it is *cro*$^+$. Thus, each of these proteins exerts an effect in the absence of the other, a finding that rules out the possibility that the sole function of either is to inhibit or reverse the action of the other. This conclusion is limited to the extent that an "N^-" phage operationally means an unsuppressed double amber mutant, in which a small amount of gp*N* function might conceivably be present.

Various measurements (Franklin, 1971; Adhya *et al.*, 1977) indicate that bacterial genes directly fused to p_L, with elimination of t_L, are subject to negative regulation by gp*cro*. This suggests a direct effect of gp*cro* either on initiation or on some unrecognized termination signal very early in the operon. Support for the latter possibility comes from data, such as those presented in Table II, that show that *trp* transcription is reduced to 35% in the absence of gp*N*, even though all known termination sites have been removed. Whether or not gp*cro* in fact influences termination, the suggestion that it does so raises the general possibility that termination may be controlled by counterbalancing regulatory proteins that respectively promote or inhibit termination within specific operons.

One consequence of the ability of gp*N* to extend transcription is that derepression of a lysogen induces some transcription of genes, such as *gal*, that are near the prophage (Table II). The *gal* induction seems to represent extension of transcription from p_L. As with the *trp* operon in λ*trp* phages, the polar effect of suppressible mutations in *gal* disappears when *gal* is being transcribed from p_L under gp*N* control. Thus, gp*N* must allow readthrough not only of the mutant site of t_{L1}, but also of all other termination sites that lie between λ and *gal*. The effect of gp*N* on RNA polymerase is to convert it into a juggernaut that continues along the DNA and ignores most of the normal stop signals that would otherwise arrest it (Adhya *et al.*, 1974).

Extension of short messages into longer ones is a process that may pervade λ regulation rather than be confined to the early operons. We have already (Section 2.1.2) mentioned that gp*cII* and gp*cIII* may promote repressor and integrase synthesis by extension of preexisting messages. The other major operon of λ, which encodes all the structural proteins of the virion as well as the enzymes that dissolve the cell envelope, is turned on by gp*Q* at a point between genes *Q* and *S* (Herskowitz and Signer, 1970*b*). A small message, 200 nucleotides long, formed *in vitro* by the action of RNA polymerase on λ DNA, also comes from between *Q* and *S*, at such a place that late transcription could result from extension of this message by gp*Q*. Extension by gp*Q* must differ from that by gp*N* at least by the fact that *in vitro* termination of the 200-nucleotide transcript does not require rho (Roberts, 1975).

What benefit does the phage derive from the device of making short messages initially and extended messages later on? A definitive answer cannot yet be given. Perhaps the best educated guess is that the natural role of gp*N* is to reinforce the stability of the switching mechanism that governs the transition between lytic and lysogenic development (Franklin, 1971). The fact that lytic development requires initiation both leftward and rightward, followed by gp*N* synthesis, protects the lysogenic cell from the hazards that might attend a momentary lapse in repressor synthesis. In an infected cell, the phage gene arrangement provides for initial synthesis of gp*N* and gp*cro*, regulatory proteins that to some extent counterbalance each other, prior to transcription of genes whose products function directly in viral development. This may increase the opportunity for controlled monitoring of conditions prior to commitment to one pathway or the other.

Is antitermination unique to viruses, whose lifestyle entails a major reorganization of overall regulatory patterns, or is specific antitermination exploited as a control device in cellular operons as well? It is now clear that several operons contain internal sites at which transcription frequently terminates, and that the proportion of transcripts that extends beyond these sites is regulated. Because termination at such sites is seldom 100% effective *in vivo,* they have been termed attenuators rather than termination sites, but it is not certain that any rigorous distinction is possible.

One attenuator that has been the object of intensive analysis lies in the *trp* operon of *Escherichia coli,* upstream from all five of the structural genes (Fig. 1). Sequence analysis shows that some 162 nucleotides at the 5′ end of the *trp* message (leader region) precede the portion of the message transcribed from the first structural gene. Jackson and Yanofsky (1973) observed that, among deletions internal to the *trp* operon, some deletions that entered the leader region had an apparent "promoter" effect on transcription of distal genes, as though deletion removed from the DNA a site that had impeded downstream transcription. Further work showed that in $trpR^-$ mutants growing in excess tryptophan, about 90% of the transcripts initiated at the *trp* promoter terminate at this (attenuator) site (Bertrand *et al.,* 1976). Transcripts can terminate at any one of the five uracil residues located 137 to 141 nucleotides from the 5′ end of the message (Bertrand *et al.,* 1977).

The frequency of termination at the attenuator is regulated by the degree of charging of tryptophanyl tRNA: It is reduced by tryptophan starvation and in mutants (*trpS*) deficient in tryptophanyl tRNA synthetase. Attenuation is not influenced by starvation for another amino acid, isoleucine (Bertrand and Yanofsky, 1976). Whether the effect of tryptophan starvation is specific to the *trp* attenuator is less certain, because tryptophan starvation can reduce termination at other stop signals, such as those exposed by polar amber mutations (Bertrand *et al.,* 1977). However, operon specificity is indicated by results with another biosynthetic operon, the histidine operon of *Salmonella typhimurium,* in which an attenuator similarly situated upstream from the known structural genes responds to charged *his* tRNA (Kasai, 1974; Artz and Broach, 1975). Attenuation in *trp* is reduced in certain rho^- mutants and hence is presumed to require the rho protein (Korn and Yanofsky, 1976). It does not require the *trp* repressor, and has customarily been studied in $trpR^-$ mutants, from which repressor is absent.

It is not known whether *trp* attenuation or its reversal requires any protein analogues to gp*N*. There is sufficient DNA between the *trp* promoter and the attenuator to permit coding of a short peptide, and the appropriate signals to initiate and terminate translation are present. Formation of such a peptide *in vivo* and any regulatory effect that it may have are as yet undemonstrated.

As in the case of the λ operons controlled by gp*N*, the introduction of a specific control point prior to the structural genes of the *trp* operon increases its responsiveness to ambient conditions. Any more profound assessment of the significance of this control system probably will await definite knowledge as to whether the attenuator proximal DNA codes for a functional product—either the RNA itself or a polypeptide encoded by it.

If the attenuated message does have functional consequences, the inclusion of the *trp* attenuator in the present chapter is clearly appropriate. If, on the other hand, the short message transcribed from the beginning of an operon is an inactive by-product of premature termination, then the attenuator effectively becomes an element of a complex operator rather than of a complex operon, even though the termination itself is an important component of normal regulation.

In the original formulation of the operon concept, the operator was implicitly defined as a site near the beginning of an operon that controls the response of messenger synthesis to specific regulatory molecules. Nothing in the theory or in the genetic operations available to Jacob and Monod (1961) distinguished controlled inhibition of initiation from controlled termination proximal to the first structural gene. Subsequent molecular studies have demonstrated that repression of *trp* message by the *trpR* product and of *lac* message by the *lacI* product, among others, entails actual interference by repressor with polymerase binding. We can, if we like, generalize this finding by redefining an operator as a site at which repressor binds and interferes with polymerase binding and/or initiation of transcription. Such a restricted definition, which seems to be currently accepted, emphasizes distinctions based on biochemical mechanisms with a concomitant deemphasis of functional equivalence. It would be fruitless to try to change the natural evolution of terminology. But we may ultimately have use for a language in which the primary distinction is made between those processes that influence the rate of transcription of the first structural gene of an operon, and those that affect the coordination of genes within the operon—a language that relegates to a secondary position the details of how these transcriptional effects are realized.

In any event, regulation of tryptophan biosynthesis clearly involves regulation at two sites, operator and attenuator, that function by different mechanisms and respond to somewhat different controls. Controlled attenuation in *trp* has some mechanistic similarity to controlled termination in the early operons of λ.

2.2.3 *Recognition Sites in Antitermination*

One of the more interesting features of antitermination in λ is the distance that separates the site of recognition for gp*N* (close to promoters p_{L} and p_{R}) from the sites of gp*N* action (t_{L}, t_{R1}, t_{R2}, and stop signals in bacterial operons that have been fused to λ promoters). It is this fact, deduced from genetic results, that focuses our thinking on polymerase modification as the most plausible mechanism of gp*N* action.

Available information on recognition sites for other regulators of internal termination is limited. The properties of certain promoter mutations in the *his* operon of *S. typhimurium* indicate that recognition of some element critical to the terminating influence of charged *his* tRNA takes place at a site upstream from the actual site of termination (Kasai, 1974). In the *E. coli trp* operon, the recognition site for *trp* tRNA could in principle lie anywhere upstream from the termination site itself.

Intriguing effects of 3′,5′-cyclic AMP on coordination of enzyme synthesis within the *gal* operon of *E. coli* (and perhaps in the *lac* operon as well) have been reported (Petit-Koskas and Contesse, 1976; Contesse *et al.,* 1970). As shown in Table III, added cAMP elevates the concentration of the transferase (product of *galT,* the second gene of the operon, Fig. 7) relative to epimerase (product of *galE,*

TABLE III. Differential Effect of Cyclic AMP Genes of E. coli

Added cAMP	Enzyme concentrations[a]	
	Epimerase	Transferase
+	57.9	0.14
−	79.5	0.48

[a]Cells were grown under conditions of mild catabolite repression (30°C, tryptone broth) with added fucose to induce the *gal* operon. Data fron Petit-Koskas and Contesse (1976).

the first gene). No information on mechanism is available. A rho-dependent stop signal between these two genes is observable *in vitro* (de Crombrugghe *et al.,* 1973). Data on messenger degradation *in vivo* suggest that a site for initiation of endonucleolytic attack lies in this position as well (Achord and Kennell, 1974). Despite the natural polarity that might be anticipated from these two findings, the *galE* and *galK* proteins are in fact synthesized in equimolar amounts under conditions in which cAMP is presumably in excess (Wilson and Hogness, 1969).

The only known recognition site for the CRP protein–cyclic AMP complex in the *gal* operon lies close to the promoter (Musso *et al.,* 1977). It is not known whether this site plays any role in regulating the differential expression of *galE* and *galT* or, in fact, whether such differential expression involves any direct interaction between the cAMP–CRP complex and *gal* DNA at all. The message whose initiation is promoted by cAMP starts at a site six bases upstream from the cAMP-dependent message. As the two messages are physically different, they could differ in the stability or rate of translation of their *galE* transcripts.*

There is no reason to imagine that all antitermination proceeds by a common mechanism. However, if polymerase modification mechanisms are at all widespread, then polymerase action *in vivo* is substantially more sophisticated than current concepts might indicate. It seems useful to examine all effectors influencing termination for similarities to the gp*N* system. A complementary approach is to examine regulatory proteins, such as gp*cro* and cAMP–CRP, that are known to be specific for sites early in the operon, to see whether they influence termination farther downstream. While encouraging explorations along these lines, I must stress that currently there is no unambiguous evidence to indicate that either gp*cro* or cAMP–CRP influences termination directly.

2.3 *Divergent Transcription*

2.3.1 *The Biotin Cluster of Escherichia coli*

In a typical *E. coli* operon, such as *lac* or *gal,* transcription initiated at a promoter site proceeds in a fixed direction. In some gene clusters, such as the

*The author is grateful to S. Adhya for calling his attention to this possibility.

arginine biosynthetic genes *argECBH* and the biotin genes *bioABFCD,* although all genes of the cluster are under coordinate control, transcription starts at a position internal to the cluster and proceeds both leftward and rightward from it (Elseviers *et al.,* 1972; Guha *et al.,* 1971). Thus, the leftmost gene of the biotin cluster *(bioA)* is transcribed leftward, whereas the other four are transcribed rightward. Guha *et al.* (1971) observed that leftward message from the biotin genes hybridized to DNA of the *bioA* gene, whereas rightward message hybridized to *bioBFCD.* These findings were soon corroborated by genetic studies of polar mutations and deletions penetrating the *bio* cluster that showed that the *bioBFCD* genes behaved as a single, biotin-repressible operon, even when most of the *bioA* gene had been removed by deletion (Cleary *et al.,* 1972). Leftward transcription of *bioA* was also confirmed by isolation of a deletion mutation that fuses the distal part of the galactose operon to the proximal segment of the *bioA* gene (Ketner and Campbell, 1974) (see Fig. 8).

None of these results demonstrates rigorously that the biotin cluster comprises a single, structurally complex operon. From these facts alone, it would be equally possible to imagine that the leftward and rightward transcripts represent two adjacent but distinct operons, each with its own operator and promoter, that respond to the same controls either because their operators are similar or because one of the operons is regulated by a product of the other.

The distinction between two separate operons and one complex operon is not only instructive as to genetic methodology, but is also of real concern because there are precedents for closely linked, metabolically related operons that interact

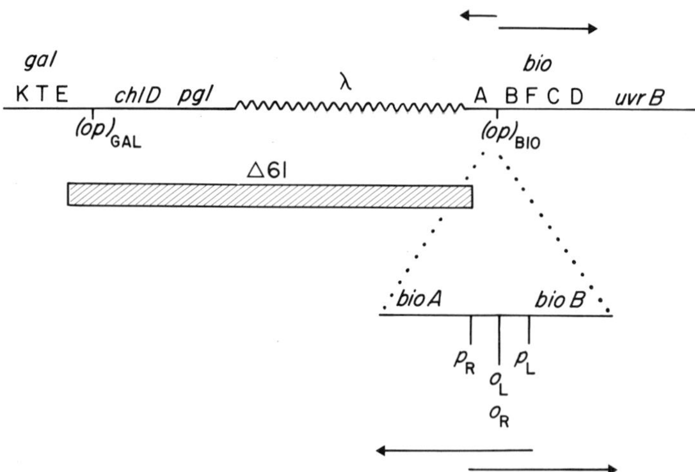

Figure 8. The biotin operon and neighboring genes on the *E. coli* chromosome. *bioA,* 7,8-diaminopelargonic acid aminotransferase; *bioB,* synthesis of biotin from dethiobiotin; *bioF,* 7-keto-8-aminopelargonic acid synthetase; *bioC,* early step in biotin biosynthesis; *bioD,* dethiobiotin synthetase. Other gene symbols as in Fig. 7. Δ61 is a deletion that fuses the *galKT* genes onto the *bioA* promoter p_L. The possible structure of the operator–promoter region is shown on an expanded scale below. p_L, promoter for leftward transcription; p_R, promoter for rightward transcription; o_L, operator for leftward transcription; o_R, operator for rightward transcription.

by virtue of the regulatory effects of their diffusible products (Smith and Magasanik, 1971). In evaluating the *bio* experiments to be described next, the reader should bear in mind what analogous results would be obtained if the same methodology were applied to the two early operons of λ by an investigator who had discovered the existence of genes *redX, O, P,* and *Q* but who was still unaware of genes *N* and *cI.* Not only are *redX* and *OPQ* transcribed in opposite directions, but also their transcription is more or less coordinate (because both are controlled by the same repressor). Furthermore, *OPQ* expression depends on gp*N,* so that promoter mutations reducing leftward transcription indirectly curtail rightward transcription as well. Nevertheless, these are two separate operons, not one operon with divergent transcription.

The evidence that transcription of all the biotin genes emanates from a common control region rests on the study of operator and promoter mutations in the biotin cluster (Ketner and Campbell, 1975). The mutations were isolated in strains carrying the Δ61 deletion (Fig. 8) that fuses the *galK* gene onto the biotin promoter. Use of this fusion strain permits selection of mutants with altered regulation of leftward transcription on the basis of their Gal phenotype, without selecting for any alteration in expression of the known genes of the rightward transcript. In fact, all mutations selected on this basis were found to affect both leftward *and* rightward transcription. These included both operator constitutive mutations and polar insertions of an IS element (probably IS*1*) (T. Otsuka and J. Abelson, personal communication).

The fact that a single insertion shuts off transcription in both directions suggests either that the two transcripts derive from a common promoter or that they overlap each other physically (Fig. 8). Alternatively, the finding could be explained by a regulatory effect of one operon on the other. For example, some gene of the rightward operon might make a product needed to turn on leftward transcription, just as gp*N* extends transcription in the rightward operon of λ. The critical test in this case is to construct a strain that is genetically diploid for the biotin cluster, in which the two clusters differ from each other in an identifiable manner (Table IV). Diploids can then be made that are heterozygous for regulatory mutations of either the operator or the promoter type. In principle the "one operon" and "two operon" models are also distinguishable by studying transcription or translation *in vitro,* using DNA from a regulatory mutant and all other components from a wild type strain. Since no definitive answer is yet available at that level, the *in vivo* tests will be discussed here in some detail.

Table IV lists the expected phenotypic properties of the two possible diploids heterozygous for operator or promoter mutants, under the one-operon hypothesis and the two-operon hypothesis, respectively. The two-operon hypothesis requires that the leftward and rightward transcripts have separate promoters and that the two transcripts not overlap. To explain how a single mutation may shut off both transcripts, it must be assumed that some product of one transcript (e.g., the rightward one) is needed to turn on the other, just as gp*N* of λ is needed to turn on most of the early rightward transcript of λ. The mutation would then have a direct effect only on the rightward promoter, its leftward effect being an indirect consequence of shutting off production of the regulatory protein. We would then expect that, in any strain diploid for *bio,* the two leftward transcripts (one from each *bio* operon) would behave in an identical manner. Either both would be

TABLE IV. Experimental Distinction between One-Operon and Two-Operon Models for the Biotin Cluster[a]

Diploid structure:

*gal*KTE$^{\Delta}$(Δ61)$^{\Delta}$*bioA*(op)^{I}B λ *bioA*(op)^{II}BFCD[b]

Regulatory mutations		Biotin concentration used to score phenotype	Predicted phenotype[c]		Observed phenotypes
(op)^{I}	(op)^{II}		One-operon	Two-operon	
Wild type	Polar insertion	Low[d]	Gal^{S}Bio^{-}	Gal^{S}Bio^{-}	Gal^{S}Bio^{-} and
Polar insertion	Wild type	Low[d]	Gal^{R}Bio^{+}	Gal^{S}Bio^{+}	Gal^{R}Bio^{+}
Wild type	Constitutive	High[e]	Gal^{R}Bio^{+}	Gal^{R}Bio^{+}	Gal^{R}Bio^{+} and
Constitutive	Wild type	High[e]	Gal^{S}Bio^{+}	Gal^{R}Bio^{+}	Gal^{S}Bio^{+}

[a] The one-operon model assumes that operator and promoter mutations directly affect both transcripts. In the two-operon model, the effect on leftward transcription is the indirect result of a regulatory effect on leftward transcription of some gene product of the *bioB*-distal portion of the rightward transcript. Based on Ketner and Campbell (1975).

[b] Constructed by lysogenization of Δ61 strain carrying appropriate regulatory mutation by λ*bio*1.

[c] Gal^{S}, sensitive to galactose, therefore transcribing *galK;* Bio^{+}, able to synthesize biotin, therefore transcribing all five *bio* genes.

[d] At high biotin, all strains are Gal^{R}.

[e] At low biotin, all strains are expected to be Gal^{S}. However, Gal^{R}Bio^{+} strains are in fact Gal^{R} also at low concentration of added biotin, presumably because of endogenous biotin generated constitutively.

turned off or both would be turned on. This is a necessary consequence of the fact that both lie in the same cell and are exposed to the same concentration of regulatory protein.* [Since in the particular partial diploid under discussion here one of the two rightward transcripts is incomplete, the exact predictions are different depending on whether the regulatory gene is distal or proximal to *bioB*. Independent evidence (J. Kotval and A. Campbell, unpublished data) indicates that neither polar mutations in *bioB* nor deletions removing much of *bioB* and DNA to the right of it change the rate of leftward transcription. Therefore in Table IV we list only the predictions for a regulatory gene upstream from *bioB*.]

The one-operon hypothesis, on the other hand, assumes either a single promoter or a physical overlap of the two transcripts, so that a single mutation can directly affect both transcriptions. In a diploid, we would then expect that each leftward transcript is controlled by its own operator–promoter, which should also control the adjacent rightward transcript.

As shown in the last column of Table IV, the observed phenotypes are those expected on the basis of the one-operon model. The most critical phenotype is Bio^{+}Gal^{R}, a strain that is able to synthesize biotin (able to express *bioA*), yet is resistant to galactose (does not express *galK*). The existence of strains with this phenotype indicates that the two leftward transcripts in a cell that is diploid for *bio* are under different control, and therefore that the mutation (in this case a polar insertion) affects transcription directly. The possibility that the results shown in Table IV arise from some technical artifact inherent to *bio* genetics is reduced by the existence of similar cis-specific regulatory mutations in two other divergent

*However, a regulatory protein could affect the two promoters differentially if its concentration within the cell were nonuniform. See Section 4.

gene clusters of *E. coli: argECBH* (Jacoby, 1972) and *malEFKlamB* (Hofnung, 1974).

Although these data strongly favor a one-operon model, there are some results indicating that control of the two biotin operon transcripts is not completely coordinate. Hybridization studies show that the biotin analog homobiotin, when added to a culture of the wild type strain, represses leftward but not rightward transcription (Vrancic and Guha, 1973). On the other hand, the same analog failed to repress the production of either the *bioA* protein or the *bioD* protein when added to a culture of a *bioB* mutant that had been grown in the presence of biotin (Eisenberg, 1975). Thus, it is possible that homobiotin exerts its repressive effect indirectly, perhaps by influencing some enzyme(s) of the biotin pathway. At any rate, the fact that there are conditions under which leftward transcription is differentially repressed indicates that, in addition to the common region controlling transcription in both directions, there are other control points specific for the individual transcripts. The possibility for superimposing specific controls onto common controls could in fact be one reason for the existence of bidirectional transcription.

Common control of the two transcripts is also indicated by the properties of regulatory mutations unlinked to the *bio* cluster (Campbell *et al.*, 1972; Eisenberg *et al.*, 1975; Pai, 1972). Investigation of spontaneous *ts* mutants selected for their ability to transcribe *galK* in the presence of biotin (in the Δ 61strain) permits grouping such mutations into two loci, *birA* and *birB* (R. Chang, G. Ketner, D. Barker, and A. Campbell, unpublished data). The *birA* mutations map near *rpo* at 88 min (Bachmann *et al.*, 1976) and influence both repression and uptake to various degrees. Extracts from *birA* mutants are deficient in catalyzing covalent binding of biotin to protein in the presence of ATP (holoenzyme synthetase). The *birB* mutations map near *metE* (84 min) and may be simple transport mutations like those selected for resistance to the analog α-dehydrobiotin (Eisenberg *et al.*, 1975). A third group of mutants (called *birC*) differs from wild type in at least two loci, one of which is indistinguishable by genetic mapping from *birB*, and the other of which lies 1–2 min away from it. Mutations in all three groups affect leftward and rightward transcription equally.

One model for the biotin operon compatible with the genetic results would have leftward and rightward transcripts overlap, with operator elements lying in the common region (Fig. 8). However, *in vitro* transcription studies indicate that the transcripts do not overlap, but start within a few bases of each other (T. Otsuka and J. Abelson, personal communication). Hence, what actually overlap must be the promoters rather than the transcripts.

2.3.2 Bidirectional Initiation in Nucleic Acid Biosynthesis

It not necessary to conceive of bidirectional transcription as inherently more complicated than unidirectional transcription. Comparing replication with transcription, one finds that in living cells replication is frequently bidirectional and occasionally unidirectional, whereas the proportions seem to be reversed for transcription. It is appropriate to wonder whether divergent transcription is truly unidirectional in the same sense that replication is, that is, whether the same event triggers transcription in both directions at once.

2.4 Overlapping Transcription

In contrast to the *bio* story are situations in which a physical overlap of two oppositely oriented transcripts is well documented, although the two transcripts in question are subject to diverse controls and are formed at different times. We have already met one example in that segment of λ DNA including the *cro* and *cII* genes (Section 2.1.2a). Following infection, the *cro* gene is transcribed rightward from p_R as part of the *cro cII O P Q* transcript. During lysogenization, under the influence of gp*cII* and gp*cIII*, transcription proceeds leftward through this region, perhaps as an extension of the 81-nucleotide *oop* transcript. The *oop* transcription itself is stimulated by gp*O* and gp*P* (Hayes and Szybalski, 1973) and may serve both as a primer for DNA replication and as a leader for *cI* message (Honigman *et al.*, 1976). At any rate, it is clear that the same DNA segment is transcribed under different circumstances in different directions, but the functional significance of the overlap itself is unknown.

3 Evolution of Complexity

In the preceding pages I have documented some cases of operon complexity and mentioned some ideas concerning their possible biological functions. Most of these ideas suffer from their extreme generality. It is clear, for instance, that divergent transcription can allow an additional degree of freedom by superimposing specific and common controls; but that statement in itself gives no clue as to why *bio* and *arg* are divergent whereas *trp* and *his* are not.

Since its original formulation by Jacob and Monod, the operon concept has been refined by many data on molecular mechanisms and has been expanded with variations on the original theme. We now recognize positive as well as negative regulation, translational as well as transcriptional control, divergent transcription and transcription extension as well as unidirectionality and coordinacy. Whereas each of these developments has expanded our horizons and has provided useful tools for exploring common mechanisms, one may fairly ask at this stage to what extent biologists may have been collectively engaged in a kind of molecular stamp-collecting, with each variation deriving its main interest from mere novelty.

One problem in this area is that experimentation has in some ways considerably outdistanced theory. Documentation of still other types of complexity may be less helpful than the formulation of a theory that attempts to explain in a more than anecdotal or *ad hoc* fashion why a given operon serving a particular function has the structure that it does, a theory that correlates operon structure with physiological or ecological information on operon function. The limited success that classical biologists have had in formulating a theory of natural selection with predictive as well as explanatory value does not encourage great optimism in this direction; on the other hand, it does not justify our neglecting such an attempt, either. I will not try to develop a theory fully here, but will briefly consider some aspects that it may embrace. One approach involves consideration of the function of a given operon in the context of the whole organism, and the way in which the products of genes within the operon interact to perform that function. This generates theories that stress the maximization of adaptive value toward which organisms and their component genetic elements are obviously driven by natural selection (see Chapter 3).

Another relevant consideration is that the properties of any biological system reflect its history as well as its present function. Biological systems typically represent local rather than absolute maxima in adaptive value, and the particular local maximum selected in a given system depends strongly on previous as well as present selective forces. Campbell *et al.* (1977) have discussed the dual promoters of the integrase system in this light. Their discussion presupposes some hypotheses (Campbell, 1972) about evolution of viruses in general, and λ in particular. One striking feature of λ is the extreme degree to which genes are clustered according to function (Fig. 2). Thus, the whole λ genome can be divided into segments concerned with insertion, recombination, regulation, replication, lysis, and virion structure; the last segment can be further divided into a head structure portion and a tail structure portion, and even within those subgroupings genes whose products interact tend to neighbor each other. Some degree of functional clustering is observed in the genomes of all organisms, but clustering is most apparent in certain viruses and prokaryotes. A possible adaptive value of this clustered arrangement is to facilitate coordinate expression of related functions. A more historically oriented explanation (emphasized here) is that the lambda virus evolved by the progressive association of genetic segments derived from diverse sources, and that each of its segments already served a function similar to its present one before it became part of the complex.

For example, one of the many possible pathways for λ evolution would involve first a nonconjugal, noninserting, self-replicating plasmid, perhaps derived from a bacterial replication origin that had broken away from the rest of the chromosome; and independently, the development in the host chromosome of other genes whose products catalyze autolysis and still others concerned with DNA packaging. The accidental translocation of these chromosomal genes onto the plasmid would then allow emergence of a complex that could both transfer its DNA into other cells and replicate independently once it arrived there. This plasmid would then constitute a virus, whose later acquisition of preexisting regulation genes and insertion genes would result in a temperate phage such as λ.

The exact order by which the various elements of the viral genome become associated with each other is unimportant. Indeed, to start with one component and to add others *seriatim,* as described here, perhaps helps to foster or perpetuate a misconception as to the significance of the whole process. In my experience, most prokaryote geneticists will accept a sequence such as the one just described as plausible or even likely, but fail to accept its full implications. It therefore may help to state explicitly what I consider the right and the wrong lessons the scheme is intended to illustrate.

The wrong lesson is that we can trace the viral lineage by starting with a simple object such as a plasmid, and have it evolve into something more complex by progressive acquisition of functions. That viewpoint deemphasizes the precise mechanism by which functions are acquired (mutation of DNA within the plasmid, or incorporation of genes from other sources). The right lesson is that complex genetic elements have developed by associations of simpler ones, that each element of the complex has an equally ancient history, and that if we try to trace the pedigree of the complex element backward in time, it is entirely arbitrary which of the various components we follow. Thus, if a primitive virus indeed evolved by association of replication genes with lysis genes, it is a matter of taste whether we describe the situation by saying that a plasmid acquired the ability to lyse cells, or that some lysis genes acquired the ability to replicate autonomously. A well-informed observer might summarize the same facts in either manner, depending on where his own interests are focused.

This point deserves emphasis because it is in fact far easier to imagine the acquisition of new functions by incorporation of preexisting units than by evolving new ones from scratch. The genes within a given pathway, such as those coding for different components of the viral capsid, may well have arisen by duplication and differentiation within a single DNA segment; but genes in very distinct pathways, such as the capsid genes and the replication genes, are much more likely to be derived from separate sources and to have been modified to fit into a new context. One consequence of this viewpoint is that the various segments of λ may have closer phylogenetic relationships to functionally similar segments found elsewhere in prokaryotes than they do to each other. For example, the segment of λ DNA specifically concerned with insertion and excision is about 1300 base pairs long (L. W. Enquist and R. A. Weisberg, personal communication), and thus similar in size to other elements (the insertion sequences IS*2* and IS*3*) that can insert at various sites of the *E. coli* genome either by themselves or in association with drug resistance determinants (Bukhari *et al.*, 1977). We would then visualize that at some stage in the history of λ, an element similar to one of these had become covalently linked to determinants of other viral functions.* The important point here is that, if an ancestor that resembled IS*2* or IS*3* did become part of λ at some time, then that sequence very likely already contained an insertion site, genes for insertion and excision, and a promoter from which those genes were transcribed. We might then consider the p_{I} promoter of present-day λ as a lineal descendant of this primordial promoter for insertion genes, whose properties have been modified in order to integrate the insertion functions into the overall scheme of λ biology. Some of the more bizarre features of this promoter, such as its location within a structural gene rather than between genes, may prove more readily understandable if viewed in terms of ancestral as well as present function.

The data most relevant to such evolutionary hypotheses comprise comparative studies of the regulatory structures of analogous elements in different species or strains. Information on λ and related coliphages has been tabulated in various places (Hershey and Dove, 1971; Campbell, 1976). These data indicate a high degree of similarity in overall organization in the face of considerable variation in the specificity of regulatory components. Repressor, gp*N*, and gp*Q* all vary sufficiently to be ineffective on the heterologous target sites of related phages; yet the locations of these genes and of their target sites are constant. The specificity of integrase and its target site *att* also varies among λ-related phages. The existence and properties of p_{I} are not documented for phages other than λ itself. In any case, the available information suggests that much of the regulatory circuitry of λ is of sufficient adaptive value to have been preserved at least during recent

*Whether IS*2* and IS*3* themselves are ancestral to or descended from temperate phages is unspecified and irrelevant here.

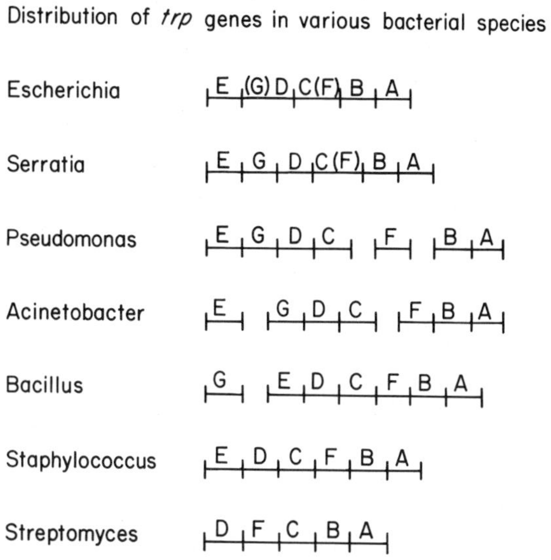

Distribution of *trp* genes in various bacterial species

Escherichia	E (G) D C(F) B A
Serratia	E G D C(F) B A
Pseudomonas	E G D C F B A
Acinetobacter	E G D C F B A
Bacillus	G E D C F B A
Staphylococcus	E D C F B A
Streptomyces	D F C B A

Figure 9. Distribution of tryptophan biosynthetic genes into one or more clusters in various bacterial species. *trpEBA,* as in Fig. 1. The *trpD* gene of *E. coli* (Fig. 1) combines two functions that are separate in some species: phosphoribosyltransferase (*D*) and glutamine amidotransferase subunit of anthranilate synthetase (*G*). Similarly, the *trpC* of *E. coli* combines InGPS (*C*) and PRAI (*F*) functions. (Redrawn from Crawford, 1975.)

evolution. For the ideas discussed here, comparative data on promoter locations in other λ-related phages and on functionally similar elements such as IS2 should be very helpful. Whether phages λ and P2 are homologous in this respect depends on which of the models for split operon control described in Section 2.1.3 turns out to be correct.

We may also ask whether the *trp* internal promoter originated within the operon or, instead, traces its ancestry back to a time when the distal portion of the cluster was an independent operon with its own promoter and later became fused to the proximal portion. The arrangement of the *trp* biosynthetic genes has been examined in various prokaryotic and eukaryotic species (Fig. 9) (Crawford, 1975). The data indicate that translocation of components of the *trp* operon must have occurred early in evolution and must have survived in different taxonomic groups to generate the diversity currently observed. Translocation can either split a single operon into two or fuse two operons into one. Whereas the data do not permit construction of a unique family tree for the *trp* operons of different species, it is noteworthy that the protein product of the *trpD* gene of *E. coli* combines two biochemical functions that are catalyzed by separate proteins in other species. Thus, it is very likely that a fusion event took place within the *trpD* gene at some time during the evolution of the *E. coli trp* operon to connect the two proteins into one polypeptide chain.*

Unfortunately, the internal promoter lies downstream from the polypeptide fusion point, so that is cannot simply mark the location of a promoter that once governed the phosphoribosyltransferase gene. If it is descended from the promoter of a once independent operon that became connected to the *trp* promoter, an additional fusion event (beyond that responsible for the known polypeptide fusion) must be postulated. Perhaps the fact most directly relevant here is that the internal promoter seems to be ubiquitous throughout the Enterobacteriaceae

*Such multifunctional proteins, presumably the consequence of gene fusion, may be fairly common in eukaryotes (Calvo and Frank, 1971). With respect to evolution and natural selection, gene fusion may be considered a minor variation on gene clustering.

(Largen and Belser, 1973). Such conservation suggests that this promoter occurred early and continues to be of adaptive value.

4 Methodological Implications for Studies of Genome Organization

4.1 Use of the Cis/Trans Test

I mentioned at the outset an intention to stress the methodology of prokaryotic genetics. Throughout the examples discussed here, the primary tool used to probe operon structure has been some variation of the cis/trans test, a test that reveals that the effects of certain regulatory mutations are restricted to genes on the same molecule of DNA (rather than just in the same cell) as the mutation itself. Such regulatory mutations modify or delete sites of interaction either with specific components (repressor, gp*N*, CRP) or with nonspecific elements of the regulated process (RNA polymerase, rho factor). Consistent application of this methodology has revealed the variations on operon structure described here.

None of these studies addresses the question (approachable in principle by similar techniques) of whether even the complex units discussed here function completely independently or whether they instead exert any cis-specific influence on each other. Thirty years ago enzymologists were occasionally criticized for treating the living cell as a mere "bag of enzymes" rather than as an integrated whole. Modern molecular biologists might likewise be charged with having conceived of genomes as "bags of operons"—that is, insofar as operons behave as independent units that relate to each other only through the effects of their diffusible products, the disposition of operons within the genome should have no regulatory consequences. To a first approximation, that prediction seems to be fulfilled in the material (mostly prokaryotic) that has been examined in depth.

4.2 Cis-Acting Proteins

Notable exceptions to this rule are provided by several examples involving "cis-acting" proteins. In these examples, a mutation know to exert its effect by altering the structure of a regulatory protein influences the expression of the genes it regulates only when the regulated genes and the mutations are in the cis configuration, that is, when they lie nearby on the same DNA molecule, not just when they are present in the same cell. Cis specificity can be virtually complete, as with the *A* proteins of phages ϕX174 (Francke and Ray, 1972) and P2 (Lindahl, 1970), or partial, as with p*Q* of λ (Echols *et al.*, 1976). Cis specificity can be attributed to limited diffusibility of active protein *in vivo*, so that the adjacent target site is exposed to a higher local concentration of regulatory protein than a distant one would be, a possibility that seems especially plausible for regulatory proteins that find their targets by unidimensional diffusion along the DNA (Riggs *et al.*, 1970). Echols *et al.* (1976) have suggested specifically that failure of a protein to reach distant target sites may be due to competitive weak binding to sites of lower affinity along the DNA.

The existence of cis-specific proteins might seem to complicate the operon concept itself, inasmuch as the operational distinction between mutations of the operator and promoter on the one hand and of regulatory genes on the other assumed strictly cis and trans behavior, respectively. We can circumvent the problem by accepting as a matter of definition that a group of genes that behave as a unit in regulation constitute an operon, regardless of the basis of its unitary behavior. In phage P2, for example, we would then consider the positive regulator gene *A* and the regulated gene *B* (Fig. 6) to lie in a single, albeit complex operon. This tactic allows us to retain a crisp operational distinction between one-operon and two-operon models (Table IV). The crispness disappears for proteins, such as p*Q* of λ, that are clearly active in trans but work preferentially in cis. Such quantitative differences between cis and trans can easily go undetected in qualitative tests, and may be quite common. Formally, the *cro-Q* genes and the *S-J* genes of λ behave not quite as a single complex operon nor yet as two separate operons, but as something in between. We may anticipate that more refined analysis will reveal other examples of quantitative differences between cis and trans configurations, generated by various mechanisms. Convenience and taste will dictate whether a gene block such as *cro-J* is treated as one complex operon or as two separate operons with weak cis-specific interactions. The important point is that the same methodology that has been used effectively to probe the structure of regulatory units of the dimensions we have discussed here may also be useful in exploring higher levels of chromosome organization.

Acknowledgments

I thank Naomi Franklin and Charles Yanofsky for many helpful and informative comments on the manuscript. Research results from my laboratory were supported by Grant AI08573 from the National Institutes of Health.

References

Achord, D., and Kennell, D., 1974, Metabolism of messenger RNA from the *gal* operon of *Escherichia coli*, *J. Mol. Biol.* **90**:581.

Adhya, S., Gottesman, M., and Court, D., 1977, Independence of *N* and tof functions of bacteriophage λ, *J. Mol. Biol.* **112**:657.

Adhya, S., Gottesman, M., and de Crombrugghe, B., 1974, Release of polarity in *Escherichia coli* by gene *N* of phage λ: Termination and antitermination of transcription, *Proc. Natl. Acad. Sci. U.S.A.* **71**:2534.

Artz, S. W., and Broach, J. R., 1975, Histidine regulation in *Salmonella typhimurium;* An activator–attenuator model of gene regulation, *Proc. Natl. Acad. Sci. U.S.A.* **72**:3453.

Atkins, J. F., and Loper, J. C., 1970, Transcription initiation in the histidine operon of *Salmonella typhimurium*, *Proc. Natl. Acad. Sci. U.S.A.* **65**:925.

Bachman, B. J., Low, K. B., and Taylor, A. C., 1976, Recalibrated linkage map of *Escherichia coli* K-12, *Bacteriol. Rev.* **40**:116.

Barrell, B. G., Air, G. M., and Hutchinson, C. A., III, 1976, Overlapping genes in bacteriophage φX174, *Nature* **264**:34.

Barrett, K., Barclay, S., Calendar, T., Lindqvist, B., and Six, E., 1975, Reciprocal trans activation in a two-chromosome system, in: *Mechanisms of Virus Disease,* ICN-UCLA Symposia on Molecular Biology, Vol. 1 (W. S. Robinson and C. F. Fox, eds.), pp. 385–401, W. A. Benjamin, Menlo Park, California.

Berget, S. M., Moore, C., and Sharp, P. A., 1976, Spliced segments at the 5′ terminus of adenovirus late mRNA, *Proc. Natl. Acad. Sci., U.S.A.* **74**:317.

Bertani, L. E., 1970, Split-operon control of a prophage gene, *Proc. Natl. Acad. Sci., U.S.A.* **65**:331.

Bertani, L. E., 1971, Stabilization of P2 tandem double lysogens by *int* mutations in the prophage, *Virology* **46**:426.

Bertrand, K., and Yanofsky, C., 1976, Regulation of transcription termination in the leader region of the tryptophan operon of *Escherichia coli* involves tryptophan or its metabolic product, *J. Mol. Biol.* **103**:339.

Bertrand, D., Squires, C., and Yanofsky, C., 1976, Transcription termination *in vivo* in the leader region of the tryptophan operon of *Escherichia coli, J. Mol. Biol.* **103**:319.

Bertrand, K., Korn, L. J., Lee, F., and Yanofsky, C., 1977, The attenuator of the tryptophan operon of *Escherichia coli.* a. Heterogeneous 3'-OH-termini *in vivo.* b. Deletion mapping of functions, *J. Mol. Biol.* **117**:227.

Bukhari, A. I., Adhya, S., and Shapiro, J., eds., 1977, *DNA Insertion Elements, Plasmids and Episomes,* Cold Spring Harbor Lab., Cold Spring Harbor, New York.

Butler, B., and Echols, H., 1970, Regulation of bacteriophage λ development by gene *N:* Properties of a mutation that bypasses *N* control of late protein synthesis, *Virology* **40**:212.

Buttin, G., Jacob, F., and Monod, J., 1960, Synthèse consititutive de galactokinase consécutive au développement du bactériophage λ chez *Escherichia coli* K12, *C. R. Acad. Sci. Paris* **250**:2471.

Calvo, J. M., and Fink, G. R., 1971, Regulation of biosynthetic pathways in bacteria and fungi, *Annu. Rev. Biochem.* **40**:943.

Campbell, A., 1972, Episomes in evolution, in: *Evolution of Genetic Systems* (H. H. Smith, ed.), Brookhaven Symp. Biol., Vol. 23, pp. 546–562, Gordon and Breach, London.

Campbell, A., 1976, Significance of constitutive integrase synthesis, *Proc. Natl. Acad. Sci. U.S.A.* **73**:887.

Campbell, A., del Campillo-Campbell, A., and Chang, R., 1972, A mutant of *Escherichia coli* that requires high concentrations of biotin, *Proc. Natl. Acad. Sci. U.S.A.* **69**:676.

Campbell, A., Heffernan, L., Hu, S.-L., and Szybalski, W., 1977, The integrase promoter of bacteriophage λ, in: *DNA Insertion Elements, Plasmids and Episomes* (A. I. Bukhari, S. Adhya, and J. Shapiro, eds.), Cold Spring Harbor Lab., Cold Spring Harbor, New York.

Cleary, P. P., Campbell, A., and Chang, R., 1972, Location of promoter and operator sites in the biotin gene cluster of *Escherichia coli, Proc. Natl. Acad. Sci. U.S.A.* **69**:2219.

Contesse, G., Crépin, M., and Gros, F., 1970, Transcription of the lactose operon in *E. coli,* in: *The Lactose Operon* (J. R. Beckwith and D. Zipser, eds.), pp. 111–142, Cold Spring Harbor Lab., Cold Spring Harbor, New York.

Contesse, G., Bracone-Mahlie, A., and Gros, F., 1973, Interaction between λ or ϕ80 prophage and *gal* operon expression, *J. Mol. Biol.* **73**:527.

Court, D., and Campbell, A., 1972, Gene regulation in *N* mutants of phage lambda, *J. Virol.* **9**:938.

Court, D., and Sato, K., 1969, Studies of novel transducing variants of lambda. Dispensability of genes *N* and *Q, Virology* **39**:348.

Court, D., Adhya, S., Nash, H., and Enquist, L., 1977, The phage λ integration protein (Int) is subject to control by the *cII* and *cIII* gene products, in: *DNA Insertion Elements, Plasmids and Episomes* (A. I. Bukhari, S. Adhya, and J. Shapiro, eds.), Cold Spring Harbor Lab., Cold Spring Harbor, New York.

Crawford, I. P., 1975, Gene rearrangements in the evolution of the tryptophan pathway, *Bacteriol. Rev.* **39**:87.

de Crombrugghe, B., Adhya, S., Gottesman, M., and Pastan, I., 1973, Effect of rho on transcription of bacterial operons, *Nature New Biol.* **241**:260.

Denney, R. M., and Yanofsky, C., 1972, Detection of tryptophan messenger RNA in several bacterial species and examination of the properties of heterologous DNA-RNA hybrids, *J. Mol. Biol.* **64**:319.

Dickson, R. C., Abelson, J., Barnes, W. M., and Reznikoff, W. S., 1975, Genetic regulation: The *lac* control region, *Science* **187**:27.

Dove, W., 1971, Biological inferences, in: *The Bacteriophage Lambda* (A. D. Hershey, ed.), pp. 297–312, Cold Spring Harbor Lab., Cold Spring Harbor, New York.

Echols, H., 1972, Developmental pathways for the temperate phage: Lysis vs. lysogeny, *Annu. Rev. Genet.* **6**:257.

Echols, H., Court, D., and Green, L., 1976, On the nature of cis-acting regulatory proteins and genetic organization in bacteriophage: The example of gene *Q* of bacteriophage λ, *Genetics* **83**:5.

Eisenberg, M. A., 1975, Mode of action of α-dehydrobiotin, a biotin analogue, *J. Bacteriol.* **123**:248.

Eisenberg, M. A., Mee, B., Prakash, O., and Eisenberg, M. R., 1975, Properties of α-dehydrobiotin-resistant mutants of *Escherichia coli* K-12, *J. Bacteriol.* **122**:66.

Elseviers, D., Cunin, R., Glansdorff, N., Baumberg, S., and Ashcraft E., 1972, Control regions within the *argECBH* gene cluster of *Escherichia coli* K-12, *Mol. Gen. Genet.* **117**:349.

Fiers, W., Contreras, R., Duerinck, F., Haegeman, G., Merregaert, J., Minjou, W., Raeymaeker, A., Remant, E., Volchaert, G., and Yselbaert, M., 1975, Bacteriophage MS2 RNA: The complete nucleotide sequence of a viral genome. Secondary structure and biological functions, *Abstract Third International Congress on Virology*, p. 17.

Francke, B., and Ray, D. S., 1972, Cis-limited action of the gene-*A* product of bacteriophage ϕX174 and the essential bacterial site, *Proc. Natl. Acad. Sci. U.S.A.* **69**:475.

Franklin, N., 1971, The *N* operon of lambda: Extent and regulation as observed in fusions to the tryptophan operon of *Escherichia coli*, in: *The Bacteriophage Lambda* (A. D. Hershey, ed.), pp. 621–638, Cold Spring Harbor Lab., Cold Spring Harbor, New York.

Franklin, N. C., 1974, Altered reading of genetic signals fused to the *N* operon of bacteriophage λ: Genetic evidence for modification of polymerase by the protein product of the *N* gene, *J. Mol. Biol.* **89**:33.

Friedman, D. I., Jolly, C. A., Mural, R. J., Ponce-Campos, R., and Baumann, M. F., 1976, Growth of λ variants with added or altered promoters in *N*-limiting bacterial mutants. Evidence that an *N*-recognition site lies in the p_R promoter, *Virology* **71**:61.

Galland, P., Cortini, R., and Calef, E., 1975, Control of gene expression in bacteriophage λ: Suppression of *N* mutants by mutations of the antirepressor, *Mol. Gen. Genet.* **142**:155.

Golomb, M., and Chamberlin, M., 1974, A preliminary map of the major transcription units read by T7 RNA polymerase on the T7 and T3 bacteriophage chromosomes, *Proc. Natl. Acad. Sci. U.S.A.* **71**:760.

Guha, A., Saturen, Y., and Szybalski, W., 1971, Divergent orientation of transcription from the biotin locus of *Escherichia coli*, *J. Mol. Biol.* **56**:53.

Hayes, S., and Szybalski, W., 1973, Control of short leftward transcripts from the immunity and *ori* regions in induced coliphage lambda, *Mol. Gen. Genet.* **7**:289.

Hershey, A. D., and Dove, W., 1971, Introduction to lambda, in: *The Bacteriophage Lambda* (A. D. Hershey, ed.), pp. 3–12, Cold Spring Harbor Laboratories, Cold Spring Harbor, N.Y.

Herskowitz, I., and Signer, E., 1970*a*, Control of transcription from the *r* strand of bacteriophage lambda, *Cold Spring Harbor Symp. Quant. Biol.* **35**:365.

Herskowitz, I., and Signer, E. R., 1970*b*, A site essential for expression of all late genes in bacteriophage λ, *J. Mol. Biol.* **47**:545.

Hofnung, M., 1974, Divergent operons and the genetic structure of the maltose B region in *Escherichia coli* K12, *Genetics* **76**:169.

Honigman, A., Oppenheim, A., Oppenheim, A. B., and Stevens, W. F., 1975, A pleiotropic regulatory mutation in λ bacteriophage, *Mol. Gen. Genet.* **138**:85.

Honigman, A., Hu, S.-L., Chase, R., and Szybalski, W., 1976, 4s *oop* RNA is a leader sequence for the immunity-establishment transcription in coliphage λ, *Nature* **262**:112.

Honigman, A., Hu, S.-L., and Szybalski, W., 1977, Unpublished results quoted by Campbell *et al.* (1977).

Jackson, E. N., and Yanofsky, C., 1972, Internal promoter of the tryptophan operon of *Escherichia coli* is located in a structural gene, *J. Mol. Biol.* **69**:307.

Jackson, E. N., and Yanofsky, C., 1973, The region between the operator and the first structural gene of the tryptophan operon of *Escherichia coli* may have a regulatory function, *J. Mol. Biol.* **76**: 89.

Jacob, F., and Monod, J., 1961, Genetic regulatory mechanisms in the synthesis of proteins, *J. Mol. Biol.* **3**:318.

Jacob, F., and Monod, J., 1965, Genetic mapping of the elements of the lactose region in *Escherichia coli*, *Biochem, Biophys, Res. Commun.* **18**:513.

Jacoby, C. A., 1972, Control of the *argECBH* cluster in *Escherichia coli*, *Mol. Gen. Genet.* **117**:337.

Kasai, T., 1974, Regulation of the expression of the histidine operon in *Salmonella typhimurium, Nature* **249**:523.

Katzir, N., Oppenheim, A., Belfort, M., and Oppenheim, A. B., 1976, Activation of the lambda *int* gene by the *cII* and *cIII* gene products, *Virology* **74**:327.

Ketner, G., and Campbell, A., 1974, A deletion mutation placing the galactokinase gene of *Escherichia coli* under control of the biotin promoter, *Proc. Natl. Acad. Sci. U.S.A.* **71**:2698.

Ketner, G., and Campbell, A., 1975, Operator and promoter mutations affecting divergent transcription in the *bio* gene cluster of *Escherichia coli, J. Mol. Biol.* **96**:13.

Korn, L. W., and Yanofsky, C., 1976, Rho factor mediates transcription termination at the attenuator of the tryptophan operon of *Escherichia coli, J. Mol. Biol.* **106**:231.

Largen, M., and Belser, W., 1973, The apparent conservation of the internal low efficiency promoter of the tryptophan operons of several species of *Enterobacteriaceae, Genetics* **75**:19.

Levin, B., 1974, *Gene Expression—1. Bacterial Genomes,* Wiley, New York.

Lindahl, G., 1970, Bacteriophage P2: Replication of the chromosome requires a protein which acts only on the genome which codes for it, *Virology* **42**:522.

Lozeron, H. A., Dahlberg, J. A., and Szybalski, W., 1976, Processing of the major leftward mRNA of coliphage lambda, *Virology* **71**:262.

Luzzati, D., 1970, Regulation of λ exonuclease synthesis: Role of the *N* gene product and λ repressor, *J. Mol. Biol.* **49**:515.

Margolin, P., and Bauerle, R. H., 1966, Determinants for regulation and initiation of expression of tryptophan genes, *Cold Spring Harbor Symp. Quant. Biol.* **31**:311.

McDermit, M., Pierce, M., Staley, D., Shimaji, M., Shaw, R., and Wulff, D., 1976, Mutations masking the lambda *cin-1* mutation, *Genetics* **82**:417.

Monod, J., and Jacob, F., 1961, General conclusions: Teleonomic mechanisms in cellular metabolism, growth and differentiation, *Cold Spring Harbor Symp. Quant. Biol.* **26**:389.

Musso, R., Di Lauro, R., Rosenberg, M., and de Crombrugghe, B., 1977, Nucleotide sequence of the operator–promoter region of the galactose operon of *Escherichia coli, Proc. Natl. Acad. Sci. U.S.A.* **74**:106.

Nakamura, Y., and Yura, T., 1976, Induction of sigma factor synthesis in *Escherichia coli* by the *N* gene product of bacteriophage lambda, *Proc. Natl. Acad. Sci. U.S.A.* **73**:-4405.

Nash, H. A., 1975, Integrative recombination of bacteriophage lambda DNA *in vitro, Proc. Natl. Acad. Sci. U.S.A.* **72**:1072.

Pachl, C. A., and Young, E. G., 1976, Detection of polycistronic and overlapping bacteriophage T7 late transcripts by *in vitro* translation *Proc. Natl. Acad. Sci. U.S.A.* **73**:312.

Pai, C. H., 1972, Mutant of *Escherichia coli* with derepressed levels of the biotin biosynthetic enzymes, *J. Bacteriol.* **112**:1280.

Petit-Koskas, E., and Contesse, G., 1976, Stimulation in trans of synthesis of *E. coli gal* operon enzymes by lambdoid phages during low catabolite repression, *Mol. Gen. Genet.* **143**:203.

Pribnow, D., 1975, Nucleotide sequence of an RNA polymerase binding site at an early T7 promoter, *Proc. Natl. Acad. Sci. U.S.A.* **72**:784.

Ptashne, M., 1971, Repressor and its action, in: *The Bacteriophage Lambda* (A. D. Hershey, ed.), pp.221–238, Cold Spring Harbor Lab., Cold Spring Harbor, New York.

Ptashne, M., Bachman, K., Humazun, M. Z., Jeffrey, A., Maurer, R., Meyer, B., and Sauer, R. T., 1976, Autoregulation and function of a repressor in bacteriophage lambda, *Science* **194**:156.

Richardson, J. P., Grimley, C., and Lowery, C., 1975, Transcription termination factor rho activity is altered in *Escherichia coli* with *suA* gene mutations, *Proc. Natl. Acad. Sci. U.S.A.* **72**:1725.

Riggs, A. D., Bourgeois, S., and Cohn, M., 1970, The *lac* repressor–operator interaction. III. Kinetic studies, *J. Mol. Biol.* **53**:401.

Roberts, J. W., 1970, The ρ factor: Termination and anti-termination in lambda, *Cold Spring Harbor Symp. Quant. Biol.* **35**:121.

Roberts, J. W., 1975, Transcription termination and late control in phage lambda, *Proc. Natl. Acad. Sci. U.S.A.* **72**:2300.

Shimada, K., and Campbell, A., 1974, *Int*-constitutive mutants of bacteriophage lambda, *Proc. Natl. Acad. Sci. U.S.A.* **71**:237.

Shimada, K., Weisberg, R., and Gottesman, M. E., 1973, Prophage lambda at unusual chromosomal locations. II. Mutations induced by bacteriophage lambda in *Escherichia coli* K12, *J. Mol. Biol.* **80**: 297.

Smith, G. R., and Magasanik, B., 1971, Nature and self-regulated synthesis of the repressor of the *hut* operons in *Salmonella typhimurium, Proc. Natl. Acad. Sci. U.S.A.* **68**:1493.

Spiegelman, W. G., Reichardt, L. F., Yaniv, M., Heineman, S. F., Kaiser, A. D., and Eisen, H., 1972, Bidirectional transcription and the regulation of phage λ repressor synthesis, *Proc. Natl. Acad. Sci. U.S.A.* **69**:3156.

Szybalski, W., Bøvre, K., Fiandt, M., Hayes, S., Hradecna, Z., Kumar, S., Lozeron, H. A., Nijkamp, H. J. J., and Stevens, W. F., 1970, Transcriptional units and their controls in *Escherichia coli* phage λ: Operons and scriptons, *Cold Spring Harbor Symp. Quant. Biol.* **35**:341.

Thomas, R., 1970, Control of development in temperate bacteriophage. III. Which prophage genes are and which are not transactivable in the presence of immunity? *J. Mol. Biol.* **49**:393.

Vrancic, A., and Guha, A., 1973, Evidence of two operators in the biotin locus of *Escherichia coli, Nature New Biol.* **245**:106.

Wilson, D. B., and Hogness, D. S., 1969, The enzymes of the galactose operon in *Escherichia coli.* IV. The frequencies of translation of the terminal cistrons in the operon, *J. Biol. Chem.* **244**:2143.

Wulff, D. L., 1976, Lambda *cin-1,* a new mutation which enhances lysogenization by bacteriophage lambda, and the genetic structure of the lambda *cy* region, *Genetics* **82**:401.

Yarmolinsky, M. B., and Wiesmeyer, H., 1960, Regulation by coliphage lambda of the expression of the capacity to synthesize a sequence of host enzymes, *Proc. Natl. Acad. Sci. U.S.A.* **46**:1626.

3

Autogenous and Classical Regulation of Gene Expression: A General Theory and Experimental Evidence

MICHAEL A. SAVAGEAU

Quis custodiet ipsos custodes?

1 *Introduction*

Control of genetic expression by a constitutively synthesized repressor protein is a central feature of the classical operon model proposed in 1961 by Jacob and Monod (see also Chapter 2). Although this model was originally developed for a specific set of genes in a particular bacterium, its influence quickly spread to all of biology. Jacob and Monod knew well the attractions of their model and warned against its indiscriminant use, but their prophetic words were largely ignored as it became the dominant paradigm for biologists concerned with the normal processes of differentiation, growth, and homeostasis, and such pathological manifestations as infectious diseases, metabolic disorders, and cancer. By any number of criteria the Jacob–Monod model has been one of the most seminal ideas in modern biology.

Inducible operons in enteric bacteria were the first to be understood at the

MICHAEL A. SAVAGEAU • Department of Microbiology, The University of Michigan, Ann Arbor, Michigan

level of molecular detail required for testing this model. In several of the best studied examples, such as the systems for the catabolism of lactose (Beckwith and Zipser, 1970) and galactose (Nakanishi *et al.*, 1973), the classical Jacob–Monod machanism appears to be operative. Control of repressible operons has remained more obscure, but there is now good evidence that this same mechanism is involved in regulation of the tryptophan (Zubay *et al.*, 1972; Squires *et al.*, 1973; Rose *et al.*, 1973; McGeoch *et al.*, 1973; Zalkin *et al.*, 1974; Shimizu *et al.*, 1974) and arginine (Urm *et al.*, 1973; Cunin *et al.*, 1976) biosynthetic systems in enteric bacteria. In recent years alternatives to this classical model have also been discovered.

Not all regulators are synthesized constitutively. For example, control of the inducible system for histidine utilization in certain enteric bacteria involves a repressor protein that is not synthesized constitutively; it is directly involved in modulating expression of its own structural gene (Smith and Magasanik, 1971). This feature of the system for histidine utilization provides the most thoroughly documented example of a more general class of phenomena that has been called *autogenous regulation* (Goldberger, 1974).

Autogenous regulation was not well recognized as a general phenomenon until the subject was reviewed in 1974 by Robert F. Goldberger, although, as he pointed out, the basic idea is not new. In fact, it was proposed in one form more than 25 years ago and has since been rediscovered many times. Goldberger's review has been instrumental in bringing about widespread recognition of autogenous regulation for two reasons. First, he clearly defined the concept of autogenous regulation and provided unambiguous criteria for experimentally identifying the phenomenon. Second, he gathered together many sporadic and disconnected reports and impressively documented the occurrence of autogenous regulation throughout the phylogenetic spectrum: prokaryotes, the bacteria and their phages, lower eukaryotes, especially fungi, and higher eukaryotes, including the human.

Another feature of the Jacob–Monod model that has proved to be true only for some systems and not for others is the repressive function of the regulator protein; in certain control systems the regulator has been found to be an activator, or positive element.* Regulation by an activator was first (and most thoroughly) demonstrated for the inducible arabinose operon of *Escherichia coli* (see Englesberg and Wilcox, 1974). Another well-documented example of a protein that acts as a positive element in gene expression is the catabolite activator protein that, together with cyclic AMP, stimulates expression of many inducible catabolic operons in bacteria (Nissley *et al.*, 1971; Eron and Block, 1971; deCrombrugghe *et al.*, 1971; Lee *et al.*, 1974; Tyler *et al.*, 1974).

As molecular biologists have realized that not all regulator proteins are synthesized constitutively and that not all regulator proteins are repressors, they have had to discard the notion that the model of Jacob and Monod is applicable to all systems of regulating gene expression. In certain respects, this has had a

*The difference between control exerted by a repressor and that exerted by an activator may be likened to the two types of genii depicted in Arabian folk tales. For one, activity is initiated by removing the seal or stopper (repressor), thus allowing the genie to escape from the lamp. For the second type, activity commences only after a suitable password (activator) has been supplied.

liberating effect; as new systems come to light, they now may be studied with an awareness of the rich possibilities that exist for regulating gene expression. On the other hand, recognition that the Jacob–Monod paradigm does not suffice to explain all of genetic regulation has caused many scientists to despair of finding unifying principles that could provide an understanding of the diversity of regulated systems, principles that could guide the investigation of new systems.

In this chapter I show that the two well-established modes of regulation, autogenous and classical, can be understood within a unifying theoretical framework based on function of the intact system and that, because of its predictive value, this theory provides a useful guide for the investigation of new systems. Whether a system is autogenously or classically regulated depends in part on the nature of the regulator. Therefore, I begin by considering the functional differences between repressor- and activator-controlled systems, and by showing that the nature of the regulator is correlated with demand for expression of the regulated structural genes. The principal comparison of autogenous and classical regulation then is presented for several classes of systems including a degenerate class of "unregulated" systems that I have called *autonomous*. For each class, predictions based on functional effectiveness are deduced and recent experimental evidence is reviewed within this theoretical framework.

2 *Repressors and Activators*

Molecular mechanisms involving repressor and activator control of gene expression are represented schematically in Figs. 1 and 2, respectively. In this section I shall focus on their functional differences, which determine in part whether a system is autogenously or classically regulated.

Are these different mechanisms simply historical accidents that represent functionally equivalent solutions to the same regulatory problem? Alternatively, have they been selected to meet specific needs and, if so, can we determine the functional implications inherent in each molecular design and the nature of the selective forces that have given rise to them?

Questions of this type are difficult to answer using only the direct experimental approach. For example, one cannot draw conclusions about the differences between repressor and activator mechanisms by comparing directly two representative systems such as the inducible lactose (repressor-controlled) and arabinose (activator-controlled) operons because there may be other (unknown) elements involved in their control, and because the systems differ in many ways that are irrelevant to the comparison of the two modes of regulation *per se*. Ideally, one would like a controlled comparison in which the two systems are identical in every respect except one: a difference in the type of control mechanism utilized. Although this is difficult to obtain experimentally, it can be simulated by appropriate mathematical analysis (Savageau, 1976). The results of such an analysis show that a repressor-controlled mechanism is selected for regulation in an environment in which there is low demand for expression of the operon, whereas an activator-controlled mechanism is selected for regulation in an environment in which there is high demand for expression of the operon. This general conclusion

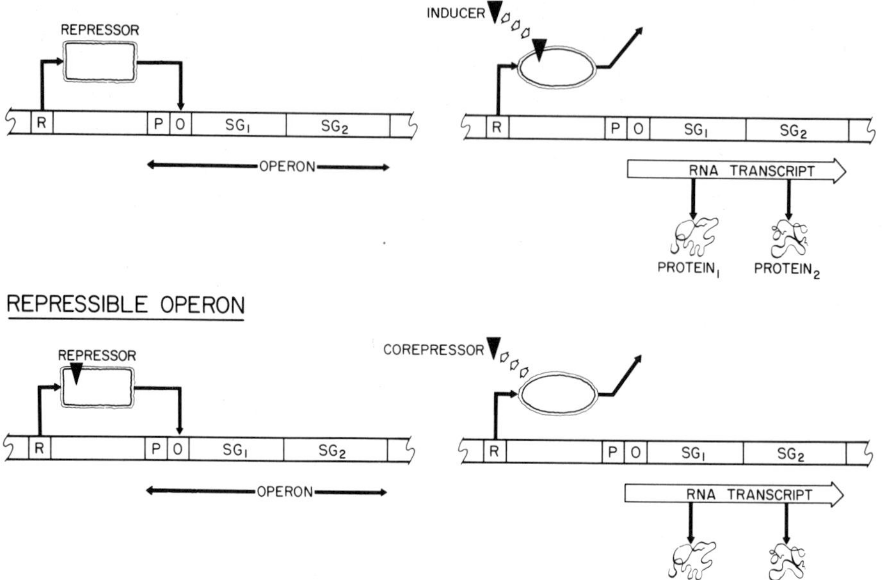

Figure 1. Control of gene expression by a repressor. A regulatory gene (R) directs the synthesis of a repressor protein, which in its "active" state binds tightly to the DNA at the promoter–operator region (P–O) and prevents the structural genes (SG) of the operon from being expressed. The structural genes of an inducible operon are switched on in the presence of an inducer that "inactivates" the repressor, whereas in a repressible operon they are switched on only when a corepressor, which normally "activates" the repressor, is removed. In either case, an RNA transcript of the structural genes is produced and subsequently translated into the corresponding protein products.

is useful in relating physiological function to underlying molecular determinants in a wide variety of systems. Perhaps its validity is best demonstrated with examples.

2.1 *Inducible Catabolic Systems*

The best studied examples of systems from this class are found in enteric bacteria. In response to an environmentally supplied nutrient, the organism, under appropriate conditions, is able to induce specific enzymes needed for utilization of that nutrient. In the present context, repressor control, which correlates with low demand for expression, is expected for inducible systems whose substrate is seldom present at high concentrations in the organism's environment, whereas activator control, which correlates with high demand for expression, is expected for inducible systems whose substrate is often available at high concentrations in the organism's environment.

In enteric bacteria there are more than half a dozen inducible catabolic systems for which the nature of the regulator is known. For example, the systems for the utilization of galactose (Nakanishi *et al.*, 1973), glycerol (Cozzarelli *et al.*, 1968), histidine (Smith and Magasanik, 1971), and lactose (Beckwith and Zipser,

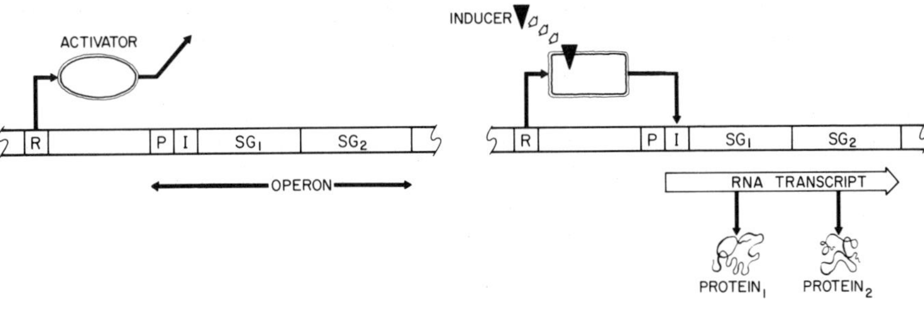

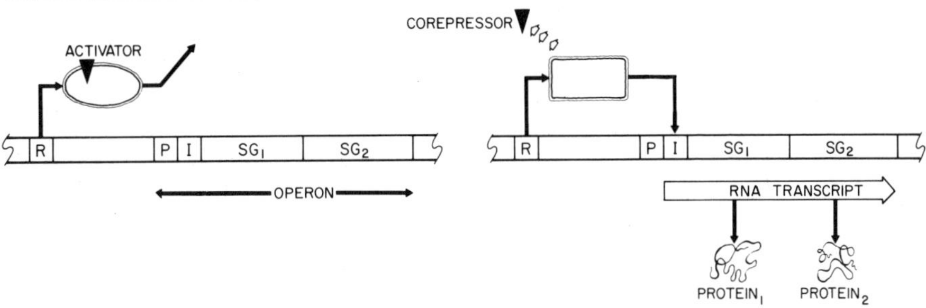

Figure 2. Control of gene expression by an activator. A regulatory gene (R) directs the synthesis of an activator protein, which in its "inactive" state is unable to facilitate initiation of transcription of the structural genes (SG) at the promoter–initiator site (P–I). The structural genes of an inducible operon are switched on in the presence of an inducer that converts the regulator protein into its "active" state; in a repressible operon these genes are switched on only when a corepressor, which normally "inactivates" the activator, is removed.

1970) are under the control of repressors, whereas those for the utilization of arabinose* (Wilcox *et al.*, 1971), maltose (Hofnung and Schwartz, 1971), and rhamnose (Power, 1967) are under the control of activators. From this information one would predict that the first group of substrates is seldom present at high concentrations in the colon, where the enteric bacteria are localized. Conversely, the second group of substrates is expected to be there frequently at high concentrations.

Although the local environment in the animal colon is complex and largely undefined, the relative concentrations of specific nutrients in the colon can be estimated indirectly from their abundance in the diet and their absorption patterns in the small intestine. The results of such studies have been reviewed elsewhere (Savageau, 1974*a*) and are in agreement with the predictions in the preceding paragraph (see Table I). The disaccharides lactose and maltose are enzymatically split into their constituent sugars early and late, respectively, during

*In the absence of the inducer arabinose, the regulator protein acts as a repressor of the structural genes that code for the enzymes that catabolize arabinose (see Englesberg and Wilcox, 1974). Thus, the regulator protein of the *ara* system can function both as an activator and as a repressor. However, the predominant functional properties are those of an activator, and for the present purposes the *ara* system can be considered a (modified) activator-controlled system (Savageau, 1976).

transit through the intestines. Glycerol, the sugar galactose, and the amino acid histidine are all absorbed effectively at the beginning of the small intestine and are unlikely to reach the colon in high concentrations; the sugars arabinose and rhamnose are poorly absorbed and probably reach the colon without extensive attenuation in concentration.

From such absorption data one can also make predictions about molecular control mechanisms that have yet to be fully characterized in enteric bacteria. For example, the sugars mannose and xylose appear to be absorbed slowly by the small intestine (Bihler, 1969) and therefore may be present at relatively high concentrations in the colon. In the bacteria one would expect the inducible catabolic systems for the utilization of mannose and xylose to be activator-controlled. Similarly, the

TABLE I. Nature of Regulator Correlates with Demand for Expression of Regulated Genes[a]

	Nature of regulator		Demand for expression	
System[b]	Observed[c]	Predicted	Predicted	Observed[c]
Inducible catabolic pathways:				
Arabinose	Activator →		→ High	High
Galactose	Repressor →		→ Low	Low
Glycerol	Repressor →		→ Low	Low
Histidine	Repressor →		→ Low	Low
Lactose	Repressor →		→ Low	Low
Maltose	Activator →		→ High	High
Rhamnose	Activator →		→ High	High
Mannose	?	Activator ←	←	High
Tryptophan	?	Activator ←	←	High
Xylose	?	Activator ←	←	High
Repressible biosynthetic pathways:				
Arginine	Repressor →		→ Low	Low
Cysteine	Activator →		→ High	High
Isoleucine–valine[d]	Activator →		→ High	High
Lysine	Repressor →		→ Low	Low
Tryptophan	Repressor →		→ Low	Low
Histidine	?	Activator ←	←	High
Isoleucine–valine	?	Activator ←	←	High
Inducible drug resistance:				
Penicillin[e,f]	Repressor →		→ Low	Low
Tetracycline	Repressor →		→ Low	Low
Chloramphenicol[e]	?	Repressor ←	←	Low
Erythromycin[e]	?	Repressor ←	←	Low
Inducible prophages:				
Lambda	Repressor →		→ Low	Low
P1	Repressor →		→ Low	Low
P2	Repressor →		→ Low	Low
P22	Repressor →		→ Low	Low

[a]An arrow indicates direction of inference. An "observed" column adjacent to a "predicted" column represents the results of independent observations that are used to test the predictions.
[b]Enteric bacteria unless indicated otherwise.
[c]Evidence of various types was used in constructing this table. The evidence is not equally conclusive in each case.
[d]*Saccharomyces cerevisiae.*
[e]*Staphylococcus aureus.*
[f]*Bacillus licheniformis.*

relative abundance of tryptophan in the colon (see Section 2.2) suggests that the inducible tryptophanase system involves an activator-controlled mechanism. These predictions are summarized in Table I, although they remain to be tested experimentally.

2.2 *Repressible Biosynthetic Systems**

The enzymatic machinery required for endogenous biosynthesis of amino acids is considerable (Davis, 1961). It makes economic sense for an organism to repress synthesis of the enzymes in a biosynthetic pathway whose end product is available preformed in the environment; the selective advantage of such repression mechanisms has also been shown experimentally (Zamenhof and Eichhorn, 1967). The general principle, restated for this class of systems, is the following: Repressor control, which correlates with low demand for expression, is expected for repressible systems whose end product is often present at high concentrations in the organism's natural environment; activator control, which correlates with high demand for expression, is expected for repressible systems whose end product is seldom present at high concentrations in the natural environment.

There is now evidence concerning the nature of the regulator for many repressible biosynthetic systems in microorganisms. Perhaps the clearest examples, in which a repressor element in the control has been demonstrated *in vitro,* are the tryptophan (Zubay *et al.*, 1972; Squires *et al.*, 1973; Rose *et al.*, 1973; McGeoch *et al.*, 1973; Zalkin *et al.*, 1974; Shimizu *et al.*, 1974) and arginine (Urm *et al.*, 1973; Cunin *et al.*, 1976) biosynthetic systems in *Escherichia coli.* Less direct but strong genetic evidence suggests control by an activator for the cysteine biosynthetic system in enteric bacteria (Jones-Mortimer, 1968; Kredich, 1971) and the isoleucine–valine biosynthetic system in *Saccharomyces cerevisiae* (Bollon and Magee, 1971, 1973; Bollon, 1974, 1975). Kelleher and Heggeness (1976) have demonstrated "escape synthesis" of diaminopimilate decarboxylase, the last enzyme in the lysine biosynthetic pathway of *E. coli,* and have suggested repressor control for this system. From this information one would predict that arginine, lysine, and tryptophan are often present and cysteine seldom present at high concentrations in the colon, where the enteric bacteria are localized. Similarly, one would expect isoleucine and valine to occur infrequently at high concentrations in the natural environment of yeast.

Although the evidence is indirect, in each case it appears to agree with these predictions (see Table I). Relative to other amino acids tryptophan is poorly absorbed by the small intestine (Adibi *et al.*, 1967; Adibi and Gray, 1967) and would be likely to reach the colon. Furthermore, the ability to catabolize exogenous tryptophan appears to occur only among those microorganisms that are able to inhabit the gut (DeMoss and Moser, 1969), and among the intestinal flora of virtually all the animals they examined there was at least one species of microorganism that was able to catabolize exogenous tryptophan. Arginine by itself is slowly absorbed by the intestine (Wilson, 1962), but in a mixture with other amino

*Portions of this material appeared originally in *Biochemical Systems Analysis* by M. A. Savageau and are reproduced here with permission of the publishers, Addison-Wesley/W. A. Benjamin, Inc., Advanced Book Program, Reading, Massachusetts.

acids its absorption appears to be accelerated (Adibi *et al.*, 1967). The most direct observations show that the concentration of free arginine reaching the distal end of the small intestine is the third highest of all amino acids under a variety of dietary conditions (Nixon and Mawer, 1970). In this last study, the concentration of lysine was the highest of all amino acids at the distal end of the small intestine, whereas that of cysteine was the third lowest.

In the case of *Saccharomyces cerevisiae,* a eukaryotic organism, the expected infrequent occurrence of isoleucine and valine at high concentrations in the natural environment is consistent with the evolution of this organism in sugar-rich, nitrogen-poor environments. This is further supported by the observation of Gross (1969) that biosynthetic pathways of fungi generally are "set" higher than those of enteric bacteria.

The correlation just supported can also be used to predict the nature of the regulator protein rather than the nature of the organism's environment. Indirect evidence shows that histidine can be absorbed readily by the small intestine (Gibson and Wiseman, 1951) although in mixtures of amino acids its rate of absorption is lower (Adibi *et al.*, 1967). By more direct measurements, the concentration of free histidine reaching the distal end of the small intestine is among the lowest of all amino acids under a variety of dietary conditions (Nixon and Mawer, 1970). Thus, one would expect that the system for histidine biosynthesis is under the control of an activator protein, which is consistent with the results of Artz and Broach (1975). A similar prediction can be made regarding isoleucine and valine. Each of these amino acids is absorbed rapidly by the small intestine when present alone (Finch and Hird, 1960) or in mixtures of amino acids (Adibi *et al.*, 1967; Adibi and Gray, 1967). Furthermore, the concentrations of free isoleucine and valine that reach the colon are among the lowest of all amino acids under a variety of dietary conditions (Nixon and Mawer, 1970). Thus, one would expect the system for biosynthesis of isoleucine and valine in *E. coli* to be under control of an activator protein, which is consistent with the results of Levinthal *et al.* (1973). These predictions are also summarized in Table I, although experimental support cannot be claimed because there is still conflicting evidence concerning the histidine and isoleucine–valine biosynthetic systems in enteric bacteria (see Sections 4.6 and 4.7.1).

2.3 *Inducible Drug Resistance*

Natural selection of drug-resistant microorganisms has occurred with expanded clinical and agricultural use of antibiotics during the past two decades (for examples see Anderson, 1968; Watanabe, 1971; Davies and Rownd, 1972). Simultaneous resistance to several drugs, which is rapidly transmitted to other microorganisms via extrachromosomal elements (Davies and Rownd, 1972), is of particular interest because of the obvious implications for clinical practice but also because of the fundamental molecular mechanisms involved. Although resistance is expressed constitutively in most instances (Mitsuhashi, 1971; Davies and Rownd, 1972), there are well-documented examples in which resistance to high levels of a drug occurs in microorganisms following their exposure to low (or subeffective) concentrations of that drug. In some cases this increased resistance is the result of gene amplification among the resistance determinants (Rownd and Mickel, 1971); in others, it is the result of increased gene expression. Examples of the latter type, which are of primary concern in the present context, are inducible resistance to tetracycline (Yang *et al.*, 1976), penicillin (Kelly and Brammar, 1973*b;* Imsande and Lilleholm, 1976), erythromycin (Weisblum *et al.*, 1971), and chloramphenicol (Winshell and Shaw, 1969). Repressor control, which correlates with low demand for expression, is expected for such inducible systems when the corresponding antibiotic is seldom present at high concentrations in the organism's natural environment, whereas activator control, which correlates with high demand for expression, is expected for systems of this type when the corresponding antibiotic is often present at high concentrations in the natural environment.

Recent studies *in vitro* of inducible tetracycline resistance in *E. coli* strongly suggest that this system is under the control of a repressor protein (Yang *et al.*, 1976). The same conclusion for inducible penicillin resistance in *Bacillus licheniformis* (Kelly and Brammar, 1973*b*) and *Staphylococcus aureus* (Imsande and Lilleholm, 1976) has been supported by genetic and biochemical studies. From the molecular evidence in these cases one would expect that tetracycline and penicillin are seldom present at relatively high concentrations in the natural environment of these bacteria (see Table I).

Again there is only indirect evidence to support the prediction. Although use of such antibiotics has increased in the past two decades, the fraction of time that microorganisms are exposed to high doses of these drugs may still be considered relatively small (see Watanabe, 1971). This exposure, while certainly significant for selection of drug-resistant organisms, is probably not sufficient for selection of activator mechanisms controlling expression of such resistance.

These arguments could also be used to predict that induction of chloramphenicol resistance and induction of erythromycin resistance are under the control of repressor proteins (Table I).

2.4 *Inducible Prophages*

Temperate phages or bacterial viruses have two alternative modes of replication. In the lysogenic mode, the virus exists as a prophage or viral genome that replicates along with its host but is otherwise almost totally quiescent. In the lytic mode, the virus grows vegetatively and eventually progeny are released from the infected cell. The phage can switch from the lysogenic to the lytic mode of replication in response to signals indicating the physiological state of the host cell. This process is called induction.

One can conceive of an inducible prophage under the control of an activator protein. The prophage synthesizes an activator that is normally unable to stimulate transcription of the other viral genes, but in response to the appropriate host signal(s), the activator is converted to a conformation that facilitates expression of viral genes required for lytic growth. Induction could also be under the control of a repressor protein. In this case, the prophage synthesizes a repressor that blocks expression of almost all viral genes. In response to the appropriate host signal(s), the repressor is converted to a form that no longer is able to block transcription of viral genes.

The same physiological function—induction—is realized in each case, but the molecular mechanisms are different. Activator control, which correlates with high demand for expression, is expected for a prophage whose induction is a frequent event, whereas repressor control, which correlates with low demand for expression of the regulated genes, is expected for a prophage whose induction is an infrequent event.

Bacteriophage lambda has been studied most thoroughly and in this case induction is under the control of a repressor protein (see review by Ptashne *et al.*, 1976). Among other temperate phages that have been well studied at the molecular level, repressor control also appears to be the rule (P1: Scott, 1970; P2: Bertani and Bertani, 1971; P22: Levine, 1972). From the preceding deductions and the known molecular nature of the regulator protein for several bacteriophages, one would predict that induction of these prophages is relatively rare in nature.

Although good experimental evidence concerning the frequency of phage induction in nature is not available, there is evidence that tends to support this prediction. Spontaneous induction is relatively rare under a variety of conditions in which it has been examined (see Barksdale and Arden, 1974) and most if not all natural isolates of coliform organisms are lysogenic for one or more phages (see Reanney, 1976; and Chapter 2 of this volume).

2.5 *Other Systems*

As stated at the beginning of Section 2, the correlation between the molecular nature of the regulator protein and demand for expression of the regulated genes may be of quite general importance. This is indicated by the deductions from which this correlation was predicted, but it also is evident in the large number of examples, representing systems of four different types, that have been considered in this section. In almost all cases, the best evidence is available for systems in enteric bacteria and their phages. There are many more examples among prokaryotes and lower eukaryotes, in which there is evidence for either the molecular mechanism or the demand for the physiological function, but not both. Thus, additional cases should soon be available for testing this correlation.

Control of gene expression in differentiated cells of higher eukaryotes has been studied by somatic cell hybridization and the evidence suggests control by a repressor in some cases and by an activator in others (see Davis and Adelberg, 1973; Davidson, 1974; Bernhard, 1976). When these mechanisms are confirmed at the molecular level and more is known about the function of the regulated genes, it will be possible to test the correlation between molecular control mechanism and demand for expression of the regulated genes in eukaryotic systems.

3 *Inducible Systems*

As just demonstrated, whether the regulator is a repressor or an activator is determined in large part by *qualitative* differences in behavior of the alternative mechanisms. These differences are inherent in the component mechanisms

CLASSICAL REGULATION (REPRESSOR)

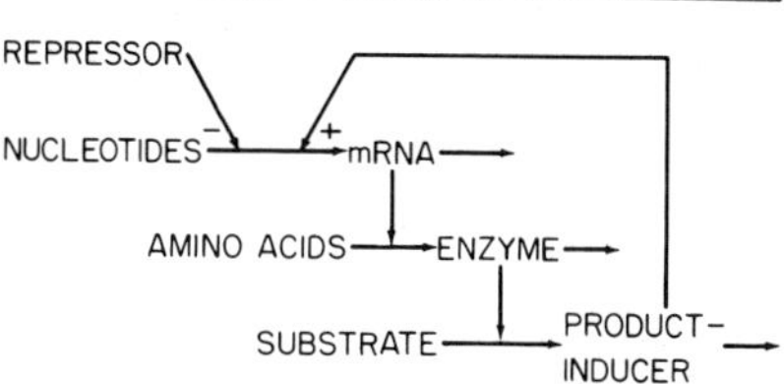

CLASSICAL REGULATION (ACTIVATOR)

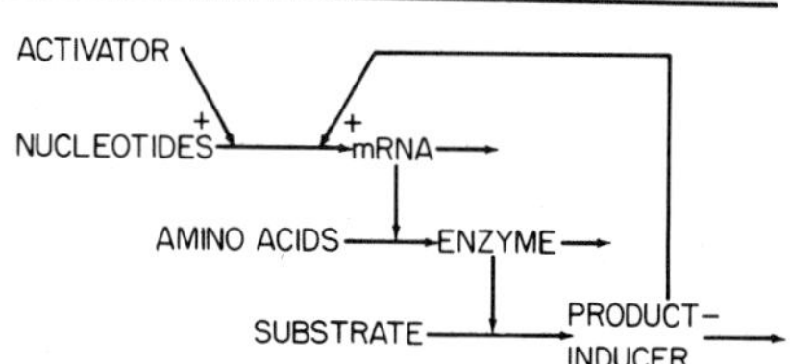

AUTOGENOUS REGULATION (REPRESSOR)

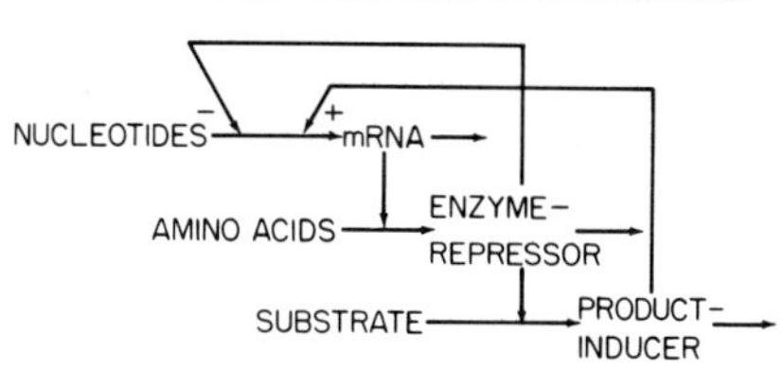

AUTOGENOUS REGULATION (ACTIVATOR)

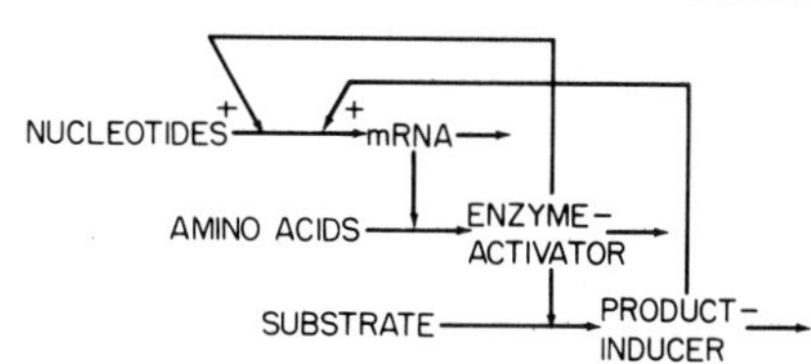

Figure 3. Autogenous and classical regulation of inducible catabolic systems. In an autogenously regulated system the structural gene of the regulator is located within the transcriptional unit that the regulator controls; in a classically regulated system the structural gene of the regulator is located outside the operon and the rate of synthesis of the regulator is unresponsive to inducer. In this schematic representation the horizontal arrows indicate chemical transformations at the messenger RNA, enzyme, and metabolite levels; the vertical arrows represent modifier or catalytic influences (the sense of the influence is positive unless otherwise indicated). The "product" is identified as the true inducer, since this is the usual case (see Savageau, 1976).

depicted in Figs. 1 and 2. In contrast, the type of regulation governing regulator proteins themselves is influenced more by *quantitative* considerations, such as the relative strengths of regulatory interactions, and the function of the intact system. The models in Fig. 3 have been drawn to represent this emphasis on circuitry of the integrated system rather than its component mechanisms.

I shall focus first on inducible catabolic systems in bacteria because these are the systems for which the most information is available. Nevertheless, the results can be generalized readily to inducible systems of other types, as is shown in Sections 3.8 to 3.10. Repressible systems are considered separately in Section 4.

3.1 Criteria for Functional Effectiveness

The physiological properties listed in Table II are reasonable requirements for any inducible catabolic system to function effectively.* Criterion 1 protects the organism from wasteful synthesis of inducible enzymes when the level of the corresponding substrate is so low that insufficient benefit would be gained from induction. Criterion 2 benefits the organism whenever the substrate of the system is the nutrient limiting growth. Criterion 3, stability, is obviously essential for a

*These criteria apply to inducible systems of other types, although their interpretation then may be slightly different. See Sections 3.8 through 3.10.

TABLE II. Criteria for Functional Effectiveness of Inducible Catabolic Systems

1. A sharp threshold in the concentration of the substrate necessary for induction
2. The ability to make the most product available to the organism from a given suprathreshold increment in the concentration of the substrate
3. Stability
4. Temporal responsiveness
5. Insensitivity to perturbations in the system's component parts

system to function properly. Criterion 4 implies rapid induction, which is an advantage when substrate levels change abruptly. Criterion 5 ensures that the system will continue to function in spite of the continual perturbations it experiences in any real environment. These perturbations result from nonlethal mutations, errors in transcription and translation of the genetic information, and physical influences such as temperature shifts. Note that none of these criteria assumes anything about the type of regulatory mechanism involved.

3.2 Autogenous and Classical Regulation

These two possibilities are represented schematically in Fig. 3 for repressor- and activator-controlled systems. Except for the differences in mode of regulation these models are assumed to be equivalent (for a detailed discussion of equivalence see Savageau, 1976). The criteria given in Table II for functional effectiveness of these systems can be quantified readily and then by mathematical analysis alternative systems represented in Fig. 3 can be compared systematically (Savageau, 1974*b*, 1976). The important results to consider here are the following: Repressor-controlled systems with autogenous regulation are superior to the corresponding system with classical regulation according to all five criteria for functional effectiveness. Just the opposite is true for activator-controlled systems: According to all these same criteria, the classically regulated system is superior to the corresponding autogenously regulated systems.

3.3 Predictions

The functional differences discussed in Section 3.2 have obvious implications for the natural selection of classical and autogenous regulation in simple inducible systems of the type represented in Fig. 3. Systems governed by an activator protein are not expected to utilize autogenous regulation. On the other hand, systems governed by a repressor protein can be predicted to involve autogenous regulation as well. Experimental evidence from several inducible systems are now reviewed in terms of these general predictions. The results are summarized in Table III.

3.4 Arabinose

There is no known instance of autogenous regulation in an activator-controlled inducible system. This is consistent with the predictions of the preceding section, but it must be noted that there are few examples in which such systems are demonstrated to be classically regulated. The arabinose (*ara*) system in *E. coli* is the most thoroughly documented case of activator control in such systems (Englesberg and Wilcox, 1974), and in this case the system appears to be classically regulated.

The activator's structural gene (*araC*) is not located in a common transcriptional unit with any structural gene known to be under the control of the arabinose activator, but is located in a separate transcriptional unit (Wilcox *et al.*, 1974*a*). Thus, a simple autogenously regulated operon is excluded. One could, of course, postulate more complex models of autogenous regulation in which the activator itself is autogenously regulated and its level varies with the expression of structural genes in other transcriptional units under its control. Functionally, a model of this type would be indistinguishable in most respects from the simple model shown in Fig. 3. There is evidence, however, against the more complex model.

Casadaban (1976*a*) has found that expression of the *araC* gene does not vary with induction of the arabinose system. In effect, the activator is constitutively synthesized and the arabinose system is classically regulated. These observations were made *in vivo* using strains of *E. coli* in which most of the *araC* structural gene is missing and in which the promoter of the *araC* gene is fused to the structural genes of the lactose operon. In such strains, transcription from the *araC* promoter

TABLE III. Nature of Regulator Correlates with Type of Regulatory Circuit in the Intact System[a]

	Nature of regulator		Type of regulatory circuit	
Inducible system[b]	Observed[c]	Predicted	Predicted	Observed[c]
Catabolic pathways:				
Arabinose	Activator	⟶	Classical	Classical
Histidine	Repressor	⟶	Autogenous	Autogenous
Lactose	Repressor	⟶	Autogenous	Classical
Nitrate[d]	Repressor	⟶	Autogenous	Autogenous
Maltose	Activator	⟶	Classical	?
D-Serine	Activator	⟶	Classical	?
Biosynthetic pathways:				
Tryptophan[e]	Repressor	⟶	Autogenous	Autogenous
Isoleucine–valine	Activator	⟶	Classical	?
Drug resistance:				
Penicillin[f]	Repressor	⟶	Autogenous	Autogenous
Penicillin[g]	Repressor	⟶	Autogenous	?
Prophages:				
Lambda	Repressor	⟶	Autogenous	Autogenous
P1	Repressor	⟶	Autogenous	?
P2	Repressor	⟶	Autogenous	?
P22	Repressor	⟶	Autogenous	?

[a]An arrow indicates direction of inference. An "observed" column adjacent to a "predicted" column represents the results of independent observations that are used to test the predictions.
[b]Enteric bacteria unless indicated otherwise.
[c]Evidence of various types was used in constructing this table. The evidence is not equally conclusive in each case.
[d]*Aspergillus nidulans.*
[e]*Pseudomonas putida.*
[f]*Bacillus licheniformis.*
[g]*Staphylococcus aureus.*

can be monitored easily by assaying for β-galactosidase activity. This enzyme activity responds to the level of wild type *araC* gene product; it decreased five- to tenfold when a wild type *araC* gene was introduced in the trans position. (This phenomenon is discussed more fully in Section 5.2.) More important in the present context, expression of genes fused to the *araC* promoter was not influenced in the merodiploid strains by the presence or absence of arabinose, the natural inducer of the arabinose system.

Since most of the *araC* gene on the chromosome is deleted in these fusion strains, a site (at the level of transcription or translation) required for arabinose-modulated expression could have been eliminated. However, this seems unlikely. The wild type *araC* gene on the episome would have preserved such a site, and if an increase in level of wild type *araC* gene product had occurred in the presence of arabinose, then this increase would have been reflected in a reduced expression of genes fused to the *araC* promoter on the chromosome. In fact, the data of Casadaban (1976*a*) show that if a reduction in expression occurred, its magnitude was insignificant. Thus, in functional terms, the arabinose system is classically regulated.

3.5 *Other Activator-Controlled Inducible Catabolic Systems*

There is no evidence for autogenous regulation in other well-studied inducible catabolic systems that are under the control of an activator, but conclusive evidence of classical regulation also is lacking.

3.5.1 D-*Serine*

The D-serine deaminase (*dsd*) system consists of two known genes closely linked on the *E. coli* chromosome. One of these codes for D-serine deaminase, the enzyme that catabolizes D-serine; the other is a regulatory gene. In an earlier study, McFall and Bloom (1971) suggested that this system might be autogenously regulated and repressor-controlled. More recent studies (Bloom *et al.*, 1975) have shown, however, that this system is in fact activator-controlled. Mutations in the regulatory gene that lead to superrepressed expression of the gene for D-serine deaminase are cis and trans recessive to the wild type allele in merodiploid strains. Integration of phage lambda into the regulatory gene also leads to superrepressed expression of the gene for D-serine deaminase. The behavior of lambda lysogens that are D-serine deaminase negative, and other findings suggest that the initiator site for the structural gene of D-serine deaminase is located between this gene and the regulatory gene. Since the two known structural genes are not within the same operon, simple autogenous regulation can be ruled out, as Bloom *et al.* (1975) have suggested. However, a more complex regulon model in which expression of the two genes is regulated coordinately cannot be rejected until the level of regulator has been assayed and shown not to vary with induction of D-serine deaminase.

3.5.2 *Maltose*

The maltose (*mal*) system is another well-studied inducible catabolic system in which the structural gene for the activator is located outside the transcriptional units known to be under the control of this activator in *E. coli* (Hofnung, 1974). Again, simple autogenous regulation can be ruled out, but studies that could demonstrate classical regulation have yet to be done. The general approach of Casadaban (1976*b*) ought to be particularly helpful in resolving this question. The structural gene of β-galactosidase recently has been fused to structural genes of the maltose system (Silhavy *et al.*, 1976; Casadaban *et al.*, 1977). By using strains in which the fusion involves the *malP-Q* operon, new regulatory mutations in the structural gene (*malT*) for the activator have been obtained. These mutations result in a constitutive phenotype and are dominant over the wild type allele in merodiploid strains, adding further support to the previous proposal (Hofnung and Schwartz, 1971) that the maltose regulator has no repressor function (M. Schwartz, personal communication).

3.6 Histidine Utilization

Although the concept of autogenous regulation has been discussed in hypothetical terms for years (see Goldberger, 1974), it has come into prominence only recently with the discovery of such mechanisms at the molecular level. The histidine utilization (*hut*) system in *Salmonella typhimurium* was the first well-documented example of autogenous regulation, and the work of Magasanik and his associates, on this system in *S. typhimurium* and related organisms, has had considerable influence on subsequent thought in this field.

The four enzymes of histidine degradation in *S. typhimurium* are encoded in two closely linked transcriptional units. The first two enzymes of the pathway have their structural genes in the "right-hand" operon; the second two have theirs in the "left-hand" operon, along with the structural gene for the repressor of this system. The repressor blocks expression of both operons and induction occurs when the natural inducer urocanate, the first intermediate of the pathway, binds the repressor, causing its removal from the two operator sites on the DNA. This model of the histidine utilization system, and the evidence supporting it, was presented by Smith and Magasanik in 1971.

More recent studies have confirmed this picture and extended our understanding at the molecular level (see Hagen and Magasanik, 1973, 1976; Tyler *et al.*, 1974; Parada and Magasanik, 1975). The studies *in vitro* by Hagen and Magasanik (1973, 1976) are most relevant in the present context. These authors developed an assay for *hut* repressor that utilizes lambda transducing phages carrying the *hut* genes and the ability of phosphocellulose to bind the repressor. With this assay they determined the level of repressor under various conditions, partially purified the repressor, and examined repressor binding to DNA *in vitro*.

The first of these studies (Hagen and Magasanik, 1973) demonstrated that the intracellular concentrations of repressor and of the enzymes encoded with it in the left-hand operon are coordinately regulated, and thus that synthesis of repressor itself is subject to induction, catabolite repression, and mutational changes in the promoter of this operon. The *hut* repressor was also shown to bind specifically to *hut* DNA in the absence of the inducer urocanate but not in its presence. In the second study (Hagen and Magasanik, 1976), repressor binding was studied independently for each of the two *hut* operators. It was shown that binding of repressor to the right-hand operator is three to four times stronger than is binding of repressor to its own operator, that of the left-hand operon. Repressors from mutant strains carrying "superrepressor" alleles of the regulatory gene were also examined for their binding to the two *hut* operators. The ratio of affinities for the two operators could be increased or decreased by mutation, which tends to emphasize further the independence of repressor interaction with the two operators.

In some respects the *hut* system has become the paradigm for autogenous regulation. It is therefore important to examine the thoughts of Hagen and Magasanik concerning the functional significance of autogenous regulation in this system. These authors state that the *hut* system is buffered strongly against premature induction and quick to shut off during deinduction. "The *hut* operons are thus very cautious in their expression" (Hagen and Magasanik, 1973). But what does cautious mean in this context? There are faster responding systems and slower responding systems, but obviously this difference can be due to many things other than autogenous regulation. The comparison these authors had in mind is presumably with a hypothetical *hut* system that lacks autogenous regulation of its repressor. If the concentration of this (hypothetical) constitutively synthesized repressor is assumed to be identical to the uninduced concentration of the autogenously regulated one, then the statements of Hagen and Magasanik (1973) make sense. Both systems would begin induction under the same conditions; but since the repressor is inducible in the autogenously regulated system, further induction of that system becomes more difficult. Once histidine has been exhausted from the medium, the concentration of the inducer urocanate drops, and the higher concentration of repressor in the autogenously regulated system more rapidly shuts off synthesis of *hut* enzymes.

However, if the concentration of the constitutively synthesized repressor is assumed identical to the fully expressed concentration of the autogenously regulated repressor, which is equally reasonable, then the conclusions are just opposite to those reached by Hagen and Magasanik (1973). When precautions are taken to exclude factors irrelevant to the comparison, such as the constitutive concentration of repressor in the present case, one finds that autogenously regulated repressor-controlled systems are inherently superior to classically regulated systems according to the criteria in Table II. In particular, the autogenously regulated systems are quicker to induce.

Thus, the experimental evidence presented by Magasanik and co-workers for autogenous regulation of the *hut* system is consistent with the predictions based on function given in Section 3.3; the statements of Hagen and Magasanik (1973) concerning the significance of autogenous regulation in the *hut* system, however, are inconsistent with these predictions.

3.7 *Other Repressor-Controlled Inducible Catabolic Systems*

There are many systems in this class, but for most of them there is no conclusive evidence for autogenous or classical regulation. I will discuss the lactose (*lac*) system in *E. coli*, for which there is evidence against autogenous regulation, and the nitrate reductase system in *Aspergillus nidulans,* which illustrates a system that is in many respects functionally equivalent to simple autogenous regulation.

3.7.1 Lactose

This system consists of one transcriptional unit with three structural genes under the control of a repressor protein whose structural gene, although closely linked to the others, lies within a second transcriptional unit (see Beckwith and Zipser, 1970; Reznikoff, 1972). Simple autogenous regulation is excluded, since the structural gene for the *lac* repressor lies outside the only transcriptional unit known to be under its control.

Early studies suggested that *lac* repressor could not be induced more than twofold (Novick *et al.*, 1965; Gilbert and Müller-Hill, 1966).

Additional evidence suggesting constitutive synthesis of *lac* repressor comes from the study of mutant strains in which the genes for the *lac* repressor and β-galactosidase have been fused (Müller-Hill and Kania, 1974). Expression of the β-galactosidase gene under the control of the repressor's promoter is not inducible by isopropyl-β, D-thiogalactoside and is not reduced in merodiploid strains containing a wild type repressor gene in trans; it is a function of the promoter to which the β-galactosidase gene is fused. All these fusion strains contain mutations in (or near) the promoter, which are believed to be of the "up-promoter" type. Such mutations could be of the operator-constitutive type. Binding studies with *lac* DNA and repressor have not revealed a strong binding site in the region immediately preceding the regulatory gene (Reznikoff *et al.*, 1974), but one might expect that any binding of repressor to an operator in this region would be weak if it were to allow proper function of the entire *lac* system (see Section 5.8).

Recent evidence concerning operator-like sequences that bind repressor indicates one site within the structural gene for β-galactosidase (Reznikoff *et al.*, 1974; Gilbert *et al.*, 1975), and another within the structural gene for the repressor (Bourgeois and Pfahl, 1976). The function of these sites is unknown.

Thus, all the evidence concerning regulation of the *lac* system suggests that the system is not regulated autogenously. There could be two reasons for this disagreement with the predictions in Section 3.3. The *lac* system might be regulated in some way that is functionally equivalent to simple autogenous regulation. Alternatively, the *lac* system might have additional, as yet unknown, functions that are not reflected in the criteria for functional effectiveness and that require the repressor not to be regulated autogenously.

3.7.2 Nitrate

The nitrate reductase system in the fungus *Aspergillus nidulans* does not involve autogenous regulation in its simplest form, but it appears to represent one mechanism that can be functionally equivalent to this type of regulation (see Section 6). In the absence of the inducer nitrate, nitrate reductase participates in blocking expression of its own structural gene as well as those of nitrite reductase and hydroxylamine reductase (Pateman *et al.*, 1964, 1967). Nitrate reductase is not a traditional repressor, but appears to function in a complex with another protein (or to modify another protein), thereby causing repression (Cove and Pateman, 1969; Cove, 1970). In any case, expression is effectively blocked in the absence of inducer. Upon addition of nitrate, nitrate reductase binds this inducer and is converted to a form that no longer blocks expression of its own structural gene, and induction is initiated. As induction proceeds, more nitrate reductase is produced and hence the level of potential "repressor" increases along with expression of the regulated genes. Thus, nitrate reductase is involved as a negative element in the control of its own synthesis, in keeping with the predictions in Section 3.3. There are other features of this complex system that are interesting and important in themselves (see Arst, 1976, and references cited therein), but they are outside the scope of this discussion, which is intended only to show the functional similarity of this system and a conventional autogenously regulated system, such as *hut*.

3.8 Inducible Biosynthetic Systems

All examples considered thus far in Section 3 have involved inducible *catabolic* enzymes. There also are inducible enzymes in *biosynthetic* pathways. Criteria for the functional effectiveness of such inducible biosynthetic systems are identical to those given in Table II, and the occurrence of autogenous regulation with repressor-controlled, but not activator-controlled, operons can be predicted as before.

3.8.1 Isoleucine–Valine

Acetohydroxy acid isomeroreductase, one of the enzymes shared for the biosynthesis of isoleucine and valine in *E. coli*, is subject to induction by its substrates, either α-acetolactate or α-acetohydroxybutyrate (Ratzkin *et al.*, 1972). Based on the predicted high demand for expression of the isoleucine–valine (*ilv*) system (Section 2.2), one would expect the induction of isomeroreductase to be under the control of an activator whose synthesis does not vary with induction.

Pledger and Umbarger (1973) have described a mutation in *E. coli* that eliminates inducibility of isomeroreductase and is recessive in merodiploid strains carrying the wild type allele in the trans position. On the basis of these and other observations, Pledger and Umbarger have suggested a model involving an activator, or positive element, in the control of isomeroreductase induction. Whether the level of this activator varies with expression of the regulated gene is unknown. One suggestion is that threonine deaminase, the first enzyme in the pathway for the biosynthesis of isoleucine, is the activator (Pledger and Umbarger, 1973). If this were the case, then isomeroreductase would not be autogenously regulated. Since induction of this enzyme can occur while threonine deaminase remains repressed (Ratzkin *et al.*, 1972), the regulation need not be functionally equivalent to autogenous regulation either. Normally, however, induction of isomeroreductase and threonine deaminase is likely to be coordinate and the regulation then would be functionally equivalent to autogenous regulation.

Thus, there is agreement between predictions and observations concerning involvement of an activator in the induction of isomeroreductase, but the identity of the activator and the nature of its regulation remain to be determined.

3.8.2 Tryptophan

Tryptophan synthase, the last enzyme in the pathway for biosynthesis of tryptophan, is induced by its substrate indole-3-glycerol phosphate (InGP) in

Pseudomonas putida (Crawford and Gunsalus, 1966; Maurer and Crawford, 1971) and in *Pseudomonas aeruginosa* (Calhoun *et al.*, 1973). This enzyme is composed of α and β chains, encoded in adjoining cistrons with coordinate expression. The proximal enzymes of this pathway appear to have their levels regulated (Crawford and Gunsalus, 1966; Maurer and Crawford, 1971; Calhoun *et al.*, 1973) in a manner similar to that of the equivalent enzymes in *E. coli* (see Section 2.2). This suggests that in these two microorganisms demand for expression of the tryptophan (*trp*) system also is low and that one can expect induction of tryptophan synthase to be under repressor control. The level of this repressor can also be expected to vary coordinately with that of tryptophan synthase.

Proctor and Crawford (1975) described a mutant of *Pseudomonas putida* harboring a lesion in the structural gene for the α chain of tryptophan synthase that allows induction of this enzyme in response to indole, which normally is not an inducer. These authors further demonstrated that indole is not converted to the natural inducer InGP in this mutant. This recognition of a new effector suggests that a regulator protein has been modified in the mutant strain, and since the mutation that causes the modification is located in the structural gene for the α chain, it can be concluded that the α chain (or a protein acting with it) is a regulator in the induction of tryptophan synthase (Proctor and Crawford, 1975).

In a more recent study, Proctor and Crawford (1976) have described another point mutation in the structural gene of the α chain that causes pleiotropic effects in *P. putida*. In addition to supplying further evidence for autogenous regulation in this system, these results indicate that the regulator might be a repressor. All these results are consistent with the predictions made in Section 3.3.

3.9 Inducible Drug Resistance

Four examples of drug resistance in bacteria, in which the level of resistance varies as a result of regulated gene expression, were discussed in Section 2.3. Resistance to tetracycline in *E. coli* involves decreased cellular permeability (Franklin, 1967; Franklin and Higginson, 1970) whereas resistance to erythromycin in *S. aureus* is caused by a modification of its cellular target, the 50 S ribosomal subunit (Weisblum and Demohn, 1969; Lai and Weisblum, 1971; Lai *et al.*, 1973*a,b*). These two systems are considered in Section 4.8. In the cases of penicillin (Ambler and Meadway, 1969; Kelly and Brammar, 1973*a*) and chloramphenicol (Winshell and Shaw, 1969), resistance results from induction of an enzyme that degrades the drug. It is assumed that these latter systems are formally identical to inducible catabolic systems. Thus, criteria for their functional effectiveness are given in Table II and the results in Section 3.2 allow one to predict that repressor-controlled systems will involve autogenous regulation whereas activator-controlled systems will not.

The system in *Bacillus licheniformis* conferring resistance to penicillin consists of two known genes, the structural gene for penicillinase and a closely linked regulatory gene (Sherratt and Collins, 1973). Constitutive synthesis of penicillinase is produced readily by mutations in the regulatory gene (Dubnau and Pollack, 1965; Kelly and Brammar, 1973*b*), suggesting that the regulator may be a repressor. A mutation in what appears to be the promoter–operator region suggests the gene order: (promoter–operator)–penicillinase gene–repressor gene

(Sherratt and Collins, 1973). Kelly and Brammar (1973*b*) have shown that a frameshift mutation affecting the carboxy terminus of the penicillinase protein has a strong polar effect on expression of the regulatory gene, strongly suggesting that these genes code for a polycistronic mRNA and confirming the gene order just given. This mutation, which has a mild effect on molecular activity of the enzyme (Kelly and Brammar, 1973*a*), results in high-level constitutive synthesis of penicillinase. Thus, the penicillinase system in *B. licheniformis* appears to be repressor-controlled and autogenously regulated in agreement with the predictions in the preceding paragraph.

The structural genes of penicillinase and its regulator also are closely linked in *Staphylococcus aureus* (Richmond, 1965; Lindberg and Novick, 1973). Incorporation of 5-methyltryptophan, a tryptophan analog, into the regulator protein of the penicillinase operon appears to inactivate this protein and cause constitutive synthesis of penicillinase, which contains no tryptophan. The same result occurs both in the wild type organism (Imsande *et al.*, 1972; Imsande, 1973) and in superrepressed mutants with a regulator that appears to have lost its ability to bind inducer (Imsande and Lilleholm, 1976). These results strongly suggest that this system is under the control of a repressor; autogenous regulation is also expected according to these predictions.

The evidence concerning the penicillinase system appears to be consistent with the above predictions. However, it should be noted that the natural inducer of this system is unknown and, indeed, that the natural function(s) of this system have yet to be clearly established (Sachithanandam *et al.*, 1974; Sykes and Matthew, 1976). Thus, the models in Fig. 3 and/or the criteria for functional effectiveness might require modification when more is known about these systems. A reanalysis, which takes these modifications into account, would then be necessary to determine the relative functional advantages of autogenous and classical circuits of regulation.

The system in *S. aureus* conferring resistance to chloramphenicol is regulated (Winshell and Shaw, 1969), but the multiple-copy nature of the plasmid involved has made it difficult to isolate regulatory mutants. Indirect approaches and recombinant DNA methodology are being used currently to investigate the regulation of this system (Shaw, 1974; W. V. Shaw, personal communication), and it seems only a matter of time before these predictions can be tested in the case of chloramphenicol.

3.10 *Inducible Prophage Lambda*

Temperate phages, such as lambda, exist in one of two different states: the prophage state, in which the viral genome is quiescent and replicated along with the host chromosome, or the vegetative state, in which the phage actively directs its own replication at the expense of the host. The ability to respond quickly to appropriate host signals and switch from the prophage to the vegetative state is critical for temperate phages and they have evolved elaborate control systems to ensure that this transition is orderly (see, for example, review by Herskowitz, 1973). Criteria for the functional effectiveness of this induction process are essentially identical to those given in Table II with the exception of criterion 2, which is irrelevant here. Again, functional analysis of autogenous and classical regulation allows one to predict their occurrence in repressor- and activator-controlled systems, respectively.

As described in Section 2.4, coliphage lambda is the most thoroughly studied of the temperate viruses; its induction is under the control of a repressor (Ptashne, 1971; Herskowitz, 1973). Lambda repressor has been shown to exert negative control or repression over its own structural gene *in vivo* (Tomizawa and Ogawa, 1967), in a crude transcription–translation system *in vitro* (Dottin *et al.*, 1975), and in a transcription system with purified components (Meyer *et al.*, 1975). It is clear that this mode of regulation is operative in the lysogenic state and that there is agreement with the foregoing predictions.

There is also evidence that at low concentrations lambda repressor is an activator for transcription of its own structural gene (Reichardt, 1975; Dottin *et al.*, 1975; Meyer *et al.*, 1975). The significance of this activation is not clearly understood. It has been suggested (Ptashne *et al.*, 1976) that in the lysogen a constant concentration of repressor is maintained by combining positive and negative autogenous regulation. However, it can be shown that positive control only makes matters worse; such a combination of positive and negative autogenous regulation is less effective in maintaining a constant repressor concentration than is negative autogenous regulation alone. It seems more likely that the activator function is involved, together with the product of the *tof* gene, in turning off repressor synthesis once induction has occurred, that is, in making the induction process irreversible.

Thus, autogenous regulation of lambda induction appears to be entirely consistent with the predictions based on function. In the lysogen, in the absence of "inducer", the repressor system may be thought of as autonomous (see Section 5) and again autogenous regulation is in accord with the predictions for systems of that class.

4 *Repressible Systems*

As noted in Section 3, the type of regulation governing regulator proteins themselves is influenced primarily by quantitative considerations, such as the relative strengths of regulatory interactions, and the functions that the intact systems must fulfill. Because of the latter factor, inducible systems have been considered separately from repressible ones. Most experimental evidence relating to repressible systems comes from the study of systems for the biosynthesis of amino acids in bacteria. The models in Fig. 4 have been drawn to represent this emphasis on circuitry of the integrated systems; generalizations to repressible systems of other types are given in Section 4.8.

4.1 *Criteria for Functional Effectiveness*

The physiological function of repressible biosynthetic pathways, such as those for the amino acids, is reflected in the criteria for functional effectiveness listed in

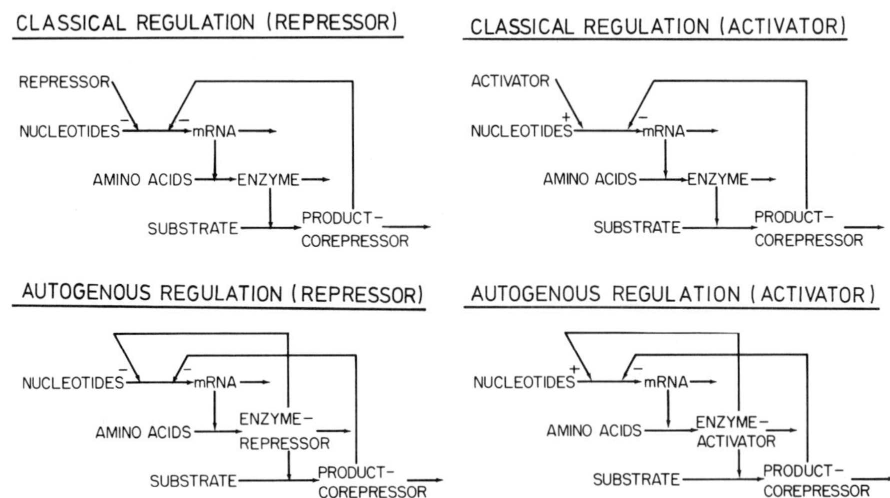

Figure 4. Autogenous and classical regulation of repressible biosynthetic systems. In a classically regulated system the rate of synthesis of regulator is unresponsive to corepressor. These models are analogous to those for inducible catabolic systems (see Fig. 3).

Table IV. Criterion 1 implies that the end product will not be so depleted that its rate of utilization is limited when the demand is increased. Criterion 2 is required for redistribution of common intermediate metabolites to occur in response to a changing environment. Since biosynthetic enzymes become superfluous when the end product is available preformed in the environment, the importance of criterion 3 for cellular economy is obvious. Criterion 4, stability, is essential. Presumably one of the prime functions of pathways synthesizing amino acids is to provide a relatively constant supply of their end products for protein synthesis. An effective system would not oscillate wildly, starving the organism for end product during one phase and overproducing and wasting it in the next. The impaired growth of mutants exhibiting such instability is readily apparent. Criterion 5 implies that the regulatory mechanism should enable the system to respond quickly to changes in its environment. Criterion 6 ensures that the system will continue to function in spite of perturbations in its own structure. Small changes in the system may result from mutations, errors in transcription or translation, or physical influences, such as temperature shifts. These changes tend to exert a deleterious effect on the cell and cells "buffered" against such harmful effects are likely to have a selective advantage.

TABLE IV. Criteria for Functional Effectiveness of Repressible Biosynthetic Systems

1. Minimization of the change in the concentration of end product as it shifts from one steady state to another in response to a change in demand for the end product
2. Maximization of the change in the concentration of end product as it shifts from one steady state to another in response to a change in the availability of the initial substrate
3. Reduction in the rate of enzyme synthesis when the end product is supplied exogenously
4. Stability
5. Temporal responsiveness
6. Insensitivity to perturbations in the system's component parts

These two possibilities are represented schematically in Fig. 4 for repressor- and activator-controlled systems. Again, except for the differences in mode of regulation these models are assumed to be equivalent. These systems have been compared according to the criteria for functional effectiveness given in Table IV (Savageau, 1975, 1976) and the important results are the following: If the first and third criteria for functional effectiveness are more important than the second, as seems reasonable for biosynthetic systems, then systems with constitutive synthesis of repressor are superior to those with autogenously regulated synthesis of repressor. The opposite is true for activator-controlled systems: Systems with autogenously regulated synthesis of activator are superior to systems with constitutive synthesis of activator.

4.3 Predictions

The functional differences given in Section 4.2 have clear implications for the natural selection of autogenous and classical mechanisms of regulation in simple repressible systems of the type represented in Fig. 4. Systems under the control of a repressor protein are not expected to utilize autogenous regulation, whereas those governed by an activator protein can be predicted to have such regulation. Experimental evidence from several repressible systems is now considered within the framework provided by these general predictions. The results are summarized in Table V.

4.4 Tryptophan

There is now good evidence for repressor control in several repressible biosynthetic systems. There is no evidence for autogenous regulation of repressor

TABLE V. Nature of Regulator Correlates with Type of Regulatory Circuit in the Intact System[a]

	Nature of regulator		Type of regulatory circuit	
Repressible system[b]	Observed[c]	Predicted	Predicted	Observed[c]
Biosynthetic pathways:				
Isoleucine–valine[d]	Activator →		→ Autogenous	Autogenous
Tryptophan	Repressor →		→ Classical	Classical
Arginine	Repressor →		→ Classical	?
Histidine	?	Activator ←	←	Autogenous
Isoleucine–valine	?	Activator ←	←	Autogenous
Drug sensitivity:				
Tetracycline	Repressor →		→ Classical	Classical
Erythromycin[e]	Repressor →		→ Classical	?

[a]An arrow indicates direction of inference. An "observed" column adjacent to a "predicted" column represents the results of independent observations that are used to test the predictions.
[b]Enteric bacteria unless indicated otherwise.
[c]Evidence of various types was used in constructing this table. The evidence is not equally conclusive in each case.
[d]*Saccharomyces cerevisiae.*
[e]*Staphylococcus aureus.*

in any of these systems, although studies specifically examining regulation of their regulators have not been reported. The tryptophan (*trp*) operon in *E. coli* is the best documented example of a repressible system involving control by a repressor.

The enzymes of the tryptophan biosynthetic pathway in *E. coli* are formed from five polypeptides encoded in a single cluster of genes, the *trp* operon. Early genetic and physiological studies by Yanofsky and his colleagues and by others showed that expression of this operon is controlled by a repressor protein, whose gene is far removed from the operon itself. The repressor acts at an operator site preceding the structural genes of the operon. This mechanism has been confirmed by recent studies *in vitro.* Tryptophan, the end product of the pathway, is a corepressor that converts the aporepressor into a repressor that in turn binds the operator and prevents transcription (Zubay *et al.*, 1972; Squires *et al.*, 1973; Rose *et al.*, 1973; McGeoch *et al.*, 1973; Zalkin *et al.*, 1974; Shimizu *et al.*, 1974).

There is no evidence for autogenous regulation of the *trp* operon involving repressor. The structural gene for the tryptophan repressor is located outside the known transcriptional units under its control (Bachmann *et al.*, 1976) so the simplest form of autogenous regulation is excluded. More complex models, functionally equivalent to autogenous regulation, in which the concentration of repressor varies in parallel with expression of the *trp* operon, cannot be excluded, since a careful study of repressor concentrations under these conditions has not been reported. However, such variation of repressor concentrations seems unlikely when one compares the data of Zubay *et al.* (1972) and Zalkin *et al.* (1974). The same value apparently was obtained in titration measurements of repressor from cells grown with tryptophan (Zalkin *et al.*, 1974) and those grown without it (Zubay *et al.,* 1972), although methodological differences did exist.

A second mode of regulation for the *trp* operon recently has been identified. At a site between the *trp* operator and the first structural gene of the operon a fraction of the polymerase molecules transcribing this region terminates transcription, and this fraction is dependent on tryptophan availability. Regulation at such a site was suggested by the elevated expression of operator-distal genes in strains with *trp* deletions leaving the operator intact and removing the operator-proximal end of the first structural gene (Jackson and Yanofsky, 1973). This site, which, according to the terminology of Kasai (1974), is called the *attenuator* (Bertrand *et al.*, 1975), has been located approximately 20 to 30 nucleotide pairs before the first structural gene and 130 to 140 nucleotide pairs following the point at which transcription of the *trp* operon is initiated (Pannekoek *et al.*, 1975; Lee *et al.*, 1976). When cells are grown in excess tryptophan about seven of every eight transcripts initiated at the *trp* promoter are terminated at the attenuator (Bertrand *et al.*, 1976). Termination at the *trp* attenuator is relieved when the intracellular concentration of tryptophan is lowered. This does not occur when the intracellular concentrations of other amino acids are lowered (Bertrand and Yanofsky, 1976; Morse and Morse, 1976). Although the extent to which termination is relieved depends on the availability of tryptophan, the actual effector may be $tRNA^{Trp}$, the cognate synthetase, or some product or complex derived from these (Morse and Morse, 1976). The protein or proteins involved in mediating regulation at the *trp* attenuator are unknown, but Pouwels and van Rotterdam (1975) have described a partially purified protein that acts as a positive element to relieve termination at this attenuator. Whether autogenous regulation is involved is unknown.

The well-documented existence of attenuator control in the *trp* system raises the question of its functional significance. Bertrand *et al.* (1975) and Bertrand and Yanofsky (1976) have suggested that it expands the range of regulated expression from 70-fold (repressor control alone) to 700-fold (repressor plus attenuator control). However, the range of regulated gene expression may be greater than 1000-fold in other systems, such as the lactose operon, where only repressor control is involved. Thus, attenuator control is not *necessary* to achieve a wide range of regulated gene expression and is likely to serve additional functions.

Anthranilate synthetase, the first enzyme of the pathway for the biosynthesis of tryptophan, has been suggested for a role in regulation of the *trp* operon (Somerville and Yanofsky, 1965). Stetson and Somerville (1971) and Somerville and Stetson (1974) have reported additional evidence for this view and suggested, on the basis of cis/trans tests, that the enzyme acts as a positive element in the regulation. Such an autogenously regulated activator would be consistent with the predictions discussed in Section 4.3.

As an activator, anthranilate synthetase might be expected to influence the derepressed or repressed frequency at which transcription is initiated in the promoter–operator region. However, normal values of these frequencies are found in strains possessing repressor control but lacking anthranilate synthetase (Hiraga and Yanofsky, 1972). Moreover, in strains (*trpR*) lacking repressor control the normal derepressed frequency exists whether anthranilate synthetase is present or absent (Bertrand and Yanofsky, 1976; Morse and Morse, 1976). Thus, if anthranilate synthetase has a significant role in regulating the initiation of transcription, it must modulate the control by repressor in a way that is apparent only at intermediate frequencies of initiation.

Alternatively, anthranilate synthetase might be expected to influence the deattenuated or attenuated fraction of polymerase molecules that continue transcription beyond the attenuator region. Bertrand and Yanofsky (1976) and Morse and Morse (1976) have shown that these fractions are normal in strains with attenuator control but lacking anthranilate synthetase. However, the converse situation, in which the attenuator control (demonstrated in the absence of anthranilate synthetase) is eliminated by a suitable mutation in the "antiattenuator" and in which anthranilate synthetase is present, has not been examined. This leaves open possibilities such as the following: Anthranilate synthetase and the putative antiattenuator are both activators. The attenuated fraction of polymerase molecules that continues transcription beyond the attenuator is determined by the physical properties of the attenuator in the absence of these activators. The deattenuated fraction is determined by the physical properties of the attenuator in the presence of activator; these properties are assumed to be the same regardless of which activator is present. In such a model, removing one or the other activator (but not both) would leave the attenuated and deattenuated expression of the operon unchanged, since each of the two activators could substitute for the other in this regard. Which of the two normally has the major role in regulating the termination of transcription would be determined by the effective concentration of each activator, and this influence on control would be apparent only when the fraction of polymerase molecules terminated is intermediate between the attenuated and deattenuated fractions. A modulator role for anthranilate synthetase, as discussed earlier with regard to the control of initiation, is also possible in the control of termination.

In spite of the multiple controls that potentially operate in the *trp* system, control by repressor appears to be the dominant mode. Repressor regulates expression of the *trp* operon over a 70-fold range, while expression at the attenuator is regulated over an eight-fold range (Bertrand and Yanofsky, 1976). Bertrand *et al.*, (1975) have suggested that repressor control operates primarily under conditions near tryptophan excess, while attenuator control predominates near tryptophan starvation. Since tryptophan seems to be frequently present in the natural environment of *E. coli* (see Section 2.2), the suggestion of Bertrand *et al.* (1975) is in agreement with the predicted correlation of repressor control and frequent presence of end product in the environment (Section 2.2).

4.5 *Arginine*

The structural genes of the arginine (*arg*) system in *E. coli* are distributed among six separate sites on the chromosome, the only cluster being the bipolar *argECBH* operon. Expression of this system is regulated mainly by the product of the *argR* regulatory gene (see Vogel *et al.*, 1971), which maps separately from the structural genes (Bachmann *et al.*, 1976). Earlier studies suggesting that the *argR* product is an aporepressor (Urm *et al.*, 1973) have been confirmed and extended (Cunin *et al.*, 1976). Neither the arginine aporepressor, partially purified and free of tRNA and arginyl-tRNA synthetase (Kelker *et al.*, 1976), nor the corepressor arginine blocks expression of the *argECBH* operon *in vitro,* but together they effectively repress its transcription (Cunin *et al.*, 1976).

In recent studies, Kelker and Eckhardt (1977) have shown that expression of the structural gene for β-galactosidase, which has been fused to the promoter of the *argA* gene, is repressed by partially purified repressor plus arginine in a system with transcription and translation coupled *in vitro.* These results suggest that the mechanism of regulation governing the *argA* gene and that governing the *arg-ECBH* genes may be essentially identical.

There is no evidence for autogenous regulation of the *arg* system involving repressor. The structural gene for the arginine repressor is not within any of the transcriptional units known to be under its control, a finding that excludes the simplest form of autogenous regulation. Whether repressor plus arginine can block expression of the *argR* gene, or whether the concentration of repressor is modulated in parallel with expression of the regulated structural genes in a less direct manner, must be determined before a form of regulation functionally equivalent to autogenous regulation can be excluded.

There is evidence for significant regulation of the bipolar *argECBH* operon at a point following the initiation of transcription (see Cunin *et al.,* 1975; Krzyzek and Rogers, 1976*a,b,* and references cited therein), but the molecular nature of this control is still uncertain. Whether autogenous regulation is involved is of course also unknown.

Since arginine appears to be frequently present at high concentrations in the natural environment of *E. coli,* the primary importance of control by repressor in the *arg* system is consistent with the predicted correlation of repressor control and frequent presence of end product at high concentrations in the environment (Section 2.2).

The ten enzymes of the pathway of histidine biosynthesis in *Salmonella typhimurium* are encoded in a single gene cluster, the *his* operon. In contrast to the tryptophan and arginine systems, there is no evidence of a traditional aporepressor for the *his* operon, although this system has also been studied extensively. Histidyl-tRNA, rather than histidine itself, has been clearly implicated as the corepressor in repression control of the *his* operon in *S. typhimurium* (see Chapter 10, this volume). The inability to find a traditional regulator protein that interacts with this corepressor has led to the suggestion that the putative regulator must have some additional, essential function under the conditions used in selecting regulatory mutants. The possibilities include histidyl-tRNA synthetase (*negative element:* Brenner and Ames, 1971; *positive element:* Wyche *et al.*, 1974) and the first enzyme of the pathway for histidine biosynthesis (*negative element*: Goldberger and Kovach, 1972; *positive element*: Rothman-Denes and Martin, 1971). These and other models are now being tested *in vitro* with purified (or partially purified) components.

Goldberger (1974) has reviewed evidence implicating the first enzyme of histidine biosynthesis as a negative element in repression control, and additional biochemical (Smith *et al.*, 1974; Meyers *et al.*, 1975*a*) and genetic (Meyers *et al.*, 1975*b*) evidence has since been reported. Such autogenous regulation is inconsistent with repressor control according to the predictions in Section 4.3. Repressor control is also inconsistent with the infrequent presence of histidine at high concentrations in the natural environment of *S. typhimurium* (Section 2.2) and difficult to reconcile with the evidence for positive control *in vitro* discussed later.

Coupled transcription–translation *in vitro* does not exhibit escape synthesis; instead, it reaches a plateau with added *his* DNA, suggesting that a positive factor(s) is required for expression of the *his* operon (Artz and Broach, 1975). A positive factor ought to stimulate its own synthesis (Section 4.3). The uncoupling experiments of Artz and Broach (1975), in which transcription first is carried out in the absence of translation and then accumulated mRNA is translated in the absence of transcription, show that transcription of the *his* operon requires translation of its specific mRNA. This result is consistent with the earlier prediction based on function of the intact system, although other interpretations of the experimental data are possible (Artz and Broach, 1975). With this coupled system repression and derepression can be mimicked *in vitro* without adding the first enzyme, which, taken together with the results of the uncoupling experiment described previously, implies that as a negative element the first enzyme cannot be the sole control involved in repression of the *his* operon (Artz and Broach, 1975).

The results of Artz and Broach (1975) do not exclude the possibility that the first enzyme is a positive element in repression control. A biregulator role for this enzyme might resolve some but not all of the conflicting evidence, and, as discussed elsewhere (Savageau, 1975, 1976), such a role is justified by analysis demonstrating the advantages of biregulator control and its essential similarity to activator control.

The first enzyme cannot be the only element, positive or negative, controlling expression of the *his* operon because normal repressed and derepressed levels of expression have been found in strains of *S. typhimurium* from which essentially all of the structural gene for the first enzyme has been deleted (Scott *et al.,* 1975). However, the extent to which the first enzyme normally might be involved in repression control of the *his* operon, or if it is involved at all, cannot be determined from these results. Until the control exhibited by such deletion mutants is characterized and appropriately eliminated the possibility of control by the first enzyme cannot be examined unambiguously. The situation is analogous to that described for the *trp* operon (Section 4.4).

The absence of dominant control by a traditional aporepressor and the indications of activator control in the *his* operon are consistent with the predictions based on demand for expression of the operon (Section 2.2). The prediction of an autogenously regulated activator (Section 4.3) is supported by experimental evidence, but not proven.

4.7 *Isoleucine–Valine*

Regulation of enzyme levels in the pathway common to the biosynthesis of isoleucine and valine is considered in the prokaryote *E. coli* and the eukaryote *Saccharomyces cerevisiae,* which, despite wide differences in their genetic organization, have some common themes in their regulation of the isoleucine–valine (*ilv*) system.

4.7.1 *ilv in Escherichia coli*

This system consists of three (and possibly more) transcriptional units that code for the enzymes of isoleucine and valine biosynthesis. With the exception of an inducible enzyme (Section 3.8.1) and one multivalently repressed by leucine plus valine, the system is multivalently repressed by isoleucine, leucine, and valine (see Smith *et al.,* 1976).

Although the *ilv* system is more complex than is the *his* operon, there are many similarities: (1) The corepressors appear to be isoleucyl-, valyl-, and leucyl-tRNA rather than the amino acids themselves (see Chapter 10, this volume); (2) in spite of intensive investigation for many years there is no evidence for a traditional regulator (aporepressor), such as that found in the *trp* and *arg* systems; (3) the first enzyme of the biosynthetic pathway for isoleucine, threonine deaminase, has been suggested as a negative (Hatfield and Burns, 1970) and a positive (Levinthal *et al.,* 1973) element in repression control of this system; and (4) repression control appears to be normal in a strain from which the structural gene for threonine deaminase has been deleted (Kline *et al.,* 1974).

The behavior of the strain missing the structural gene for threonine deaminase shows that this enzyme cannot be the sole element in repression control of the *ilv* system, Whether it is normally involved and, if so, what its relative contribution might be remain to be clearly demonstrated. A major role for threonine deaminase as a negative element in the control (Hatfield and Burns, 1970) is inconsistent with the prediction of activator control, based on demand for expression of the *ilv* operon (Section 2.2), and with the predicted correlation of autogenous regulation with activator control (Section 4.3). A major role for threonine deaminase as a positive element in the control (Levinthal *et al.,* 1973) is consistent with these same predictions. A biregulator role for this enzyme might resolve some of the conflicting evidence. Once again the situation is analogous to that described in Section 4.6 for the *his* operon. The essential similarity of activator and biregulator control implies that the latter would also be consistent with the predictions just discussed.

Smith *et al.* (1976) have recently reinterpreted the genetic organization of the *ilv* system in light of polarity effects induced by phage mu and have presented a model of *ilv* regulation that resolves some of the previously existing conflicts. This most recent model includes threonine deaminase as an antiattenuator or positive element in the (autogenous) control of the transcriptional unit harboring its own structural gene. It also includes an undetermined control element(s) affecting initiation of transcription at the corresponding promoter region. Although far from complete, this model includes nothing that is inconsistent with the predictions in Sections 2.2 and 4.3.

4.7.2 ilv in Saccharomyces cerevisiae

The five enzymes of the *ilv* system are located on several different chromosomes in *Saccharomyces cerevisiae* (see deRobichon-Szulmajster and Surdin-Kerjan, 1971). Threonine deaminase, the first enzyme in the biosynthetic pathway for isoleucine, is repressed in response to isoleucine alone, whereas at least three of the remaining enzymes in this system are multivalently repressed by isoleucine, leucine, and valine (Magee and Hereford, 1969; Bussey and Umbarger, 1969; Bollon, 1975). As in *E. coli,* the aminoacyl-tRNAs, rather than the free amino acids, may be the effectors (corepressors) in this system (McLaughlin *et al.,* 1969).

Mutants of three types have implicated threonine deaminase in repression control of the *ilv* system in *S. cerevisiae.* A missense mutation in the structural gene for threonine deaminase that decreases the enzyme's sensitivity to the feedback inhibitor isoleucine by 100-fold alters repression control of most (if not all) enzymes of the *ilv* system with the exception of threonine deaminase (Bollon and Magee, 1971, 1973; Bollon, 1974, 1975). Some nonsense mutations mapping within the structural gene for threonine deaminase simultaneously eliminate the catalytic activity of threonine deaminase and render the remaining enzymes of the system superrepressed, suggesting that threonine deaminase is a positive element in repression control of the latter enzymes (Bollon and Magee, 1971, 1973; Bollon, 1974, 1975). Two missense mutations in the structural gene for threonine deaminase, each of which eliminates its catalytic activity but leaves repression control of the other enzymes of the system unaffected, result in polypeptide products that complement each other in diploid strains to yield partial restoration of threonine deaminase activity, reduced sensitivity of this activity to the feedback inhibitor isoleucine, and altered repression control of the *ilv* enzymes (Bollon, 1974, 1975). Although not specifically noted in these studies, the data of Bollon (1975) indicate that repression control of threonine deaminase has also been altered by this intragenic complementation.

The involvement of threonine deaminase as a positive element in the control is consistent with the predictions in Section 2.2. The predictions in Section 4.3 indicate that the *ilv* system also ought to be autogenously regulated. Since the genes for this system do not lie in a single operon, the simplest form of autogenous regulation is excluded. However, under conditions in which the system is responding to isoleucine (that is, with isoleucyl-tRNA acting as corepressor), the regulation appears to be functionally equivalent to autogenous regulation because all the enzyme concentrations vary in parallel (Magee and Hereford, 1969; Bollon, 1975). From another point of view, whenever valine and leucine are present in relative excess, the *ilv* system effectively functions as an unbranched biosynthetic pathway whose regulation is responsive to its end product isoleucine. Under these conditions regulation of the system is functionally equivalent to autogenous regulation. Under conditions in which regulation of the *ilv* system is responsive to valine or leucine, the regulation may not be functionally equivalent to autogenous regulation (Magee and Hereford, 1969; Bollon, 1975).

Whether threonine deaminase is involved in autogenous regulation of its own structural gene (as opposed to autogenous regulation of the entire system) is unclear. Bollon (1975) concludes that it is not, but there is no evidence in his paper to support this conclusion and, in fact, the results of intragenic complementation (Bollon, 1975) suggest the opposite (see earlier comment).

There are several more enzymes in biosynthetic pathways for which autogenous regulation has been implicated (for examples see Goldberger, 1974), but in most cases neither the nature of the regulator—repressor or activator—nor the type of effector—inducer or corepressor—is clearly defined.* Therefore, these examples cannot be used to test the predictions in Sections 3.3 and 4.3 until additional information becomes available.

4.8 *Repressible Drug Sensitivity*

The last examples in Section 4 were chosen to illustrate repressible systems other than biosynthetic pathways. Although inducible drug resistance, which was used in Section 3.9, and repressible drug sensitivity might be considered synonymous terms, the latter was chosen for the systems in this section because it more accurately reflects the function of these intact systems. Whereas the systems described in Section 3.9 produce resistance by *induction* of an alternate fate for the drug—detoxification—the systems in this section produce resistance by *repression* of some step(s) in the normal pathway for drug action: transport into the cell, in the case of tetracycline; binding to the ribosome target, in the case of erythromycin.

Criteria for the functional effectiveness of these repressible systems are similar to those in Table IV. Criteria 4, 5, and 6 may be retained as they are, and criteria 1 and 3 can be ignored, since they do not apply here. However, criteron 2 must be reinterpreted slightly: The end effect, cell viability or growth, should be unresponsive to change in availability of the drug. In other words, criterion 2,

*Glutamine synthetase from *Klebsiella aerogenes* is an interesting example. It may be considered to be part of an inducible system with α-ketoglutarate being the inducer and/or it may be considered to be part of a repressible system with glutamine being the corepressor (Foor *et al.*, 1975). There are also several other functions of this complex protein to consider (see Ginsburg and Stadtman, 1973; Wohlhueter *et al.*, 1973; Magasanik *et al.*, 1974; Foor *et al.*, 1975). Additional experimentation and analysis will be required before the complex, multifunctional physiology of the system(s) involving this protein can be satisfactorily understood.

which was stated in terms of responsiveness, should be minimized rather than maximized. In fact, this is precisely what occurred during the previous analysis of repressible biosynthetic systems because of the emphasis placed on criteria 1 and 3 (Section 4.2). Thus, the same analysis as that described in Section 4.2 can be applied to repressible drug sensitivity and the results allow one to predict that activator-controlled systems will involve autogenous regulation whereas repressor-controlled systems will not.

4.8.1 *Tetracycline*

Plasmid-determined resistance of *E. coli* to tetracycline is associated with repression of drug transport into the cell (Franklin, 1967; Franklin and Higginson, 1970). This repression of transport also correlates with the presence of a membrane protein specified by the tetracycline resistance determinant of the plasmid (Levy and McMurry, 1974).

Yang *et al.* (1976) have studied the synthesis of this membrane protein *in vitro* by using a coupled transcription–translation system and have identified another plasmid-specified protein that acts as repressor. Partially purified preparations of this repressor specifically block synthesis of the membrane protein and this repression is reversed by addition of tetracycline, or a tetracycline analog, to the system (Yang *et al.*, 1976).

According to the predictions given in Section 4.8, one would expect the repressor to be synthesized constitutively, that is, the same amount of repressor is expected for resistant cells as for sensitive cells. The following evidence suggests that this may be the case. A large amount of membrane protein was synthesized *in vitro* upon addition of tetracycline at subinhibitory concentrations. This was true whether the cell-free system was prepared from cells grown in the presence (resistant cells) or absence (sensitive cells) of tetracycline (Yang *et al.*, 1976). However, more careful measurements of repressor concentration will have to be made by appropriate titrations before one can test adequately the prediction of constitutive synthesis.

4.8.2 *Erythromycin*

The resistance of *S. aureus* to erythromycin and antibiotics of two other classes is due to methylation of a specific adenine residue in the 23 S ribosomal ribonucleic acid (Lai and Weisblum, 1971; Lai *et al.*, 1973*a,b*), which in turn blocks binding of these drugs to the 50 S ribosomal subunit (Weisblum and Demohn, 1969). The molecular basis for regulation of this modification, which occurs in response to erythromycin in the environment, has not been established. *De novo* synthesis of RNA and protein is required for repression of sensitivity (Weisblum *et al.*, 1971) and mutants with constitutive expression of resistance can be obtained readily (Weisblum *et al.*, 1971), providing a hint that this system might be repressor-controlled. If such a mechanism were confirmed by further investigation, then one could predict that the repressor is synthesized constitutively.

TABLE VI. *Experimental Tests for the Predicted Correlation of Autonomous Systems, Autogenous Regulation, and Repressor Control*

System[a]	Autonomous system[b]	Autogenous regulation[b]	Repressor control[b]
Regulator of the arabinose operon	+	+	+
Regulator of phage λ*dv* replication	+	+	+
6-Phosphogluconate dehydrogenase	+	?	?
T antigen of simian virus 40	+	+	+
Termination factor rho	+	+	+
RNA polymerase	?	+	+
Histones of HeLa cells	+	+	+
Unwinding protein of phage T4	+	+	+
Scaffolding protein of phage P22	+	+	+

[a]The organism is *Escherichia coli* unless indicated otherwise.
[b]Evidence of various types was used in constructing this table. The evidence is not equally conclusive in each case.

5 Autonomous Systems

A special case of the systems considered in Sections 3 and 4 occurs when the regulator protein modulating gene expression does not require participation of an effector molecule (inducer or corepressor) in its regulatory function. Such regulators are functionally autonomous (Savageau, 1975, 1976) and systems governed by them are called *autonomous systems.*

5.1 Functional Implications and Predictions

Analysis of autonomous systems indicates that those with an autogenously regulated repressor are optimal with respect to: (1) maintenance of a constant intracellular concentration of gene product; (2) stability; (3) temporal responsiveness to change; and (4) insensitivity to perturbations in the structure of the system itself (Savageau, 1975, 1976). On the basis of these results one could predict that many genes or operons previously thought to be unregulated, or constitutively expressed,* might actually be subject to an autogenously regulated repressor. Goldberger and Deeley (1976) have reviewed several proposed mechanisms for constitutive gene expression and have suggested a similar conclusion.

Experimental evidence that can be used as a test for the predicted correlation of autonomous systems, autogenous regulation, and repressor control is discussed in Sections 5.2 through 5.10. The results are summarized in Table VI.

*I will continue to use the term constitutive in an operational sense to indicate a system with apparently unregulated or invariant expression in response to effector molecules. However, when it is clear that a system involves autogenous regulation, and the term constitutive would be contradictory, the system will be called autonomous.

The classical model of Jacob and Monod (1961) involves a constitutively expressed regulatory gene. The results described in Sections 3 and 4 require that expression of certain regulatory genes not vary in response to particular effectors. Such genes would be obvious candidates for testing the predictions in Section 5.1.

Of the regulatory genes required by theory to have constitutive expression, only the *araC* gene of *E. coli* has been examined carefully for regulation at the molecular level. As predicted, Casadaban (1976*a*) found that expression of the *araC* gene promoter, which was fused to the structural gene for β-galactosidase and monitored by the activity of this enzyme, was not influenced by the inducer arabinose (see Section 3.4). Furthermore, expression of the *araC* gene promoter decreased five- to ten-fold when the product of a wild type *araC* gene was introduced via an episome, a finding that strongly suggests autogenous regulation by a repressor protein. The putative operator for the *araC* operon may have already been detected in the filter-binding studies of Wilcox *et al.* (1974*b*) and the high-resolution electron microscopic studies of Hirsh and Schleif (1976). Thus, with respect to arabinose-specific regulation, the *araC* operon appears to be autonomous, in agreement with the predictions in Section 5.1.

The genetic methods employed by Casadaban to examine regulation of the *araC* gene are quite general (for details see Casadaban, 1976*b*). As he has pointed out, it should be possible to fuse the structural gene of β-galactosidase to many if not all promoters and in this way facilitate the study of gene expression, particularly in those systems with gene products that are difficult to assay. (Regulatory genes are a prime example in the latter category.)

5.3 Regulator of DNA Replication

Another example of an autonomous regulator is that predicted for a replicon, the minimal genetic structure exhibiting autonomous replication, as defined by Jacob *et al.* (1963). The identification of replicons is most advanced in the case of certain small plasmids. For example, plasmid λ*dv* is a fragment one-ninth the length of the coliphage lambda genome and consisting almost entirely of genes that function in regulation and replication; the genes specifying the structural elements of the phage particle have been deleted (Matsubara and Kaiser, 1968). The use of new techniques for obtaining λ*dv* plasmids with mutations in specific genes has reduced further the number of genes that might be involved in this replicon. From such studies the minimum essential elements of the λ*dv* replicon have been identified (Berg, 1974; Matsubara, 1976). It consists of a circular DNA molecule with a single transcriptional unit specifying repressor protein (*tof* gene product) and initiator proteins (*O* and *P* gene products). The repressor acts at the operator (O_R) of this unit to block its transcription (Folkmanis *et al.*, 1976). Between the *tof* gene and the *O* and *P* genes is a site (*ori*) at which DNA replication is initiated when the site has been transcribed and then acted upon by the products of genes *O* and *P*. The number of plasmid copies per cell and repressor concentration are maintained at a rather constant level by this mechanism. If they should be in excess for any reason, repressor shuts off repressor synthesis and plasmid replication; if they should fall below the required level for any reason, repressor synthesis is derepressed and DNA replication initiated. Thus, this system appears to be autonomous, repressor-controlled, and autogenously regulated, in agreement with the predictions in Section 5.1. See Chapter 6 for a more detailed treatment concerning regulation of DNA replication.

5.4 6-Phosphogluconate Dehydrogenase

The intracellular concentrations of many enzymes concerned with the metabolism of central intermediates are relatively invariant (see review by Fraenkel and Vinopal, 1973) and these traditionally have been regarded as some of the best examples of proteins specified by constitutively expressed genes. Thus, one might expect an enzyme such as 6-phosphogluconate dehydrogenase, belonging to the hexose monophosphate shunt in *E. coli,* to be autonomous according to the predictions in Section 5.1.

The structural gene for this enzyme has been obtained on a transducing phage, which has been used recently to program the synthesis of 6-phosphogluconate dehydrogenase in a coupled transcription–translation system *in vitro* (Isturiz and Wolf, 1975). With this system, it now is possible to determine whether some form of the enzyme itself, or some protein encoded in the same transcriptional unit, is involved in repressing expression of the structural gene for 6-phosphogluconate dehydrogenase.

5.5 T Antigen

There is a massive literature on simian virus 40, or SV40 (see *Cold Spring Harbor Symp. Quant. Biol.* **39**: Part 1, 1974), but I shall focus only on recent reports suggesting that synthesis of the *A* gene product, which is believed to be the intranuclear tumor antigen, or T antigen, is autogenously regulated.

There are at least two functions identified with the *A* gene that might require its product to be an autonomous regulator: (1) The *A* gene function is required to establish the transformed state and to maintain the characteristics of that state in some transformed cell lines (Brugge and Butel, 1975; Kimura and Itagaki, 1975; Martin and Chou, 1975; Osborn and Weber, 1975; Tegtmeyer, 1975). This function may be somewhat analogous to that of λ repressor (Section 3.10), which is essentially an autonomous regulator during lysogeny. (2) The *A* gene function is continually required for initiation of DNA replication in a productive infection (Tegtmeyer, 1972; Chow *et al.,* 1974; Manteuil and Girard, 1974). This function may be analogous to that of initiator protein(s) in a replicon.

That the *A* gene might also be autogenously regulated was a proposal made by Tegtmeyer *et al.* (1975) based on the behavior of cells infected with SV40 harboring temperature-sensitive mutations in the *A* gene. Under restrictive conditions these mutants elicit overproduction of a 100,000-dalton protein in both productively infected and transformed cells. The mutant protein was degraded more rapidly *in vivo* and its binding to intranuclear components was reduced in comparison to immunologically cross-reacting protein from cells infected with wild type SV40. These and other results (Tegtmeyer *et al.*, 1975) strongly suggest that the 100,000-dalton protein is the viral tumor antigen encoded in the *A* gene.

The elevated expression of the mutant *A* gene during a productive infection, demonstrated by Tegtmeyer *et al.* (1975), was confirmed and extended by Reed *et al.* (1976), who showed that the accumulation and rate of synthesis of early RNA are also elevated under the same conditions. A difference in expression between the mutant and wild type SV40 exists at permissive temperatures and is accentuated at nonpermissive temperatures. Thus, control of the *A* gene is exerted at the level of transcription.

The product of the *A* gene may be the only viral-coded protein synthesized in the transformed state and early in a productive infection, since it can account for the entire coding capacity of the early region of the SV40 genome (Khoury *et al.*, 1975). Unless one is willing to assign an enormous number of functions to this single protein, one must assume that its profound influence results from an ability to elicit an array of secondary responses specified by the host. In most, if not all, instances it is currently impossible to determine whether a given response is primary or secondary. This is true for regulation of the *A* gene as well, as Tegtmeyer *et al.* (1975) have noted.

Conclusive evidence for autogenous regulation in this system will have to await the purification of the *A* gene product and refinement of transcription systems *in vitro* for SV40. Nevertheless, an *A* gene product that is an autonomous regulator, a repressor, and autogenously regulated would be consistent with the predictions in Section 5.1.

5.6 Transcription Termination Factor Rho

Early transcripts of coliphage lambda are terminated at specific sites by the action of a host protein called rho unless rho's action is antagonized by a phage-specified protein, the product of lambda gene *N* (Roberts, 1969, 1976). Since its discovery (Roberts, 1969) and the elucidation of its role in regulating transcription (see Roberts, 1976), rho has been implicated in attenuator control of the *trp* (Korn and Yanofsky, 1976; Pouwels and Pannekoek, 1976) and *ilv* (Smith *et al.*, 1976) systems in *E. coli.* It seems likely that rho (or rholike proteins) will be an important regulator of other systems involving attenuator control (see Section 4).

The studies of Blumenthal *et al.* (1976), showing that the level of rho varies little when cultures of *E. coli* are grown in media of different composition, lead to the prediction (Section 5.1) that rho is repressor-controlled and autogenously regulated. Ratner (1976) had previously reported results that are consistent with these expectations. He found expression of the structural gene for rho to be elevated in cells having mutations that alter the structure of rho and noted the possibility of autogenous regulation. It remains to be seen whether rho affects its own regulation directly and as a repressor.

5.7 *RNA Polymerase*

DNA-dependent RNA polymerase, apparently the only enzyme responsible for RNA synthesis in *E coli,* is composed of at least four subunits (α, β, β', and σ) arranged according to the formula $\alpha_2\beta\beta'\sigma$ (see Chamberlin and Losick, 1976). the β and β' subunits are encoded in a single operon with the cistron for the β subunit located operator proximal to that for the β' subunit (Errington *et al.,* 1974; Kirschbaum and Scaife, 1974). Evidence of three types suggests that some form of the RNA polymerase molecule may be involved in regulating expression of this $\beta\beta'$ operon:

1. Partial suppression of nonsense mutations in the structural gene for the β subunit leads to steady state rates of β' synthesis that are elevated to levels that depend on the degree of polarity of the amber mutations, whereas the rates of synthesis of the β subunit are nearly normal in these mutants (Glass *et al.,* 1975).

2. Temperature-sensitive mutants with a lesion in the structural gene for the β' subunit synthesize β and β' subunits at an elevated rate when grown at the nonpermissive temperature (Kirschbaum *et al.,* 1975; Taketo *et al.,* 1976). These differences persist for more than 2 h after a shift to the nonpermissive temperature, although there are some differences in detail between the responses of different alleles. The stability of both β and β' subunits in these mutants is greatly diminished (Kirschbaum *et al.,* 1975; Taketo *et al.,* 1976), as it is when they are produced in excess by strains diploid for the $\beta\beta'$ operon (Hayward *et al.,* 1974). Degradation of these subunits may be the normal response when they cannot be assembled into functional polymerase molecules. It has been shown for one of the temperature-sensitive β' mutants that subunit assembly, but not activity of the purified polymerase, is temperature sensitive (Ishihama *et al.,* 1976; Taketo and Ishihama, 1976). The effect of this lesion on expression of the $\beta\beta'$ operon is trans recessive to the wild type allele in merodiploid strains (see Taketo *et al.,* 1976). Recent studies involving another mutant (strain XH110) with a temperature-sensitive β' subunit have shown a striking decrease in polymerase activity at the nonpermissive temperature *in vivo,* whereas assembly of the enzyme from subunits is largely unaffected (J. B. Kirschbaum, personal communication). Of all the temperature-sensitive alleles examined by Kirschbaum and his colleagues, this mutation causes the most dramatic increase in the rate of synthesis of β and β' subunits; there is also an elevation in the rate of synthesis of mRNA from the $\beta\beta'$ operon, but not from the *lac* operon, suggesting specific transcriptional control of the $\beta\beta'$ operon (J. B. Kirschbaum, personal communication). A double mutant, harboring lesions in the structural gene of the β subunit that confer resistance to rifampicin and temperature sensitivity, synthesizes β and β' subunits at rates lower than wild type when grown at the nonpermissive temperature (Kirschbaum *et al.,* 1975; Taketo *et al.,* 1976). However, when the mutant is maintained under these conditions, the rates of β and β' synthesis tend to increase with time.

3. In merodiploid strains containing both rifampicin-sensitive and -resistant β subunits, addition of the drug causes a marked increase in the initial rate of β and β' synthesis (Hayward *et al.,* 1973; Tittawella and Hayward, 1974; Tittawella, 1976). A similar response has been observed when rifampicin was added to some haploid strains (see Nakamura and Yura, 1975). This response was not seen in

experiments with merodiploid strains containing β subunits both sensitive and resistant to streptolydigin, another inhibitor of RNA synthesis that affects the β subunit, demonstrating that the response is not due simply to inhibition of RNA synthesis (Tittawella and Hayward, 1974). In the continued presence of rifampicin some additional mechanism(s) apparently comes into play, because the rates of β and β' synthesis decline from their high initial values until the levels that existed before drug addition are reached (Hayward *et al.*, 1973; Tittawella and Hayward, 1974; Tittawella, 1976). It is unclear why the elevated expression of the $\beta\beta'$ operon does not persist, as it does in the first two types of studies described. However, once the rates of synthesis begin to decline, this trend continues whether or not the drug is removed (Tittawella, 1976). Somewhat analogous observations and interpretations have been reported by Somers *et al.* (1975) for RNA polymerase II activity in a myoblast cell line that is partially resistant to the drug α-amanitin, a specific inhibitor of RNA polymerase II activity.

This evidence is consistent with involvement of autogenous regulation in the control of the $\beta\beta'$ operon in *E. coli*, although in no case is the evidence conclusive. Transcription studies *in vitro* with purified components should soon help to clarify the regulatory interactions governing the synthesis of this important enzyme.

5.8 *Histones*

The histones are basic proteins, rather homogeneous, consisting of five major fractions, and virtually unchanged during evolution of eukaryotic cells (Phillips, 1971). These proteins bind to DNA in performing their function, although their role appears to be more concerned with structural determination of the chromatin than with regulation of gene expression *per se* (Kornberg and Thomas, 1974; Thomas and Kornberg, 1975; Felsenfeld, 1975; Whitlock and Simpson, 1976; Finch and Klug, 1976). Histones generally are found in a 1:1 (weight:weight) ratio with DNA, a finding that can be interpreted as constitutive expression of the histone genes, and from this one can predict an autonomous system governing histone synthesis.

Evidence that (1) newly synthesized DNA is rapidly bound with histones, (2) histones accumulate and their synthesis rapidly declines when DNA synthesis is inhibited, (3) histone synthesis is more resistant than general protein synthesis to partial inhibition by cycloheximide, and (4) the initial rate of histone synthesis is greatly enhanced after removal of this inhibitor of protein synthesis—together with other evidence—led Weintraub (1973) and Butler and Mueller (1973) to propose that free histones shut off expression of their own structural genes in an autonomous fashion.

It was suggested in these studies that newly synthesized histones preferentially bind to regions on the DNA near replicating forks. Recent studies, however, have presented evidence that newly synthesized histones do not become *specifically* associated with the DNA near replicating forks or with any other region of the DNA, but appear to be dispersed randomly throughout the DNA (Jackson *et al.*, 1976; Seale, 1976). Previously bound histones migrate toward regions of DNA active in replication and leave behind regions of unbound DNA, which are then filled in with newly synthesized histones. These observations do not change in any essential way the model for autonomous expression of histone genes.

The function of such a regulatory system, however, would be improved if "nascent DNA" were to have a higher affinity for histones and "operators of the histone genes" a lower affinity for histones than has bulk DNA. It would seem particularly important for this model that the putative operator sites for histone genes have a relatively lower affinity for histones. Since a relatively small fraction of total DNA is free of histones, the migration of histones from other sites to the region of replication would leave the putative operator sites open for transcription an equally small fraction of time if their affinity for histones were identical to that of bulk DNA. This could be insufficient to meet the need for histone synthesis. However, if the affinity of histones for these operators were less than it is for bulk DNA, then histones could migrate preferentially away from the operators and leave them open for transcription a high fraction of time. This higher potential for expression of histone genes would be reduced of course to an appropriate level by accumulation of a free-histone pool. In effect, by having the operators of the histone genes the last to be covered by histones ensures a sufficiently rapid rate of histone synthesis and complete coverage of the DNA when synthesis ceases.

Recent experiments examining the effect of chromosomal proteins on transcription *in vitro* of histone genes from HeLa cells are relevant here. Stein *et al.* (1976) report that "histones by themselves inhibit transcription from DNA, including transcription of histone genes, in a dose-dependent, nonspecific manner." These results indicate that the putative operators of the histone genes have the same affinity for histones as does bulk DNA. Thus, the histone/DNA ratio of chromatin would have to fall by one-half before the rate of histone synthesis would derepress to 50% of its full capacity. These authors also showed that nonhistone proteins by themselves neither activate nor inhibit transcription of histone genes.

However, Park *et al.* (1976) have shown that dissociated chromatin reconstituted in the presence of nonhistone proteins obtained from cells undergoing DNA replication is an active template for transcription of histone genes. Furthermore, the active component appears to be a phosphoprotein from the nonhistone fraction (Thompson *et al.*, 1976) and chromatin reconstituted with enzymatically dephosphorylated nonhistone proteins is less active as a template for transcription of histone genes (Kleinsmith *et al.*, 1976). Phosphorylated nonhistone protein appears specifically to counteract histone-mediated repression of transcription from histone genes. The phosphoprotein(s) might, for example, reduce the "off time" for histone binding at the putative operators for the histone genes, but regardless of the exact mechanism, which remains to be elucidated, the end result may be the same. Namely, the effective affinity of histones for the putative operators of their own structural genes would be lowered, as seems required for autogenous regulation of histone gene expression.

5.9 *Unwinding Protein*

Like many of the proteins discussed thus far in Section 5, *unwinding protein* (the product of gene 32) from coliphage T4 is a protein whose principal function involves binding DNA and, as convincingly documented by Gold and his colleagues (Gold *et al.*, 1976; Russel *et al.*, 1976), it is autonomous. However, unlike previously discussed systems for which the evidence either clearly indicates transcriptional control or is entirely consistent with the predominance of such control, the control of gene 32 expression occurs at the level of translation (Russel *et al.*, 1976).

The protein product of gene 32 is required throughout T4 infection; it is essential for both DNA synthesis and genetic recombination; and it plays an important role in DNA repair and probably other processes by protecting or stabilizing single-stranded DNA. *In vitro,* this protein binds strongly to single-stranded DNA and stimulates DNA synthesis catalyzed by the T4-specified DNA polymerase (see Gold *et al.,* 1976, and references cited therein).

That gene 32 might be autogenously regulated was suggested by the work of Krisch *et al.* (1974). These authors showed that strains with a nonsense mutation in gene 32 overproduce the protein fragment encoded in that portion of the gene proximal to the mutation. Similarly, the temperature-sensitive products resulting from missense mutations in gene 32 were overproduced at the nonpermissive temperature. The overproduction of altered protein by these mutants is recessive in mixed infections with wild type phage. Finally, Krisch *et al.* showed that overproduction of the normal product of gene 32 occurred when normal DNA synthesis or expression of late genes was altered by mutations that nevertheless permit DNA synthesis.

Gold and his colleagues demonstrated that expression of gene 32 is controlled directly by its protein product and that the concentration of this protein is in turn responsive to the amount of single-stranded DNA present during infection. Direct control of gene 32 expression is indicated because many different mutations in gene 32 all lead to overproduction of the altered protein but have no detectable effect on other proteins in the system (Gold *et al.,* 1976). An indirect effect would have to be attributed to a change in some minor protein or to some small metabolite accumulated or depleted as a result of mutation in gene 32. These possibilities are made extremely unlikely because the translational control observed in infected cells is also seen *in vitro*. Translation of mRNA from gene 32 is repressed efficiently and specifically *in vitro* by addition of the protein product (see Russel *et al.,* 1976). The indirect effect of single-stranded DNA on expression of gene 32 was demonstrated with T4 phage wild type for gene 32. The amount and state of intracellular DNA were systematically varied by infecting cells with phage mutated in one or more of 16 different genes. The expression of gene 32 in all cases was correlated with the amount of single-stranded DNA, as determined from the known properties of the mutants used (Gold *et al.,* 1976). Thus, gene 32 in phage T4 represents an autonomous system, repressor-controlled and autogenously regulated.

Russel *et al.* (1976) have demonstrated that this control occurs at the translational level: (1) The functional half-life of mRNA from gene 32 is the longest of any mRNA in the infected cell; (2) expression of gene 32, normally repressed late during infection, can be fully derepressed without any increase in mRNA synthesis; and (3) the amount of mRNA from gene 32 that can direct the synthesis of protein *in vitro* does not decrease when expression of gene 32 is repressed. Furthermore, unpublished results indicate that translation of this mRNA is repressed efficiently and specifically *in vitro* by addition of the protein product (see Russel *et al.,* 1976).

Gold and his colleagues have suggested, as the simplest model consistent with all of their observations and the known DNA- and RNA-binding properties of gene 32 protein, that this protein binds *nonspecifically* to its own mRNA and blocks translation. This repression is made *specific* for gene 32 expression by virtue of a uniquely designed mRNA (Russel *et al.*, 1976).

These authors have shown experimentally or by carefully reasoned deductions the advantages of their model according to several functional criteria. Their conclusions are in complete agreement with an independent analysis showing that such an autonomous model represents an optimal design by several of the same criteria (Savageau, 1975, 1976). Furthermore, Gold and his colleagues have pointed out unique attributes of their model that are inherent in its molecular detail. Among the most interesting is that mutations in the initiation region cannot lead to constitutive expression, as might occur if repressor binding required recognition of a specific RNA sequence (Russel *et al.*, 1976). The feature of nonspecific binding in this model may, however, exclude its widespread use in other systems, since only one such mechanism at a time can function effectively in a given cell.

5.10 Scaffolding Protein

Proper assembly of the capsid structure for phage P22 requires a protein, *scaffolding protein,* the product of phage gene 8, but this protein is absent from the mature phage particle. Scaffolding protein interacts with coat protein during assembly of a precursor or "prohead" structure, which contains these proteins in a numerical ratio of about 250:420, but at the time of DNA encapsulation scaffolding protein is removed and reused for the assembly of additional capsid structures (King and Casjens, 1974).

Scaffolding protein normally is synthesized at a relatively constant rate throughout late infection. However, in mutant phages that cannot encapsulate DNA the scaffolding protein becomes trapped in prohead structures and the rate of synthesis of scaffolding protein increases. Conversely, in mutants that cannot form the prohead structure, unassembled scaffolding protein accumulates and its rate of synthesis declines. Thus, scaffolding protein appears to repress expression of its own structural gene (King *et al.*, 1976). Once again, these results are consistent with the predicted (Section 5.1) concurrence of apparent constitutive expression, an autonomous system, repressor control, and autogenous regulation. However, whether the repressive effect of scaffolding protein on expression of its own structural gene is direct (autogenous regulation) and whether this occurs at the level of transcription or translation are questions that remain to be answered.

6 Discussion

This is an appropriate place to review and reevaluate the working definitions with which this chapter began. Autogenous regulation, as defined by Goldberger (1974), is direct involvement of a protein in regulating expression of its own structural gene. This might be called a "structural" definition of autogenous regulation. Autogenous regulation in this sense is excluded by the definitions of classical repressor-controlled (Jacob and Monod, 1961) and activator-controlled (Englesberg and Wilcox, 1974) operons. Quite rigorous experimental conditions must be satisfied before autogenous regulation of a system can be claimed (Goldberger, 1974; Goldberger and Deeley, 1976). Therefore, it is not surprising that although this form of regulation appears to be very widespread, there are still relatively few well-documented examples.

Many examples of autogenous regulation involve a regulator protein that also has a catalytic or structural role, and is, therefore, bifunctional or multifunctional. Cove (1974) has suggested that autogenous regulation is a special case of catalytic or structural proteins that also have a regulatory role, whether or not they regulate the expression of their own structural gene. In at least one other case, the two phenomena have been considered synonymous (Graf *et al.*, 1976). However, there are autogenously regulated proteins that are not bifunctional, and bifunctional regulator proteins that are not involved in autogenous regulation. Furthermore, the functional implications of these two phenomena can be distinguished by analysis of the intact regulatory systems. It is appropriate, therefore, to keep a clear distinction between the concepts of autogenous regulation and bifunctionality of proteins.

Autogenous regulation also might be defined in a "functional" sense as regulation in which the intracellular concentration of regulator protein varies in parallel with expression of the regulated structural genes according to the degree of repression or induction. (Autogenous regulation in the structural sense then would be a special case.) Classical regulation, by the corresponding functional definition, is that in which the intracellular concentration of regulator remains unchanged, whereas expression of the regulated structural genes varies according to the degree of repression or induction. If one accepts this functional perspective, then the need to distinguish between direct and indirect involvement of the regulator in modulating expression of its own structural gene is less pressing and autogenous regulation can be understood in terms of a general theory with predictive value. In spite of these advantages, there is a possibility for confusion if this terminology was used now. Therefore, I have continued to use the term autogenous regulation as originally defined by Goldberger (1974) and have referred to the larger class of mechanisms as functionally equivalent to autogenous regulation. The need to maintain this distinction may disappear naturally if the functional perspective becomes widely adopted.

In dealing with more complex systems of regulation, Goldberger (1974) has suggested a corollary to his definition of autogenous regulation: "If any one of the proteins that influences expression of a gene is autogenously regulated, then regulation of the entire system is autogenous." In light of recent findings this corollary should be modified. In the case of the histidine utilization system (Section 3.6), one operon, but not the other, is regulated autogenously according to the structural definition. However, both are regulated autogenously by the functional definition. In the case of the arabinose system (Sections 3.4 and 5.2), the regulator protein itself is regulated autogenously, but the other operons governed by this regulator are not regulated autogenously, by either the structural or functional definition. I propose that the modes of control for each transcriptional unit of a system be specified separately, and that only if all transcriptional units of a system are regulated autogenously, or regulated by a mechanism functionally equivalent to autogenous regulation, should the entire system be called autogenously regulated.

An autonomous system, by definition, lacks effector molecules that can alter its regulated expression. This definition has the drawback of involving "negative" criteria. Thus, effectors may be found for some of the systems considered in Section 5 and, in those cases, the system would have to be reclassified as inducible. It should also be pointed out that in the absence of their effectors, inducible systems involving an autogenously regulated repressor are equivalent to autonomous systems.

The theory described in this chapter is not formulated in terms of any particular system, but is abstract and general. It is based on a mathematical analysis of inducible, repressible, and autonomous systems involving repressor and activator mechanisms integrated in autogenous and classical networks of regulation. This theory has yielded the first general insight into the functional significance of repressor and activator control mechanisms, and the autogenous and classical networks of regulation that include these mechanisms. The principal implications of this theory may be stated in terms of four theorems of regulation.

1. *Repressor control occurs in systems for which there is a low demand for expression of the regulated structural genes.*
2. *Activator control occurs in systems for which there is a high demand for expression of the regulated structural genes.*
3. *Autogenous regulation occurs with repressor control in inducible and autonomous systems and with activator control in repressible systems.*
4. *Classical regulation occurs with activator control in inducible systems and with repressor control in repressible systems.*

The evidence examined within the framework of this theory covers a spectrum of quality. In a few cases the evidence is rigorous and conclusive; in many other cases it is substantial though not definitive; and in some cases the evidence is only suggestive. Although agreement between theory and experimental evidence in any single case cannot be weighted heavily, uniform agreement with evidence taken from a large number and variety of systems is unlikely to be fortuitous. Of all the cases examined, only that of the lactose operon (Sections 2.1 and 3.7.1) appears to be inconsistent with the theory and this inconsistency is only partial. Since there is currently no other theory of regulation with comparable predictive value, this theory seems to be the best available guide for the investigation of new systems until it can be replaced by a better one.

Aside from its use in testing the theory just summarized, the experimental evidence itself suggests additional generalizations. First, the assembly of complex structures appears to be facilitated by having the structural elements synthesized in a carefully regulated fashion. For a structural protein, which may or may not have enzymatic activity, the protein itself would seem to be a natural regulator for expression of its own structural gene. Structures with components exhibiting such autogenous regulation include phage capsid (Section 5.10), chromatin (Section 5.8), and RNA polymerase (Section 5.7). Autogenous regulation may also be involved in coordinating the synthesis of ribosomes (Dennis and Nomura, 1975*a,b*) (see also Chapter 12, this volume) and the outer cell envelope of *E. coli* (Datta *et al.,* 1976).

Second, proteins whose principal function involves binding to nucleic acid (see Chapter 8, this volume) appear frequently to be regulated autogenously. Examples include repressor of the *hut* system (Section 3.6), repressor of phage lambda (Section 3.10), *tof* gene product of phage lambda (Section 5.3), regulator of the *ara* system (Section 5.2), *A* gene product of SV40 (Section 5.5), termination factor rho (Section 5.6), RNA polymerase (Section 5.7), histones (Section 5.8), unwinding protein of phage T4 (Section 5.9), and DNA polymerase of phage T4 (Russel, 1973; reviewed by Goldberger, 1974). Russel (1973) had noted earlier that autogenous regulation in phage T4 seems to be restricted to two (or possibly three) early proteins that are essential for DNA synthesis. Krisch *et al.* (1974) conjectured that since these two proteins already had one site for the recognition of DNA, acquisition of a second but related specificity, that of a repressor, could have occurred with minimal alteration of the proteins. The frequency with which autogenous regulation occurs among DNA-binding proteins tends to confirm this conjecture and suggests that it may be an even more general phenomenon.

The generalizations in the last two paragraphs may provide a clue to the early evolution of mechanisms regulating gene expression. It is widely accepted that the earliest form of life as we know it had to include a self-replicating structure of nucleic acid and, if not initially then soon after, protein (Miller and Orgel, 1974). The earliest of proteins, which were encoded by and facilitated the replication of nucleic acid, undoubtedly were quickly selected for efficient assembly and function within the replicating structure, and thus were also the first to have their synthesis regulated. Since autogenous regulation appears to be favored among such nucleic acid-binding proteins, autogenous regulation was probably the first mechanism governing specific gene expression.

Later, when more specialized proteins for biosynthetic and catabolic pathways were evolving, their regulatory mechanisms no longer could be autonomous but had to respond to specific environmental signals or effectors. The regulator proteins for such systems require two specificities: one for a specific effector and the other for nucleic acid. This dual specificity could be obtained most readily by evolution starting with a protein that already possessed one of the two specificities: (1) a protein that already had a binding site for the effector, such as a previously evolved enzyme in the pathway to be regulated, or (2) a nucleic acid-binding protein. Cove (1974) and Goldberger and Deeley (1976) have argued, according to the first possibility, that autogenous regulation occurred early in the evolution of regulatory mechanisms for biosynthetic and catabolic pathways. However, the evolution of regulation in such pathways according to the second possibility might have been more likely. Based on the reasoning given earlier, proteins that bind nucleic acid were probably the first to evolve. Later, there undoubtedly would have been many more of these, in comparison to the number of proteins that bind a specific effector, from which the protein with both specificities could have evolved. Furthermore, it would seem no more difficult for a protein with both specificities to evolve from one that binds nucleic acid than from one that binds a specific effector. Whether systems that evolved in this second way would initially have been autogenously regulated or classically regulated would depend on the nucleic acid-binding protein from which their regulator evolved and the specific mutational events that occurred during this evolution.

ACKNOWLEDGMENTS

My own research was supported in part by grants from the National Science Foundation. I thank all those who made comments on earlier versions of the manuscript; Dr. M. J. Casadaban, Dr. N. E. Kelker, Dr. J. B. Kirschbaum, Dr. W. K. Maas, and Dr. M. Schwartz for communicating unpublished results to me; Dr. M. Osborn and Dr. K. Weber for helpful discussions; and Dr. R. F. Goldberger for detailed criticisms. This chapter was written in 1976 while I was on sabbatical at the Max-Planck-Institut für biophysikalische Chemie in Göttingen. I acknowledge the facilities provided by Dr. M. Eigen, the hospitality of the Fulbright Kommission, and the generosity of the John Simon Guggenheim Memorial Foundation.

References

Adibi, S. A., and Gray, S. J., 1967, Intestinal absorption of essential amino acids in man, *Gastroenterology* **52**:837–845.

Adibi, S. A., Gray, S. J., and Menden, E., 1967, The kinetics of amino acid absorption and alteration of plasma composition of free amino acids after intestinal perfusion of amino acid mixtures, *Am. J. Clin. Nutr.* **20**:24–33.

Ambler, R. P., and Meadway, R. J., 1969, Chemical structure of bacterial penicillinases, *Nature* **222**:24–26.

Anderson, E. S., 1968, The ecology of transferable drug resistance in the enterobacteria, *Annu. Rev. Microbiol.* **22**:131–180.

Arst, H. N., 1976, Integrator gene in *Aspergillus nidulans, Nature* **262**:231–234.

Artz, S. W., and Broach, J. R., 1975, Histidine regulation in *Salmonella typhimurium:* An activator–attenuator model of gene regulation, *Proc. Natl. Acad. Sci. U.S.A.* **72**:3453–3457.

Bachmann, B. J., Low, K. B., and Taylor, A. L., 1976, Recalibrated linkage map of *Escherichia coli* K-12, *Bacteriol. Rev.* **40**;116–167.

Barksdale, L., and Arden, S. B., 1974, Persisting bacteriophage infections, lysogeny and phage conversions, *Annu. Rev. Microbiol.* **28**:265–299.

Beckwith, J. R., and Zipser, D., eds., 1970, *The Lactose Operon,* Cold Spring Harbor Lab., Cold Spring Harbor, New York.

Berg, D. E., 1974, Genes of phage λ essential for λ*dv* plasmids, *Virology* **62**:224–233.

Bernhard, H. P., 1976, The control of gene expression in somatic cell hybrids, *Int. Rev. Cytol.* **47**:289–325.

Bertani, L. E., and Bertani, G., 1971, Genetics of P2 and related phages, *Adv. Genet.* **16**:199–237.

Bertrand, K., and Yanofsky, C., 1976, Regulation of transcription termination in the leader region of the tryptophan operon of *Escherichia coli* involves tryptophan or its metabolic product, *J. Mol. Biol.* **103**:339–340.

Bertrand, K., Korn, L., Lee, F., Platt, T., Squires, C. L., Squires, C., and Yanofsky, C., 1975, New features of the regulation of the tryptophan operon, *Science* **189**:22–26.

Bertrand, K., Squires, C., and Yanofsky, C., 1976, Transcription termination *in vivo* in the leader region of the tryptophan operon of *Escherichia coli, J. Mol. Biol.* **103**:319–337.

Bihler, I., 1969, Intestinal sugar transport: Ionic activation and chemical specificity, *Biochim. Biophys. Acta* **183**:169–181.

Bloom, F. R., McFall, E., Young, M. C., and Carothers, A. M., 1975, Positive control in the D-serine deaminase system of *Escherichia coli* K-12, *J. Bacteriol.* **121**:1092–1101.

Blumenthal, R. M., Reeh, S., and Pedersen, S., 1976, Regulation of transcription factor ρ and the α subunit of RNA polymerase in *Escherichia coli* B/r, *Proc. Natl. Acad. Sci. U.S.A.* **73**:2285–2288.

Bollon, A. P., 1974, Fine structure analysis of a eukaryotic multifunctional gene, *Nature* **250**:630–634.

Bollon, A. P., 1975, Regulation of the *ilv* 1 multifunctional gene in *Saccharomyces cerevisiae, Mol. Gen. Genet.* **142**:1–12.

Bollon, A. P., and Magee, P. T., 1971, Involvement of threonine deaminase in multivalent repression of the isoleucine–valine pathway in *Saccharomyces cerevisiae, Proc. Natl. Acad. Sci. U.S.A.* **68**:2169–2172.

Bollon, A. P., and Magee, P. T., 1973, Involvement of threonine deaminase in repression of the isoleucine–valine and leucine pathways in *Saccharomyces cerevisiae, J. Bacteriol.* **113**:1333–1344.

Bourgeois, S., and Pfahl, M., 1976, Repressors, *Adv. Protein Chem.* **30**:1–99.

Brenner, M., and Ames, B. N., 1971, The histidine operon and its regulation, in: *Metabolic Regulation,* Vol. V (H. J. Vogel, ed.), pp. 349–387, Academic Press, New York.

Brugge, J. S., and Butel, J. S., 1975, Involvement of the simian virus 40 gene *A* function in the maintenance of transformation, *J. Virol.* **15**:619–635.

Bussey, H., and Umbarger, H. E., 1969, Biosynthesis of branched-chain amino acids in yeast: Regulation of synthesis of the enzymes of isoleucine and valine biosynthesis, *J. Bacteriol.* **98**:623–628.

Butler, W. B., and Mueller, G. C., 1973, Control of histone synthesis in HeLa cells, *Biochim. Biophys. Acta* **294**:481–496.

Calhoun, D. H., Pierson, D. L., and Jensen, R. A., 1973, The regulation of tryptophan biosynthesis in *Pseudomonas aeruginosa, Mol. Gen. Genet.* **121**:117–132.

Casadaban, M. J., 1976*a,* Regulation of the regulatory gene for the arabinose pathway, *araC, J. Mol. Biol.* **104**:557–566.

Casadaban, M. J., 1976*b,* Transposition and fusion of the *lac* genes to selected promoters in *Escherichia coli* using bacteriophage lambda and mu, *J. Mol. Biol.* **104**:541–555.

Casadaban, M. J., Silhavy, T. J., Berman, M. L., Shuman, H. A., Sarty, A. V., and Beckwith, J. R., 1977, Construction and use of gene fusions directed by bacteriophage mu insertions, in: *DNA Insertion Elements, Plasmids, and Episomes* (A. I. Bukhari, J. A. Shapiro, and S. L. Adhya, eds.), pp. 531–535, Cold Spring Harbor Lab., Cold Spring Harbor, New York.

Chamberlin, M., and Losick, R., eds., 1976, *RNA Polymerases,* Cold Spring Harbor Lab., Cold Spring Harbor, New York.

Chow, J. Y., Avila, J., and Martin, R. G., 1974, Viral DNA synthesis in cells infected by temperature-sensitive mutant of simian virus 40, *J. Virol.* **14**:116–124.

Cove, D. J., 1970, Control of gene action in *Aspergillus nidulans, Proc. Roy. Soc. Lond. B* **176**:267–275.

Cove, D. J., 1974, Evolutionary significance of autogenous regulation, *Nature* **251**:256.

Cove, D. J., and Pateman, J. A., 1969, Autoregulation of the synthesis of nitrate reductase in *Aspergillus nidulans, J. Bacteriol.* **97**:1374–1378.

Cozzarelli, N. R., Freedberg, W. B., and Lin, E. C. C., 1968, Genetic control of the L-α-glycerophosphate system in *Escherichia coli, J. Mol. Biol.* **31**:371–387.

Crawford, I. P., and Gunsalus, I. C., 1966, Inducibility of tryptophan synthase in *Pseudomonas putida, Proc. Natl. Acad. Sci. U.S.A.* **56**:717–724.

Cunin, R., Boyen, A., Pouwels, P., Glansdorff, N., and Crabeel, M., 1975, Parameters of gene expression in the bipolar *argECBH* operon of *E. coli* K12: The question of translational control, *Mol. Gen. Genet.* **140**:51–60.

Cunin, R., Kelker, N., Boyen, A., Yang, H.-L., Zubay, G., Glansdorff, N., and Maas, W. K., 1976, Involvement of arginine in *in vitro* repression of transcription of arginine genes C, B and H in *Escherichia coli* K12, *Biochem. Biophys. Res. Commun.* **69**:377–382.

Datta, D. B., Krämer, C., and Henning, U., 1976, Diploidy for a structural gene specifying a major protein of the outer cell envelope membrane from *Escherichia coli* K-12, *J. Bacteriol.* **128**:834–841.

Davidson, R. L., 1974, Gene expression in somatic cell hybrids, *Annu. Rev. Genet.* **8**:195–218.

Davies, J. E., and Rownd, R., 1972, Transmissible multiple drug resistance in Enterobacteriaceae, *Science* **176**:758–768.

Davis, B. D., 1961, The teleonomic significance of biosynthetic control mechanisms, *Cold Spring Harbor Symp. Quant. Biol.* **26**:1–10.

Davis, F. M., and Adelberg, E. A., 1973, Use of somatic cell hybrids for analysis of the differentiated state, *Bacteriol. Rev.* **37**:197–214.

deCrombrugghe, B., Chen, B., Anderson, W., Nissley, P., Gottesman, M., and Pastan, I., 1971, *Lac* DNA, RNA polymerase and cyclic AMP receptor protein, cyclic AMP, *lac* repressor and inducer are the essential elements for controlled *lac* transcription, *Nature New Biol.* **231**:139–142.

DeMoss, R. D., and Moser, K., 1969, Tryptophanase in diverse bacterial species, *J. Bacteriol.* **98**:167–171.

Dennis, P. P., and Nomura, M., 1975*a*, Stringent control of the transcriptional activities of ribosomal protein genes in *E. coli, Nature* **255**:460–465.

Dennis, P. P., and Nomura, M., 1975*b*, Regulation of the expression of ribosomal protein genes in *Escherichia coli, J. Mol. Biol.* **97**:61–76.

deRobichon-Szulmajster, H., and Surdin-Kerjan, Y., 1971, Nucleic acid and protein synthesis in yeasts: Regulation of synthesis and activity, in: *The Yeasts,* Vol. 2 (A. H. Rose and J. S. Harrison, eds.), pp. 335–418, Academic Press, New York.

Dottin, R. P., Cutler, L. S., and Pearson, M. L., 1975, Repression and autogenous stimulation *in vitro* by bacteriophage lambda repressor, *Proc. Natl. Acad. Sci. U.S.A.* **72**:804–808.

Dubnau, D. A., and Pollock, M. R., 1965, Genetics of *Bacillus licheniformis* penicillinase: A preliminary analysis from studies on mutation and inter-strain and intra-strain transformations, *J. Gen. Microbiol.* **41**:7–21.

Englesberg, E., and Wilcox, G., 1974, Regulation: Positive control, *Annu. Rev. Genet.* **8**:219–242.

Eron, L., and Block, R., 1971, Mechanism of initiation and repression of *in vitro* transcription of the *lac* operon of *Escherichia coli, Proc. Natl. Acad. Sci. U.S.A.* **68**:1828–1832.

Errington, L., Glass, R., Hayward, R., and Scaife, J., 1974, Structure and orientation of an RNA operon in *Escherichia coli, Nature* **249**:519–522.

Felsenfeld, G., 1975, String of pearls, *Nature* **257**:177–178.

Finch, J. T., and Klug, A., 1976, Solenoidal model for superstructure in chromatin, *Proc. Natl. Acad. Sci. U.S.A.* **73**:1897–1901.

Finch, L. R., and Hird, F. J. R., 1960, The uptake of amino acids by isolated segments of rat intestine. II. A survey of affinity for uptake from rates of uptake and competition for uptake, *Biochim. Biophys. Acta* **43**:278–287.

Folkmanis, A., Takeda, Y., Simuth, J., Gussin, G., and Echols, H., 1976, Purification and properties of a DNA-binding protein with characteristics expected for the Cro protein of bacteriophage λ, a repressor essential for lytic growth, *Proc. Natl. Acad. Sci. U.S.A.* **73**:2249–2253.

Foor, F., Janssen, K. A., and Magasanik, B., 1975, Regulation of synthesis of glutamine synthetase by adenylylated glutamine synthetase, *Proc. Natl. Acad. Sci. U.S.A.* **72**:4844–4848.

Fraenkel, D. G., and Vinopal, R. T., 1973, Carbohydrate metabolism in bacteria, *Annu. Rev. Microbiol.* **27**:69–100.

Franklin, T. J., 1967, Resistance of *Escherichia coli* to tetracyclines: Changes in permeability to tetracyclines in *Escherichia coli* bearing transferable resistance factors, *Biochem. J.* **105**:371–378.

Franklin, T. J., and Higginson, B., 1970, Active accumulation of tetracycline by *Escherichia coli, Biochem. J.* **116**:287–297.

Gibson, Q. H., and Wiseman, G., 1951, Selective absorption of stereo-isomers of amino acids from loops of the small intestine of the rat, *Biochem. J.* **48**:426–429.

Gilbert, W., and Müller-Hill, B., 1966, Isolation of the *lac* repressor, *Proc. Natl. Acad. Sci. U.S.A.* **56**:1891–1898.

Gilbert, W., Gralla, J., Majors, J., and Maxam, A., 1975, Lactose operator sequences and the action of lac repressor, in: *Protein–Ligand Interactions* (H. Sund and G. Blauer, eds.), pp. 193–206, de Gruyter, Berlin.

Ginsburg, A., and Stadtman, E. R., 1973, Regulation of glutamine synthetase in *Escherichia coli,* in: *The Enzymes of Glutamine Metabolism* (S. Prusiner and E. R. Stadtman, eds.), pp. 9–43, Academic Press, New York.

Glass, R. E., Goman, M., Errington, L., and Scaife, J., 1975, Induction of RNA polymerase synthesis in *Escherichia coli. Mol. Gen. Genet.* **143**:79–83.

Gold, L., O'Farrell, P. Z., and Russel, M., 1976, Regulation of gene 32 expression during bacteriophage T4 infection of *Escherichia coli, J. Biol. Chem.* **251**:7251–7262.

Goldberger, R. F., 1974, Autogenous regulation of gene expression, *Science* **183**:810–816.

Goldberger, R. F., and Deeley, R. G., 1976, Autogenous regulation of gene expression, in: *Regulatory Biology* (J. C. Copeland and G. A. Marzluf, eds.), pp. 178–195, Ohio State Univ. Press, Columbus.

Goldberger, R. F., and Kovach, J. S., 1972, Regulation of histidine biosynthesis in *Salmonella typhimurium, Curr. Top. Cell. Regul.* **5**:285–308.

Graf, L. H., Jr., McRoberts, J. A., Harrison, T. M., and Martin, D. W., Jr., 1976, Increased PRPP synthetase activity in cultured rat hepatoma cells containing mutations in the hypoxanthine-guanine phosphoribosyltransferase gene, *J. Cell. Physiol.* **88**:331–342.

Gross, S. R., 1969, Genetic regulatory mechanisms in the fungi, *Annu. Rev. Genet.* **3**:395–424.

Hagen, D. C., and Magasanik, B., 1973, Isolation of the self-regulating repressor protein of the *hut* operons of *Salmonella typhimurium, Proc. Natl. Acad. Sci. U.S.A.* **70**:808–812.

Hagen, D. C., and Magasanik, B., 1976, Deoxyribonucleic acid-binding studies on the *hut* repressor and mutant forms of the *hut* repressor of *Salmonella typhimurium, J. Bacteriol.* **127**:837–847.

Hatfield, G. W., and Burns, R. O., 1970, The specific binding of leucyl transfer RNA to an immature form of L-threonine deaminase: Its implication in repression, *Proc. Natl. Acad. Sci. U.S.A.* **66**:1027–1035.

Hayward, R. S., Tittawella, I. P. B., and Scaife, J. G., 1973, Evidence for specific control of RNA polymerase synthesis in *Escherichia coli, Nature New Biol.* **243**:6–9.

Hayward, R. S., Sustin, S. J., and Scaife, J. G., 1974, The effect of gene dosage on the synthesis and stability of RNA polymerase subunits in *Escherichia coli. Mol. Gen. Genet.* **131**:173–180.

Herskowitz, I., 1973, Control of gene expression in bacteriophage lambda, *Annu. Rev. Genet.* **7**:289–324.

Hiraga, S., and Yanofsky, C., 1972, Normal repression in a deletion mutant lacking almost the entire operator-proximal gene of the tryptophan operon of *E. coli, Nature New Biol.* **237**:47–49.

Hirsh, J., and Schleif, R., 1976, Electron microscopy of gene regulation: The L-arabinose operon, *Proc. Natl. Acad. Sci. U.S.A.* **73**:1518–1522.

Hofnung, M., 1974, Divergent operons and the genetic structure of the maltose B region in *Escherichia coli* K12, *Genetics* **76**:169–184.

Hofnung, M., and Schwartz, M., 1971, Mutations allowing growth on maltose of *Escherichia coli* K12 strains with a deleted *malT* gene, *Mol. Gen. Genet.* **112**:117–132.

Imsande, J., 1973, Repressor and antirepressor in the regulation of staphylococcal penicillinase synthesis, *Genetics* **75**:1–17.

Imsande, J., and Lilleholm, J. L., 1976, Characterization of mutations in the penicillinase operon of *Staphylococcus aureus, Mol. Gen. Genet.* **147**:23–27.

Imsande, J., Zyskind, J. W., and Mile, I., 1972, Regulation of staphylococcal penicillinase synthesis, *J. Bacteriol.* **109**:122–133.

Ishihama, A., Taketo, M., Saitoh, T., and Fukuda, R., 1976, Control of formation of RNA polymerase in *Escherichia coli,* in: *RNA Polymerases* (M. Chamberlin and R. Losick, eds.), pp. 485–502, Cold Spring Harbor Lab., Cold Spring Harbor, New York.

Isturiz, T., and Wolf, R. E., Jr., 1975, *In vitro* synthesis of a constitutive enzyme of *Escherichia coli,* 6-phosphogluconate dehydrogenase, *Proc. Natl. Acad. Sci. U.S.A.* **72**:4381–4384.

Jackson, E. N., and Yanofsky, C., 1973, The region between the operator and the first structural gene of the tryptophan operon of *Escherichia coli* may have a regulatory function, *J. Mol. Biol.* **76**:89–101.

Jackson, V., Granner, D., and Chalkley, R., 1976, Deposition of histone onto the replicating chromosome: Newly synthesized histone is not found near the replication fork, *Proc. Natl. Acad. Sci. U.S.A.* **73**:2266–2269.

Jacob, F., and Monod, J., 1961, Genetic regulatory mechanisms in the synthesis of proteins, *J. Mol. Biol.* **3**:318–356.

Jacob, F., Brenner, S., and Cuzin, F., 1963, On the regulation of DNA replication in bacteria, *Cold Spring Harbor Symp. Quant. Biol.* **28**:329–348.

Jones-Mortimer, M. C., 1968, Positive control of sulphate reduction in *Escherichia coli:* The nature of the pleiotropic cysteineless mutants of *E. coli* K12, *Biochem. J.* **110**:597–602.

Kasai, T., 1974, Regulation of the expression of the histidine operon in *Salmonella typhimurium, Nature* **249**:523–527.

Kelker, N., and Eckhardt, T., 1977, Regulation of *argA* operon expression in *Escherichia coli* K-12: Cell-free synthesis of beta-galactosidase under *argA* control, *J. Bacteriol.* **132**:67–72.

Kelker, N. E., Maas, W. K., Yang, H.-L., and Zubay, G., 1976, *In vitro* synthesis and repression of arginino-succinase in *Escherichia coli* K12; partial purification of the arginine repressor, *Mol. Gen. Genet.* **144**:17–20.

Kelleher, R. J., and Heggeness, M., 1976, Repression of diaminopimelic acid decarboxylase in *Escherichia coli:* Gene dosage effects and escape synthesis, *J. Bacteriol.* **125**:376–378.

Kelly, L. E., and Brammar, W. J., 1973*a,* A frameshift mutation that elongates the penicillinase protein of *Bacillus licheniformis, J. Mol. Biol.* **80**:135–147.

Kelly, L. E., and Brammar, W. J., 1973*b,* The polycistronic nature of the penicillinase structural and regulatory genes in *Bacillus licheniformis, J. Mol. Biol.* **80**:149–154.

Khoury, G., Howley, P., Nathans, D., and Martin, M., 1975, Post-transcriptional selection of simian virus 40-specific RNA, *J. Virol.* **15**:433–437.

Kimura, G., and Itagaki, A., 1975, Initiation and maintenance of cell transformation by simian virus 40: A viral genetic property, *Proc. Natl. Acad. Sci. U.S.A.* **72**:673–677.

King, J., and Casjens, S., 1974, Catalytic head assembling protein in virus morphogenesis, *Nature* **251**:112–119.

King, J., Botstein, D., Casjens, S., Earnshaw, W., Harrison, S., and Lenk, E., 1976, Structure and assembly of the capsid of bacteriophage P22, *Phil. Trans. Roy. Soc. Lond. B* **276**:37–49.

Kirschbaum, J., and Scaife, J., 1974, Evidence for a transducing phage carrying the genes for the β and β' subunits of *Escherichia coli* RNA polymerase, *Mol. Gen. Genet.* **132**:193–201.

Kirschbaum, J. B., Claeys, I. V., Nasi, S., Molholt, B., and Miller, J. H., 1975, Temperature-sensitive RNA polymerase mutants with altered subunit synthesis and degradation, *Proc. Natl. Acad. Sci. U.S.A.* **72**:2375–2379.

Kleinsmith, L. J., Stein, J. L., and Stein, G. S., 1976, Dephosphorylation of nonhistone proteins specifically alters the pattern of gene transcription in reconstituted chromatin, *Proc. Natl. Acad. Sci. U.S.A.* **73**:1174–1178.

Kline, E. L., Brown, C. S., Coleman, W. G., Jr., and Umbarger, H. E., 1974, Regulation of isoleucine-valine biosynthesis in an *ilvDAC* deletion strain of *Escherichia coli* K-12, *Biochem. Biophys. Res. Commun.* **57**:1144–1151.

Korn, L. J., and Yanofsky, C., 1976, Polarity suppressors increase expression of the wild-type tryptophan operon of *Escherichia coli, J. Mol. Biol.* **103**:395–409.

Kornberg, R. D., and Thomas, J. O., 1974, Chromatin structure: Oligomers of the histones, *Science* **184**:865–868.

Kredich, N. M., 1971, Regulation of L-cysteine biosynthesis in *Salmonella typhimurium* I. Effects of growth on varying sulfur sources and *O*-acetyl-L-serine on gene expression, *J. Biol. Chem.* **246**:3474–3484.

Krisch, H. M., Bolle, A., and Epstein, R. H., 1974, Regulation of the synthesis of bacteriophage T4 gene 32 protein, *J. Mol. Biol.* **88**:89–104.

Krzyzek, R. A., and Rogers, P., 1976*a,* Dual regulation by arginine of the expression of the *Escherichia coli argECBH* operon, *J. Bacteriol.* **126**:348–364.

Krzyzek, R. A., and Rogers, P., 1976*b,* Effect of arginine on the stability and size of *argECBH* messenger ribonucleic acid in *Escherichia coli, J. Bacteriol.* **126**:365–376.

Lai, C.-J., and Weisblum, B., 1971, Altered methylation of ribosomal RNA in an erythromycin-resistant strain of *Staphylococcus aureus, Proc. Natl. Acad. Sci. U.S.A.* **68**:856–860.

Lai, C.-J. Weisblum, B., Fahnestock, S. R., and Nomura, M., 1973*a,* Alteration of 23S ribosomal RNA and erythromycin-induced resistance to lincomycin and spiramycin in *Staphylococcus aureus, J. Mol. Biol.* **74**:67–72.

Lai, C.-J., Dahlberg, J. E., and Weisblum, B., 1973*b,* Structure of an inducibly methylatable nucleotide sequence in 23S ribosomal ribonucleic acid from erythromycin-resistant *Staphylococcus aureus, Biochemistry* **12**:457–460.

Lee, F., Squires, C. L., Squires, C., and Yanofsky, C., 1976, Termination of transcription *in vitro* in the *Escherichia coli* tryptophan operon leder region, *J. Mol. Biol.* **103**:383–393.

Lee, N., Wilcox, G., Gielow, W., Arnold, J., Cleary, P., and Englesberg, E., 1974, *In vitro* activation of the transcription of *araBAD* operon by *araC* activator, *Proc. Natl. Acad. Sci. U.S.A.* **71**:634–638.

Levine, M., 1972, Replication and lysogeny with phage P22 in *Salmonella typhimurium, Curr. Top. Microbiol. Immunol.* **58**:135–156.

Levinthal, M., Williams, L. S., Levinthal, M., and Umbarger, H. E., 1973, Role of threonine deaminase in the regulation of isoleucine and valine biosynthesis, *Nature New Biol.* **246**:65–68.

Levy, S. B., and McMurry, L., 1974, Detection of an inducible membrane protein associated with R-factor-mediated tetracycline resistance, *Biochem. Biophys. Res. Commun.* **56**:1060–1068.

Lindberg, M., and Novick, R. P., 1973, Plasmid-specific transformation in *Staphylococcus aureus, J. Bacteriol.* **115**:139–145.

Magasanik, B., Prival, M. J., Brenchley, J. E., Tyler, B. M., DeLeo, A. B., Streicher, S. L., Bender, R. A., and Paris, C. G., 1974, Glutamine synthetase as a regulator of enzyme synthesis, *Curr. Top. Cell. Regul.* **8**:119–138.

Magee, P. T., and Hereford, L. M., 1969, Multivalent repression of isoleucine–valine biosynthesis in *Saccharomyces cerevisiae, J. Bacteriol.* **98**:857–862.

Manteuil, S., and Girard, M., 1974, Inhibitors of DNA synthesis: Their influence on the replication and transcription of simian virus 40 DNA, *Virology* **60**:438–454.

Martin, R. G., and Chou, J. Y., 1975, Simian virus 40 functions required for the establishment and maintenance of malignant transformation, *J. Virol.* **15**:599–612.

Matsubara, K., 1976, Genetic structure and regulation of a replicon of plasmid λ*dv*, *J. Mol. Biol.* **102**:427–439.

Matsubara, K., and Kaiser, A. D., 1968, λ*dv:* An autonomously replicating DNA fragment, *Cold Spring Harbor Symp. Quant. Biol.* **33**:769–775.

Maurer, R., and Crawford, I. P., 1971, New regulatory mutation affecting some of the tryptophan genes in *Pseudomonas putida, J. Bacteriol.* **106**:331–338.

McFall, E., and Bloom, F. R., 1971, Catabolite repression in the D-serine deaminase system of *Escherichia coli* K-12, *J. Bacteriol.* **105**:241–248.

McGeoch, D., McGeoch, J., and Morse, D., 1973, Synthesis of tryptophan operon RNA in a cell-free system, *Nature New Biol.* **245**:137–140.

McLaughlin, C. A., Magee, P. T., and Hartwell, L. H., 1969, Role of isoleucyl-transfer ribonucleic acid synthetase in ribonucleic acid synthesis and enzyme repression in yeast, *J. Bacteriol.* **100**:579–584.

Meyer, B. J., Kleid, D. G., and Ptashne, M., 1975, λ repressor turns off transcription of its own gene, *Proc. Natl. Acad. Sci. U.S.A.* **72**:4785–4789.

Meyers, M., Blasi, F., Bruni, C. B., Deeley, R. G., Kovach, J. S., Levinthal, M., Mullinix, K. P., Vogel, T., and Goldberger, R. F., 1975*a,* Specific binding of the first enzyme for histidine biosynthesis to the DNA of the histidine operon, *Nucleic Acids Res.* **2**:2021–2036.

Meyers, M., Levinthal, M., and Goldberger, R. F., 1975*b,* Trans-recessive mutation in the first structural gene of the histidine operon that results in constitutive expression of the operon, *J. Bacteriol.* **124**:1227–1235.

Miller, S. L., and Orgel, L. E., 1974, *The Origins of Life on the Earth,* Prentice-Hall, Englewood Cliffs, New Jersey.

Mitsuhashi, S., 1971, *Transferable Drug Resistance Factor R,* University Park Press, Baltimore, Maryland.

Morse, D. E., and Morse, A. N. C., 1976, Dual-control of the tryptophan operon is mediated by both tryptophanyl-tRNA synthetase and the repressor, *J. Mol. Biol.* **103**:209–226.

Müller-Hill, B., and Kania, J., 1974, *Lac* repressor can be fused to β-galactosidase, *Nature* **249**:561–563.

Nakamura, Y., and Yura, T., 1975, Evidence for a positive regulation of RNA polymerase synthesis in *Escherichia coli, J. Mol. Biol.* **97**:621–642.

Nakanishi, S., Adhya, S., Gottesman, M. E., and Pastan, I., 1973, *In vitro* repression of the transcription of *gal* operon by purified *gal* repressor, *Proc. Natl. Acad. Sci. U.S.A.* **70**:334–338.

Nissley, S. P., Anderson, W. B., Gottesman, M. E., Perlman, R. L., and Pastan, I., 1971, *In vitro* transcription of the *gal* operon requires cyclic adenosine monophosphate and cyclic adenosine monophosphate receptor protein, *J. Biol. Chem.* **246**:4671–4678.

Nixon, S. E., and Mawer, G. E., 1970, The digestion and absorption of protein in man 2. The form in which digested protein is absorbed, *Br. J. Nutr.* **24**:241–258.

Novick, A., McCoy, J. M., and Sadler, J. R., 1965, The noninducibility of repressor formation, *J. Mol. Biol.* **12**:328–330.

Osborn, M., and Weber, K., 1975, Simian virus 40 gene *A* function and maintenance of transformation, *J. Virol.* **15**:636–644.

Pannekoek, H., Brammar, W. J., and Pouwels, P. H., 1975, Punctuation of transcription *in vitro* of the tryptophan operon of *Escherichia coli:* A novel type of control of transcription, *Mol. Gen. Genet.* **136**:199–214.

Parada, J. L., and Magasanik, B., 1975, Expression of the *hut* operons of *Salmonella typhimurium* in *Klebsiella aerogenes* and in *Escherichia coli, J. Bacteriol.* **124**:1263–1268.

Park, W. D., Stein, J. L., and Stein, G. S., 1976, Activation of *in vitro* histone gene transcription from HeLa S_3 chromatin by S-phase nonhistone chromosomal proteins, *Biochemistry* **15**:3296–3300.

Pateman, J. A., Cove, D. J., Rever, B. M., and Roberts, D. B., 1964, A common co-factor for nitrate reductase and xanthine dehydrogenase which also regulates the synthesis of nitrate reductase, *Nature* **201**:58–60.

Pateman, J. A., Rever, B. M., and Cove, D. J., 1967, Genetic and biochemical studies of nitrate reduction in *Aspergillus nidulans, Biochem. J.* **104**:103–111.

Phillips, D. M. P., ed., 1971, *Histones and Nucleohistones,* Plenum Press, New York.

Pledger, W. J., and Umbarger, H. E., 1973, Isoleucine and valine metabolism in *Escherichia coli.* XXII. A pleiotropic mutation affecting induction of isomeroreductase activity, *J. Bacteriol.* **114**:195–207.

Pouwels, P. H., and Pannekoek, H., 1976, A transcriptional barrier in the regulatory region of the tryptophan operon of *Escherichia coli:* Its role in the regulation of repressor-independent RNA synthesis, *Mol. Gen. Genet.* **149**:255–265.

Pouwels, P. H., and van Rotterdam, J., 1975, *In vitro* synthesis of enzymes of the tryptophan operon of *Escherichia coli:* Evidence for positive control of transcription, *Mol. Gen. Genet.* **136**:215–226.

Power, J., 1967, The L-rhamnose genetic system in *Escherichia coli* K-12, *Genetics* **55**:557–568.

Proctor, A. R., and Crawford, I. P., 1975, Autogenous regulation of the inducible tryptophan synthase of *Pseudomonas putida, Proc. Natl. Acad. Sci. U.S.A.* **72**:1249–1253.

Proctor, A. R., and Crawford, I. P., 1976, Evidence for autogenous regulation of *Pseudomonas putida* tryptophan synthase, *J. Bacteriol.* **126**:547–549.

Ptashne, M., 1971, Repressor and its action, in: *The Bacteriophage Lambda* (A. D. Hershey, ed.), pp. 221–237, Cold Spring Harbor Lab., Cold Spring Harbor, New York.

Ptashne, M., Backman, K., Humayun, M. Z., Jeffrey, A., Maurer, R., Meyer, B., and Sauer, R. T., 1976, Autoregulation and function of a repressor in bacteriophage lambda, *Science* **194**:156–161.

Ratner, D., 1976, Evidence that mutations in the *suA* polarity suppressing gene directly affect termination factor rho, *Nature* **259**:151–153.

Ratzkin, B., Arfin, S. M., and Umbarger, H. E., 1972, Isoleucine and valine metabolism in *Escherichia coli.* XVIII. The induction of acetohydroxy acid isomeroreductase, *J. Bacteriol.* **112**:131–141.

Reanney, D., 1976, Extrachromosomal elements as possible agents of adaptation and development, *Bacteriol. Rev.* **40**:552–590.

Reed, S. I., Stark, G. R., and Alwine, J. C., 1976, Autoregulation of simian virus 40 gene *A* by T antigen, *Proc. Natl. Acad. Sci. U.S.A.* **73**:3083–3087.

Reichardt, L. F., 1975, Control of bacteriophage lambda repressor synthesis: Regulation of the maintenance pathway by the *cro* and *cI* products, *J. Mol. Biol.* **93**:289–309.

Reznikoff, W. S., 1972, The operon revisited, *Annu. Rev. Genet.* **6**:133–156.

Reznikoff, W. S., Winter, R. B., and Hurley, C. K., 1974, The location of the repressor binding sites in the *lac* operon, *Proc. Natl. Acad. Sci. U.S.A.* **71**:2314–2318.

Richmond, M. H., 1965, Dominance of the inducible state in strains of *Staphylococcus aureus* containing two distinct penicillinase plasmids, *J. Bacteriol.* **90**:370–374.

Roberts, J. W., 1969, Termination factor for RNA synthesis, *Nature* **224**:1168–1174.

Roberts, J. W., 1976, Transcription termination and its control in *E. coli,* in: *RNA Polymerases* (M. Chamberlin and R. Losick, eds.), pp. 247–271, Cold Spring Harbor Lab., Cold Spring Harbor, New York.

Rose, J. K., Squires, C. L., Yanofsky, C., Yang, H.-L., and Zubay, G., 1973, Regulation of *in vitro* transcription of the tryptophan operon by purified RNA polymerase in the presence of partially purified repressor and tryptophan, *Nature New Biol.* **245**:133–137.

Rothman-Denes, L., and Martin, R. G., 1971, Two mutations in the first gene of the histidine operon of *Salmonella typhimurium* affecting control, *J. Bacteriol.* **106**:227–237.

Rownd, R., and Mickel, S., 1971, Dissociation and reassociation of RTF and r-determinants of the R-factor *NR1* in *Proteus mirabilis, Nature New Biol.* **234**:40–43.

Russel, M., 1973, Control of bacteriophage T4 DNA polymerase synthesis, *J. Mol. Biol.* **79**:83–94.

Russel, M., Gold, L., Morrissett, H., and O'Farrell, P. Z., 1976, Translational, autogenous regulation of gene 32 expression during bacteriophage T4 infection, *J. Biol. Chem.* **251**:7263–7270.

Sachithanandam, S., Lowery, D. L., and Saz, A. K., 1974, Endogeneous, spontaneous formation of beta-lactamase in *Staphylococcus aureus, Antimicrob. Ag. Chemother.* **6**:763–769.

Savageau, M. A., 1974*a,* Genetic regulatory mechanisms and the ecological niche of *Escherichia coli, Proc. Natl. Acad. Sci. U.S.A.* **71**:2453–2455.

Savageau, M. A., 1974*b,* Comparison of classical and autogenous systems of regulation in inducible operons, *Nature* **252**:546–549.

Savageau, M. A., 1975, Significance of autogenously regulated and constitutive synthesis of regulatory proteins in repressible biosynthetic systems, *Nature* **258**:208–214.

Savageau, M. A., 1976, *Biochemical Systems Analysis: A Study of Function and Design in Molecular Biology,* Addison-Wesley, Reading, Massachusetts.

Scott, J. R., 1970, Clear plaque mutants of phage P1, *Virology* **41**:66–71.

Scott, J. F., Roth, J. R., and Artz, S. W., 1975, Regulation of histidine operon does not require *hisG* enzyme, *Proc. Natl. Acad. Sci. U.S.A.* **72**:5021–5025.

Seale, R. L., 1976, Temporal relationships of chromatin protein synthesis, DNA synthesis, and assembly of deoxyribonucleoprotein, *Proc. Natl. Acad. Sci. U.S.A.* **73**:2270–2274.

Shaw, W. V., 1974, Genetics and enzymology of chloramphenicol resistance, *Biochem. Soc. Trans.* **2**:834–838.

Sherratt, D. J., and Collins, J. F., 1973, Analysis by transformation of penicillinase system in *Bacillus licheniformis, J. Gen. Microbiol.* **76**:217–230.

Shimizu, N., Shimizu, Y., Fujimura, F. K., and Hayashi, M., 1974, Repression of tryptophan operon RNA synthesis by *trp* repressor in an *in vitro* coupled transcription–translation system, *FEBS Lett.* **40**:80–83.

Silhavy, T. J., Casadaban, M. J., Shuman, H. A., and Beckwith, J. R., 1976, Conversion of β-galactosidase to a membrane-bound state by gene fusion, *Proc. Natl. Acad. Sci. U.S.A.* **73**:3423–3427.

Smith, G. R., and Magasanik, B., 1971, Nature and self-regulated synthesis of the repressor of the *hut* operons in *Salmonella typhimurium, Proc. Natl. Acad. Sci. U.S.A.* **68**:1493–1497.

Smith, J. M., Smolin, D. E., and Umbarger, H. E., 1976, Polarity and the regulation of the *ilv* gene cluster in *Escherichia coli* strain K-12, *Mol. Gen. Genet.* **148**:111–124.

Smith, O., Meyers, M. M., Vogel, T., Deeley, R. D., and Goldberger, R., 1974, Defective *in vitro* binding of histidyl-transfer ribonucleic acid to feedback resistant phosphoribosyl transferase of *Salmonella typhimurium, Nucleic Acids Res.* **1**:881–888.

Somers, D. G., Pearson, M. L., and Ingles, C. J., 1975, Regulation of RNA polymerase II activity in a mutant rat myoblast cell line resistant to α amanitin, *Nature* **253**:372–374.

Somerville, R. L., and Stetson, H., 1974, Expression of the tryptophan operon in merodiploids of *Escherichia coli.* II. Effects of polar mutations in the *trpE* gene, *Mol. Gen. Genet.* **131**:247–261.

Somerville, R. L., and Yanofsky, C., 1965, Studies on the regulation of tryptophan biosynthesis in *Escherichia coli, J. Mol. Biol.* **11**:747–759.

Squires, C. L., Rose, J. K., Yanofsky, C., Yang, H.-L., and Zubay, G., 1973, Tryptophanyl-tRNA and tryptophanyl-tRNA synthetase are not required for *in vitro* repression of the tryptophan operon, *Nature New Biol.* **245**:131–133.

Stein, J. L., Reed, K., and Stein, G. S., 1976, Effect of histones and nonhistone chromosomal proteins on the transcription of histone genes from HeLa S_3 cell DNA, *Biochemistry* **15**:3291–3295.

Stetson, H., and Somerville, R. L., 1971, Expression of the tryptophan operon in merodiploids of *Escherichia coli* I. Gene dosage, gene position and marker effects, *Mol. Gen. Genet.* **111**:342–351.

Sykes, R. B., and Matthew, M., 1976, The β-lactamases of gram-negative bacteria and their role in resistance to β-lactam antibiotics, *J. Antimicrob. Chemother.* **2**:115–157.

Taketo, M., and Ishihama, A., 1976, Biosynthesis of RNA polymerase in *Escherichia coli.* IV. Accumulation of intermediates in mutants defective in the subunit assembly, *J. Mol. Biol.* **102**:297–310.

Taketo, M., Ishihama, A., and Kirschbaum, J. B., 1976, Altered synthesis and stability of RNA polymerase holoenzyme subunits in mutants of *Escherichia coli* with mutations in the β or β′ subunit genes, *Mol. Gen. Genet.* **147**:139–143.

Tegtmeyer, P., 1972, Simian virus 40 deoxyribonucleic acid synthesis: The viral replicon, *J. Virol.* **10**:591–598.

Tegtmeyer, P., 1975, Function of simian virus 40 gene *A* in transforming infection, *J. Virol* **15**:613–618.

Tegtmeyer, P., Schwartz, M., Collins, J. K., and Rundell, K., 1975, Regulation of tumor antigen synthesis by simian virus 40 gene *A, J. Virol.* **16**:168–178.

Thomas, J. O., and Kornberg, R. D., 1975, An octamer of histones in chromatin and free in solution, *Proc. Natl. Acad. Sci. U.S.A.* **72**:2626–2630.

Thompson, J. A., Stein, J. L., Kleinsmith, L. J., and Stein, G. S., 1976, Activation of histone gene transcription by nonhistone chromosomal phosphoproteins, *Science* **194**:428–431.

Tittawella, I. P. B., 1976, Evidence against autorepression of the ββ′ operon in *Escherichia coli, Mol. Gen. Genet.* **145**:223–226.

Tittawella, I. P. B., and Hayward, R. S., 1974, Different effects of rifampicin and streptolydigin on the control of RNA polymerase subunit synthesis in *Escherichia coli, Mol. Gen. Genet.* **134**:181–186.

Tomizawa, J., and Ogawa, T., 1967, Effect of ultraviolet irradiation on bacteriophage lambda immunity, *J. Mol. Biol.* **23**:247–263.

Tyler, B., Deleo, A. B., and Magasanik, B., 1974, Activation of transcription of *hut* DNA by glutamine synthetase, *Proc. Natl. Acad. Sci. U.S.A.* **71**:225–229.

Urm, E., Yang, H., Zubay, G., Kelker, N., and Maas, W., 1973, *In vitro* repression of *N*-α-acetyl-L-ornithinase synthesis in *Escherichia coli, Mol. Gen. Genet.* **121**:1–7.

Vogel, R. H., McLellan, W. L., Hirvonen, A. P., and Vogel, H. J., 1971, The arginine biosynthetic system and its regulation, in: *Metabolic Regulation V* (H. J. Vogel, ed.), pp. 463–488, Academic Press, New York.

Watanabe, T., 1971, Infectious drug resistance in bacteria, *Curr. Top. Microbiol. Immunol.* **56**:43–98.

Weintraub, H., 1973, The assembly of newly replicated DNA into chromatin, *Cold Spring Harbor Symp. Quant. Biol.* **38**:247–256.

Weisblum, B., and Demohn, V., 1969, Erythromycin-inducible resistance in *Staphylococcus aureus:* Survey of antibiotic classes involved, *J. Bacteriol.* **98**:447–452.

Weisblum, B., Siddhikol, C., Lai, C. J., and Demohn, V., 1971, Erythromycin-inducible resistance in *Staphylococcus aureus:* Requirements for induction, *J. Bacteriol.* **106**:835–847.

Whitlock, J. P., and Simpson, R. T., 1976, Removal of histone H1 exposes a fifty base pair DNA segment between nucleosomes, *Biochemistry* **15**:3307–3314.

Wilcox, G., Clemetson, K. J., Santi, D. V., and Englesberg, E., 1971, Purification of the *araC* protein, *Proc. Natl. Acad. Sci. U.S.A.* **68**:2145–2148.

Wilcox, G., Boulter, J., and Lee, N., 1974*a,* Direction of transcription of the regulatory gene *araC* in *Escherichia coli* B/r, *Proc. Natl. Acad. Sci. U.S.A.* **71**:3635–3639.

Wilcox, G., Clemetson, K. J., Cleary, P., and Englesberg, E., 1974*b,* Interaction of the regulatory gene product with the operator site in the L-arabinose operon of *Escherichia coli, J. Mol. Biol.* **85**:589–602.

Wilson, T. H., 1962, *Intestinal Absorption,* Saunders, Philadelphia.

Winshell, E., and Shaw, W. V., 1969, Kinetics of induction and purification of chloramphenicol acetyltransferase from chloramphenicol-resistant *Staphylococcus aureus, J. Bacteriol.* **98**:1248–1257.

Wohlhueter, R. M., Schutt, H., and Holzer, H., 1973, Regulation of glutamine synthetase *in vivo* in *E. coli,* in: *The Enzymes of Glutamine Metabolism* (S. Prusiner and E. R. Stadtman, eds.), pp. 45–64, Academic Press, New York.

Wyche, J. H., Ely, B., Cebula, T. A., Snead, M. C., and Hartman, P. E., 1974, Histidyl-transfer ribonucleic acid synthetase in positive control of the histidine operon in *Salmonella typhimurium, J. Bacteriol.* **117**:708–716.

Yang, H.-L., Zubay, G., and Levy, S. B., 1976, Synthesis of an R plasmid protein associated with tetracycline resistance is negatively regulated, *Proc. Natl. Acad. Sci. U.S.A.* **73**:1509–1512.

Zalkin, H., Yanofsky, C., and Squires, C. L., 1974, Regulation *in vitro* synthesis of *Escherichia coli* tryptophan operon messenger ribonucleic acid and enzymes, *J. Biol. Chem.* **249**:465–475.

Zamenhof, S., and Eichhorn, H. H., 1967, Study of microbial evolution through loss of biosynthetic functions: Establishment of "defective" mutants, *Nature* **216**:456–458.

Zubay, G., Morse, D. E., Schrenk, W. J., and Miller, J. H. M., 1972, Detection and isolation of the repressor protein for the tryptophan operon of *Escherichia coli, Proc. Natl. Acad. Sci. U.S.A.* **69**:1100–1103.

4

Regulation of Enzyme Synthesis in the Bacteria: A Comparative and Evolutionary Study

PATRICIA H. CLARKE

1 *Introduction*

It is a truism of biology that all the parts must fit together to make a viable organism. In terms of cellular metabolic activities this means that the rates at which enzymes act must be compatible with one another, and the appropriate amounts of enzymes must be synthesized at the time they are required. The regulatory systems have evolved to do this.

The bacteria are characterized by rapid growth rates and many can survive and grow in a wide range of different physiological conditions and at the expense of a variety of growth substrates. The enzymes synthesized, and the amounts made of each of them, may vary greatly under these different growth conditions. The bacteria have therefore been the main targets of studies on regulation of the rates of enzyme synthesis and of investigations of factors controlling gene expression. Detailed studies have been restricted to a relatively small number of bacterial species and much of current theory has been developed from investigations of *Escherichia coli* K12. There are less detailed studies with some other species of

PATRICIA H. CLARKE • Department of Biochemistry, University College London, London, England

prokaryotes and eukaryotes which complement the findings with *E. coli* and there is a very large number of fragmentary observations still to be fitted into the jigsaw puzzle. Even so, we can see that different regulatory patterns have evolved in different groups of bacteria and similarities can be seen between certain types of regulation in prokaryotes and eukaryotes.

The questions before us are as follows: (1) how did the different ways of regulating enzyme activity evolve and (2) how did the different ways of controlling rates of enzyme synthesis evolve? We need to take into account that although many metabolic pathways take a similar course in bacteria belonging to different genera, both enzyme activity and enzyme synthesis may be regulated in very different ways. One interpretation of this is that enzymes evolved first and are therefore more likely to be similar to one another, and that regulatory systems evolved later to fit the particular ecological niches that were being colonized. This interpretation fits also with the view that primitive enzymes were probably not very efficient and became more effective catalysts as successive modifications were made to their structures. Comparative sequence studies have indicated that internal duplication has occurred for many proteins during their long evolutionary history. One consequence of internal gene duplication would be that a region of the protein could then have been available for adaptation to an additional function, such as forming a binding site for an inhibitor or activator.

Duplication of entire genes is thought to have been followed by mutational divergence to produce families with related catalytic activities. It could also have led to the evolution of genes specifying proteins with regulatory functions. In many cases there are close chemical resemblances between enzyme substrates and effector molecules that regulate induction and repression. It is possible that duplication of the structural gene for an enzyme could give rise to the regulator gene which now controls the expression of the enzyme structural gene. In other cases, the regulator gene might have arisen by duplication of an unrelated gene. As a secondary, and later, development, and this may have occurred for genes of related catabolic pathways, both structural genes and regulator genes may have been duplicated and taken on new functions (see Section 7).

Duplication and divergence are not the only options open. Gene fusion can bring together segments of DNA specifying different protein functions and this could have played a part in the origin of proteins with multiple catalytic functions, of proteins with catalytic and regulatory regions, and of proteins regulating gene expression. Gene fusion is feasible only if genes can be brought to adjacent sites. Many instances are now known of gene transposition and the role of insertion sequences may be very important in these gene rearrangements (see Section 2.2).

This chapter gives a brief account of the ways in which ideas about regulatory mechanisms have changed and developed as experiments have become more detailed. This will be essentially an account of the evolution of ideas and the evolution of experimental techniques. Since much of the comparative information is fragmentary, attention is directed particularly to studies in which it has been possible to combine biochemical and genetic investigations. This is used as a basis for a wider examination of other regulatory systems in an attempt to see how some of the regulatory mechanisms for various enzymes and metabolic pathways might have evolved in different organisms. The pathways to be discussed were chosen because they are complex and present many interesting problems of regulation.

Sections 4 and 5 are concerned with the biosynthesis and degradation of aromatic compounds. The branched biosynthetic pathways exemplify, in a simplified way, the problems of integrating the whole of cellular metabolism so that each of the metabolic building blocks is available for macromolecular synthesis. A feedback system in which one of the end products of a branched pathway has total control of the activity, or the rate of synthesis, of early enzymes would be nonviable and a few of the ways in which this has been overcome are described. The catabolic pathways to be discussed will also be related mainly to aromatic compounds. In the natural environment the biodegradation of aromatic compounds is carried out by fungi and bacteria which are present in various mixed associations. Laboratory studies of the regulation of these pathways are still mostly concerned with the biochemical or physiological approach but for a few species genetic evidence is rapidly accumulating and these are the ones discussed in Section 5.

One way of studying evolution with microorganisms is to set up experimental systems in the laboratory. During the last decade several research groups have been successful in selecting strains with new growth phenotypes and many of these experiments are relevant to our attempts to understand the evolution of regulatory systems. Some of these are discussed in Section 7. Most of the sections of this chapter are concerned with the regulation of enzyme synthesis, but in the living organism this is only one aspect of cellular integration. At the same time as certain enzymes are being inhibited and activated, some enzymes are separated from each other by compartmentalization and others are very intimately associated as aggregates or polymeric proteins. The interactions of regulation of enzyme activity with regulation of enzyme synthesis is particularly striking with respect to biosynthetic pathways and this is discussed in Section 4.

Another control system that is of importance to a cell is that of exclusion or concentration of metabolites; alterations in transport systems are associated with resistance to toxic analogs and ability to utilize novel substrates, but these topics are considered only briefly and in reference to those metabolic pathways selected for discussion.

2 *The Nature of the Evidence*

2.1 *Theories of Regulation of Enzyme Synthesis*

Although enzyme regulation has been studied in a variety of organisms there is remarkably little that can be said with certainty about how the different regulatory systems have evolved. Such ideas as we might have on this subject are of course dependent on our knowledge of the regulatory systems themselves. This is necessarily incomplete and we have no way of knowing whether we are still lacking systems that might be the key ones for interpreting general evolutionary relationships. What we can say is that *ideas* about the mechanisms by which rates of enzyme synthesis are regulated have evolved considerably since the primordial concept, usually ascribed to Karstrom (1930), that some enzymes are constitutive and others are induced by ingredients of the growth media. In the main branch of evolutionary descent of theories about regulation are those derived from studies on the enzymes of lactose metabolism. With the publication of the classic paper of Jacob and Monod (1961) on the *lac* operon, it appeared to many, at least for a time, that in principle all problems of regulation were almost solved. Many new and important developments have occurred since then and have produced surprises. The idea of negative control with a repressor molecule, acting by blocking gene expression until it could be removed from the operator-binding site by interaction with an inducer, was easy to comprehend and so intellectually satisfying that it was rather difficult for alternative hypotheses to evolve. However, the elegant and detailed work of Englesberg and his collaborators on the enzymes of L-arabinose metabolism made it clear that regulation of enzyme synthesis could also occur by a positive control system with a regulator molecule *required* for gene transcription. Several positive control systems have now been well defined (Englesberg and Wilcox, 1974).

One of the more puzzling aspects of regulation was the "glucose effect," also known as *catabolite repression*. Several of the more intricate theories concerning the mechanism of this rather general aspect of regulation have now petered out. The finding that, at least in *E. coli*, the main effect of glucose in repressing enzyme synthesis was due to lowering the intracellular level of cyclic AMP cleared away much of the confusion and gave rise to one of the most powerful of current regulatory models. There is now good evidence indicating that transcription of many catabolic genes requires a positive regulator protein known as the *cyclic AMP receptor protein*, CRP (also known as the *catabolic gene activator protein*, CAP). To be effective in activating transcription by RNA polymerase, the CRP requires cyclic AMP and thus the glucose effect can be fitted comfortably into the other regulatory mechanisms of regulator proteins and low-molecular-weight effector molecules acting at the level of gene transcription (Pastan and Adhya, 1976). There are still gaps to be filled in the theories of general metabolic regulation to account for all the facts. For example, synthesis of some catabolic enzymes which are subject to catabolite repression is not affected by cyclic AMP. In addition, Ullman *et al.* (1976) report that even in *E. coli* there is yet another catabolite modulator factor which exerts very strong effects on catabolite repression-sensitive operons. *In vitro lac* transcription has been found to be stimulated to higher levels by a low-molecular-weight protein, the H_1 factor (Crepin *et al.*, 1975), and by a protein factor together with guanosine tetraphosphate (Aboud and Pastan, 1975).

Ideas on the nature of regulator molecules have also undergone a number of changes. The first outline of the operon theory (Jacob and Monod, 1961) allowed that the regulator gene could control the structural genes equally well via a protein molecule or an RNA molecule, and it was only later that the idea of the repressor as a protein became firmly established. One of the most valuable pieces of evidence which went into the development of the concepts of the *lac* operon was the finding that inducer specificity was not identical to substrate specificity. This completely disposed of the mass action theory of enzyme induction which had very plausibly suggested that by binding the substrate, the equilibrium between the enzyme and its precursors was disturbed and more enzyme could be made (Yudkin, 1938). The disparities in inducer and substrate specificities gave a basis for the concept of a regulator gene specifying a regulator protein that was quite distinct from the enzyme (or enzymes) specified by the structural gene(s). This neat compartmentalization no longer holds, and indeed was not an essential tenet of the original operon theory. Some enzymes act as regulator proteins for their own synthesis and some enzymes act as regulator proteins for the synthesis of other enzymes (see Sections 5 and 6). Whereas for the *lac* operon it was very early shown that the *lacI* gene (the gene that specifies the repressor of the *lac* operon) was not inducible (Novick *et al.,* 1965), and later that constitutive synthesis allowed the production of a very small amount of the repressor during each generation (Gilbert and Müller-Hill, 1966), it is now known that some regulator proteins that are not enzymes also regulate their own synthesis. The process whereby a protein may regulate its own synthesis is known as *autogenous regulation* and has been shown to apply to a variety of systems (Goldberger, 1974; Calhoun and Hatfield, 1975). Savageau (see Chapter 3 of this volume) discusses the different requirements that may have led to the development of autogenously regulated systems compared with the classically regulated systems with separate and distinct regulatory genes.

There are many variations on the detailed interactions for both classical and autogenously regulated gene–enzyme systems. Once it is accepted that an enzyme can be involved in its own regulation it is possible to contemplate that it could be involved in regulation of a number of other enzymes, as appears to be the case for glutamine synthetase (Magasanik *et al.,* 1974). It may be difficult to guess the identity of the low-molecular-weight effector that interacts with the regulatory protein and it may not be the obvious candidate. Despite the widely known generalization that the effector (inducer) for an inducible catabolic enzyme is the substrate of that enzyme, exceptions are readily found. For example, the inducer for histidase in *Klebsiella aerogenes* and *Pseudomonas aeruginosa* (although not for *Bacillus subtilis*) is the product and not the substrate. The natural inducer of β-galactosidase is now known to be allolactose and not lactose (Jobe and Bourgeois, 1972). More complex interactions are suggested for the repressible systems for several amino acid biosynthetic pathways where transfer RNAs and transfer RNA synthetases have been implicated (Calvo and Fink, 1971; Jackson *et al.,* 1974). Regulation of synthesis of enzymes for the catabolism of aromatic compounds has evolved in such a way that some intermediates of the pathways are product inducers of earlier enzymes and substrate inducers of later enzymes (see Section 5).

One of the most complex of the regulatory systems to have been analyzed is that for the regulation of the histidine biosynthetic enzymes in *Salmonella typhimurium.* The early experimental evidence showed clearly that the first enzyme of the pathway was subject to feedback inhibition by histidine and that this amino acid also repressed synthesis of all the enzymes of the pathway. The *his* genes in this species are arranged in a single operon and are transcribed into a single giant RNA messenger (Martin, 1963). The difficulties came with the isolation of regulator mutants. It was possible to isolate mutants which were derepressed for the *his* enzymes but instead of finding a single regulator gene it was found that mutations in any one of several quite distinct genes could lead to derepression (Roth and Ames, 1966; Roth *et al.,* 1966; Silbert *et al.,* 1966). Mutations in the structural genes for histidinyl-tRNA and for histidinyl-tRNA synthetase and also in the genes concerned with modification of histidinyl-tRNA can all result in derepressed synthesis of the histidine enzymes (see Brenner and Ames, 1971). These findings can be explained by the fact that it is histidyl-tRNA, not histidine itself, that plays a role in repression of the histidine operon.

The evolution of theories of regulation of the *his* operon has been an epic in itself. The finding that some feedback inhibition-resistant mutants were also altered in regulation implicated the *hisG* product, the first enzyme of the pathway, as another component in the regulation of the *his* operon enzymes (Kovack *et al.*, 1973). Some of the contradictions in experimental findings with the complex regulatory system for the histidine operon remain to be resolved (see Chapters 3 and 10 of this volume for more detailed discussions). It is clear that during the evolution of the regulatory control for histidine biosynthesis mutations have occurred which allow the products of a number of genes to act in concert in regulating the expression of the *his* genes. This is a long way from the neat simplicity of the earliest picture of the regulation of the *lac* operon by the product of the *lacI* regulator gene. Actually, as we have already seen, the *lac* system itself is not so simple; positive control by the catabolic gene activator system is superimposed on the *lac* repressor operon-specific control and other general regulatory systems may also affect the synthesis of the lactose operon proteins.

Regulation of enzyme synthesis is fundamentally a problem in gene expression while regulation of enzyme activity enables the flow of metabolites through the pathway to be controlled by fluctuating levels of precursors and end products. The same molecules may be effectors in both inhibition/activation and repression/induction regulation. If the enzymes of a biosynthetic pathway are synthesized constitutively, but at a very low level, it may be a satisfactory and economical solution for the cell to regulate them entirely by feedback inhibition rather than by repression. This solution is particularly likely to have evolved in organisms that live in environments low in organic compounds where the supply of exogenous amino acids and other metabolites is seldom likely to reach levels high enough to obviate the need for biosynthesis. The regulation of enzyme activity by feedback inhibition would suffice for adjusting intracellular levels of metabolites to fit other biosynthetic demands (see Section 4).

2.2 *Gene Arrangements*

For many of the gene–enzyme systems of *E. coli* it has been possible to identify mutants and to carry out genetic mapping to determine the locations of all the genes involved. The regulator genes may or may not be closely linked to the structural genes they control and there are no differences in this respect between negatively and positively regulated systems. The *lacI* (repressor) gene is adjacent to the *lac* operon, mapping at 8 min on the *E. coli* chromosome (Bachmann *et al.*, 1976). The *gal* operon, which is also under classical negative control, maps at 17 min while the regulator (repressor) gene *galR* maps at 61 min. The regulator (positive activator) gene *dsdC* for serine deaminase is adjacent to the structural gene *dsdA* at 50 min. The regulator (positive activator) gene *araC* is closely linked to the *araBAD* operon and maps at 1 min, but also controls an arabinose permease gene mapping at 61 min. One can ask whether the relative map positions can tell us anything about the origins of these regulator genes and whether the arrangements of regulator and structural genes on the chromosome offer particular advantages.

One suggestion that has been made (Zipkas and Riley, 1975) is that the chromosome of *E. coli* has undergone two successive duplications and that related genes might be found in each quarter of the chromosome. Bachmann *et al.* (1976)

point out that the genes mapped at that date are distributed along the chromosome in a nonrandom fashion and appear to exhibit a certain symmetry of gene clustering. Gene duplication of various types is generally thought to have been the means of increasing the total genetic potential and total duplication of the *E. coli* primitive bacterial chromosome is a reasonable hypothesis. In a more detailed study, Riley *et al.* (1978) analyzed the clustering of related genes and found that there was a strong tendency for the genes governing the sequential reactions of glucose catabolism to lie in four clusters located approximately 90° to 180° from one another. These authors were unable to find the same relationship between other gene pairs tested in this way. There are several reasons why such relationships might not be apparent in present-day *E. coli.* Gene transposition could have disturbed the primitive arrangements so severely as to have obscured any symmetry that might have existed at an earlier date. It is pointed out later that the arrangements of biosynthetic genes in different organisms may have been brought about, at least in part, by gene translocation to enable particular types of transcriptional regulation to be adopted. Similarly, transposition of catabolic genes could associate or dissociate genes for sequential reactions and in some species this may be related to the evolution of catabolic plasmids. Another factor that might make it difficult to find evidence for any large-scale duplications is that some genes may have become silent. Bachmann *et al.* (1976) point out that there are some regions of the chromosome that are very much less occupied by known gene clusters and suggest that this might be related to the physicochemical structure of the chromosome. The alternative explanation is that the genes in these regions have become silent because their original functions have become superseded. In this respect it is interesting to recognize that recent work on experimental evolution has revealed a number of hitherto silent genes that have become reactivated by appropriate mutational selection (see Section 7).

The genes for enzymes of biosynthetic pathways are frequently clustered in *E. coli* and related organisms. The five *trp* genes are located at 27 min and the regulator (repressor) gene, *trpR,* is unlinked to the *trp* operon and is located at 100 min. The clustering of genes, and the common and coordinate regulation of synthesis of the enzymes determined by the linked genes, provided part of the evidence for the operon theory of Jacob and Monod (1961). To take only the classical negative control model, the product of the regulator gene is required to interact with an operator site that must be adjacent to the structural genes so controlled, but there is no necessity for the regulator gene itself to be at the same location. The essential feature of the classical model is that the product of the regulator gene, whether it is a repressor or activator, is independent of the operon it controls and acts by diffusing through the cytoplasm to the DNA-binding site. There are advantages in clustering genes for enzymes of the same pathway since they can be regulated simultaneously, and the surprise is not that clustering of genes for biosynthetic pathways occurs frequently in the enteric bacteria but that it is absent in other genera. If the biosynthetic genes are not under repression control, then gene clustering offers no advantages.

Even in *E. coli* there are exceptions to the clustering of genes for related biosynthetic enzymes. The genes for arginine biosynthesis are scattered around the chromosome and only four occur together in the *argECBH* cluster. Even this cluster is not a simple regulatory unit but consists of two separate operons transcribed in opposite directions (Elseviers *et al.,* 1972). Nevertheless, the arginine biosynthetic genes are under the common control of the *argR* gene, and in *E. coli* K12 the enzymes are all repressed by the product of this gene plus arginine. The rates of synthesis of the different enzymes in repressed and various derepressed growth conditions are coincident but not coordinate. The common regulatory control led to the suggestion that a common regulator molecule interacted at a number of different operator sites adjacent to each of the structural gene loci (Baumberg *et al.,* 1965). This requires the evolution of a regulator molecule able to recognize several DNA-binding sites which may be slightly different in sequence. Among other early findings on the regulation of arginine biosynthesis was the observation that in a mutant of *E. coli* strain W, one of the enzymes, *N*-acetylornithine aminotransferase, was not repressed by arginine but was induced (Vogel *et al.,* 1963). This was an unexpected finding for an enzyme of a biosynthetic pathway but it was later found by Jacoby and Gorini (1967) that a single mutation could change inducibility to repressibility. The arginine enzymes of *E. coli* strain B are slightly induced by arginine but a single mutation in *argR* could make them repressible by arginine. Further, it was possible to transduce *E. coli* K12 with *argR* from strain B and convert it from being arginine repressible to being arginine inducible. Thus, a single mutation in a regulator gene could change an inducible control into a repressible control with the same low-molecular-weight effector molecule, and a particular regulator gene could be expressed in the same way in two different genetic backgrounds.

Linkage of structural genes is therefore not essential for common regulation. However, in other species scattering of genes may be associated with very different types of regulatory control. The arginine genes of *Pseudomonas aeruginosa* are scattered around the chromosome and arginine auxotrophic mutants could be separated by transductional analysis into seven linkage groups (Feary *et al.,* 1969). Of the arginine biosynthetic enzymes only ornithine carbamoyltransferase is repressed by arginine to a marked extent although *N*-acetylglutamic semialdehyde dehydrogenase is slightly repressed (Isaac and Holloway, 1972; Voellmy and Leisinger, 1975). None of the other biosynthetic enzymes was repressed by arginine, but *N*-acetylornithine aminotransferase was induced. This enzyme also has a catabolic role in *P. aeruginosa* and catalyzes the transamination of ornithine in the arginine degradative pathway. The regulation of this transaminase is related to its catabolic role. During the evolution of *P. aeruginosa* the enzyme has evolved its dual function and the separation of the two roles is related to the equilibrium for the reactions with acetylated or nonacetylated substrates. With the acetylated substrates the reaction favors the synthesis of *N*-acetylornithine, whereas the equilibrium with the nonacetylated substrates favors the catabolic direction with formation of glutamic semialdehyde. The role of this enzyme in the catabolism of arginine may be the key to understanding why it is induced to higher levels in the presence of arginine. The inducibility of the aminotransferases in certain strains of *E. coli* may reflect a similar evolutionary history. Thus, in analyzing the significance of the arrangement of genes for a particular metabolic pathway it must be borne in mind that some enzymes have more than one metabolic role and this might be reflected in the regulatory pattern.

The genes for the enzymes of other biosynthetic pathways are much less clustered in *P. aeruginosa* than in *E. coli* (Fargié and Holloway, 1965). The histidine genes, for example, are separated into five distinct transductional linkage groups (Mee and Lee, 1967) while the tryptophan genes are found in three linkage groups. This scattering of biosynthetic genes is probably more widespread than is the tight clustering of the enteric bacteria, and among eukaryotes, genes for the sequential enzymes of a pathway may be found on different chromosomes. In Section 4 are discussed the regulation and the arrangements of the genes for the enzymes of the pathways for the biosynthesis of the aromatic amino acids. Crawford (1975) reviewed the gene arrangements of the *trp* genes.

Although the classical model for gene regulation does not demand any contiguity of the regulator gene and the structural genes it controls, there are conditions for which this arrangement is desirable. If organisms are able to acquire new metabolic activities by the transfer of genes from other organisms, then it is essential that they acquire also any elements needed for the expression of these genes. Transductional transfer of chromosomal fragments carrying *lac* or *dsd* genes of *E. coli* will include both the structural and regulator genes and if the genes can be accepted and other necessary conditions are satisfied, the presence of the inducer will result in enzyme synthesis. If a negatively controlled structural gene is transferred to a cell that does not possess the related regulator gene, then it may result in constitutive enzyme synthesis and drain off as much as 10% of the cell's resources for *protein synthesis,* whereas with a positively regulated system in the absence of the regulator gene no enzyme synthesis would occur. The genes for catabolic enzymes may be carried on transmissible plasmids rather than chromosomes (see Section 5) and it seems probable that the plasmids also carry the relevant regulator genes. In this respect it is notable that although the biosynthetic genes of pseudomonads are scattered around the chromosome (Holloway, 1975), some of the catabolic genes are grouped in supraoperonic clusters. This may be an evolutionary advantage in that if transfer can occur either by transduction or by transfer by plasmids, then the related genes of the pathway will be transferred and will allow growth on the organic compounds available (Wheelis, 1975). Plasmids of *Streptomyces coelicolor* carry the genes for synthesis of an antibiotic, methylenomycin (Kirby and Hopwood, 1977), and in this case also it would be essential for gene expression for the plasmid to carry all the structural genes and any regulator genes that might be required for the synthesis of the enzymes needed to make the antibiotic.

2.3 The Experimental Approach

Current theories of regulation have developed from studies of the relatively few enzyme systems that have been examined by biochemical and genetic experiments. Without such intensive investigations we would have no basis for constructing general theories of regulation and no point from which to begin to speculate how these regulatory systems might have arisen. However, some of the most interesting of the regulatory systems that have evolved are concerned with complicated biochemical pathways that occur in organisms for which only limited genetic systems are available, if any at all. Apart from the difficulties of genetic studies there may be very great technical limitations on the types of biochemical experiments that can be carried out. The catabolic pathways for aromatic compounds in *Pseudomonas* species (see Section 5) provide a good example of the problems presented. The pathway intermediates are often chemically unstable, or difficult to prepare, and some of the enzyme assays may present technical difficulties. Certain strains may be suitable for biochemical investigations but may not be the strains for which even limited genetic analysis can be undertaken. And yet, no general theories of evolution are of value unless they take into account the findings from all available sources. Many bacteria can utilize a much wider range of organic compounds as growth substrates than can *E. coli.* Investigations of the regulation of metabolic pathways for the dissimilation of organic compounds can be carried out only with those species with the appropriate genetic makeup. In addition, for biosynthetic pathways it is very important to make comparative studies to test the general validity of regulatory models and to explore the extent to which regulatory systems have diverged or converged during evolution.

For these reasons we need to take into account information derived from a variety of different experiments. Measurements of the amounts of enzymes produced after batch growth for set periods of time provide at least qualitative data and are often the first step in a new investigation. This will be an obligatory step in a study of a series of potential growth substrates and can provide preliminary data about specific inducing and repressing compounds and indicate which carbon compounds are likely to produce catabolite repression. More exact information comes from studies with cultures growing exponentially in batch culture, but if growth is slow and enzyme assays are difficult, this type of experiment may not be possible. Chemostat culture can be used even when growth rates are low and substrates have unusual properties (for example, the hydrocarbon homologous series ranging from gases to nonpolar solids). The advantages of continuous culture for regulatory studies have been insufficiently recognized although there are a few important exceptions. A culture can be maintained in a steady state for a long period of time; growth can be limited by carbon or nitrogen or other essential nutrients; perturbations can be introduced to disturb the steady state and accentuate regulatory oscillations.

A most important contribution of continuous culture to understanding regulation was that of Meers *et al.* (1970) who found that glutamate dehydrogenase was not involved in ammonia assimilation by *Klebsiella aerogenes* in nitrogen-limited growth. This observation led to the discovery of the role of glutamine synthetase and glutamine: α-oxoglutarate aminotransferase (see Section 6). Later studies by Senior (1975) compared *K. aerogenes* and *E. coli* and demonstrated very clearly the differences in the regulation of nitrogen assimilation in these two organisms. Measurements of enzyme synthesis over a range of growth rates with an inducing substrate used for growth limitation can provide quantitative data on the balance between induction and catabolite repression. This is discussed by Clarke and Lilly (1969) who examined the effects of altering growth rates on the induction and repression of enzymes of *P. aeruginosa.* In the natural environment, changes in growth rates, as potential substrates come and go, are very common and it is also probable that cultures will be presented with more than one compound at a time. The experiments of Higgins and Mandelstam (1972) were concerned with the growth of *P. putida* in chemostat culture with pairs of potential growth substrates. This organism can utilize mandelate and *p*-hydroxymandelate which are converted to benzoate and *p*-hydroxybenzoate, respectively. One of the interesting features of the regulation of this pathway is that some of the intermediates are able to repress the synthesis of earlier enzymes and induce the later enzymes. In chemostat culture the severity of this inhibition was apparent with a mutant blocked later in the *p*-hydroxybenzoate (protocatechuate) pathway (see Section 5). This mutant could grow on mandelate (metabolized via catechol) but when both mandelate and *p*-hydroxybenzoate were presented in the chemostat the repression by *p*-hydroxybenzoate of the mandelate enzymes was so severe that growth was inhibited. This type of investigation sheds some light on the ways in which regulatory systems interact in growing cultures.

Mutant strains are invaluable for regulatory studies but are not always obtainable. Even when mutants are available a considerable amount of evidence will depend on physiological experiments. If a group of enzymes belongs to a common pathway, the rates of synthesis will be coincident. Such evidence does not necessarily mean that all enzymes under coincident regulation belong to the same pathway, since other factors may result in the gratuitous synthesis of unrelated enzymes. Enzymes under common regulation (controlled by the same regulator gene) will be synthesized *coordinately* and to assess whether enzyme synthesis is coordinate or merely coincident it is essential to compare enzyme levels over as wide a range of experimental conditions as possible. Physiological experiments of this sort made it possible to identify the inducing intermediates of the ortho pathway for catabolism of catechol (see review by Ornston and Parke, 1977). Catechol is not the inducer of 1,2-oxygenase and this enzyme is induced by the product *cis,cis*-muconate which also induces the next two enzymes of the pathway. The three enzymes induced by *cis,cis*-muconate in *P. putida* are induced coincidently, but only the last two are under coordinate regulation. Although under most conditions the three enzymes vary in a similar manner, the synthesis of catechol 1,2-oxygenase is much more sensitive than the other two to catabolite repression by glucose, and this finding indicated that there were two distinct regulatory units (see Section 5).

Some of the inferences that can be drawn from investigations with a variety of different experimental methods are discussed in later sections of this chapter. It is encouraging that not only are genetic transfer systems being established for species which have interesting biochemical characteristics but that gene transfer between genera has be successful by several different techniques. An example of this is the use of the P class plasmids to pick up genes from *P. aeruginosa* that can then be maintained in *E. coli* (Hedges *et al.*, 1977).

3 The Molecular Basis of Regulation of Gene Expression

3.1 Binding Domains

A regulator protein, exerting its effect by negative control of transcription, is thought to bind to DNA at a specific site in such a way that this interaction prevents the initiation of transcription by RNA polymerase. A positive regulator protein must do quite the opposite and make it easier for RNA polymerase to start transcribing. Most of the detailed studies on how this might come about are inevitably concerned with the *lac* operon.

From the analysis of many mutants of the *lacI* gene it has been possible to identify regions of the *lac* repressor responsible for operator binding and those responsible for inducer binding (Müller-Hill, 1975; Miller *et al.*, 1977; Schmeissner *et al.*, 1977). Mutants defective in the inducer-binding region are phenotypically lactose negative and were designated by Willson *et al.* (1964) as *lacI*^{S} or superrepressed. Mutants lacking, or defective in, the operator-binding site are constitutive and Müller-Hill and colleagues have selected a number of these which bind inducer normally and are dominant constitutives, *lacI*$^{-d}$. In addition to the inducer- and operator-binding regions, Müller-Hill (1975) defines the transmitter region which links the two. From one point of view the *lac* repressor is an example of a DNA-binding protein, many of which are now known. Among these are the restriction endonucleases which have been shown to bind to very specific DNA sequences characterized by possessing palindromic sequences (Murray and Old, 1974). The *lac* operator also contains such sequences as do those promoter sequences that have been determined (Dickson *et al.*, 1975; Musso *et al.*, 1977).

The association of polypeptides and polynucleotides is a very ancient one and is likely to have preceded the evolution of cellular organisms. Although the characteristics of such ancestral associations are not known it is clear that nucleotide binding of polypeptides has always been of fundamental importance. Haldane (1965) suggested that the primitive system required only one enzyme which could be classified as a general phosphokinase to account for the synthesis (albeit inefficient) of all polynucleotides and polypeptides. Rossman has compared the tertiary structures of a number of nucleotide-binding enzymes and finds remarkable similarities (Rossman *et al.*, 1974; Eventoff and Rossman, 1975). He proposes that the nucleotide-binding dehydrogenases evolved from a common general-purpose ancestral dehydrogenase. It would be possible to envisage a similar process for the evolution of the array of regulator proteins, both positive and negative, either from a single protein or from two archetypal DNA-binding proteins, one of which stabilized the DNA structure and the other of which destabilized it. If this were so, it would be predicted that there would be considerable homology of sequence between regulator proteins and similarities of tertiary structure. This approach leaves out the consideration of autogenously regulated systems.

If we start from the other end of the molecule, then the argument goes that the inducer of the gene for a catabolic enzyme is very like the substrate and that the regulator protein could very reasonably have evolved from another copy of the enzyme it controls by gene duplication. This fits in much better with autogenously controlled systems since it can be argued for these that they have acquired regulator function without disposing of their original catalytic activities. Duplication of the autogenous regulator could then be followed by loss of catalytic activity for one copy which became solely regulatory in character. Cove (1974) pointed out that if the inducer (or corepressor) is taken as the starting point, then the protein need not have been the enzyme itself, but could have been any suitable protein which could react with a related metabolite likely to vary in concentration in the same way as the enzyme substrate or product. He suggests, for example, that the *lac* repressor might have evolved from an enzyme which bound allolactose rather than from β-galactosidase, which binds lactose itself. An obvious way of assessing whether a regulatory protein might have evolved from an enzyme it regulates (or from any other candidate) is to compare the sequence of the two proteins. For the *lac* repressor and β-galactosidase there are no significant similarities (Beyreuther *et al.,* 1973). Another β-galactoside-binding protein is the product of the *ebg* gene (Hartl and Hall, 1974). This too is under negative control by another β-galactoside-binding protein, the product of the *ebgR* gene (Hall and Hartl, 1975). There may be some ancestral connection between the *ebg* and the *lac* genes although in the present-day *E. coli* only the latter appear to be functional. The inducer of the *ebg* gene is lactose itself, but it is possible that this was originally the case for the *lac* genes as well and that the role of allolactose as inducer is a later refinement.

A different way to look at the evolution of regulator molecules is to consider that they could have arisen by a complex series of events involving gene translocation, partial gene duplication, and gene fusion. If so, a regulator molecule would have mixed ancestry and homologies would be expected, not for large sequences of the molecule, but for each of the binding domains. There is evidence that translocation of genes and of small insertion sequences of DNA can occur with high frequency (Saedler *et al.,* 1972; Saedler and Heiss, 1973; Cohen, 1976). Gene duplication has been considered to have been of fundamental importance in the evolutionary divergence of proteins (see Ohno, 1970; Koch, 1972). Fusion of genes to give proteins with more than one functional group can be observed in experimental conditions (Yourno *et al.,* 1970) and there is evidence that it has occurred during the evolution of the genes of certain metabolic pathways (see Crawford, 1975). These three processes are likely to have played a significant part in the evolution of the genes for regulator proteins as well as the genes for the enzymes themselves. A regulator protein that evolved in this way would be expected to contain a regulatory region corresponding to the DNA-binding domain, a region recognizing and binding the effector molecule and such other regions as it had retained from its ancestral protein.

The discussion will be limited to regulator proteins concerned with the synthesis of specific groups of enzymes and not those of more general function such as the cyclic AMP–CRP systems. It is being assumed that the specific regulator proteins all must interact at a DNA site and that the clues for the origins of part or all of these proteins will be found by identifying the effector molecules with which they interact and tracking down the gene clusters that they control. Interactions with RNA polymerase are of great importance in controlled transcription, but this is outside the scope of this chapter. The complexities of some regulatory systems, such as that for the histidine operon, only became apparent from the analysis of many mutants and detailed genetic studies. Some of the catabolic and biosynthetic pathways now being analyzed will certainly turn out to have more complicated regulatory systems than we now recognize.

3.2 *Origins of Regulatory Genes*

A complete account of the evolution of proteins that act as regulators of gene transcription requires knowledge of the DNA sequences with which they interact as well as the three-dimensional structures of the proteins themselves. There is evidence to suggest that new efficient promoter sites can evolve within structural genes (Bruenn and Hollingsworth, 1973) and some operons have additional internal promoters even though the genes are under common regulation. The second low-level promoter for the *trp* operon of *E. coli* is within a structural gene (Jackson and Yanofsky, 1972). Promoters, in the sense of RNA polymerase-binding sites, must be associated with any gene region in order for it to be transcribed, and specific promoters could evolve from existing DNA sequences. Insertion sequences may confer new regulator properties. In addition to the strong polar effect of IS sequences within genes it is reported that IS2 contains a stop signal for transcription when it is inserted in one orientation, and provides a promoter which can initiate constitutive transcription when it is inserted in the opposite orientation (Saedler *et al.,* 1974). Translocation of structural genes can bring them under the control of other operators and promoters. It was shown by Jacob *et al.* (1965) that the *pur* genes could be fused to *lac* genes by deletion of the intervening region so that the *lacY* and *lacA* genes were brought under the control of the *pur* operator and were regulated by repression by adenine and not by induction by β-galactosides. Such gene fusions have been particularly useful in experimental studies on regulation since assays for β-galactosidase are more convenient than for many other enzymes. During the course of evolution, translocation of genes could have brought unregulated genes together prior to the evolution of regulatory systems, or added an unregulated gene to a site at which a regulatory system had already evolved. The known processes of mutation and selection together with translocation can probably account for the evolution of the DNA regulatory regions.

If we attempt to go back to very primitive organisms, there may have been a time when most or all of the genes for enzymes were transcribed constitutively and at the same rate. A very simple device to increase the amount of any one enzyme would have been to duplicate that gene. If the second copy were readily lost and regained, such a mechanism could provide a very simple way of regulating enzyme synthesis. Theories of enzyme evolution have postulated the necessity of a gene duplication to allow for divergence and improved enzyme activities but some have then implicated the idea that one gene copy had to become silent for new activities to evolve (Koch, 1972). Gene duplication is a common regulatory occurrence in selecting for "fitter" mutants in continuous culture (Hartley, 1974). A duplicate gene can be retained as long as the growth conditions require it and can be a mechanism for compensating for a less efficient enzyme. Folk and Berg (1971) found that a mutation giving a defective glycyl-tRNA synthetase could be suppressed by duplication of the defective gene. Gene duplication may not only have been the way in which the total genetic potential was increased but could have been the way in which enzyme levels were originally regulated. There is evidence that gene dosage may still play some part in enzyme regulation. The position of genes on the chromosome and the distance from the origin of replication determine how many copies are likely to be present at different growth rates. In accounting for the arrangement of genes on chromosomes the relationship between distance from the origin and enzyme regulation is not often considered. The differences in enzyme levels that can be achieved by this means are trivial compared with regulation by induction and repression but may be important for the regulation of the constitutive enzymes of a complex metabolic pathway. Tribe *et al.* (1976) have shown that the level of the constitutive enzymes of the pathway for the biosynthesis of the aromatic amino acids is related to chromosomal position (see also Section 4).

Fusion of structural genes can give rise to multifunctional enzymes. One of the first multifunctional enzymes recognized was that determined by the *hisB* gene of *E. coli* (see Brenner and Ames, 1971). This enzyme carried both histidinol phosphatase and imidazole glycerol phosphate dehydratase activities. Among the enzymes of the *trp* operon are several produced by *E. coli* that show evidence of previous gene fusion although this does not appear to have occurred in other species (Crawford, 1975). The evolution of multifunctional proteins by gene fusion can also account for the origin of regulator proteins with binding sites for both DNA and low-molecular-weight effector molecules.

Gene fusion has also been produced experimentally. A series of genetic events led to the fusion of the *hisC* and *hisD* genes to produce a protein carrying both transaminase and histidinol dehydrogenase activities (Yourno *et al.,* 1970). A remarkable series of chimaeras have been formed by the fusion of the *lacI* and *lacZ* genes (Müller-Hill *et al.,* 1976). The resulting proteins are hybrids, some of which have both repressor and β-galactosidase activities. The success in obtaining gene fusions in which both functions were retained was high. Small changes in protein structure as a result of mutation can make enzymes inactive but it is also known that some changes in amino acids do not have this effect since homologous proteins from different organisms can differ in sequence but be equally active. Some amino acid substitutions can confer new and favorable properties on enzymes (see Section 7). The results of the *lac* gene fusion experiments show that even with the highly evolved proteins of present-day organisms both catalytic activity and the inducer- and operator-binding domains of the repressor could be functional in the hybrid protein.

Another type of duplication that has played a part in protein evolution is internal duplication within a gene. This is evidence by repeated sequences of amino acids. Engel (1973) suggested that internal duplications could allow a protein to evolve a regulatory site in addition to a substrate-binding site. From the sequence of bovine glutamate dehydrogenase he suggested that the nucleotide-binding regulatory site could have arisen by a duplication of the sequence containing the binding site for the NADH substrate. Regulation of biosynthetic pathway enzymes is based on a combination of feedback inhibition and repression by end products or intermediates (see Section 4). Frequently the same molecule acts both as feedback inhibitor and as effector for the repressor. If the inhibitor has some structural resemblance to the enzyme substrate, it could be envisaged that the inhibitor-binding site arose by an internal duplication of this sort. The alternative possibility would be that the inhibitor-binding sequence had been acquired by partial gene fusion. In either event, the enzyme that now carried both substrate- and inhibitor-binding sites would be a potential ancestor of the protein regulating enzyme synthesis (see Chapter 3 of Volume 2 of this treatise).

4 *The Biosynthesis of Aromatic Amino Acids*

The metabolic pathways for the biosynthesis of aromatic amino acids are essentially similar in all species studied but it is possible to find wide variations in the ways in which both enzyme synthesis and enzyme activities are regulated. The patterns of regulation may differ considerably from one microbial genus to another but the patterns are quite similar within the species belonging to a single genus. This suggests that the metabolic pathways evolved before the regulatory systems but that once established the regulatory patterns that had developed were strongly conserved.

In looking at the possible evolutionary origins of the regulatory systems for enzyme synthesis it will be useful to look at the same time at the controls relating to enzyme activity, since in some organisms these play a more important role in the overall regulatory pattern. The most detailed studies have been made with the enteric bacteria but some other species have been examined in sufficient detail to highlight the way in which different solutions have evolved to solve the same biological problem—that of synthesizing the right amount of each of the aromatic amino acids for balanced cell growth.

4.1 *Aromatic Pathway Enzymes and Regulation in Escherichia coli*

In *E. coli* and related organisms the system for aromatic acid biosynthesis is characterized by the findings that (1) there are isoenzymes at several points in the pathway; (2) there are a limited number of regulatory genes; (3) enzymes may occur as aggregates; and (4) some enzymes have multiple functions. Most of the genes are widely scattered on the chromosome with the notable exception of the five *trp* genes and a small group of *tyr* genes (Fig. 1). The first reaction of the common pathway is catalyzed by three DAHP (3-deoxy-D-arabinoheptulosonate-7-phosphate) synthetases, each of which is independently regulated (Fig. 2). One is subject to feedback inhibition and repression by tyrosine, the second to inhibition and repression by phenylalanine, and the third to repression by tryptophan. This pattern with multiple isoenzymes for the first reaction is similar to that for the control by end products of the three aspartokinases which carry out the first step of the branching pathway leading from aspartate to the biosynthesis of lysine, threonine, and methionine (Cohen, 1969). There are two isoenzymes for shikimate kinase and this is the only other point at which the common enzymes are specifically regulated (Tribe *et al.*, 1976).

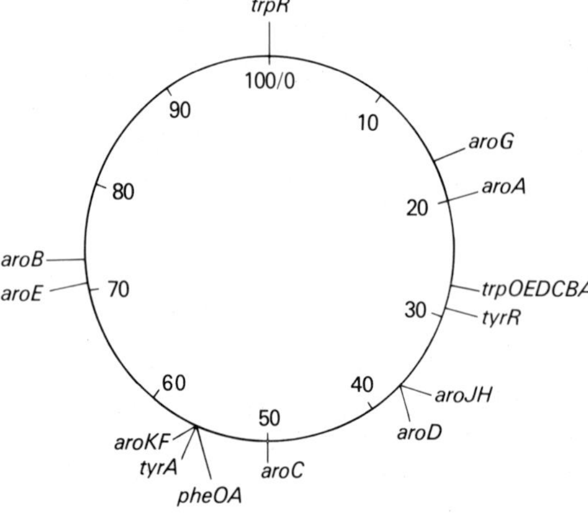

Figure 1. Chromosome of *E. coli* with locations of genes for the biosynthesis of the aromatic amino acids. *aroA*, EPSP (3-enolpyruvylshikimate-5-phosphate) synthetase; *aroB*, DHQ (dehydroquinate) synthetase; *aroC*, chorismate synthetase; *aroD*, DHQ dehydratase; *aroE*, dehydroshikimate reductase; *aroF*, DAHP (3-deoxy-D-arabinoheptulosonate-7-phosphate) synthetase (tyr); *aroG*, DAHP synthetase (phe); *aroH*, DAHP synthetase (trp); *pheA*, chorismate mutase-prephenate dehydratase; *tryA*, chorismate mutase-prephenate dehydrogenase. Numbers indicate map positions given by Bachmann *et al.* (1976).

At a later stage there are two isoenzymes for chorismate mutase, one of which is related to phenylalanine synthesis and the other to tyrosine synthesis. These are both bifunctional enzymes and one also carries prephenate dehydratase activity and is sensitive to feedback inhibition by phenylalanine. The second enzyme carries both chorismate mutase and prephenate dehydrogenase activities and only the latter activity is inhibited by tyrosine. The two enzymes are similar in physico-chemical properties and they carry out one common enzymatic reaction in each case. Davidson *et al.* (1972) suggest that these enzymes may have arisen from a common ancestor by gene duplication.

Two regulatory genes have been identified. The gene *trpR* determines the apopressor which, with tryptophan, represses the synthesis of DAHP synthetase (trp) and the five terminal enzymes of tryptophan biosynthesis. The *tyrR* gene determines the regulator protein which binds with tyrosine to repress the synthesis of DAHP synthetase (tyr), one of the two shikimate kinases and the terminal enzymes for tyrosine biosynthesis. A particularly interesting characteristic of this regulatory protein is that it can also accept phenylalanine and in this form it represses the synthesis of the DAHP synthetase (phe). Thus, this regulator gene product responds to two different effector molecules to regulate two separate genes in an independent manner.

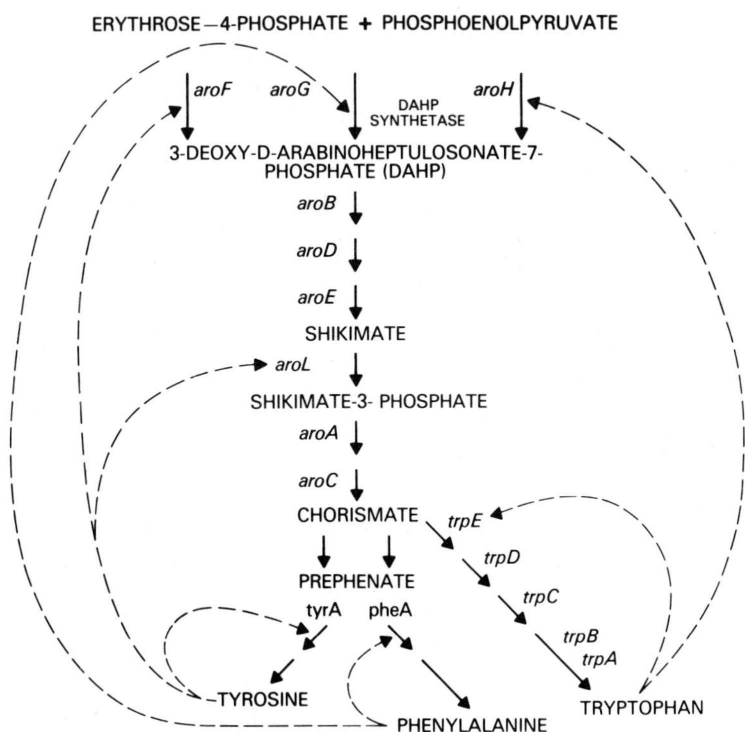

Figure 2. Outline of the pathway for the biosynthesis of the aromatic amino acids in *E. coli.* Genes and enzymes are designated as described in the legend for Fig. 1. The dashed lines indicate the main sites of regulation by feedback inhibition and/or repression. The regulator gene *tyrR* is concerned with repression of DAHP synthetase (phe) and DAHP synthetase (tyr). The regulator gene *trpR* is concerned with the regulation of DAHP synthetase (trp) and the *trp* operon. Details in text.

The two regulatory genes *trpR* and *tyrR* may have a common origin or may have evolved independently. The product of the *tyrR* gene has evolved in such a way that when combined with tyrosine or phenylalanine it effectively controls synthesis of DAHP synthetase (tyr) or DAHP synthetase (phe), respectively. Were it to be duplicated now, it is easy to see that two distinct regulatory genes could become more highly specific—one for tyrosine alone and one for phenylalanine alone. However, the *tyrR* product is very interesting in that when tyrosine is bound it switches off DAHP synthetase (tyr) by interacting at *aroK,* the operator for the small gene cluster that includes in the same operon *aroF* for DAHP synthetase (tyr) and *tyrA* for the multifunctional enzyme chorismate mutase (tyr)-prephenate dehydrogenase. When phenylalanine is bound to the *tyrR* product, there is no effect on the *aroF tyrA* operon but *aroG,* DAHP synthetase (phe) is switched off. Thus, a single protein can change its conformation in different ways as a result of binding each of these two effector molecules, and the two active forms of the *tyrR* repressor can recognize different operator sites.

There are several candidates for the role of ancestor of the regulator genes *trpR* and *tyrR*. One possibility is the gene for the first enzyme of the pathway. The arrangements for the DAHP synthetase genes do not suggest that the genes evolved by a recent duplication since the three genes *aroG* (phe), *aroH* (trp), and *aroF* (tyr) are located quite separately at 17, 37, and 56 min on the *E. coli* chromosome (Bachman *et al.,* 1976). However, if the suggestion of Zipkas and Riley (1975) is correct, such an arrangement would be compatible with two successive duplications of the primitive *E. coli* chromosome. Before the first such duplication, the organism could already have evolved a regulatory protein for a single DAHP synthetase gene, perhaps partially repressed by any one of the three aromatic amino acids. Later (after the duplications), divergence could have allowed specificity of repression to develop. Two of the DAHP synthetase enzymes are also subject to feedback inhibition and therefore carry binding sites for phenylalanine or tyrosine. These two functional domains might have arisen by partial fusion of a gene for unregulated DAHP synthetase with part of a gene for a protein with a binding site for both tyrosine and phenylalanine. This fusion could give rise to an enzyme whose activity could be regulated by inhibition by both phenylalanine and tyrosine. Duplication of this gene, followed by fusion to a DNA sequence specifying a regulatory function, could provide a regulatory protein. On this basis, regulation of the activity of other enzymes by phenylalanine or tyrosine could have been a later development. On the other hand, a case could be made for one of the other duplicate enzymes, also regulated by feedback inhibition by tyrosine or phenylalanine, to have provided the core of the ancestor protein to which was added, by fusion, a tyrosine- or phenylalanine-binding domain and then a regulatory function sequence. At this stage, too little is known about the structures of the proteins concerned to judge which is the most likely route for the regulatory protein to have evolved. It is, however, reasonable to suggest that phenylalanine regulation and tyrosine regulation are so intimately bound up that, at least as far as feedback inhibition is concerned, a common sequence may have been transposed and fused to the genes for the individual enzymes. Most of the enzymes of the common pathway are constitutive but shikimate kinase is an exception and the regulation at this step is again bound up with the existence of isofunctional enzymes (Tribe *et al.,* 1976). Clearly gene duplication has frequently been an evolutionary step to more efficient regulation of both activity and control of synthesis in this pathway.

The *trpR* gene does not seem to be involved in the regulation of any of the common pathway enzymes with the exception of DAHP synthetase (trp) (Tribe *et al.,* 1976). This might be due to a mutation in the region near the duplicated gene *aroH* or the operator of the *trp* operon could have been translocated to the *aroH* site. The *trpABCDE* cluster is one of the tightly regulated operons of *E. coli* and tryptophan itself is thought to be the corepressor (see Crawford, 1975). Manson and Yanofsky (1976) transferred the *trp* operon genes and the $trpR^+$ repressor gene, in separate experiments, from *E. coli* into other species of the enterobacteria and found that, with minor quantitative variations, the regulation was identical to that in *E. coli.* This provides good evidence for the conservation of repressor–operator recognition during the divergence of this genus. Manson and Yanofsky (1976) point out that the additional requirement for regulation at the *aroH* operator site would tend to conserve these sequences once they had evolved.

Studies on the *trp* operon have also provided evidence for gene duplication and translocation. Jackson and Yanofsky (1973) showed that a single mutational event that relieved the polar effect of a *trpE* mutation and released *trpB* from tryptophan repression, consisted of duplication of part of the *trp* operon and translocation of the duplicated part to another site in the chromosome. Among seven mutants studied, the length of the transposed segment and the site to which it was transferred differed in each case. The duplication–translocation event observed in these experiments provided a new independent operon, not controlled by tryptophan, for one or more of the *trp* genes without the loss of the original operon and supports the view that these events were important in the evolution of regulatory controls.

Another important feature of the enzymes of the aromatic pathways is the tendency to form aggregates. This is not confined to the enterobacteria but occurs in other species as well and may have regulatory significance (see Crawford, 1975). The ultimate in protein association is provided by gene fusion and there are many examples of multifunctional enzymes, such as chorismate mutase (tyr)-prephenate dehydrogenase and chorismate mutase (phe)-prephenate dehydratase in *E. coli.* The variations on aggregation of separate enzymes and gene fusion leading to multifunctional proteins among the final enzymes of the tryptophan pathway are considered in Section 4.4.

4.2 *Aromatic Pathway Enzymes and Regulation in Bacillus*

Jensen and Nester (1965) found that feedback inhibition of the aromatic amino acid biosynthetic enzymes in *Bacillus subtilis* follows a different pattern from that in *E. coli.* The bacilli have only a single DAHP synthetase and it is subject to feedback inhibition by intermediates, not by the amino acid end products. Prephenate and chorismate each produces partial inhibition and at low concentrations the effects of the two compounds are additive. At higher concentrations, when inhibition by one of the compounds has reached the maximum level, there is no increased inhibition when the other intermediate is added. Jensen and Nester suggested that the concentrations of prephenate and chorismate reflect the concentrations of the final products of the pathway. Nester *et al.* (1969) showed that regulation of synthesis of the enzymes falls into several distinct groups. The first enzyme, DAHP synthetase, is repressed by tyrosine and this repression is increased if phenylalanine is also present. Tryptophan has no repressing effect. Shikimate kinase and chorismate mutase are repressed to the same extent as DAHP synthetase, and in the same manner, by tyrosine and phenylalanine. The common enzymes of the pathway are thus regulated mainly by the concentration of tyrosine. The enzymes belonging to the tryptophan terminal sequence are all repressed by tryptophan (see also Crawford, 1975), while tyrosine represses the enzymes of its own terminal pathway.

Most of the genes for the enzymes required for the synthesis of the aromatic amino acids were found to be grouped together in a small segment of the chromosome of *B. subtilis* (Nester *et al.*, 1963; Hoch and Nester, 1973), as shown in Fig. 3. One of the odd features of this clustering of the *trp–tyr* and *aro* genes was the occurrence of *hisH* among the 12 gene loci concerned with the synthesis of the aromatic amino acids. In *B. subtilis hisH* is the gene specifying histidinol phosphate transaminase whereas in *E. coli* this activity has been assigned to *hisC* (Nester and Montoya, 1976*a,b*). Experimental evidence had previously indicated that the biosynthetic pathways for histidine and the aromatic amino acids interlocked in some way and Nester and Montoya (1976*a,b*) have now shown that the *hisH* enzyme also acts in the pathways for phenylalanine and tyrosine biosynthesis. Mutants with *hisH* lesions have about 60% less tyrosine and phenylalanine transaminase activity than the wild type. Mutants totally lacking transaminase activity had not turned up among the tyrosine and phenylalanine auxotrophs. This problem was resolved by finding that extracts of the wild type strain gave two peaks of tyrosine and phenylalanine transaminase activity and that one of the peaks coincided with histidinol phosphate transaminase. Mutants were subsequently isolated which lacked either or both activities.

Thus, the *hisH* gene appears to play an active role in both histidine and aromatic amino acid synthesis. The expression of *hisH* is under the combined control of histidine, tryptophan, and tyrosine, and this reflects the importance of the *hisH* enzyme for both pathways. Single-site mutants can be isolated that are depressed for most enzymes of histidine and aromatic amino acid biosynthesis. Roth and Nester (1971) showed that the *hisH* gene is expressed together with the *trp* genes under conditions of tryptophan limitation, but histidine limitation results in depression of *hisH,* but not the *trp* genes. On the other hand, tyrosine represses the synthesis of the *hisH* transaminase and the prephenate dehydrogenase determined by the adjacent gene.

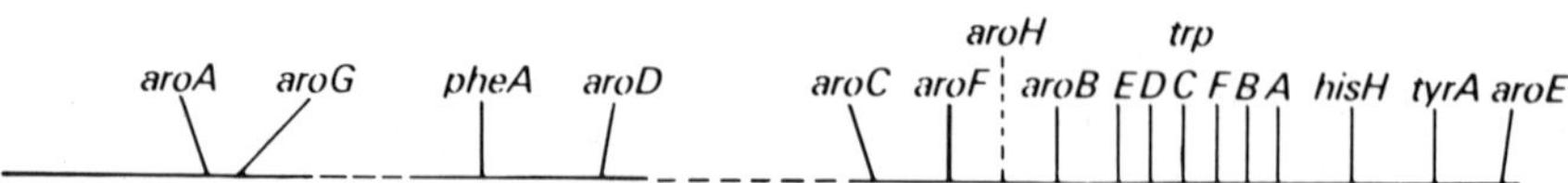

Figure 3. Arrangements of genes for the enzymes of the aromatic amino acid biosynthetic pathway in *Bacillus subtilis. aroA,* DAHP synthetase (a single enzyme only); *aroB,* DHQ synthetase; *aroC,* DHQ dehydratase; *aroD,* dehydroshikimate reductase; *aroE,* EPSP synthetase; *aroF,* chorismate synthetase; *aroG* and *aroH,* chorismate mutases; *pheA,* prephenate dehydratase; *tyrA,* prephenate dehydrogenase; *hisH,* histidinol phosphate, tyrosine, phenylalanine, transaminase; *trpEDCFBA,* tryptophan enzymes (Nester *et al.*, 1963; Hoch and Nester, 1973; Nester and Montoya, 1976*a,b*).

The rest of the *his* genes of *B. subtilis* are not tightly clustered together as they are in *E. coli* but are located in several linkage groups (Chapman and Nester, 1969). The attachment of *hisH* to some of the aromatic pathway genes appears to offer no obvious advantage and may be a purely fortuitous arrangement consequent on the transposition of a gene segment. For a long time the overlap in regulation of the histidine and aromatic amino acid pathways was extremely puzzling and the identification of a transaminase with dual functions has provided a most interesting solution. Regulation of the expression of *hisH* is elaborate but the other transaminase, which is devoid of histidinol phosphate transaminase activity but is active with tyrosine and phenylalanine, is not derepressed by starvation for phenylalanine or tyrosine and there is no evidence to suggest that it is regulated by any of the aromatic amino acids (Nester and Montoya, 1976*b*). The gene for the aromatic transaminase has not yet been mapped.

4.3 Aromatic Pathway Enzymes and Regulation in Other Genera

Jensen *et al.* (1967) examined microorganisms belonging to 32 genera to find out whether the regulation of the first enzyme of the pathway, DAHP synthetase, resembled that of *E. coli* or *B. subtilis,* or exhibited yet other patterns. Bacilli of a number of different species all resembled *B. subtilis* in having the sequential feedback control of a single DAHP synthetase, while other members of the enterobacteria resembled *E. coli* in possessing isoenzymes for the first step. Several other patterns of control by feedback inhibition were found for other genera and were similar for related species. The *Pseudomonas* species all appeared to possess a single DAHP synthetase and Jensen *et al.* (1973) showed that the enzyme from *P. aeruginosa* was sensitive to inhibition by tyrosine, tryptophan, and phenylpyruvate. A combination of the three inhibitors produced a cumulative, or less than cumulative, effect. It was interesting that phenylpyruvate, rather than phenylalanine, was the feedback inhibitor. The inhibition was complex in nature, and although inhibition by tyrosine or tryptophan appeared to be competitive with respect to phosphoenolpyruvate, it was found that phenylpyruvate inhibition was competitive with respect to the other substrate of the pathway, erythrose-4-phosphate. Inhibition by either tyrosine or phenylpyruvate could reach 90% and these appear to be the only controls operating on DAHP synthetase since Jensen *et al.* (1973) were unable to find any evidence for repression of this enzyme. Calhoun *et al.* (1973) concluded that the synthesis of DAHP synthetase in *P. aeruginosa* is set at a fairly high level since although tyrosine is a potent feedback inhibitor, it does not inhibit growth by cutting down the supply of intermediates to the other branches of the pathway. Calhoun *et al.* (1973) found a complicated control, which they termed the *channel-shuttle model,* at the tyrosine–phenylalanine branch point. Prephenate dehydratase is a bifunctional enzyme that also catalyzes the chorismate mutase reaction and is sensitive to phenylalanine inhibition. With high tyrosine and limited flow of intermediates through the pathway, the single chorismate mutase would be directed toward phenylalanine synthesis. With excess phenyalanine, on the other hand, the prephenate dehydratase activity is severely inhibited whereas inhibition of chorismate mutase is only partial. This could allow chorismate to be diverted to tyrosine synthesis. If both amino acids were present, then tyrosine would inhibit DAHP synthetase and phenylalanine would inhibit chorismate mutase and prephenate dehydratase, and possibly chorismate mutase would also be inhibited by accumulated prephenate. Under normal growth conditions in minimal medium these interlocking controls would work together to give balanced growth.

4.4 *Genes and Enzymes of Tryptophan Biosynthesis*

The gene–enzyme relationships of the tryptophan biosynthetic pathway (Fig. 4) have been more thoroughly explored in different genera than the common pathway enzymes. A recent review by Crawford (1975) gives an excellent account of current knowledge. Some of the features of enzyme aggregates and possible gene fusions have already been mentioned. What is particularly interesting is that the arrangements of genes and regulatory controls have now been compared in very closely related and distantly related groups. It is convenient to consider the *trp* group of enzymes separately but it is important to bear in mind that the rates of tryptophan biosynthesis depend not only on the specific controls of the *trp* pathway but on the supply of precursors from the common aromatic pathway and on the rates at which intermediates are drained away for the synthesis of phenylalanine, tyrosine, and other cell metabolites.

The tryptophan genes of *E. coli* are clustered in an operon and are thus regulated coordinately (Squires *et al.,* 1973, 1975). This clustering of genes and coordinate repression is well suited to the ecological niche of *E. coli* in which tryptophan may be expected to be found in fluctuating amounts. More subtle types of regulation have recently been found. In addition to the main *trp* pro-

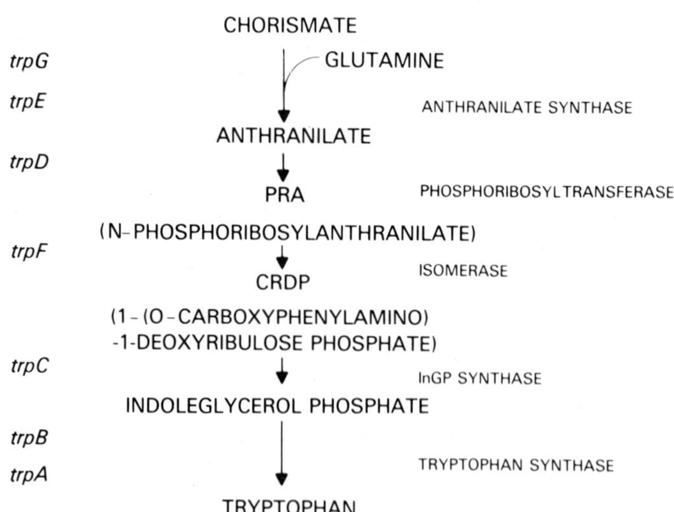

Figure 4. Genes and enzymes of the pathway for the biosynthesis of tryptophan. In *E. coli* the protein determined by *trpD* carries phosphoribosyltransferase activity and also the activity of the smaller subunit of anthranilate synthase determined by *trpG* in *B. subtilis* and *P. aeruginosa.* The *trpC* protein of *E. coli* carries PRA isomerase and InGP synthase activities which are determined by *trpC* and *trpF* in *B. subtilis* and *P. aeruginosa,* respectively.

moter adjacent to gene *trpE,* there is a secondary promoter. Under fully repressing conditions, initiation of transcription from the second promoter becomes significant and the last three enzymes are produced at about five times the basal rate of the first two.

In addition to this modification of *trp* operon control, it has been found that there is a region (*leader* sequence) preceding the *trpE* gene which can affect transcription. Bertrand *et al.* (1975) reported that under conditions of tryptophan excess there is "attenuation" of transcription in this leader sequence that prevents the RNA polymerase from progressing into the structural genes. At minimal tryptophan concentrations this control does not operate and thus more enzymes can be made. Both the second promoter and the attenuator control can be regarded as later refinements added to the *trp* operon. From the comparative studies it can be suggested that *trpD* and *trpC,* which determine multifunctional proteins, have arisen by gene fusion. However, tryptophan synthetase, which has two nonidentical subunits, is coded by two separate genes. This arrangement holds for all the prokaryotes examined but is not true for *Neurospora,* a finding that led Bonner *et al.* (1965) to suggest that gene fusion, which produced the single *Neurospora* tryptophan synthetase gene, might be a more general indication of the direction of evolution. Other fungi resemble *Neurospora* in this respect but the widespread occurrence of other gene fusions in prokaryotes makes it unlikely that gene fusion *per se* indicates an evolutionary direction (Crawford, 1975).

Regulation of the activity of the tryptophan enzymes in all prokaryotes is by feedback inhibition of the first enzyme and is related to the larger subunit of the multimeric enzymes. This is therefore likely to have been evolved as the first stage of regulation of this pathway. Although repression of synthesis of all the *trp* enzymes is common in prokaryotes, even for species in which the genes do not form a single operon, it is by no means universal. The greater variation in control of rates of synthesis indicates that these may have arisen later in evolution than the enzymes themselves. One variation on the theme is found in *Acinetobacter calcoaceticus* in which the *trp* genes are found in three separate linkage groups. Mutants can be obtained that are derepressed for all the enzymes, suggesting that there may be a single *trpR* gene (Crawford, 1975). The *trp* genes of *Bacillus subtilis* are repressed by tryptophan, including *trpG,* which is not linked to the cluster containing all the other *trp* genes (Fig. 5).

Regulation of the tryptophan enzymes in *Pseudomonas aeruginosa* and

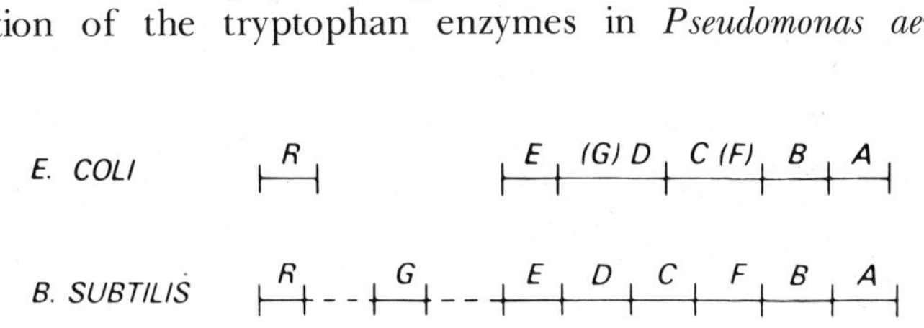

Figure 5. Arrangements of genes for enzymes of tryptophan biosynthesis (see Fig. 4). The *trpR* regulator gene is involved in repression by tryptophan of all the enzymes in *E. coli* and *B. subtilis.* In *P. aeruginosa* only the *trpEGDC* group of enzymes is repressed, *trpBA,* tryptophan synthase is induced by indoleglycerol phosphate, and the *trpF* enzyme is constitutive (Crawford, 1975).

P. putida is quite different. The *trp* genes are found in three separate linkage groups widely dispersed on the chromosome. The earlier studies on the regulation of the aromatic biosynthetic enzymes were hampered by the lack of response to amino acid analogs (Waltho and Holloway, 1966). However, Calhoun and Jensen (1972) found that 4-fluorotryptophan was a potent growth inhibitor and mutants were isolated that were derepressed for the enzymes of the *trpEGDC* group (Fig. 5). These are the only genes of the pathway that are under tryptophan repression control and affect anthranilate synthase (large and small subunits), anthranilate phosphoribosyltransferase, and indoleglycerol phosphate synthase. The single gene *trpF,* phosphoribosylanthranilate isomerase, appears to be unregulated while the genes *trpAB,* determining the two subunits of tryptophan synthetase, are induced by the substrate indoleglycerol phosphate. It can be seen that as in the *Bacillus* and *Acinetobacter* species there have been none of the gene fusions to be found in the enteric bacteria. The *trpAB* regulation is of particular interest in that it involves the induction of a biosynthetic enzyme. Moreover, this appears to be an example of a very neat autogenous regulation with the tryptophan synthetase A subunit acting as the repressor of its own synthesis and of the linked *trpB* gene (Proctor and Crawford, 1975). The evolution of this regulation could result very easily from the fusion of a regulatory sequence to the ancestral *trpA* gene. Partial sequence comparisons have been made of the N-terminal regions of the *P. putida* tryptophan synthetase α chain with that of *E. coli* and it would be most interesting to identify the regulatory domain of this protein.

Although regulation of the *trp* genes is essentially the same in *P. aeruginosa* and *P. putida,* the level set for the solitary *trpF* appears to be different. Calhoun *et al.* (1973) found that a derepressed mutant of *P. aeruginosa* excreted tryptophan. The amount of indoleglycerol phosphate produced by the earlier enzymes was sufficient to induce tryptophan synthetase and amounts of the precursors were sufficient for excess tryptophan to be synthesized. In *P. putida* Crawford and Gunsalus (1966) had found the same type of regulation but Maurer and Crawford (1971) found that the analog-resistant mutants accumulated anthranilate and not tryptophan. The derepressed enzymes were synthesized in 20-fold excess of the level of the wild type so that the flow of intermediates through the pathway seemed to have been controlled by the activity of the *trpF* product. The differences in the *trpF* genes of *P. aeruginosa* and *P. putida,* combined with similar mutations in the regulatory gene *trpR* leading to derepression of the early enzymes of the pathway, account for the excretion of tryptophan by *trpR* mutants of one species and anthranilate by the other species. In some metabolic pathways, piling up an intermediate such as anthranilate in this way could have a dramatic effect on the activities or the synthesis of enzymes of another pathway. One advantage of evolving more highly specific regulatory systems is that casual effects of mutations in the regulation of other pathways can be minimized.

The variations in the gene–enzyme arrangements for the pathways for the biosynthesis of the aromatic amino acids, and the different patterns of regulation that have evolved, illustrate very well that among the prokaryotes alternative solutions to these regulatory problems are feasible and efficient. Although the special arrangements of the enterobacteria may be interpreted in relation to their particular ecological niche this seems inadequate to account for the difference between the various organisms usually found in soil and water.

5.1 Induction and Repression

The regulation of catabolic enzymes is often easier to study than that of biosynthetic enzymes. Many catabolic enzymes are synthesized only in response to the presence of the inducer (or a related compound) in the growth medium and this makes it possible to compare inducer and substrate specificities. The amounts of catabolic enzymes synthesized also depend on the presence or absence of other carbon or nitrogen compounds in the growth medium that may produce catabolite repression. The increase in the rate of synthesis of catabolic enzymes in the presence of inducer, above the basal rate, may be a 1000-fold. Such large increases in rates of synthesis of catabolic enzymes in response to environmental changes depend on the properties of the regulator proteins and of the regulatory DNA regions (promoters, initiators, operators). To understand how the regulatory systems for catabolic enzymes might have evolved it is important to bear in mind the large incremental changes that are possible between expressed and nonexpressed states of catabolic genes. A further point is the complexity of some of the catabolic pathways. The utilization of complex organic compounds may require the *de novo* synthesis of ten or more enzymes that convert the initial growth substrate into an intermediate of the central metabolic pathways. Further, intermediates of the pathway may occur in nature and in order to make use of these compounds the regulatory systems must provide for induction of some of the enzymes by pathway intermediates. In nature, organic compounds seldom occur in pure solutions so that organisms are presented with a mixture of potential growth substrates. It is usually found that compounds that support rapid growth rates repress the synthesis of inducible enzymes required for compounds that support lower growth rates. Although glucose is a preferred growth substrate for the enterobacteria it takes a secondary role among the more biochemically versatile pseudomonads. Hamlin *et al.* (1967) showed that citrate repressed the synthesis of glucose catabolic enzymes in *P. aeruginosa* and in general the intermediates of the tricarboxylic acid cycle are the most important effectors of catabolite repression in this genus.

5.2 Catabolism of Aromatic Compounds

The study of catabolic pathways for the degradation of aromatic compounds has provided problems of great biochemical interest. For a long time, most of the available information concerned the possible chemical transformations and the intermediates formed. There are several different routes for the aerobic dissimilation of aromatic compounds; recent reviews in this field include those of Dagley (1971, 1975), Stanier and Ornston (1973), and Ornston and Parke (1977). Aerobic degradation is carried out by bacteria and fungi and for certain reactions molecular oxygen is required. There are also anaerobic degradation pathways known (Dutton and Evans, 1969; Guyer and Hegeman, 1969). Many of the strains used for biochemical studies were isolated by elective subculture and this has allowed a wealth of biochemical information to be collected. The diversity of species and strains isolated in this manner means that in many respects the evidence is fragmentary. Only a few species have been subject to genetic analysis and information about the physiological inducers and repressors is often scanty. However, sufficient information is now available to compare the regulatory patterns that have evolved for some of these pathways. The interpretation of the physiological experiments on rates of enzyme synthesis has recently been augmented by genetic studies with *P. aeruginosa* and *P. putida* (Holloway, 1975) and transformation (Juni, 1972) and conjugation systems (Towner and Vivian, 1976) have been established for *Acinetobacter calcoaceticus,* an organism that has also been used for detailed biochemical studies by several research groups.

5.3 The β-Ketoadipate Pathway

The best understood of the pathways for the dissimilation of aromatic compounds is that converging on β-ketoadipate (Fig. 6). It is present in many species of bacteria and fungi (Cain *et al.,* 1968; Ornston, 1971; Stanier and Ornston, 1973) and allows many naturally occurring aromatic compounds (and some of the novel chemical compounds produced by the chemical industry) to be completely degraded and thereby to provide carbon and energy for growth. Whereas the enzymatic steps of the pathway follow similar routes (although the ring fusion products in bacteria and fungi are different), there is a diversity of regulatory mechanisms any of which is seemingly as efficient in its particular organism as is any other. Elucidation of the regulation of this pathway in *P. putida* and *Acinetobacter calcoaceticus* owes much to the work of Ornston and Stanier (1966). Ornston and

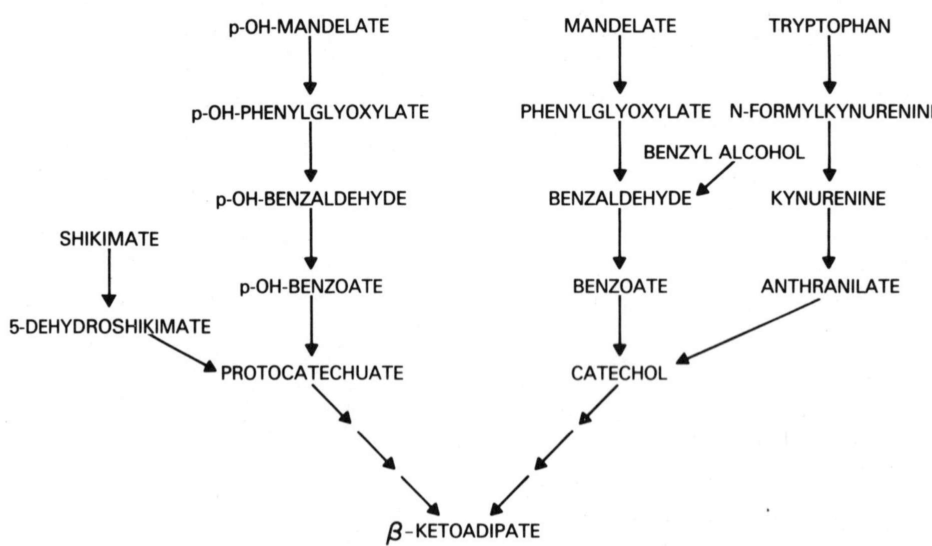

Figure 6. Outline of convergent catabolic pathways for aromatic compounds leading to β-ketoadipate. Enzymes are sequentially induced by metabolites of the pathway and constitute a number of separate regulation groups (see text for details).

Parke (1977) review "the evolution of induction mechanisms in bacteria" mainly on the basis of the results obtained from the studies on the β-ketoadipate pathway. Some of their conclusions are discussed here but the review should be consulted for a full interpretation of this extremely interesting metabolic pathway and its regulation. Figure 6 outlines the main intermediates of the β-ketoadipate pathway and gives examples of the converging pathways that allow various compounds to be metabolized by this route. The key metabolites are catechol and protocatechuate and these are cleaved by oxygenases that attack the bond between the adjacent hydroxyls. This is also termed the ortho cleavage pathway in contradistinction to an alternative fission by meta cleavage which is discussed later. The peripheral enzymes convert a variety of aromatic compounds into catechols, or substituted catechols, which can then be subjected to ring cleavage. In earlier studies on tryptophan metabolism, Palleroni and Stanier (1964) found that the enzymes were induced sequentially, either singly or in groups. The result is that intermediates induce enzymes for their own metabolism without triggering the production of all the enzymes of the pathway. The inducer is frequently a product, not the first substrate, and in the catabolism of tryptophan the first two enzymes are induced by their product, kynurenine, which also induces the next enzyme converting kynurenine to anthranilate. The enzymes converting anthranilate to catechol form the third regulation group and are induced by anthranilate (Fig. 6).

5.3.1 Mandelate Enzymes in Pseudomonas

Regulation of the mandelate and *p*-hydroxymandelate group of enzymes in *P. putida* was elucidated by Hegeman (1966*a–c*) and Mandelstam and Jacoby (1965). The group of enzymes (Fig. 7) is able to metabolize *p*-hydroxymandelate and its products as well as mandelate. The enzymes are coordinately synthesized, indicating that they form a single regulation group. The specificity of induction is relatively low; both L- and D-mandelate and several intermediates can induce all the enzymes, as can the *p*-hydroxy compounds. Hegeman (1966*c*) identified nonsubstrate inducers, including phenoxyacetate, and isolated constitutive mutants that synthesize the entire group of enzymes in the absence of an inducer. There is therefore circumstantial evidence for operon control by a single regulator gene. One interesting feature of this part of the pathway in *P. putida* is the presence of two benzaldehyde dehydrogenases. This appears to be an example of functional gene duplication. One of the enzymes requires NAD and the other NADP, although both are thought to act primarily for mandelate metabolism (another dehydrogenase is associated with the metabolism of benzyl alcohol).

Rosenberg and Hegeman (1969) made similar observations with *P. aeruginosa,* although in this species the racemase is absent and only L-mandelate is utilized. It should be recognized that in most cases detailed studies on these complex metabolic pathways have been carried out with single strains so that species comparisons may not always be valid. An alternative arrangement to a racemase converting D-mandelate to L-mandelate is the presence of two dehydrogenases, one of which acts on L-mandelate while the other acts on D-mandelate, and both producing phenylglyoxylate (benzoylformate). The latter arrangement is found in *Acinetobacter calcoaceticus* and suggests an earlier gene duplication.

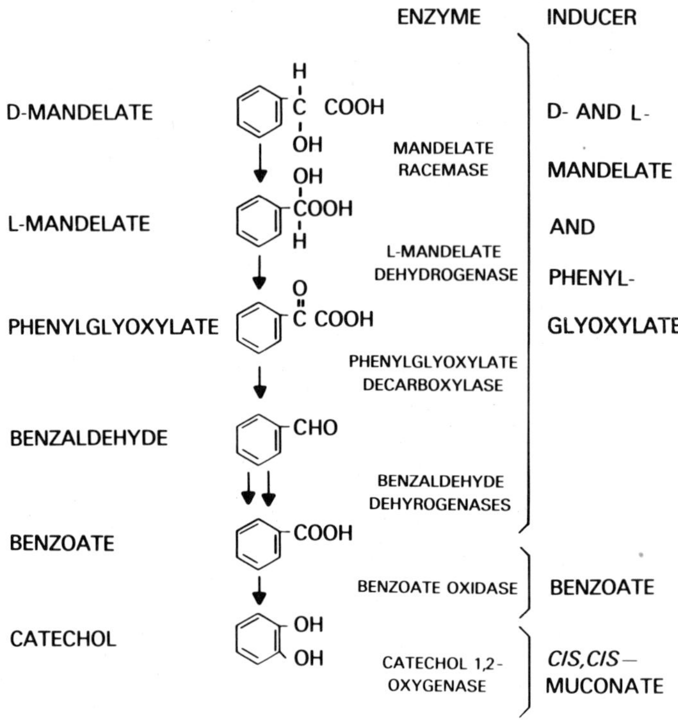

Figure 7. Regulation of the mandelate group of enzymes in *Pseudomonas putida*. The *p*-hydroxy compounds are metabolized by the same enzymes. Benzoate and *p*-hydroxybenzoate repress the synthesis of the mandelate group of enzymes and benzoate represses the *p*-hydroxybenzoate oxidase.

There is no direct information available on the molecular basis of the regulation of the mandelate group of enzymes but some interesting observations were made by Higgins and Mandelstam (1972) on the utilization of one of a pair of potential growth substrates by cultures provided with a mixture of two aromatic compounds. It had been observed earlier by Mandelstam and Jacoby (1965) that the synthesis of the mandelate enzymes was repressed by the pathway intermediates, benzoate, *p*-hydroxybenzoate, and catechol, and also by the terminal products of the pathway, succinate and acetate. The repression by the latter compounds is probably due to general catabolite repression, but the earlier intermediates have more specific effects. In nitrogen-limited continuous culture, both benzoate and *p*-hydroxybenzoate severely repressed the mandelate enzymes. The bacteria were fully induced for the latter enzymes and the alternative substrates (either benzoate or *p*-hydroxybenzoate) were utilized in preference to mandelate. With a mixture of benzoate and hydroxybenzoate, the enzymes required for *p*-hydroxybenzoate metabolism were repressed and benzoate was preferentially utilized. The specificity of the inducer-binding site of the regulator protein controlling the mandelate enzymes may allow compounds of similar structure to bind at the inducer-binding site and prevent transcription so that benzoate and *p*-hydroxybenzoate could be regarded as analog repressors of the mandelate enzymes. Similarly, benzoate could act as an analog repressor with respect to the *p*-hydroxybenzoate enzymes.

The preferential utilization of the intermediates does not result in more rapid growth since mandelate supports a faster growth of *P. putida* than does either benzoate or *p*-hydroxybenzoate. The only obvious advantage in this type of control seems to be that fewer enzymes are required, but in the natural environment the concentrations and the fluctuations of these and related compounds may be very different from those in laboratory experiments and there may be hidden growth advantages. It is clear that regulation of the synthesis of the enzymes forming this pathway is dependent on the synthesis of blocks of enzymes in response to sequential induction by pathway intermediates and that imposed upon the induction control is a second system involving repression of the synthesis of earlier enzymes by pathway intermediates. In addition to these pathway-specific regulation controls all the enzymes are subject to more general catabolite repression, which is expressed in the preferential utilization of compounds such as succinate.

5.3.2 *Mandelate Enzymes in Acinetobacter*

Acinetobacter calcoaceticus can grow on either L-mandelate or benzyl alcohol but the specific enzymes required for benzyl alcohol catabolism are repressed by L-mandelate. Rather surprisingly, benzyl alcohol, when present as the sole carbon source, can support a faster growth rate and give a higher growth yield than can L-mandelate (Beggs *et al.,* 1976; Cook *et al.,* 1975; Moyes and Fewson, 1976). The two enzymes specific for benzyl alcohol metabolism, benzyl alcohol dehydrogenase and benzaldehyde dehydrogenase II, are coordinately induced by either substrate. The product is benzoate which is also an intermediate of the L-mandelate pathway (Fig. 8). Beggs and Fewson (1974, 1977) have suggested that the regulator

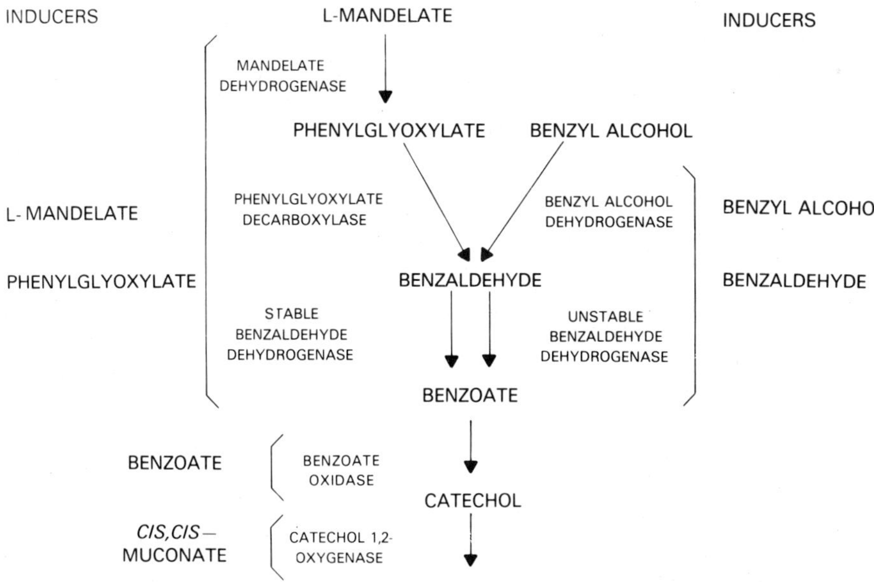

Figure 8. Regulation of the mandelate and benzyl alcohol pathway enzymes in *Acinetobacter calcoaceticus.* Mandelate and phenylglyoxylate repress the synthesis of the benzyl alcohol dehydrogenase and the unstable benzaldehyde dehydrogenase (Beggs and Fewson, 1974, 1977).

molecule responsible for repression of benzyl alcohol enzymes is phenylglyoxylate carboxy-lyase. These authors showed that both phenylglyoxylate and mandelate repressed synthesis of both dehydrogenases in the wild type strain but that mutants lacking mandelate dehydrogenase, and therefore unable to convert mandelate to phenylglyoxylate, were not repressed by mandelate. Mutants lacking phenylglyoxylate carboxy-lyase were not at all repressed by phenylglyoxylate whereas a mutant blocked later in the pathway, but still possessing an inducible phenylglyoxylate carboxy-lyase, was repressed by phenylglyoxylate. These results suggested that both phenylglyoxylate and the carboxy-lyase were required for repression of the benzyl alcohol pathway enzymes. Cultures that had been preinduced by the gratuitous inducer, thiophenoxyacetate, in the absence of phenylglyoxylate, were repressed for benzyl alcohol dehydrogenase, a finding that indicated that phenylglyoxylate was not directly concerned. Further evidence came from the study of constitutive mutants. Beggs and Fewson (1977) found that mutants that were constitutive for the mandelate enzymes were severely repressed for benzyl alcohol dehydrogenase, providing additional evidence that the enzyme phenylglyoxylate carboxy-lyase was itself the regulator molecule. If this enzyme is involved in this way in the repression of a pair of enzymes synthesized coordinately, then it might be expected to have a DNA-binding site which would allow it to prevent transcription of the relevant genes. Since benzyl alcohol and benzaldehyde dehydrogenases are induced by their substrates there must also be another regulator molecule involved in that part of the control system for these enzymes. Beggs and Fewson (1977) point out that *Acinetobacter acetocalcoaceticus* appears to have evolved an evolutionarily disadvantageous type of control since it is capable of growing faster and better on benzyl alcohol than on L-mandelate and that possibly the natural substrates of the enzymes are substituted molecules for which the relative growth advantages might be reversed or that the control by phenylglyoxylate carboxy-lyase is an accidental feature of some other regulatory system. Benzyl alcohol dehydrogenase is also repressed during rapid growth or by the addition of compounds such as succinate to the growth medium, suggesting that a catabolite repression type of control may also exist.

5.3.3 Catechol and Protocatechuate Pathways

In *P. putida* the regulatory patterns for the two converging branches of the β-ketoadipate pathway differ in several respects (Fig. 9). Benzoate oxidase is induced by benzoate but catechol 1,2-oxygenase is induced by the product *cis,cis*-muconate. This requires a basal level of the 1,2-oxygenase sufficient to convert catechol to inducer. However, it also has the consequence that in strains possessing the meta pathway enzyme, 2,3-oxygenase, which is induced by earlier substrates such as phenol and substituted catechols and toluates, catechol is removed too rapidly to allow *cis,cis*-muconate to build up to the level required for induction of the 1,2-oxygenase. These two different regulatory systems for the two oxygenases make it possible for benzoate to be metabolized via the ortho pathway and phenol to be metabolized mainly via the meta pathway, even though the common intermediate is catechol (Feist and Hegeman, 1969). The next two enzymes of the β-ketoadipate pathway are also induced by *cis,cis*-muconate but form a separate regulation group. Along the other branch of the pathway, protocatechuate induces its own oxygenase while the next three enzymes are induced by β-ketoadipate. As can be seen from Fig. 9 the enzyme β-ketoadipate enol lactone hydrolase is common to both pathways and during growth on mandelate or benzoate the production of β-ketoadipate leads to the gratuitous induction of two of the enzymes of the protocatechuate branch (Ornston, 1966).

This regulatory pattern also occurs in *P. aeruginosa* (Kemp and Hegeman, 1968) and is common to all the fluorescent pseudomonads. It has been retained during subsequent divergence of these species in spite of its apparent inefficiency. The regulatory genes for this pathway clearly evolved later than the structural genes but at the present time the only clue to their origins could be the congruity of the inducer specificities with some of the enzymes. The regulatory pattern is quite otherwise in *Acinetobacter calcoaceticus.* The final steps of the pathway are carried out by isoenzymes so that there are two enol lactone hydrolases and two transferases. Protocatechuate induces all the enzymes of its own pathway including one of each of the pairs of isoenzymes. *cis,cis*-Muconate induces catechol 1,2-oxygenase and the remaining enzymes of the catechol branch are also induced by *cis,cis*-muconate (Fig. 10). Thus, this organism, unlike *P. putida,* has evolved a high degree of regulatory coordination. Protocatechuate also induces the synthesis of enzymes that convert another precursor, shikimate, to protocatechuate. The concentration of endogenous shikimate produced in biosynthesis is not sufficient for induction to occur but growth with either *p*-hydroxybenzoate or protocatechuate will trigger the gratuitous induction of the enzymes of shikimate catabolism (Canovas *et al.,* 1968). This does not appear to deplete the provision of shikimate for the biosynthesis of the aromatic amino acids unduly. There is some evidence for product inhibition by 5-dehydroshikimate, and there has been some suggestion of compartmentalization of the catabolic enzymes (Tresguerres *et al.,* 1970).

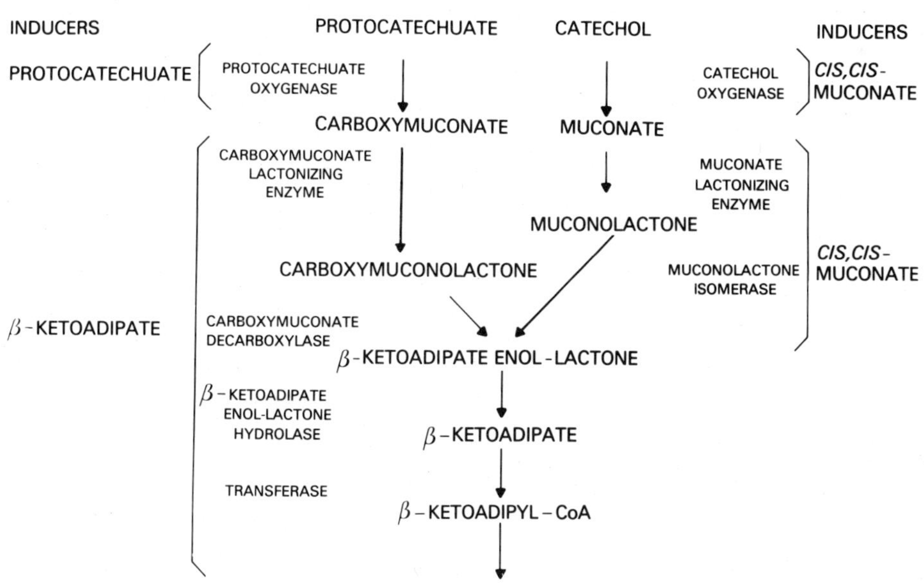

Figure 9. Regulation of the converging branches of the β-ketoadipate pathway in *Pseudomonas putida.*

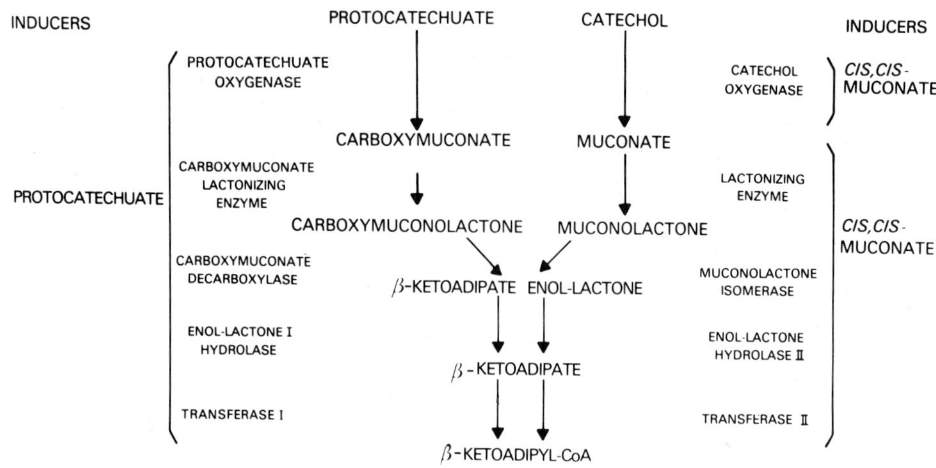

Figure 10. Regulation of the converging branches of the β-ketoadipate pathway in *Acinetobacter calcoaceticus.*

From the available data on regulatory patterns it appears that *Acinetobacter* has evolved less regulatory genes than has *P. putida,* but that these regulatory genes are able to control larger gene clusters than those of *P. putida.* In those cases in which a single metabolite is the inducing molecule for more than one group of structural genes it could bind to a single regulatory molecule with a dual regulatory role or it could bind to two different regulatory genes. The suggestion of Beggs and Fewson (1977) that phenylglyoxylate carboxy-lyase is itself a regulator molecule opens very interesting possibilities. The very large number of inducible catabolic enzymes possessed by many of these organisms would make it reasonable in terms of economy of gentic material if other enzymes were also able to act as regulator proteins.

The enzymes of the parallel catechol and protocatechuate pathways of *P. putida* and *A. calcoaceticus* show many similarities and Ornston and Parke (1977) discuss the evidence from physicochemical comparisons and partial sequence studies of some of the enzymes, that the β-ketoadipate pathway had a common origin in these two bacterial species. There are, however, sufficient differences between the bacteria and the fungal pathways to suggest that the β-ketoadipate pathway might have evolved independently in the bacteria and fungi. There are some similarities in regulation between *Pseudomonas* and *Acinetobacter* in that similar compounds have been selected as inducers, but this may be less significant than appears at first. The choice of inducers is limited by the intermediates of the pathway and may originally have been less specific in both organisms.

In *A. calcoaceticus* and *P. putida* only three of the intermediates in the reactions from catechol and protocatechuate to β-ketoadipate act as inducers (Fig. 10). In *Acinetobacter* all the enzymes of the catechol branch, including one of the enol lactone hydrolases and transferases, are induced by *cis,cis*-muconate. The same compound is the inducer of the first three enzymes of the pathway in *P. putida* while the single enol lactone hydrolase and transferase are induced by β-ketoadipate. The two inducing compounds are not dissimilar in structure. For the protocatechuate branch of the pathway the inducer of the first enzyme is protocatechuate in both species but, whereas it is also the inducer of the rest of the enzymes of the pathway in *Acinetobacter,* it does not act in this way in *P. putida.* β-Ketoadipate induces the carboxymuconate lactonizing enzyme and carboxymuconate decarboxylase so that these two enzymes are induced gratuitously during growth on catechol. It is possible that at an earlier stage of development of this pathway other intermediates could also serve as inducers. There are suggestions from experiments with other strains of *P. putida* that this might be so. Bayly and McKenzie (1976) working with *P. putida* strain PsU and mutants derived from it thought that benzoate or an intermediate before catechol, might be another inducer of catechol 1,2-oxygenase.

5.4 Meta Pathway Enzymes

The enzymes of the meta cleavage pathway are less specific than are those of the ortho pathway. Aromatic compounds with various substituent groups can be accepted for cleavage and their products can be metabolized further. Figure 11 shows the general outline of the pathway for catechol and methyl catechols. From the regulatory aspect perhaps the most significant finding is that the enzymes of

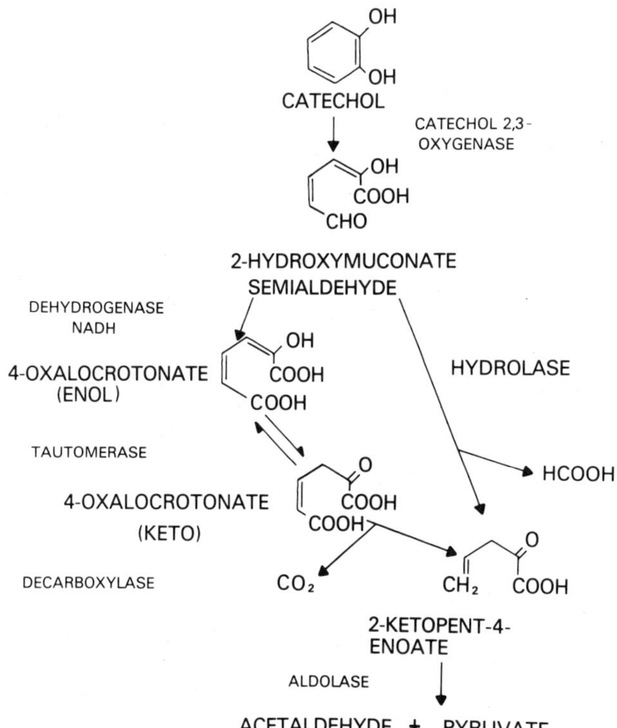

Figure 11. Outline of the meta pathway for the catabolism of aromatic compounds that can be converted to catechol or its methyl derivatives. The enzymes of this pathway are induced by a large number of aromatic compounds and the enzymes have fairly broad specificities. Details in text.

the entire meta pathway are induced from above (that is, by early substrates of the pathway) and not by the intricate system of intermediates evinced in the regulation of the enzymes of the ortho cleavage pathway. This very broad specificity for substrates and inducers allows the meta pathway to be used for the breakdown of a wide variety of aromatic compounds. In some strains of *P. putida,* genes for the meta pathway enzymes, as well as the genes for a group of peripheral enzymes, are carried on plasmids. Thus, by plasmid transfer a bacterium may acquire all the genetic information needed to catabolize compounds such as toluene and *m*- and *p*-xylene (Williams and Murray, 1974).

The pivotal position of catechol, the common intermediate of ortho and meta cleavage pathways, was first clarified in regulatory terms by Feist and Hegeman (1969), who examined a strain of *P. putida* capable of synthesizing the enzymes of both pathways. These authors showed that phenol and its methyl derivatives directly induced all of the meta enzymes. This included catechol 2,3-oxygenase, so that catechol would have been removed too fast to allow an adequate conversion to the ortho pathway inducer, *cis,cis*-muconate.

The meta pathway has two sets of reactions available for the conversion of 2-hydroxymuconic semialdehyde to 2-ketopent-4-enoate. Both sets of enzymes are present at the same time but they have different roles in the metabolism of catechol and the methyl catechols. The hydrolase route is significant for the metabolism of *o*- and *m*-cresol and the dehydrogenase route for the metabolism of *p*-cresol and phenol (Bayly and Wigmore, 1973; Murray and Williams, 1974).

The PsU strain of *P. putida,* which produces ortho pathway enzymes when grown on benzoate (Feist and Hegeman, 1969) but meta pathway enzymes when grown on phenol, has now been shown also to produce low levels of catechol 2,3-oxygenase and other meta pathway enzymes when grown with catechol. The enzymes of the two pathways may therefore both be present although one pathway will predominate. Feist and Hegeman (1969) showed that phenol could be metabolized via the ortho pathway if the 2,3-oxygenase were removed by mutation. Bayly and McKenzie (1976) showed that catechol could be metabolized via the meta pathway if the 1,2-oxygenase were absent. This freedom to be metabolized by the enzymes of either the ortho or meta pathway enzymes does not extend to the substituted catechols and the compounds that give rise to them. Wigmore and Bayly (1974) obtained mutants of *P. putida* in which the specificity of regulation had been altered and concluded that a single regulator protein controlled all the enzymes of the meta pathway. The first enzyme, phenol hydroxylase, was induced by phenol but its synthesis was not coordinate with the synthesis of the subsequent enzymes of the pathway. Bayly *et al.* (1977) suggest that the phenol hydroxylase gene is regulated independently and that the genes for catechol 2,3-oxygenase together with those for the two groups of meta pathway enzymes—(1) hydrolase, hydratase, and dehydrogenase, (2) tautomerase and decarboxylase (Fig. 11)—form a single operon under unified control. Some regulatory mutants produced all the later group of enzymes constitutively but were defective in phenol hydroxylase activity. Revertants regained inducibility for enzymes of both operons. Wigmore *et al.* (1977) suggest that there is a single regulator gene that exerts positive control on the expression of the phenol hydroxylase gene, and negative control on the expression of the genes for the other meta pathway enzymes. The authors argue against a role for phenol hydroxylase as an autogenous regulator since some mutants were defective in phenol hydroxylase but remained inducible for the other enzymes. This is insufficient evidence to rule out the possibility that phenol hydroxylase acts as a regulator protein; further investigation of this system may clarify the situation.

The differences in regulation between the ortho and meta pathways for the catabolism of aromatic compounds are related to the respective roles that these pathways play in the microbial environment. It appears at first sight that there is considerable genetic and metabolic redundancy in the provision of two major pathways for catechol in a single organism. However, the ortho pathway is very specific in terms of the compounds it serves to metabolize. This is evidenced by the rather high substrate and inducer specificities of the sequential groups of enzymes. Ornston and Parke (1977) suggest that the ortho pathway has a role not only in the catabolism of aromatic compounds but also in providing a pathway for the utilization of the intermediates as growth substrates. They suggest that *cis,cis*-muconate and β-ketoadipate may be present in the environment as a result of the degradative activities of other bacteria and fungi and that this might account for the inducing activities of these compounds. Most strains of *P. putida* are impermeable to *cis,cis*-muconate but mutants with increased permeability have been isolated. Ornston and Parke (1977) have discovered an inducible transport system for the uptake of β-ketoadipate. If β-ketoadipate were a primary substrate for the ortho pathway, then its role as an inducer of other enzymes could be a secondary development.

The meta pathway, on the other hand, is well adapted to the dissimilation of a wide variety of substituted compounds. These may be produced in complex mixtures as the result of the initial degradation of natural products and in some places will be found as mixtures of manufactured chemicals. The lower efficiency of the meta pathway in terms of growth on benzoate of some strains could be balanced by its greater value in the catabolism of a mixed bag of aromatic compounds.

5.5 *Arrangements of Genes of Aromatic Pathway Enzymes*

Transductional analysis in *P. aeruginosa* and *P. putida* has shown tight linkage of the genes for enzymes that are coordinately regulated, suggesting that they are arranged in operons under common transcriptional control (Chakrabarty *et al.*, 1968; Kemp and Hegeman, 1968; Wheelis and Stanier, 1970). The linkage of some of the regulatory mutations has been explored and the *catBC* genes (muconate lactonizing enzyme and muconolactone isomerase) were found to be linked to the regulatory gene *catR*. This may be a positive control system (Wheelis and Ornston, 1972; Wu *et al.*, 1972). Mutations in the regulatory gene *pcaR* confer constitutivity on the *pcaBCD* gene cluster. This determines the carboxymuconate lactonizing enzyme, carboxymuconate decarboxylase, enol lactone hydrolase, and also the β-ketoadipate transport system (Ornston and Parke, 1977). Wigmore *et al.* (1977) suggest that a single regulatory gene has a dual role as activator of the phenol hydroxylase gene and repressor of the five meta pathway enzymes. The evidence for this idea comes from the properties of mutants but the genes have not been mapped.

One of the unexpected findings about the genes for the degradation of aromatic compounds was that many genes occurred very near each other on the chromosome, although they were not under common regulation. This was apparent from the cotransduction of genes for enzymes known to be independently regulated. For example, *pcaA,* which is regulated by induction by protocatechuate, is linked to the *pcaBCD* cluster, regulated by induction by β-ketoadipate. This clustering of catabolic genes includes others that are not related to the degradation of aromatic compounds. Leidigh and Wheelis (1973) showed that in *P. putida,* genes for histidine and nicotinic acid catabolism were cotransduced with some of the aromatic pathway genes. The authors suggested that the supraoperonic clustering of genes for unrelated enzymes was not concerned with regulation but had evolved to enable simultaneous transfer of genes for several catabolic pathways. It is not true, however, that all catabolic genes are to be found in a small segment of the chromosome of all *Pseudomonas* species. The genes for the catabolism of histidine, arginine, lysine, and acetamide are scattered around the chromosome of *P. aeruginosa* (Day *et al.,* 1975; Rahman and Clarke, 1975). Canovas and his colleagues (de Torrontegui *et al.,* 1976) have been working on the enzymes of glucose catabolism in *P. putida* and have found that five genes comprising at least three separate operons are cotransduced. This gives another example of a supraoperonic cluster, but these genes were not linked to the *aro–his* segment. The previous studies had shown that in *P. putida* the *aro–his* segment included 21 structural genes belonging to at least 14 operons and comprising about 10% of the chromosome. It was important to establish whether or not all catabolic genes were located in the same region (de Torrontegui *et al.,* 1976).

The supraoperonic clustering of catabolic genes in *P. putida* may represent one aspect of the evolution of the very high biochemical potential of the *Pseudomonas* species. Transfer of genes for whole pathways and also genes for segments of pathways, together with all essential regulator genes, could allow rapid evolution of catabolic pathways for novel substrates. An advantageous mutation could be spread very rapidly through a microbial population if the gene segment could be added to a transmissible plasmid.

5.6 *Plasmids and Regulation*

A genetic device that is commonplace in bacteria is the maintenance of dispensable genes on an extrachromosomal element or plasmid (Wheelis, 1975; Chakrabarty, 1976; Williams and Worsey, 1976*b*). This can be regarded as a regulation of a rather extreme type and for the peripheral genes of complex catabolic pathways it has obvious advantages. In some strains the genes for certain aromatic pathway enzymes are plasmid-borne. These include enzymes for metabolizing naphthalene (NAH: Dunn and Gunsalus, 1973), salicylate (SAL: Chakrabarty, 1972), and the xylenes and toluates (XYL, XAL, TOL: Williams and Murray, 1974; Wong and Dunn, 1974). Some chromosomal genes code for enzymes for which there are alternative genes found on plasmids (Grund *et al.,* 1975). With the evidence that drug resistance may be carried by transposons which may transfer between plasmids and chromosomes and from one plasmid to another (Cohen, 1976), it becomes possible to envisage a similar route for the evolution of the catabolic plasmids. Thus, a natural bacterial population may contain within it several minority groups carrying different catabolic plasmids. The influx of particular compounds would then not only give a growth advantage to those cells harboring the appropriate plasmid but would result in rapid spread of the plasmid through the population. There are some suggestions that the plasmid transfer rate is higher when the catabolic genes are being expressed.

Williams and Murray (1974) found that the TOL plasmid of *P. putida* (*arvilla*) was lost if the cultures were grown on benzoate. Further, whereas in the parent strain aromatic compounds were metabolized primarily by the meta pathway, the cured strains metabolized benzoate via the ortho pathway. Benzoate is capable of supporting a faster growth rate than toluate and the bacteria which have lost the plasmid rapidly outgrow the rest. Williams and Worsey (1976*a*) have obtained a number of different natural isolates which are capable of utilizing toluene and the xylenes; some contain plasmids apparently identical with that of the *P. putida* (*arvilla*) strain so that this plasmid gene pool may have importance in nature. From studies of strains harboring defective plasmids, Worsey and Williams (1977) have suggested a control system for the TOL groups of enzymes involving at least two regulator genes.

Pseudomonas putida MT20 is one of the strains carrying a TOL plasmid that codes for the enzymes oxidizing toluene and *m*- and *p*-xylene to benzoate and *m*- and *p*-toluate, followed by meta cleavage of the aromatic ring. The plasmid can be lost completely by selecting on benzoate medium since in the absence of the plasmid the ortho pathway enzymes can be synthesized and this pathway supports a faster growth rate. Occasionally, mutants appear that retain the ability to grow on toluene and *m*-xylene but have lost the ability to grow on *m*-toluate; most of these have also lost the ability to transfer the plasmid. Worsey and Williams (1977) suggest that some genes have been lost from the plasmid including a gene for a regulator protein. These authors suggest that the genes for the early enzymes of this pathway are arranged in at least two groups under the control of two regulator genes, one of which responds to the hydrocarbons (possibly the alcohols as well) and the other of which responds to benzoate and *m*-toluate. The loss of the second regulator gene could explain the loss of inducibility in strains that are still able to metabolize *m*-toluate if the enzymes have been induced by *m*-xylene. This effect would occur if the enzymes were under positive regulation. Plasmids may have very profound effects on the strains that harbor them and in this instance the plasmid carries the set of structural and regulatory genes that masks the chromosomal genes of the ortho cleavage pathway. Studies on the interactions between the regulation of plasmid and chromosomal genes have not been extended far enough to say whether such plasmid dominance occurs for other catabolic pathways.

6 *Nitrogen Metabolism and Regulation*

6.1 *Glutamine Synthetase*

Until recently, regulation of the enzymes of nitrogen metabolism has seldom been treated as a separate problem. In the analysis of degradative pathways many nitrogen-containing compounds, including amino acids such as tryptophan, have been regarded in the same way as compounds of related structure that do not contain nitrogen. Most of the bacteria used for regulatory studies can grow in ammonium salt media with a range of carbon compounds and for a long time it was assumed that nitrogen assimilation took place by means of one of the well-known amination reactions. Ammonia was seldom considered to present any regulatory problems, although it was known that enzymes of nitrogen fixation were not active if either ammonium salts or complex organic nitrogen compounds were available. One of the key enzymes for ammonia assimilation was considered to be glutamate dehydrogenase, and undoubtedly it does play this role. However, the observations of Meers *et al.* (1970) that glutamate dehydrogenase activity was very low in chemostat cultures grown under nitrogen limitation led to the finding that there was an alternative route. When ammonium salt concentration is low, the low-affinity glutamate dehydrogenase (GDH) is inadequate and ammonia assimilation takes place via the two enzymes, glutamine synthetase (GS) and glutamine: α-oxoglutarate aminotransferase (GOGAT).

Glutamate occupies a key position in metabolism since it is the precursor of several amino acids and both glutamate and glutamine are required for the biosynthesis of a number of other amino acids and essential cell metabolites. It is also the product of several amino acid degradative pathways and plays an essential role in transamination reactions. It is not surprising that complex regulatory mechanisms have evolved to control the level of enzyme activities for glutamine and glutamate metabolism and to control the amounts of glutamine and glutamate synthesized. In some of these regulatory systems glutamine synthetase itself is the regulatory molecule. This is discussed in detail elsewhere (Magasanik *et al.,* 1974). The discussion here is concerned with some observations that may be relevant to the evolution of this complex regulatory system. In *Klebsiella aerogenes* the concentration of glutamate dehydrogenase is high and that of glutamine synthetase is low in cultures grown with excess ammonia, whereas in nitrogen-limited media the concentration of glutamine synthetase is high and that of glutamate dehydrogenase is low (Brenchley *et al.,* 1973). From physiological studies of the enzyme levels of cultures grown in various media and the properties of glutamine synthetase mutants, Magasanik and colleagues concluded that glutamine synthetase regulated both its own synthesis and the synthesis of glutamate dehydrogenase (Magasanik *et al.,* 1974). Glutamine synthetase occurs in an active form and also in an inactive form in which each of the 12 subunits may be covalently modified by adenylylation with AMP. The adenylylated form is predominant in ammonia-rich media and is considered to act as the autogenous repressing molecule.

In *E. coli* the level of glutamate dehydrogenase is not inversely related to the level of glutamine synthetase and is not severely repressed in nitrogen-limited media (Cole *et al.,* 1974; Newman and Cole, 1977; Senior, 1975; Striecher *et al.,* 1976). However, the structural gene for glutamine synthetase, *glnA,* could be transferred to *K. aerogenes* from *E. coli.* Normal repression of glutamate dehydrogenase occurred so that the absence of repression in *E. coli* was due to the evolutionary divergence of regulation of its glutamate dehydrogenase. One of the most interesting of the regulatory functions of glutamine synthetase lies in its relation to catabolite repression of certain amino acid pathways. Magasanik *et al.* (1974) described positive regulation by glutamine synthetase of tryptophan permease and the enzymes of histidine catabolism in *K. aerogenes.* The synthesis of the *hut* (histidine utilization) enzymes of *K. aerogenes* is induced by urocanate, the product of the first enzyme, histidase, and is repressed by glucose in ammonium salt media (Prival *et al.,* 1973). The *hut* genes are under the operon-specific control of a regulator gene, *hutC,* and the general CRP–cAMP positive regulation of catabolic genes. Under nitrogen limitation there is no repression by glucose, and glutamine synthetase takes over the role of positive regulator (Prival and Magasanik, 1971; Tyler *et al.,* 1974). *Salmonella typhimurium* can also catabolize histidine and is subject to catabolite repression by glucose, but in this case there is no relief from catabolite repression in nitrogen-limited growth. However, the *hut* genes of *S. typhimurium* (hut_S) can be transferred to *K. aerogenes* and they are then able to escape catabolite repression during nitrogen limitation. Goldberg *et al.* (1976) also found that it was possible to transfer the *hut* genes of *K. aerogenes* (hut_K) to *E. coli* which does not normally catabolize histidine. The *hut* genes of *K. aerogenes* and *S. typhimurium* could then be compared in three different genetic backgrounds. In *E. coli* hut_S histidase was more sensitive to glucose repression and less relief was observed under nitrogen-limited growth than with the hut_K histidase. When the hut_K genes were transferred to *S. typhimurium* it was found that they were not at all repressed by glucose. Thus, the regulation of the *hut* enzymes and the amounts of histidase synthesized in different growth media were dependent on the interactions of other regulatory systems of these organisms including, presumably, the CRP–cAMP regulatory control.

In *P. aeruginosa* histidase is also subject to strong catabolite repression but in this case succinate and not glucose is the most effective repressing compound. With a strain carrying a defect for histidine uptake it was possible to show that uner nitrogen-limited growth in pyruvate–histidine medium there was no repression by succinate (Potts and Clarke, 1976). Thus, in this unrelated genus it seems probable that glutamine synthetase plays a similar role in the activation of the *hut* genes.

Earlier it was suggested that autogenous regulation (control by an enzyme of its own structural gene) could be considered a primitive model for control of gene transcription. The evidence for autogenous control of glutamine synthetase indicates that it is very far from being a simple system (Foor *et al.,* 1975). For its role in the regulation of transcription, glutamine synthetase requires a specific nucleotide-binding domain as discussed earlier. For activation of the *hut* genes it is required to be present in the nonadenylylated form (Tyler *et al.,* 1974). In the repression of its own gene, *glnA,* glutamine synthetase appears to be in the adenylylated state although a high level of adenylylation does not always reduce *glnA* expression (Streicher *et al.,* 1976). The enzymatic reactions by which glutamine synthetase is adenylylated and deadenylylated are complex and involve the products of several other genes, *glnB, glnD,* and *glnE* (Foor *et al.,* 1975).

The state of adenylylation of glutamine synthetase affects its enzymatic activity as well as its regulatory activity, and in addition to this covalent modification the enzyme is subject to multiple feedback inhibition by a number of different end products of metabolism (Ginsburg and Stadtman, 1973). Such complexity of binding sites and the conformational changes possible in the molecule indicate clearly the evolutionary sophistication of this enzyme. Glutamine synthetase has evolved to become both a key metabolic enzyme and a key regulator of enzyme synthesis.

Tronick *et al.* (1973) compared glutamine synthetases from a number of gram-negative and gram-positive bacteria and found significant immunological cross-reactions between many from gram-negative bacteria. The enzymes giving cross-reactions gave increased activities after treatment with snake venom phosphodiesterase, indicating that they were capable of the same adenylylation modification as those of the enterobacteria. This did not occur with glutamine synthetases from the gram-positive species, suggesting that glutamine synthetase activity might not be regulated in the same way in these bacteria.

6.2 *Nitrogen Regulation in Fungi*

Fungal enzymes do not vary in response to environmental changes to the same extent as those of bacteria but regulation by induction and repression of a number of enzymes has been studied, particularly in *Saccharomyces cerevisiae, Aspergillus nidulans,* and *Neurospora crassa.* Many fungi can use a variety of organic compounds and their activities in nature are important in biodegradation. The enzymes required for the dissimilation of nitrogenous compounds may be inducible and also subject to catabolite repression by carbon compounds and to repression by other nitrogen compounds (for review see Arst and Cove, 1973). Among the enzyme systems shown to be regulated by the three controls of *induction, carbon repression,* and *nitrogen repression* is that for proline metabolism. Arst and MacDonald (1975) showed that mutations in a regulator gene, linked to two genes for proline enzymes, gave relief from carbon repression and were cis dominant.

Regulation of amidases in *Aspergillus nidulans* has been investigated in some detail. Hynes (1970, 1972, 1974) has obtained a series of mutants for the acetamidase that is induced by acetamide and some amide analogs and repressed by glucose and ammonium salts. One mutant produced higher levels of the normal acetamidase under inducing conditions and the mutation was cis dominant to the wild type regulation. Hynes (1975) suggested that the mutation might be in the regulatory site that accepted a positive regulatory molecule.

A gene exerting a general control of enzymes for nitrogen metabolism was described by Arst and Cove (1973). Mutations in this gene, *areA,* led to the inability to grow on various amides, amino acids, purines, nitrate, and nitrite but did not affect growth in ammonium salt media. The role suggested for the *areA* product was that of a positive regulator protein that was required for the synthesis of many of the enzymes of nitrogen metabolism although a negative role was not excluded. The interactions observed between carbon and nitrogen metabolism were complex.

Arst and Scazzocchio (1975) described a mutation leading to a constitutive uric acid-xanthine permease that was tightly linked to the structural gene it controlled and was cis dominant. This had been isolated by selecting mutants able to grow on uric acid from a strain carrying an *areA* mutation that reduced the level of the permease without affecting its catabolism once inside the cell. The expression of the permease gene requires a specific regulator gene and the general nitrogen regulation gene *areA.* The constitutive mutations were considered to be in an initiator site that accepted the two positive regulator proteins. More complex interactions of regulator gene products were suggested by Arst (1976) to account for some multiple effects of certain regulator mutations. Nitrate reductase has been suggested to act in the repression of its own synthesis and the synthesis of other enzymes of nitrogen metabolism (Cove and Pateman, 1969).

7 *Experimental Evolution*

7.1 *Growth on Novel Substrates*

In recent years experiments have been designed to study the evolution of enzymes and metabolic pathways under controlled conditions in the laboratory. Success has led to the isolation of mutant strains, and the end results, in some instances, have been mutants that might well have been isolated by procedures designed for the analysis of metabolic pathways or regulatory systems. Frequently, only regulatory mutants were isolated, although the selection procedure had been designed to obtain strains producing new enzymes. This might indicate that regulatory genes are inherently less stable and have greater evolutionary potential than structural genes. It might be that any of a number of different mutations could change the properties of a regulator protein in an appropriate way but that a structural gene can undergo only very limited changes without losing enzyme activity, so that there would be fewer mutational sites available. Another factor that must be taken into account is that inducer and substrate specificities are often similar and that a change in the substrate specificity of an inducible enzyme would not be at all useful if the new enzyme could not be induced.

Some of the systems studied required only one enzyme reaction to convert the novel substrate into a compound that could enter one of the usual metabolic pathways. More complex problems are introduced when a number of enzymes are needed for the early reactions, as in the catabolic pathways for aromatic compounds. The inducer may be a later metabolite of the pathway, as in the case of induction of two of the protocatechuate pathway enzymes of *P. putida* by β-ketoadipate (see Section 5.4). An unusual substrate might be attacked by an existing enzyme but the resulting metabolites, even if they were acted on by the subsequent enzymes of the original pathway, would not necessarily give rise to the normal inducing molecule. In some cases it might be necessary for several new enzyme activities to arise for the novel growth substrate to be metabolized at a significant rate. At the same time, more than one regulatory mutation might be required to enable these new enzymes to be synthesized in response to the novel compound. It is widely known that some synthetic chemicals, such as the chlorinated hydrocarbons, are very resistant to biodegradation, whereas others are broken down fairly readily. Prolonged selection, often carried out in continuous culture, can give rise to mutants that can degrade some of these persistent chemicals and analysis of the mutations that have occurred will eventually be valuable in understanding how regulatory changes may arise. At the present time, most of the information that can be interpreted in detail comes from the study of simpler systems in which only one or two changes have occurred in regulatory genes and structural genes.

The new growth phenotypes of mutant strains that can grow on novel substrates are in some cases entirely due to regulatory alterations and include mutations to constitutivity, altered inducibility, resistance to analog or catabolite repression, and duplication of genes specifying enzymes that already had some activity for the novel substrate. Studies on pentose and pentitol metabolism have shown that mutations in regulatory genes can allow enzymes from different metabolic pathways to be assembled into new pathways for the utilization of pentoses and pentitols not normally found in the natural environment (Mortlock, 1976).

7.2 *Amidase*

Pseudomonas aeruginosa and *P. putida* can grow on low-molecular-weight aliphatic amides that induce an amidase with high activity for acetamide and propionamide but very low activity for butyramide. The enzymes in the two species are very similar (Clarke, 1972), but the growth phenotypes differ in one important respect. The enzyme of *P. putida* is induced by butyramide and, although the specific activity is only about 3% of that for acetamide the fully induced level of the enzyme is quite sufficient for growth. With *P. aeruginosa,* on the other hand, butyramide is unable to induce the enzyme and, indeed, competes with inducing amides and prevents enzyme synthesis. Butyramide also prevents amidase synthesis in some constitutive mutants (Brown and Clarke, 1970). These findings suggest that butyramide competes with inducing amides at the inducer-binding site of the regulator protein. Such competition by inducer analogs provides an additional constraint on the enzyme system. The wild type strain is restricted in its growth on aliphatic amides by the amidase substrate specificity, inducer specificity, and the effects of acetamide analogs. The class of constitutive mutants that retains sensitivity to repression by butyramide appears to have kept the amide-binding site of the regulator protein. Those constitutive mutants that are resistant to butyramide repression may have mutations affecting the amide-binding site or mutations that affect the conformation of the regulator protein.

Regulation of amidase has been studied in more detail than that of other *Pseudomonas* enzymes. The substrate and inducer specificities of wild type strains are strictly limited. Many amides are available for experimental studies, and amides that can be utilized by mutant strains may provide carbon or nitrogen sources for growth. Various classes of constitutive mutants can be isolated from plate media containing succinate as carbon source and formamide as nitrogen source. This medium also allows the selection of mutants that have a much higher rate of induction by formamide than does the wild type (Brammar *et al.,* 1967). There are few reports of mutants with altered inducibility (see Section 7.5) and yet it is important for a strain with a new growth phenotype to avoid wasteful protein synthesis. A constitutive mutant may obtain a temporary growth advantage but would be likely to be outgrown by inducible strains when the selection pressure is removed. Constitutive mutants can also be isolated from butyramide plates and these, of necessity, must be resistant to butyramide repression.

Growth on a novel substrate can be achieved either by an increase in the amount of a preexisting enzyme that has low activity for the new substrate or by a mutation resulting in an enzyme with increased activity (Clarke, 1974). This can be illustrated by considering the butyramide-utilizing mutants of *P. aeruginosa.* The class of butyramide-utilizing mutants that are constitutive and resistant to repression can be isolated directly from the wild type strain on butyramide medium (or indirectly from a constitutive mutant unable to grow on butyramide). Enzyme activity toward butyramide is low and the growth rate in medium containing butyramide as the carbon source is limited by the rate of butyramide hydrolysis (Brown and Clarke, 1970; Clarke and Lilly, 1969). An amidase mutant with altered substrate specificity was isolated on butyramide medium from a constitutive mutant that was sensitive to butyramide repression (Brown *et al.,* 1969). It was not possible to test the sensitivity of this strain to butyramide repression, since it was hydrolyzed too rapidly, but the sensitivity to repression by other amide analogs, such as cyanoacetamide, remained the same as the parent strain. It is interesting that none of the selection media tested so far has given rise to butyramide-inducible mutants although this would appear to be a possible solution to the problem of growth on butyramide. There is no obvious reason why *P. putida* should have evolved a regulator gene that allows butyramide to be an inducer whereas *P. aeruginosa* has evolved a less permissive phenotype, with growth on butyramide prevented by repression as well as by the limitation of low substrate affinity.

Amidases with a range of substrate specificities have been evolved by selection on media containing some of the higher amides. Many of these amides, including valeramide, hexanoamide, and phenylacetamide, were able to repress amidase induction. It is uncertain that combinations of these particular amides would ever be encountered in the natural environment and these properties may be fortuitous. However, competitive repression of enzyme induction by analogs of the inducing substrate does illustrate an important principle—namely, that it acts as a double barrier to the evolution of new growth phenotypes since mutations have to circumvent the repression effect as well as change the enzyme specificity (Betz *et al.,* 1974).

The amidase regulator gene, *amiR,* is closely linked to the structural gene, *amiE,* and is cotransduced at a very high frequency. There is good evidence that the *amiR* gene exerts positive control on amidase synthesis. There are indications that some of the genes specifying enzymes of the pathways for the catabolism of aromatic compounds may also be under positive control (see Sections 5.5 and 5.6), although the evidence is less direct than for amidase. The classical test for distinguishing between positive and negative control of gene expression is based on analyzing cis–trans dominance with partial diploids, and such experiments led Englesberg *et al.* (1965) to suggest that the L-arabinose operon was under positive control. Although dominance tests with diploids can be carried out only with systems for which genetic analysis has become fairly sophisticated, it is possible to learn a great deal from physiological experiments with mutants. In order to assess the significance of positive versus negative control of genes for various metabolic pathways, it is important to find out how frequently each type of control occurs in bacteria other than the enterobacteria.

If amidase were under negative (repressor) control, it would have been expected that among the regulatory mutants isolated would be some that were inducible at lower temperatures and constitutive at higher temperatures because they produced a temperature-sensitive repressor (see Sadler and Novick, 1965, on the *lac* operon). None had been isolated and attention was directed to mutants that were negative at higher temperatures and amidase positive at lower temperatures that might have temperature-sensitive activator–regulator molecules (see Irr and Englesberg, 1971, on the *ara* operon). Mutants producing temperature-sensitive enzymes could be eliminated, since amidase is a very thermostable enzyme, and only mutants producing enzymes with the same thermal stability as the wild type were considered. Farin and Clarke (1978) found that some mutants isolated from one of the constitutive butyramide-sensitive strains were amidase negative at 42°C and amidase positive at 28°C and produced the normal thermostable enzyme at the permissive temperature. A temperature shift from the permissive to the nonpermissive temperature resulted in the cessation of amidase synthesis with kinetics similar to that of other treatments thought to act directly, or indirectly, on the transcription process. Further evidence for positive control was obtained from an examination of the regulatory patterns of revertants of amidase-negative strains, derived from a mutant carrying a structural gene for an altered enzyme and thought to be defective in the regulatory gene. This is similar to the analysis used by Bloom *et al.* (1975) for serine deaminase.

Another question that needs to be asked is whether catabolite repression in other bacterial genera is the same as in the enterobacteria. Although catabolite repression is known to affect *Bacillus* enzymes there is no evidence of any cyclic AMP involvement (Bernlohr *et al.*, 1974; Setlow, 1973). Gilboa-Garber *et al.* (1973) found that catabolite repression of acetylcholine esterase by glucose in a strain of *P. aeruginosa* could be relieved by cyclic AMP. Smyth and Clarke (1975*a,b*) found that partial relief of catabolite repression of amidase synthesis could be obtained by adding cyclic AMP but that high concentrations were needed. These authors found that a presumed promoter mutant had a very high rate of amidase synthesis and was insensitive to concentrations of cyclic AMP that stimulated the rates of synthesis of inducible and constitutive strains. There is not yet any conclusive evidence on the type of catabolite repression control that has evolved in *Pseudomonas* species for amidase and other enzymes although there appear to be some similarities with *E. coli* (Pastan and Adhya, 1976).

7.3 Evolved β-Galactosidase

The many studies on the *E. coli lac* operon β-galactosidase have provided a vast range of regulator mutants. Some of these exhibit altered response to β-galactosides and others are constitutive, and new growth phenotypes have evolved as a result of regulatory mutations. Despite the many methods that have been used to select mutations in the *lacZ* gene, no mutant has been found that produces a more efficient β-galactosidase (although many were found that produce an enzyme with impaired activity). Campbell *et al.* (1973) obtained a strain producing another, and different, β-galactosidase by selecting for lactose-fermenting mutants from a strain with a *lacZ* deletion. Later, Hall and Hartl (1975) showed that strains growing on lactose in the absence of a functional *lac* operon had acquired mutations in two genes—*ebgA*, specifying the structure of the new β-galactosidase, and its regulatory gene, *ebgR*.

The wild type strain is normally capable of producing this second β-galactosidase, which is induced by lactose and not by IPTG (isopropyl-β-D-galactoside). The amount of enzyme synthesized by the wild type strain is very low and although lactose is the inducer, the affinity of the enzyme for lactose is low. The wild type strain with the *lacZ* deletion was therefore unable to utilize this second enzyme for growth on lactose (Hartl and Hall, 1974). The *ebg* genes in the wild type strain are in a latent state and can be brought into an active role by mutation if there is sufficient selective pressure. Hall and Hartl (1974) used a *lacY*$^+$ strain and added IPTG to induce the permease and found that most of the Ebg$^+$ mutants selected were inducible and only a few were constitutive. The inducible mutants all produced higher levels of enzyme in the presence of lactose. The evidence suggests that the *ebgR* gene exerts negative control on expression of the *ebgA* gene so that a mutation to improved inducibility by lactose probably results in an inducer-binding site with a higher affinity for lactose.

When the enzymes from the Ebg$^+$ mutants were purified Hall (1976*a*) found that there were several variants with lactose-hydrolyzing activity. The genes for the different enzymes were combined with regulator genes determining wild type, constitutive, or the increased inducibility type of regulation. Hall and Clarke (1977) showed that for growth to occur on lactose it was essential for the lactose-hydrolyzing activity of the strain to reach a threshold value. The *ebgR* mutation determined the rate at which the enzyme was synthesized and the *ebgA* mutation determined the specific activity of the enzyme. The wild type strain never reaches this threshold value. In a genetic background including the *lac* operon, the *ebg* genes may have drifted into a quiescent state. There is, however, a genetic potential available that is not confined to recovering, or acquiring, a second enzyme for lactose. Hall (1976*b*) found that several *ebg* mutants were able to grow on methyl-β-D-galactoside.

The two *ebg* genes may be related to the *lacI* and *lacZ* genes by an ancient duplication event. Alternatively, they might belong to an enzyme whose original function had been lost during the evolution of the present-day *E. coli*. The properties of such a hypothetical ancestral gene are of course unknown, but it might have defined a rather inefficient β-galactosidase and the present *lacZ* gene could represent the subsequent evolution of a more efficient enzyme. The *lac* operon might even have been relatively recently acquired by *E. coli* by genetic transfer from another species, since plasmids carrying *lac* genes are known among the enterobacteria (Reeve and Braithwaite, 1973) and the *ebg* genes could be evolutionary remnants. Another possibility is that the *ebg* genes have a definite role that has not yet been identified. From the point of view of comparative regulation it is of particular interest that both the *lac* and *ebg* genes should be under negative control.

7.4 Gene Duplications

Duplication of structural genes provides a means for increasing the size of the gene pool and, as pointed out earlier (Section 3.3), that transient gene duplication also could have been a primitive mechanism for regulation of enzyme synthesis. With experimental systems it has been found that duplication of genes can result in new growth phenotypes. Chemostat growth allows the selection of mutants that have a growth advantage over the parental culture. This method was exploited by Hartley and colleagues to select strains of *Aerobacter aerogenes* with higher growth rates on xylitol. This is an example of selection for an enzyme that can act on a compound not normally found in nature. The normal substrate for the enzyme is the naturally occurring pentitol, ribitol. The wild type enzyme has low activity toward xylitol although xylitol cannot induce it. Wu *et al.* (1968) had previously isolated a mutant able to grow on xylitol in which the ribitol dehydrogenase had become constitutive, and this strain was used for chemostat selection experiments. The mutants obtained in these experiments had much higher xylitol dehydrogenase activities but this was due to gene amplification (Rigby *et al.,* 1974). The multiplication of the ribitol dehydrogenase gene is a regulatory device which gave an evolutionary advantage to the mutants, but the extra gene copies were lost when the selection pressure was removed. The temporary growth advantage and the genetic instability of the gene duplications bear some resemblance to the acquisition and loss of catabolic plasmids. In principle, if duplicate genes could be transferred to plasmids, they might be less easily lost.

If two copies of the ribitol dehydrogenase gene could be retained, it could be envisaged that one could retain the function of ribitol dehydrogenase and the other become a xylitol dehydrogenase. Charnetsky and Mortlock (1974) showed that in *Klebsiella aerogenes* the genes for the ribitol (*rbtD, rbtK*) and D-arabitol (*dalD* and *dalK*) enzymes were closely linked. The similarities of the enzyme reactions make it reasonable to suggest that they have common origins and have evolved by gene duplications. In the evolution of the catabolic pathways for these two pentitols, structural genes and regulator genes have evolved with the appropriate specificities. Interlied and Mortlock (1977) also selected xylitol mutants of *K. aerogenes* in continuous cultures. Within 48 generations they selected xylitol-utilizing mutants that were hyperconstitutive for ribitol dehydrogenase due to gene duplication. These authors found that if the dilution rate of the culture was raised, the level of ribitol dehydrogenase synthesized by these mutants was diminished, and ascribed this to the onset of catabolite repression. They suggested that this could put additional selective pressure on the system and perhaps make it easier to select altered enzyme mutants. The duplication of the ribitol dehydrogenase gene (*rbtD*) did not extend to the closely linked structural genes (*rbtK, dalD,* and *dalK*). This gene duplication was therefore restricted to the gene subjected to the selective pressure and involved only a very limited DNA segment. The occurrence of such limited duplications is of great interest in view of the role that duplications of specific gene segments might play in the evolution of regulatory systems.

Escherichia coli strain C, unlike K12, is able to metabolize both ribitol and D-arabitol. The genes are carried on the chromosome but Reiner (1975) has suggested that they might have been acquired relatively recently from *K. aerogenes* by plasmid transfer. Among the plasmids carried by strains of *K. aerogenes* are some that carry genes for lactose fermentation (Reeve and Braithwaite, 1973). These are frequently lost during laboratory culture and further studies with fresh isolates may reveal many more plasmids carrying other catabolic genes concerned with the metabolism of carbohydrates.

7.5 *New Metabolic Pathways*

Some potential growth substrates require only a single enzyme to convert them into a compound that can be metabolized by existing enzymes while others require several enzymes to do this. Mortlock (1976) describes the various ways in which "unnatural carbohydrates" can be metabolized by microorganisms and in cases in which organisms have adapted to grow on these compounds mutations have always involved changes in regulation. The way in which this can happen is illustrated by the isolation of mutants of *K. aerogenes* (in earlier work strains were described as *Aerobacter*) and of *E. coli* with the ability to grow on D-arabinose. *Aerobacter aerogenes* PRL-R3 can grow on L-fucose, which induces the synthesis of fucose isomerase, and this enzyme has some activity for D-arabinose although it is not induced by it. Mortlock *et al.* (1965) showed that extracts of the D-arabinose-utilizing mutant could isomerize L-fucose as well as D-arabinose and the mutants were shown to be constitutive for this enzyme. The mutation allowing growth on D-arabinose affected a regulatory gene and the required isomerase was released from its normal control. By selecting for faster growth on D-arabinose, Oliver and Mortlock (1971) were able to obtain a mutant with an isomerase possessing a higher affinity for D-arabinose and the regulatory mutation had paved the way to obtaining a more effective enzyme. An alternative solution to the problem of unlocking the isomerase is to change the specificity of induction. St. Martin and Mortlock (1976) isolated a series of D-arabinose-utilizing mutants from *K. aerogenes* W70 which included constitutives, producing various levels of enzyme, and some strains in which the isomerase was inducible by D-arabinose. The product of the reaction with D-arabinose is D-ribulose and the further metabolism of this compound is ensured since it is the inducer of D-ribulokinase as well as ribitol dehydrogenase, both enzymes required for ribitol catabolism. *E. coli* K12 does not grow on ribitol so that this pathway is not available for the further metabolism of D-ribulose produced from D-arabinose. Leblanc and Mortlock (1971*a,b*) isolated D-arabinose-utilizing mutants and found that they were altered in the specificity of induction of L-fucose isomerase. This enzyme was inducible in the mutants both by L-fucose and D-arabinose. The subsequent catabolism of D-arabinose was via the other enzymes of the L-fucose pathway, and these were also induced by growth with D-arabinose. The new metabolic pathway for D-arabinose in the *E. coli* mutant involved at least three enzymes belonging to the L-fucose pathway (isomerase, kinase, and aldolase), and the regulation of all of these was altered. This suggested that the mutational event had occurred in a regulator gene that controlled the synthesis of the three related enzymes. This is not a new metabolic pathway for the organism but represents the capture of an existing pathway by a change in the specificity of induction. For the enzymes to be used in this way it is necessary that they have sufficiently broad specificities to be able to accept D-arabinose and its metabolites as well as L-fucose and its metabolites.

The possibility also exists that *K. aerogenes* could use the L-fucose pathway enzymes. The mutants originally isolated were shown to metabolize D-arabinose by the constitutive L-fucose isomerase "borrowed" from the L-fucose pathway, and D-ribulokinase "borrowed" from the ribitol pathway converted D-ribulose to D-ribulose-5-phosphate. If the L-fucose pathway enzymes were used, the product would be D-ribulose-1-phosphate, which would then be subjected to aldolase cleavage to produce dihydroxyacetone phosphate and glycolaldehyde (Fig. 12).

St. Martin and Mortlock (1977) constructed strains with mutations in the D-ribulokinase gene that blocked the ribitol catabolic pathway. Some of these were made constitutive for the L-fucose enzymes and others were made inducible, and both classes were able to grow with D-arabinose. The main disadvantage of the D-ribulose-1-phosphate route in *K. aerogenes* is that one of the products is glycolaldehyde which is not further metabolized. The rate of flow of intermediates through the pathway depends on the specificities of the L-fucose enzymes for the D-arabinose metabolites and it appeared that D-ribulose could be produced more rapidly than it could be phosphorylated and cleaved by the later enzymes. In the D-arabinose-positive mutants (inducible or constitutive for L-fucose isomerase) and carrying the normal ribitol pathway genes, there would be a tendency for D-ribulose to build up, and this would induce D-ribulokinase which would divert the metabolism of D-arabinose and its products to the D-ribulose-5-phosphate pathway (Fig. 12). When the mutation has occurred in *K. aerogenes* to make L-fucose isomerase constitutive (or inducible by D-arabinose) there is an apparent choice of routes. The predominance of the D-ribulose-5-phosphate route is due in part to the low affinity of the L-fuculokinase for D-ribulose, but the more important factor is that D-ribulose is the actual inducer of the ribitol pathway enzymes. St. Martin and Mortlock (1976) found that the most probable inducer of the L-fucose pathway enzymes was L-fucose-1-phosphate and that in the inducible mutants the enzymes were inducible by D-ribulose-1-phosphate as well. D-Ribulose-1-phosphate would be unlikely to build up in sufficient amount to divert the pathway and the dominant induction would be by D-ribulose.

A novel pathway for the utilization of L-1,2-propanediol has been shown to be dependent solely on regulatory mutations (Lin *et al.,* 1976). After repeated subculture a strain of *E. coli* K12 was isolated that could grow on L-1,2-propanediol. The origin of the enzyme activity was obscure until it was observed that the mutant could not grow on L-fucose and that the gene for the new activity mapped at or very near the genes for the L-fucose enzymes. One of the products of L-fucose metabolism is lactaldehyde, and during aerobic growth this is oxidized by an NAD-linked dehydrogenase. During anaerobic growth the lactaldehyde is reduced by an NADH-coupled oxidoreductase to propanediol which is not metabolized further but is excreted into the medium. In the mutant this enzyme is now produced constitutively under aerobic conditions and acts in the reverse direction converting L-1,2-propanediol to lactaldehyde which is oxidized to lactate by the lactaldehyde dehydrogenase of the L-fucose pathway. The lactate is then oxidized

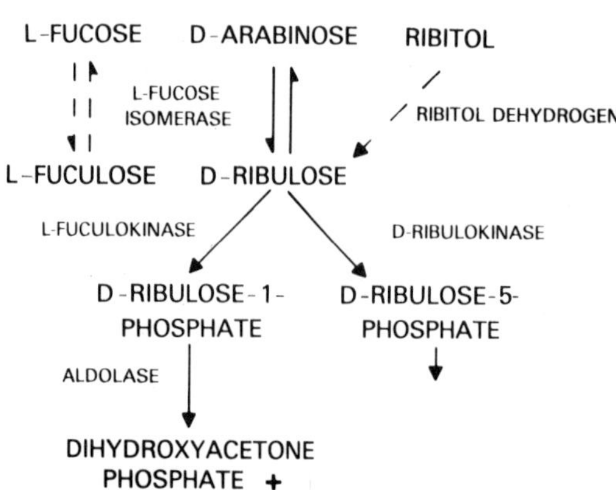

Figure 12. Novel pathways for the utilization of D-arabinose in *E. coli* and *K. aerogenes.* The initial reaction involves L-fucose isomerase which becomes inducible in *E. coli* mutants and may be inducible or constitutive in *K. aerogenes* mutants. Subsequent metabolism may involve either the L-fucose pathway enzymes or the ribitol pathway enzymes in *K. aerogenes. E. coli* K12 does not utilize ribitol. Details in text.

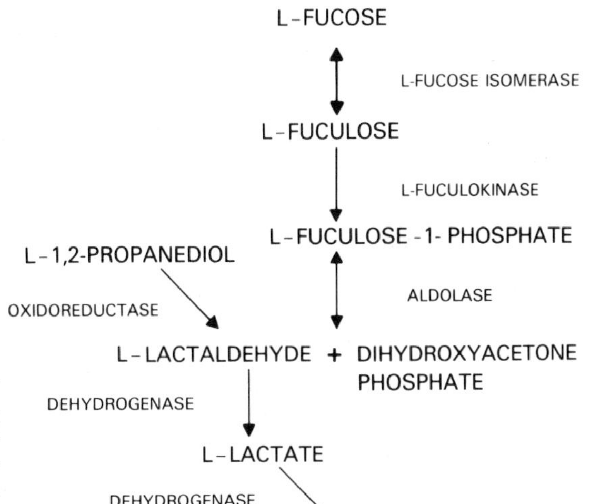

Figure 13. Novel pathway for the utilization of L-1,2-propanediol in *E. coli* K12. Propanediol oxidoreductase is a normal enzyme of the L-fucose pathway and during anaerobic growth reduces lactaldehyde to L-1,2-propanediol. The mutant synthesizes this enzyme constitutively.

to pyruvate by a flavoprotein dehydrogenase induced by lactate (Fig. 13). During the emergence of the propanediol-positive strain several mutational events had taken place. An early mutation resulted in an increased level of propanediol oxidoreductase under aerobic growth conditions. Two of the L-fucose pathway enzymes, the isomerase and kinase, and also L-fucose permease became noninducible so that the strain became L-fucose negative. Mutations to regain the L-fucose-positive phenotype resulted in constitutivity, indicating that this had been a regulatory mutation. Later selection produced mutants which were constitutive for the aldolase and another appeared to produce lactaldehyde dehydrogenase constitutively but had lost the aldolase (Cocks *et al.,* 1974; Hacking and Lin, 1976). Lin *et al.* (1976) point out that five different phenotypes have been identified among the propanediol-utilizing mutants and that all of these affect regulatory genes and not the structural genes for enzymes. The new propanediol catabolic pathway built up as a result of these mutations is already present in *E. coli.* The enzymes are there and apparently possess all the catalytic activities needed. The pathway is cryptic in the parent strain because the enzymes are not synthesized in response to the presence of propanediol in the growth medium. The significance of the various mutations in the L-fucose pathway genes is not clear but these are all regulatory mutations. It was pointed out earlier that the choice of the ribitol pathway as the major route for D-arabinose catabolism in the *K. aerogenes* mutants could be ascribed to the inducer specificities and the buildup of pathway intermediates. The blocks in the L-fucose pathway resulting from regulatory mutations in the propanediol mutant may be important in reducing the level of certain intermediates.

The propanediol mutant is of particular interest in that it was used as the parent strain to select xylitol- and D-arabitol-utilizing strains of *E. coli* K12. This stepwise selection of strains with new growth capacities has shown how important the regulatory systems are not only for controlling the amount of enzymes produced for a particular sequence of reactions but also in determining the potential growth phenotype. Wu (1976*a,b*) was successful in obtaining mutants

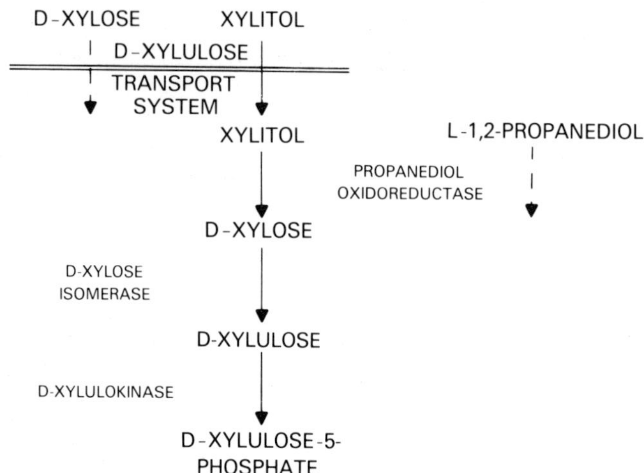

Figure. 14. Novel pathway for the catabolism of xylitol in *E. coli* K12. Xylitol is taken up by the constitutive D-xylose transport system and oxidized by the constitutive L-1,2-propanediol oxidoreductase (Wu, 1976*a*).

able to grow on both these pentitols but the stages in the adaptations were not the same.

The extract of the propanediol-utilizing strain was found to contain xylitol dehydrogenase activity and this was shown to be due to the propanediol oxidoreductase. Wu (1976*a*) found that whereas the K_m of the enzyme for propanediol was about 2 mM, the K_m for xylitol was 100-fold greater, but this is similar to the affinity of ribitol dehydrogenase of *K. aerogenes* for xylitol (Section 7.4). The enzyme is produced constitutively in the *E. coli* propanediol mutant but even so, it is unable to grow on xylitol. The reason for this is that xylitol is not taken up into the bacterial cells. However, xylitol can be transported by D-xylose permease, and a mutant was isolated that was constitutive for this permease. The K12 xylitol-utilizing mutant was constitutive for both the permease able to transport xylitol (D-xylose permease) and for the dehydrogenase that could convert xylitol into D-xylose (L-1,2-propanediol oxidoreductase). The subsequent metabolism takes place by the normal D-xylose catabolic enzymes (Fig. 14). The mutants grew rather slowly on xylitol; faster growth might require some improvement in the enzymes. The pathway is nevertheless functional and uses only enzymes that are already available. It would probably be advantageous to impose some regulation by xylitol. If xylitol could become the inducer for both the permease and the dehydrogenase, this would be similar to other established pentitol pathways.

The L-1,2-propanediol mutant was also used by Wu (1976*b*) to select D-arabitol-utilizing mutants of *E. coli* K12. These were obtained but the oxidoreductase was not involved this time. The important feature of the mutant was that it also possessed a constitutive L-1,2-propanediol uptake system and this could transport D-arabitol. Another mutation was required before D-arabitol could be metabolized. The D-arabitol-utilizing mutant synthesized a dehydrogenase that was not found in the parent strain. Wu (1976*b*) suggests that the normal substrate of this dehydrogenase is D-galactose and that it might be an evolutionary remnant of a pathway that is no longer required by *E. coli*. The dehydrogenase could not be induced by D-galactose or any of the sugars or polyols tested and the gene was therefore silent. The new pathway for D-arabitol is shown in Fig. 15. It involves transport of D-arabitol by the constitutive propanediol permease followed by the dehydrogenation of D-arabitol to D-xylulose by the novel constitutive dehydrogenase. D-Xylulose is a normal intermediate of D-xylose metabolism in *E. coli* and both D-xylose isomerase and D-xylulokinase were induced when the mutant was grown on D-arabitol. This pathway for D-arabitol is another example of the patchwork assembly of a new catabolic pathway by assembling genes from other catabolic pathways that have been released from their normal regulatory constraints.

8 *Discussion*

The regulatory systems which have evolved in bacteria to control the rates of synthesis of amino acids to maintain balanced growth must obviously be efficient enough for the environment in which these microorganisms find themselves. Yet, in respect to both feedback inhibition and repression, the patterns of regulation are quite varied even among the small number of genera considered in this chapter. The rapid growth rates that occur in favorable media lead one to suppose that bacteria will tend to evolve the most efficient systems possible but a balance will have to be struck between the regulation of one metabolic pathway and another, and between biosynthetic and catabolic pathways. For any one metabolic pathway the regulatory system might not be the most efficient one, but the one that is efficient enough, given the other requirements of the organism. The

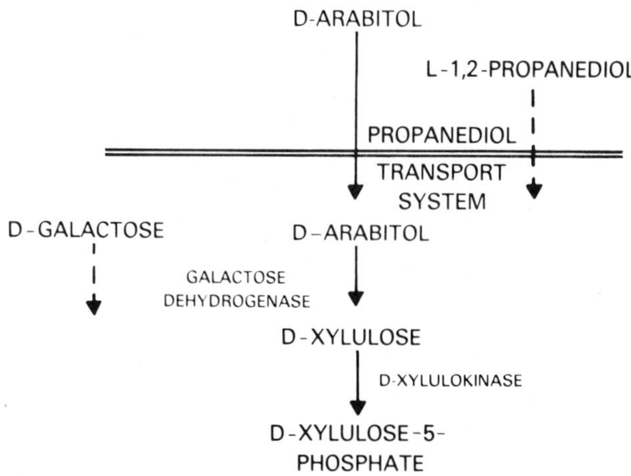

Figure 15. Novel pathway for the catabolism of D-arabitol in *E. coli* K12. D-Arabitol is transported by the constitutive L-1,2-propanediol transport system and oxidized by the constitutive D-galactose dehydrogenase which is nonfunctional in the parent strain. D-Xylulose induces subsequent enzymes (Wu, 1976*b*).

pathways for the biosynthesis of the aromatic amino acids may be regulated by inhibition and repression of isoenzymes at the branch points or by multivalent inhibition and repression by various combinations of amino acid end products or intermediates. Neither of these arrangements could be thought of as more primitive than the other and both are effective in the species that have adopted them. The similarities in the properties of the isoenzymes suggest that they arose by duplication of the structural genes for the enzymes. In Section 4 some of the enzymes of the pathway were put forward as possible ancestors of the effector-binding domains of the regulator proteins and it was suggested that these could have arisen by partial duplication and gene fusion combined with gene translocations. At the present time it is not possible to give unequivocal assignments to the origins of any of these regulatory genes.

In considering the evolution of the enzymes of the catabolic pathways and their regulatory proteins it may be necessary to take into account other microbial companions present in the ecosystem. Degradation of biological material in nature is seldom the responsibility of pure cultures. Mixtures of bacteria and fungi break down complex polymers, sometimes with the assistance of protozoa, worms, and insects. Even in the degradation of relatively simple organic molecules a mixture of microorganisms may be more effective than a single strain. For biochemical and genetic investigations it is essential to use pure cultures to elucidate pathways, to identify intermediates, and to analyze the nature of the regulatory molecules, but growth studies can be made with mixed cultures as well as pure cultures. Bull and Slater and their colleagues found that selection in a chemostat for cultures able to grow at the expense of the herbicide Dalapon resulted in a steady state mixed culture (Senior *et al.,* 1976). Some of the organisms were primary utilizers and could grow with Dalapon as the sole carbon source, although more slowly than in mixed culture, whereas others were secondary utilizers and could not grow on Dalapon in pure culture. Interactions of this sort are to be expected in the natural environment and the regulatory pattern of a single species may be related to its association in the ecosystem with other species. This interpretation has been put forward by Ornston and Parke (1977) to account for the evolutionary choice of *cis,cis*-muconate and β-ketoadipate as inducers of the ortho pathway enzymes in *P. putida.* Other organisms in the microenvironment may produce these intermediates of the degradation of aromatic compounds in amounts in excess of their own metabolic needs. A stable association of microorganisms might include the release of metabolic intermediates by one organism and the destruction of toxic compounds by the other partner so that both derived benefit from the association. In circumstances such as these the evolution of the regulatory system of one species in the partnership would not have taken place in isolation.

Examples were given of regulatory interactions that resulted in the repression of catabolic enzymes for a substrate that could support a higher rate of growth than the compound used preferentially (Sections 5.3.1 and 5.3.2). The laboratory conditions may not reflect the distribution and quantities of the potential growth substrates encountered in the natural environment. The extreme case of this is when inhibition or repression of an enzyme by a normal metabolite inhibits growth. This can be seen in the inhibition of growth by one of the end products of the branched pathway for the biosynthesis of the amino acids derived from aspartate. The single aspartokinase of *Rhodopseudomonas capsulata* is inhibited by a combination of lysine and threonine and in minimal salt medium the addition of these two amino acids inhibits growth. This can be relieved by the addition of methionine, the other end product of the pathway (Datta, 1969). It is possible to account for the evolution of this apparently suicidal form of regulatory control by suggesting that these bacteria never encounter a high level of lysine and threonine in nature or, if they do, methionine is likely to be present as well. Another way to assess whether or not this type of regulation would be a serious hazard would be to investigate whether other organisms in the microenvironment would remove any excess lysine and threonine by using them as growth substrates. In places in which amino acids are to be found in any abundance it is to be expected that microorganisms would use them for growth. This again envisages some form of peaceful coexistence between different microbial species.

The view was put forward that gene duplication and segregation provide a primitive form of regulation and that is is an important step in evolving both new enzymes and new regulator proteins. It was also suggested that constitutive mutants may have provided an intermediate stage in the evolution of novel inducible (or repressible) enzyme systems. One of the objections to gene duplications, especially of unregulated genes, has always been that the gratuitous synthesis of proteins during periods when they were not required would put a burden on the cell and cause a serious growth disadvantage. This view is supported by the observations that the multiple gene copies of hyperstrains selected in chemostats segregate when the selection pressure is removed (Section 7.3). The competition experiments of Zamenhof and Eichhorn (1967) showed that a derepressed $trpR^-$ mutant of *Bacillus subtilis* was soon outgrown when it was put in chemostat culture with the corresponding $trpR^+$ strain. This seems patently obvious since constitutive and derepressed mutants are not normally found among natural isolates. There are, however, very many constitutive and derepressed mutants in laboratory collections that are not always maintained under appropriate selection pressures and these are not invariably lost. The growth disadvantage of gratuitous synthesis of one or two enzymes may not be so significant as has been thought.

Andrews and Hegeman (1976) reexamined the question of the selective disadvantage of nonfunctional protein synthesis by comparing the growth rates of strains of *E. coli* that produced various levels of β-galactosidase but were otherwise isogenic. These authors found that there was a measurable reduction of the growth rates of strains with very high levels of nonfunctional protein synthesis, but for strains producing rather low levels of the nonfunctional protein the differences in growth rates were very small and not proportional to the extent of wasteful protein synthesis. Andrews and Hegeman suggest that variants carrying more than one gene copy would not necessarily be strongly selected against if these variants had only a low level of excess protein synthesis. If gene duplication can occur and persist for some time in this way, mutations giving new inducer and enzyme specificities present few difficulties.

The experimental evolution of bacteria with new metabolic capacities illustrates how gene duplications and mutations in regulatory genes can allow enzymes to be used for new purposes. The new D-arabinose pathway evolved in *E. coli* K12 represents a takeover of the L-fucose pathway enzymes. Since D-arabinose is now an inducer for this pathway it only requires mutations in one or more of the enzymes to give higher affinities for D-arabinose metabolites for the original function of the pathway to begin to disappear. The xylitol-utilizing mutant isolated by Wu (1976*a*) might in time change the regulation of the constitutive propanediol oxidoreductase to become xylitol inducible and the constitutive D-xylose permease and kinase to become xylitol inducible. If this happened, it would be extremely difficult to hit on the right candidate as the ancestor of the regulator genes.

The experimental evolution results suggest that regulatory systems can be potent agents of change and that the derepression of one or more enzymes can lead to new growth phenotypes that might in time become more permanently imprinted. These events are observed to occur when organisms are subjected to very strong selection pressures. We need to set against this the stabilizing effects of regulatory systems in the absence of any demand for new and unusual activities. The presence as sole growth substrate of a compound that occurs seldom, if at all, in nature is certainly not a normal condition of microbial life. That such adaptations can occur is fortunate for the future prospects of biodegradation of man-made materials. Most of the time we would expect the interlock of inducer (or corepressor) specificities with enzyme specificities to tend to keep things much the same. This stability is increased if additional controls are present, such as repression by analogs of inducers or by other cell metabolites. The peripheral enzymes and their regulatory systems have the most freedom to change and it is among these that variations among strains are most apparent. Enzymes that are central to metabolism are under more complex controls, and these are the ones that are least likely to change in response to the unfair demands of the experimenter, and remain most stable amid the fluctuating fortunes of the microbial world.

References

Aboud, M., and Pastan, I., 1975, Activation of transcription by guanosine 5′-diphosphate, 3′-diphosphate, transfer ribonucleic acid, and a novel protein from *Escherichia coli, J. Biol. Chem.* **250**:2189.

Andrews, K. J., and Hegeman, G. D., 1976, Selective disadvantage of non-functional protein synthesis in *Escherichia coli, J. Mol. Evol.* **8**:371.

Arst, H. N., 1976, Integrator gene in *Aspergillus nidulans, Nature* **262**:231.

Arst, H. N., and Cove, D. J., 1973, Nitrogen metabolite repression in *Aspergillus nidulans, Mol. Gen. Genet.* **126**:111.

Arst, H. N., and MacDonald, D. W., 1975, A gene cluster in *Aspergillus nidulans* with an internally located cis-acting regulatory region, *Nature* **254**:26.

Arst, H. N., and Scazzochio, C., 1975, Initiator constitutive mutation with an "up-promoter" effect in *Aspergillus nidulans, Nature* **254**:31.

Bachmann, B. J., Low, K. B., and Taylor, A., 1976, Recalibrated linkage map of *Escherichia coli* K-12, *Bacteriol. Rev.* **40**:116.

Baumberg, S., Bacon, D. F., and Vogel, H. J. 1965, Individually repressible enzymes specified by clustered genes of arginine synthesis, *Proc. Natl. Acad. Sci. U.S.A.* **53**:1029.

Bayly, R. C., and McKenzie, D. I., 1976, Catechol oxygenases of *Pseudomonas putida* mutant strains, *J. Bacteriol.* **127**:1098.

Bayly, R. C., and Wigmore, G. J., 1973, Metabolism of phenol and cresols by mutants of *Pseudomonas putida, J. Bacteriol.* **113**:1112.

Bayly, R. C., Wigmore, G. J., and McKenzie, D. I., 1977, Regulation of the enzymes of the meta-cleavage pathway of *Pseudomonas putida:* The regulon is composed of two operons, *J. Gen. Microbiol.* **100**:71.

Beggs, J. D., and Fewson, C. A., 1974, Repression of the enzymes converting benzyl alcohol into benzoate in *Acinetobacter calcoaceticus, Biochem. Soc. Trans.* **2**:924.

Beggs, J. D., and Fewson, C. A., 1977, Regulation of synthesis of benzyl alcohol dehyrogenase in *Acinetobacter calcoacetions, J. Gen. Microbiol.* **103**:127.

Beggs, J. D., Cook, A. M., and Fewson, C. A., 1976, Regulation of growth of *Acinetobacter calcoaceticus* NC1B8250 on benzyl alcohol in batch culture, *J. Gen. Microbiol.* **96**:365.

Bernlohr, R. W., Haddox, M. K., and Goldberg, N. D., 1974, Cyclic guanosine-3′,5′-monophosphate in *Escherichia coli* and *Bacillus licheniformis, J. Biol. Chem.* **249**:4329.

Bertrand, K., Korn, L., Lee, F., Platt, T., Squires, C. L., Squires, C., and Yanofsky, C., 1975, New features of the tryptophan operon, *Science* **189**:22.

Betz, J. L., Brown, P. R., Smyth, M. J., and Clarke, P. H., 1974, Evolution in action, *Nature* **247**:261.

Beyreuther, K., Adler, K., Geisler, N., and Klemm, A., 1973, The amino-acid sequence of the *lac* repressor, *Proc. Natl. Acad. Sci. U.S.A.* **70**:3576.

Bloom, F. R., McFall, E., Young, M. C., and Carothers, A. M., 1975, Positive control in the D-serine deaminase system of *Escherichia coli, J. Bacteriol.* **121**:1092.

Bonner, D. M., DeMoss, J. A., and Mills, S. E., 1965, The evolution of an enzyme, in: *Evolving Genes and Proteins* (V. Bryson and H. J. Vogel, eds.), pp. 305–318, Academic Press, New York.

Brammar, W. J., Clarke, P. H., and Skinner, A. J., 1967, Biochemical and genetic studies with regulator mutants of the *Pseudomonas aeruginosa* 8602 amidase system, *J. Gen. Microbiol.* **47**:87.

Brenchley, J. E., Prival, M. J., and Magasanik, B., 1973, Regulation of the synthesis of enzymes responsible for glutamate formation in *Klebsiella aerogenes, J. Biol. Chem.* **248**:6122.

Brenner, M., and Ames, B. N., 1971, The histidine operon and its regulation, in: *Metabolic Pathways* (H. J. Vogel, ed.), pp. 349–387, Academic Press, New York.

Brown, J. E., and Clarke, P. H., 1970, Mutations in a regulator gene allowing *Pseudomonas aeruginosa* 8602 to grow on butyramide, *J. Gen. Microbiol.* **64**:329.

Bruenn, J., and Hollingsworth, H., 1973, A mutant of *Escherichia coli* with a new highly efficient promoter for the lactose operon, *Proc. Natl. Acad. Sci. U.S.A.* **70**:3693.

Cain, R. B., Bilton, R. F., and Darrah, J. A., 1968, The metabolism of aromatic acids by micro-organisms. Metabolic pathways in the fungi, *Biochem. J.* **108**:797.

Calhoun, D. H., and Hatfield, G. W., 1975, Autoregulation of gene expression, *Annu. Rev. Microbiol.* **29**:275.

Calhoun, D. H., and Jensen, R. A., 1972, Significance of altered carbon flow in aromatic amino acid synthesis: An approach to the isolation of regulatory mutants in *Pseudomonas aeruginosa, J. Bacteriol.* **109**:365.

Calhoun, D. H., Pierson, D. L., and Jensen, R. A., 1973, Channel-shuttle mechanism for the regulation of phenylalanine and tyrosine synthesis at a metabolic branch point in *Pseudomonas aeruginosa, J. Bacteriol.* **113**:241.

Calvo, J. M., and Fink, G. R., 1971, Regulation of biosynthetic pathways in bacteria and fungi, *Annu. Rev. Biochem.* **40**:943.

Campbell, J. H., Lengyel, J. A., and Langridge, J., 1973, Evolution of a second gene for β-galactosidase in *Escherichia coli, Proc. Natl. Acad. Sci. U.S.A.* **70**:1841.

Canovas, J. L., Wheelis, M. L., and Stanier, R. Y., 1968, Regulation of the enzymes of the β-ketoadipate pathway in *Moraxella calcoacetica.* 2. The role of protocatechuate as inducer, *Eur. J. Biochem.* **3**:293.

Chakrabarty, A. M., 1972, Genetic basis of the biodegradation of salicylate in *Pseudomonas, J. Bacteriol.* **112**:815.

Chakrabarty, A. M., 1976, Plasmids in *Pseudomonas, Annu. Rev. Genet.* **10**:7.

Chakrabarty, A. M., Gunsalus, C. F., and Gunsalus, I. C., 1968, Transduction and the clustering of genes in fluorescent pseudomonads, *Proc. Natl. Acad. Sci. U.S.A.* **60**:168.

Chapman, L. F., and Nester, E. W., 1969, Gene–enzyme relationships in histidine biosynthesis in *Bacillus subtilis, J. Bacteriol.* **97**:1444.

Charnetsky, W. T., and Mortlock, R. P., 1974, Close linkage of the determinants of the ribitol and D-arabitol catabolic pathways in *Klebsiella aerogenes, J. Bacteriol.* **119**:176.

Clarke, P. H., 1972, Biochemical and immunological comparisons of aliphatic amidases produced by *Pseudomonas* species, *J. Gen. Microbiol.* **71**:241.

Clarke, P. H., 1974, The evolution of enzymes for the utilisation of novel substrates, in: *Evolution in the Microbial World,* Vol. 24 (M. J. Carlile and J. J. Skehel, eds.), *Symp. Soc. Gen. Microbiol.,* pp. 183–217, Cambridge Univ. Press, London and New York.

Clarke, P. H., and Lilly, M. D., 1969, The regulation of enzyme synthesis during growth, in: *Microbial Growth,* Vol. 19 (P. M. Meadow and S. J. Pirt, eds.), *Symp. Soc. Gen. Microbiol.,* pp. 113–159, Cambridge Univ. Press, London and New York.

Cocks, G. T., Aguilar, J., and Lin, E. C. C., 1974, Evolution of L-1,2-propanediol catabolism in *Escherichia coli* by recruitment of enzymes for L-fucose and L-lactate metabolism, *J. Bacteriol.* **118**:83.

Cohen, G. N., 1969, The aspartokinases and homoserine dehydrogenases of *Escherichia coli, Curr. Top. Cell. Regul.* **1**:183.

Cohen, S. N., 1976, Transposable genetic elements and plasmic evolution, *Nature* **263**:731.

Cole, J. A., Coleman, K. J., Compton, B. E., Kavanagh, B. M., and Keevil, C. W., 1974, Nitrite and ammonia assimilation by anaerobic continuous cultures of *Escherichia coli, J. Gen. Microbiol.* **85**: 11.

Cook, A. M., Beggs, J. D., and Fewson, C. A., 1975, Regulation of growth of *Acinetobacter calcoaceticus* NC1B8250 on L-mandelate in batch culture, *J. Gen. Microbiol.* **91**:325.

Cove, D., 1974, Evolutionary significance of autogenous regulation, *Nature* **251**:256.

Cove, D. J., and Pateman, J. A., 1969, Autoregulation of the synthesis of nitrate reductase in *Aspergillus nidulans, J. Bacteriol.* **97**:1374.

Crawford, I. P., 1975, Gene arrangements in the evolution of the tryptophan synthetic pathway, *Bacteriol. Rev.* **39**:87.

Crawford, I. P., and Gunsalus, I. C., 1966, Inducibility of tryptophan synthetase in *Pseudomonas putida, Proc. Natl. Acad. Sci. U.S.A.* **56**:717.

Crepin, M., Cukier-Kahn, R., and Gros, F., 1975, Effect of a low-molecular-weight DNA-binding protein, H_1 factor, on the *in vitro* transcription of the lactose operon in *Escherichia coli, Proc. Natl. Acad. Sci. U.S.A.* **72**:333.

Dagley, S., 1971, Catabolism of aromatic compounds by micro-organisms, *Adv. Microbiol. Physiol.* **6**:1.

Dagley, S., 1975, A biochemical approach to some problems of environmental pollution, in: *Essays in Biochemistry,* Vol. 11 (P. N. Campbell and W. N. Aldridge, eds.), pp. 81–138, Academic Press, New York.

Datta, P., 1969, Regulation of branched biosynthetic pathways in bacteria, *Science* **165**:556.

Davidson, B. E., Blackburn, E. H., and Dopheide, T. A. A., 1972, Chorismate mutase-prephenate dehydratase from *Escherichia coli* K-12, *J. Biol. Chem.* **247**:4441.

Day, M., Potts, J. R., and Clarke, P. H., 1975, Location of genes for the utilization of acetamide, histidine and proline on the chromosome of *Pseudomonas aeruginosa, Genet. Res.* **25**:71.

de Torrontegui, G., Diaz, R., Wheelis, M. L., and Canovas, J. L., 1976, Supra-operonic clustering of genes specifying glucose dissimilation in *Pseudomona putida, Mol. Gen. Genet.* **144**:307.

Dickson, R. C., Abelson, J., Barnes, W. M., and Reznikoff, W. S., 1975, Genetic regulation: The *lac* control region, *Science* **187**:27.

Dunn, N. W., and Gunsalus, I. C., 1973, Transmissible plasmid coding early enzymes of naphthalene oxidation in *Pseudomonas putida, J. Bacteriol.* **114**:974.

Dutton, P. L., and Evans, W. C., 1969, The metabolism of aromatic compounds by *Rhodopseudomonas palustris:* A new, reductive, method of aromatic ring metabolism, *Biochem. J.* **113**:525.

Elseviers, D., Cunin, R., Glansdorff, N., Baumberg, S., and Ashcroft, E., 1972, Control regions within the *argECBH* gene cluster of *Escherichia coli* K12, *Mol. Gen. Genet.* **117**:349.

Engel, P. S., 1973, Evolution of enzyme regulator sites: Evidence for partial gene duplication from amino-acid sequence of bovine glutamate dehydrogenase, *Nature* **241**:118.

Englesberg, E., and Wilcox, G., 1974, Regulation: Positive control, *Annu. Rev. Genet.* **8**:219.

Englesberg, E., Irr, J., Power, J., and Lee, N., 1965, Positive control of enzyme synthesis by gene C in the L-arabinose system, *J. Bacteriol.* **90**:946.

Eventoff, W., and Rossman, M. G., 1975, The evolution of dehydrogenases and kinases, *Crit. Rev. Biochem.* **3**:111.

Fargie, B., and Holloway, B. H., 1965, Absence of clustering of functionally related genes in *Pseudomonas aeruginosa, Genet. Res.* **6**:284.

Farin, F., and Clarke, P. H., 1978, Positive regulation of amidase synthesis in *Pseudomonas aeruginosa, J. Bacteriol.* **135**.

Feary, T. W., Williams, B., Calhoun, D. H., and Walker, T. A., 1969, An analysis of arginine requiring mutants in *Pseudomonas aeruginosa Genetics* **62**:673.

Feist, C. F., and Hegeman, G. D., 1969, Phenol and benzoate metabolism by *Pseudomonas putida:* Regulation of tangential pathways, *J. Bacteriol.* **100**:869.

Folk, W. R., and Berg, P., 1971, Duplication of the structural gene for the glycyl-transfer RNA synthetase in *Escherichia coli, J. Mol. Biol.* **58**:595.

Foor, F., Janssen, K. A., and Magasanik, B., 1975, Regulation of synthesis of glutamine synthetase by adenylylated glutamine synthetase, *Proc. Natl. Acad. Sci. U.S.A.* **72**:4844.

Gilbert, W., and Müller-Hill, B., 1966, Isolation of the lac repressor, *Proc. Natl. Acad. Sci. U.S.A.* **56**:1891.

Gilboa-Garber, N., Zakut, V., and Mizrahi, L., 1973, Production of choline esterase by *Pseudomonas aeruginosa,* its regulation by glucose and cyclic AMP and inhibition by antiserum, *Biochim. Biophys. Acta* **297**:120.

Ginsburg, A., and Stadtman, E. R., 1973, Regulation of glutamine synthetase in *Escherichia coli,* in: *The Enzymes of Glutamine Metabolism* (S. Prusiner and E. R. Stadtman, eds.), pp. 9–43, Academic Press, New York.

Goldberg, R. B., Bloom, F. R., and Magasanik, B., 1976, Regulation of histidase synthesis in intrageneric hybrids of enteric bacteria, *J. Bacteriol.* **127**:114.

Goldberger, R. F., 1974, Autogenous regulation of gene expression, *Science* **183**:810.

Grund, A., Shapiro, J., Fennewald, M., Bacha, P., Leahy, J., Markbreiter, K., Nieder, M., and Toepfer, M., 1975, Regulation of alkane oxidation in *Pseudomonas putida, J. Bacteriol.* **123**:546.

Guyer, M., and Hegeman, G. D., 1969, Evidence for a reductive pathway in the anaerobic metabolism of benzoate, *J. Bacteriol.* **99**:906.

Hacking, A. J., and Lin, E. C. C., 1976, Disruption of the fucose pathway was a consequence of genetic adaptation to propanediol as a carbon source in *Escherichia coli, J. Bacteriol.* **126**:1166.

Haldane, J. B. S., 1965, Data needed for a blueprint of the first organism, in: *Origins of Prebiological Systems and of Their Molecular Matrices* (S. W. Fox, ed.), pp. 11–18, Academic Press, New York.

Hall, B., 1976*a,* Experimental evolution of a new enzymatic function. Kinetic analysis of the ancestral (*ebg*°) and evolved (*ebg*$^+$) enzymes, *J. Mol. Biol.* **107**:71.

Hall, B., 1976*b,* Methyl galactosidase activity: An alternative evolutionary destination for the *ebgA*° gene, *J. Bacteriol.* **126**:536.

Hall, B., and Clarke, N. D., 1977, Experimental evolution of a new enzymatic function. II. An obligatory pathway for the evolution of lactobionate utilization by *Escherichia coli, Genetics* **85**:193.

Hall, B., and Hart, D. L., 1974, Regulation of newly evolved enzymes. I. Selection of a novel lactase regulated by lactose in *Escherichia coli, Genetics* **76**:391.

Hall, B., and Hartl, D. L., 1975, Regulation of newly evolved enzymes. II. The *ebg* repressor, *Genetics* **81**:427.

Hamlin, B. T., Ng, F. M.-W., and Dawes, E. A., 1967, Regulation of enzymes of glucose metabolism in *Pseudomonas aeruginosa* by citrate, in: *Microbial Physiology and Continuous Culture* (E. O. Powell, C. G. T. Evans, R. E. Strange, and D. W. Tempest, eds.), p. 211, HM Stationery Officer London.

Hartl, D. L., and Hall, B., 1974, Second naturally occurring β-galactosidase in *E. coli, Nature* **248**:152.

Hartley, B. S., 1974, Enzyme familes, in: *Evolution in the Microbial World,* Vol. 24 (M. J. Carlile and J. J. Skehel, eds.), *Symp. Soc. Gen. Microbiol.,* pp. 151–182, Cambridge Univ. Press, London and New York.

Hedges, R. W., Jacob, A. E., and Crawford, I. P., 1977, Wide ranging plasmid bearing the *Pseudomonas aeruginosa* tryptophan synthase genes, *Nature* **267**:283.

Hegeman, G. D., 1966*a,* Synthesis of the enzymes of the mandalate pathway by *Pseudomonas putida.* I. Synthesis of enzymes by the wild type, *J. Bacteriol.* **91**:1140.

Hegeman, G. D., 1966*b,* Synthesis of the enzymes of the mandelate pathway by *Pseudomonas putida.* II. Isolation and properties of blocked mutants. *J. Bacteriol.* **91**:1155.

Hegeman, G. D., 1966*c,* Synthesis of the enzymes of the mandelate pathway by *Pseudomonas putida.* III. Isolation and properties of constitutive mutants, *J. Bacteriol.* **91**:1161.

Higgins, S. J., and Mandelstam, J., 1972, Regulation of pathways degrading aromatic substrates in *Psuedomonas putida.* Enzymic response to binary mixtures of substrates, *Biochem. J.* **126**:901.

Hoch, J. A., and Nester, E. W., 1973, Gene–enzyme relationships of aromatic acid biosynthesis in *Bacillus subtilis, J. Bacteriol.* **116**:59.

Holloway, B. W., 1975, Genetic organization of Pseudomonas, in: *Genetics and Biochemistry of Pseudomonas* (P. H. Clarke and M. H. Richmond, eds.), pp. 133–161, Wiley, New York.

Hynes, M. J., 1970, Induction and repression of amidase enzymes in *Aspergillus nidulans, J. Bacteriol.* **103**:482.

Hynes, M. J., 1972, Mutants with altered glucose repression of amidase enzymes in *Aspergillus nidulans, J. Bacteriol.* **111**:717.

Hynes, M. J., 1974, Effect of ammonium, L-glutamate and L-glutamine on nitrogen catabolism in *Aspergillus nidulans, J. Bacteriol.* **120**:1116.

Hynes, M. J., 1975, A cis-dominant regulatory mutation affecting enzyme induction in the eukaryote *Aspergillus nidulans, Nature* **253**:210.

Interlied, C. B., and Mortlock, R. P., 1977, Growth of *Klebsiella aerogenes* on xylitol: Implications for bacterial enzyme evolution, *J. Mol. Evol.* **9**:181.

Irr, J., and Englesberg, E., 1971, Control of expression of the L-arabinose operon in temperature-sensitive mutants of gene *araC* in *Escherichia coli* B/r, *J. Bacteriol.* **105**:136.

Isaac, J. H., and Holloway, B. W., 1972, Control of arginine synthesis in *Pseudomonas aeruginosa, J. Gen. Microbiol.* **73**:427.

Jackson, E. N., and Yanofsky, C., 1972, Internal promoter of the tryptophan operon of *Escherichia coli* is located in a structural gene, *J. Mol. Biol.* **69**:307.

Jackson, E. N., and Yanofsky, C., 1973, Duplication-translocations of tryptophan operon genes in *Escherichia coli, J. Bacteriol.* **116**:33.

Jackson, J., Williams, L. S., and Umbarger, H. E., 1974, Regulation of synthesis of the branched-chain amino acids and cognate aminoacyl transfer ribonucleic acid synthetases of *Escherichia coli:* A common regulatory element, *J. Bacteriol.* **120**:1380.

Jacob, F., and Monod, J., 1961, Genetic regulatory mechanisms in the synthesis of proteins, *J. Mol. Biol.* **3**:318.

Jacob, F., Ullman, A., and Monod, J., 1965, Délétions fusionnant l'operon lactose et un opéron purine chez *Escherichia coli, J. Mol. Biol.* **13**:704.

Jacoby, G. A., and Gorini, L., 1967, Genetics of control of the arginine pathway in *Escherichia coli* B and K, *J. Mol. Biol.* **24**:41.

Jensen, R. A., and Nester, E. W., 1965, The regulatory significance of intermediary metabolites: Control of aromatic acid biosynthesis by feedback inhibition in *Bacillus subtilis. J. Mol. Biol.* **12**: 468.

Jensen, R. A., Nasser, D. S., and Nester, E. W., 1967, Comparative control of a branch-point enzyme in microorganisms, *J. Bacteriol.* **94**:1582.

Jensen, R. A., Calhoun, D. H., and Stenmark, S. L., 1973, Allosteric inhibition of 3-deoxy-D-arabino-heptulosonate-7-phosphate synthetase by tyrosine, tryptophan and phenylpyruvate in *Pseudomonas aeruginosa, Biochim. Biophys. Acta* **293**:256.

Jobe, A., and Bourgeois, S., 1972, *lac* repressor–operator interaction VI. The natural inducer of the *lac* operon, *J. Mol. Biol.* **69**:397.

Juni, E., 1972, Interspecies transformation of *Acinetobacter:* Genetic evidence for a ubiquitous genus, *J. Bacteriol.* **112**:917.

Karstrom, H., 1930, Uber die Enzymbildung in Bakterien, Thesis, Helsingfors.

Kemp, M. B., and Hegeman, G. D., 1968, Genetic control of the β-ketoadipate pathway in *Pseudomonas aeruginosa, J. Bacteriol.* **96**:1488.

Kirby, R., and Hopwood, D. A., 1977, Genetic determination of methylenomycin synthesis by the SCP1 plasmid of *Streptomyces coelicolor* A3(2), *J. Gen. Microbiol.* **98**:239.

Koch, A. L., 1972, Enzyme evolution: I. The importance of untranslatable intermediates, *Genetics* **72**:297.

Kovach, J. S., Ballesteros, A. O., Meyers, M., Soria, M., and Goldberger, R. F., 1973, A cis/trans test of the effect of the first enzyme for histidine biosynthesis on regulation of the histidine operon, *J. Bacteriol.* **114**:351.

Leblanc, D. J., and Mortlock, R. P., 1971*a,* Metabolism of D-arabinose: Origin of a D-ribulokinase activity in *Escherichia coli, J. Bacteriol.* **106**:82.

Leblanc, D. J., and Mortlock, R. P., 1971*b,* Metabolism of D-arabinose: A new pathway in *Escherichia coli, J. Bacteriol.* **106**:90.

Leidigh, B. J., and Wheelis, M. L., 1973, The clustering on the *Pseudomonas putida* chromosome of genes specifying dissimilatory functions, *J. Mol. Evol.* **2**:235.

Lin, E. C. C., Hacking, A. J., and Aguilar, J., 1976, Experimental models of acquisitive evolution, *Bioscience* **26**:548.

Magasanik, B., Prival, M. J., Brenchley, J. E., Tyler, B. M., Deleo, A. B., Streicher, S. L., Bender, R. A., and Paris, C. G., 1974, Glutamine synthetase as a regulator of enzyme synthesis, *Curr. Top. Cell. Regul.* **8**:119.

Mandelstam, J., and Jacoby, G. A., 1965, Induction and multi-sensitive end product repression in the enzymic pathway degrading mandelate in *Pseudomonas fluorescens, Biochem. J.* **94**:569.

Manson, M. D., and Yanofsky, C., 1976, Tryptophan operon regulation in interspecific hybrids of enteric bacteria, *J. Bacteriol.* **126**:679.

Martin, R. G., 1963, The one operon–one messenger theory of transcription, *Cold Spring Harbor Symp. Quant. Biol.* **28**:357.

Maurer, R., and Crawford, I. P., 1971, New regulatory mutation affecting some of the tryptophan genes in *Pseudomonas putida, J. Bacteriol.* **106**:331.

Mee, B. J., and Lee, B. T. O., 1967, An analysis of histidine requiring mutants in *Pseudomonas aeruginosa, Genetics* **55**:709.

Meers, J. L., Tempest, D. W., and Brown, C. M., 1970, "Glutamine (amide):2-oxoglutarate aminotransferase oxido-reductase (NADP)," an enzyme involved in the synthesis of glutamate by some bacteria, *J. Gen. Microbiol.* **64**:187.

Miller, J., Ganem, D., Lu, P., and Schmitz, A., 1977, Genetic studies of the *lac* repressor, I. Correlation of mutational sites with specific amino acid residues: Construction of colinear gene-protein map, *J. Mol. Biol.* **109**:275.

Mortlock, R. P., 1976, Catabolism of unnatural carbohydrates by micro-organisms, *Adv. Microb. Physiol.* **13**:1.

Mortlock, R. P., Fossitt, D. D., and Wood, W. A., 1965, A basis for the utilization of unnatural pentoses and pentitols by *Aerobacter aerogenes, Proc. Natl. Acad. Sci. U.S.A.* **54**:572.

Moyes, H. M., and Fewson, C. A., 1976, Constitutivity of the mandelate enzymes in *Acinetobacter calcoaceticus* NC1B 8250 and its effect on the synthesis of benzyl alcohol dehydrogenase, *Biochem. Soc. Trans.* **4**:1105.

Müller-Hill, B., 1975, Lac repressor and lac operator, *Prog. Biophys, Mol. Biol.* **30**:227.

Müller-Hill, B., Heidecker, G., and Kania, J., 1976, Repressor-galactosidase-chimaeras, in: *Structure–Function Relationships of Proteins, Third John Innes Symposium* (R. Markham and R. W. Horne, eds), pp. 167–179, Elsevier/North-Holland, Amsterdam.

Murray, K., and Old, R. W., 1974, The primary structure of DNA, *Prog. Nucleic Acid Res. Mol. Biol.* **14**:117.

Murray, K., and Willaims, P. A., 1974, Role of catechol and methyl-catechols as inducers of aromatic metabolism in *Pseudomonas putida, J. Bacteriol.* **117**:1153.

Musso, R., Di Lauro, R., Rosenberg, M., and de Crombrugghe, B., 1977, Nucleotide sequence of the operator-promoter region of the galactose operon of *Escherichia coli, Proc. Natl. Acad. Sci. U.S.A.* **74**:106.

Nester, E. W., and Montoya, A. L., 1976*a,* An enzyme common to histidine and aromatic amino acid biosynthesis in *Bacillus subtilis, J. Bacteriol.* **126**:699.

Nester, E. W., and Montoya, A., 1976*b,* Involvement of a histidine locus in tyrosine and phenylalanine synthesis in Bacillus subtilis, in: *Microbiology—1976* (D. Schlessinger, ed.), pp. 141–144, American Society for Microbiology, Washington, D.C.

Nester, E. W., Schafer, M., and Lederberg, J., 1963, Gene linkage in DNA transfer: A cluster of genes concerned with aromatic acid biosynthesis in *Bacillus subtilis, Genetics* **48**:592.

Nester, E. W., Jensen, R. A., and Nasser, D. S. 1969, Regulation of enzyme synthesis in the aromatic amino acid pathway of *Bacillus subtilis, J. Bacteriol.* **97**:83.

Newman, B. M., and Cole, J. A., 1977, Lack of a regulatory function for glutamine synthetase protein in the synthesis of glutamate dehydrogenase and nitrate reductase in *Escherichia coli* K12, *J. Gen. Microbiol.* **98**:369.

Novick, A., McCoy, J. M., and Sadler, J. R., 1965, The noninducibility of repressor formation, *J. Mol. Biol.* **12**:328.

Ohno, S., 1970, "Evolution by Gene Duplication," Allen & Unwin, London.

Oliver, E. J., and Mortlock, R. P., 1971, Metabolism of D-arabinose by *Aerobacter aerogenes.* Purification of the isomerase, *J. Bacteriol.* **108**:293.

Ornston, L. N., 1966, The conversion of catechol and protocatechuate to β-ketoadipate by *Pseudomonas putida.* IV. Regulation, *J. Biol. Chem.* **241**:3800.

Ornston, L. N., 1971, Regulation of catabolic pathways in *Pseudomonas, Bacteriol. Rev.* **35**:87.

Ornston, L. N., and Parke, D., 1977, The evolution of induction mechanisms in bacteria; Insights derived from the study of the β-ketoadipate pathway, *Curr. Top. Cell. Regul.* **12**:210.

Ornston, L. N., and Stanier, R. Y., 1966, The conversion of catechol and protocatechuate to β-ketoadipate by *Pseudomonas putida.* I. Biochemistry, *J. Biol. Chem.* **241**:3776.

Palleroni, N. J., and Stanier, R. Y., 1964, Regulatory mechanisms governing synthesis of the enzymes of tryptophan oxidation by *Pseudomonas fluorescens, J. Gen. Microbiol.* **35**:319.

Pastan, I., and Adhya, S., 1976, Cyclic adenosine 5′-monophosphate in *Escherichia coli, Bacteriol. Rev.* **40**:527.

Potts, J. R., and Clarke, P. H., 1976, The effect of nitrogen limitation on catabolite repression of amidase, histidase and urocanase in *Pseudomonas aeruginosa, J. Gen. Microbiol.* **93**:377.

Prival, M. J., and Magasanik, B., 1971, Resistance to catabolite repression of histidase and proline oxidase during nitrogen-limited growth of *Klebsiella aerogenes, J. Biol. Chem.* **246**:6288.

Prival, M. J., Brenchley, J. E., and Magasanik, B., 1973, Glutamine synthetase and the regulation of histidase formation in *Klebsiella aerogenes, J. Biol. Chem.* **248**:4334.

Proctor, A. R., and Crawford, I. P., 1975, Autogenous regulation of the inducible tryptophan synthase of *Pseudomonas putida, Proc. Natl. Acad. Sci. U.S.A.* **72**:1249.

Rahman, M., and Clarke, P. H. 1975, Mapping of genes for catabolic enzymes on the chromosome of *Pseudomonas aeruginosa, Proc. Soc. Gen. Microbiol.* **3**:52.

Reeve, E. C. R., and Braithwaite, J. A., 1973, Lac$^+$ plasmids are responsible for the strong lactose-positive phenotype found in many strains of *Klebsiella* species, *Genet. Res.* **22**:329.

Reiner, A. B., 1975, Genes for ribitol and D-arabitol catabolism in *Escherichia coli:* Their loci in C strains and absence in K-12 and B strains, *J. Bacteriol.* **123**:530.

Rigby, P. W. J., Burleigh, B. D., and Hartley, B. S., 1974, Gene duplication in experimental enzyme evolution, *Nature* **251**:200.

Riley, M., Solomon, J., and Zipkas, D., 1978, Relationship between gene function and gene location in *Escherichia coli, J. Mol. Evol.* **11**:47.

Rosenberg, S. L., and Hegeman, G. D., 1969, Clustering of functionally related genes in *Pseudomonas aeruginosa, J. Bacteriol.* **99**:353.

Rossmann, M. G., Moras, D., and Olsen, K. W., 1974, Chemical and biological evolution of a nucleotide-binding protein, *Nature* **250**:194.

Roth, C. W., and Nester, E. W., 1971, Co-ordinate control of tryptophan, histidine and tyrosine enzyme synthesis in *Bacillus subtilis, J. Mol. Biol.* **62**:577.

Roth, J. R., and Ames, B. N., 1966, Histidine regulatory mutants in *Salmonella typhimurium.* II. Histidine regulatory mutants having altered histidinyl-tRNA synthetase, *J. Mol. Biol.* **22**:325.

Roth, J. R., Anton, D. N., and Hartman, P. E., 1966, Histidine regulatory mutants in *Salmonella typhimurium.* I. Isolation and general properties, *J. Mol. Biol.* **22**:305.

Sadler, J. R., and Novick, A., 1965, The properties of repressor and the kinetics of its action, *J. Mol. Biol.* **12**:305.

Saedler, H., and Heiss, B., 1973, Multiple copies of the insertion-DNA sequences IS1 and IS2 in the chromosome of *E. coli* K12, *Mol. Gen. Genet.* **122**:267.

Saedler, H., Besemer, J., Kemper, B., Rosenwirth, B., and Starlinger, P., 1972, Insertion mutations in the control region of the *gal* operon of *E. coli.* I. Biological characterization of the mutations, *Mol. Gen. Genet.* **115**:258.

Saedler, H., Reif, H. J., Hu, S., and Davidson, N., 1974, IS2, a genetic element for turn-off and turn-on of gene activity in *E. coli, Mol. Gen. Genet.* **132**:265.

St. Martin, E. J., and Mortlock, R. P., 1976, Natural and altered induction of the L-fucose catabolic enzymes in *Klebsiella aerogenes J. Bacteriol.* **127**:91.

St. Martin, E. J., and Mortlock, R. P., 1977, A comparison of alternate metabolic strategies for the utilization of D-arabinose, *J. Mol. Evol.* **10**:111.

Schmeissner, U., Ganem, D., and Miller, J. H., 1977, Genetic studies of the *lac* repressor, II. Fine structure deletion map of the *lac-I* gene, and its correlation with the physical map, *J. Mol. Biol.* **109**:303.

Senior, E., Bull, A. T., and Slater, J. H., 1976, Enzyme evolution in a microbial community growing on the herbicide Dalapon, *Nature* **263**:476.

Senior, P. J., 1975, Regulation of nitrogen metabolism in *Escherichia coli* and *Klebsiella aerogenes:* Studies with the continuous culture technique, *J. Bacteriol.* **123**:407.

Setlow, P., 1973, Inability to detect cyclic AMP in vegetative or sporulating cells or dormant spores of *Bacillus megaterium, Biochem. Biophys. Res. Commun.* **52**:365.

Silbert, D. F., Fink, G. R., and Ames, B. N., 1966, Histidine regulatory mutants in *Salmonella typhimurium.* III. A class of regulatory mutants deficient in tRNA for histidine, *J. Mol. Biol.* **22**:335.

Smyth, P. F., and Clarke, P. H., 1975*a,* Catabolite repression of *Pseudomonas aeruginosa* amidase: The effect of carbon source on amidase synthesis, *J. Gen. Microbiol.* **90**:81.

Smyth, P. F., and Clarke, P. H., 1975*b,* Catabolite repression of *Pseudomonas aeruginosa* amidase: Isolation of promoter mutants, *J. Gen. Microbiol.* **90**:91.

Squires, C. L., Rose, J. K., Yanofsky, C., Yang, H.-L., and Zubay, G., 1973, Tryptophanyl-tRNA and tryptophanyl-tRNA synthetase are not required for *in vitro* repression of the tryptophan operon, *Nature New Biol.* **245**:131.

Squires, C. L., Lee, F. D., and Yanofsky, C., 1975, Interaction of the *trp* repressor and RNA polymerase with the *trp* operon. *J. Mol. Biol.* **92**:93.

Stanier, R. Y., and Ornston, L. N., 1973, The β-keto adipate pathway, *Adv. Microbiol. Physiol.* **9**:89.

Streicher, S. L., Deleo, A. B., and Magasanik, B., 1976, Regulation of enzyme formation in *Klebsiella aerogenes* by episomal glutamine synthetase of *Escherichia coli, J. Bacteriol.* **127**:184.

Thomas, K. A., and Schechter, A., 1978, Protein folding: Structural and evolutionary approach, in: *Biological Regulation and Development,* Vol. 2, *Molecular Organization and Cell Function* (R. F. Goldberger, ed.), Plenum Press, New York, in press.

Towner, K. J., and Vivian, A., 1976, RP4-mediated conjugation in *Acinetobacter calcoaceticus, J. Gen. Microbiol.* **93**:355.

Tresguerres, M. E. F., de Torrentegui, G., Ingledew, W. M., and Canovas, J. L., 1970, Regulation of the enzymes of the β-ketoadipate pathway in *Moraxella.* Control of quinate oxidation by protocatechuate, *Eur. J. Biochem.* **14**:445.

Tribe, D. E., Camakaris, H., and Pittard, J., 1976, Constitutive and repressible enzymes of the common pathway of aromatic biosynthesis in *Escherichia coli* K-12: Regulation of enzyme synthesis at different growth rates, *J. Bacteriol.* **127**:1085.

Tronick, S. R., Ciardi, J. E., and Stadtman, E. R., 1973, Comparative biochemical and immunological studies of bacterial glutamine synthetases, *J. Bacteriol.* **115**:858.

Tyler, B., Deleo, A. B., and Magasanik, B., 1974, Activation of transcription of *hut* DNA by glutamine synthetase, *Proc. Natl. Acad. Sci. U.S.A.* **71**:225.

Ullman, A., Tillier, F., and Monod, J., 1976, Catabolite modulator factor: A possible mediator of catabolite repression in bacteria, *Proc. Natl. Acad. Sci. U.S.A.* **73**:3476.

Voellmy, R., and Leisinger, T., 1975, Dual role for *N*-acetylornithine-5-aminotransferase from *Pseudomonas aeruginosa* in arginine biosynthesis and arginine catabolism, *J. Bacteriol.* **122**:799.

Vogel, H. J., Bacon, D. F., and Baich, A., 1963, Induction of acetylornithine δ-transaminase during pathway-wide repression, in: *Informational Macromolecules* (H. J. Vogel, V. Bryson, and J. O. Lampen, eds.), pp. 293–300, Academic Press, New York.

Waltho, J., and Holloway, B. W., 1966, Suppression of fluorophenylalanine resistance by mutation to streptomycin resistance in *Pseudomonas aeruginosa, J. Bacteriol.* **92**:35.

Wheelis, M. L., 1975, The genetics of dissimilatory pathways in *Pseudomonas, Annu. Rev. Microbiol.* **29**:505.

Wheelis, M. L., and Ornston, L. N., 1972, Genetic control of enzyme induction in the β-ketoadipate pathway of *Pseudomonas putida:* Deletion mapping of *cat* mutations, *J. Bacteriol.* **109**:790.

Wheelis, M. L., and Stanier, R. Y., 1970, The genetic control of dissimilatory pathways in *Pseudomonas putida, Genetics* **66**:245.

Wigmore, G. J., and Bayly, R. C., 1974, A mutant of *Pseudomonas putida* with altered regulation of the enzymes for degradation of phenol and cresols, *Biochem. Biophys. Res. Commun.* **60**:48.

Wigmore, G. J., Di Berardino, D., and Bayly, R. C., 1977, Regulation of the enzymes of the meta-cleavage pathway of *Pseudomonas putida:* A regulatory model, *J. Gen. Microbiol.* **100**:81.

Williams, P. A., and Murray, K., 1974, Metabolism of benzoate and the methylbenzoates by *Pseudomonas putida* (*arvilla*) mt-2: Evidence for the existence of a TOL plasmid, *J. Bacteriol.* **120**:416.

Williams, P. A., and Worsey, M. J., 1976*a,* Ubiquity of plasmids in coding for toluene and xylene metabolism in soil bacteria: Evidence for the existence of new TOL plasmids, *J. Bacteriol.* **125**: 818.

Williams, P. A., and Worsey, M. J., 1976*b,* Plasmids and catabolism, *Biochem. Soc. Trans.* **4**:466.

Willson, C., Perrin, D., Cohn, M., Jacob, F., and Monod, J., 1964, Non-inducible mutants of the regulator gene in the "Lactose" system of *Escherichia coli, J. Mol. Biol.* **8**:582.

Wong, C. L., and Dunn, N. W., 1974, Transmissible plasmid coding for the degradation of benzoate and *m*-toluate in *Pseudomonas arvilla* mt-2, *Genet. Res.* **23**:227.

Worsey, M. J., and Williams, P. A., 1977, Characterization of a spontaneously occurring mutant of the TOL20 plasmid in *Pseudomonas putida* MT20: Possible regulatory implications, *J. Bacteriol.* **130**:1149.

Wu, C.-H., Ornston, M. K., and Ornston, L. N., 1972, Genetic control of enzyme induction in the β-ketoadipate pathway of *Pseudomons putida:* Two-point crosses with a regulatory mutant strain, *J. Bacteriol.* **109**:796.

Wu, T. T., 1976*a,* Growth of a mutant of *Escherichia coli* K-12 on xylitol by recruiting enzymes for D-xylose and L-1,2-propanediol metabolism, *Biochem. Biophys. Acta* **428**:656.

Wu, T. T., 1976*b,* Growth on D-arabitol of a mutant strain of *Escherichia coli* K-12 using a novel dehydrogenase and enzymes related to L-1,2-propanediol and D-xylose metabolism, *J. Gen. Microbiol.* **94**:246.

Wu, T. T., Lin, E. C. C., and Tanaka, S., 1968, Mutants of *Aerobacter aerogenes* capable of utilizing xylitol as a novel carbon source, *J. Bacteriol.* **96**:447.

Yourno, J., Kohno, T., and Roth, J. R., 1970, Enzyme evolution: Generation of a bifunctional enzyme by fusion of adjacent genes, *Nature* **228**:820.

Yudkin, J., 1938, Enzyme variation in micro-organisms, *Biol. Rev.* **13**:93.

Zamenhof, S., and Eichhorn, H. H., 1967, Study of microbial evolution through loss of biosynthetic function: Establishment of "defective" mutants, *Nature* **216**:456.

Zipkas, D., and Riley, M., 1975, Proposal concerning mechanism of evolution of the genome of *Escherichia coli, Proc. Natl. Acad. Sci. U.S.A.* **72**:1354.

5

Importance of Symmetry and Conformational Flexibility in DNA Structure for Understanding Protein–DNA Interactions

HENRY M. SOBELL

1 *Introduction*

During the past few years, there has been a growing realization that two aspects of DNA structure, namely, its symmetry and its conformational flexibility, play important roles in protein–DNA interactions involved in genetic regulation. This chapter reviews evidence along these lines, beginning with our structural studies of drug intercalation (these studies have given direct structural information concerning the roles of symmetry and conformational flexibility in DNA structure in a variety of drug–DNA interactions), enlarging on this theme with a discussion of the nature of DNA breathing and DNA denaturation, and ending with discussions of specific protein–DNA interactions that involve various aspects of DNA symmetry and flexibility—that is, histone–DNA interactions in chromatin, RNA polymerase–promoter recognition, operator–repressor recognition, and so on. No attempt is made to review the wealth of nucleotide sequence information

---

HENRY M. SOBELL • Department of Chemistry, and Department of Radiation Biology and Biophysics, University of Rochester School of Medicine and Dentistry, Rochester, New York

currently available; rather, specific examples of protein–DNA interactions are discussed in detail and these will serve to illustrate the general principles involved.

2 *Symmetry in DNA Structure*

To understand protein–DNA interactions, it is important to realize that DNA has *two* distinct dyad axes perpendicular to its helix axis (see Fig. 1). These axes fall at the level of each base pair and between two adjacent base pairs, and this gives rise to (potential) symmetric recognition that spans either an odd or an even number of base pairs. Specific examples of protein–DNA recognition of this type are found in the interaction of the restriction nucleases with DNA, which results in DNA molecules that possess single-stranded ends having either an odd or an even number of nucleotide bases (see, for example, Hedgpeth *et al.*, 1972; Bigger *et al.*, 1973). Another example may be the interactions between proteins that contain

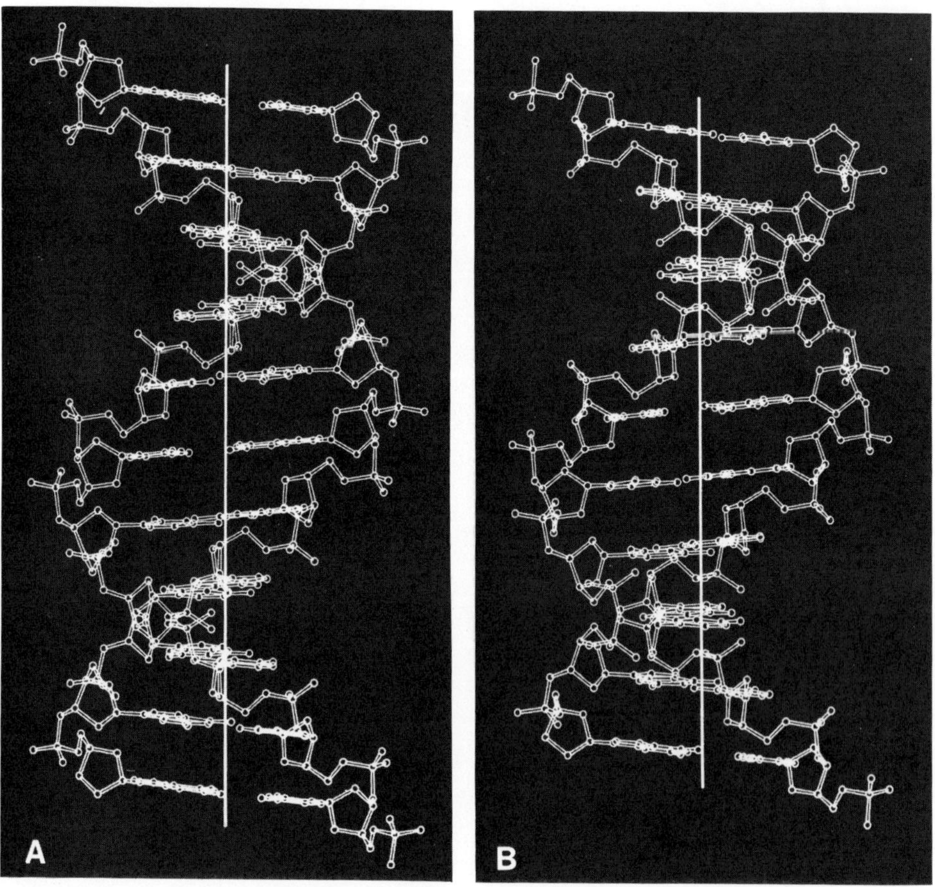

Figure 1. Computer-drawn illustrations showing dyad axes of two types in DNA. (A) Dyad axis falling at the level of a base pair; (B) dyad axis falling between two adjacent base pairs. See text for discussion.

localized domains of antiparallel β-pleated sheet structure and the minor groove of the DNA double helix (Carter and Kraut, 1974; Kim *et al.*, 1977); interactions of this type involve localized recognition of both dyad axes in DNA. They do not, however, require the presence of base sequence symmetries.

It is important to note that symmetry arguments make no specific prediction about the detailed nature of protein–DNA interactions. For example, a DNA sequence involved in symmetric recognition could be in denatured, double helical, kinked, or hairpin form (Crick, 1971; Kelly and Smith, 1970; Crick and Klug, 1975; Sobell *et al.*, 1976, 1977*a,b;* Gierer, 1966); symmetry arguments say nothing about this. Furthermore, atomic interactions between proteins and DNA may not always require exact sequence symmetries; approximately symmetric sequences can be utilized in the interactions. By approximately symmetric sequences, we mean guanine–adenine or cytosine–thymine nucleotide bases related across a twofold axis; such sequences possess common functional groups, such as the N-7 imidazole nitrogen on guanine and adenine, and the O-2 keto oxygen on cytosine and thymine, that can be utilized in symmetric recognition. Such sequences may be intermixed with blocks of exactly symmetric sequences involved in protein–DNA interactions.

Perfect crystallographic point group symmetries rarely (if ever) occur in biological systems. This probably reflects the size and complexity of subunits in oligomeric proteins and close stereochemical contacts between subunits across local twofold axes (see, for example, *Structure and Function of Proteins at the Three-Dimensional Level*). Departures from symmetry that relates protein subunits have important consequences in understanding certain examples of protein–DNA recognition (for example, the *lac* operator–repressor interaction). These result in asymmetric recognition near the axis of symmetry of a polynucleotide base sequence, and symmetric recognition of sequences (and other aspects of DNA structure) farther away. A particularly important model system to understand various aspects of protein–DNA interactions is the actinomycin–DNA binding reaction. We will discuss the detailed structure and supporting evidence for this miniature protein–DNA interaction shortly.

3 *Flexibility in DNA Structure—The Kink*

An important aspect of DNA structure is its flexibility. This is shown in Fig. 2. DNA has natural flexibility down its dyad axis—that is, the dyad axis falling between adjacent base pairs. This flexibility reflects the ability of neighboring base pairs to "roll" upon each other's van der Waals surfaces, as well as the occurrence of small alterations of bond distances and angles from their equilibrium values in both sugar-phosphate chains. Conceptually, it is possible to bend DNA in one of two directions. If we bend DNA into the narrow groove, negatively charged phosphate groups on opposite chains approach each other and this introduces a repulsive term into the total free energy. Furthermore, strain energy in both sugar-phosphate chains appears to be evenly distributed over four or so nucleotide base pairs. There is no tendency for DNA to kink so as to relieve this strain energy. If we bend DNA into the wide groove, however, negatively charged phosphate groups do not approach each other significantly, even after bending DNA 20° to 30°. Moreover, the strain energy in both sugar-phosphate chains remains highly localized to the central nucleotide base-paired region. This strain energy can be relieved rather dramatically by simultaneously altering the puckering of deoxyribose sugar residues across the dyad axis to give a C3′ *endo* (3′–5′) C2′ *endo* sugar puckering pattern. This results in a kinked DNA structure in which base pairs are partially unstacked (base pairs form an angle of about 40° to one another) and helical axes for B DNA sections above and below the kink are displaced by about −1.0 Å (see Figs. 3A and 3B). Kinking DNA causes it to unwind. We estimate that adjacent base pairs at the kink are twisted by about 26° (this has been calculated as described later for ethidium intercalation, using projectional geometry); this gives rise to an effective angular unwinding by about −10°.

A conformational change such as this would require minimal stereochemical rearrangement and would involve small energies. DNA may, therefore, be undergoing bending and kinking conformational fluctuations in solution due to normal modes in the polymer excited at thermal energies (these modes probably contain a mixture of bending, stretching, unwinding, and sliding motions, all of which are tightly coupled). Once formed, a kink then becomes a flexible hinge in DNA. This hinge can straighten and base pairs separate to create the space for drug intercalation (see Figs. 4A and 4B). These motions would again involve bending, stretching, unwinding, and sliding motions associated with the normal modes in DNA.

One can imagine three classes of intercalative drugs or dyes, each distinguished by its direction of entrance into DNA (see Fig. 5). Compounds of the first class (class A) intercalate exclusively from the narrow groove. These include actinomycin and ethidium (irehdiamine A may also be included in this class, although strictly speaking it is not an intercalative drug; see later discussions). These drugs may first bind to the kink and then subsequently "slip into" the interior of the double helix. Numerous van der Waals contacts may be made between the planar chromophore and base pairs at the kink. Some could be retained and others added during the insertion process, an effect that would help to facilitate the entrance of the drug into DNA. Compounds of the second class (class B) intercalate exclusively from the wide groove. Examples of drugs in this class are daunomycin, proflavin, and acridine orange (Pigram *et al.*, 1972; Reddy *et al.*, 1977). These drugs could first bind to the sugar-phosphate chains at the kink and then intercalate as DNA straightens and base pairs separate. Finally, there may be intercalative drugs of a third class (class C) that can intercalate from either the wide or the narrow grooves. One such drug may be 9-aminoacridine (Sakore *et al.*, 1977*a*, *b*). It is possible that the biological activities of these drugs and the kinetics of their binding reactions reflect their directional entrance into DNA, as defined by these three classes.

4 *Detailed Models for Drug–DNA Binding*

The postulate that DNA kinking is a spontaneous process induced at thermal energies by normal modes in the polymer and that the phenomenon of drug

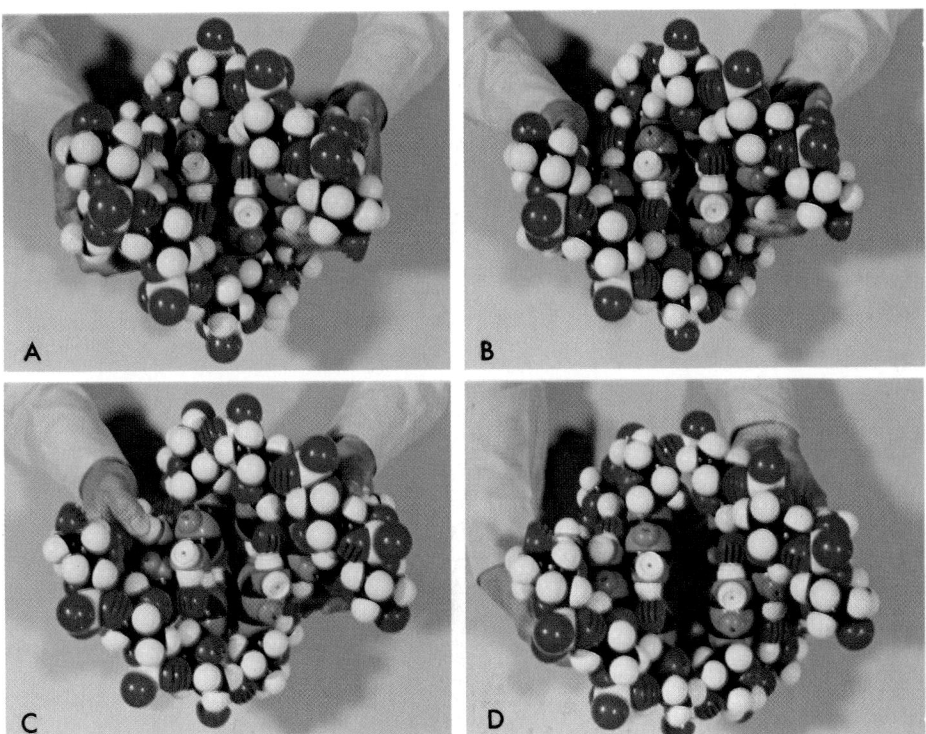

Figure 2. Space-filling Corey–Pauling–Koltun (CPK) molecular models of DNA that demonstrate its conformational flexibility. (A) B DNA viewed down its dyad axis; (B) B DNA flexed about 15° into the major groove of the helix; (C) the kink, viewed from the minor groove; and (D) the flexible hinge that results, allowing the molecule to straighten and base pairs to separate for drug intercalation.

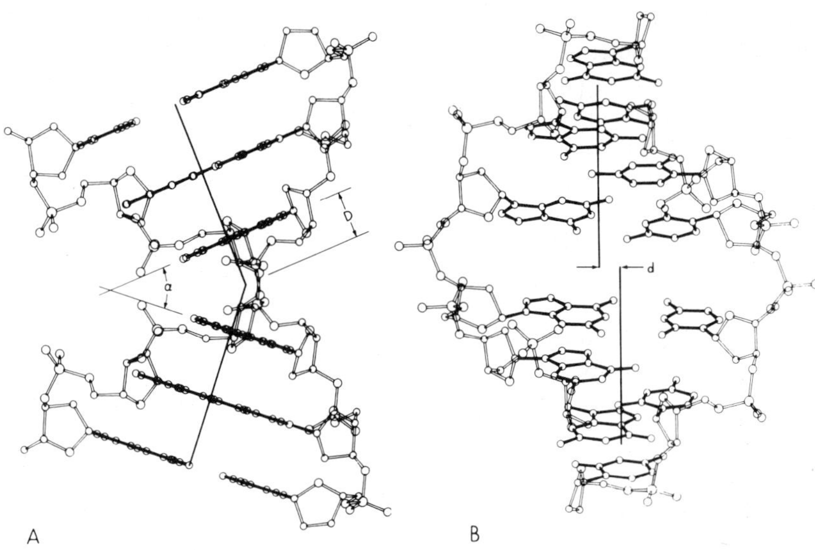

Figure 3. Detailed illustration of the kink in DNA. (A) The kink, viewed from a sidewise direction to show the bending (α) and the stretching (D) of DNA; and (B) the kink, viewed down the dyad axis of DNA to show the displacement or dislocation (d) of helix axes of DNA above and below the kink.

intercalation results from this conformational flexibility in DNA allows us to understand many of the kinetic features of drug–DNA binding. Moreover, it allows us to understand more fully the detailed nature of DNA binding by ethidium, actinomycin, and irehdiamine, three molecules that utilize intercalative and kinked type geometries in their interactions with DNA.

Over the past few years, we have been able to form a large number of drug–nucleic acid crystalline complexes with mononucleosides and self-complementary dinucleoside monophosphates, and to solve their three-dimensional structures to atomic resolution by X-ray crystallography (Sobell *et al.*, 1971, 1977*a;* Jain and Sobell, 1972; Sobell and Jain, 1972; Tsai *et al.*, 1977; Jain *et al.*, 1977; Sakore *et al.*, 1977*a,b;* Reddy *et al.*,, 1977).

Figure 4. Conversion of the kink (A) to the intercalative structure (B) through alterations in the glycosidic torsional angles of the nucleotides. It seems likely that a certain amount of strain energy could develop in the sugar-phosphate backbone due to small departures from the normal bond distances and angles in the immediate region. This could allow DNA to "bite down" on an intercalative drug or dye and provide additional energy to stabilize the drug–DNA interaction. This conformational change reflects bending, stretching, unwinding, and sliding motions that are associated with the normal modes in DNA.

Figure 6 shows the chemical structures of drugs used in these studies. All structures have demonstrated sandwich type association or intercalative binding between the planar drug chromophore and the nucleic acid component. In the ethidium– and aminoacridine–dinucleoside monophosphate studies, common features in the stereochemistry of the sugar-phosphate chains include a sugar-puckering pattern of the type C3′ *endo* (3′–5′) C2′ *endo* in the immediate intercalation region. This change, as well as additional small but systematic changes in torsional angles that describe the sugar-phosphate chains, allow base pairs to separate 6.8 Å and give rise to the observed twist angle between base pairs above and below the intercalative drug or dye. We have been able to use this information to understand the general nature of intercalative drug binding to DNA. This is shown in Figs. 7A and 7B.

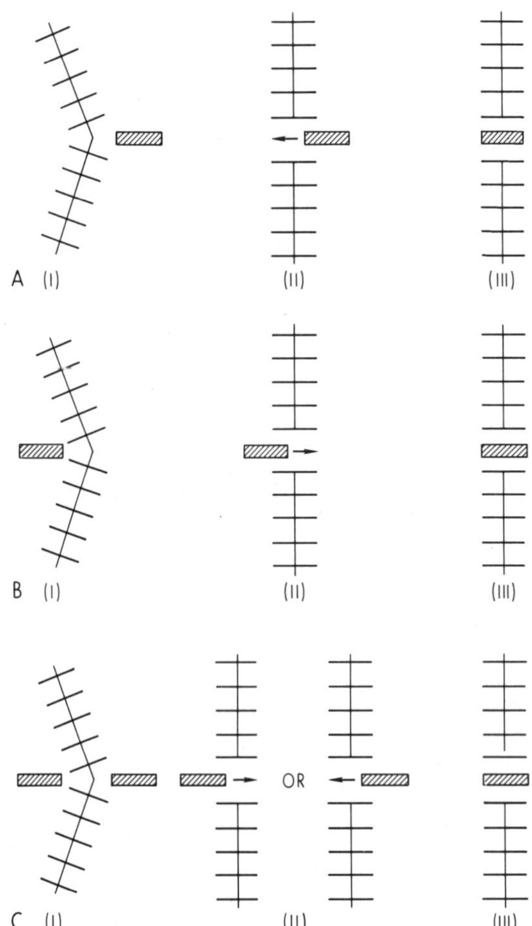

Figure 5. Schematic figure to show the three classes of intercalative drugs and dyes as defined by their directional entrance into DNA. (A) Drugs entering from the narrow groove (class A); (B) drugs entering from the wide groove (class B); and (C) drugs entering from either the narrow or wide groove (class C).

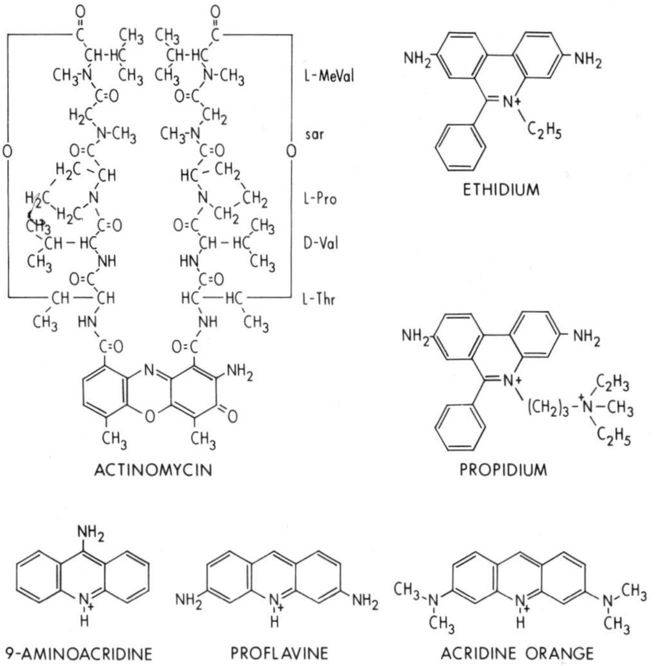

L-MeVal
sar
L-Pro
D-Val
L-Thr
ACTINOMYCIN
ETHIDIUM
PROPIDIUM
9-AMINOACRIDINE
PROFLAVINE
ACRIDINE ORANGE

Figure 6. Chemical structures of drugs and dyes that form crystalline complexes with nucleic acid components and that have been studied by X-ray crystallography.

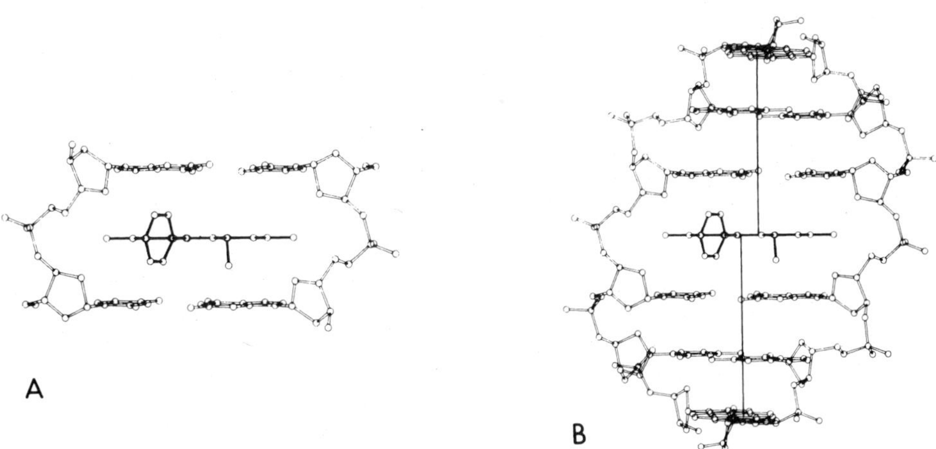

Figure 7. Steps in assembly of the ethidium–DNA binding model. (A) Idealized (deoxyribose-containing) configuration common to both the ethidium–iodoUpA and ethidium–iodoCpG crystalline complexes; and (B) B DNA added above and below the intercalated deoxyribodinucleotide monophosphate to demonstrate the helical screw axis dislocation in B DNA accompanying ethidium intercalation.

4.1 *Ethidium*

To construct the ethidium–DNA binding model, we have added B DNA to either side of an idealized (deoxyribose-containing) configuration common to both the ethidium–iodoUpA and ethidium–iodoCpG crystal structures (Tsai *et al.*, 1977; Jain *et al.*, 1977). This is done easily and without steric difficulty. An important realization that immediately emerges is the concept that drug intercalation into DNA is accompanied by a helical screw axis displacement (or dislocation) in its structure. (For ethidium intercalation, we estimate that helical axes for B DNA on either side of the phenanthridinium ring system are displaced approximately +1.0 Å.) Base pairs in the immediate region are twisted by 10° (a value that has been estimated by projecting the interglycosidic carbon vectors on a plane passing midway between base pairs, and then measuring the angle between them). This gives rise to an angular unwinding of −26° at the immediate site of drug intercalation. We have also observed that intercalated base pairs are tilted relative to one another by about 8° in both ethidium crystal structures; this results in a small residual "kink" of 8° at the intercalation site, and has been included in our ethidium–DNA binding model.

The magnitude of angular unwinding predicted by our ethidium–DNA binding model is in good agreement with Wang's estimate of ethidium–DNA angular unwinding based on alkaline titration studies of superhelical DNA in cesium chloride density gradients (Wang, 1974). Moreover, the C3′ *endo* (3′–5′) C2′ *endo* mixed sugar puckering necessarily predicts that intercalation be limited to every *other* base pair at maximal drug–nucleic acid binding ratios (that is, a neighbor exclusion structure) (Crothers, 1968). We have examined the stereochemistry of this model carefully. The effect of having a helical screw axis displacement every other base pair, combined with an 8° kink, is to give rise to a maximally unwound DNA structure possessing a slow right-handed superhelical writhe. This is shown in Fig. 8.

4.2 *Actinomycin*

Actinomycin forms a 1:2 stoichiometric complex with deoxyguanosine, and the stereochemical information provided by this structure has led us to propose a model for actinomycin–DNA binding (Sobell *et al.*, 1971; Jain and Sobell, 1972; Sobell and Jain, 1972). The model involves intercalation of the phenoxazone ring system between base pairs in the DNA double helix and the utilization of specific hydrogen bonds, van der Waals forces, and hydrophobic interactions between the pentapeptide chains on actinomycin and chemical groups in the minor groove of the DNA helix. Important elements in the recognition of actinomycin for DNA are the guanine specificity and the use of symmetry in the interaction. This predicts a base sequence binding preference of the type d-GpC in actinomycin–DNA binding, and this has been verified in model dinucleoside monophosphate solution-binding studies and in synthetic polymer-binding studies (Krugh, 1972; Wells and Larson, 1970).

Although we believe that the general features of our actinomycin–DNA binding model are correct, we have presented a slightly modified version of this

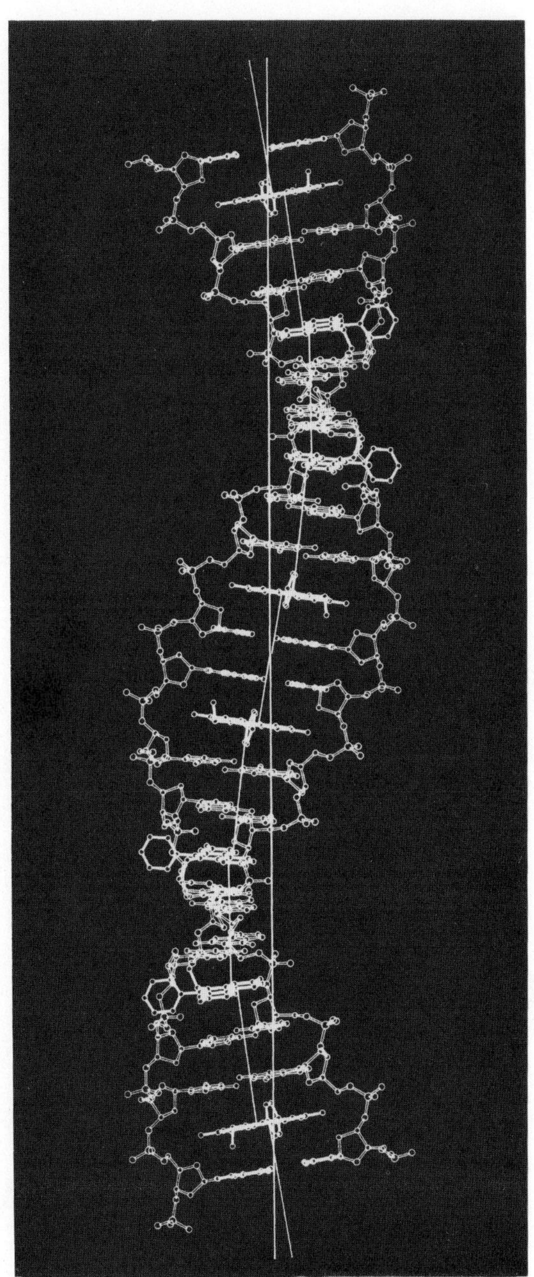

Figure 8. Computer graphics illustration of the neighbor exclusion model for ethidium–DNA binding. This structure is a maximally elongated and unwound DNA molecule that possesses a slow right-handed superhelical writhe. It contains a perfectly alternating pattern of sugar puckering [that is, C3′ endo (3′–5′) C2′ endo (3′–5′) C3′ endo, and so on down the polynucleotide sugar-phosphate backbone. See text for additional discussion.

model that utilizes many of the insights afforded by the ethidium– and aminoacridine–dinucleoside monophosphate crystal studies (Sobell *et al.*, 1977*a*). See Fig. 9. Major features of our revised actinomycin–DNA model are as follows: (1) Intercalation of the phenoxazone ring system between base-paired d-GpC sequences accompanied by a helical screw axis dislocation of about −0.4 Å. (This value was about −1.5 Å in our previous model, since we attempted to use the precise

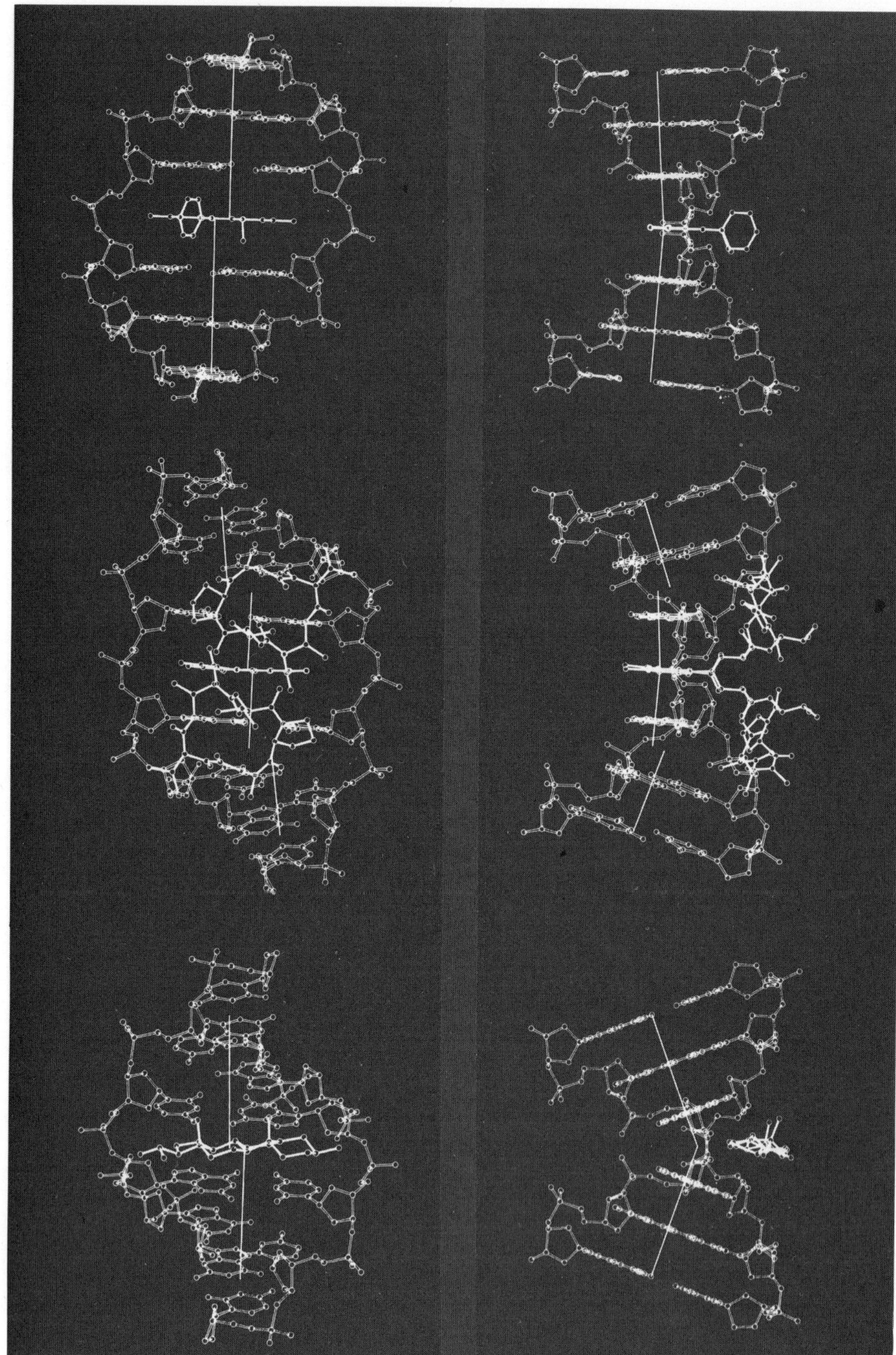

Figure 9. Detailed stereochemical models for drug–DNA interactions. Top, ethidium–DNA binding model viewed down the dyad axis (left) and from a sidewise direction (right); middle, actinomycin–DNA binding model viewed down the dyad axis (left) and from a sidewise direction (right); and bottom, irehdiamine–DNA binding model viewed down the dyad axis (left) and from a sidewise direction (right). See text for discussion.

configuration observed in the actinomycin–deoxyguanosine crystalline complex; subsequent study has shown, however, that the polymer backbone imposes stereochemical constraints which limit this dislocation to about −0.4 Å.) (2) A departure from *exact* twofold symmetry in the complex; this reflects the asymmetry of the phenoxazone ring system and the inability of both pentapeptide chains to interact simultaneously with (or span) guanine residues on opposite chains due to the smaller magnitude of this helical screw axis dislocation. (3) An additional bending distortion in DNA on both sides of the intercalation site to accommodate the steric bulk of the pentapeptide chains on actinomycin; this gives rise to a kinked type DNA structure (helical axes for B DNA sections on either side of this bend form an angle of about 40° and are displaced by about −1.6 Å).

Important questions often raised in discussions of actinomycin–DNA binding concern the origin of the guanine specificity (does this reflect specific stacking interactions between the phenoxazone ring system, or is this due to specific hydrogen-bonding interactions between the pentapeptide chains and the 2-amino group of guanine, as predicted by our model?) and of the importance of symmetry in the interaction. These questions have recently been resolved in an interesting way with implications in understanding the broader nature of protein–DNA interactions.

Ethidium is known to possess a sequence-dependent pyrimidine (3′–5′) purine binding preference for double-helical dinucleotides and dinucleoside monophosphates in solution and in the solid state (Krugh *et al.,* 1975; Krugh and Reinhardt, 1975; Tsai *et al.,* 1975*a,b*). Actinomycin, on the other hand, has a strong preference for purine (3′–5′) pyrimidine sequences in similar studies with self-complementary dinucleotides and oligonucleotides containing guanine and cytosine residues (Krugh, 1972; Patel, 1976) as well as with mixtures of complementary dinucleotides in which at least one purine residue is guanine (Krugh, 1974; Chiao and Krugh, 1977). These sequence binding preferences are at least partly maintained when these drugs bind to DNA, as evidenced by early studies of noncompetitive binding of ethidium to DNA in the presence of actinomycin (LePecq and Paoletti, 1967). Reinhardt (1976) has recently investigated the binding of ethidium to DNA in the presence of actinomycin and, also, actinomine, an analog of actinomycin that lacks the pentapeptide chains. His studies (shown as fluorescence Scatchard plots in Figs. 10A and 10B) clearly indicate that, whereas actinomycin binds noncooperatively with ethidium for sites on DNA, actinomine *competes* with ethidium for these sites. These data, as well as additional data (see Sobell, 1973), indicate that the pentapeptide chains on actinomycin play a key role in determining its sequence-dependent guanine specificity in the DNA-binding reaction.* The twofold symmetry that relates the pentapeptide chains on actinomycin approximately coincides with the twofold axis of DNA that relates the sugar-phosphate chains and the d-GpC base sequence. The pattern of recognition in this miniature protein–DNA interaction and the known mode of action of this

*The sequence binding preferences demonstrated by ethidium most likely reflect the relative energies required to unstack pyrimidine–purine versus purine–pyrimidine sequences in the double helix, rather than differences in stacking energies between ethidium and base pairs in these sequences (see Pack and Loew, 1977; Young and Krugh, 1975). However, the guanine specificity and the purine–pyrimidine sequence specificity demonstrated by actinomycin in binding DNA most probably reflect hydrogen bonding between the 2-amino group of guanine and the carbonyl oxygen of the L-threonine residues on the pentapeptide chains (see Sobell, 1973).

antibiotic (that is, to inhibit passage of the RNA polymerase enzyme along DNA through direct stereochemical blockage) have suggested this to be a model system to understand the larger naturally occurring operator–repressor interactions (Sobell *et al.*, 1971). This is discussed further in Section 8.

4.3 *Irehdiamine*

Irehdiamine A is a steroidal diamine that binds tightly to DNA (Mahler *et al.*, 1968). Although the details of the interaction are unknown, many aspects of the irehdiamine–DNA binding reaction suggest an intercalative binding mode. Thus, for example, irehdiamine A unwinds superhelical DNA, as evidenced by its ability to produce a fall and rise in the sedimentation coefficient of covalently circular supercoiled DNA molecules (Waring, 1970; Waring and Chisholm, 1972; Waring and Henley, 1975). These studies have also indicated that, on a molar basis, irehdiamine A unwinds DNA roughly half as much as does ethidium. Moreover, at maximal drug–nucleic acid binding ratios, the molecule demonstrates an upper binding limit of one irehdiamine to every four nucleotides (Mahler *et al.*, 1968). Finally, although irehdiamine produces different hydrodynamic effects in DNA solutions than most intercalative drugs or dyes (that is, there are no clear viscosity or sedimentation changes associated with irehdiamine–DNA binding), it is possible that irehdiamine mimics intercalative drug binding by *binding to the kink in DNA* (Sobell *et al.*, 1977*a,b*). This is shown in Fig. 9. The van der Waals surfaces of the irehdiamine molecule resemble a triangular wedge whose overall shape allows it to fit into the kink in DNA. Forces stabilizing the complex include electrostatic, van der Waals, and hydrophobic forces. Our model predicts an effective unwinding of $-10°$ associated with irehdiamine–DNA binding. It also predicts a neighbor exclusion structure at saturating concentrations of irehdiamine whose axial repeat

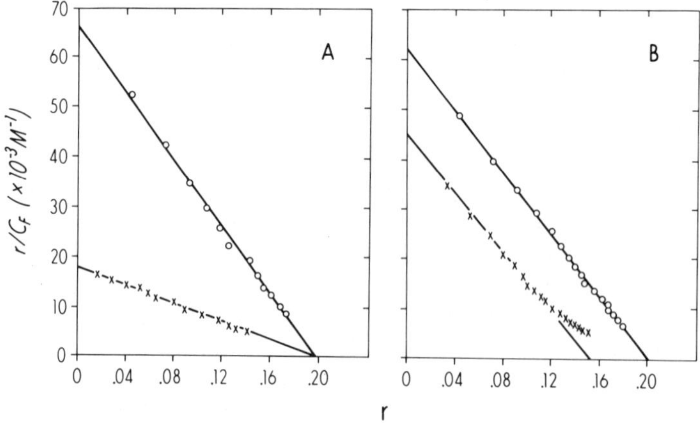

Figure 10. Fluorescence Scatchard plots of ethidium bromide binding to DNA in the presence of actinomine and actinomycin. (A) Competition of actinomine with ethidium bromide for binding DNA; and (B) competition of actinomycin D with ethidium for binding DNA. Ethidium concentrations range from 1.0 to 28.3 μM, DNA concentration 3.5 M, 50 mM Tris, pH 7.5, 0.2 M NaCl. [DNA-P]/[actinomine] = [DNA-P]/[actinomycin D] = 5.0.

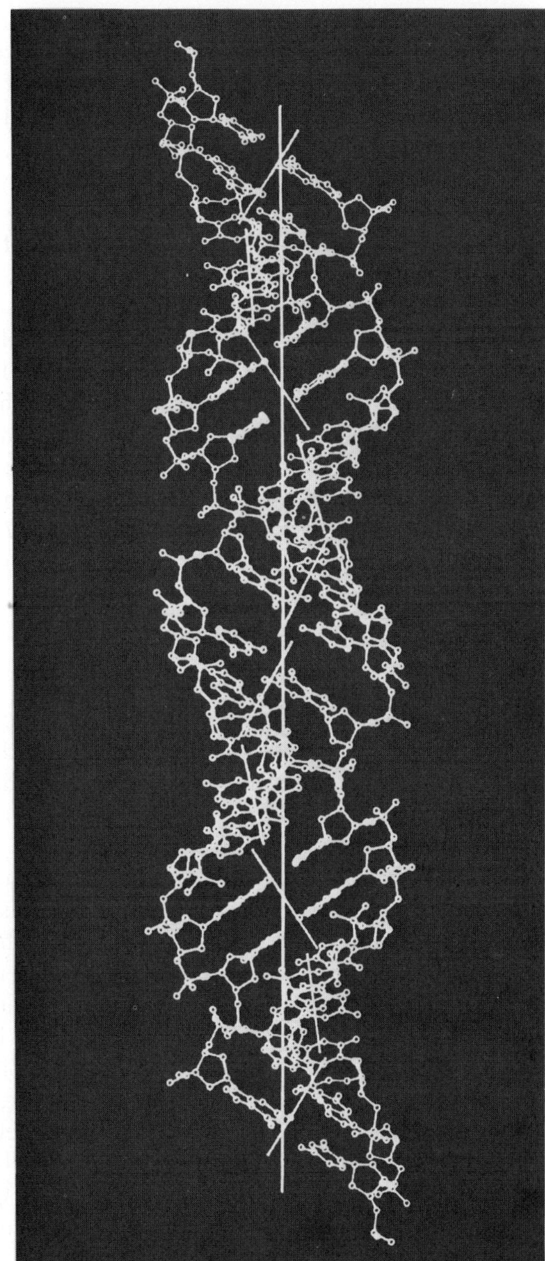

Figure 11. Computer-graphic-drawn β-kinked DNA structure. This structure is a right-handed superhelix with a diameter of about 20 Å and a pitch of 34 Å, and contains 9.6 base pairs per turn. Irehdiamine A could induce the formation of this structure at maximal drug/DNA binding ratios (that is, this would correspond to a neighbor exclusion structure for irehdiamine–DNA binding). Localized domains of this structure could exist immediately prior to thermal DNA denaturation, exposing base pairs to interactions with water molecules. A structure such as this could be recognized by the RNA polymerase enzyme when binding to (and denaturing) the promoter. See text for additional discussion.

length is similar to that of B DNA. This structure (denoted β-kinked B DNA) is shown in Fig. 11.

Pohl and Jovin (1972) have described an extremely interesting salt-induced cooperative conformational change in poly(dG-dC) that is accompanied by large changes in the circular dichroism and optical dispersion spectra of this polymer. Although the molecular nature of the transition is unknown, thermodynamic measurements have indicated that the conversion from the low salt form (denoted

R) to the high salt form (denoted L) is almost entirely entropy-driven, while kinetic studies have shown that the process is first order, requiring many minutes for completion, and having an activation energy of about 22 kcal/mol. An important finding concerns the nature of ethidium bromide binding to these two different conformational states of the polymer (Pohl *et al.,* 1972). At low salt, ethidium binds to the R form without detectable cooperativity, whereas at high salt, ethidium binds to the L form with *extremely high cooperativity.* It is of interest that proflavin and quinacrine (drugs that may intercalate from the wide groove) show no comparable effect.

We have wondered whether the L form of poly(dG-dC) is, in fact, the β-kinked DNA structure. Such a structure would be expected to bind ethidium with high cooperativity to promote the formation of expanding domains of neighbor exclusion structure (shown in Fig. 8). We have therefore proposed this mechanism to explain the allosteric transition that ethidium induces in poly(dG-dC) and, by extension, to understand an important feature of several protein–DNA interactions (for example, the mechanism that gives rise to RNA polymerase–promoter tight binding; see Section 7) (Sobell *et al.,* 1977*a,b*).

5 *Nature of DNA Breathing*

If DNA kinks spontaneously to form a flexibly jointed (or hinged) structure and if drugs can utilize this flexibility by intercalating into DNA, what other molecules could enter into the interior of the double helix by virtue of this flexibility?

It has been known for many years that DNA undergoes conformational fluctuations in its structure that allow deuterium and tritium to exchange with protons normally involved in Watson–Crick base pairing (Printz and von Hippel, 1965; Teitelbaum and Englander, 1975*a,b*). The ability of DNA to "breathe" has also been studied with formaldehyde, a reagent that reacts with both exocyclic amino and endocyclic imino groups on the purine and pyrimidine bases (McGhee and von Hippel, 1975*a,b*). These observations have suggested that (at any given time) a small number of base pairs in the double helix are transiently denatured, even at temperatures well below the thermal transition temperature. This would allow bases to exchange their interior hydrogens from endocyclic imino groups with solvent, and would enhance the reactivity of these groups to chemical attack by formaldehyde.

We have recently proposed a structural mechanism that interrelates DNA breathing with drug intercalation (Sobell *et al.,* 1977*a,b;* Lozansky *et al.,* 1978*b*); this is shown schematically in Fig. 12. We envision DNA to be a dynamic structure in solution, continuously undergoing coupled motions that involve bending, stretching, unwinding, and shearing components excited through Brownian motion of solvent molecules at normal thermal energies. To understand the origin of DNA kinking, it is important to discuss the formation of a multiple-kinked DNA structure, β-kinked DNA (i.e., DNA kinked every other base pair). We postulate this structure to form as the result of structural distortions in DNA associated with travelling wave forms (these forms could resemble shock waves that act to unwind the helix and to promote right-handed superhelical writhe, maintaining a constant linkage). In other words, if the energy of a wave exceeds some elastic limit (that is, due to an occasional collision by a "hot" solvent molecule from the tail of the Maxwell–Boltzmann distribution and/or due to constructive interference between elastic traveling waves), DNA will crack (or kink) to form short regions of β-kinked DNA. Once formed, β-kinked DNA can undergo further elastic deformations to form the neighbor exclusion intercalative structure. This structure is the maximally unwound form possible for double-stranded DNA. Further unwinding results in weakening, and, eventually, in disruption of base pairing. We envision a combination of these activated DNA forms to give rise to DNA breathing and to drug intercalation.

Our concepts of drug–DNA binding now take on a more general meaning with regard to the dynamic nature of DNA structure: *drugs such as ethidium and irehdiamine act to immobilize (or freeze out) specific structural intermediates in DNA unwinding that give rise to DNA breathing.* The subject of drug intercalation is intimately related to a wide variety of protein–DNA interactions that require DNA flexibility, and we will now describe these.

6 *Organization of DNA in Chromatin*

Conformational flexibility in DNA structure is an important concept in understanding the organization of DNA in chromatin (Crick and Klug, 1975). We have recently described a particularly interesting superhelical DNA structure obtained by kinking DNA every ten base pairs, a structure we have called κ-kinked

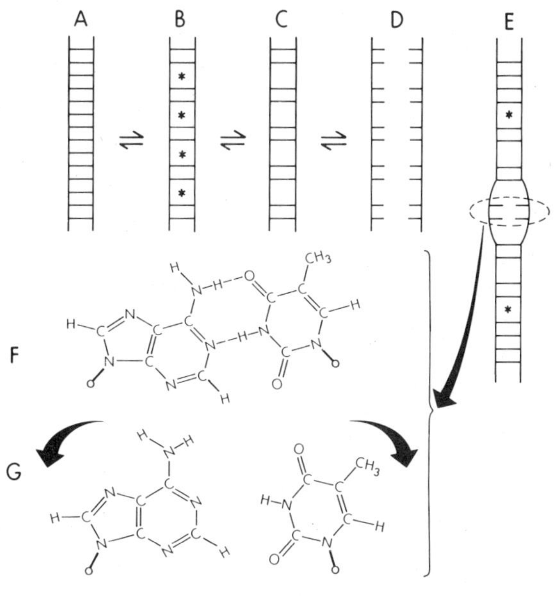

Figure 12. Schematic illustration summarizing structural concepts of DNA breathing and DNA melting. The conformational intermediates that are shown are highly schematic and it is suggested that CPK models be utilized to investigate their structures. (A) B DNA structure, gently unwound to possess a slow right-handed superhelical writhe; (B) β-kinked B DNA structure, further unwound to possess greater right-handed superhelical writhe; (C) neighbor exclusion intercalative structure, still further unwound and now possessing less superhelical writhe; (D) denatured DNA structure that is completely unwound (notice bases stacked in groups of two); (E) figure to show conformations A–D existing as a continuum within a given region of DNA undergoing DNA breathing; (F) Watson-Crick base pairing between adenine and thymine in the "closed" state; and (G) transient disruption of adenine–thymine Watson–Crick base pairing by torsional motions just described, leading to the "open" state. This structure could give rise to DNA breathing. See text for additional discussion.

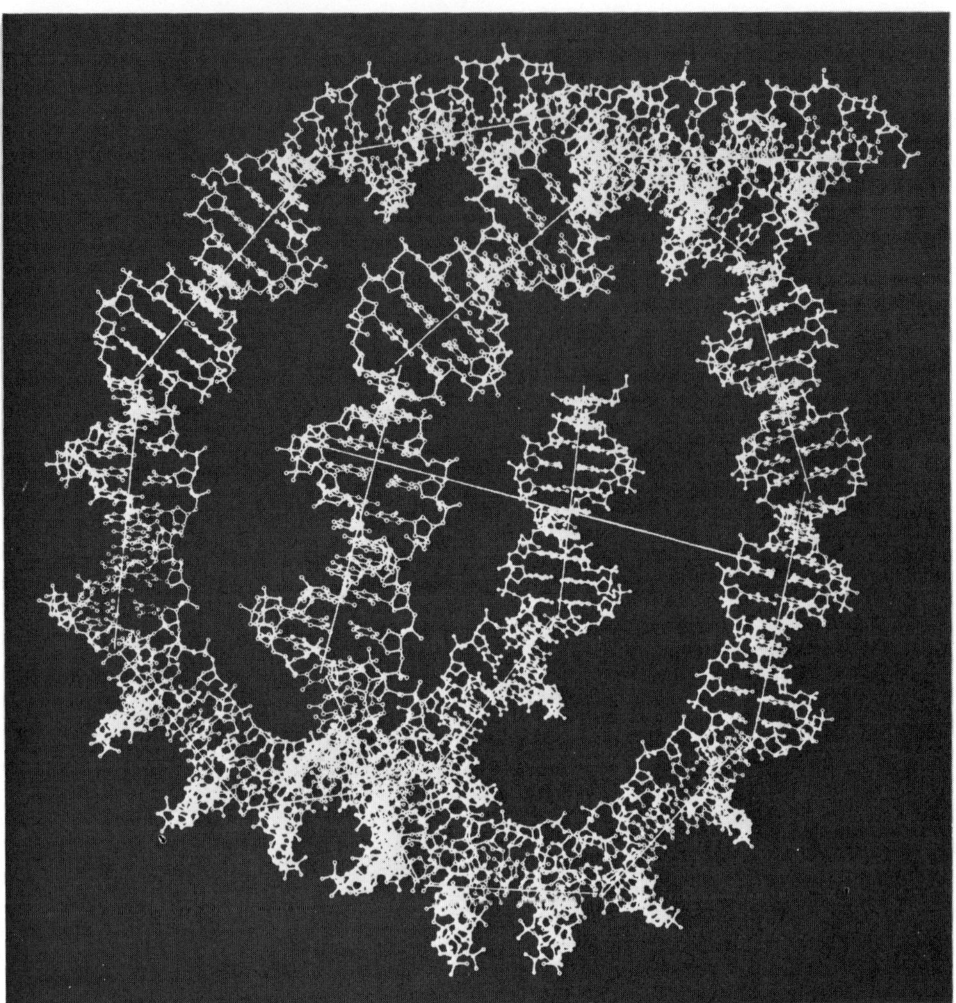

Figure 13. Perspective illustration of κ-kinked B DNA drawn by computer graphics. The structure is a left-handed kinked superhelix with a diameter of about 100 Å and contains about 1.5 turns per 140 base pairs. The long central line indicates the superhelix axis. This basic structure may be utilized in the organization of DNA within the nucleosome in chromatin.

B DNA (Sobell *et al.*, 1976, 1977*a*,*b*, 1978*a*). This structure (shown in Fig. 13) is a left-handed superhelix with a diameter of approximately 100 Å. Each residue of the helix contains 10 base pairs; the helix is generated from this residue by a twist of $-41.1°$ and a translation of 5.26Å. This gives rise to a structure that contains about 1½ turns per 140 base pairs and has an axial length of about 80 Å.* The dimensions of this structure are in reasonable agreement with current estimates of

*Although the pitch of our helix is 47 Å, this value is expected to be extremely sensitive to both the magnitude of angular unwinding at the kink and the presence of exactly B DNA between the kinks. We estimate, for example, that if B DNA has 9.95 (rather than 10.0) base pairs per turn, this would decrease the pitch of the κ helix to about 35 Å. This value is in good agreement with preliminary estimates of the pitch from X-ray crystallographic studies of nucleosomes (Finch and Klug, 1977).

the size of nucleosomes (Olins and Olins, 1974). Moreover, its twofold symmetry could allow it to interact with the octameric substructure of chromatin (Thomas and Kornberg, 1975) along a common twofold axis. We have therefore proposed this structure as a model to understand the organization of DNA within the nucleosome in chromatin.

Finch and Klug (1976) have recently presented evidence for a higher order solenoidal organization of nucleosomes in chromatin. This solenoidal structure may correspond to the "thick" 250- to 300-Å fiber observed in metaphase chromosomes and requires the presence of H-1 histones and trace amounts of divalent cations for its formation. Worcel (1978) and Worcel and Benyajati (1977) have proposed a detailed model to explain this higher order coiling of DNA in chromatin. Their model involves a distortion of the 60-base-pair internucleosomal DNA region (most easily described as a continuous toroidal bending of our κ helix) to give rise to a higher order superhelix with a pitch of 110 Å. Important postulates of their model are that this 60-base-pair connecting region lies on the outside of the solenoid (the 140-base-paired superhelical DNA region within each nucleosome is not distorted in their model) and that twofold symmetry be retained in the overall structure.

We have calculated a variety of different solenoids obtained by distorting our basic κ helix either continuously or discontinuously throughout the 200-base-pair repeat (Sobell *et al.*, 1977*b*). Three classes of models can be distinguished to describe the path of DNA through the solenoid (refer to Fig. 14).

The first class, shown in Fig. 14A, is composed of solenoids obtained by (semi-) continuously deforming our κ helix (best described as a continuous toroidal bending of the κ helix) to an outer diameter of about 280 Å by adjusting dihedral angles between planes subtended by B DNA helix axes θ, keeping the kink angle α constant (Crick and Klug, 1975). Although it is possible to distort a 9.8- or 9.9-fold κ helix in this way to give a left-handed superhelix with a pitch of 110 Å and a diameter of 280 Å, it is not possible to form an acceptable structure with an 8.8- or 9.0-fold κ helix (these structures have pitches in excess of 500 Å when constrained to the same diameter). Furthermore, numerous close contacts (below 20 Å) are observed between DNA molecules in the center of these structures (that is, the so-called wedge effect) and this makes this class of models seem less likely.

The second and third classes of models (shown in Figs. 14B and 14C) are obtained by limiting the distortion of the κ helix to the 60-base-pair connecting region between nucleosomes. Distortion of this type is accomplished by varying *both* θ and α, and is most easily described as a *helical screw axis dislocation* in the basic κ helix. These two classes of models are distinguished by the connecting DNA being on the inside or the outside of the solenoid. When the connecting DNA is on the inside of the solenoid, the solenoid is most naturally *right-handed* (although it is possible to form left-handed variants if one unwinds DNA between the kinks sufficiently). The diameters of these structures, however, are between 350 and 400 Å and, for this reason, we consider these to be less likely. When the connecting DNA is on the outside of the solenoid, the solenoid is most naturally *left-handed* (although, again, it is possible to convert these structures to right-handedness by winding DNA between the kinks sufficiently). Structures of this class can readily be formed with the right pitch (110 Å) and the right diameter (150–300 Å) and pose few, if any, stereochemical problems. We therefore propose this class of models as

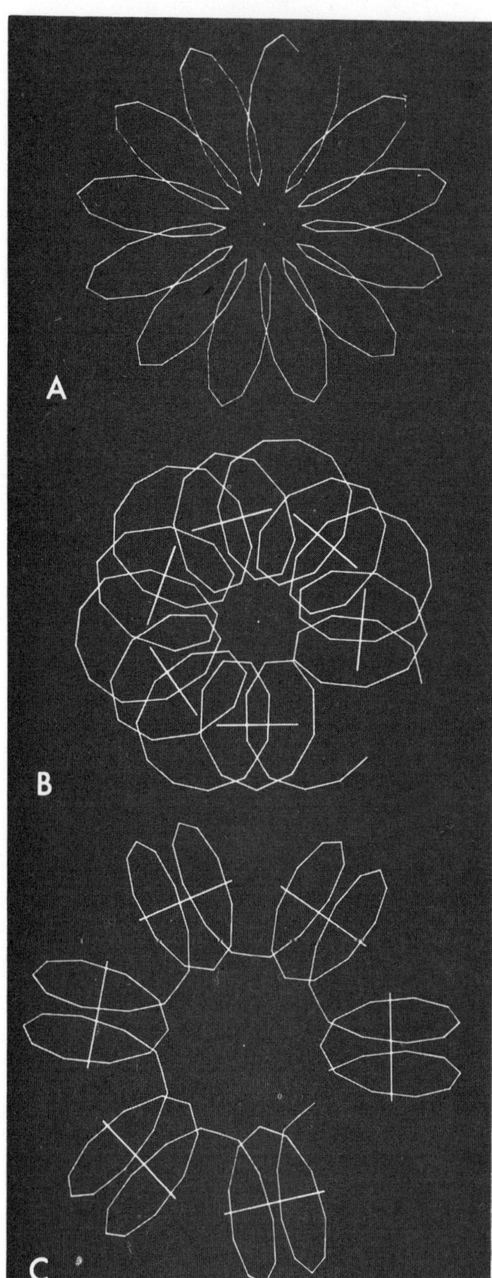

Figure 14. Three classes of solenoid models obtained by distorting the basic κ helix either continuously or discontinuously throughout the 200-base-pair repeat. (A) (Semi-)continuous distortion of the κ helix to give a left-handed superhelix with a pitch of 110 Å and a diameter of 280 Å. This can be achieved only by distorting a 9.8- or 9.9-fold κ helix; 8.8- or 9.0-fold κ helices have pitches in excess of 500 Å when constrained to the same diameter. (B) discontinuous distortion of the κ helix by a helical screw axis dislocation accomplished by varying θ and α in the 60-base-pair connecting region. The connecting region lies on the outside of the solenoid. Solenoids in this class form left-handed superhelices with the right pitch (110 Å) and the right diameter (250–300 Å), using the 8.8-fold κ helix, and pose few, if any, stereochemical problems. We consider this class of models most likely in describing the path of DNA within the Finch–Klug solenoid. (C) Same as B, except that the connecting region lies on the inside of the solenoid. Solenoids in this class form right-handed superhelices with the right pitch (110 Å); however, their diameters are too large (350–400 Å) to fit the Finch–Klug solenoid structure.

most likely to describe the path of DNA within the Finch–Klug type of solenoidal structure.

Figure 15 shows one such (5.8-fold) solenoid in which the connecting DNA between nucleosomes lies on the outside and twofold symmetry is maintained. An important prediction that can be made from models of this general class concerns the ability of intercalative drugs and dyes to bind tightly to their internucleosomal DNA regions. This follows from the molecular nature of the helical screw axis dislocation that has been postulated every 200 base pairs in the κ helix: This

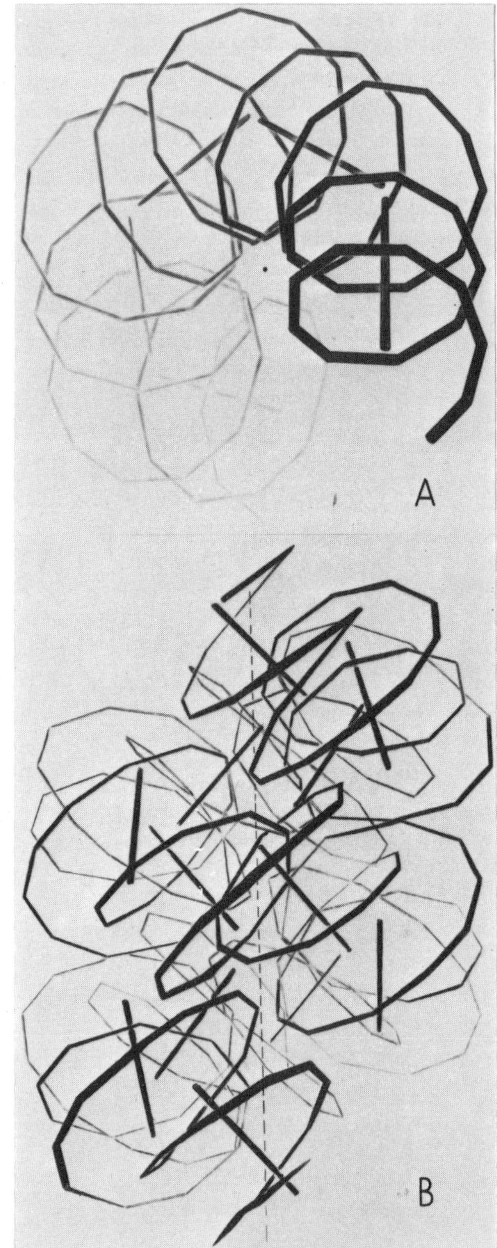

Figure 15. A 5.8-fold solenoid obtained by a discontinuous distortion of the κ helix by a helical screw axis dislocation accomplished by varying θ and α in the 60-base-pair connecting region. The connecting region lies on the outside of the solenoid. An important prediction that this class of models makes concerns the ability of these structures to bind intercalative drugs and dyes rightly in the internucleosomal DNA regions. This follows from the molecular nature of the helical screw axis dislocation postulated for the κ helix; this involves opening three kinks (each to a different extent) to intercalative type of geometries in the connecting region. The most central of these most nearly corresponds to the geometry of DNA when it undergoes drug intercalation. This leads us to speculate that, whereas pancreatic DNase recognizes the kink in chromatin and cleaves every ten base pairs, micrococcal nuclease recognizes (and cleaves) the intercalative geometry present every 200 base pairs. See text for additional discussion.

involves opening three kinks (each to a different extent) to geometries of the intercalative type in the connecting region. The geometry of the most central of these corresponds closely to that of DNA when it undergoes drug intercalation. This finding leads us to speculate that, whereas the pancreatic DNase recognizes the kink in chromatin and cleaves every 10 base pairs, micrococcal nuclease recognizes (and cleaves) the intercalative geometry that is present every 200 base pairs.

Both pancreatic DNase and micrococcal nuclease may, therefore, have features in common when binding to DNA (that is, pancreatic DNase could partially intercalate, while micrococcal nuclease could completely intercalate aromatic side chains into DNA). Their interactions with DNA may be directly pertinent to the whole subject of drug intercalation into DNA (see Section 4).

Lawrence and Duane (1976) have recently described the existence of a limited number of binding sites for ethidium in native chromatin with binding constants two orders of magnitude greater than in naked DNA. These sites disappear when H-1 histones are removed, suggesting a correlation with the solenoid described here. Of further interest, Hearst (1977) describes psoralin-sensitive sites in internucleosomal DNA regions. Our model would predict that these sites occur at integral multiples of ten base pairs and are highly clustered in the 60-base-pair connecting region.

7 *Active Form of DNA in Transcription, Replication, and Recombination*

The concept that DNA has a dynamic structure in solution and that drug intercalation is a phenomenon that reflects this dynamic structure leads to a structural hypothesis concerning the form of DNA that is active in RNA transcription, DNA replication, and genetic recombination.

Transcription, replication, and recombination all involve unwinding and melting of DNA. The question is, how does DNA unwind? From previous discussion (see Section 5), a key intermediate in DNA breathing (a process that necessitates DNA unwinding and local melting) is the multiple-kinked DNA structure, β-kinked DNA. One can understand the B DNA $\rightarrow$ β-kinked DNA structural transition most easily through examination of space filling Corey–Pauling–Koltun (CPK) molecular models of DNA. As one unwinds the double helix, two things tend to happen: base pairs unstack and strain energy accumulates in the two sugar–phosphate chains. Both are energetically unfavorable and can be minimized if one allows DNA to assume a right-handed superhelical writhe [part of the twist in the helix, T, is converted into superhelical writhe, W, to maintain a constant linkage, L, according to the Fuller (1971) relation, $W = L - T$]. The effect of introducing further unwinding causes DNA to attain still greater superhelical writhe; eventually, however, one finally overcomes the elastic properties of DNA and DNA kinks (or "cracks") to form localized regions of β-kinked DNA structure. This structure has right-handed superhelical writhe of sufficient magnitude to offset the unwinding localized at the kinks, and this allows it to maintain a linkage that is similar to B DNA (Sobell *et al.*, 1977*a,b*). Introducing still further unwinding converts this structure to the neighbor-exclusion intercalative structure. This structure is the maximally unwound form possible for double-helical DNA; further unwinding results in weakening, and, eventually, in the disruption of base pairing in DNA. *We therefore postulate these conformational changes to accompany DNA unwinding and denaturation, and speculate that β-kinked DNA plays a central role in transcription, replication, and recombination, processes that all involve unwinding the double helix.*

Energy is necessary to unstack base pairs when converting the β-kinked DNA structure to the neighbor exclusion-intercalative structure. For DNA breathing (and for heat-induced DNA denaturation) this energy is supplied by heat. Proteins could capitalize on intercalation between bases by aromatic amino acids during solvent denaturation. This could reduce the activation energy necessary for base unstacking and hydrogen bond breakage, and would enhance the binding affinities these proteins have to single-stranded DNA. For this reason, we have postulated that proteins such as the *Escherichia coli* RNA polymerase recognize the presence of β-kinked DNA structure (that is, at the promoter) and then intercalate aromatic side chains into DNA to catalyze the process of DNA denaturation.

Coleman *et al.* (1976) have studied the interactions between a variety of single-stranded oligonucleotides of defined sequence and the fd gene 5 protein, using high-resolution nuclear magnetic resonance. Pronounced upfield chemical shifts are observed for protons on three tyrosine residues in the presence of these oligonucleotides; similar chemical shifts are observed with protons on the purine and pyrimidine rings. These data are consistent with stacking interactions between tyrosine residues and the purine and pyrimidine bases, most likely by a mechanism of the intercalative type. Recent success in forming a crystalline complex between the gene 5 protein and a variety of single-stranded oligonucleotides of defined sequence heralds an understanding of this protein–DNA interaction at atomic resolution by X-ray crystallography (A. Rich, personal communication). Our model predicts that tyrosine residues intercalate and also predicts the presence of an alternating pattern of sugar puckering in these single-stranded oligonucleotides. These structural predictions are direct consequences of our models for DNA breathing and DNA denaturation, and can be readily tested through these X-ray crystallographic studies.

The mechanism of recognition and attachment by the RNA polymerase enzyme to the promoter is a subject that has attracted wide interest in recent years (for reviews, see Travers, 1974; Sobell, 1976) (see Chapter 7 of this volume). Several lines of evidence point to the possibility that recognition, entry, tight binding, and, finally, messenger RNA initiation are all different elements of the promoter–polymerase interaction. It is possible that recognition and entry may occur outside the tight binding site for RNA polymerase attachment. Evidence for this idea has been the discovery of promoter mutations that fall outside the region protected from nuclease attack by the tight binding of RNA polymerase (Dickson *et al.,* 1975; Blattner and Dahlberg, 1972; Walz and Pirrotta, 1975). Comparison between the *lac* and λ promoter regions has revealed two features that may be common to both: the presence of one or more AAAT sequence(s), and regions of very high guanine–cytosine content flanking a central region of very high adenine–thymine content. These sequence features lie outside the tight binding site, and it is possible that DNA must first "breathe" in order for RNA polymerase to attach to it. Presumably, this would occur at A–T-rich regions and the ease of "breathing" may reflect the binding of other proteins (such as the CAP-binding protein) in the immediate neighborhood. Model studies with a duplex block oligonucleotide, d($C_{15}A_{15}$)d($T_{15}G_{15}$), have demonstrated the influence of sequence-specific drugs (actinomycin, GC; netropsin, AT) on thermal denaturation and nuclease susceptibility of DNA regions remote from the drug-binding site (Burd *et al.,* 1975). Studies such as these have suggested the possibility that protein binding at one location of DNA can transmit thermal stability (or instability) to other regions of DNA (a phenomenon termed telestability). In terms of our concepts of DNA kinking, telestability could refer to an altered probability for DNA kinking distal to some protein attachment site. This could arise from constructive interference between incident and reflected waveforms that propagate along DNA and experience a barrier (this could be either a drug–DNA complex or a protein–DNA complex). Related concepts have been described in detail elsewhere (Lozansky *et al.,* 1978; Sobell *et al.,* 1978*b*).

Is there a nucleotide sequence common to all promoters that is used to initiate messenger RNA synthesis? Pribnow (1975) has advanced the hypothesis that there exists such a sequence homology between all promoters studied thus far, and that it exists -7 (or -6) to -13 (or -12) base pairs to the left of messenger RNA initiation. The possible sequence homology between promoters is

5′ T A T Pu A T G 3′
3′ A T A Py T A C 5′

It is possible that this sequence plays a key role in determining the tight binding of RNA polymerase to these promoters and in fixing the initiation point for messenger RNA synthesis. Further sequence studies of other promoters will be of particular interest in this regard (see Chapter 7).

8 *Operator–Repressor Interactions*

A wealth of information has become available in recent years about nucleotide sequences involved in positive and negative control of RNA transcription. This reflects the growing arsenal of site-specific restriction endonucleases available for cleaving DNA into defined oligonucleotide segments and important recent advances in chemical methods for nucleotide sequencing. This discussion, however, will limit itself to studies of the *lac* operator and its interactions with the *lac* repressor. These studies have indicated the utilization of (an approximate) twofold axis of symmetry in this protein–DNA interaction and have suggested, in addition, the possibility that DNA kinking or intercalation accompanies the binding reaction. We will first review evidence indicating pseudosymmetry in the *lac* operator–repressor interaction. We will then discuss the question of whether DNA kinking accompanies the binding reaction.

The *lac* operator region has been delineated by three basic techniques. The first, based on repressor protection of a double-helical DNA fragment against pancreatic deoxyribonuclease digestion followed by transcription of this fragment into RNA using RNA oligonucleotide primers (Gilbert and Maxam, 1973), has yielded a nucleotide sequence 24 base pairs long with the following sequence:

5′ T G G A A T T G T G A G C G G A T A A C A A T T 3′
3′ A C C T T A A C A C T C G C C T A T T G T T A A 5′

Of these 24 base pairs, a total of 16 base pairs fits a twofold symmetric pattern, and these have been boxed. The presence of symmetry in this sequence immediately suggests the possible utilization of symmetry in *lac* repressor–operator interaction. The *lac* repressor is known to be a tetrameric protein, molecular weight 150,000, containing four identical subunits that are probably related by approximate 222 symmetry (Steitz *et al.,* 1974). One can visualize repressor binding to operator using a common twofold axis of symmetry in which either two or four subunits are involved in the interaction; however, it is clear that the sequence information by itself is not adequate to define precisely the nature of the repressor–operator interaction. It is important, in addition, to establish which bases in this sequence are involved in protein–nucleic acid recognition. Two other techniques have given information concerning this.

One of these has been to determine the change in nucleotide sequence associated with a specific operator constitutive mutation and to note whether or not it falls within a symmetric region of the *lac* operator structure (Gilbert *et al.*, 1975). Another method has been to react DNA with dimethyl sulfate in the presence and absence of repressor, and to note which guanine or adenine bases are altered in their chemical reactivity under these conditions (Gilbert and Maxam, 1976). The results of these studies are summarized in the accompanying display:

```
                               -     - + +         -           -
5'  T G T G T G G A A T T G T G A G C G G A T A A C A A T T T C A C A C A  3'
3'  A C A C A C C T T A A C A C T C G C C T A T T G T T A A A G T G T G T  5'
                     - -             +             -

                          ↓   ↓ ↓ ↓ ↓ ↓     ↓     ↓
     oc mutations         A   T G T T A     C     T             operator
                          T   A C A A T     G     A

  homology           A A      G T G A G C G     T A A C A A          C     A
                     T T      C A C T C G C     A T T G T T          G     T

                               -     - + +       -
5'  C A A C A T T A A A T G T G A G C G A G T A A C A A C C C G T C G G A  3'
3'  G T T G T A A T T T A C A C T C G C T C A T T G T T G G G C A G C C T  5'
                                                   -
                                                              pseudooperator
```

Exactly symmetric regions have been boxed, while vertical arrows show position of mutational changes; + denotes methylation enhanced by repressor, – denotes methylation blocked by repressor.

It is seen that operator constitutive mutations change base pairs that fall both within and outside of symmetric sequences. Most alterations are to the left of the symmetry axis; this indicates an element of asymmetry in operator–repressor interaction, a fact further confirmed by observing the pattern of chemical reactivity in the presence of repressor in the dimethyl sulfate methylation studies. Two operator constitutive mutations act to increase symmetry in the sequence (these fall to the left of the symmetry axis and are transversions of the type G–C → T–A). These same G–C base pairs are observed to have altered chemical reactivities in the methylation studies. These data, taken together, strongly suggest a key role for these G–C base pairs (which fall outside a symmetry region) in repressor–operator binding. Similar evidence suggests that the central G–C base pair falling on the symmetry axis is involved in repressor binding.

An important additional piece of information has been provided by the detection and nucleotide sequencing of a region of DNA that falls within the *lacZ* structural gene and binds repressor. This DNA region has no known physiological function; however, since it binds to repressor with high affinity (the binding constant is estimated to be only a few orders of magnitude smaller than that of the real operator–repressor interaction), it has been termed the *pseudooperator.* The sequence of the pseudooperator (see page 193) is important since it has provided direct information about sequence homologies with the real (or primary) operator and, by inference, has assisted in understanding which nucleotide bases are important for repressor binding. Inspection of this homology confirms the importance of the central nucleotide sequences located immediately to the left of (and including) the axis of symmetry in the sequence.

In addition to these nucleotide sequence studies, Gilbert and Maxam have used the organic chemical probe, ethyl nitrosourea, to study the pattern of altered reactivity of phosphate groups in the operator region in the presence of repressor (A. Maxam and W. Gilbert, personal communication). The results of these studies, shown in the accompanying display, indicate that the repressor binds to DNA from the side and aligns itself along the pseudotwofold axis of the operator

```
         * *             *
5′ pApApTpTpGpTpGpApGpCpGpGpApTpApApCpApApTpT  3′
3′  TpTpApApCpApCpTpCpGpCpCpTpApTpTpGpTpTpApAp 5′
                   * *               * *
```

(asterisks indicate those phosphate groups whose chemical reactivity with ethyl nitrosourea decreases in the presence of repressor).

How do these observations relate to understanding the role of symmetry in this protein–DNA interaction? The repressor is a tetrameric protein molecule that possesses pseudo-222 symmetry. One twofold axis in the molecule is used in the protein–DNA interaction. This dyad need not be an exact one (it is commonly found in protein crystallography that proteins are related by approximate noncrystallographic symmetry, departures from symmetry being especially prevalent near an axis of symmetry, due to close stereochemical contacts between subunits or between neighboring molecules within the unit cell). This being the case, the binding sites need not have identical stereochemistry and DNA-binding affinity. (Note that, for this argument, it is not important to know whether the binding site is composed of one protein subunit or if it is formed along an interface between two adjoining subunits. What is important is the prediction this explanation affords—that departures from symmetry should occur most frequently near the axis of symmetry in the operator nucleotide base sequence.) With this in mind, the data are consistent with the following pattern of protein–DNA interaction:

```
5′ [A A T T G T G A G C] G [G] A [T] A [A C A A T T] 3′
3′ [T T A A C A C T C G] C [C] T [A] T [T G T T A A] 5′
```

This predicts that tightest binding occurs to the left of the symmetry axis. Although not boxed, the central G–C base pair may be involved in protein–nucleic acid contacts with both protein-binding sites. The departure from symmetry in the interaction could reflect the departure from symmetry in the three-dimensional structure of the *lac* repressor, and a detailed understanding of this protein–DNA interaction must await an X-ray crystallographic examination of the *lac* repressor in future years.

Does DNA kinking accompany the *lac* operator–repressor interaction? Some preliminary evidence suggests this to be the case. Chan *et al.* (1977) have described the presence of mung bean nuclease-sensitive sites in the *lac* operator. Once cleaved (to give single-strand breaks), the repressor no longer binds operator. This suggests some conformational heterogeneity in the DNA (such as a kink) that is recognized by this nuclease and is important in *lac* operator–repressor binding. Evidence for partial intercalation by tyrosine residues has been provided by Fanning (1975), who has iodinated repressor with KI_3 to modify three tyrosine residues in its structure. This iodinated repressor appears to retain properties similar to those of the native molecule (that is, it is still a tetramer, molecular weight 150,000, and binds IPTG, isopropylthiogalactoside, efficiently); however, it has lost its affinity for operator. Control experiments, in which repressor is iodinated in the presence of operator DNA (as well as with nonoperator DNA), indicate that operator–repressor complexing protects tyrosine residues against iodination and against inactivation. These experiments suggest (but do not prove) the involvement of tyrosine residues in *lac* operator–repressor binding. It is possible that these tyrosine residues intercalate (either partly or completely) into DNA. Evidence for partial intercalation by small oligopeptides containing tyrosine, phenylalanine, and tryptophan has, in addition, been provided by Gabbay and his colleagues (Gabbay *et al.*, 1976*a,b*) and by Helene and his colleagues (Brun *et al.*, 1975; Durand *et al.*, 1975). It is possible that these model peptide–DNA interactions relate to understanding the *lac* repressor–operator interaction and to other protein–DNA interactions as well.

9 *Concluding Remarks*

This chapter has reviewed concepts of symmetry and flexibility in DNA, two aspects of DNA structure that play important roles in protein–DNA interactions. An important feature of DNA is its dynamic structure in solution (this is due to the presence of normal modes in the polymer excited at thermal energies). This results in the kinetic phenomena of drug intercalation and DNA breathing. Structural studies of drug intercalation have revealed a conformational change in the DNA sugar-phosphate chains of the type C3′ *endo* (3′–5′) C2′ *endo*. This conformational change is best explained by assuming that DNA kinks spontaneously in solution prior to drug intercalation. Once formed, the kink can be likened to a flexible hinge in DNA—this hinge can straighten and base pairs separate to provide the space for drug intercalation. The concept of the kink also leads to an understanding of the molecular nature of DNA breathing. A key intermediate in our mechanism to explain DNA breathing is the multiply kinked DNA structure, β-kinked DNA. This structure is one of a series of structural intermediates that lead to DNA melting. We therefore postulate this structure to be the active form for DNA during RNA transcription, DNA replication, and genetic recombination, processes that all involve double-helix unwinding and localized regions of DNA denaturation. Proteins such as the *E. coli* RNA polymerase, gene 32 protein of bacteriophage T4, and gene 5 protein of bacteriophage fd may recognize the transient formation of β-kinked DNA structure and then intercalate aromatic amino acids between every other base in the process of protein-catalyzed DNA denaturation. In addition to these dynamic aspects of

DNA structure, the kink allows us to understand how DNA is organized within the nucleosome in chromatin. DNA that is kinked every ten base pairs (a structure denoted κ-kinked B DNA) forms a left-handed superhelix with a diameter of about 100 Å. This structure could interact with the octameric histone substructure of chromatin (the nucleosome) to give rise to the 100 Å fiber. Higher order coiling in chromatin can be achieved by opening the kink to the intercalative geometry every 200 base pairs and by allowing nucleosome disks to slip on each others' flat surfaces to give a left-handed solenoidal structure with a pitch of about 110 Å. This structure (or better, this class of structures) may occur in the 250-to 300-Å-thick chromatin fiber. These models for the organization of DNA in chromatin lead us to speculate that, whereas pancreatic DNase recognizes the kink in chromatin and cleaves every ten base pairs, micrococcal nuclease recognizes (and cleaves) the intercalative geometry that is present every 200 base pairs. Their interactions with DNA, therefore, may be directly related to the whole subject of drug intercalation into DNA.

Acknowledgments

This work has been supported by grants from the National Institutes of Health, the American Cancer Society, and the Office of Energy (OEA). This paper has been assigned report number UR-3490-1301 at the OEA, the University of Rochester.

References

Bigger, C. H., Murray, K., and Murray, N. E., 1973, Recognition sites in phage λ for a restriction endonuclease from *Excherichia coli* fi⁻ R factor, *Nature New Biol.* **244**:7.

Blattner, F. R., and Dahlberg, J. E., 1972, Distance from a promoter mutation to an RNA synthesis startpoint, *Nature New Biol.* **237**:232.

Brun, F. B., Toulme, J. J., and Helene, C., 1975, Interactions of aromatic residues of proteins with nucleic acids. Fluorescence studies of the binding of oligopeptides containing tryptophan and tyrosine residues to polynucleotides, *J. Biol. Chem.* **250**:5109.

Burd, J. F., Wartell, R. M., Dogson, J. B., and Wells, R. D., 1975, Transmission of stability (telestability) in deoxyribonucleic acid, *J. Biol. Chem.* **250**:5109.

Carter, C. W., and Kraut, J., 1974, A proposed model for interaction of polypeptides with RNA, *Proc. Natl. Acad. Sci. U.S.A.* **71**:283.

Chan, H. W., Dodgson, J. B., and Wells, R. D., 1977, Influence of DNA structure on the lactose operator–repressor interaction, *Biochemistry* **16**:2356.

Chiao, Y. C. C., and Krugh, T. R., 1977, Actinomycin D complexes with oligonucleotides as models for the binding of the drug to DNA. Paramagnetic-induced relaxation experiments on drug–nucleic acid complexes, *Biochemistry* **16**:747.

Coleman, J. E., Anderson, R. A., Ratcliffe, R. G., and Armitage, I. M., 1976, Structure of gene 5 protein–oligodeoxynucleotide complexes as determined by ^{1}H, ^{19}F and ^{31}P nuclear magnetic resonance, *Biochemistry* **15**:5419.

Crick, F. H. C., 1971, General model for the chromosomes of higher organisms, *Nature* **234**:25.

Crick, F. H. C., and Klug, A., 1975, Kinky helix, *Nature* **255**:531.

Crothers, D. M., 1968, Calculation of binding isotherms for heterogeneous polymers, *Biopolymers* **6**:575.

Dickson, R. C., Abelson, J., Barnes, W. M., and Reznikoff, W. S., 1975, Genetic regulation: The *lac* control region, *Science* **187**:27.

Durance, M., Maurizot, J. C., Borazan, H. N., and Helene, C., 1975, Interaction of aromatic residues of proteins with nucleic acids. Circular dichroism studies of the binding of oligopeptides to polyadenylic acid, *Biochemistry* **14**:563.

Fanning, J. G., 1975, Iodination of *Escherichia coli lac* repressor. Effect of tyrosine modification on repressor activity, *Biochemistry* **14**:2512.

Finch, J. T., and Klug, A., 1976, Solenoidal model for superstructure in chromatin, *Proc. Natl. Acad. Sci. U.S.A.* **73**:1897.

Finch, J. T., and Klug, A., 1977, X-ray and electron microscope analyses of crystals of nucleosome cores, *Cold Spring Harbor Symp. Quant. Biol.* **42**:1–9.

Fuller, F. B., 1971, The writhing number of a space curve, *Proc. Natl. Acad. Sci. U.S.A.* **86**:815.

Gabbay, E. J., Adawadkar, P. D., Kapicak, L., Pearce, S., and Wilson, W. D., 1976*a*, The interaction specificity of peptides with DNA. Evidence for peptide β-sheet–DNA binding, *Biochemistry* **15**:152.

Gabbay, E. J., Adawadkar, P. D., and Wilson, W. D., 1976*b*, Stereospecific binding of distereomeric peptides to salmon sperm DNA, *Biochemistry* **15**:146.

Gierer, A., 1966, Model for DNA and protein interactions and the function of the operator, *Nature* **212**:1480.

Gilbert, W., and Maxam, A., 1973, The nucleotide sequence of the *lac* operator, *Proc. Natl. Acad. Sci. U.S.A.* **70**:3581.

Gilbert, W., Gralla, J., Majors, J., and Maxam, A., 1975, Lactose operator sequences and the action of *lac* repressor, in: *Protein–Ligand Interactions,* pp. 193–210, de Gruyter, Berlin.

Hearst, J., 1977, *Cold Spring Harbor Symp. Quant. Biol.* **42**.

Hedgpeth, J., Goodman, H. M., and Boyer, H. W., 1972, The DNA nucleotide sequence restricted by the RI endonuclease, *Proc. Natl. Acad. Sci. U.S.A.* **69**:3448.

Jain, S. C., and Sobell, H. M., 1972, The stereochemistry of actinomycin binding to DNA. I. Refinement and further structural details of the actinomycin–deoxyguanosine crystalline complex, *J. Mol. Biol.* **68**:1.

Jain, S. C., Tsai, C.-C., and Sobell, H. M., 1977, Visualization of drug–nucleic acid interactions at atomic resolution. II. Structure of an ethidium/dinucleoside monophosphate crystalline complex, ethidium:5-iodocytidylyl(3′-5′)guanosine, *J. Mol. Biol.* **114**:317.

Kelly, T. J., and Smith, H. O., 1970, A restriction enzyme from *Hemophillus influenzae.* II. Base sequence of the recognition site, *J. Mol. Biol.* **51**:393.

Kim, S.-H., Church, G. M., and Sussman, J. L., 1977, Secondary structural complementarity between DNA and proteins, *Proc. Natl. Acad. Sci. U.S.A.* **74**:1458.

Krugh, T. R., 1974, Sequence specificity in the interaction at actinomycin D with deoxydinucleotides as a model for the binding of the drug to DNA, in: *Molecular and Quantum Pharmacology* (E. Bergman and B. Pullman, eds.), pp. 465–471, Reidel, Dordrecht, The Netherlands.

Krugh, T. R., 1974, in: *Molecular and Quantum Pharmacology* (E. Bergman and B. Pullman, eds.), pp. 465–471, Reidel, Dordrecht, The Netherlands.

Krugh, T. R., and Reinhardt, C. G., 1975, Evidence for sequence preferences in the intercalative binding of ethidium bromide to dinucleoside monophosphates, *J. Mol. Biol.* **97**:133.

Krugh, T. R., Wittlin, F. N., and Cramer, S. P., 1975, Ethidium bromide–dinucleotide complexes: Evidence for intercalation and sequence preferences in binding to double-stranded nucleic acids, *Biopolymers* **14**:197.

Lawrence, J. J., and Daune, M., 1976, Ethidium bromide as a probe of conformational heterogeneity of DNA in chromatin. The role of histone H1, *Biochemistry* **15**:3301.

LePecq, J. B., and Paoletti, C. J., 1967, A fluorescent complex between ethidium bromide and nucleic acids, *J. Mol. Biol.* **27**:87.

Lozansky, E. D., Sobell, H. M., and Lessen, M., 1978, Wave propagation in double-helical deoxyribonucleic acid, *Proc. Natl. Acad. Sci. USA* (in press).

Mahler, H. R., Green, G., Goutarel, R., and Khuong-Huu, Q., 1968, Nucleic acid–small molecule interactions. VII. Further characterization of deoxyribonucleic acid–diamino steroid complexes, *Biochemistry* **7**:1568.

McGhee, J. D., and von Hippel, P. H., 1975*a*, Formaldehyde as a probe of DNA structure. I. Reaction with exocylic amino groups of DNA bases, *Biochemistry* **14**:1281.

McGhee, J. D., and von Hippel, P. H., 1975*b*, Formaldehyde as a probe of DNA structure. II. Reaction with endocyclic imino groups of DNA bases, *Biochemistry* **14**:1297.

Olins, D. E., and Olins, A. L., 1974, Spheroid chromatin unit (ν bodies), *Science* **183**:330.

Pack, G. R., and Loew, G. H., 1977, The origins of sequence specificity of ethidium nucleic acid intercalation, *Int. J. Quant. Chem. Quant. Biol. Symp.* **4**:87.

Patel, D. J., 1976, Proton and phosphorous NMR studies of *d*-CpG(pCpG)$_n$ duplexes in solution. Helix coil transition and complex formation with actinomycin D, *Biopolymers* **15**:533.

Pigram, W. J., Fuller, W., and Hamilton, L. D., 1972, Stereochemistry of intercalation: Interaction of daunomycin with DNA, *Nature New Biol.* **235**:17.

Pohl, F. M., and Jovin, T. M., 1972, Salt-induced cooperative conformational change of a synthetic DNA: Equilibrium and kinetic studies with poly(dG-dC), *J. Mol. Biol.* **67**:375.

Pohl, F. M., Jovin, T. M., Baehr, W., and Holbrook, J. J., 1972, Ethidium bromide as a cooperative effector of a DNA structure, *Proc. Natl. Acad. Sci. U.S.A.* **69**:3805.

Pribnow, D., 1975, Nucleotide sequence of an RNA polymerase binding site of an early T7 promoter, *Proc. Natl. Acad. Sci. U.S.A.* **72**:784.

Printz, M. P., and von Hippel, P. H., 1965, Hydrogen exchange studies of DNA structure, *Proc. Natl. Acad. Sci. U.S.A.* **53**:363.

Reddy, B. S., Seshadri, T. P., Sakore, T. D., and Sobell, H. M., 1977, Visualization of drug–nucleic acid interactions at atomic resolution. V. Structures of two aminoacridine-dinucleoside monophosphate crystalline complexes, proflavin:5-iodocytidylyl(3′-5′)guanosine and acridine orange:5-iodocytidylyl(3′-5′)guanosine, *J. Mol. Biol.* (in press).

Reinhardt, C. G., 1976, Ph.D. Thesis, Univ. of Rochester, Rochester, New York.

Sakore, T. D., Reddy, B. S., and Sobell, H. M., 1977*a*, Visualization of drug–nucleic acid interactions at atomic resolution. IV. Structure of an aminoacridine-dinucleoside monophosphate crystalline complex, 9-aminoacridine:5-iodocytidylyl(3′-5′)guanosine, *J. Mol. Biol* (in press).

Sakore, T. D., Jain, S. C., Tsai, C.-C., and Sobell, H. M., 1977*b*, Mutagen–nucleic acid intercalative binding: Structure of a 9-aminoacridine:5-iodocytidylyl-(3′-5′)guanosine crystalline complex, *Proc. Natl. Acad. Sci. U.S.A.* **74**:188.

Sobell, H. M., 1973, Stereochemistry of actinomycin binding to DNA and its implications in molecular biology, in: *Progress in Nucleic Acid Research and Molecular Biology,* Vol. 13 (W. E. Cohn and J. N. Davidson, eds.), pp. 153–190, Academic Press, New York.

Sobell, H. M., 1976, Symmetry in nucleic acid structure and its role in protein–nucleic acid interactions, *Annu. Rev. Biophys. Bioeng.* **5**:307.

Sobell, H. M., and Jain, S. C., 1972, Stereochemistry of actinomycin binding to DNA. II. Detailed molecular model for actinomycin–DNA complex and its implications, *J. Mol. Biol.* **68**:21.

Sobell, H. M., Jain, S. C., Sakore, T. D., and Nordman, C. E., 1971, Stereochemistry of actinomycin–DNA binding, *Nature* **231**:200.

Sobell, H. M., Tsai, C.-C., Gilbert, S. G., Jain, S. C., and Sakore, T. D., 1976, Organization of DNA in chromatin, *Proc. Natl. Acad. Sci. U.S.A.* **73**:3068.

Sobell, H. M., Tsai, C.-C., Jain, S. C., and Gilbert, S. G., 1977*a*, Visualization of drug–nucleic acid interactions at atomic resolution. III. Unifying structural concepts in understanding drug–DNA interactions and their broader implications in understanding protein–DNA interactions, *J. Mol. Biol.* **114**:333.

Sobell, H. M., Reddy, B. S., Bhandary, K. K., Jain, S. C., Sakore, T. D., and Seshadri, T. P., 1977*b*, Conformational flexibility in DNA structure as revealed by structural studies of drug intercalation and its implications in understanding the organization of DNA in chromatin, *Cold Spring Harbor Symp. Quant. Biol.* **42**: 87–102.

Sobell, H. M., Tsai, C.-C., Jain, S. C., and Sakore, T. D., 1978*a*, Conformational flexibility in DNA structure and its implications in understanding the organization of DNA in chromatin, *Phil. Trans. Roy. Soc. Lond. B* **283**:295–298.

Sobell, H. M., Lozansky, E. D., and Lessen, M., 1978*b*, Structural and energetic considerations of wave propagation in DNA, *Cold Spring Harbor Symp. Quant. Biol.* **43** (in press).

Steitz, T. A., Richmond, T. J., Wise, D., and Engelman, D., 1974, The *lac* repressor protein: Molecular shape, subunit structure and proposed model for operator interaction based on structural studies of microcrystals, *Proc. Natl. Acad. Sci. U.S.A.* **71**:593.

Structure and Function of Proteins at the Three-dimensional Level, Cold Spring Harbor Symp. Quant. Biol. **36**.

Teitelbaum, H., and Englander, S. W., 1975*a*, Open states in native polynucleotides. I. Hydrogen-exchange study of adenine-thymine containing double helices, *J. Mol. Biol.* **92**:55.

Teitelbaum, H., and Englander, S. W., 1975*b*, Open states in native polynucleotides. II. Hydrogen exchange study of guanine-cytosine containing double helices, *J. Mol. Biol.* **92**:79.

Thomas, J. O., and Kornberg, R. D., 1975, An octamer of histones in chromatin and free in solution, *Proc. Natl. Acad. Sci. U.S.A.* **72**:2626.

Travers, A., 1974, RNA polymerase–promoter interactions: Some general principles, *Cell* **3**:97.

Tsai, C.-C., Jain, S. C., and Sobell, H. M., 1975*a*, X-ray crystallographic visualization of drug–nucleic acid intercalative binding: Structure of an ethidium-dinucleoside monophosphate crystalline complex, ethidium:5-iodouridylyl(3′-5′)adenosine, *Proc. Natl. Acad. Sci. U.S.A.* **72**:628.

Tsai, C.-C., Jain, S. C., and Sobell, H. M., 1975*b*, Drug–nucleic acid interaction: X-ray crystallographic determination of an ethidium-dinucleoside monophosphate crystalline complex, ethidium:5-iodouridylyl(3′-5′)adenosine, *Phil. Trans. Roy. Soc. Lond. B* **272**:137.

Tsai, C.-C., Jain, S. C., and Sobell, H. M., 1977, Visualization of drug–nucleic acid interactions at atomic resolution. I. Structure of an ethidium-dinucleoside monophosphate crystalline complex, ethidium:5-iodouridylyl(3′-5′)adenosine, *J. Mol. Biol.* **114**:301.

Walz, A., and Pirrotta, V., 1975, Sequence of the P_R promoter of phage λ, *Nature* **254**:118.

Wang, J. C., 1974, The degree of unwinding of the DNA helix by ethidium. I. Titration of twisted PN2 DNA molecules in alkaline cesium chloride density gradients, *J. Mol. Biol.* **89**:783.

Waring, M. J., 1970, Variation of supercoils in closed circular DNA by binding of antibiotics and drugs: Evidence for molecular models involving intercalation, *J. Mol. Biol.* **54**:247.

Waring, M. J., and Chisholm, J. W., 1972, Uncoiling of bacteriophage PM2 DNA by binding of steroidal diamines, *Biochim. Biophys. Acta* **262**:18.

Waring, M. J., and Henley, S. M., 1975, Stereochemical aspects of the interaction between steroidal diamines and DNA, *Nucleic Acids Res.* **2**:567.

Wells, R. D., and Larson, J. E., 1970, Studies on the binding of actinomycin D to DNA and DNA model polymers, *J. Mol. Biol.* **49**:319.

Worcel, A., 1978, Molecular architecture of the chromatin fiber, *Cold Spring Harbor Symp. Quant. Biol.* **42**:313–324.

Worcel, A., and Benyajati, C., 1977, Higher order coiling of DNA in chromatin, *Cell* **2**:83.

Young, M. A., and Krugh, T. R., 1975, Proton magnetic resonance studies of double helical oligonucleotides. The effect of base sequence on the stability of deoxydinucleotide dimers, *Biochemistry* **14:** 4841–4847.

6

Some Aspects of the Regulation of DNA Replication in *Escherichia coli*

KARL G. LARK

1 Introduction

The purpose of this chapter is to discuss recent results in the field of DNA replication which provide insight into a regulatory mechanism of an unusual type: regulation by protein–protein interaction. DNA replication represents one example of this regulatory mechanism. However, regulation of this type may govern many other aspects of cell division and differentiation.

The reader is referred to reviews (Lark, 1969; Bonhoeffer and Messer, 1969; Cairns, 1972; Kornberg, 1974; Gefter, 1975; Dressler, 1975; Alberts and Sternglanz, 1977; Davern, 1978; Johnston *et al.*, 1978) and to proceedings of meetings in this field (Cairns and Frisch, 1968; Wells and Inman, 1973; Goulian *et al.*, 1975; Molineux and Kohiyama, 1978) for more comprehensive treatment of various aspects of DNA replication.

2 DNA Replication in *Escherichia coli*

Most of the data discussed herein concern *E. coli* because of the relative ease with which biochemistry and genetics can be related in this organism. The *E. coli*

KARL G. LARK • Department of Biology, University of Utah, Salt Lake City, Utah

chromosome is a single replicating unit which is replicated bidirectionally from a unique origin (Cairns, 1963; Masters and Broda, 1971; Bird *et al.*, 1972; Prescott and Kuempel, 1972; Rodriguez *et al.*, 1973). Replication ceases in a region (opposite to the origin) (Rodriguez *et al.*, 1973) that appears to contain a termination site (Kuempel *et al.*, 1977). Replication, once initiated, proceeds by chain elongation in which one nascent strand of the DNA appears to elongate rapidly in a $5' \rightarrow 3'$ direction, copying a template of the opposite polarity (Kornberg, 1974; Alberts and Sternglanz, 1977). The other strand is synthesized discontinuously as small pieces *(Okazaki pieces)* which themselves are also synthesized in a $5' \rightarrow 3'$ direction. These small pieces are joined together by polynucleotide ligase (Kornberg, 1974; Alberts and Sternglanz, 1977).

A number of temperature-sensitive, conditionally lethal mutants have been isolated which are unable to replicate DNA at the nonpermissive temperature (Wechsler, 1978). Several of these mutants cease DNA synthesis immediately *(immediate stop mutants)* when the temperature is raised. Mutants of this class include *dnaB,* certain alleles of *dnaC, dnaE, dnaG, dnaL, dnaM, dnaZ,* and a DNA gyrase mutant. The functions of only a few of these are known: *dnaE* specifies the structural gene of polymerase III, which is active in replication when combined with several elongation factors, one of which is specified by *dnaZ; dnaG* is required for synthesis of Okazaki pieces and appears to be responsible for synthesizing a polynucleotide primer which is subsequently elongated. DNA gyrase supercoils DNA (Gellert *et al.*, 1976). The functions controlled by the other immediate stop mutations are not known. However, all of them appear to be necessary for chain elongation.

Mutants of another class *(initiation mutants)* cease replication only after a period at the nonpermissive temperature. It has been shown that these can complete a cycle of replication at the nonpermissive temperature. However, they are unable to initiate new cycles unless the temperature is lowered. These include *dnaA, dnaC, dnaI,* and *dnaP* mutants. None of these gene functions has been identified with a biochemical activity, although the function of *dnaC* is required to replicate DNA in some *in vitro* systems.

One should note here that the list of genes affecting replication is probably not complete, since many biochemical functions exist which, although essential for *in vitro* replication systems, are not correlated with any genetic locus. Examples are elongation factors which interact with polymerase III (Wickner and Hurwitz, 1976); DNA-binding proteins (Alberts *et al.*, 1968; Alberts and Sternglanz, 1977); and proteins i and n, which participate in forming the ϕX174 *in vitro* replication complex (Scheckman *et al.*, 1975). The number of genes essential for DNA replication indicates that replication of DNA is not a simple process and must require the cooperative participation of many components. To date, each aspect that has been closely examined appears *unduly* complicated.

Because the DNA double helix is antiparallel and because both templates are replicated in the same direction and at the same time (Cairns, 1963), the two nascent chains must grow with different polarities ($5' \rightarrow 3'$ and $3' \rightarrow 5'$). Three bacterial DNA polymerases are known, all of which extend the elongating nascent chain only by addition to the $3'$ end. Yet only one of these polymerases (Pol III), the product of an essential gene, *dnaE* (in which impairment of function is a lethal event), is required for DNA replication (Kornberg, 1974). Why should the cell have several polymerases?

Instead of a polymerase that can extend the elongating nascent chain by addition at the 5′ position, the cell has solved the problem of 3′ → 5′ elongation by constructing small Okazaki pieces which are initiated and then elongated backward in a 5′ → 3′ direction. These are then joined together (Kornberg, 1974; Alberts and Sternglanz, 1977). Such intermediates are essential for replication since the gene that regulates their initiation, *dnaG,* is also an essential gene. An elegant explanation for the elongation of DNA by addition of triphosphates only to the 3′-OH group has been proposed by Alberts and Sternglanz (1977). These authors have suggested that the restrictions on chain elongation have been evolved as a consequence of the need to maintain sequence fidelity during replication.

Although replication could, in principle, proceed most simply by continuous synthesis of one strand 5′ → 3′ and discontinuous synthesis (plus joining) of the other, it does not do so. There is evidence that in bacteria and in some bacterial plasmids and viruses a very small intermediate (200–400 nucleotides) is involved in the elongation of the 5′ → 3′ nascent chain (Lark and Wechsler, 1975). Why?

Deoxyuridine triphosphate (dUTP) is present in bacteria and is incorporated into DNA in place of thymidine triphosphate (TTP). This is subsequently excised, creating single-strand breaks (Duncan and Warner, 1976; Olivera, 1978; Tye *et al.,* 1977). For what purpose?

It is important to remember that *not all of the restriction on replication that we observe need be essential to replication per se.* Because the role of DNA is central to all of the activities of the cell, many properties of replication may be the result of adapting the replication process to improve or accommodate some other essential function such as transcription, recombination, or the condensation of the chromosome. Some restrictions may be related to the basic process of simple chain elongation—for example, the interactive use of *dnaE* and *dnaG* gene products. Other restrictions could evolve from pressure to increase the cells' DNA content (usable information). Thus, the condensed supercoiled chromosome must be unwound and again rewound as it is replicated. This probably explains the involvement of proteins such as gyrase (Gellert *et al.,* 1976) and possibly omega (Wang, 1971) in replication. Still other restrictions must be imposed on replication by the participation of DNA in transcription and in genetic events (such as recombination).

3 *Stoichiometry of DNA Replication*

The synthesis, or replication, of DNA is unique in that an exact number of copies of the genetic material is made. An example from bacteria can serve to illustrate the importance of proper genome dosage to the survival of a cell population. Bacteria can vary their content of DNA at different growth rates (Schaechter *et al.,* 1958; Cooper and Helmstetter, 1968). They do this by increasing the rate at which chromosome replication is initiated (Yoshikawa *et al.,* 1964; Helmstetter and Cooper, 1968). As a consequence, several replication forks may be active simultaneously. When this occurs, genes located near the chromosomal origin of replication are present in the greatest frequency (Yoshikawa *et al.,* 1964; Bird *et al.,* 1976). The frequency of such genes increases as the number of active forks per chromosome increases. Both *E. coli* and *Salmonella* have evolved in similar, almost identical environments. In both organisms the order of a number of genetic loci on the chromosomes is almost identical (Taylor and Trotter, 1972; Sanderson, 1972), yet the base sequence homology is so different that little recombination can occur between the two DNAs (Middleton and Mojica-a, 1971; Sanderson, 1972; J. R. Roth, personal communication). It is possible that this particular order of genes on the chromosome may have evolved so that these loci, such as those for ribosomal components (Bachmann *et al.,* 1976), for which extra copies are useful at rapid growth rates, are near the origin.

4 Regulation Is at the Level of Initiating Chromosome (or Replicon) Replication

In 1963, Jacob, Brenner, and Cuzin outlined a replicon hypothesis for regulation of DNA synthesis in bacteria (Jacob *et al.,* 1963). The most useful aspect of this model was the definition of the replicon as the regulatory unit of replication. The entire *E. coli* chromosome was considered a single unit or replicon. According to this hypothesis, replication of this unit was independent of the replication of other such units and, once initiated, would proceed until replication of the entire unit was completed.

Important among the data leading up to this model was the finding that inhibition of protein and/or RNA synthesis in *E. coli* eventually inhibits DNA synthesis but not until replication of the chromosome is completed (Maaløe and Hanawalt, 1961; Lark *et al.,* 1963). Thus, new proteins must be synthesized to initiate chromosome replication but are not needed for continued replication. This model was supported by the isolation of mutants of two types in which there was a temperature-sensitive (conditional lethal) defect in DNA replication (Jacob *et al.,* 1963; Wechsler and Gross, 1971). In those of one type *(immediate stop)*, replication ceased immediately when cells were placed at the nonpermissive temperature. In those of the other type *(delayed stop)*, replication ceased only after an amount of DNA was synthesized that corresponded to the completion of the chromosome. Moreover, when two different replication units exist in a bacterium (such as a plasmid or episome in addition to the chromosome) each is controlled independently as determined by either genetic or physiological analysis (Jacob *et al.,* 1963; Helinski *et al.,* 1975; Rownd *et al.,* 1975).

As further tests were applied to the model, its support was strengthened. A series of experiments demonstrated that the bacterial chromosome was replicated from a unique origin and that replication from that origin proceeded sequentially (Lark *et al.,* 1963; Yoshikawa and Sueoka, 1963; Eberle and Lark, 1969; Bird *et al.,* 1972). When such experiments were carried out on delayed stop mutants, it was possible to demonstrate that, after DNA synthesis had been stopped at the nonpermissive temperature, it was reinitiated at this origin of replication (Hirota *et al.,* 1970; Monk and Gross, 1971; Schubach *et al.,* 1973; Bird *et al.,* 1976).

Replication is regulated by regulating initiation on individual replicon units. When bacteria grow rapidly, extra initiations occur at the unique origin of replication (Yoshikawa *et al.,* 1964; Helmstetter and Cooper, 1968; Bird and Lark, 1968). If replication is inhibited by inhibiting growth, the replication cycles are always completed before DNA synthesis ceases. Mutants which stop replication immediately, or cells in which replication is inhibited by physiological means, initiate new cycles of replication *from their chromosome origin* without awaiting the completion of replication cycles already underway (Pritchard and Lark, 1964; Kogoma and Lark, 1970; Worcel, 1970).

Certain extrachromosomal genetic units, F factors, are capable of suppressing delayed stop DNA mutants when integrated into the chromosome, a phenomenon referred to as *integrative suppression* (Wechsler and Gross, 1971; Nishimura *et al.*, 1971; Tresguerres *et al.*, 1975). Placing temperature-sensitive *dnaA* mutants containing an integrated F factor at the restrictive temperature does not inhibit DNA synthesis. A variety of other plasmids or episomes have been shown to be capable of integrative suppression as well as F (Lindahl *et al.*, 1971; Bird *et al.*, 1976). DNA replication in these strains is not initiated at the *E. coli* chromosome origin but at a new locus corresponding to the site of the integrated F factor (Nishimura *et al.*, 1971; Bird *et al.*, 1976). Presumably these replicons are capable of donating a gene product necessary for initiation but this product must act on its own site or origin.

The universality of regulation by initiating replicon replication is demonstrated by the existence in all eukaryotes of multiple replicons arranged in tandem on the same DNA fiber (Cairns, 1966; Hubermann and Riggs, 1968; Hori and Lark, 1973; Callan, 1973; Blumenthal *et al.*, 1973). These are not replicated simultaneously and evidence exists suggesting that they can be bounded by termination points (Hori and Lark, 1973, 1974). In at least two systems, it has been possible to show that the spacing of the units which initiate replication can be varied, depending on the developmental or physiological conditions of the cells (Callan, 1973; Blumenthal *et al.*, 1973). These situations are very similar to the integrative suppression observed in bacteria in which tandem replicons are present but not always active.

5 *The Replication Complex*

Several years ago, the author proposed, as a working hypothesis, that replication is carried out by a complex of interacting proteins (much like a ribosome) (Kogoma and Lark, 1970; Lark, 1972). The characteristics of replication, determined by the interaction of these proteins, might be established during formation of the complex. This image, and an increasing confidence in its usefulness, developed from the results of *in vivo* experiments on replication in *E. coli* (Kjeldgaard *et al.*, 1958; Bird and Lark, 1970; Urban and Lark, 1971; Kogoma and Lark, 1970). The data from these experiments suggest that the conditions (such as temperature or nutritional state) under which the cell is growing at the time that the replication complex is formed determine the characteristics of the complex during the subsequent replication cycle (Bird and Lark, 1970; Urban and Lark, 1971). Thus, the rate of chain elongation continues to reflect the rate at the previous growth condition until a new cycle of replication is initiated.

A characteristic of replication in *E. coli* is the inability to initiate a new cycle of replication without protein synthesis (that is, the proteins necessary for initiation cannot be recycled). In the absence of protein or RNA synthesis, replication cycles that are underway are completed, but new ones are not initiated (Maaløe and Hanawalt, 1961; Lark *et al.,* 1963; Bird and Lark, 1968; Lark, 1972).

A striking example of the effect of growth conditions upon the characteristics of replication is the formation of a replication apparatus which can be reutilized (Kogoma and Lark, 1970, 1975). This occurs as a result of growth in which the synthesis of DNA lags behind the synthesis of protein (such as growth of a thymine auxotroph in low thymine, or a shift up from poor to rich medium). Once formed, this replication complex can be reused again and again, even in the absence of protein or RNA synthesis. The conditions that lead to its formation suggest that its special attributes are irreversibly determined early in the assembly of the replication apparatus (Kogoma and Lark, 1975). During subsequent growth, these special "reusable" replication complexes are diluted out of the population in proportion to the extent of growth, indicating that their formation was transient, corresponding to the transient change in growth conditions that gave rise to them.

These are observations that support the idea that the properties of the replication complex are determined when it is formed. But what evidence exists that DNA replication is accomplished by a complex of interacting proteins? In *E. coli,* the evidence for interaction between replication proteins has centered around the *dnaB* protein. The earliest evidence which suggested that the *dnaB* protein was altering its functional state came from studies of conditional forms of DNA replication.

Studies of DNA replication mutants of λ phage showed that products of the λ *P* gene and of the *E. coli dnaB* gene were required for λ DNA replication (Georgopoulos and Herskowitz, 1971). Mutations in the viral gene could suppress mutations in the gene of the bacterial host and demonstrated that the two replication genes functioned as external suppressors or modifiers of each other's activities. In a different study, it was shown that a mutant *dnaB* gene product was altered during replication. Its activity was *irreversibly* lost at high temperature if cells were heated before replication began, but, once replication was underway, raising the temperature resulted only in *reversible* inactivation (Kogoma and Lark, 1970; Kogoma, 1976). From both of these studies we conclude that the *dnaB* gene product was capable of altering its conformation during replication.

The second study just cited utilized a system of two replication complexes, one of which was reusable, the other not. It was possible to conclude that the active *dnaB* protein could not be transferred between complexes at the end of a replication cycle (Kogoma and Lark, 1970).

More recent experiments indicated that *dnaB* proteins interacted with each other *in vivo* (Lark and Wechsler, 1975). In partial diploids, containing both a mutant and a wild type allele (F′*dnaB*$^+$/*dnaBts*) negative complementation was observed. The results could be explained by making the following assumptions: The active form of *dnaB* protein was a tetramer containing four identical subunits; the mutant and wild type protein subunits could associate in an active form at the permissive temperature; raising the temperature irreversibly inactivated any *dnaB* tetramer that contained at least one mutant ts subunit; and inactive tetramers *could not* be replaced by active ones.

The *dnaB* protein has now been purified to homogeneity and its active form is a tetramer (Wickner *et al.,* 1974; Ueda *et al.,* 1978). In an *in vitro* replication system, it was shown that replication is carried out by a complex of proteins (Wickner *et al.,* 1974; Scheckman *et al.,* 1975) in which the *dnaB* protein appears to act stoichiometrically instead of catalytically (Weiner *et al.,* 1976). An estimate of the cellular content of *dnaB* protein suggests that there are about 20 tetramers per rapidly growing cell (Ueda *et al.,* 1978). Depending on the number of chromosomes and replication forks per chromosome, this would suggest that between three and six are available for each replication fork. In an *in vitro* system in which ϕX174 single-strand DNA is converted to the double-strand replicative form, only one *dnaB* tetramer is required (Ueda *et al.,* 1978). It seems likely that for the replication of double-strand DNA, at least two tetramers may be required per replication fork. Thus, the amount of *dnaB* product available is close to the stoichiometric amount required for replication.

In this *in vitro* system, *dnaB* and *dnaC* proteins interact, altering the properties of both proteins (Wickner and Hurwitz, 1975). This alteration appears to be stoichiometric and to require the participation of other protein factors. An important aspect of the *dnaB–dnaC* interaction is that the proteins form a nucleoprotein complex before replication begins and this association, at least in the ϕX174 *in vitro* system, is stable and extremely tight (Weiner *et al.,* 1976). The *dnaB* and *dnaC* proteins show different rate dependencies on concentration for their participation in the ϕX nucleoprotein complex. The rate increases linearly as *dnaB* protein concentration is increased. However, the rate of complex formation when *dnaC* protein is limiting increases in a manner suggestive of multiple interactions or possibly a cooperative interaction. In summary, *dnaB* and *dnaC* proteins interact in the ϕX *in vitro* system to participate in a nucleoprotein complex composed of one tetramer of *dnaB* and several *dnaC* protein molecules.

The interaction of *dnaC* and *dnaB* proteins explains the phenotypes of certain mutations at these loci. Most *dnaC* mutants belong to the initiation class, completing replication of the chromosome before cessation of DNA synthesis (Carl, 1970; Schubach *et al.,* 1973). However, certain mutants cease chain elongation when the temperature is raised sufficiently (Wechsler, 1975). These alleles may contain mutations that change the *dnaC* protein so that it alters *dnaB* function. The *dnaB* mutation is an immediate stop mutation affecting chain elongation. One mutant has been isolated, which behaves as an initiator mutation: Replication ceases only when it reaches the end of the chromosome (Zyskind and Smith, 1977). This mutant could be one in which changes in the *dnaB* protein alter *dnaC* function. Another example of *in vivo dnaB–dnaC* interaction is the observation that in *dnaB–dnaC* double mutants the effect of an external suppressor of *dnaB* (P1 *bac* prophage) is modified, such that only partial suppression is observed. Either the phenotype of *dnaC* or that of the suppressed *dnaB* is altered in the double mutant (Schuster *et al.,* 1977).

Examples of other possible replication protein interactions have been observed in *in vivo* experiments. As mentioned earlier, under certain growth conditions, the cell forms replication complexes that can continue DNA synthesis in the absence of protein synthesis. A mutation, *dnaT,* has been isolated which causes these stable complexes to terminate (see later). Alleles at this locus are epistatic to *dnaC,* increasing the growth rate of *dnaCts* mutants and completely suppressing the temperature sensitivity of some *dnaCts* mutants (K. G. Lark and C. Lark, unpublished data). Another example of this phenotype (not able to recycle) occurs in a revertant of *dnaBts* (Kogoma, 1976). This revertant maps at the *dnaB* locus. A possible explanation of the phenotype is that an alteration in *dnaB* protein changes the behavior of the *dnaT* product either directly, by interacting with the *dnaT* product itself, or indirectly, by interacting with the *dnaC* protein in concert with the *dnaT* product.

Mutants that do not permit growth of λ phage have been isolated that also affect host DNA replication. Although these suppress the activity of λ *P* protein, they do not map at the *dnaB* locus. Mutations in the λ *P* gene overcome the suppression (Sunshine *et al.*, 1977; Georgopoulos, 1977). These findings also suggest an interaction with *dnaB*, since λ *P* protein interacts with *dnaB* protein.

The *dnaA* and *dnaC* mutants are affected by changes in the lipid content of the host cells. When unsaturated fatty acid synthesis is blocked, the temperature transition of mutants at these loci is altered and reversible temperature effects on initiation of replication become irreversible (Fralick and Lark, 1973). The effects are most easily explained by an association between *dnaA* or *dnaC* products and lipid in which the association cannot be reversed at high temperature. In this connection, several suppressors of *dnaA* have been isolated which may be mutations affecting membrane integrity (Wechsler and Zdzienicka, 1975).

It has been shown that a mutation in the β subunit of RNA polymerase *(rif)* can result in suppression of the *dnaA* phenotype (Bagdasarian *et al.*, 1977). Since both the *dnaA* gene product and RNA synthesis are required for initiation of replication, interactions between these gene products are not unexpected. The temperature sensitivity of the double mutant is different from that of the original *dnaA* mutant, suggesting that there is a structural interaction.

A mutant of *dnaE* (W. Kurtz and J. A. Wechsler, personal communication) also behaves as if its gene product were altered by the replication process. Upon raising the temperature, the kinetics of DNA synthesis become linear (continuing for many replication cycles), as if any new gene product were inactivated, but enzyme (polymerase III) which is already in use were not. Some of the effects of rifampin on DNA replication suggest that raising the temperature alters the replication process in this mutant. One explanation for the data may be an association, between polymerase III (*dnaE* gene product) and parts of the replication complex, in which the activity of polymerase III is stabilized.

It is not surprising that the replication of DNA should be accomplished by a structure composed of many protein subunits. Such a structure can easily lead to the regulation of the stoichiometry of replication by requiring that several subunits of each protein participate in the complex. When this occurs, the assembly will be of a cooperative nature, such that formation of the complex will be exceedingly improbable below a certain subunit concentration and almost certain at higher concentrations. Moreover, during assembly, the population of subunits will again be reduced below that necessary for initiating a new assembly. The precision of this process will be greatly enhanced by utilizing different proteins and combining the probabilities of their interactions. In this way, the vagaries of statistical processes can be avoided.

6 *The Destruction of the Replication Complex*

The stoichiometry of replication restricts the number of replication complexes that are active. An important aspect of this stoichiometry is that replication cannot be reinitiated in the absence of protein synthesis. When protein synthesis is restored, replication is reinitiated after a lag during which the capacity to initiate replication is recovered in stages. At least three stages can be distinguished by measuring the increase in capacity to initiate a cycle of replication in the presence of an inhibitor. Two of these (Lark and Renger, 1969; Messer, 1972) are distinguished by sensitivity to different concentrations of chloramphenicol (stages of protein synthesis). The third is sensitive to inhibition of only RNA synthesis (Lark, 1972; Messer, 1972). For each of these different types of inhibition, the capacity to initiate replication, in the presence of inhibitor, is acquired at characteristic, but different, times before replication is initiated. However, in all cases, irrespective of the inhibitor treatment used, *the capacity is lost when replication is initiated* (Lark, 1969). If initiation is delayed, the capacities gained through protein synthesis are maintained until initiation of replication occurs (Lark, 1972). These results suggest the formation of several components whose usefulness in initiating replication is destroyed by the act of initiation. This is a description of the formation of a "disposable" complex which cannot be "recycled."

As mentioned earlier, under certain growth conditions the cell forms a replication complex that can be reused (Kogoma and Lark, 1975). Studies of these growth conditions led to the following conclusions: The condition of growth irreversibly affects an early stage in the formation of the complex (probably by a failure to alter some part of the complex). Other replication proteins, which are subsequently synthesized and added to the complex, "lock in" this configuration. These data suggest that termination of a replication cycle and the loss of the replication complex involve an active component, a *destructor* (Kogoma and Lark, 1975).

It was hypothesized that the destructor was usually activated by a second component, and that the activation had to occur before the inactive destructor was buried in the complex and inaccessible. Treatments that altered the growth pattern were believed to result in the asynchronous synthesis of these two components, resulting in the synthesis of the replication complex around an inactive destructor. (Because the partial inhibition of RNA synthesis led to "recycling" DNA replication it was believed that the reuse of the complex resulted from *inactivation* of a destructor rather than from *synthesis* or activation of a preservator.)

A prediction of this hypothesis was that mutants should exist that are never able to recycle the replication complex—that is, in which the destructor is always active. Moreover, these mutants might be expected to be dominant. Such obligatory termination mutants *(dnaT)* have been isolated (C. Lark, J. Riazi, and K. G. Lark, unpublished data). The two *dnaT* alleles which have been studied are both dominant over wild type, and are conditional lethal mutations, defective in DNA replication. Another prediction of the original hypothesis was that the destructor should interact with at least one or more of the critical—that is, stoichiometric—replication proteins. I have already discussed the evidence that the *dnaT* product interacts with the *dnaC* protein, fulfilling the prediction.

It should also be possible to isolate termination defective mutants in which the replication complex is always recycled during balanced growth. Attempts to isolate such mutants have been only partially successful to date. One mutant with this phenotype (Sdr^c) has been isolated (Kogoma, 1978). Although it has not been mapped precisely, it appears to be located in a different quadrant of the chromosome from *dnaT*. This mutant expresses only a fraction of the possible number of recyclable replication forks which it is capable of producing when it is also subjected to thymine starvation. This (together with the difficulty in obtaining such mutants) may indicate that full expression of this phenotype is lethal.

7 *Repair Replication (a Possible Example of Regulative Assembly)*

A large amount of experimental evidence indicates that the mutagenic effect of ultraviolet light is due to the expression of an inducible error-prone repair system referred to as S.O.S. repair (Witkin, 1976; Radman, 1975). Protein synthesis is necessary for mutant formation after exposure to ultraviolet light and it is hypothesized that DNA synthesis is altered in some way that allows replication to proceed past pyrimidine dimers or other lesions in DNA, resulting in inaccuracies. Repair of this type requires functional *recA* and *lex* genes and appears to have several steps in common with λ prophage induction.

Many of the critical experiments that deal with S.O.S. repair involve the *recA* locus. Different alleles of this locus exist (such as *tif, zab, lexB*) which alter the spontaneous expression of error-prone repair and which affect, as does *recA,* the induction of phage λ (Castellozzi *et al.,* 1977). Recent experiments strongly implicate a protease as the agent which, in destroying λ repressor, mediates phage induction (Roberts *et al.,* 1977).

The precise mechanism by which DNA replication can proceed past lesions producing errors is not known. Suggested possibilities are that proofreading mechanisms built into known polymerases, such as the 3′,5′-exonuclease of polymerase I, are altered or lacking or that a new error-prone polymerase is induced (Caillet-Fauquet *et al.,* 1977). An alternative hypothesis can be formulated on the basis of the results discussed here. Namely, that error-prone repair is the result of replication by an altered replication apparatus, assembled in a modified form such that it can replicate past lesions in DNA. According to this hypothesis, protein synthesis is necessary to synthesize the normal replication proteins (instead of, or in addition to, some new induced function). The assembly of these proteins would then be altered by *recA* function which, itself, could be modified by ultraviolet treatment of the cell.

Several of the treatments that give rise to nonterminating replication are also known to induce S.O.S. repair. These include thymine starvation and incubation of certain *dnaBts* mutants at 42°C (Witkin, 1976). In a *uvrA* (excision$^-$) strain of *E. coli,* nonterminating DNA replication appears to be less sensitive to ultraviolet irradiation than is normal DNA replication (T. Kogoma, personal communication). Moreover, nonterminating DNA replication complexes cannot be induced in *recA* cells (K. G. Lark and C. Lark, unpublished data; T. Kogoma, unpublished data). Evidence also exists which indicates that the *recA*-dependence of DNA replication continues in the absence of protein synthesis (K. G. Lark and C. Lark, unpublished data): It is possible to induce recycling replication at the permissive temperature using a *recAts* strain. This can be maintained in the absence of protein synthesis (that is, in the presence of chloramphenicol), but ceases when the strain is placed at a nonpermissive temperature. Synthesis (still in chloramphenicol) resumes immediately when the culture is restored to a permissive temperature.

These last data (which are almost identical with the earlier data on the effects of *dnaB* mutations which led to the conclusion that *dnaB* protein was part of a replication complex) raise the possibility that the *recA* product could alter the replication complex or even be an integral part of an altered replication apparatus.

8 Regulation of the Quality of DNA Replication

Alteration of the replication complex represents a regulation of the quality of DNA replication. All of the instances presented earlier (alteration in the rate of chain elongation, the temperature sensitivity of replication, or the ability of replication to continue in the absence of protein synthesis) constitute examples of changes in the nature of the replication process rather than in the amount of replication that is accomplished.

Bacteria live in variable environments, exposed to extreme fluctuations in the conditions that restrict their growth. If each replication complex bears an imprint of the growth pattern experienced during the previous replication cycle which relates the growth of the cell to the replication currently in progress, the cell has a system for matching replication to growth condition. Since in most instances the replication complex is destroyed at the end of each cycle, replication and growth conditions are compared and adjusted by a short-term memory system (one replication cycle delay).

The most interesting regulation of replication quality could be the occasional formation of an error-prone replication complex in response to drastic alteration in growth conditions. This response would lead to a wave of mutagenesis which, although repeated in a single cell for many cycles, will be diluted by normal cell growth. In this way, the bacterial population could receive an increase in genetic variability which is directly correlated to the extent of variation encountered in its growth.

9 RNA and the Initiation of Replication

Integrative suppression occurs when a replicon with its own origin of replication is integrated into the chromosome of a *dnaA* mutant (Lindahl *et al.*, 1971; Nishimura *et al.*, 1971; Tresguerres *et al.*, 1975; Bird *et al.*, 1976). At high temperatures, replication occurs, but proceeds from the new origin of replication. Only *dnaA* mutations are integratively suppressed, suggesting that this function is specific for initiating replication at the normal origin of the *E. coli* chromosome. There is evidence that suggests that *dnaA* function is related to the RNA (I-RNA) required to initiate replication.

After the synthesis of the protein components required for DNA replication, RNA still must be synthesized before a cycle can commence (Lark, 1972) (a requirement originally shown for replication of phage λ DNA) (Dove *et al.*, 1971). If requisite RNA synthesis is delayed, protein components that already have been synthesized will retain their potential activity until RNA synthesis occurs (Lark, 1972). This indicates that RNA synthesis provides a controlling element in the utilization of the replication complex.

Attempts to link gene function with required RNA synthesis have met with mixed success (Saitoh and Hiroga, 1975; Hanna and Carl, 1975; Messer *et al.*, 1975; Zyskind *et al.*, 1977; Bagdasarian *et al.*, 1977). However, at least three laboratories have succeeded in obtaining some evidence linking *dnaA* function with RNA synthesis (Messer *et al.*, 1975; Zyskind *et al.*, 1977; Bagdasarian *et al.*, 1977). The most promising link has been the demonstration that mutations in the gene for the β subunit of RNA polymerase, *rpoB,* suppress the temperature sensitivity of *dnaA* mutants (Bagdasarian *et al.*, 1977).

Another correlation has been obtained from studies of RNA covalently inserted into the *E. coli* chromosome (Messer *et al.*, 1975). Such a structure has been well established in replicating ColEl plasmid DNA. In this system, circles of ColEl DNA containing a uniquely located RNA segment can be accumulated in chloramphenicol (Helinski *et al.,* 1975). Although nothing is known about the location of the *E. coli* RNA segment, it has been shown that very little of this RNA accumulates in *dnaA* mutants under conditions in which it is demonstrable in wild type cells or *dnaC* mutants (Messer *et al.,* 1975).

The synthesis of various components of the replication complex represents a positive control over the stoichiometry of DNA replication. Regulation of the synthesis of I-RNA could allow the operation of a negative regulatory control, several models of which have been proposed (Rosenberg *et al.*, 1969; Pritchard *et al.,* 1969; Blau and Mordoh, 1972; Sompayrac and Maaløe, 1973; Messer *et al.*, 1975). So far, no direct evidence has been obtained for a repressor system. A suggestion was made that *dnaA* mutants might be dominant over wild type (Beyersmann *et al.*, 1974). However, at least one *dnaA* mutant is clearly recessive to wild type (Gotfried and Wechsler, 1977) and the previous suggestion of dominance may, in reality, be negated by the possibility of negative complementation of subunit multimers such as that described for *dnaB* (Lark and Wechsler, 1975).

Whether a negative or positive control is exercised over the formation of I-RNA, it is clear that this step presents a second type of regulation of replication, one in which replication is switched on provided that the necessary machinery is already assembled.

10 Conclusion

Two types of regulation of DNA replication in *E. coli* have been discussed: One is the on–off switch involving the *dnaA* protein and synthesis of some type of initiation RNA. The other regulates the stoichiometry of DNA by regulating the synthesis and assembly of the replication complex. By altering the interactions of the subunits, the characteristics of replication can be changed. (Although this chapter has concentrated on DNA replication in *E. coli,* similar interactions of DNA replication proteins have been demonstrated for bacteriophage T4.) The reader is referred to the elegant experiments of Mosig and Breschkin (1975) and Breschkin and Mosig (1977*a,b*) and to the review of Alberts and Sternglanz (1977).

The regulative assembly of this type is not unique, but may occur whenever proteins interact in a complex in order to accomplish a particular task. Assembly of phage head proteins is known to be modified by gene product interaction and cell membranes bear the imprint of the conditions (temperature or nutrition) that existed at the time of their assembly.

Can we expect our conclusions to apply to eukaryotic systems?

Unicellular eukaryotes which are subject to varying growth conditions could have evolved replication proteins with a similar set of plastic interactions. Thus, DNA replication in yeast may have variable characteristics as it does in *E. coli.* Because higher plants evolve in variable environments, plant cells also may be capable of responding to changing growth conditions by altering the assembly of their DNA replication apparatus. However, in such multicellular organisms we might expect to find populations of cells that respond differently during growth according to the growth conditions prevailing when the replication proteins were synthesized. Mammalian cells grow in a very constant environment and therefore many of these considerations may not apply to them. However, the aspects of replication discussed in this chapter may be useful in explaining some properties of eukaryotic chromosome replication and structure which are not subject to environmental influence.

A characteristic of eukaryotic cells is that the genome is composed of many (often thousands) replicating units. Thus, it is possible to assemble different replication complexes by changing the gene products that are present during different stages of DNA replication (S phase). There is some evidence that replication units that are replicated early in S control the replication of later units (Hori and Lark, 1973). Thus, late replication complexes could be altered by the incorporation of special gene products not available early in S. It should be possible to alter the replication sequence (using an RNA-controlled on–off switch) such that special replication complexes synthesized early or late in S will only be concerned with the replication of certain genes. For example, replication may be ordered in such a way that some late replicating units are subject to repeated replication (leading to gene amplification or recombination).

The investigation of DNA replication in higher eukaryotes is just beginning. We may expect to find that the control of replication plays a central role in the regulation of growth and differentiation.

Acknowledgments

The author is grateful to Cynthia Lark for helpful suggestions and extensive revisions of the manuscript. Unpublished research referred to in this chapter was supported by a grant (AI10056) from the National Institutes of Health.

References

Alberts, B., and Sternglanz, R., 1977, Recent excitement in the DNA replication problem, *Nature* **269**:655.

Alberts, B., Amodio, F., Jenkins, M., Gutmann, E., and Ferris, F. L., 1968, Studies with DNA-cellulase chromatography. I. DNA binding proteins from *Escherichia coli, Cold Spring Harbor Symp. Quant. Biol.* **33**:289.

Bachmann, B. J., Low, K. B., and Taylor, A. L., 1976, Recalibrated linkage map of *Escherichia coli* K-12, *Bacteriol. Rev.* **40**:116.

Bagdasarian, M. M., Izakowska, M., and Bagdasarian, M., 1977, Suppression of the *DnaA* phenotype by mutations in the *rpoB* cistron of ribonucleic acid polymerase in *Salmonella typhemurium* and *Escherichia coli, J Bacteriol.* **130**:577.

Beyersmann, D., Messer, W., and Schlicht, M., 1974, Mutants of *Escherichia coli* B/r defective in deoxyribosenucleic acid initiation: *dnaI,* a new gene for replication, *J. Bacteriol.* **118**:783.

Bird, R., and Lark, K. G., 1968, Initiation and termination of DNA replication after amino acid starvation of *E. coli 15T⁻, Cold Spring Harbor Symp. Quant. Biol.* **33**:799.

Bird, R., and Lark, K. G., 1970, Chromosome replication in *Escherichia coli* at different growth rates: Rate of replication of the chromosome and the rate of formation of small pieces, *J. Mol. Biol.* **49**:343.

Bird, R., Louarn, J., Martuscelli, J., and Caro, L., 1972, Origin and sequence of chromosome replication in *Escherichia coli, J. Mol. Biol.* **70**:549.

Bird, R., Chandler, M., and Caro, L., 1976, Suppression of an *Escherichia coli dnaA* mutation by the integrated R factor R.-100.1: Change of chromosome replication origin in synchronized cultures, *J. Bacteriol.* **126**:1215.

Blau, S., and Mordoh, J., 1972, A new element in the control of DNA initiation in *Escherichia coli, Proc. Natl. Acad. Sci. U.S.A.* **69**:2895.

Blumenthal, A. B., Kriegstein, H. J., and Hogness, D. S.,` 1973, The units of DNA replication in *Drosophila melanogaster* chromosomes, *Cold Spring Harbor Symp. Quant. Biol.* **38**:205.

Bonhoeffer, F., and Messer, W., 1969, Replication of the bacterial chromosome, *Annu. Rev. Genet.* **3**:233.

Breschkin, A. M., and Mosig, G., 1977*a,* Multiple interactions of a DNA-binding protein *in vivo.* I. Gene *32* mutations of phage T4 inactivate different steps in DNA replication and recombination, *J. Mol. Biol.* **112**:279.

Breschkin, A. M., and Mosig, G., 1977*b,* Multiple interactions of a DNA-binding protein *in vivo.* II. Effects of host mutations on DNA replication of phage T4 gene *32* mutants, *J. Mol. Biol.* **112**:295.

Caillet-Fauquet, P., Defais, M., and Radman, M., 1977, Molecular mechanisms of induced mutagenesis. Replication *in vivo* of bacteriophage ϕX174 single-stranded ultraviolet light-irradiated DNA in intact and irradiated host cells, *J. Mol. Biol.* **25**:95.

Cairns, J., 1963, The chromosome of *Escherichia coli, Cold Spring Harbor Symp, Quant. Biol.* **28**:43.

Cairns, J., 1966, Autoradiography of HeLa cell DNA, *J. Mol. Biol.* **15**:372.

Cairns, J., 1972, DNA synthesis, *Harvey Lect.* **1970–1971**:1.

Cairns, J., and Frisch, L., 1968, Replication of DNA in Microorganisms, *Cold Spring Harbor Symp. Quant Biol.* **33**.

Callan, H. G., 1973, DNA replication in the chromosomes of eukaryotes, *Cold Spring Harbor Symp. Quant. Biol.* **38**:195.

Carl, P. L., 1970, *Escherichia coli* mutants with temperature-sensitive synthesis of DNA, *Mol. Gen. Genet.* **109**:107.

Castellozzi, M., Morand, P., George, J., and Buttin, G., 1977, Prophage induction and cell division in *E. coli.* V. Dominance and complementing analysis in partial diploids with pleiotropic mutations (*tif, recA, zab,* & *lexB*) at the *recA* locus, *Mol. Gen. Genet.* **153**:297.

Cooper, S., and Helmstetter, C. E., 1968, Chromosome replication and the division cycle of *Escherichia coli* B/r, *J. Mol. Biol.* **31**:519.

Davern, C. I., 1978, Replication of the prokaryotic chromosome with emphasis on the bacterial chromosome replication in relation to the cell cycle, in: *Cell Biology: A Comprehensive Treatise 2* (D. Prescott, ed.), Academic Press, New York (in press).

Dove, W. F., Inokuchi, H., and Stevens, W. F., 1971, Replication control in phage lambda, in: *The Bacteriophage Lambda* (A. D. Hershey, ed.), pp. 747–771, Cold Spring Harbor Lab., Cold Spring Harbor, New York.

Dressler, D., 1975, The recent excitement in the DNA growing point problem, *Annu. Rev. Microbiol.* **29**:525.

Duncan, B., and Warner, R., 1976, Metabolism of uracil containing DNA by DNA *N*-glycosidase, *Fed. Proc.* **35**:1493.

Eberle, H., and Lark, K. G., 1969, Relation of the segregative origin of chromosome replication to the origin of replication after amino acid starvation, *J. Bacteriol.* **398**:536.

Fralick, J. A., and Lark, K. G., 1973, Evidence for the involvement of unsaturated fatty acids in initiating chromosome replication in *Escherichia coli, J. Mol. Biol.* **80**:459.

Gefter, M. L., 1975, DNA replication, *Annu. Rev. Biochem.* **44**:45.

Gellert, M. Mizuuchi, K., O'Dea, M. H., and Nash, H. A., 1976, DNA gyrase: An enzyme that introduces superhelical turns into DNA, *Proc. Natl. Acad. Sci. U.S.A.* **73**:3872.

Georgopoulos, C. P., 1977, A new bacterial gene *(groPC)* which affects λ DNA replication, *Mol. Gen. Genet.* **151**:35.

Georgopoulos, C. P., and Herskowitz, I., 1971, *Escherichia coli* mutants blocked in lambda DNA synthesis, in: *The Bacteriophage Lambda* (A. D. Hershey, ed.), pp. 553–564, Cold Spring Harbor Lab., Cold Spring Harbor, New York.

Gotfried, F., and Wechsler, J. A., 1977, Dominance of $dnaA^+$ to *dnaA* in *Escherichia coli, J. Bacteriol.* **130**:963.

Goulian, M., Hanawalt, P., and Fox, C. F., 1975, *DNA Synthesis and Its Regulation,* Benjamin, Menlo Park, California.

Hanna, M. H., and Carl, P. L., 1975, Reinitiation of deoxyribonucleic acid synthesis by deoxyribonucleic acid initiation mutants of *Escherichia coli:* Role of ribonucleic acid synthesis, protein synthesis and cell division, *J. Bacteriol.* **121**:219.

Helinski, D. R., Lovett, M. A., Williams, P. H., Katz, L., Collins, J., Kupersztock-Portnoy, Y., Sato, S., Leavitt, R. W., Sparks, R., Hershfield, V., Guiney, D. G., and Blair, D. G., 1975, Modes of plasmid DNA replication in *Escherichia coli,* in: *DNA Synthesis and Its Regulation* (M. Goulian and P. Hanawalt, eds.), pp. 514–536, Benjamin, Menlo Park, California.

Helmstetter, C. E., and Cooper, S., 1968, DNA synthesis during the division cycle of rapidly growing *Escherichia coli, J. Mol. Biol.* **31**:507.

Hirota, Y., Mordoh, J., and Jacob, F., 1970, On the process of cellular division in *Escherichia coli.* III. Thermosensitive mutants of *E. coli* altered in the process of DNA initiation, *J. Mol. Biol.* **53**:369.

Hori, T.-A., and Lark, K. G., 1973, Effect of puromycin on DNA replication in Chinese hamster cells. *J. Mol. Biol.* **77**:391.

Hori, T.-A., and Lark, K. G., 1974, Autoradiographic studies of the replication of satellite DNA in the kangaroo rat, *J. Mol. Biol.* **88**:221.

Hubermann, J. A., and Riggs, A. D., 1968, On the mechanism of DNA replication in mammalian chromosomes, *J. Mol. Biol.* **32**:327.

Jacob, F., Brenner, S., and Cuzin, F., 1963, On the regulation of DNA replication in bacteria, *Cold Spring Harbor Symp. Quant. Biol.* **28**:329.

Johnston, L. H., Bonhoeffer, F., and Symons, P., 1978, The molecular principles and the enzymatic machinery of DNA replication, in: *Cell Biology: A Comprehensive Treatise 2* (D. Prescott, ed.), Academic Press, New York (in press).

Kjeldgaard, N. O., Maaløe, O., and Schaechter, M., 1958. The transition between different physiological states during balanced growth of *Salmonella typhimurium, J. Gen. Microbiol.* **19**:607.

Kogoma, T., 1976, Two types of temperature sensitivity in DNA replication of an *Escherichia coli dnaB* mutant, *J. Mol. Biol.* **103**:191.

Kogoma, T., 1978, A novel *Escherichia coli* mutant capable of DNA replication in the absence of protein synthesis, *J. Mol. Biol.* **121**:55.

Kogoma, T., and Lark, K. G., 1970, DNA replication in *Escherichia coli:* Replication in absence of protein synthesis after replication inhibition, *J. Mol. Biol.* **52**:143.

Kogoma, T., and Lark, K. G., 1975, Characterization of the replication of *Escherichia coli* DNA in the absence of protein synthesis: Stable DNA replication, *J. Mol. Biol.* **94**:243.

Kornberg, A., 1974, *DNA Synthesis,* Freeman, San Francisco, California.

Kuempel, P. L., Duerr, S. A., and Seeley, N. R., 1977, Terminus region of the chromosome in *Escherichia coli* inhibits replication forks, *Proc. Natl. Acad. Sci. U.S.A.* **74**:3927.

Lark, K. G., 1969, Initiation and control of DNA synthesis, *Annu. Rev. Biochem.* **38**:569.

Lark, K. G., 1972, Evidence for the direct involvement of RNA in the initiation of DNA replication in *Escherichia coli* $15T^-$, *J. Mol. Biol.* **64**:47.

Lark, K. G., and Renger, E. H., 1969, Initiation of DNA replication in $15T^-$: Chronological dissection of three physiological processes required for initiation, *J. Mol. Biol.* **42**:221.

Lark, K. G., and Wechsler, J. A., 1975, DNA replication in *dnaB* mutants of *Escherichia coli:* Gene product interaction and synthesis of 4S pieces, *J. Mol. Biol.* **92**:145.

Lark, K. G., Repko, T., and Hoffman, E. J., 1963, The effect of amino acid deprivation and subsequent DNA replication, *Biochim. Biophys. Acta* **76**:9.

Lindahl, G., Hirota, Y., and Jacob, F., 1971, On the process of cellular division in *Escherichia coli:* Replication of the bacterial chromosome under control of prophage P2, *Proc. Natl. Acad. Sci. U.S.A.* **68**:2407.

Maaløe, O., and Hanawalt, P., 1961, Thymine deficiency and the normal DNA replication cycle, *J. Mol. Biol.* **3**:144.

Masters, M., and Broda, P., 1971, Evidence for the bidirectional replication of the *Escherichia* chromosome, *Nature New Biol.* **232**:137.

Messer, W., 1972, Initiation of deoxyribonucleic acid replication in *Escherichia coli* B/r: Chronology of events and transcriptional control of initiation, *J. Bacteriol.* **112**:7.

Messer, W., Dankworth, L., Tippe-Schindler, R., Womack, J. E., and Zahn, G., 1975, Regulation of the initiation of DNA replication in *E. coli.* Isolation of I-RNA and the control of I-RNA synthesis, in: *DNA Synthesis and Its Regulation* (M. Goulian and P. Hanawalt, eds.), pp. 602–617, Benjamin, Menlo Park, California.

Middleton, R. B., and Mojica-a, T., 1971, Homology in the enterobacteraceae based on intercrosses between species, *Adv. Genet.* **16**:53.

Molineux, I., and Kohiyama, M., 1978, *NATO ASI: DNA Synthesis Present and Future,* Plenum Press, New York.

Monk, M., and Gross, J., 1971, Induction of prophage λ in a mutant of *E. coli* K-12 defective in initiation of DNA replication at high temperatures, *Mol. Gen. Genet.* **110**:299.

Mosig, G., and Breschkin, A. M., 1975, Genetic evidence for an additional function of page T4 gene 32 protein: Interaction with ligase, *Proc. Natl. Acad. Sci. U.S.A.* **72**:1226.

Nishimura, Y., Caro, L. G., Berg, C. M., and Hirota, Y., 1971, Chromosome replication in *Escherichia coli.* IV. Control of chromosome replication and cell division by an integrated episome, *J. Mol. Biol.* **55**:441.

Olivera, B. M., 1978, DNA intermediates of the *E. coli* replication fork: Effect of dUTP, *Proc. Natl. Acad. Sci. U.S.A.* **75**:238.

Prescott, D. M., and Kuempel, P. L., 1972, Bidirectional replication of the chromosome in *Escherichia coli, Proc. Natl. Acad. Sci. U.S.A.* **69**:2842.

Pritchard, R. H., and Lark, K. G., 1964, Induction of replication by thymine starvation at the chromosome origin in *Escherichia coli, J. Mol. Biol.* **9**:288.

Pritchard, R. H., Barth, P. T., and Collins, J., 1969, Control of DNA synthesis in bacteria, *J. Symp. Soc. Gen. Microbiol.* **19**:263.

Radman, M., 1975, S.O.S. hypothesis: Phenomenology of an inducible DNA repair which is accompanied by mutagenesis, in: *Molecular Mechanisms for Repair of DNA,* Part A (P. Hanawalt and R. Setlow, eds.), pp. 355–367, Plenum Press, New York.

Roberts, J. W., Roberts, C. W., and Mount, D. W., 1977, Inactivation and proteolytic cleavage of phage λ repressor *in vitro* in an ATP-dependent reaction, *Proc. Natl. Acad. Sci. U.S.A.* **74**:2283.

Rodriguez, R. L., Dalbey, M. S., and Davern, C. I., 1973, Autoradiographic evidence for bidirectional DNA replication in Escherichia coli, *J. Mol. Biol.* **74**:599.

Rosenberg, B. H., Cavalieri, L. F., and Ungers, G., 1969, The negative control mechanism for *E. coli* DNA replication, *Proc. Natl. Acad. Sci. U.S.A.* **63**:1410.

Rownd, R. H., Pulman, D., Goto, N., and Appelbaum, E. R., 1975, Control of RTF and r-determinants replication in composite R plasmids, in: *DNA Synthesis and Its Regulation* (M. Goulian and P. Hanawalt, eds.), pp. 537–559, Benjamin, Menlo Park, California.

Saitoh, T., and Hiroga, S., 1975, Initiation of DNA replication in *Escherichia coli.* III. Genetic analysis of the DNA mutant exhibiting rifampicin-sensitive resumption of replication, *Mol. Gen. Genet.* **137**:249.

Sanderson, K. E., 1972, Linkage map of *Salmonella typhimurium, Bacteriol. Rev.* **36**:558.

Schaechter, M., Maaløe, O., and Kjeldgaard, N. O., 1958, Dependency on medium and temperature of cell size and chemical composition during balanced growth of *Salmonella typhimurium, J. Gen Microbiol.* **19**:592.

Scheckman, R., Weiner, J. H., Weiner, A., and Kornberg, A., 1975, Ten proteins required for conversion of φX174 single-stranded DNA to duplex form *in vitro, J. Biol. Chem.* **250**:5859.

Schubach, W. H., Whitmer, J. D., and Davern, C. I., 1973, Genetic control of DNA initiation in *Escherichia coli, J. Mol. Biol.* **74**:205.

Schuster, H., Schlicht, M., Lanka, E., Mikolajczyk, M., and Edelbluth, C., 1977, DNA synthesis in an *Escherichia coli dnaB dnaC* mutant, *Mol. Gen. Genet.* **151**:11.

Sompayrac, L., and Maaløe, O., 1973, Autorepressor model for control of DNA replication, *Nature New Biol.* **241**:133.

Sunshine, M., Feiss, M., Stuart, J., and Yochem, J., 1977, A new host gene *(groPC)* necessary for lambda DNA replication, *Mol. Gen. Genet.* **151**:27.

Taylor, A. L., and Trotter, C. D., 1972, Linkage map of *Escherichia coli* strain K-12, *Bacteriol. Rev.* **36**:504.

Tresguerres, E. F., Nandadasa, H. G., and Pritchard, R. H., 1975, Suppression of initiation negative strains of *Escherichia coli* by integration of the sex factor F, *J. Bacteriol.* **121**:554.

Tye, B. K., Nyman, P. O., Lehman, I. R., Hochhauser, S., and Weiss, B., 1977, Transient accumulation of Okazaki fragments as a result of uracil incorporation into DNA, *Proc. Natl. Acad. Sci. U.S.A.* **74**:154.

Ueda, K., McMacken, R., and Kornberg, A., 1978, *dnaB* protein of *Escherichia coli:* Purification and role in the replication of ϕX174 DNA, *J. Biol. Chem.* **253**:261.

Urban, J. E., and Lark, K. G., 1971, DNA replication in *Escherichia coli* 15T$^-$ growing at 20°C, *J. Mol. Biol.* **58**:711.

Wang, J. C., 1971, Interaction between DNA and an *Escherichia coli* protein ω, *J. Mol. Biol.* **55**:523.

Wechsler, J. A., 1975, Genetic and phenotypic characterization of *dnaC* mutations, *J. Bacteriol.* **121**:594.

Wechsler, J. A., 1978, The genetics of *E. coli* DNA replication, in: *NATO ASI: DNA Synthesis Present and Future* (I. Molineux and M. Kohiyama, eds.), Plenum Press, New York.

Wechsler, J. A., and Gross, J. D., 1971, *Escherichia coli* mutants temperature-sensitive for DNA synthesis, *Mol. Gen. Genet.* **113**: 273.

Wechsler, J. A., and Zdzienicka, M., 1975, Cryolethal suppressors of thermosensitive *dnaA* mutations, in: *DNA Synthesis and Its Regulation* (M. Goulian and P. Hanawalt, eds.), pp. 624–639, Benjamin, Menlo Park, California.

Weiner, J. H., McMacken, R., and Kornberg, A., 1976, Isolation of an intermediate which precedes *dnaG* RNA polymerase participation in enzymatic replication of bacteriophage ϕX174 DNA, *Proc. Natl. Acad. Sci. U.S.A.* **73**:752.

Wells, R. D., and Inman, R. B., 1973, *DNA Synthesis in Vitro,* University Park Press, Baltimore, Maryland.

Wickner, S., and Hurwitz, J., 1975, Interaction of *Escherichia coli dnaB* and *dnaC*(D) gene products *in vitro, Proc. Natl. Acad. Sci. U.S.A.* **72**:921.

Wickner, S., and Hurwitz, J., 1976, Involvement of *Escherichia coli dnaZ* gene product in DNA elongation *in vitro, Proc. Natl. Acad. Sci. U.S.A.* **73**:1053.

Wickner, S., Wright, M., and Hurwitz, J., 1974, Association of DNA-dependent and -independent ribonucleoside triphosphatase activities with *dnaB* gene product of *Escherichia coli, Proc. Natl. Acad. Sci. U.S.A.* **71**:783.

Witkin, E. M., 1976, Ultraviolet mutagenesis and inducible DNA repair in *Escherichia coli, Bacteriol. Rev.* **40**:869.

Worcel, A., 1970, Induction of chromosome reinitiation in a thermosensitive DNA mutant of *E. coli, J. Mol. Biol.* **52**:371.

Yoshikawa, H., and Sueoka, N., 1963, Sequential replication of the *Bacillus subtilis* chromosome. I. Comparison of marker frequencies in exponential and stationary growth phases, *Proc. Natl. Acad. Sci. U.S.A.* **49**:559.

Yoshikawa, H., O'Sullivan, A., and Sueoka, N., 1964, Sequential replication of the *Bacillus subtilis* chromosome. III. Regulation of initiation, *Proc. Natl. Acad. Sci. U.S.A.* **52**:973.

Zyskind, J. W., and Smith, D. W., 1977, Novel *Escherichia coli dnaB* mutant: Direct involvement of the *dnaB 252* gene product in the synthesis of an origin ribonucleic acid species during initiation of a round of deoxyribonucleic acid replication, *J. Bacteriol.* **129**:1476.

Zyskind, J. W., Deen, L. T., and Smith, D. W., 1977, Temporal sequence of events during the initiation process in *Escherichia coli* deoxyribonucleic acid replication: Roles of the *dnaA* and *dnaC* gene products and ribonucleic acid polymerase, *J. Bacteriol.* **129**:1466.

Note Added in Proof. More recent studies on the regulation of DNA replication will be published in Volume 43 of the *Cold Spring Harbor Symposium of Quantitative Biology,* entitled *DNA: Replication and Recombination.*

7

Genetic Control Signals in DNA

DAVID PRIBNOW

1 DNA Control Signals

1.1 Introduction

All organisms control the expression of the genetic information stored in their DNA. The DNA in every cell of every organism contains a vast number of genes. The gene products can carry out an equally vast number of biological and biochemical functions. Utilization of this potential must be coordinated and energetically efficient so that any organism can survive the selective pressures of its environment and reproduce in numbers sufficient to avoid extinction. Overall genetic regulation has two primary objectives: first, to allow for biologically favorable adaptive responses to changes in environment, and second, to coordinate an effectively irreversible program of development leading to reproduction. Control signals, encoded in an organism's DNA, are primary elements of this genetic regulation.

Gene regulation takes place mainly at the level of *transcription.* Transcription is the process whereby the enzyme *RNA polymerase* faithfully transcribes selected units of genetic information stored in DNA into RNA. It is the first step in gene expression and is therefore the logical target for efficient regulation. Most RNA transcripts are genetic *messages* that are translated by ribosomes to produce

DAVID PRIBNOW • Department of Molecular, Cellular, and Developmental Biology, University of Colorado, Boulder, Colorado

proteins; some other transcripts (namely, rRNAs and tRNAs) are not translated but are incorporated directly into the cellular biochemical machinery. Without transcription DNA is "silent"—nothing is produced and little that is of use to a cell can take place. At least theoretically, everything that goes on inside a cell can be controlled if this one process is controlled. So how *is* transcription controlled?

Transcription is controlled by specific base pair sequences encoded in DNA and by gene products that interact with the DNA sequences and with the RNA polymerase. Particular control sequences exist in DNA that tell the RNA polymerase where and how often to start and to stop RNA synthesis. Other control sequences in DNA are recognized by regulatory proteins that affect the interactions between the RNA polymerase and its start and stop signals and thereby regulate transcription. On top of this, some regulatory proteins can alter the RNA polymerase directly, changing its response to different start and stop signals in the DNA. A carefully organized array of "starts" and "stops" coupled with an appropriate set of modulating signals provides the DNA-directed transcriptional flexibility and coordination required to carry out an organism's life support and reproduction functions. But signals that are encoded in DNA are not utilized in transcriptional regulation alone.

DNA-encoded control signals govern the three processes that are the "guts" of molecular biology: transcription, translation, and DNA replication. Translation signals are "read"—by ribosomes—as they finally appear in RNA, not DNA, and so do not particularly concern us here. Like transcription signals, replication-related signals are read at the DNA level.

Although its regulation is probably far less complex, DNA replication can be viewed much as can transcription. This is especially true given that RNA synthesis is involved in the initiation of DNA replication in many biological systems (Kornberg, 1976) (see also Chapter 6 of this volume). Replication must start somewhere on a DNA molecule and stop somewhere on a DNA molecule; but unlike transcription, DNA replication must take place over the *entire* DNA complement of a cell. Thus, the replication process is not compartmentalized spatially, as is transcription. In addition, DNA replication can be pictured generally as an ongoing process relative to transcription (not in the lifetime of a cell, however). The crux of these differences is that replication is probably governed by only a few specific DNA control signals.

Little is known about replication control signals in DNA, whereas much is known about transcription control signals. In this chapter we deal exclusively with transcription control signals, with the hope that what we learn will provide clues about the probable nature of signals that control DNA replication.

Most of our present knowledge about transcription control signals in DNA and how they operate derives from genetic and biochemical studies of gene regulation in bacterial systems, and particularly in the bacterium *E. coli* and its associated viruses (coliphages). This chapter therefore focuses on the bacterial systems. The aim of the chapter is to formulate working models for the operation of various kinds of DNA-encoded transcription control signals. Implicit in the model building is the underlying assumption that the regulation of gene expression in higher cells has been "learned" substantially from bacteria and other prokaryotes.

1.2 *DNA Sequence*

In DNA, sequence determines specificity—it is *the* source of information. A genetic control signal is a specific region on a DNA molecule, and it must be described ultimately in terms of the sequence of nucleotides that comprises the particular signal. Control signals in DNA that are recognized by the same cellular enzyme or regulatory protein are necessarily similar, if not identical, in DNA sequence. Thus, an attempt must be made to define particular kinds of control signals in terms of prototype sequences. Of course DNA sequences alone can tell us little or nothing about the *functions* of control signals.

A control signal must be recognized and interpreted in order to be of any use. Accordingly, it is necessary to determine how proteins can discriminate among DNA sequences. Further, it must be determined how sequence recognition accomplishes a regulatory event.

The first task ahead is to formulate a general model for sequence-specific interactions between proteins and DNA. (For a somewhat different approach, see Chapter 8 of this volume.) In constructing this model, several important conditions must be met in keeping with our knowledge of the system with which we are dealing.

1.3 *Sequence-Specific Protein–DNA Interactions*

1.3.1 *Multiple Contacts*

A protein must make multiple sequence-specific contacts with DNA in order to interact selectively with a particular control signal. An often cited example illustrates this point: About 4×10^6 base pairs comprise the single *E. coli* DNA molecule. Assuming a *random* base pair sequence, a minimum of 12 nucleotide pairs defines a unique sequence on a genome of this size. A protein that recognizes only one control signal on *E. coli* DNA must therefore make at least 12 sequence-specific contacts with that control signal. If there is redundancy in DNA sequence information, then the number of required contacts increases accordingly.

1.3.2 *Specific versus Nonspecific Interactions*

The binding of proteins to control sequences must be several orders of magnitude tighter than the binding to nonspecific sequences. Consider a DNA molecule that contains 10^5 base pairs and ten specific binding sequences for a particular protein. The equilibrium binding constant for the specific binding sites (where $K_B = k_d/k_a$) must be a factor of 10^4 lower than that for the average nonspecific site, just to ensure that half of the specific sites are filled at any time. The minimum specific versus nonspecific binding requirements for a given protein are dictated in principle by the amount of DNA to which the protein has access and by the number of specific binding sites that are present on that DNA. In this context, we must keep an important fact in mind: For every unique 12-base-pair DNA sequence in an *E. coli*-sized DNA molecule there are, on the average, nearly 20 other sequences that differ by only one base pair. We must determine ultimately how a single base pair difference can effect a considerable change in a protein–DNA binding constant.

1.3.3 Finding the Signal

As a rule, proteins make initial contact with the DNA on the *outside* of the double helix. In a physiological environment, DNA is essentially completely native (see von Hippel and McGhee, 1972). At any instant, a given segment of the DNA helix—including a control signal—is closed by Watson–Crick base pairing. It is probably also in the B-helix conformation state. Although DNA does undergo local structural perturbations that include melting of base pairs and, probably, helix conformation changes, such perturbations are apparently rare. From a kinetic standpoint it would be disastrous to require that a protein first encounter its specific binding sequence when that sequence is coincidentally assuming an abnormal configuration.

Control proteins find their DNA signals very rapidly. In fact, measured or inferred association constants for particular protein–DNA interactions are close to and even greater than what is expected for a diffusion-limited process. This suggests that specific DNA-binding proteins spend most of their time *on* DNA rather than wandering off through the solvent. Various modes of rapid "search" along and among DNA molecules have been suggested, but it is not yet clear how sequence-specific interactions are accomplished with such speed. Surely a very low relative affinity for nonsignal sequences helps a protein search the DNA rapidly (see chapter 8 of this volume).

1.3.4 Tight Complexes, Effectors, and Reversibility

Some proteins elicit their regulatory functions by simply binding at a specific control sequence or by not binding at the control sequence, the biochemical consequences of the bound and unbound states being different. In some cases, the binding is unregulated and is therefore irreversible, in a sense. The protein has a high intrinsic affinity for its signal, and regulation occurs only at the level of expression of the gene that encodes the protein. In other cases, the binding is regulated directly and is reversible. The protein can exist in two distinct states (conformations). In one state it has a high affinity for its control signal and a low affinity for other DNA sequences; in the other state, the protein has a low affinity for all sequences (or it could have acquired a high affinity for some *different* sequence). The choice of state is made by an *effector* molecule that interacts specifically within the protein. Effector molecules are usually small molecules, and their interactions with regulatory proteins usually are by way of concentration-dependent equilibria (see, for example, Friedman *et al.*, 1977).

A protein–effector complex acquires either increased or reduced affinity for a specific DNA control sequence relative to the affinity of the protein alone for that sequence. For example, a *repressor* protein can block the transcription of some gene(s) by binding within a DNA region that is a start signal for the RNA polymerase (see Section 3.11). A specific effector molecule alters its repressor "target," assuring either the signal-bound state that blocks transcription or the signal-free state that allows transcription to proceed. When the effector concentration is low enough, most repressor molecules are free of the effector molecule and the situation is reversed.

In this example the repressor protein cannot block transcription unless it is bound at the transcription start signal at exactly the time when the RNA polymerase attempts to use the signal. Since it cannot anticipate the polymerase, the repressor must either bind very tightly to its control signal—and thus for a relatively long time—or it must be present in the cell at such a high concentration that mass action keeps the site filled most of the time. Providing a huge reservoir of a regulatory protein that elicits control at only one or a few DNA sites probably does not make biological sense (energetically and spatially) in most instances. Very tight binding is the better alternative.

As a rule, we expect regulatory proteins whose control sequences are rare to be present in cells at low concentrations and to have high affinities for those sequences. Available information shows that at least some regulatory proteins are indeed present at low intracellular concentrations and that they have characteristic sequence-specific binding constants in the range of 10^{-10} to 10^{-13} M (Riggs *et al.*, 1970*a,c;* Chadwick *et al.*, 1970; Brack and Pirrotta, 1975; Parks *et al.*, 1971*a;* Rose and Yanofsky, 1974). Off-rates for complex dissociation are on the order of minutes to hours! In principle, the previously mentioned time constraints do not apply necessarily for the transient interactions between the RNA polymerase and its start signals in DNA. Nevertheless, the RNA polymerase does bind very tightly to at least some of its start signals ($K_B \sim 10^{-10}$–10^{-11} M), and the complexes dissociate very slowly (measured in the absence of transcription initiation) (see Chamberlin, 1976). In general, then, we can say that a regulatory protein has a very high affinity for its particular DNA control signal(s), and that in many cases the binding interaction is reversibly mediated by protein–effector molecule interactions. Tight protein–DNA binding and the reversibility of tight binding have profound implications for the energetics of protein–DNA interactions.

1.3.5 Binding Energy and Ionic Bonds

Protein–DNA equilibrium binding constants in the range of $K_B = 10^{-10}$ to 10^{-13} M reflect binding free energies of between −15 and −20 kcal/mol. Much of this energy is provided by the binding enthalpy. The overall entropy of the system probably does not change too dramatically when a protein engages DNA (but see Record *et al.*, 1976). What is the source of this large binding energy?

Ionic bonds are probably the major source of binding free energy in protein–DNA interactions. Hydrogen bonds can provide little energy, since the net energetic gain in going from the free state (where both protein and DNA make H bonds with the aqueous solvent) to the bound state is essentially zero. van der Waals interactions between transient dipoles can provide little more than about 1 kcal/mol of bonds formed. (Hydrophobic interactions are probably the source of most nonionic binding energy.) This leaves ionic bonds, which can provide up to about 6 kcal/mol of bonds formed. Several ionic pairs with an average bond energy of 3 to 4 kcal/mol easily provide the large binding enthalpy that we know is required in "tight" protein–DNA interactions. We should note that some amount of binding energy (entropic) is provided by the release of monovalent counterions from the DNA and perhaps from the protein when protein–DNA ionic pairs are formed (see Record *et al.,* 1976).

That ionic bonding occurs between regulatory proteins (or the RNA polymerase) and DNA has been demonstrated in a variety of ways (see von Hippel and McGhee, 1972; Record *et al.,* 1976; Chamberlin, 1976). For example, regulatory proteins in general bind to polyanions. DNA is a polyanion. Specific and nonspecific protein–DNA interactions are weakened at high ionic strength—more specifically, at elevated concentrations of monovalent ions—showing that ionic bonds are formed between the proteins and the DNA. Perhaps most convincing is the fact that three different *E. coli* proteins, the *lac* repressor (A. Maxam and W. Gilbert, personal communication), the catabolite activator protein (CAP) (Majors, 1978), and the RNA polymerase (L. Johnsrud, personal communication), have been shown directly to make *several* ionic contacts each with particular DNA phosphates at their respective binding sequences in the DNA of the *E. coli* lactose operon. For a regulatory protein to make several ionic contacts with DNA phosphates, it must have a stretch of basic amino acid residues lining the binding site. This has been shown for several proteins (Adler *et al.,* 1972; Beyreuther and Gronenborn, 1976; Ptashne *et al.,* 1976; Krakow *et al.,* 1976; see also von Hippel and McGhee, 1972; Record *et al.,* 1976; Zillig *et al.,* 1976).

Energetically, just one ionic bond can make the difference between specific and nonspecific protein–DNA interactions. For a given protein, a ratio of sequence-unspecific to sequence-specific binding constants of 10^4 to 10^5 reflects a difference in binding energy of 5 to 7 kcal/mol—the energy potentially available in a single ionic pair. This makes sequence discrimination by regulatory proteins easier to understand, as we will see shortly. It also has implications for reversible, tight protein–DNA interactions.

1.3.6 Energetics of Reversible Binding

Energetically, true reversibility can be effected by the making or breaking of a single ionic pair between a control protein and its DNA-binding site. An effector molecule that interacts with the protein (or, possibly, with the DNA) need induce only a conformational change that creates or eliminates an ionic bond.

In general, the objective of an effector molecule is either to put the regulatory protein on its control signal or to take—and keep—the protein off its signal. In the first case, the effector must stabilize the protein–DNA interaction so that the protein spends a relatively long time on the control signal—that is, the effector molecule must (at least) reduce the complex off-rate. Further, removal of the effector should return the system to its original state—that is, the off-rate must be reelevated by a favored, reverse change in protein conformation. Overall, these conditions require that the energy input to induce the effector-mediated conformational change be greater than the free energy gain from formation of the extra protein–DNA ionic pair. In the second case, effector binding must destabilize the protein–DNA interaction so as to leave the control signal open. The energy input of effector binding must be greater than the *sum* of the energies required to induce the conformational change and to compensate the lost ionic pair—that is, taking a protein off the DNA is harder than putting it on. In both cases, the system is responsive energetically to the effector molecule concentration. This gives the regulatory system some flexibility, particularly in the concentration range of the effector–protein equilibrium binding constant.

1.3.7 The Main Condition

All of the foregoing considerations lead to this conclusion: In sequence-specific protein–DNA interactions, energy provided by ionic bonding and information provided by many sequence-specific contacts must be coupled. This coupling must allow for strong discrimination among many DNA sequences sometimes differing by only a single base pair. A brief examination of DNA structure will show where the points of potential ionic and sequence-specific contacts reside on native DNA. With this information, we can construct a model for energy–information coupling in specific protein–DNA interactions.

1.3.8 DNA Structure

1.3.8a Ionic Contacts. The phosphate groups along the outsides of both helices of the double-stranded DNA molecule are potential sites of ionic contact. Each phosphate group has a net negative charge (-1) shared between the two oxygen atoms that point away from the helix. The positions of these two oxygens are fixed. In fact, we can consider the positions of all atoms comprising the DNA molecule to be *fixed* relative to one another, which makes the job of a protein a lot easier than it might otherwise be.

Although ionic pairs with the DNA phosphates can provide the energy we need, these contacts are probably completely *sequence unspecific.* The phosphates are distributed in an entirely regular fashion from base pair to base pair along the DNA backbone. Are there any other points of likely ionic contact, particularly ones that reflect the DNA sequence? Apparently not. [Base-pair-specific chelation of ions has been suggested as a limited source of information-linked ionic pairing (Seeman *et al.,* 1976).] There are no other ionic moieties anywhere on DNA. So, we must now identify the potential sources of sequence-specific contacts on DNA and then determine how these contacts can be specifically coupled with protein–DNA phosphate ionic contacts.

1.3.8b H-Bond Contacts. Hydrogen bonding to the heterocyclic bases in the major and minor grooves of DNA is probably the main source of sequence information in protein–DNA interactions. (van der Waals interactions are also a possible source of sequence discrimination, but we shall concentrate on the more obvious H-bond possibilities.) In native DNA, the bases are stacked and paired *à la* Watson–Crick so that only the edges of the bases are visible in the DNA grooves—actually, each base pair is seen edgewise in each groove. Each base (and, consequently, each of the four base pairs) is potentially distinguishable from the other three bases via H bonding directed into the DNA grooves. Although the thymine methyl group—protruding conspicuously into the major groove—cannot participate in H-bond formation, it provides a highly discriminatory "block" in the same position in which H bonds could be formed with any other base. It therefore provides H-bond information. The thymine methyl group is also a source of specific hydrophobic interactions (Goeddel *et al.,* 1977). [Superimposition of the base pairs in four sets of two allows for easy comparison of different H-bond potential, and this is done for us in a recent article by Seeman *et al.* (1976).] Most of the base-specific H-bond donors and acceptors reside in the major groove. Few reside in the minor groove. Therefore, *most* of the information that enables proteins to discriminate DNA sequence will be sought in the major groove.

As just implied, there are points of equivalent or near-equivalent contact with some of the bases, especially in the minor groove (see Seeman *et al.,* 1976). This means that there is some potential for redundancy in sequence information. Unfortunately, we cannot indulge here in an analysis of what base pair substitutions are allowable given specified points of H-bond contact; however, the fact of probable information redundancy should be kept in mind.

Overall, then, sequence-specific H-bond information resides in the DNA grooves, the majority in the major groove. Energetically strong points of ionic contact, the DNA phosphates, are situated along the outside of the DNA backbone. How are H-bond information and ionic bond energy *coupled?* Ionic and H-bond contacts on DNA must be coupled *spatially* through "measured" interactions with protein.

1.3.9 The Ruler Concept

A protein can be usefully viewed as a three-dimensional "ruler" (thanks to P. von Hippel). A protein has the potential for making many different noncovalent interactions, both along its polypeptide backbone and through its various amino acid side chains. Intraprotein interactions can make local regions of a protein assume a relatively rigid three-dimensional conformation in space, thereby holding points of potential DNA contact in reasonably precise relative position. (This situation obtains in the active sites of enzymes.) Allosteric effects show that protein structure can be quite rigid over large intraprotein and even intersubunit distances. Thus, a protein has the potential to make many site-specific contacts with DNA by measuring the distances between its own spatially fixed contact points and those available to it on the DNA.

1.3.10 The Model

Here is our working model for the selective interaction of a (control) protein with its specific DNA (control) sequence: The protein uses a spatially fixed array of bonding moieties to measure the distances from phosphates on the DNA backbone to base-/base-pair-specific H-bond donors and/or acceptors in the major and minor grooves of the DNA. Probably two or more sequence-specific H bonds are utilized in order to specify each ionic pair. (Specific hydrophobic interactions could be coupled similarly to H-bond contacts.) *All* contacts are coupled to each other physically through the protein structure.

Blocking of H-bond contacts is the obvious source of DNA sequence discrimination in this model. Whenever a hydrogen donor or acceptor on the protein cannot find an appropriate partner on the DNA, it will interact preferentially with the aqueous solvent so as to conserve H-bond energy. For example, if an H-bond donor on the protein is confronted with another donor in a DNA groove, it will be held at bay. This situation prevents close ionic contact between the protein and the

DNA phosphate that is coupled to the protein hydrogen donor. In this respect, the thymine methyl group is probably very important in blocking close ionic contact at nonspecific sequences along a DNA molecule. (Also, mutations that substitute A–T base pairs in control sequences could be very damaging to major groove interactions.) H-bond blocking has direct implications for effector-induced reversibility of tight binding interactions.

Altered protein H-bond potential could create or destroy the energetically stronger ionic bonding potential of a protein. An effector molecule need only induce a protein conformational change that reorients one or more of the protein H donors or acceptors, so that blocking is either effected or eliminated at the conjugate DNA signal. Effector molecule binding, by rearranging protein H-bonding potential even slightly, could actually mediate tight binding of a protein at two different, closely related DNA sequences. An effector could, of course, induce a protein rearrangement that effectively buries or exposes an important basic residue, aside from considerations of H-bond perturbations.

Whereas we have concentrated on tight protein–DNA complexes so far, the model accommodates weak complexes as well. The overall strength of the interaction reflects primarily the number and strength of the ionic pairs between the protein and the DNA. An equally large number of H-bond contacts could specify a tight complex or a weak complex, depending on the number and strength of ionic pairs involved.

This model was constructed to provide a useful conceptual framework for analyzing and understanding the interactions between regulatory proteins and their conjugate DNA control signals. For detailed information about investigations of protein–DNA interactions, the reader is directed to illuminating review articles by Yarus (1969) and by von Hippel and McGhee (1972). See also Chapter 8 of this volume.

1.4 Genetics of Control Signals

Genetics was, and usually still is, the first source of information about DNA control signals. In fact, genetic studies led directly to the operon concept of gene regulation (Jacob and Monod, 1961*a*).

Control signals do not code for diffusible gene products. Consequently, they are distinctive genetically. Mutations that alter DNA control signals are cis dominant (Jacob and Monod, 1961*a,b;* Epstein and Beckwith, 1968; Beckwith and Rossow, 1974). A cis-dominant mutation permanently alters the expression of some function *on the DNA molecule harboring the mutation.* Its effect on the DNA molecule cannot be overcome by the wild type allele residing on another DNA molecule—that is, it cannot be complemented in trans. For example, a mutation that destroys an RNA start signal on a DNA molecule absolutely prevents normal gene transcription from that signal. Providing a "good" start signal on another DNA molecule does not help the mutated molecule to be transcribed. Mutations leading to altered protein or RNA products can, however, be complemented in trans.

A molecular geneticist can exploit the unique genetics of control sequences. One can select and screen for cis-dominant mutations that affect the expression of one or more assayable gene products in a cell. A well-designed selection scheme may ultimately provide several distinct mutations in one particular control signal. Genetic mapping can then be employed to locate the mutations—and thereby the signal—relative to other mutations in the DNA. Biochemical tests can give information about the mutation-induced abnormality. In principle, secondary mutations that compensate for the primary control signal mutation(s) can be selected (see Beckwith and Rossow, 1974), providing still more information about the signal and the cellular components that interact with the signal. Consider, for instance, a mutation in a control signal that prevents a regulatory protein from recognizing the signal. One might select for a compensatory mutation in the regulatory protein (amino acid substitution) that partially restores signal recognition and therefore signal function (perhaps a mutation that rearranges protein H-bonding potential to fit the new configuration in the control signal sequence). Note that our model for protein–DNA interactions generally dictates *against* compensatory mutations in the control signal itself. A second change would only make things *worse*—except in the case of a compound signal, as we will see later.

Mutations change DNA sequence. For a particular control signal, determination of wild type and mutant DNA sequences can tell us which base pair substitutions block specific protein–signal recognition.

1.5 DNA Sequence Analysis

DNA sequencing has become rather easy, owing to recent advances in technology (Sanger and Coulson, 1975; Maxam and Gilbert, 1977; Sanger *et al.*, 1977, and references cited therein). The main problem, then, is getting the right piece of DNA to sequence. Briefly, two approaches have proved most successful. The first takes advantage of the high affinity that a regulatory protein often has for its conjugate control sequence. DNA that is complexed tightly to protein *in vitro* is generally protected from DNase digestion, while uncomplexed DNA is degraded. Thus, using a purified protein, an intact signal can be isolated and then sequenced. The second approach involves isolating a larger DNA fragment that includes the desired control signal. Restriction endonucleases that cut double-stranded DNA at specific sequences are used to generate a specific set of DNA fragments. The signal-containing fragment is then isolated either by virtue of its mutual affinity for the purified control protein or by virtue of its size. (Isolation by size requires that a linear map of the DNA restriction fragments be correlated with a genetic map locating the control signal and that the size of the desired fragment be known relative to the sizes of the other fragments.) The entire fragment is then subjected to sequence analysis. (Besides their usefulness in sequence analysis, such DNA fragments are extremely valuable for *in vitro* investigations of the function(s) of particular control signals.) Of course, it is entirely possible to know the DNA sequence of a region containing a control protein binding site and yet not know the base pairs with which the protein actually interacts.

Mutations define important base pair contacts in control signals, and so sequence analysis of mutant signals is extremely important. If feasible, saturation of a control signal with all possible mutations coupled with DNA sequence analysis identifies all base pair contacts that the conjugate protein makes with the control signal. In some special cases (where signals overlap with genes or other signals), information about *redundant* protein–DNA contacts can be obtained by noting base pair substitutions that do not impair control signal function significantly.

Again, the DNA sequences of several control signals that are recognized by the same protein must be related. If the sequences of many related signals are known, then they can be compared. In such a comparison, conservation of sequence probably reflects points of sequence-specific protein contact; and sequence variation probably reflects points of nonessential or nonspecific contact with the control protein. Mutations, of course, eliminate the guesswork; similarly, chemical probes can be used to determine the sequence(s) with which a protein actually interacts.

1.6 Chemical Probes

Certain chemical probes can be used in a purified *in vitro* system to identify points of sequence-specific protein–DNA contact. A powerful new technology has been developed that employs reagents that bond covalently to particular atoms on DNA (Gilbert *et al.,* 1976; A. Maxam and W. Gilbert, personal communication). The technology is tied directly to a rapid DNA sequencing method (Maxam and Gilbert, 1977). A particular reagent is reacted with a homogeneous population of DNA fragments containing a control signal; it is reacted separately in a mixture containing the DNA signal *and* the purified control protein, conditions being chosen so that tight protein–signal complexes are favored. Comparative analysis of reacted DNA sites subsequently identifies those sites whose reactivity was altered (blocked or enhanced) by the presence of the protein. Presumably, these are points of protein contact at the control signal. This method has now been used to determine backbone phosphate contacts and base-specific contacts between several different proteins and their conjugate DNA-binding signals (Gilbert *et al.,* 1976; Majors, 1978; Ptashne *et al.,* 1976; A. Maxam and W. Gilbert, personal communication; L. Johnsrud, personal communication).

Another very promising approach involves substitution of base analogs for specific bases in control signal DNA—a sort of "*in vitro* mutagenesis" (Stahl and Chamberlin, 1977; Goeddel *et al.,* 1977). Substitutions at particular sites followed by *in vitro* analysis of protein–signal binding parameters can show directly what contacts are or are not made at the normal signal (Goeddel *et al.,* 1977).

1.7 General Information in DNA

Proteins can probably take advantage of some *general* features of the DNA sequence in recognizing and/or utilizing a control signal. To illustrate this point, consider the initiation of RNA synthesis by the RNA polymerase. The polymerase must gain access to the *inside* of the double helix in order to polymerize ribonucleotides that are complementary to the transcribed DNA strand—that is, the polymerase must melt the DNA locally (see Chamberlin, 1976). A–T-rich regions of the DNA "melt" more readily than do other regions. Incorporation of an A–T-rich DNA segment within the domain of polymerase–DNA contacts at a start signal might facilitate the overall initiation process. Similarly, there are indications that local regions of DNA may vary in conformation from the standard B-helix form (von Hippel and McGhee, 1972) (see also Chapter 8 of this volume). This means that the relative disposition of DNA phosphates, and so on, might vary somewhat depending on the general underlying DNA sequence in a particular region. Proteins might accordingly take advantage of particular DNA helix conformations in signal recognition. In any case, it is probably useful sometimes to view control signals as being comprised of both specific and general sequence information.

2 *Transcription Control Signals*

2.1 *The Transcription Unit*

The molecular picture of transcription and its regulation in *E. coli,* and to some degree in other bacteria, is rapidly becoming clear, if somewhat elaborate. Many details of this picture derive from studies of the relatively uncomplicated and easily manipulable bacteriophages. Our understanding of the coordinated transcription process comes primarily from two sources: (1) genetic analysis of mutations affecting the transcription of specific genes; and (2) *in vitro* studies of the purified components of the transcription system and their physical and functional interactions. Combined genetic and biochemical studies have proven to be very informative. The integrated lesson from all of these studies is that transcription and its regulation can be understood best in terms of the DNA-encoded control sequences that define the basic unit of transcription.

Very simply, a transcription unit is a stretch of DNA base pairs bounded on one end by a "start sequence" or *promoter* (Jacob *et al.,* 1964; Epstein and Beckwith, 1968) and at the other end by a "stop sequence" or *terminator* (Reznikoff, 1972). The RNA polymerase initiates RNA synthesis at the promoter, it copies one strand of the DNA template synthesizing a complementary RNA molecule, and it terminates the RNA chain and disengages from the DNA when it reaches the terminator. Transcription units sometimes overlap with each other. Usually at least one gene is contained in a transcription unit; and when there are two or more genes in a particular transcription unit, they are expressed, and controlled, *coordinately.*

2.2 *Development versus Maintenance*

The transcription of two or more different transcription units is sometimes linked absolutely in time. That is, some gene product or products arising from one transcription unit are essential for the transcription of one or more other units, and the first transcription unit is not expressed all of the time. Such interdependence of transcription relies upon transcriptional regulation and constitutes a system of *transcriptional development.* Transcriptional development becomes organismic development when it is essentially irreversible.

Many transcription units are relatively independent of all other transcription units, whether or not they are regulated. These transcription units generally provide for life-sustaining functions, independent of time.

2.3 *Control of Transcription*

An unregulated transcription unit is expressed constitutively (all of the time and at an essentially constant rate). The rate of transcription is determined primarily by the rate of RNA chain initiation at the promoter and so depends on the relative promoter strength. (It is also tied to the availability of energy useful to the cell, reflected by the cell growth rate.) Control of a regulated transcription unit is exerted at the *promoter* and/or at the *terminator*—that is, by regulation of promoter and/or terminator strength. Some promoters can be activated or repressed; and, similarly, some terminators can be effectively turned on or turned off.

Here, we explore what is known genetically and biochemically about promoters, terminators, associated regulatory sequences, and the proteins that interact with all of these signals to bring about or to control the transcription of specific genes. We define each type of genetic signal, and we construct a model for the function of each type of signal. Keeping in mind the fact that it is the *organization* of the signals around selected genes that turns biochemistry into real biology, we begin at the beginning—with the promoter.

3 *The Promoter and Its Regulation*

3.1 *The Promoter*

The basic promoter (Jacob *et al.,* 1964; Epstein and Beckwith, 1968) is only that particular DNA sequence that is recognized directly and used by the RNA polymerase as a start signal for transcription. The basic promoter can be strong, weak, or somewhere in between. Oftentimes a promoter is more complex: In addition to the polymerase start signal, it contains a sequence that is the binding site for a specific *activator* protein (Gilbert, 1976; Majors, 1977; Musso *et al.,* 1977; see Englesberg and Wilcox, 1974). By complexing with its binding sequence in a promoter, an activator protein can stimulate the initiation of transcription at the promoter, probably through secondary interactions with the RNA polymerase. Operationally, then, a promoter is probably defined best as the DNA sequence or sequences at the beginning of a transcription unit that are required for the maximum observable rate of initiation of transcripts across that unit under normal *in vivo* conditions. By this definition, the promoter includes the DNA sequence recognized directly by the RNA polymerase plus any auxiliary sequence that is an activator-binding site. The definition intentionally excludes negative regulatory elements of the DNA.

A promoter is characterized genetically by cis-dominant mutations that affect the level of expression of a given transcription unit, that map at the beginning of the transcription unit, and that are not operator (see later) mutations (Jacob and

Monod, 1961*b;* Epstein and Beckwith, 1968; Reznikoff, 1972; Beckwith and Rossow, 1974). One obvious type of promoter mutation changes the basic promoter start sequence so that the direct interaction between the polymerase and the promoter is altered, leading to an increase or, more likely, a decrease in the rate of initiation at the promoter. A mutation in the binding site for an activator protein is different, though also cis dominant. Such a mutation does not (usually) affect the efficiency of the promoter in the absence of the activator protein, but it does block or hinder the activator from stimulating transcription to its maximum level (Miller *et al.,* 1968; Englesberg and Wilcox, 1974). A mutation that alters a specific activator protein is not cis dominant. It can be pleiotropic, affecting the expression of more than one transcription unit, and it can be genetically distant from, or not linked to, the affected transcription unit(s) (see Englesberg and Wilcox, 1974).

3.2 Transcription Initiation

The initiation of transcription is a complex, multistep process (see Chamberlin, 1976; Krakow *et al.,* 1976). The RNA polymerase must locate a promoter, open the DNA strands so as to accommodate ribonucleotide base pairing with the transcribed DNA strand, bind to the initiating and subsequent ribonucleotides, and catalyze the formation of the first (and subsequent) phosphodiester bond(s). The polymerase must also move along the DNA, away from the promoter. How do all of the initiation events ensue from polymerase–promoter interactions?

We know that the *E. coli* RNA polymerase binds to a promoter (usually tightly) in what is aptly called a *rapid-start* (RS) *complex* (Mangel and Chamberlin, 1974*a,c;* see Chamberlin, 1976). An RS complex initiates RNA synthesis with very fast kinetics in the presence of RNA precursors—the DNA strands are open and the RNA polymerase is primed to go. For our purposes, establishing the RS complex is the primary objective of polymerase–promoter interactions. And so how is the RS complex formed? Examination of RNA polymerase structure and of detailed promoter structure provides the basis for constructing an answer.

3.3 The RNA Polymerase

The *E. coli* RNA polymerase is made up of five noncovalently joined proteins: one β subunit, one β' subunit, two identical α subunits, and a σ subunit, with a total mass of approximately 500,000 daltons (see Burgess, 1976; Zillig *et al.,* 1976). In structure, and probably in function, the *E. coli* RNA polymerase is representative of prokaryotic RNA polymerases in general (see Burgess, 1976). β, β', and α_2 comprise the RNA polymerase *core,* which is solely responsible for RNA chain *elongation* (Travers and Burgess, 1969; Pettijohn *et al.,* 1970; Krakow *et al.,* 1969; Zillig *et al.,* 1976). The σ subunit (sigma factor) is loosely associated with the core polymerase, and it dissociates from the core after RNA chain initiation.

The sigma factor is responsible for successful promoter *recognition* (see Losick, 1972; Chamberlin, 1974), and thereby promoter *utilization.* The core polymerase is deficient in the initiation of RNA chains on natural DNA molecules, and it does not utilize promoter sequences to a significant extent (Burgess *et al.,* 1969;

Goff and Minkley, 1970; Zillig *et al.*, 1970*a;* Burgess, 1971; Losick, 1972; Chamberlin, 1974). The sigma factor reduces RNA polymerase affinity for nonpromoter sequences (Hinkle and Chamberlin, 1972). This is probably a direct consequence of the fact that sigma mediates sequence-specific polymerase contacts with promoter DNA (see later). In association with the sigma factor, the core polymerase initiates RNA synthesis very efficiently on natural DNA molecules and shows a high degree of specificity for initiation at promoter sequences (see Losick, 1972; Chamberlin, 1974). What distinguishes a promoter from all other DNA sequences?

3.4 The Basic Promoter

The basic promoter contains three sets of sequence information that are organized spatially along the DNA (Pribnow, 1975*a,b;* Gilbert, 1976). An idealized promoter sequence which is a composite of over 20 known promoter sequences is presented in Fig. 1 together with a small collection of coliphage and *E. coli* promoter sequences (only the nontranscribed DNA strand is presented). Transcription proceeds to the right from **I,** the initiation point(s) of incorporation of the initiating (5′) ribonucleotide(s). There are two sets of DNA sequence that are largely conserved among promoter sequences, designated $\mathbf{R}_\sigma$ (for sigma "recognition") and $\mathbf{R}_c$ (for core polymerase recognition), as indicated in the figure. Some base pairs within these two sequences are more variable than others. $\mathbf{R}_\sigma$ and $\mathbf{R}_c$ undoubtedly reflect regions of sequence-specific polymerase promoter contact, with a potential for about 20 such contacts at an ideal promoter. $\mathbf{R}_\sigma$, $\mathbf{R}_c$, and **I** are uniformly distributed in all promoters, as indicated in Fig. 1 by the small variation in the number of base pairs intervening between the promoter elements. Spatial organization of the sequences is clearly important. Although these sequences probably do not represent *all* of the specific contacts that the RNA polymerase makes at a promoter, they probably *do* represent the most important contacts for promoter function.

A simple model for promoter function is presented in the following section and is schematized in Fig. 2. The model derives from the foregoing considerations of RNA polymerase structure (and function) and promoter structure—and from a large assortment of genetic and biochemical studies of promoter structure and function, several of which are discussed shortly using the model as an interpretive framework. Many features of this model have been put forth before, based on much of the same evidence (Walter *et al.*, 1967; Burgess, 1971; Losick, 1972; Chamberlin, 1974; Pribnow, 1975*b;* Gilbert, 1976).

3.5 Promoter Function—The Model

Promoter utilization by RNA polymerase occurs in three general steps: recognition, melting in, and initiation.

Recognition: The RNA polymerase first recognizes the promoter by making sequence-specific contacts at $\mathbf{R}_\sigma$ and at $\mathbf{R}_c$. Both sequences are in the native (undenatured) state. The sigma factor makes the contacts at $\mathbf{R}_\sigma$, and the core polymerase makes the contacts at $\mathbf{R}_c$ (as pictured in Fig. 2). Both sets of specific interactions enable promoter sequence discrimination. They also orient the polymerase in the proper direction for transcription from the promoter. Core contacts at $\mathbf{R}_c$ ensure that the RNA polymerase is poised over the region of the promoter that has to be opened and then held open so that the rapid-start (RS) complex is formed. The sigma interactions are necessary to stabilize the core interactions at $\mathbf{R}_c$ and beyond.

Melting in: The RNA polymerase core establishes a set of strong, probably ionic, contacts with the two DNA strands that actually pull the strands apart. These melting interactions, largely sequence unspecific, are localized around the initiation point, **I**. The melted complex is the RS complex. Core contacts at $\mathbf{R}_c$ are maintained in the RS complex; similarly, the sigma factor maintains specific contacts at $\mathbf{R}_\sigma$, and it stays bound to the core enzyme. This whole set of interactions stabilizes the catalytic center of the core polymerase at the melted initiation point **I**, so that a complete active site is established. The RNA polymerase does not

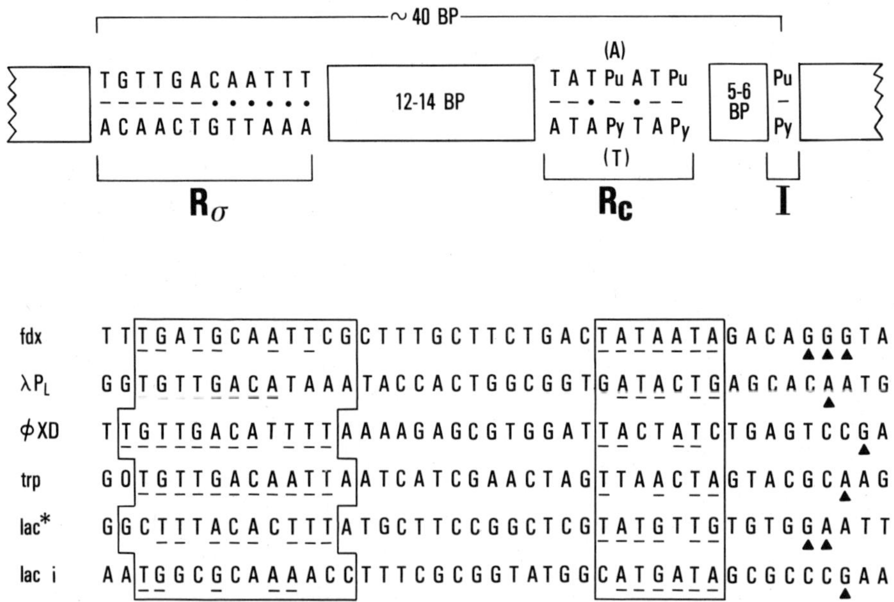

Figure 1. Promoter sequences. An ideal promoter sequence is presented at the top of the figure. It contains the three essential promoter elements, $\mathbf{R}_\sigma$, $\mathbf{R}_c$, and **I**, in which the sequences show those base pairs that are most common among over 20 known promoter sequences. The spacers between $\mathbf{R}_\sigma$ and $\mathbf{R}_c$ and between $\mathbf{R}_c$ and **I** usually contain 12 to 14 base pairs and 5 to 6 base pairs, respectively. The bottom DNA strand is the transcribed strand, and it always contains a pyrimidine nucleotide (Py) at the initiation point, **I**. The middle base pair in the $\mathbf{R}_c$ sequence is always a purine→ pyrimidine pair (Pu→ Py), but evidence noted in the text suggests that an A → T pair is best. (The 40 base pairs that are indicated from $\mathbf{R}_\sigma$ to **I** are not the same as the 40 base pairs that are protected by the polymerase from digestion by DNase I.) Dashes between paired bases indicate the most invariant base pairs in $\mathbf{R}_\sigma$ and in $\mathbf{R}_c$. The lower portion of the figure shows six known promoter sequences; fdX (Schaller *et al.*, 1975; Sugimoto *et al.*, 1975); λP_L (Maniatis *et al.*, 1974); ϕXD (Sanger *et al.*, 1977; J. Sims, unpublished data); *trp (E. coli)* (Bennett *et al.*, 1976); *lac** (ZYA) (Dickson *et al.*, 1975); and *lacI* (M. Calos, manuscript in preparation). Other known sequences are too numerous to include here. Only the untranscribed strand is presented. In each case, the $\mathbf{R}_\sigma$ and $\mathbf{R}_c$ sequences are boxed in, and the ideal base pairs are underlined in each sequence. Triangles indicate the initiation points.

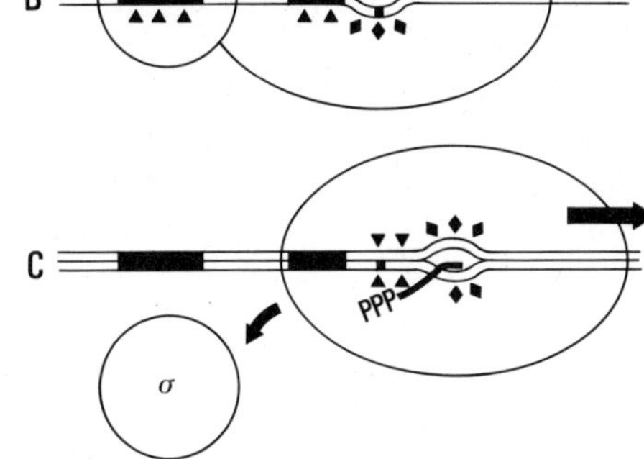

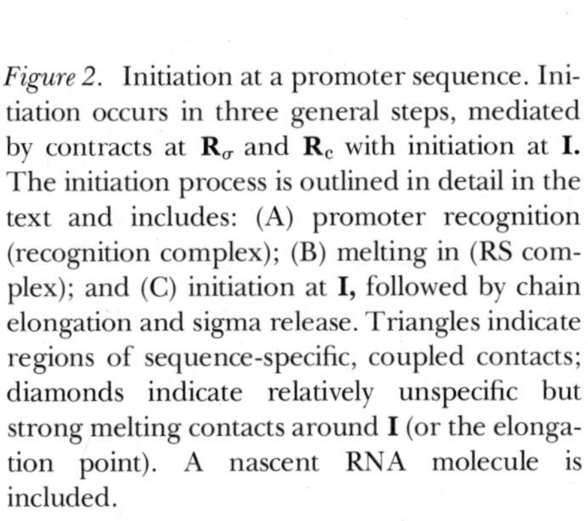

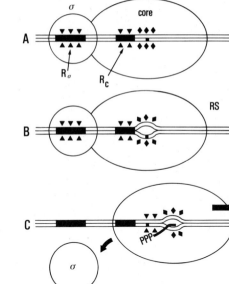

Figure 2. Initiation at a promoter sequence. Initiation occurs in three general steps, mediated by contracts at $\mathbf{R}_\sigma$ and $\mathbf{R}_c$ with initiation at **I.** The initiation process is outlined in detail in the text and includes: (A) promoter recognition (recognition complex); (B) melting in (RS complex); and (C) initiation at **I,** followed by chain elongation and sigma release. Triangles indicate regions of sequence-specific, coupled contacts; diamonds indicate relatively unspecific but strong melting contacts around **I** (or the elongation point). A nascent RNA molecule is included.

move laterally along the DNA helix during the transition from the recognition complex to the melted RS complex.

Initiation: Once the RS complex is established the RNA polymerase core accepts the first and second template-complementary ribonucleotides at **I** (and at **I** + 1) and catalyzes formation of the first phosphodiester bond. The *E. coli* RNA polymerase initiates natural transcripts with either ATP or GTP (Maitra *et al.*, 1967), so **I** can be defined precisely as the dT or dC (*purine* nucleotide) in the transcribed DNA strand that pairs with the initiating (5′) triphosphate. At this point initiation is effectively complete, and the polymerase chugs away from the promoter elongating the nascent RNA chain. Soon after initiation, the sigma factor dissociates from the core enzyme, perhaps because it is dragged onto DNA sequences that it does not interact with favorably.

In this model, the spatial distribution of functional sites *on the RNA polymerase* dictates the spatial distribution of sequence elements ($\mathbf{R}_\sigma$, $\mathbf{R}_c$, **I**) in the promoter DNA. The polymerase measures the distance from $\mathbf{R}_\sigma$ and $\mathbf{R}_c$ to **I**, ultimately establishing strong contacts at **I**. How does this model fit our notions about coupling of energy and information in sequence-specific protein–DNA interactions?

3.6 Energy–Information Coupling

Sequence discrimination in promoter *recognition* involves many (potentially about 20) specific contacts. Since the recognition complex is relatively weak (see later), few strong, ionic contacts are made within the immediate domain of $\mathbf{R}_\sigma$ and $\mathbf{R}_c$ contacts. However, more ionic contacts are coupled indirectly to interactions at $\mathbf{R}_\sigma$ and $\mathbf{R}_c$. In the RS complex, the polymerase makes relatively unspecific ionic contacts with the separated DNA strands. Since recognition contacts at $\mathbf{R}_\sigma$ and $\mathbf{R}_c$

are essential for RS complex formation, the set of unspecific contacts around **I** is coupled to the specific recognition contacts.

3.7 Binding Energy and Kinetics

The totality of polymerase–promoter interactions in the RS complex is quite strong (see Chamberlin, 1976), and for good reason. First, energy input is required to hold the two DNA strands apart against the forces that stabilize base pairing. More importantly, however, polymerase dissociation from the promoter must be slow compared to the rate of RNA chain initiation so that transcription can actually take place. A polymerase molecule in the recognition complex has two choices: it can dissociate, or it can undergo the transition to the RS complex. Strong interactions with the melted promoter DNA ensure that the transition is favored energetically over dissociation. These strong interactions then hold the polymerase in the melted RS complex long enough for initiation to take place. This competition between polymerase off-rate and initiation rate has a great deal to do with relative promoter *strength.*

Overall, our promoter model is quite pleasing, at least intuitively. It suggests how sequence variations might influence promoter strengths, and it is readily adaptable to mechanisms of promoter regulation. Before exploring these possibilities, though, we had better justify, and to some extent refine, our model.

3.8 Justifying the Model

3.8.1 Simultaneous Contacts

The *E. coli* RNA polymerase covers about 65 base pairs of DNA when it is bound at a promoter, roughly 20 base pairs downstream from the initiation point and 45 base pairs preceding the initiation point (as pictured in Fig. 2). This was shown directly in an experiment in which RNA polymerase bound to a coliphage lambda promoter protected the approximately 65 base pairs of DNA from digestion by a combination of lambda exonuclease and S1 nuclease (B. Meyer, unpublished observations). The polymerase therefore situates itself upon a promoter so that it can contact the recognition sequences $\mathbf{R}_\sigma$ and $\mathbf{R}_c$ *simultaneously,* as required by the model. In fact, recent chemical probe studies show that RNA polymerase bound stably to the *E. coli lac* (UV5) promoter makes simultaneous contacts at $\mathbf{R}_\sigma$, $\mathbf{R}_c$, and **I** (L. Johnsrud, personal communication). (Some contacts are also made in the intervening space.) The polymerase actually contacts those base pairs in the *lac* (UV5) $\mathbf{R}_\sigma$ and $\mathbf{R}_c$ sequences that are the ideal base pairs shown in Fig. 1, thus attaching functional significance to the sequence homology among promoters.

3.8.2 Sigma Recognition

The recognition sequence $\mathbf{R}_\sigma$ is required for the RNA polymerase to locate a promoter and to form the RS complex. It is not, however, required to *maintain* the RS complex once formed; nor is it required for accurate initiation to take place.

Truncated promoters that are missing most or all of the $\mathbf{R}_\sigma$ recognition sequence have been produced and studied *in vitro* (Heyden *et al.*, 1975; Walz and Pirrotta, 1975; Pribnow, 1975*b;* Maurer *et al.*, 1974; Sekiya *et al.*, 1976; Takanami *et al.*, 1976). They do *not* function as promoters for the RNA polymerase. Neither tight binding that is characteristic of most RS complexes nor initiation of RNA synthesis can be achieved on these half-promoters. Both can be achieved on the whole promoters—that is, where the DNA extends beyond the $\mathbf{R}_\sigma$ region. In addition, mutations that either reduce or enhance promoter utilization *in vivo* and *in vitro* have been located in $\mathbf{R}_\sigma$ (Meyer *et al.*, 1975; Kleid *et al.*, 1976; Dickson *et al.*, 1977; M. Calos, personal communication). These mutations show unambiguously that sequence-specific information is contained in $\mathbf{R}_\sigma$, and that shortness is not the only failing of the truncated promoters. If the RS complex is formed *before* the promoter is truncated at $\mathbf{R}_\sigma$, then the complex is maintained, and accurate initiation occurs.

An RS complex on a truncated promoter can be produced by DNase I digestion of DNA containing the preformed RS complex (Schaller *et al.*, 1975; Pribnow, 1976*b;* Walz and Pirrotta, 1975; Sugimoto *et al.*, 1975). The DNA that is protected by the polymerase from DNase I digestion includes about 20 base pairs on either side of the initiation point. It does not contain *any* of the $\mathbf{R}_\sigma$ sequence. [It is not known why DNase I digestion generates a shorter protected fragment (40 base pairs) than the lambda exonuclease–S1 nuclease combined digestion. Presumably, the DNase I fragment is the *core* polymerase-protected DNA.] The mere fact that the DNA *is protected* by the polymerase demonstrates maintenance of the RS complex in the absence of $\mathbf{R}_\sigma$. However, it has been shown directly that the stability of the complex on a short promoter is approximately the same as the stability of the complex on the whole promoter (Heyden *et al.*, 1975). Thus, polymerase–promoter contacts that maintain the RS complex are contained on the short fragment, and they are probably coupled with specific contacts at the core recognition sequence $\mathbf{R}_c$. When ribonucleotides are fed to the DNase-generated complex, the RNA polymerase initiates transcription accurately and runs off the end of the fragment, copying the 18 to 20 template nucleotides that remain (Schaller *et al.*, 1975; Heyden *et al.*, 1975; Walz and Pirrotta, 1975; Sugimoto *et al.*, 1975). So the $\mathbf{R}_\sigma$ sequence is not required for initiation, either; and the complex that remains after DNase I treatment appears to be a true RS complex. All of these facts establish that the $\mathbf{R}_\sigma$ recognition sequence is important only for the *formation* of the RS complex—even though sigma contacts are apparently maintained in the RS complex.

Direct interaction of the sigma factor with promoter DNA has not yet been demonstrated. The reasons for assigning the recognition function (at $\mathbf{R}_\sigma$) to the sigma factor will become clear as the whole promoter scenario develops.

3.8.3 Core Polymerase Recognition

There is direct genetic evidence that promoter utilization involves specific contacts in the $\mathbf{R}_c$ promoter sequence. Two *E. coli gal* promoters overlap each other—the $\mathbf{R}_c$ sequences actually overlap (Musso *et al.*, 1977). A single mutation changes both $\mathbf{R}_c$ sequences and blocks utilization of both *gal* promoters (R. di Lauro, personal communication). Also, two "up" promoter mutations in the *E. coli lac* (ZYA) promoter, both located in the $\mathbf{R}_c$ sequence (see Fig. 1), increase *lac* promoter utilization substantially (J. Gralla, unpublished observations; J. Majors and E. Lacy, unpublished observations). [*lac* (ZYA) gene transcription is normally activated by the catabolite activator protein, CAP. The up mutations were selected using an intrinsically *weak lac* promoter for which activation was precluded via a mutation in the activator or its binding site (de Crumbrugghe *et al.*, 1971*a;* Arditti *et al.*, 1968).] These mutations all change the $\mathbf{R}_c$ sequence(s) so as to decrease or to increase promoter sequence homology in $\mathbf{R}_c$. [Another *lac* up promoter mutation apparently lies just outside of the $\mathbf{R}_c$ sequence (Dickson *et al.*, 1977). It is not as effective as the other *lac* mutations, but it does show that at least some contacts between $\mathbf{R}_\sigma$ and $\mathbf{R}_c$ are likely to be important.] Sequence conservation among promoters in $\mathbf{R}_c$ is important functionally.

Sequence-specific contacts at $\mathbf{R}_\sigma$ and $\mathbf{R}_c$ together establish the *recognition complex.* This proposed complex has been eluding biochemists for a long time (see Chamberlin, 1976), suggesting that it is a relatively *weak* complex. However, the promoter-specific recognition complex has been demonstrated at long last utilizing electron microscopy (Williams and Chamberlin, 1977). The complex is observed on coliphage T7 promoters at low temperature (0°C)—that is, on *native* promoter DNA. The sequence-specific contacts (at $\mathbf{R}_\sigma$ and $\mathbf{R}_c$) in this recognition complex position the polymerase so that it can "melt" the promoter DNA.

3.8.4 Promoter Melting

During initiation a promoter must be held in a melted condition so that RNA precursors can base pair with the DNA template at and beyond the initiation point, **I**. Melting is probably the main event in promoter utilization; it is also a strong barrier to promoter utilization (Stahl and Chamberlin, 1977; Seeburg *et al.*, 1977; see also Chamberlin, 1976; Krakow *et al.*, 1976).

That the RNA polymerase melts promoter DNA has been shown directly by measuring the resulting DNA overwinding (Saucier and Wang, 1972). The melting does not occur at 0°C, and the core polymerase cannot induce melting at any temperature. Melting is also inferred from the temperature dependence of RS complex formation and/or transcription initiation (Zillig *et al.*, 1970*b;* Mangel and Chamberlin, 1974*b,c;* Richardson, 1975; Seeburg *et al.*, 1977; Stahl and Chamberlin, 1977). Formation of the RS complex is cooperative with temperature, and every promoter has a characteristic transition temperature (usually between 15 and 25°C) at which half of the maximum observable number of RS complexes are formed at the promoter *in vitro.* A few degrees below the promoter transition temperature very few RS complexes form—that is, the transition is quite *sharp.* The sharpness of the transition probably reflects the intrinsic rate of "spontaneous" melting of each particular promoter sequence at **I**, a rate that is increased effectively by the RNA polymerase.

The RNA polymerase core actually "pulls" the DNA strands apart. Several different prokaryotic RNA polymerases can utilize the same *(E. coli)* promoters on coliphage T7 DNA. The transition temperature for a given T7 promoter varies, depending on the particular RNA polymerase that is using the promoter (J. Wiggs, J. Bush and M. J. Chamberlin, personal communication). That each polymerase exerts a characteristic pull on the promoter DNA strands is the simplest interpretation. The transition temperature is lowest for the *E. coli* RNA polymerase. This is consistent with the proposal that "recognition" contacts mediate the "melting" contacts. It also hints that melting contacts are somewhat sequence-specific (we return to this notion much later).

The RNA polymerase holds at least a few base pairs on either side of the initiation point "open." Dinucleotide (diribonucleoside monophosphate: NpN-OH) primers are used by the RNA polymerase to initiate transcription *in vitro* (Downey and So, 1970; Downey *et al.*, 1971). These primers are incorporated at or close to the normal (ATP or GTP) initiation point, and they are utilized only when complementary to the promoter DNA in this short region (Minkley and Pribnow, 1973; Maizels, 1973; Pribnow, 1975*a,b;* Dausse *et al.*, 1975; Küpper *et al.*, 1976). Experiments with dinucleotides show directly that at least five or six base pairs, including the initiation point, are held open by the polymerase at a promoter. The actual extent of promoter melting is not known precisely.

Direct physical measurements of polymerase-induced DNA "overwinding" (RS complex) suggest that five to ten base pairs of DNA are melted in a promoter (Saucier and Wang, 1972). Estimates from the temperature dependence of promoter melting/utilization are in the same range (Mangel and Chamberlin, 1974*c*). Overall, it looks as though melting is localized immediately around **I** and probably does not extend into the region of core recognition contacts at $\mathbf{R}_c$ (implying that $\mathbf{R}_c$ contacts are the *same* in the recognition and RS complexes).

The rate-limiting step in the kinetics of transcription initiation precedes binding of the first ribonucleotide (ATP or GTP) (see Krakow *et al.*, 1976), which occurs in the RS complex. Presumably, promoter melting is the rate-limiting step (Mangel and Chamberlin, 1974*c;* Seeburg *et al.*, 1977; Stahl and Chamberlin, 1977). As noted previously, it would be unfeasible kinetically to require *coincidental* polymerase–promoter contact and local promoter melting. The role of the recognition complex is to stabilize the polymerase upon the promoter transiently so that it can melt the DNA directly. However, complexes resembling those of the RS type can apparently be formed *in vitro without* a prior recognition step.

RNA polymerase initiates RNA synthesis at some A–T-rich and apparently *nonpromoter* sequences on negatively supercoiled DNA (Richardson, 1975; Botchan, 1976; H. Echols, personal communication). Negatively supercoiled DNA melts more readily than does nonsupercoiled or positively supercoiled DNA, and it does so at A–T-rich regions preferentially (Wang, 1974; Botchan, 1976). These regions spend a considerable amount of time in a melted state *in vitro* (Dean and Lebowitz, 1971; Beard *et al.*, 1973; see Botchan, 1976). Presumably, the RNA polymerase can bind to DNA sequences that are "already" melted, and it can do so without making the recognition contacts normally required for RS complex formation (Botchan, 1976). Some complexes on negatively supercoiled DNA are nonproductive for transcription. (Richardson, 1975; H. Echols, personal communication), implying that some sequence specificity is essential even at the "abnormal" promoter sites. This whole line of evidence suggests strongly that the recognition complex is used normally to overcome the kinetic barrier of promoter melting.

3.8.5 Specific and Unspecific Contacts in the RS Complex

In the RS complex, polymerase–promoter interactions that hold the DNA strands apart are largely, but not completely, sequence unspecific. As noted earlier, "nonpromoter" initiations show some degree of sequence selectivity. The core polymerase makes the relatively unspecific (ionic) contacts that melt the DNA.

Again, the core enzyme is solely responsible for RNA chain elongation. In this process, the core must maintain strong interactions that are largely sequence unspecific, since it must transcribe through a very wide variety of DNA sequences without being perturbed substantially (*except* at a terminator, as we see later). Abnormal starts at A–T-rich sequences (supercoiled DNA) occur with the rapid kinetics characteristic of RS complexes (Richardson, 1975). The sequences at these start points are not known, but it is improbable that they are very promoter-like, since they are not true promoters. Nevertheless, the core recognition sequence ($\mathbf{R}_c$, Fig. 1) is almost completely A $\longleftrightarrow$ T base pairs*; so core recognition contacts—maintained in the RS complex—may actually occur at the "abnormal" transcription start sites.

What is the purpose of core contacts at $\mathbf{R}_c$ in the RS complex? (We do not consider $\mathbf{R}_\sigma$ contacts, since they are apparently important only for establishing the RS complex.) $\mathbf{R}_c$ contacts probably stabilize the complex to some extent; they probably prevent the polymerase from "wandering" in the DNA; but more importantly, they could stabilize a polymerase conformation in which a functional active site is formed at the initiation point (see Section 4.6).

The set of melting contacts that the core polymerase makes (at **I**) in our promoter model does not change in the transition from initiation to elongation. This accommodates the relatively unspecific nature of melting interactions made during RNA chain extension. However, the core polymerase probably continues to seek out specific contacts of the $\mathbf{R}_c$ type all along the transcribed DNA, suggesting that certain sequences perturb the elongation process (see Section 4.6.1).

3.9 Promoter Strength

3.9.1 Forming the RS Complex

Different promoters are used more or less frequently by the RNA polymerase as start signals for transcription. What determines differences in relative promoter strengths?

Transcription initiation is a transient process, so our attention focuses on initiation kinetics. Comparative kinetic studies on the promoters of coliphage fd (Seeburg *et al.*, 1977) and coliphage T7 (Stahl and Chamberlin, 1977) show that the *rate of RS complex formation* determines promoter strength. The main kinetic barrier is melting, as we have seen. Specifically, then, our question is as follows: How do promoter sequence variations affect promoter recognition and promoter melting, and thereby the kinetics of RS complex formation?

**A $\rightarrow$ T base pair(s)* indicates specifically A(s) on one DNA strand paired with T(s) on the homologous DNA strand. *A $\longleftrightarrow$ T base pair(s)* indicates A(s) and/or T(s) on one DNA strand paired with T(s) and/ or A(s) on the homologous strand, respectively.

The contribution of promoter recognition to promoter strength is determined by the stability of the recognition complex. Recognition contacts at the promoter sequences $\mathbf{R}_\sigma$ and $\mathbf{R}_c$ (see Fig. 1) are coupled to contacts that melt the promoter DNA. If the recognition contacts are unstable, there is little chance that melting will occur. Promoter mutations in $\mathbf{R}_\sigma$ (Maurer *et al.*, 1974; Kleid *et al.*, 1976; Dickson *et al.*, 1977) that reduce promoter strength can be interpreted as base pair changes that destabilize the recognition complex substantially (specifically, contacts made by the sigma factor). Rapid dissociation of the recognition complex precludes promoter utilization. At the same time, base changes that "improve" the $\mathbf{R}_\sigma$ sequence at a promoter should stabilize the recognition complex and increase the promoter strength. There is an "up" mutation in the *lacI* gene promoter sequence (I^q) that increases promoter strength about tenfold (Müller-Hill *et al.*, 1968). The mutation is a change in the promoter $\mathbf{R}_\sigma$ sequence (M. Calos, personal communication) that makes the sequence better for sigma contacts. As seen in Fig. 1, the $\mathbf{R}_\sigma$ sequence for the *lacI* promoter is rotten, with only 5 of the 12 "ideal" base pairs (the mutation makes it 6). The *lacI* gene is probably transcribed only once or twice per *E. coli* generation (see Gilbert, 1976). It is a very weak promoter, probably mostly due to unstable sigma–$\mathbf{R}_\sigma$ interactions (the *lacI* promoter has a pretty good $\mathbf{R}_c$ sequence). We cannot be as certain about the effects of $\mathbf{R}_c$ mutations, since they presumably alter the stability of both the recognition complex and the RS complex. However, we can (guardedly) apply the same reasoning as that used for $\mathbf{R}_\sigma$.

That the strength of the recognition complex is fairly "weak" makes sense biologically. If $\mathbf{R}_\sigma$ and $\mathbf{R}_c$ recognition contacts were always *tight* interactions, then the ability of the polymerase to melt promoter DNA would be the sole determinant of promoter strength.

3.9.3 *Promoter Melting*

The DNA sequence in the immediate vicinity of the initiation point helps to determine the rate of promoter melting in two ways. First, A–T-rich sequences melt more readily than do other sequences, so A $\longleftrightarrow$ T base pairs at or near the initiation point must be easier for the polymerase to pull apart. (Remember that in the A–T-rich regions on negatively supercoiled DNA the DNA strands were apparently open already.) Second, we have seen that there is some degree of specificity to the strong interactions that the core polymerase uses to hold the DNA strands apart. The polymerase probably opens the DNA strands more readily at some initiation points than at others, depending on how specific the melting contacts are at each sequence. (We see in Section 4.6.1 that a specific arrangement of T $\rightarrow$ A base pairs is probably bad for melting contacts.) As long as the core polymerase can open the DNA strands (aided by coupled contacts at $\mathbf{R}_\sigma$ and $\mathbf{R}_c$) it is probably home free, actual initiation kinetics being so fast. Overall, we can understand relative promoter strengths in terms of *sequences*—at $\mathbf{R}_\sigma$, $\mathbf{R}_c$, and around the initiation point, **I**.

3.9.4 Sequence Organization

Intrinsic promoter strength may also be influenced by the manner in which sequence information is organized. For example, variation of the number of base pairs intervening between $\mathbf{R}_\sigma$ and $\mathbf{R}_c$ could effect changes in complex stability. This guess is based on the fact that promoter contacts made by the RNA polymerase must be restricted physically.

Some promoters, such as the *lacI* gene promoter, are intrinsically very weak. Some of these intrinsically weak promoters can be turned up or activated by interactions of specific activator proteins with the promoters.

3.10 Promoter Activation

3.10.1 Activator Model

Promoter "activation" means increasing the promoter strength. Activator proteins "turn up" the promoters that they regulate by stabilizing the recognition complex so as to ensure promoter melting. An activator protein cannot increase promoter strength by opening DNA base pairs around the initiation point, since access to these base pairs (especially the DNA phosphates) is blocked by the RNA polymerase. So how does an activator protein stabilize the recognition complex at a promoter?

An activator protein has a specific binding sequence in the promoter that is adjacent to the DNA that is "covered" by the RNA polymerase (see Gilbert, 1976). The binding site is on the "sigma side" of the promoter. (If it was on the transcribed side, the bound activator might interfere with transcription initiation and/or elongation.) When bound at the promoter, the activator is positioned so that it can contact the RNA polymerase (sigma factor) directly, but without blocking RNA polymerase access to its promoter contacts. Thus, the activator provides a "nesting site" for the polymerase, and protein–protein interactions between the bound activator and the (incoming) RNA polymerase stabilize the polymerase in the recognition complex. The melting transition to the RS complex is thereby assured. By this model, the activator acts only to compensate for bad polymerase contacts at $\mathbf{R}_\sigma$ and/or at $\mathbf{R}_c$.

Activator binding in a promoter must be quite strong. Tight binding ensures coincidence of activator and polymerase interactions at the promoter and provides the extra DNA-binding energy needed to stabilize the entire complex. In principle, tight activator binding is not necessarily reversible. (In a program of activator-mediated transcriptional development, regulation would occur at the level of activator protein synthesis.) In most cases, however, activator–promoter binding probably is reversible.

As discussed earlier, an effector molecule can mediate the choice of an activator protein (or other control protein) to be on or off its promoter-binding site. Figure 3 depicts the overall model for positive regulation of promoter utilization by an effector–activator team (effector-induced binding was chosen arbitrarily over effector-induced dissociation). The effector molecule cyclic AMP (cAMP) and the *E. coli* catabolite activator protein (CAP) regulate several *E. coli* promoters in just this way (see Englesberg and Wilcox, 1974; Gilbert, 1976). Let us take a look at this representative system.

3.10.2 cAMP–CAP Activation

Cyclic AMP induces the *E. coli* CAP protein to bind at and thereby activate several promoters in the bacterial genome—that is, CAP is a "pleiotropic" activator protein (Zubay *et al.*, 1970; Emmer *et al.*, 1970; de Crombrugghe *et al.*, 1971*a,b;* Majors, 1975*a,b;* Mitra *et al.*, 1975). We focus on CAP interaction at the *E. coli lac* promoter, which has been characterized extensively. (The transcription unit governed by the *lac* promoter contains three genes whose products are used to transport lactose into the cell and to split the disaccharide so that it can be utilized in glycolysis.)

cAMP binds to CAP, inducing a conformational change in this activator (Krakow and Pastan, 1973; Wu *et al.*, 1974). This change enables CAP to bind specifically to its conjugate control sequence in the *lac* promoter (Majors, 1975*a*, 1978). CAP probably binds to the promoter as a dimer (Anderson *et al.*, 1971), and the CAP-binding sequence is symmetric (a palindrome), so that each CAP monomer can "see" equivalent sequence information (Dickson *et al.*, 1977; Majors, 1978). The CAP dimer makes symmetric contacts with the binding site bases and with the (coupled) DNA phosphates (Majors, 1978). Two point mutations that prevent CAP activation (Beckwith *et al.*, 1972; Hopkins, 1974) are symmetrically disposed in the CAP binding sequence (Dickson *et al.*, 1977). The bound complex is quite stable (Majors, 1975*a*, 1978), but a binding constant for the interaction has not yet been determined.

Bound at the activator site, CAP can interact directly with the RNA polymerase (sigma factor). The CAP-binding sequence is centered 60.5 base pairs upstream from the *lac* initiation point—that is, about 15 base pairs from the "edge" of the RNA polymerase bound in the recognition complex (Majors, 1977; Dickson *et al.*, 1977). One CAP monomer (mol. wt. ~22,500 daltons) can easily cover 15 base pairs if it is only slightly nonspherical in shape (see Gilbert, 1976).

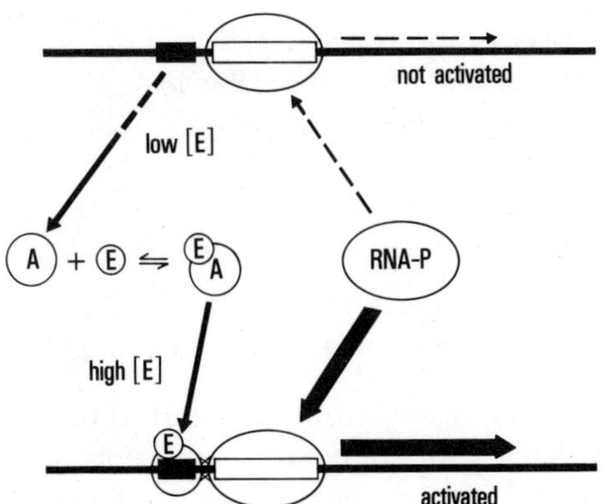

Figure 3. Effector-induced promoter activation. At low effector (E) concentrations, the activator (A) is essentially free of E, and it does not bind to the activator-binding sequence. At high effector concentrations, the equilibrium favors the effector–activator complex, which has a high affinity for the activator-binding sequence. When bound adjacent to the polymerase-binding site, the activator stabilizes the polymerase–promoter recognition complex through direct interactions with the RNA polymerase sigma factor. An increase in the rate of RS complex formation results, boosting the rate of promoter utilization. When the effector concentration drops, the activator vacates its binding site, and promoter utilization returns to the lower basal level. (The actual equilibria involved are not shown.)

Direct CAP-RNA polymerase interactions have not been demonstrated. Nevertheless, such interactions provide the simplest apparent explanation for how RNA polymerase interactions with the *lac* promoter are facilitated.

CAP binding increases *lac* transcription about 50-fold (de Crombrugghe *et al.*, 1971*a,b;* Majors, 1975*b*). We interpret this as a 50-fold increase in the stability of the RNA polymerase–*lac* promoter recognition complex. Energetically weak CAP–polymerase interactions (~2 kcal/mol binding energy) and somewhat stronger CAP–DNA interactions could accomplish this degree of stabilization. The *lac* UV5 (J. Gralla, unpublished observations) and *lac* p^s (J. Majors and E. Lacy, unpublished observations) up promoter mutations in the *lac* $\mathbf{R}_c$ sequence (see Gilbert, 1976) suggest that it is primarily bad contacts in the $\mathbf{R}_c$ sequence that are overcome energetically by CAP-mediated recognition. *A priori,* activator binding might be expected to overcome bad contacts in $\mathbf{R}_\sigma$ as well. CAP interaction with the *E. coli gal* promoter (Nissley *et al.*, 1971) probably does compensate poor $\mathbf{R}_\sigma$ contacts.

CAP is an activator of *gal* gene transcription (Miller *et al.*, 1971; Parks *et al.*, 1971*b;* Nissley *et al.*, 1971; Nakanishi *et al.*, 1973*a*). The *gal* genes are transcribed from two different, overlapping promoters; one promoter (gal_{CAP}) functions in the presence of bound cAMP–CAP, the other in its absence (see Musso *et al.*, 1977). A symmetric sequence that is very similar to the *lac* CAP sequence is centered 60.5 base pairs upstream from the gal_{CAP} promoter initiation point, and it is probably the *gal* CAP-binding sequence (Musso *et al.*, 1977; Majors, 1978). Two more base pairs separate the symmetry elements in the *gal* sequence as compared with the *lac* sequence, suggesting that the CAP dimer is "hinged" between the monomers. (CAP binding at the *gal* sequence might preclude initiation at the second *gal* promoter by excluding the polymerase from the promoter. This in turn would mean that a CAP monomer "covers" at least ten base pairs of DNA, since the two *gal* promoters are "shifted" relative to each other by only five base pairs.) Presumably, CAP activates *gal* transcription in the same way that it activates *lac* transcription—by stabilizing the recognition complex. The gal_{CAP} $\mathbf{R}_\sigma$ promoter sequence is missing one good base pair compared to the *lac* sequence (Musso *et al.*, 1977; Dickson *et al.*, 1977).

We have dealt with the mechanism of promoter activation, but not the biology of this regulatory process. Where does the effector molecule (cAMP) that mediates activation come from? Regulation at the *E. coli lac* and *gal* promoters is actually more complex than just described, since it also involves *repression* by specific repressor proteins. We first indulge in an analysis of repressor-linked negative control and then describe how all of this is biologically important in effector-driven promoter regulation.

3.11 Repressors, Operators, and Negative Control

3.11.1 Repressor–Operator Systems

Repressor proteins exert direct negative control over gene expression at specific promoters (see Bourgeois and Pfal, 1976). A transcription unit that is subject to repressor control contains an *operator* sequence that overlaps the RNA polymerase-binding site at the promoter for the transcription unit (Bennet *et al.*,

1976; Maniatis *et al.,* 1975; Ptashne *et al.,* 1976; Gilbert and Maxam, 1973; Dickson *et al.,* 1975). The operator is the binding sequence for a specific repressor protein. When it is bound to the operator, the repressor prevents the RNA polymerase from initiating RNA synthesis by sterically excluding the polymerase from its crucial promoter contacts (Majors, 1975*b;* Meyer *et al.,* 1975; Walz and Pirrotta, 1975; Squires *et al.,* 1975; Nakanishi *et al.,* 1973*b*)—that is, the repressor blocks promoter *recognition.* The relative position of the operator sequence is not fixed as it is in the case of an activator sequence, since the repressor does not interact specifically with the polymerase and therefore need not be positioned precisely with relation to it. In the case of an "activated" promoter, it is conceivable that the repressor might bind so as to preclude promoter activation only. The operator would overlap the activator sequence but not the polymerase-binding site (see Musso *et al.,* 1977).

A cis-dominant mutation in the operator sequence decreases repressor–operator binding (Rose *et al.,* 1973; Gilbert and Müller-Hill, 1967; Jobe *et al.,* 1974; see Ptashne *et al.,* 1976), thereby "relieving" repression of the transcription unit. Such a mutation is called an *operator constitutive* (O^c) mutation. When critical promoter and operator sequences overlap to some extent, one expects to find point mutations that are at once promoter and operator mutations. Mutations in the repressor protein itself—that is, in the repressor gene—that reduce repressor–operator binding affinity also relieve repression, but they are not cis dominant (see Beckwith and Rossow, 1974).

Repressor binding occurs with the native operator DNA (Wang *et al.,* 1974; Maniatis and Ptashne, 1973), and in many cases the binding is mediated by a small effector molecule. As in the case of effector-mediated promoter activation, repressor–effector interactions favor either repressor binding or repressor dissociation (nonspecific binding) (Zubay *et al.,* 1972; Rose *et al.,* 1973; Riggs *et al.,* 1970*b;* Gilbert and Müller-Hill, 1970; Nakanishi *et al.,* 1973*a*). In the bound state, the repressor makes strong, sequence-coupled interactions with the operator DNA. A stable complex ensures exclusion of the polymerase from the promoter. Measured equilibrium binding constants for specific repressor–operator pairs are very low ($K_B \sim 10^{-12}$ M) (Riggs *et al.,* 1970*a,c;* Chadwick *et al.,* 1970; Rose and Yanofsky, 1974), suggesting that several coupled ionic pairs stabilize the complexes. Dissociation rates are slow. A model for effector-induced repression (though *de*repression could have been chosen) is presented schematically in Fig. 4.

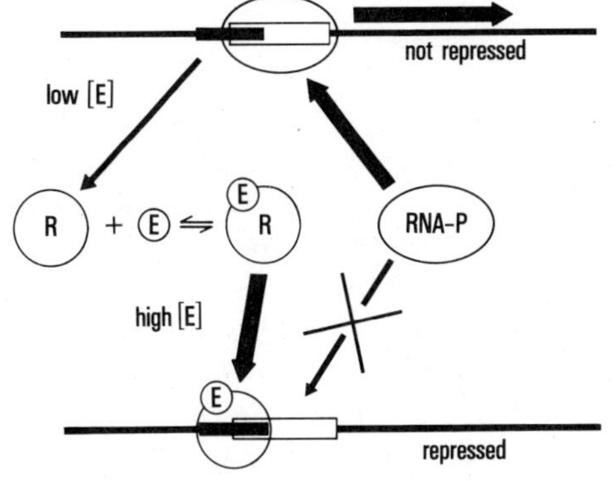

Figure 4. Effector-induced repression. At low effector (E) concentrations, the repressor (R) is essentially free of E, and it does not bind to the repressor-binding sequence. At high effector concentrations, the equilibrium favors the effector–repressor complex, which has a high affinity for the repressor-binding sequence. Since the repressor-binding sequence overlaps the polymerase-binding site, bound repressor physically prevents promoter utilization. When the effector concentration drops, the repressor vacates its binding site, and promoter utilization returns to the derepressed level. (The actual equilibria involved are not shown.)

A great deal is known about the *E. coli lac* repressor and the coliphage λ C_I repressor and the interactions with their respective operator sequences. Conceptually, the regulation effected by the two repressors is very different even though the mechanism of repression is basically the same. We look at both systems, first *lac,* with the aim of getting an overall picture of regulation of *E. coli lac* gene expression. (*lac* repressor–operator interactions are considered in some detail by von Hippel in Chapter 8 of this volume, and so our analysis is brief.)

3.11.2 lac Repressor and Operator

The *lac* operator overlaps the binding site for the RNA polymerase in the *lac* promoter (Gilbert and Maxam, 1973; Gilbert *et al.,* 1973; Dickson *et al.,* 1975; J. Gralla, unpublished observations), and bound *lac* repressor prevents the RNA polymerase from utilizing the promoter (Majors, 1975*a*). The *lac* operator sequence contains a large palindrome (Gilbert and Maxam, 1973) that was originally thought to reflect binding of the symmetric repressor tetramer. However, some *lac* operator constitutive mutations change base pairs that are not part of the symmetric sequence (Gilbert *et al.,* 1973, 1975). Further, studies using chemical probes show that the *lac* repressor makes several nonsymmetric contacts with operator DNA (Gilbert *et al.,* 1976; W. Gilbert and A. Maxam, personal communication). Thus, symmetry is apparently not used here as a way of "doubling" sequence-specific contacts, unlike the case of CAP–*lac* promoter interactions (Majors, 1978).

In vivo, repressor–operator binding is mediated by the effector molecule, allolactose (Jobe and Bourgeois, 1972), a by-product of lactose catabolism. The allolactose analog, isopropyl-thio-β-D-galactoside (IPTG), mimics the action of allolactose (see Gilbert and Müller-Hill, 1970; Contesse *et al.,* 1970). In the absence of IPTG, *lac* repressor forms an extremely stable complex with operator DNA (Gilbert and Müller-Hill, 1967; Riggs *et al.,* 1968)—a good thing, given that the average *E. coli* cell contains only about ten repressor tetramers (Gilbert and Müller-Hill, 1966) (see Chapter 8 of this volume). IPTG attaches to operator-bound (as well as free) repressor (see Friedman *et al.,* 1977), driving it off the operator (Riggs *et al.,* 1970*a,b;* Gilbert and Müller-Hill, 1970; Gilbert, 1972)—that is, it induces *lac* gene expression. [Some repressor mutations, denoted I^s, weaken effector–repressor interactions and prevent normal induction (Willson *et al.,* 1964; Bourgeois and Jobe, 1970).] Now we go on to incorporate these facts into a biologically meaningful picture of effector-mediated control of *lac* gene expression.

3.11.3 lac Control

Control of *lac* gene expression in *E. coli* is representative of the type of regulation that enables an organism to make favorable responses to changes in its environment—that is, to sustain life. (The *lac* operon is a relatively independent transcription unit.)

When *E. coli* is growing in the absence of lactose, it does not waste time, energy, or space producing the *lac* proteins needed to transport and to catabolize lactose. The *lac* genes are turned off by the *lac* repressor. The *lac* promoter is not completely silent, however, since the bacterium must be able to sense the presence of lactose in its surroundings in the event that lactose becomes the best available source of food and energy. *E. coli* senses the presence of lactose through the products of the *lac* genes that are expressed at a low basal level.

If lactose is plentiful, it is brought into the cell and catabolized (slowly, at first) giving rise to the *inducer,* allolactose. Repression of the *lac* genes is lifted by interaction of sufficient allolactose with the operator-bound and free repressor. Thus, expression of the *lac* genes (and thereby lactose utilization) is turned up, but only *slightly*. Activation of the *lac* promoter is required to increase *lac* gene transcription to its maximum level, which is about 100-fold over the basal level (see Contesse *et al.,* 1970).

Escherichia coli prefers glucose over other sugars, including lactose (see Magasanik, 1970). (Glucose is probably "preferred" energetically, since cellular glycolytic enzymes are produced constitutively.) Cellular production of cyclic AMP is reduced during growth on glucose. If, in addition to lactose, glucose is plentiful, little cAMP is produced in the cell and the *lac* genes are not activated. [Other conditions also reduce cAMP levels, preventing activation of catabolite-sensitive genes (see Magasanik, 1970).] Only when lactose is plentiful and glucose is scarce are both conditions for induction of the *lac* operon met: removal of the repressor by interaction with allolactose and a rise in cAMP concentration to a level that is sufficient to drive the activator protein, CAP, to activate *lac* gene expression fully.

3.11.4 Lambda Repressor and Operators

The coliphage λ repressor mediates the choice between *vegetative* (or "lytic") growth that leads to mass production of phage and *lysogenic* growth that involves stable, but reversible integration of λ DNA into the host genome ("prophage" state) (see Herskowitz, 1973). The λ repressor controls the expression of all λ vegetative genes by interacting with specific operator sequences at two promoters, P_L and P_R (see Ptashne, 1971; Ptashne *et al.,* 1976). [Lambda lysogeny is a remarkable and complicated process that involves several control functions (Herskowitz, 1973; Reichardt, 1975; Ptashne *et al.,* 1976).] The λ vegetative genes are included in several transcription units, but transcription of most vegetative genes is linked absolutely to prior transcription of the genes governed by the promoters P_L and P_R (see Herskowitz, 1973). Repression at P_L and P_R therefore regulates λ development. (We look further into λ transcriptional development later.)

The lambda operators, O_L and O_R, overlap the two repressible promoters, P_L and P_R (see Ptashne *et al.,* 1976), and bound repressor excludes the (host) RNA polymerase (Meyer *et al.,* 1975; Walz and Pirrotta, 1975). Repressor binding is measurably very tight at both operators (Pirrotta *et al.,* 1970; Chadwick *et al.,* 1970; Brack and Pirrotta, 1975). [Each operator is actually a compound repressor-binding site (see Ptashne *et al.,* 1976).] The two operator sequences are nearly identical (see Maniatis *et al.,* 1975; Ptashne *et al.,* 1976). When repression and hence the prophage state is established, there are only about 200 repressor monomers in the infected cell, and the functional repressor is an oligomer (Chadwick *et al.,* 1970; Brack and Pirrotta, 1975). The mechanism of derepression (induction of the integrated prophage) is not known, but appears to involve destruction of the repressor protein (Roberts and Roberts, 1975). A typical effector-mediated control system has not been implicated so far.

The *lac* and λ repressors exert negative promoter control by the same basic mechanism, but the control has very different consequences. This reflects temporal independence of *lac* gene transcription and temporal interdependence of λ vegetative gene transcription (see Section 5.3.4). Next we look at what is probably the most important mechanism that regulates the temporal development of gene transcription.

3.12 Promoters and Development

3.12.1 RNA Polymerase Modification

The RNA polymerase can be modified so that it changes the promoter sequences that it recognizes and uses as start signals (see Losick and Pero, 1976; Rabussay and Geiduschek, 1976). The result is a change in the transcription program of the cell(s) involved. The modification function links transcription of one set of genes (including the gene(s) expressing the modification function) to transcription of a new set of genetic information. A temporal sequence of one or more such modifications constitutes a program of transcriptional—and, usually, organismic—development.

In *E. coli*, the RNA polymerase sigma factor is the specificity determinant that guides the RNA polymerase to its promoter sites. Replacement of the sigma factor with one or more proteins directs the RNA polymerase core to utilize different promoters (Losick and Pero, 1976; Rabussay and Geiduschek, 1976), as was proposed long ago when the sigma factor was first characterized (Burgess *et al.*, 1969). Probably, one or more sigma type proteins direct promoter selection by a core polymerase in all prokaryotic systems. The *Bacillus subtilis* system is the most informative on the subject of sigma-mediated development.

3.12.2 Sigma Factors in Bacillus subtilis

Bacillus subtilis phage SP01 transcription occurs in three fixed stages: *early*, *middle*, and *late*. [This is an intentionally oversimplified picture (see Rabussay and Geiduschek, 1976).] Middle transcription is dependent on early transcription; similarly, late transcription depends on middle transcription (see Fujita *et al.*, 1971). Each stage of transcription involves a different set of SP01 promoters (Fujita *et al.*, 1971; Losick and Pero, 1976; Rabussay and Geiduschek, 1976; D. Shub, personal communication). [The related phage SP82 is virtually identical to SP01 in these respects (Spiegelman and Whiteley, 1974; Whiteley *et al.*, 1976).]

The product of the early SP01 gene 28 (P28) is a "sigma" protein that directs SP01 middle promoter selection. Mutations in gene 28 abort the early-to-middle transition in SP01 gene expression (Fugita *et al.*, 1971; Okubo *et al.*, 1972). RNA polymerase isolated from SP01-infected cells at the time of middle transcription contains P28 instead of the host sigma factor (Pero *et al.*, 1975*a;* Duffy and Geiduschek, 1975). The P28-modified enzyme transcribes preferentially SP01 middle genes *in vitro* (Duffy and Geiduschek, 1973, 1975; Pero *et al.*, 1975*a;* Losick and Pero, 1976). P28 confers properties on the *B. subtilis* core polymerase that define it as a true sigma protein (Petrusek *et al.*, 1976; Duffy and Geiduschek, 1977*a,b*), in the respects that the *E. coli* sigma factor was defined earlier. [A host protein apparently influences the behavior of the P28-modified enzyme (see Losick and Pero, 1976), but P28 appears to be the specificity determinant (Duffy *et al.,* 1975).] Just how P28 actually replaces the host sigma factor that directs early gene transcription is not known, but the host sigma factor is not inactivated irreversibly during phage SP01 infection (see Rabussay and Geiduschek, 1976).

Two SP01 middle gene products, P33 and P34, appear to be the combined sigma directing SP01 late transcription. Mutations in these two genes block the middle-to-late transition in SP01 development (Fujita *et al.,* 1971; Okubo *et al.,* 1972). P33 and P34 are part of the late SP01-modified RNA polymerase, and together they confer upon the core enzyme the ability to transcribe SP01 late genes specifically *in vitro* (Pero *et al.,* 1975*b;* see Losick and Pero, 1976).

In unfavorable conditions, the *B. subtilis* bacterium undergoes a reversible change to a spore-forming state. The vegetative-to-sporulation change in the *B. subtilis* transcription program involves a change in promoter selectivity and is probably directed by a sigma type factor (see Losick and Pero, 1976). That the transition is adaptive (reversible and independent of time) shows that a basically developmental mechanism can be used for (complex) life-sustaining gene regulation.

3.12.3 Sigma Replacement Model

Given what we already know about specific RNA polymerase–promoter interactions, how does replacement of a sigma factor change promoter selectivity? The answer should be obvious: Different sigma factors "see" different sigma recognition sequences ($\mathbf{R}_\sigma$).

A model for sigma-mediated transcriptional development is presented in Fig. 5. Sigma I directs the RNA polymerase core to those promoters having the appropriate recognition sequence, $\mathbf{R}_{\sigma_\mathrm{I}}$. Sigma II is produced via gene expression directed by sigma I. Sigma II replaces sigma I on the core polymerase and redirects the polymerase to those previously "silent" promoters having the recognition sequence $\mathbf{R}_{\sigma_\mathrm{II}}$. (More sigma proteins can be added to the scheme in the same way.) The rate of sigma replacement determines the rate of change of the transcriptional program. Replacement of sigma I by sigma II is assured if a gene

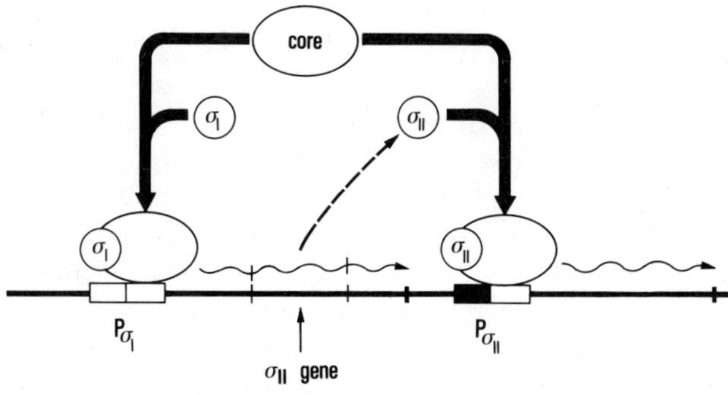

Figure 5. Transcription development by sigma replacement. The course of events here is described in the text. $\mathbf{R}_{\sigma_\mathrm{I}}$ and $\mathbf{R}_{\sigma_\mathrm{II}}$ are different sequences. $\mathbf{R}_\mathrm{c}$ is the same at each promoter.

product that is expressed together with sigma II modifies sigma I so as to prevent its association with the core polymerase (a variety of schemes can be envisioned with different overall kinetics). Although this model does not include effector molecules as mediators of sigma replacement, such a mechanism could exist and could function in adaptive or in developmental transcription control.

Any particular sigma factor stabilizes the RNA polymerase recognition complex only at those promoters that include the conjugate $\mathbf{R}_\sigma$ sequence. The model does not specify exactly the location of the $\mathbf{R}_\sigma$ sequence in a given promoter; it simply requires that direct sigma contacts with the promoter DNA (at $\mathbf{R}_\sigma$) be coupled physically through the RNA polymerase to core-specific contacts at $\mathbf{R}_c$. [$\mathbf{R}_\sigma$ in coliphage T4 late promoters may actually be an unusual, sequence-specific nick in one of the promoter DNA strands (see Rabussay and Geiduschek, 1976).] The model states explicitly that all promoters that are used as start signals by the same core RNA polymerase have the *same core recognition sequence,* $\mathbf{R}_c$. Unfortunately, we do not have comparative promoter sequence information (from *B. subtilis* phage SP01, for example) with which to test these ideas.

3.12.4 New RNA Polymerase

Synthesis of a completely new RNA polymerase that recognizes and uses a "new" set of promoters is another way of regulating transcriptional development (see Bautz, 1976). The consequences of new polymerase synthesis are virtually identical to those for the sigma replacement scheme. The obvious difference is that promoters recognized by the "new" polymerase can be completely unlike those recognized by the "old" polymerase. (In principle, though, the requirement for a recognition complex that leads to the formation of a rapid-start complex must be retained.) Coliphages T7 and T3 regulate their temporal development of gene expression by this new polymerase mode (see Bautz, 1976).

This completes our analysis of promoters and promoter regulation. Moving toward the end of this chapter, we now move to the other end of the transcription unit, the *terminator.*

4 The Terminator and Its Function

4.1 Terminators and Rho Factor

The terminator at the end of a transcription unit is a DNA sequence that induces the RNA polymerase core to stop RNA synthesis and to disengage from the DNA (Reznikoff, 1972; Chamberlin, 1974; Roberts, 1976). The completed RNA chain is released in the (normal) termination process. A mutation in a terminator is a cis-dominant lesion that allows the RNA polymerase to read through to the next DNA-encoded terminator sequence, sometimes leading to the expression of genes that are distal to the mutated terminator (see Miller, 1970; Herskowitz, 1973; Bertrand *et al.,* 1976; McDermit *et al.,* 1976). In *E. coli,* readthrough occurs pathologically in the absence of the normal termination factor, *rho* (see Roberts, 1976).

Rho factor (protein) (Roberts, 1969) maximizes the efficiency of most, if not all, terminators (see Roberts, 1976). In general, terminators vary in relative efficiency in the absence of rho, and rho boosts the efficiency of all terminators to or close to 100%. Mutations in rho are pleiotropic, resulting in widespread terminator readthrough (Richardson *et al.*, 1975; Ratner, 1976*a;* Adhya *et al.*, 1974; Korn and Yanofsky, 1976*a,b;* Das *et al.*, 1976; see Roberts, 1976). Some *rho* mutations lead to differential reductions in termination at different terminator sequences (Korn and Yanofsky, 1976*a,b;* Kiefer *et al.*, 1977; Adhya *et al.*, 1976; Bertrand *et al.*, 1977). This probably is a consequence of different intrinsic terminator strengths—that is, stronger terminators are not affected so much as weaker terminators.

4.2 Attenuators

An *attenuator* is a terminator that is not 100% efficient, even in the presence of rho factor (Kasai, 1974; Pannekoek *et al.*, 1975; Korn and Yanofsky, 1976*a,b;* Yanofsky, 1976). (This definition pertains to the *in vivo* situation; see later.) Rho factor is involved in termination at attenuators (Korn and Yanofsky, 1976*a,b*), as it is in the case of terminators. Presumably, different attenuators also have different intrinsic strengths—that is, without rho.

4.3 Rho-Dependent and Rho-Independent Terminators

Some terminators are rho-independent *in vivo* (see Chamberlin, 1976; Roberts, 1976; Adhya *et al.*, 1976; Kiefer *et al.*, 1977). At these DNA sequences all steps of the termination process occur in the absence of rho. In contrast, at least one terminator, λt_{R1}, appears to be highly rho dependent *in vivo* (M. Rosenberg, personal communication). The polymerase usually reads through λt_{R1} in the absence of normal rho factor. Other terminators lie somewhere in between these two extremes in terms of rho affects. *In vitro* transcription in the presence or absence of rho factor shows the same thing: Different terminators are more or less dependent on rho factor to induce transcription termination (see Chamberlin, 1976; Roberts, 1976).

In the analysis of terminator structure and function that follows, we attempt to discern what distinguishes weak terminators from strong terminators. We also see how rho factor might act at terminators to maximize their efficiency. Presumably, rho dependence increases as intrinsic terminator strength decreases. Since at least some completely rho-independent terminators exist, we begin the analysis leaving rho factor completely out of the picture. A word of caution is given: Some aspects of terminator function are not well understood, but we proceed as though that is not the case.

4.4 Reversing Initiation . . . Somewhat

The termination of RNA synthesis at a terminator DNA sequence is related to the initiation of RNA synthesis at a promoter DNA sequence. *In vitro,* the purified *E. coli* RNA polymerase carries out multiple rounds of transcription across defined transcription units on at least some natural DNA templates (see Chamberlin, 1974, 1976). The polymerase does everything, given only a DNA template, RNA precursors, and a "happy" environment. Termination, therefore, almost has to be a *reversal of initiation,* at least in part.

In initiation, the RNA polymerase (core) holds the two DNA strands apart; in termination, it "lets them go." In initiation, RNA synthesis is favored kinetically over polymerase dissociation; in termination, polymerase dissociation is favored kinetically over continued RNA synthesis. In initiation, RNA precursors are accommodated at the initiation point; in termination, RNA precursors are excluded at the termination point. These important aspects of initiation and termination are, of course, mutually interdependent in each process.

These considerations suggest that promoter structure and terminator structure are inversely related in terms of specific core polymerase–DNA contacts. That is, specific contacts that are made at a promoter are blocked at a terminator. This structural interdependence places obvious restrictions on how a cell such as *E. coli* can design its start and stop signals. But let us not get carried away.

There are two major differences between initiation and termination of RNA synthesis that must be considered. First, the sigma factor is not implicated in termination and probably has nothing to do with the termination process. We therefore focus on core polymerase interactions with the terminator sequence. Second, the termination complex includes some amount of newly synthesized RNA that was transcribed by the RNA polymerase. The RNA is base-paired to a short, melted stretch of the DNA template, and it also dangles out of and away from the complex. The nascent RNA chain plays an active and essential role in the termination process.

4.5 The Basic Terminator

4.5.1 Terminator Structure

All strong terminators for the *E. coli* RNA polymerase have three common sequence features (see Gilbert, 1976; Roberts, 1976). These features are illustrated in the *hypothetical* terminator sequence shown in Fig. 6. (This sequence is representative of the known 3′ terminal RNA sequences, referenced in the legend to Fig. 6.) They are (1) a set of T → A base pairs, preceded by (2) a set of G ⟷ C base pairs, overlapping with one-half of (3) a symmetric DNA sequence (inverted repeat). The symmetric DNA sequence can apparently be any sequence, so long as it includes some or all of the G ⟷ C base pairs and some of the T → A base pairs. The actual extent of each terminator element varies somewhat among sequences for strong terminators, but all are present. Figure 6 shows a segment of the terminated RNA molecule arising from our imaginary terminator DNA sequence.

4.5.2 The Termination Point(s)

Termination occurs within the set of A → T base pairs, usually at RNA position U_7-OH, as indicated in Fig. 6 (for references, see the figure legend). However, the actual termination point at a given terminator or among different terminators sometimes varies from about U_6-OH to U_8-OH, showing that the termination point is not specified exactly (Rosenberg *et al.,* 1975; Lee *et al.,* 1976; Bertrand *et al.,* 1977; Sugimoto *et al.,* 1977). In addition, the 3′ terminal (or penultimate) RNA nucleotide is sometimes a "non-U" nucleotide (usually A) that is template-encoded.

4.5.3 Terminator Symmetry and Nascent RNA

As shown in Fig. 6, the terminated RNA molecule can assume a *hairpin* RNA–RNA hybrid structure ("stem and loop" structure). This is a direct consequence of the symmetric DNA sequence in the terminator. The RNA hairpin structure is required for termination to take place, as we see shortly. First, let us place the core polymerase alone on the terminator DNA and see what we get.

4.6 Polymerase–Terminator DNA Interactions

4.6.1 Polymerase Dissociation

4.6.1a The Binary Complex. At a terminator, the core polymerase dissociates from the DNA. Temporarily ignoring the RNA chain, what are the "bad" DNA contacts that force the core enzyme off the terminator sequence?

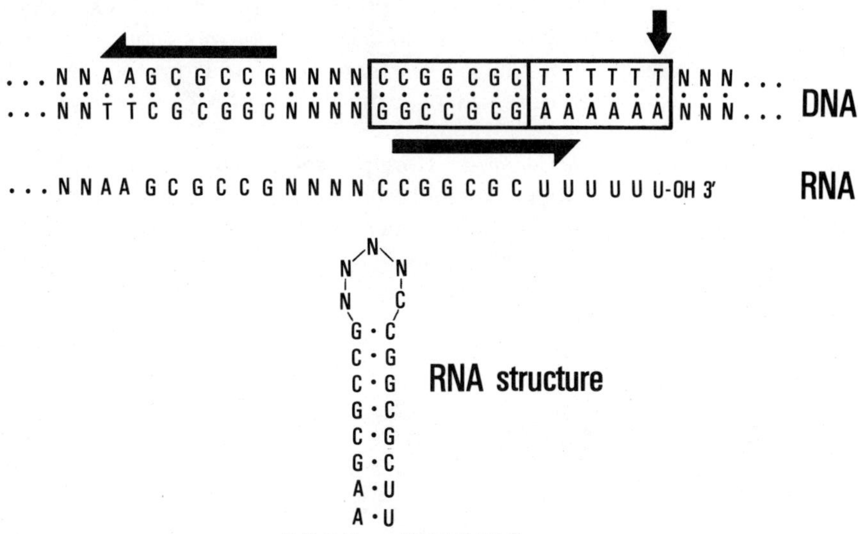

Figure 6. The strong terminator. The DNA sequence of a hypothetical strong terminator is presented at the top of the figure. The two half-arrows indicate the symmetric terminator sequence (inverted repeat). The arrow indicates the preferred, but not exclusive, termination point (3′-OH nucleotide). Below the DNA sequence is the corresponding sequence of the terminated RNA molecule. The RNA hairpin (stem and loop) structure that is essential for terminator function is included, and it is a direct consequence of the DNA inverted repeat sequence. Variations in terminator sequences and resulting RNA sequences are discussed in the text. This hypothetical terminator sequence derives from a comparison of the following known terminator (or terminated RNA) sequences: *E. coli trp* attenuator (Squires *et al.,* 1976; Lee *et al.,* 1976; Bertrand *et al.,* 1977); λ(4S) *oop* RNA (Blattner and Dahlberg, 1972); ϕ80 *oop* RNA (Pieczenick *et al.,* 1972); P22 *oop* RNA (C. Brady, S. Hilliker, and M. Rosenberg, unpublished data); λ21 *oop* RNA (C. Brady and M. Rosenberg, unpublished data); λ6S RNA (Lebowitz *et al.,* 1971); ϕX RNA (Sanger *et al.,* 1977); and fd G3 RNA (Sugimoto *et al.,* 1977). [See also Rosenberg *et al.* (1975) for λ4S and λ6S RNAs, and Rosenberg and Kramer (1977) for T7 RNA.]

Figure 7A shows the RNA polymerase core in a *binary* complex at a terminator such as the hypothetical terminator represented in Fig. 6. The enzyme is pictured as covering about 20 base pairs on either side of the termination point, as suggested by DNase I "protection" experiments with initiated promoter complexes (see Pribnow, 1975*b*). Also the core is shown holding the two DNA strands apart immediately around the termination point; and the catalytic center of the enzyme is located at the termination point. Blocked contacts in the melted T → A region and in the G ⟷ C region probably promote core dissociation with DNA strand closure.

4.6.1b T → A Sequence and Complex Stability. The DNA strands are separated around the termination point so as to accommodate the nascent RNA chain. Most of the core–DNA contacts holding the strands apart are made with poly-dT on one strand and with poly-dA on the opposite strand. This is true of contacts made during the formation of the last several poly-rU bonds of the RNA molecule. These contacts cannot all be "good"—that is, they must destabilize the binary complex so that core dissociation is favored.

We have seen that contacts holding the DNA strands apart must be relatively nonspecific so that RNA elongation proceeds smoothly. Terminator function, however, virtually demands that melting contacts with the poly-dT strand and with the poly-dA strand be relatively "bad" or weak contacts. (Perhaps the thymine methyl groups are the important blocking moieties that "uncouple" some core–terminator ionic bonding.) The core polymerase certainly encounters stretches of predominantly T → A base pairs as it elongates transcripts (not at promoter initiation points, however). Why does it not then dissociate from the DNA? The kinetics of polymerization obviously win over the kinetics of core dissociation. T → A-related destabilization is not sufficient.

4.6.1c G ⟷ C Sequence Contacts. Core polymerase interactions with the set of G ⟷ C base pairs (further) destabilize the terminator complex. At a promoter, the core polymerase makes sequence-specific recognition contacts (**R**$_c$ sequence, Fig. 1) in the region approximately 6 to 13 base pairs behind the

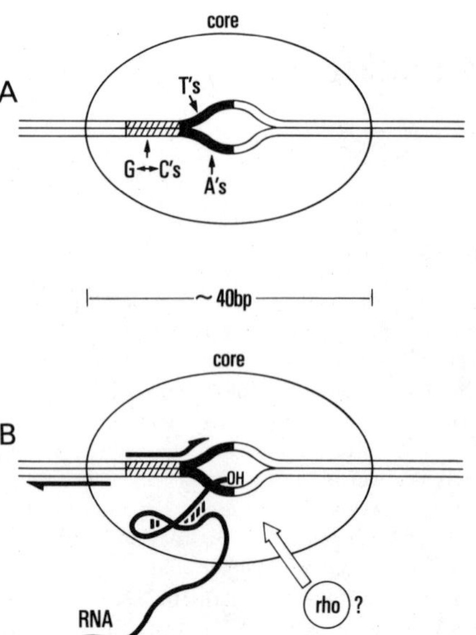

Figure 7. (A) The binary terminator complex, (B) the ternary terminator complex (see text). The terminated RNA chain is pictured as forming a short hybrid with the terminator DNA. The RNA hairpin loop structure that is required for terminator braking is also shown. The terminator DNA inverted repeat sequence is indicated by half-arrows. The probable site of rho action is shown.

initiation point(s). The core recognition sequence, $\mathbf{R}_c$, is A–T-rich, and the most important (most highly conserved) base pairs for establishing good contacts are A $\longleftrightarrow$ T base pairs. Two *lac* "up" promoter mutations both substitute an A → T base pair for a G → C base pair in the *lac* promoter $\mathbf{R}_c$ sequence, showing that a G → C (or C → G?) base pair can be bad for recognition contacts (J. Gralla, unpublished observations; Dickson *et al.,* 1977). (However, substitution of a T → A pair for an A → T pair can similarly be bad.)

We postulated that during RNA elongation the core polymerase is structurally committed to probe the DNA with its recognition fingers. At a terminator, these recognition fingers get burned by a set of G $\longleftrightarrow$ C base pairs in the region roughly 7 to 13 base pairs before the termination point, as seen in Fig. 7A. These blocked contacts must destabilize core–terminator binding.

4.6.2 *Kinetics and Terminator Contacts*

At a terminator, the kinetics of polymerase dissociation win over the kinetics of RNA chain elongation. Together, the T → A and G $\longleftrightarrow$ C bad contacts increase the rate of core dissociation at a terminator. This destabilization probably contributes to a slowdown of the elongation reaction, as well.

The core polymerase and the transcribed DNA strand together comprise the actual catalytic site for RNA synthesis. Perturbation of the close spatial relationship between core catalytic moieties and the DNA can certainly alter the catalytic site and thereby slow the rate of RNA chain growth. There is some evidence that at least the terminator G $\longleftrightarrow$ C base pairs are a weak catalytic brake.

G $\longleftrightarrow$ C base pairs that are situated at positions 8, 9, 10, and 11 (at least) behind the elongation point of a transcription complex have been shown to cause the core polymerase to pause during chain elongation. Pausing is observed *in vitro* at artificially low RNA precursor (NTP) concentrations, where the rate of RNA polymerization is slowed down approximately 50-fold relative to the normal rate (high NTP concentration) (Maizels, 1973; Gilbert *et al.,* 1973; Pribnow, 1975*b*). Each of the *lac* mutations (O^c's) in four adjacent base pairs affects the pause that occurs at one point in *lac* transcription (Gilbert *et al.,* 1973; Gilbert, 1976). The mutations are located from 8 to 11 base pairs before the pause point. An A → T to G → C transition (position −10) causes a longer pause; two G → C to T → A transversions and a C → G to T → A transition cause a shorter pause or no pause at all.

This particular kind of pause is not seen except under the artificial conditions noted earlier. Nevertheless, it is clear that G $\longleftrightarrow$ C base pairs in the core recognition domain are damaging to elongation kinetics. (Presumably, the terminator T → A sequence makes things worse.) The pause apparently does not involve the nascent RNA chain, since paused transcripts can be only a few nucleotides long. But the nascent RNA chain does induce a different and stronger kind of pause, at a terminator and under normal conditions (hereafter we refer to RNA-induced pausing as *braking*).

4.7 *RNA Structure and Braking*

The nascent RNA chain is an essential element of the basic terminator mechanism. If RNase is included in a standard *in vitro* transcription reaction, RNA synthesis (and degradation) proceeds, but termination apparently does not take place (Krakow, 1966; Darlix, 1973). What does the RNA do?

In a real terminator complex (Fig. 7B), the nascent RNA chain is the main brake of the elongation reaction. Specifically, it is the RNA hairpin structure preceding the short poly-rU tail (see Fig. 6) that transiently stops RNA chain growth. Several lines of evidence demonstrate this fact convincingly.

Coliphage ϕX174 (ϕX) DNA contains a terminator sequence that terminates transcripts initiated at one ϕX promoter but that does not terminate transcripts originating at another promoter, and both promoters *precede* the termination point (see Sanger *et al.*, 1977). How can this be? The terminated transcript starts at a promoter far upstream from the terminator (Axelrod, 1976; Hayashi *et al.*, 1976; Sanger *et al.*, 1977). The transcript can form a hairpin structure at its 3′ terminus, reflecting the symmetric terminator DNA sequence (Sanger *et al.*, 1977). The RNA hairpin structure enables successful termination. The nonterminated transcript originates *within* the symmetric terminator sequence, 20 base pairs before the probable termination point for the other transcript (Sanger *et al.*, 1977). The nonterminated RNA cannot form the hairpin structure and so the RNA polymerase does not see the terminator (Sanger *et al.*, 1977). [An abnormal transcript from a coliphage fd DNA fragment (G3′ RNA) (Sugimoto *et al.*, 1977) probably terminates because an RNA hairpin structure is formed.] This demonstrates the need for the RNA structure in terminator function, but the braking phenomenon here is inferred from other results.

The RNA polymerase normally brakes (*in vivo* and *in vitro*) at the coliphage λ terminator, designated λt_{R1}. The terminator is rho dependent but can be made relatively rho independent by a mutation (Wulff, 1976) that strengthens the stem of the RNA hairpin structure. Secondary mutations that eliminate termination at (the rho-independent) λt_{R1} (Wulff, 1976; McDermit *et al.*, 1976) are substitutions that *disrupt* RNA base pairing in the hairpin stem (the stable element of the hairpin structure). These structural alterations eliminate braking at the terminator, to the extent that braking can be measured. [The crucial biochemistry and the resulting model for the role of RNA structure in λt_{R1} function, and in terminator function in general, were graciously communicated by M. Rosenberg and coworkers (M. Rosenberg, D. Wulf, D. Court, B. de Crombrugghe, and C. Brady, manuscript in preparation).] The causal role of RNA structure in terminator braking is implicated indirectly by other evidence.

Artificial destabilization of (potential) RNA–RNA structure eliminates terminator braking. Inosine triphosphate (ITP) can be substituted for GTP in an *in vitro* transcription system; this GTP analog pairs specifically with cytosine, but it forms only *two* Watson–Crick type H bonds with cytosine. Braking that is observed normally at three different terminator sequences is reduced or abolished when transcription is carried out using ITP instead of GTP [coliphage T7 and T3 "early" terminators (Neff and Chamberlin, 1978); *E. coli trp* attenuator (C. Yanofsky, personal communication).] Similarly, 5-Br-cytosine, which pairs with guanine less tightly than does cytosine, prevents braking (Neff and Chamberlin, 1978). In each case, analog substitution in the RNA would decrease the stability of the RNA hairpin stem generated at the terminator sequence. 5-Br-Uracil, which pairs more *strongly* with adenine than does uracil, does *not* reduce braking of transcripts in which it is substituted (Neff and Chamberlin, 1978), the result expected on the basis of the stable hairpin model.

Just how the RNA hairpin effects polymerase braking at a terminator is not known, but we can guess. At different terminators, known and predicted hairpin structures are all approximately the same, if not in sequence or in stability, at least in size and in distance from the 3′-RNA terminus. (A portion of the RNA poly-rU tail is included in most structures.) The structural similarity is probably important. It suggests a structure-specific interaction between the nascent RNA molecule and the RNA polymerase core. The hairpin structure must either fit in a niche in the core–DNA complex, clogging things up, or it must bind to the polymerase via a specific arrangement of nonspecific contacts. Whatever the interactions are, they counter temporarily the energetics of the RNA chain elongation reaction. Since it is the main catalytic brake, the RNA hairpin structure probably determines indirectly the actual point(s) of termination.

4.8 Terminator DNA Melting

During transcription termination, the melted region of a terminator sequence is probably small, as indicated in Fig. 7B. If nothing else, this is a consequence of RNA hairpin formation: The RNA cannot pair with itself and with the DNA too. Where RNA is not hybridized with DNA behind the termination point, the DNA strands are probably closed. Possibly, RNA hairpin formation actually helps to dislodge the nascent RNA molecule from its complementary DNA strand, and so aids the termination process in a second way. (The actual extent of the melted terminator region during transcript elongation is unknown.) In any case, a small, melted region includes only a short (poly-rU) RNA–DNA hybrid that must be squeezed out of the DNA during chain termination.

4.9 RNA Elimination

We postulated that core polymerase interactions with terminator DNA are bad at the melted $T \rightarrow A$ sequence. This region contains the short, poly-rU tail of the RNA that must be dislodged. Since core contacts in the recognition region ($G \longleftrightarrow C$ base pairs) are coupled physically with core melting contacts, the terminator $G \longleftrightarrow C$ base pairs should add to destabilization of the melting interactions. Presumably weakened melting interactions allow for DNA strand closure, and strand closure can force the short RNA tail out of the double helix. Note that the RNA–DNA hybrid consists of relatively weak rU–dA pairs only. It is not really this simple, however.

After initiation, the RNA polymerase is complexed more stably than before initiation (in the RS complex) (see Chamberlin, 1976). Interactions with the nascent RNA chain probably supply most of the extra stability in the elongation complex. The core polymerase certainly interacts favorably with at least the 3′-OH member of the growing RNA chain, since it is one of the substrates in the polymerization reaction. Probably the core binds to one or more adjacent RNA nucleotides as well. Thus, the nascent RNA molecule works against termination not only by forming an RNA–DNA hybrid but also by interacting with the core polymerase. Nevertheless, there is a way out.

Synthesis of the poly-rU tail involves a series of reactions in which UTP (actually UMP) is linked to 3′-UMP-OH in the growing RNA chain. Apparently this is a bad combination for the RNA polymerase. The site occupied by the 3′-OH ribonucleotide during transcript elongation is probably the same site that preferentially accommodates purine nucleotides during transcript initiation. This suggests that UMP (also rCMP) does not interact so favorably with the polymerase in this substrate site. Also, the formation of a single phosphodiester bond usually stabilizes the initiated complex relative to the RS complex (against salt- or temperature-induced dissociation and attack by inhibitors of the RNA polymerase) (see Chamberlin, 1976), but this is *not* the case when the second (noninitiating) nucleotide is UTP. UTP in the incoming position can bring about abortive initiation (Johnston and McClure, 1976) or abnormal catalysis of a pyrophosphate exchange reaction between UTP and UMP (Krakow and Fronk, 1969; So and Downey, 1970). UTP has the highest ("worst") elongation K_m of the natural RNA precursors, too (see Krakow *et al.*, 1976, for a comprehensive review of kinetic parameters in the transcription process). All of this suggests that poly-rU synthesis destabilizes the RNA polymerase significantly and probably slows the rate of chain elongation somewhat. This makes elimination of the nascent RNA at a terminator sequence a bit "easier." (The AMP incorporated at the 3′-OH end of some transcripts confuses the issue. Presumably such RNA molecules are harder to eject from the termination complex, but they are ejected.)

4.10 Concerted Process

RNA chain termination at a strong terminator can be viewed as a concerted process involving the whole complicated system of interactions that we have just discussed. The terminator DNA sequence provides directly and indirectly (via RNA product) for *catalytic braking,* core polymerase *complex destabilization,* and (some) RNA–DNA *hybrid destabilization.* If braking and complex destabilization are sufficient, as they must be at a strong terminator, then the RNA polymerase dissociates and DNA strand closure expels the RNA molecule. Termination is complete.

Probably, the single most important element of the termination mechanism is the RNA hairpin structure that induces braking. (Note that we are dealing with part of a DNA control signal that is "interpreted" at the RNA level.) The main barrier to termination is the rapid and energetic extension of the RNA molecule. All terminators, regardless of strength, must encode an RNA brake. However, terminators do not have to encode anything more than an RNA brake to function in the presence of rho factor.

4.11 Rho-Dependent Termination

4.11.1 Complex Destabilization

Rho factor probably acts directly to destabilize the termination complex. More specifically, it probably destabilizes polymerase–terminator DNA interactions. The highly rho-dependent coliphage λ terminator, λt_{R1}, serves to illustrate this notion.

The λt_{R1} sequence functions poorly as a terminator *in vivo* and *in vitro* in the absence of rho factor (Roberts, 1969; and see Roberts, 1976). It accordingly bears little resemblance to strong terminators. The λt_{R1} DNA sequence does *not* have a set of G $\longleftrightarrow$ C base pairs followed by a set of T $\rightarrow$ A pairs that includes the termination point (nor does the terminated RNA have a poly-rU tail, of course). The terminator contains only the required symmetric sequence (inverted repeat) that directs synthesis of an RNA molecule that can assume a hairpin loop structure. (The RNA terminus is five, six, or seven nucleotides from the base of the hairpin stem.) Thus, λt_{R1} encodes the brake but does not have the DNA base pairs that destabilize the polymerase–terminator interactions. Braking does occur at λt_{R1}, as noted earlier. (Again, detailed information about λt_{R1} was generously provided by M. Rosenberg.) For the terminator to work, rho factor must apparently knock the terminator complex loose before the polymerase manages to escape along the DNA. How does rho factor do this? It appears to have two primary targets of action, the nascent RNA molecule and the RNA polymerase core.

4.11.2 *RNA and the Rho ATPase*

Rho factor almost certainly interacts functionally with the nascent RNA chain at a terminator. Rho has an ATPase activity (actually an *NTPase,* but ATP is the preferred substrate) (Lowery-Goldhammer and Richardson, 1974). Conversion of ATP to ADP + P_i (NTP to NDP + P_i) by the rho ATPase is obligatory for rho-dependent termination (Howard and de Crombrugghe, 1975; Galluppi *et al.,* 1976). (The stoichiometry is not known.) Further, the ATPase requires single-stranded *RNA* in order to work *in vitro* (see Galluppi *et al.,* 1976), and it is activated only by RNA that contains cytosine (Lowery-Goldhammer and Richardson, 1974). Rho has been observed to bind to single-stranded RNA, especially to poly-rC, *in vitro* (Carmichael, 1975); it also binds to the RNA polymerase.

4.11.3 *Polymerase–Rho Interactions*

Rho factor probably interacts specifically with the RNA polymerase core. It forms a reasonably stable complex with the RNA polymerase *in vitro* (Darlix *et al.,* 1971; Ratner, 1976*b*). That the observed binding is biologically meaningful is suggested by recent genetic studies. A rho^- mutation (polarity suppressor) can be compensated by a second mutation (rif^r) that alters the β subunit of the RNA polymerase (polarity restored) (A. Das, personal communication). The polymerase mutation does not simply sensitize the enzyme to terminator sequences (independent of rho), since new rho^- polarity suppressor mutations can be selected in the rif^r strain (A. Das, personal communication). This demonstrates a functional interaction between rho factor and the core polymerase at terminator sequences. Now, how does rho get at the nascent RNA chain and the polymerase at a terminator?

4.11.4 *Braking and Rho Action*

In rho-dependent termination there is a requirement for *simultaneous* events, namely, rho interaction with the nascent RNA chain *and* with the RNA polymerase while it is still sitting at the terminator. RNA hairpin structure mediates braking of transcript elongation at a terminator. This braking gives rho factor the time that it needs to locate and engage the termination complex (as suggested by Darlix and Horaist, 1975). Transcript digestion by RNase prevents rho action at terminators (Darlix, 1973), and mutations in the λt_{R1} sequence that eliminate normal braking at the terminator (by reducing RNA hairpin stability) also preclude rho action (M. Rosenberg, personal communication). So rho attacks the ternary complex when it brakes at the RNA hairpin. Now how does it bring about termination?

Using the energy of ATP hydrolysis, rho factor probably pushes the RNA polymerase off the terminator DNA, perhaps yanking the RNA out of the complex at the same time. In theory, rho could act with the single-stranded RNA either before (5′ side) or after (3′ side) the hairpin loop structure. We opt for the 3′ side, with no direct evidence to defend the choice (see Fig. 7B). There is no C-rich, single-stranded region of the nascent RNA chain that is common among 3′-terminal RNA sequences, so the rho factor ATPase is apparently activated at a termination complex by any single-stranded RNA sequence. Rho interactions with the polymerase might be required for functional expression of the ATPase activity; such interactions might also obviate the "artificial" rC requirement. The apparent specificity of rho–polymerase interactions suggests that the nascent RNA chain and the core enzyme actually do form a "compound"-binding site for rho. Once bound at this site, rho uses its ATPase to drive the polymerase off the DNA, perhaps by changing transiently the polymerase core structure. (Remember, some terminator sequences do not have the intrinsic capability to discharge the stopped polymerase.) Being bound to the RNA chain, rho probably dislodges it from the DNA. We hope to know more facts about what really happens soon. In the meantime, it is fun to toy with the possibilities.

4.12 Translation and Transcript Termination

Some transcription terminators are effectively hidden by the translation process (protein synthesis). Their existence is revealed *in vivo* by polar nonsense mutations in some genes (Franklin and Luria, 1961; Jacob and Monod, 1961*b;* Ames and Hartman, 1963; see Roberts, 1976). A nonsense mutation in a gene stops translation of the gene messenger RNA at the nonsense codon; and a nonsense mutation is polar if it reduces or eliminates the expression of distal genes in the same transcription unit.

Some *rho* mutations reduce the affects of polar nonsense mutations (Beckwith, 1963; Carter and Newton, 1971; Korn and Yanofsky, 1976*a,b;* Richardson *et al.,* 1975; Das *et al.,* 1976; Ratner, 1976*a,b*). This leads us to two important conclusions: (1) most, maybe all, polarity is caused by rho-dependent termination events; and (2) translation of RNA by ribosomes can block this rho-dependent termination. In the absence of rho, transcription termination *in vitro* seems to occur at terminators that are also used *in vivo* (see Chamberlin, 1976; Roberts, 1976). In the presence of rho, however, *in vitro* transcripts are sometimes terminated within genes (Dunn and Studier, 1973; de Crombrugghe *et al.,* 1973; Shimizu and Hayashi, 1974; D. Pribnow, unpublished data), presumably at "polar type" (rho-dependent) terminator sequences. How do ribosomes prevent termination at these signals *in vivo?*

Probably a ribosome translating the nascent RNA molecule close behind the core polymerase keeps the terminator-encoded RNA hairpin structure from forming. In this way, translation prevents braking and thereby precludes rho action. (Alternatively, the ribosome could physically exclude rho from the elongation complex, but this model is hard to picture without having the ribosome gratuitously disrupting the RNA hairpin structure at the same time.) A nonsense mutation preceding a buried terminator sequence keeps the ribosome from following the core polymerase. The RNA hairpin forms, braking occurs, and rho induces abnormal termination. A single translating ribosome gets the RNA polymerase past one of these polar terminators if it is not somehow waylaid. It would be interesting to know the sequences of several polar type terminators, since their rho dependence implies that they are intrinsically weak terminator sequences.

4.13 Terminator Strength

What is the difference between a rho-dependent or weak terminator and a rho-independent or strong terminator? We have seen that all terminators must encode an RNA hairpin structure. The details of the RNA structure, including stem (and/or loop) size and stem stability, obviously influence the extent of braking at a terminator and so influence terminator strength. The main distinction between terminators appears at the level of core polymerase–terminator DNA contacts. A strong terminator contains a spatially organized set of base pairs that discharges the polymerase from the DNA. (Poor interactions between the polymerase and the poly-rU RNA tail help the discharging process.) A weak terminator does not contain this set of base pairs and so cannot kick off the polymerase. Also, the nascent RNA molecule must be more difficult to expel from the DNA. A weak terminator depends on rho factor to effect the discharge(s).

5 Terminator Regulation

5.1 Regulation and Rho Factor

The size of a transcription unit and therefore the number of genes under the coordinate control of a single promoter can be regulated by changing the effective termination point. If termination is blocked at a particular terminator site, then transcripts are extended to the next encoded terminator sequence. Conversely, if termination is enhanced at an otherwise silent or weak terminator site, then transcripts are effectively shortened, and expression of distal genes is cut back. Various mechanisms can be envisioned whereby such terminator regulation might be accomplished (and the reader is encouraged to consider possible mechanisms). We focus upon the relationship between rho factor and terminator regulation in *E. coli.*

First, we assume that it is the most rho-dependent (intrinsically weak) terminators that are controlled. Terminator regulation, then, is a matter of either antagonizing or facilitating the interaction of rho with the elongation or termination complex. We have already seen the potential for *translational regulation* of rho-dependent termination in the case of polar type terminators. In fact, a translation-coupled mechanism of terminator *de*activation is probably used to regulate transcription in *some* prokaryotic systems. Turning things around, terminator activation could also be linked to translational control. It is just a matter of facilitating the formation of hairpin structure in the nascent RNA. The *E. coli* tryptophan (*trp*) operon attenuator is probably controlled by translation that is, in turn, controlled ultimately by the end product of *trp* enzyme action—a fascinating negative feedback loop of control.

5.2 *The trp Attenuator*

5.2.1 *trp Gene Regulation*

The *E. coli trp* gene products are needed to synthesize the amino acid tryptophan when it is not available in the cell's environment. Expression of the *trp* gene is reversibly sensitive to intracellular levels of tryptophan (see Yanofsky, 1976). The genes are turned off when tryptophan levels are high, and they are turned on when tryptophan levels are low. Two regulatory switches are involved. One is tryptophan-induced repressor binding at the *trp* gene promoter—that is, effector-mediated repressor–operator interactions (Squires *et al.*, 1975; Bennett *et al.*, 1976; see Yanofsky, 1976). The other is the *trp attenuator,* located beyond the promoter but before the structural genes of the operon.

When tryptophan is plentiful, the *trp* attenuator terminates about 85% of the transcripts initiated at the *trp* promoter (Bertrand *et al.*, 1975; Morse and Morse, 1976; Bertrand and Yanofsky, 1976). [This is observed most readily when the *trp* genes are genetically *derepressed* (*trp*R$^-$).] During tryptophan starvation, most *trp* transcripts read through the attenuator into the distal *trp* genes (Morse and Morse, 1976; Bertrand and Yanofsky, 1976). Readthrough occurs, presumably at its maximum level, when the *trp* attenuator is genetically removed from the operon (Jackson and Yanofsky, 1973; Bertrand *et al.*, 1976). The attenuator is only partially rho dependent, as determined by the effects of mutations in *rho* (Korn and Yanofsky, 1976*a,b;* Bertrand *et al.*, 1977). [There is a very interesting *trp* deletion that cuts into the attenuator T → A sequence, leaving most of the attenuator intact. The deletion reduces the maximum efficiency of the attenuator to about 50%, and it apparently changes the attenuator so that rho cannot participate in the termination process (see Bertrand *et al.*, 1977).]

5.2.2 *trp Attenuator Control*

The *trp* attenuator is regulated, indirectly, by the intracellular concentration of tryptophan. It is responsive to the extent of charging (aminoacylation) of the single tryptophan tRNA species, $tRNA^{Trp}$ (Morse and Morse, 1976; Yanofsky, 1976; Yanofsky and Soll, 1977), and charging is dependent on tryptophan availability. The involvement of charged/uncharged $tRNA^{Trp}$ suggests that translation of the *trp leader* RNA mediates termination at the attenuator.

The leader RNA is the transcript from the *trp* promoter to the attenuator (Bronson *et al.,* 1973; Squires *et al.,* 1976). The entire leader RNA sequence of (about) 140 nucleotides is known (Squires *et al.,* 1976—but see Bertrand *et al.,* 1977, for corrections). It contains a ribosome initiation site near its 5′ terminus (Platt *et al.,* 1976) that serves as a start signal for translation *in vivo* (in *trp–lac* fusions; C. Yanofsky, personal communication). In-phase translation of the leader can proceed up to a nonsense codon that is about 70 nucleotides away from the leader 3′ end. Two tandem tryptophan codons reside (in phase) in the translated segment of the leader. Where does $tRNA^{Trp}$ fit into the scheme?

In general, charged tRNA (for all amino acids) enables translation to proceed normally, while uncharged tRNA can cause ribosomes to idle at RNA codons that specify the uncharged tRNA species (see Block and Haseltine, 1974). Thus, uncharged $tRNA^{Trp}$ would cause ribosome idling at the tryptophan codons in the *trp* leader RNA. What we need is a model in which idling blocks attenuation, while efficient translation of the leader (up to nucleotide 70) allows the attenuator to work, terminating the transcripts. The following model is a candidate for translational regulation of the *trp* attenuator.

5.2.3 The Model

The RNA structure of the *trp* leader directly determines whether or not the attenuator induces termination, and *trp* leader translation mediates the RNA structure. There are (at least) two potential hairpin loop structures that include nucleotides at the 3′ terminus of the *trp* leader RNA, and both hairpin structures are observed *in vitro* (F. Lee and C. Yanofsky, manuscript in preparation). One hairpin is a standard terminator hairpin structure; the other is much larger and certainly could not be a typical braking structure. The two structures are probably mutually exclusive because the same nucleotides form part of one half-stem of each hairpin (F. Lee and C. Yanofsky, manuscript in preparation). To make things simple, we assume that the large, nonterminator hairpin is the preferred structure *in vivo.* Thus, the terminator hairpin occurs only infrequently unless the larger hairpin is somehow precluded or disrupted.

A ribosome translating the leader RNA behind the RNA polymerase in the presence of fully charged tRNAs should cover nucleotides needed to form the large hairpin stem. It should not cover nucleotides included in the terminator hairpin, since it stops at a nonsense codon that is far away. The terminator hairpin can form, inducing braking at the attenuator. Termination then occurs with some help from rho. In the presence of uncharged $tRNA^{Trp}$, however, the ribosome is halted at the leader tryptophan codon(s). It probably cannot disrupt the large hairpin structure, so the terminator hairpin does not (usually) form, and the polymerase (usually) reads past the attenuator into the *trp* genes. There is no braking and no rho action. [The reader might consider the explicit, testable predictions that this model makes. It is not without some problems (see, for example, Yanofsky and Soll, 1977).]

The model is presented here mostly because it illustrates how RNA signals (codons, secondary structure) are or could be used to control the function of a DNA signal. The model might be completely wrong, but it is instructive. Now we examine a mechanism of terminator control that operates mostly at the DNA level.

5.3 *Control by Antitermination*

5.3.1 *Antitermination in General*

An antiterminator protein can block rho-dependent termination. The antiterminator protein has a binding sequence within the transcription unit that it regulates—that is, between the promoter and the rho-dependent terminator. The antiterminator protein binds to or otherwise modifies the RNA polymerase core as the core transcribes across the unit and encounters the protein complexed at its specific signal. The modified polymerase is rendered insensitive to rho action, but it still stops transcription at the first strong, rho-independent terminator that it sees. (What if it did not?) Genetically, an antiterminator control signal (binding sequence) actually resembles a promoter for genes distal to the rho-dependent terminator, and an antiterminator protein resembles an activator. The *N* protein of coliphage λ provides the model system for this general picture of antitermination.

5.3.2 *The λ N Protein*

The λ *N* gene product (*N*) is an antiterminator protein (see Herskowitz, 1973; Roberts, 1976). It acts to block termination of transcripts originating at the two λ vegetative promoters, P_L and P_R. *In vivo* termination of P_L and P_R transcripts occurs at the rho-dependent terminators, t_{L1} and t_{R1}, respectively, in the absence of *N* protein, and these transcripts are extended to distal, rho-independent terminators in the presence of *N* (see Herskowitz, 1973; Roberts, 1976; Szybalski, 1976*a,b*). *N* function is specific for P_L and P_R transcripts (Adhya *et al.*, 1974; Franklin, 1974); rho-dependent termination of other transcripts does occur in the presence of functional *N* protein (Rosenberg *et al.*, 1975; Roberts, 1975). This situation demands that *N* recognize control signals in the *N*-regulated transcription units. Direct *N* protein–DNA interactions are implicated.

Cis-dominant mutations that prevent *N* from functioning are located within the $P_L \rightarrow t_{L1}$ and $P_R \rightarrow t_{R1}$ transcription units (Salstrom and Szybalski, 1976; W. Szybalski and M. Rosenberg, personal communications). These mutations occur in identical sequences (M. Rosenberg, personal communication) that are the presumptive *N*-protein-binding sites in the regulated transcription units. *N* protein interacts functionally with the (elongating) core polymerase, and the *N*-binding signals ensure that this interaction takes place. *N* protein has been identified as an RNA polymerase-binding protein (Epp and Pearson, 1976). The interaction is probably specific, as suggested by genetic studies (Georgopoulos, 1971; Ghysen and Pironio, 1972; see Herskowitz, 1974).

Lambda transducing phages carrying defined segments of the *E. coli* genome have been used to show that *in vivo* the *N* protein antiterminates at a variety of (properly situated) rho-dependent terminators including polar terminators, terminators at the ends of *E. coli* operons, and an attenuator (see Franklin and Yanofsky, 1976; Adhya *et al.,* 1976). More importantly, *N* can block termination sequentially at more than one rho-dependent terminator, but it cannot overcome termination at a strong, rho-independent site (see Roberts, 1976; Adhya *et al.,* 1976). This shows that *N* remains stably attached to the RNA polymerase core even after an antitermination event—that is, at least until a rho-independent terminator is encountered. Probably, termination at a strong, rho-independent terminator induces dissociation of *N* from the polymerase. [*N* appears to act catalytically—that is, it "recycles" (Radding and Echols, 1968), as do the sigma and rho factors.]

An antitermination model, using the λ *N* protein as a representative antiterminator protein, is presented in Fig. 8. In this model, *N* action overrides rho action, but how?

5.3.3 *Mechanism of Antitermination*

Antiterminator proteins such as *N* probably block rho function directly by excluding rho from the ternary terminator complex (Roberts, 1970). This guess is based on the fact that the *N*-modified polymerase does stop at strong terminator sequences, sequences that can discharge the polymerase without help from rho. If *N* somehow interfered with terminator RNA structure or prevented the polymerase from seeing the structure, then *N* would presumably enable the polymerase to bypass all terminators, weak and strong alike. When the *N* protein is scooped up by the elongating polymerase, it apparently occupies or modifies the site at which rho factor normally interacts with the polymerase (Friedman *et al.,* 1973). *N* thereby blocks rho-mediated complex dissociation at weak terminators, but it does not affect the braking that occurs at the terminators.

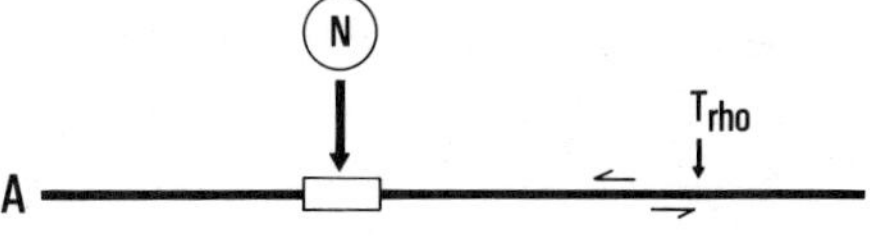

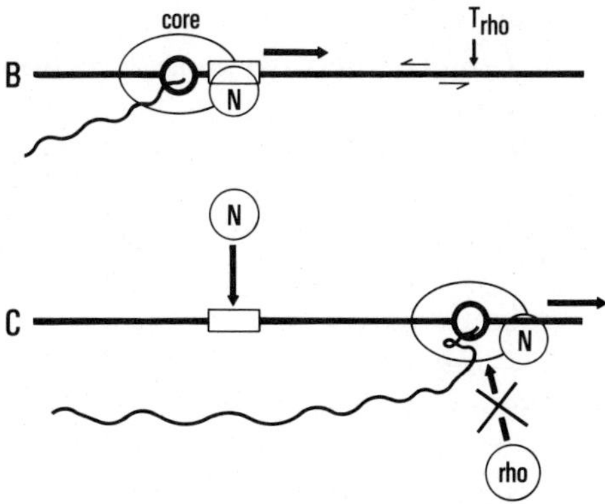

Figure 8. Antitermination by the lambda *N* protein. The series of events is described in the text: (A) *N* binds to its conjugate binding sequence within the regulated transcription unit; (B) *N* binds to the RNA polymerase core as the core elongates the RNA chain through the site that is occupied by the *N* protein; and (C) polymerase-bound *N* blocks the site of rho action at the terminator without interfering with RNA-induced terminator braking. The polymerase continues past the rho-dependent terminator, T_{rho}. (Regulated expression of the *N* gene is not shown. See text.)

Antitermination, if controlled in a reversible fashion, is an adaptive regulatory mechanism for gene expression. In λ, antitermination directs an irreversible program of transcription.

5.3.4 Antitermination and λ Development

Transcription antitermination can be a *developmental* mechanism. This is accomplished by coupling sequential extension of different transcription units with sequential expression of different antiterminator functions. Where antitermination is used as the exclusive developmental mechanism, all promoters involved are recognized and utilized as start signals by the same RNA polymerase. Prior transcription from all promoters is required for antiterminator-directed expression of genes distal to the rho-dependent terminators. Coliphage λ vegetative transcription develops according to such a scheme.

Lambda vegetative transcription occurs in three temporally distinct stages that we will call early, middle, and late. Sequential early, middle, and late gene expression makes biological sense for λ growth: Roughly speaking, early gene products establish the vegetative growth cycle (in competition with lysogeny) (see Herskowitz, 1974); middle gene products get λ DNA replication going; and late gene products provide the complex protein package for new, mature λ DNA (see Herskowitz, 1974; Szybalski, 1976*a,b*). Clearly, DNA packaging would be a disaster for λ growth if it occurred too early. (The same general pattern of gene expression obtains for all bacteriophages that utilize a developmental program of transcription, though the mechanisms vary.) Lambda switches from early to middle and from middle to late transcription via antitermination.

Lambda vegetative transcription occurs in three temporally distinct stages that we will call early, middle, and late. Sequential early, middle, and late gene expression makes biological sense for λ growth: Roughly speaking, early gene products establish the vegetative growth cycle (in competition with lysogeny) (see Herskowitz, 1974); middle gene products get λ DNA replication going; and late gene products provide the complex protein package for new, mature λ DNA (see Hersokwitz, 1974; Szybalski, 1976*a,b*). Clearly, DNA packaging would be a disaster for λ growth if it occurred too early. (The same general pattern of gene expression obtains for all bacteriophages that utilize a developmental program of transcription, though the mechanisms vary.) Lambda switches from early to middle and from middle to late transcription via antitermination.

Lambda DNA encodes two different antiterminator functions. The *N* gene product directs the early-to-middle switch (see Roberts, 1976); and the *Q* gene product directs the middle-to-late switch (Herskowitz and Signer, 1970; see Roberts, 1976). Immediately after infection (or induction of the prophage), the host RNA polymerase initiates RNA synthesis at several lambda promoters (Blattner and Dahlberg, 1972). (We focus on three vegetative gene promoters.) Short transcripts originate at the promoters P_L, P_R, and P'_R and terminate at rho-dependent terminators (see Szybalski, 1976*a,b*). P_L and P_R transcripts each include (at least) one gene. The *N* gene is expressed from P_L in this early transcription mode (see Roberts, 1976; Szybalski, 1976*a,b*). The *N* protein antiterminates the P_L and P_R transcripts, extending them into distal middle genes (see Herskowitz, 1974; Roberts, 1976; Szybalski, 1976*a,b*). Early and middle gene expressions therefore overlap. The *Q* gene is expressed from the distal portion of the extended P_R transcript (see Herskowitz, 1974; Roberts, 1975). The *Q* protein then antiterminates specifically the short P'_R transcript, extending it into the λ late genes (Roberts, 1975, 1976; see Herskowitz, 1974), and many phage are produced ultimately.

There are no real effector molecules acting directly in the process just outlined. However, there are an assortment of regulatory functions that operate in choosing lambda vegetative growth (see Herskowitz, 1974; Reichardt, 1975; Ptashne *et al.,* 1976), as noted earlier. Once the choice is made to grow, transcriptional development sets in irreversibly.

This marks the terminus of our detailed exploration of DNA control signals that regulate prokaryote *(E. coli)* and phage gene transcription.

6 Concluding Remarks

6.1 Transcription Regulatory Mechanisms

Control of the transcription process occurs at promoters and at terminators. This is a *rule.* Several different mechanisms that we have discussed operate in *E. coli* and in other prokaryotes to regulate the initiation and/or termination of RNA synthesis. There are probably other as yet undiscovered transcription regulatory mechanisms that also operate in these systems. Also, we can consider the DNA-encoded control signals and their functional interactions with control proteins described here to be the evolutionary foundations upon which other regulatory systems have been built.

In eukaryotes—and in some prokaryotes—promoters, terminators, and associated regulatory mechanisms are perhaps somewhat more complicated. [For instance, promoters and terminators need not be inversely related in sequence as they appear to be for the *E. coli* RNA polymerase (core).] Sequence specificity must be maintained at some level, but not necessarily at the level of direct polymerase–DNA or control protein–DNA interactions. Consider, for example, sequence-specific alteration of DNA primary structure. There could be two signals that are recognized sequentially by two different regulatory components: The first component, an enzyme, makes a sequence-specific modification of the DNA; then the second component, the polymerase or a control (binding) protein, recognizes the modification—not the sequence—as a control signal.

Cell–cell interactions undoubtedly play a role in the control of gene transcription in higher systems. At the DNA level, we can rely upon our arsenal of regulatory mechanisms to understand how control is exerted. But we need to know how regulatory information is conveyed from cell to cell and from the outer cell membrane to the regulatory machinery inside the cell. A variety of possibilities can be imagined. Perhaps most attractive among them is the following: (1) specific cell surface interactions between neighboring cells followed by (2) contact-mediated activation or deactivation of a membrane-bound enzyme that synthesizes a specific effector molecule. Alternatively, specific surface interactions could lead to release of membrane-bound control proteins or even to intercellular transfer of potentiating or inhibitory molecules. It is hoped that we have established a good molecular framework for understanding whatever it is that actually does go on.

6.2 *Adaptive Response versus Development*

We have seen that the distinction between *adaptive* and *developmental* regulation of transcription is largely a matter of organization. The actual regulatory mechanisms can be the same in each process, but the placement of the DNA control signals around specific genes can have very different consequences for gene expression. Effector molecules can serve to modulate either process equally well. We noted that one basically developmental scheme of transcription (*B. subtilis* sporulation via sigma replacement) is actually adaptive overall, since it is a reversible process. At the organismic level, then, reversibility is the most appropriate distinction between adaptation and development.

6.3 *A Problem in Genetic Design*

An incredibly complex transcription system can obviously be designed using the DNA control signals, proteins (or enzymes), and effector molecules of the kinds that we have examined. A careful arrangement of DNA signals interspersed among a select set of genes can make an adaptive, developing, and even living system—if it all gets going somehow. (It all got going a long, long time ago.) DNA replication machinery and control signals are needed, of course, to make the system reproduce itself.

A DNA virus is basically *DNA*—just a select arrangement of control signals and genes. The viral DNA gets going by using the biochemical machinery and energy provided by the infected cell. (The host replication machinery is often usurped for use by the infecting virus.) The design of such a simple, reproducing piece of DNA is really a trivial problem compared to the design of a petunia or a grasshopper!

Acknowledgments

Many people have contributed information and ideas that are incorporated into this chapter. I am extremely grateful to these people, both for their substantial contributions and for the generous spirit of each exchange. I particularly wish to thank Mike Chamberlin, Peter Geiduschek, Wally Gilbert, John Majors, Marty Rosenberg, Heinz Schaller, Waclaw Szybalski, Pete von Hippel, and Charley Yanofsky for providing me with a considerable amount of information and for helping me to formulate many of the concepts included in the chapter. I am also very grateful to Michele Calos, Marvin Caruthers, Asis Das, Roberto di Lauro, Hatch Echols, Jay Gralla, Lorraine Johnsrud, Alan Maxam, Barbara Meyer, David Schub, and John Sims for directly contributing unpublished data and/or good ideas. Several people constantly provided a pleasant working atmosphere and stimulating discussions on the topics included in the chapter, most notably Barry Chelm, Larry Gold, Marjorie Russell, Britta Singer, Larry Soll, and Mike Yarus. My thanks to Gail Winklemann for typing the manuscript and to Bob Goldberger for editing the manuscript carefully. Unfortunately, many more thanks must go unspecified.

This chapter was written during the tenure of National Research Service Award #5 F32 CA05194-02 from the National Cancer Institute.

References

Adhya, S., Gottesman, M., and de Crombrugghe, B., 1974, Release of polarity in *Escherichia coli* by gene *N* of phage λ: Termination and antitermination of transcription, *Proc. Natl. Acad. Sci. U.S.A.* **71**: 2534.

Adhya, S., Gottesman, M., de Crombrugghe, B., and Court, D., 1976, Transcription termination regulates gene expression, in: *RNA Polymerase* (R. Losick and M. Chamberlin, eds.), pp. 719–730, Cold Spring Harbor Lab., Cold Spring Harbor, New York.

Adler, K., Beyreuther, K., Fanning, E., Geisler, E., Gronenborn, B., Klemm, A., Müller-Hill, B., Pfahl, M., and Schmitz, A., 1972, How *lac* repressor binds to DNA, *Nature* **237**:322.

Ames, B., and Hartman, P. E., 1963, The histidine operon, *Cold Spring Harbor Symp. Quant. Biol.* **28**:349.

Anderson, W., Schneider, A., Emmer, M., Perlman, R., and Pastan, I., 1971, Purification of and properties of the cyclic adenosine 3′,5′-monophosphate receptor protein which mediates cyclic adenosine 3′,5′-monophosphate-dependent gene transcription in *Escherichia coli, J. Biol. Chem.* **246**:5929.

Arditti, R. R., Scaife, J. G., and Beckwith, J. R., 1968, The nature or mutants in the *lac* promoter region, *J. Mol. Biol.* **38**:421.

Axelrod, N., 1976, Transcription of bacteriophage φX174 *in vitro:* Selective initiation with oligonucleotides, *J. Mol. Biol.* **108**:753.

Bautz, E. K. F., 1976, Bacteriophage-induced DNA-dependent RNA polymerases, in: *RNA Polymerase* (R. Losick and M. Chamberlin, eds.), pp. 273–284, Cold Spring Harbor Lab., Cold Spring Harbor, New York.

Beard, P., Morrow, J. F., and Berg, P., 1973, Cleavage of circular, superhelical simian virus 40 DNA to a linear duplex by S1 nuclease, *J. Virol.* **12**:1303.

Beckwith, J., 1963, Restoration of operon activity by suppressors, *Biolchim. Biophys. Acta* **76**:162.

Beckwith, J., and Rossow, P., 1974, Analysis of genetic regulatory mechanisms, *Annu. Rev. Genet.* **8**:1.

Beckwith, J., Grodzicker, T., and Arditti, R., 1972, Evidence for two sites in the *lac* promoter region, *J. Mol. Biol.* **69**:155.

Bennett, G. N., Schweingruber, M. E., Brown, K. D., Squires, C., and Yanofsky, C., 1976, Nucleotide sequence of region preceding *trp* mRNA initiation site and its role in promoter and operator function, *Proc. Natl. Acad. Sci. U.S.A.* **73**:235.

Bertrand, K., and Yanofsky, C., 1976, Regulation of transcription termination in the leader region of the tryptophan operon on *Escherichia coli* involves tryptophan or its metabolic product, *J. Mol. Biol.* **103**:339.

Bertrand, K., Korn, L., Lee, F., Platt, T., Squires, C. L., Squires, C., and Yanofsky, C., 1975, New features of the regulation of the tryptophan operon, *Science* **189**:22.

Bertrand, K., Squires, C., and Yanofsky, C., 1976, Transcription termination *in vivo* in the leader region of the tryptophan operon of *Escherichia coli, J. Mol. Biol.* **103**:319.

Bertrand, K., Korn, L. J., Lee, F., and Yanofsky, C., 1977, The attenuator of the tryptophan operon of *Escherichia coli:* a. Heterogeneous 3′-OH termini *in vivo.* b. Deletion mapping of attenuator functions, *J. Mol. Biol.* **117**:227.

Beyreuther, K., and Gronenborn, B., 1976, N-terminal sequence of phage lambda repressor, *Mol. Gen. Genet.* **147**:115.

Blattner, F. R., and Dahlberg, J. E., 1972, RNA synthesis startpoints in bacteriophage λ: Are the promoter and operator transcribed? *Nature New Biol.* **237**:277.

Block, R., and Haseltine, W. A., 1974, *In vitro* synthesis of ppGpp and pppGpp, in: *Ribosomes* (M. Nomura, A. Tissieres, and P. Lengyel, eds.), pp. 747–761, Cold Spring Harbor Lab., Cold Spring Harbor, New York.

Botchan, P., 1976, An electron microscopic comparison of transcription on linear and superhelical DNA, *J. Mol. Biol.* **105**:161.

Bourgeois, S., and Jobe, A., 1970, Superrepressors of the *lac* operon, in: *The Lactose Operon* (J. R. Beckwith and D. Zipser, eds.), pp. 325–341, Cold Spring Harbor Lab., Cold Spring Harbor, New York.

Bourgeois, S., and Pfal, M., 1976, Repressors, *Adv. Protein Chem.* **30**:1.

Brack, C., and Pirrotta, V., 1975, Electron microscopic study of the repressor of bacteriophage λ and its interaction with operator DNA, *J. Mol. Biol.* **96**:139.

Bronson, M., Squires, C., and Yanofsky, C., 1973, Nucleotide sequences from tryptophan messenger RNA of *Escherichia coli:* The sequence corresponding to the amino-terminal region of the first polypeptide specified by the operon, *Proc. Natl. Acad. Sci. U.S.A.* **70**:2335.

Burgess, R. R., 1971, RNA polymerase, *Annu. Rev. Biochem.* **40**:711.

Burgess, R. R. 1976, Purification and physical properties of *E. coli* RNA polymerase, in: *RNA Polymerase* (R. Losick and M. Chamberlin, eds.), pp. 69–100, Cold Spring Harbor Lab., Cold Spring Harbor, New York.

Burgess, R. R., Travers, A. A., Dunn, J. J., and Bautz, E. K. F., 1969, Factor stimulating transcription by RNA polymerase, *Nature* **221**:43.

Carmichael, G. G., 1975, Isolation of bacterial and phage proteins by homopolymer RNA-cellulose chromatography, *J. Biol. Chem.* **250**:6160.

Carter, T., and Newton, A., 1971, New polarity suppressors in *Escherichia coli:* Suppression and messenger RNA stability, *Proc. Natl. Acad. Sci. U.S.A.* **68**:2962.

Chadwick, P., Pirrotta, V., Steinberg, R., Hopkins, N., and Ptashne, M., 1970, the λ and 434 phage repressors, *Cold Spring Harbor Symp. Quant. Biol.* **35**:283.

Chamberlin, M., 1974, The selectivity of transcription, *Annu. Rev. Biochem.* **43**:721.

Chamberlin, M. J., 1976, RNA polymerase—An overview, in: *RNA Polymerase* (R. Losick and M. Chamberlin, eds.), pp. 17–67, Cold Spring Harbor Lab., Cold Spring Harbor, New York.

Contesse, G., Crépin, M., and Gross, F., 1970, Transcription of the lactose operon in *E. coli,* in: *The Lactose Operon* (J. R. Beckwith and D. Zipser, eds.), pp. 111–142, Cold Spring Harbor Lab., Cold Spring Harbor, New York.

Darlix, J. L., 1973, The functions of *rho* in T7-DNA transcription *in vitro, Eur. J. Biochem.* **35**:517.

Darlix, J., and Horaist, M., 1975, Existence and possible roles of transcriptional barriers in T7 DNA early region as shown by electron microscopy, *Nature* **256**:288.

Darlix, J. L., Sentenac, A., and Fromageot, P., 1971, Binding of termination factor rho to RNA polymerase and DNA, *FEBS Lett.* **13**:165.

Das, A., Court, D., and Adhya, S., 1976, Isolation and characterization of conditional lethal mutants of *Escherichia coli* defective in transcription termination factor rho, *Proc. Natl. Acad. Sci. U.S.A.* **73**:1959.

Dausse, J. P., Sentenac, A., and Fromageot, P., 1975, Interaction of RNA polymerase from *Escherichia coli* with DNA. Analysis of T7 DNA early promoter sites, *Eur. J. Biochem.* **57**:569.

Dean, W., and Lebowitz, J., 1971, Alterations induced in native superhelices by formaldehyde, *Nature New Biol.* **231**:1.

de Crombrugghe, B., Chen, B., Gottesman, M., Varmus, H., Emmer, M., and Perlman, R., 1971*a,* Regulation of *lac* mRNA synthesis in a soluble cell-free system, *Nature New Biol.* **230**:37.

de Crombrugghe, B., Chen, B., Anderson, W., Nissley, S., Gottesman, M., Pastan, I., and Perlman, R., 1971*b, Lac* DNA, RNA polymerase and cyclic AMP receptor protein, cyclic AMP receptor protein, cyclic AMP, *lac* repressor and inducer are the essential elements for controlled *lac* transcription, *Nature New Biol.* **231**:139.

de Crombrugghe, B., Adhya, S., Gottesman, M., and Pastan, I., 1973, Effect of rho on transcription of bacterial operons, *Nature New Biol.* **241**:260.

Dickson, R. C., Abelson, J., Barnes, W. M., and Reznikoff, W. S., 1975, Genetic regulation: The *lac* control region, *Science* **187**:27.

Dickson, R. C., Abelson, J., Johnson, P., Reznikoff, W. S., and Barnes, W. M., 1977, Nucleotide sequence changes produced by mutations in the *lac* promoter of *Escherichia coli, J. Mol. Biol.* **111**:65.

Downey, K., and So. A., 1970, Studies on the kinetics of ribonucleic acid chain initiation and elongation, *Biochemistry* **9**:2520.

Downey, K., Jurmark, B., and So, A., 1971, Determination of nucleotide sequences at promoter regions by the use of dinucleotides, *Biochemistry* **10**:4970.

Duffy, J. J., and Geiduschek, E. P., 1973, Transcription specificity of an RNA polymerase fraction from bacteriophage SPO1-infected *B. subtilis, FEBS Lett.* **34**:172.

Duffy, J. J., and Geiduschek, E. P., 1975, RNA polymerase from phage SPO1-infected and uninfected *Bacillus subtilis, J. Biol. Chem.* **250**:4530.

Duffy, J. J., and Geiduschek, E. P., 1977*a,* The virus-specified subunits of a modified *B. subtilis* RNA polymerase are determinants of DNA binding and RNA chain initiation, *Cell* **8**:595.

Duffy, J. J., and Geiduschek, E. P., 1977*b,* Purification of a positive regulatory subunit from phage SPO1-modified RNA polymerase, *Nature* **270**:28.

Duffy, J. J., Petrusek, R. L., and Geiduschek, E. P., 1975, Conversion of *Bacillus subtilis* RNA polymerase activity *in vitro* by a protein induced by phage SPO1, *Proc. Natl. Acad. Sci. U.S.A.* **71**:2761.

Dunn, J. J., and Studier, F. W., 1973, T7 early RNAs are generated by site-specific cleavages, *Proc. Natl. Acad. Sci. U.S.A.* **70**:1559.

Emmer, M., de Crombrugghe, B., Pastan, I., and Perlman, R., 1970, Cyclic AMP receptor protein of *E. coli;* Its role in the synthesis of inducible enzymes, *Proc. Natl. Acad. Sci. U.S.A.* **66**:480.

Englesberg, E., and Wilcox, G., 1974, Regulation: Positive control, *Annu. Rev. Genet.* **8**:219.

Epp, C., and Pearson, M. L., 1976, Association of bacteriophage lambda *N* gene protein with *E. coli* RNA polymerase, in: *RNA Polymerase* (R. Losick and M. Chamberlin, eds.), pp. 667–691, Cold Spring Harbor Lab., Cold Spring Harbor, New York.

Epstein, W., and Beckwith, J., 1968, Regulation of gene expression, *Annu. Rev. Biochem.* **37**:411.

Franklin, N. C., 1974, Altered reading of genetic signals fused to the *N* operon of bacteriophage λ: Genetic evidence for modification of polymerase by the protein product of the *N* gene, *J. Mol. Biol.* **89**:33.

Franklin, N. C., and Luria, S. E., 1961, Transduction by bacteriophage P1 and the properties of the *lac* genetic region in *E. coli* and *S. dysenteriae, Virology* **15**:299.

Franklin, N. C., and Yanofsky, C., 1976, The *N* protein of λ: Evidence bearing on transcription termination, polarity and the alteration of the *E. coli* RNA polymerase, in: *RNA Polymerase* (R. Losick and M. Chamberlin, eds.), pp. 693–706, Cold Spring Harbor Lab., Cold Spring Harbor, New York.

Friedman, B. E., Olson, J. S., and Matthews, K. S., 1977, Interaction of the *lac* repressor with inducer. Kinetic and equilibrium measurements, *J. Mol. Biol.* **111**:27.

Friedman, D. I., Wilgus, G. S., and Mural, R. J., 1973, Gene *N* regulator function of phage λ *imm* 21: Evidence that a site of *N* action differs from a site of *N* recognition, *J. Mol. Biol.* **81**:505.

Fujita, D. J., Ohlsson-Wilhelm, B. M., and Geiduschek, E. P., 1971, Transcription during bacteriophage SPO1 development: Mutations affecting the program of viral transcription, *J. Mol. Biol.* **57**:301.

Galluppi, G., Lowery, C., and Richardon, J. P., 1976, Nucleoside triphosphate requirement for termination of RNA synthesis by rho factor, in: *RNA Polymerase* (R. Losick and M. Chamberlin, eds.), pp. 657–666, Cold Spring Harbor Lab., Cold Spring Harbor, New York.

Georgopoulos, C. P., 1971, Bacterial mutants in which the gene *N* function of bacteriophage lambda have an altered RNA polymerase, *Proc. Natl. Acad. Sci. U.S.A.* **68**:2977.

Ghysen, A., and Pironio, M., 1972, Relationship between the *N* function of bacteriophage λ and host RNA polymerase, *J. Mol. Biol.* **65**:259.

Gilbert, W., 1972, The *lac* repressor and the *lac* operator, *Ciba Found. Symp.* **7**:245.

Gilbert, W., 1976, Starting and stopping sequences for the RNA polymerase, in: *RNA Polymerase* (R. Losick and M. Chamberlin, eds.), pp. 193–205, Cold Spring Harbor Lab., Cold Spring Harbor, New York.

Gilbert, W., and Maxam, A., 1973, The nucleotide sequence of the *lac* operator, *Proc. Natl. Acad. Sci. U.S.A.* **70**:3581.

Gilbert, W., and Müller-Hill, B., 1966, Isolation of the *lac* repressor, *Proc. Natl. Acad. Sci. U.S.A.* **56**:1891.

Gilbert, W., and Müller-Hill, B., 1967, The *lac* operator is DNA, *Proc. Natl. Acad. Sci. U.S.A.* **58**:2415.

Gilbert, W., and Müller-Hill, B., 1970, The lactose repressor, in: *The Lactose Operon* (J. R. Beckwith and D. Zipser, eds.), pp. 93–109, Cold Spring Harbor Lab., Cold Spring Harbor, New York.

Gilbert, W., Maizels, N., and Maxam, A., 1973, Sequences of controlling regions of the lactose operon, *Cold Spring Harbor Symp. Quant. Biol.* **38**:845.

Gilbert, W., Gralla, J., Majors, J., and Maxam, A., 1975, Lactose operator sequences and the action of *Lac* repressor, in: *Symposium on Protein–Ligand Interactions* (H. Sund and G. Blauer, eds.), pp. 193–210, de Gruyter, Berlin.

Gilbert, W., Maxam, A., and Mirzebekov, A., 1976, Contacts between the *lac* repressor and DNA revealed by methylation, in: *Control of Ribosome Synthesis* (N. O. Kjelgaard and O. Maaløe, eds.), pp. 139–148, The Alfred Benzon Symposium IX, Munksgaard, Copenhagen.

Goeddel, D. V., Yansura, D. G., and Caruthers, M. H., 1977, Studies on gene control regions. VI. The 5-methyl of thymine, a *lac* repressor recognition site, *Nucleic Acids Res.* **4**:3039.

Goff, C., and Minkley, E. G., 1970, The RNA polymerase sigma factor: A specificity determinant, in: *Lepetit Colloquium on RNA Polymerase,* Vol. I (L. Silvestri, ed.), pp. 124–147, North-Holland, Amsterdam.

Hayashi, M., Fujimura, F. K., and Hayashi, M., 1976, Mapping of *in vivo* messenger RNAs for bacteriophage ϕX-174, *Proc. Natl. Acad. Sci. U.S.A.* **73**:3519.

Herskowitz, I., 1974, Control of gene expression in bacteriophage lambda, *Annu. Rev. Genet.* **7**:389.

Herskowitz, I., and Signer, E. R., 1970, A site essential for expression of all late genes in bacteriophage λ, *J. Mol. Biol.* **47**:545.

Heyden, B., Nüsslein, C., and Schaller, H., 1975, Initiation of transcription within an RNA polymerase binding site, *Eur. J. Biochem.* **55**:147.

Hinkle, D., and Chamberlin, M., 1972, Studies of the binding of *E. coli* RNA polymerase to DNA. I. The role of the sigma subunit, *J. Mol. Biol.* **70**:157.

Hopkins, J. D., 1974, A new class of promoter mutations in the lactose operon of *Escherichia coli, J. Mol. Biol.* **87**:715.

Howard, B., and de Crombrugghe, B., 1975, ATPase activity required for termination of transcription by the *E. coli* protein factor *rho, J. Biol. Chem.* **251**:2520.

Jackson, E. N., and Yanofsky, C., 1973, The region between the operator and the first structural gene of the tryptophan operon of *Escherichia coli* may have a regulatory function, *J. Mol. Biol.* **76**:89.

Jacob, F., and Monod, J., 1961*a,* Genetic regulatory mechanisms in the synthesis of proteins, *J. Mol. Biol.* **3**:318.

Jacob, F., and Monod, J., 1961*b,* On the regulation of gene activity, *Cold Spring Harbor Symp. Quant. Biol.* **26**:193.

Jacob, F., Ullman, A., and Monod, J., 1964, Le promoteur, élément génétique nécessaire à l'expression d'un operon, *C.R. Acad. Sci.* **258**:3125.

Jobe, A., and Bourgeois, S., 1972, The natural inducer of the *lac* operon, *J. Mol. Biol.* **69**:397.

Jobe, A., Sadler, J. R., and Bourgeois, S., 1974, *Lac* repressor–operator interaction. IX. The binding of *lac* repressor to operators containing O^c mutations, *J. Mol. Biol.* **85**:231.

Johnston, D. E., and McClure, W. R., 1976, Abortive initiation of *in vitro* RNA synthesis on bacteriophage λ DNA, in: *RNA Polymerase* (R. Losick and M. Chamberlin, eds.), pp. 413–428, Cold Spring Harbor Lab., Cold Spring Harbor, New York.

Kasai, T., 1974, Regulation of the expression of the histidine operon in *Salmonella typhimurium, Nature* **249**:523.

Kiefer, M., Neff, N., and Chamberlin, M. J., 1977, Transcriptional termination at the end of the early region of bacteriophages T3 and T7 is not affected by polarity suppressors, *J. Virol.* **22**:548.

Kleid, D., Humayun, Z., Jeffrey, A., and Ptashne, M., 1976, Novel properties of a restriction endonuclease isolated from *Haemophilus parahaemolyticus, Proc. Natl. Acad. Sci. U.S.A.* **73**:293.

Korn, L. J., and Yanofsky, C., 1976*a,* Polarity suppressors increase expression of the wild-type tryptophan operon of *Escherichia coli, J. Mol. Biol.* **103**:395.

Korn, L. J., and Yanofsky, C., 1976*b,* Polarity suppressors defective in transcription termination of the tryptophan operon of *Escherichia coli* have altered rho factor, *J. Mol. Biol.* **106**:231.

Kornberg, A., 1976, RNA priming of DNA replication, in: *RNA Polymerase* (R. Losick and M. Chamberlin, eds.), pp. 331–352, Cold Spring Harbor Lab., Cold Spring Harbor, New York.

Krakow, J. S., 1966, *Azobacter vinelandii* ribonucleic acid polymerase. II. Effect of ribonuclease on polymerase activity, *J. Biol. Chem.* **241**:1830.

Krakow, J. S., and Fronk, E., 1969, *Azobacter vinelandii* ribonucleic acid polymerase. VIII. Pyrophosphate exchange, *J. Biol. Chem.* **244**:5988.

Krakow, J. S., and Pastan, I., 1973, Cyclic adenosine monophosphate receptor: Loss of cAMP-dependent DNA binding activity after proteolysis in the presence of cyclic adenosine monophosphate, *Proc. Natl. Acad. Sci. U.S.A.* **70**:2529.

Krakow, J. S., Rhodes, G., and Jovin, T. M., 1976, RNA polymerase: Catalytic mechanisms and inhibitors, in: *RNA Polymerase* (R. Losick and M. Chamberlin, eds.), pp. 127–157, Cold Spring Harbor Lab., Cold Spring Harbor, New York.

Küpper, H., Contreras, R., Khorana, H. G., and Landy, A., 1976, The tyrosine tRNA promoter, in: *RNA Polymerase* (R. Losick and M. Chamberlin, eds.), pp. 473–484, Cold Spring Harbor Lab., Cold Spring Harbor, New York.

Lebowitz, P., Weissman, S. H., and Radding, C. M., 1971, Nucleotide sequence of a ribonucleic acid transcribed *in vitro* from λ phage deoxyribonucleic acid, *J. Biol. Chem.* **246**:5120.

Lee, F., Squires, C. L., Squires, C., and Yanofsky, C., 1976, Termination of transcription *in vitro* in the *Escherichia coli* tryptophan operon leader region, *J. Mol. Biol.* **103**:383.

Losick, R., 1972, *In vitro transcription, Annu. Rev. Biochem.* **41**:409.

Losick, R., and Pero, J., 1976, Regulatory subunits of RNA polymerase, in: *RNA Polymerase* (R. Losick and M. Chamberlin, eds.), pp. 227–246, Cold Spring Harbor Lab., Cold Spring Harbor, New York.

Lowery-Goldhammer, C., and Richardson, J. P., 1974, An RNA-dependent nucleoside triphosphate phosphohydrolase (ATPase) associated with rho termination factor, *Proc. Natl. Acad. Sci. U.S.A.* **71**:2003.

Magasanik, B., 1970, Glucose effects: Inducer exclusion and repression, in: *The Lactose Operon* (J. R. Beckwith and D. Zipser, eds.), pp. 189–219, Cold Spring Harbor Lab., Cold Spring Harbor, New York.

Maitra, U., Nakata, Y., and Hurwitz, J., 1967, The role of deoxyribonucleic acid in ribonucleic acid synthesis. XIV. A study of the initiation of ribonucleic acid synthesis, *J. Biol. Chem.* **242**:4908.

Maizels, N., 1973, The neucleotide sequence of the lactose messenger ribonucleic acid transcribed from the UV5 promoter mutant of *E. coli, Proc. Natl. Acad. Sci. U.S.A.* **70**:3585.

Majors, J., 1975*a,* Specific binding of CAP factor to *lac* promoter DNA, *Nature* **256**:672.

Majors, J., 1975*b,* Initiation of *in vitro* mRNA synthesis from the wild-type *lac* promoter, *Proc. Natl. Acad. Sci. U.S.A.* **72**:4394.

Majors, J., 1978, Symmetric Binding Sites for CAP Factor within the *E. coli lac* Promoter, Ph. D. Thesis, Harvard University, Cambridge, Massachusetts.

Maniatis, T., and Ptashne, M., 1973, Multiple repressor binding at the operators in bacteriophage λ, *Proc. Natl. Acad. Sci. U.S.A.* **70**:1531.

Maniatis, T., Ptashne, M., Barrell, B., and Donelson, J., 1974, Sequence of a repressor-binding site in the DNA of bacteriophage λ, *Nature* **250**:394.

Maniatis, T., Ptashne, M., Backman, K., Kleid, D., Flashman, S., Jeffrey, A., and Maurer, R., 1975, Recognition sequences of repressor and polymerase in the operators of bacteriophage lambda, *Cell* **5**:109.

Mangel, W. F., and Chamberlin, M. J., 1974*a,* Studies of RNA chain initiation by *E. coli* RNA polymerase bound to T7 DNA. I. An assay for the rate and extent of RNA chain initiation, *J. Biol. Chem.* **249**:2995.

Mangel, W. F., and Chamberlin, M. J., 1974*b,* Studies of RNA chain initiation by *E. coli* RNA polymerase bound to T7 DNA. II. The effect of alterations in ionic strength on chain intiation and on the conformation of binary complexes, *J. Biol. Chem.* **249**:3002.

Mangel, W. F., and Chamberlin, M. J., 1974*c,* Studies of RNA chain initiation by *E. coli* RNA polymerase bound to T7 DNA. III. The effect of temperature on RNA chain intiation and on the conformation of binary complexes, *J. Biol. Chem.* **249**:3007.

Maurer, R., Maniatis, T., and Ptashne, M., 1974, Promoters are in the operators in phage λ, *Nature* **249**:221.

Maxam, A. M., and Gilbert, W., 1977, A new method for sequencing DNA, *Proc. Natl. Acad. Sci. U.S.A.* **74**:560.

McDermit, M., Pierce, M., Staley, D., Shimaji, M., Shaw, R., and Wulff, D., 1976, Mutations masking the lambda *cin-1* mutation, *Genetics* **82**:417.

Meyer, B., Kleid, D., and Ptashne, M., 1975, Lambda repressor turns off transcription of its own genes, *Proc. Natl. Acad. Sci. U.S.A.* **72**:4785.

Miller, J. H., 1970, Transcription starts and stops in the *lac* operon, in: *The Lactose Operon* (J. R. Beckwith and D. Zipser, eds.), pp. 173–188, Cold Spring Harbor Lab., Cold Spring Harbor, New York.

Miller, J. H., Ippen, K., Scaife, J. G., and Beckwith, J. R., 1968, The promoter–operator region of the *lac* operon of *E. coli, J. Mol. Biol.* **38**:413.

Miller, Z., Varmus, H. E., Parks, J. S., Perlman, R. L., and Pastan, I., 1971, Regulation of *gal* messenger ribonucleic acid synthesis in *Escherichia coli* by 3′,5′-cyclic adenosine monophosphate, *J. Biol. Chem.* **246**:2898.

Minkley, E. G., Pribnow, D., 1973, Transcription of the early region of bacteriophage T7: Selective initiation with dinucleotides, *J. Mol. Biol.* **77**:255.

Mitra, S., Zubay, G., and Landy, A., 1975, Evidence for the preferential binding of the catabolite gene activator protein (CAP) to DNA containing the *lac* promoter, *Biochem. Biophys. Res. Commun.* **67**:857.

Morse, D. E., and Morse, A. N. C., 1976, Dual control of the tryptophan operon is mediated by both tryptophanyl-tRNA synthetase and the repressor, *J. Mol. Biol.* **103**:209.

Müller-Hill, B., Crapo, L., and Gilbert W., 1968, Mutants that make more *lac* repressor, *Proc. Natl. Acad. Sci. U.S.A.* **59**:1259.

Musso, R., Di Lauro, R., Rosenberg, M., and de Crombrugghe, B., 1977, Nucleotide sequence of the operator–promoter region of the galactose operon of *Escherichia coli, Proc. Natl. Acad. Sci. U.S.A.* **74**:106.

Nakanishi, S., Adhya, S., Gottesman, M. E., and Pastan, I., 1973*a, In vitro* repression of transcription of *gal* operon by purified *gal* repressor, *Proc. Natl. Acad. Sci. U.S.A.* **70**:334.

Nakanishi, S., Adhya, S., Gottesman, M. E., and Pastan, I., 1973*b,* Studies on the mechanism of action of the *gal* repressor, *J. Biol. Chem.* **248**:5937.

Neff, N., and Chamberlin, M. J., 1978, Termination of transcription by *E. coli* RNA polymerase *in vitro* is affected by ribonucleoside triphosphate base analogs, *J. Biol. Chem.* **253**:2455.

Nissley, S., Anderson, W., Gottesman, M., Perlman, R., and Pastan, I., 1971, *In vitro* transcription of the *gal* operon requires cyclic adenosine monophosphate and cyclic adenosine monophosphate receptor protein, *J. Biol. Chem.* **246**:4671.

Okubo, S., Yanagida, T., Fujita, D. J., and Ohlssen-Wilhelm, B. M., 1972, The genetics of bacteriophage SPO1, *Biken J.* **15**:81.

Pannekoek, H., Brammar, W. J., and Pouwels, P. H., 1975, Punctuation of transcription *in vitro* of the tryptophan operon of *Escherichia coli.* A novel type of control of transcription, *Mol. Gen. Genet.* **136**:199.

Parks, J. S., Gottesman, M., Shimada, K., Weisberg, A., Perlman, R. L., and Pastan, I., 1971*a,* Isolation of the *gal* repressor, *Proc. Natl. Acad. Sci. U.S.A.* **68**:1891.

Parks, J. S., Gottesman, M., Perlman, R. L., and Pastan, I., 1971*b,* Regulation of galactokinase synthesis by cyclic adenosine 3′,5′-monophosphate in cell-free extracts of *Escherichia coli, J. Biol. Chem.* **246**:2419.

Pero, J., Nelson, J., and Fox, T. D., 1975*a,* Highly asymmetric transcription by RNA polymerase containing phage-SPO1-induced polypeptides and a new host protein, *Proc. Natl. Acad. Sci. U.S.A.* **72**:1589.

Pero, J., Tijan, R., Nelson, J., and Losick, R., 1975*b, In vitro* transcription of a late class of phage SPO1 genes, *Nature* **257**:248.

Petrusek, R., Duffy, J. J., and Geiduschek, E. P., 1976, Control of gene action in phage SPO1 development: Phage-specific modifications of RNA polymerase and a mechanism of positive regulation, in: *RNA Polymerase* (R. Losick and M. Chamberlin, eds.), pp. 587–600, Cold Spring Harbor Lab., Cold Spring Harbor, New York.

Pettijohn, D., Stonington, O., and Kossman, C., 1970, Chain termination of ribosomal RNA synthesis *in vitro, Nature* **228**:235.

Pieczenik, G., Barrell, B. G., and Gefter, M. L., 1972, Bacteriophage ϕ80-induced low molecular weight RNA, *Arch. Biochem. Biophys.* **152**:152.

Pirrotta, V., Chadwick, P., and Ptashne, M., 1970, Active form of two coliphage repressors, *Nature* **227**:41.

Platt, T., Squires, C., and Yanofsky, C., 1976, Ribosome-protected regions in the leader-*trp*E sequence of *Escherichia coli* tryptophan operon messenger RNA, *J. Mol. Biol.* **103**:411.

Pribnow, D., 1975*a,* Nucleotide sequence of an RNA polymerase binding site at an early T7 promoter, *Proc. Natl. Acad. Sci. U.S.A.* **72**:784.

Pribnow, D., 1975*b,* Bacteriophage T7 early promoters: Nucleotide sequences of two RNA polymerase binding sites, *J. Mol. Biol.* **99**:419.

Ptashne, M., 1971, Repressor and its action, in: *The Bacteriophage Lambda* (A. D. Hershey, ed.), pp. 221–238, Cold Spring Harbor Lab., Cold Spring Harbor, New York.

Ptashne, M., Backman, K., Humayun, M. Z., Jeffrey, A., Maurer, R., Meyer, B., and Sauer, R. T., 1976, Autoregulation and function of a repressor in bacteriophage lambda. Interactions of a regulatory protein with sequences in DNA mediate intricate patterns of gene regulation, *Science* **194**:156.

Rabussay, D., and Geiduschek, E. P., 1976, Regulation of gene action in the development of lytic bacteriophages, *Comp. Virol.* **8**:1.

Radding, C. M., and Echols, H., 1968, The role of the *N* gene of phage λ in the synthesis of two phage-specified proteins, *Proc. Natl. Acad. Sci. U.S.A.* **60**:707.

Ratner, D., 1976*a*, The rho gene of *E. coli* maps at *suA*, in: *RNA Polymerase* (R. Losick and M. Chamberlin, eds.), pp. 645–655, Cold Spring Harbor Lab., Cold Spring Harbor, New York.

Ratner, D., 1976*b*, Evidence that mutations in the *suA* polarity suppressing gene directly affect termination factor rho, *Nature* **259**:151.

Record, M. T., Jr., Lohman, T. M., and de·Haseth, P., 1976, Ion effects on ligand–nucleic acid interactions, *J. Mol. Biol.* **107**:145.

Reichardt, L. F., 1975, Control of bacteriophage lambda repressor synthesis after phage infection: The role of the *N, cII, cIII* and *cro* products, *J. Mol. Biol.* **93**:267.

Reznikoff, W., 1972, The operon revisited, *Annu. Rev. Genet.* **6**:133.

Richardson, J. P., 1975, Initiation of transcription by *Escherichia coli* RNA polymerase from supercoiled and non-supercoiled bacteriophage PM2 DNA, *J. Mol. Biol.* **91**:477.

Richardson, J. P., Grimley, C., and Lowery, C., 1975, Transcription termination factor rho activity is altered in *E. coli* with *suA* gene mutations, *Proc. Natl. Acad. Sci. U.S.A.* **72**:1725.

Riggs, A. D., Bourgeois, S., Newby, R. F., and Cohn, M., 1968, DNA binding of the *lac* repressor, *J. Mol. Biol.* **34**:365.

Riggs, A. D., Suzuki, H., and Bourgeois, S., 1970*a*, The *lac* repressor–operator interaction. I. Equilibrium studies, *J. Mol. Biol.* **48**:67.

Riggs, A. D., Newby, R. F., and Bourgeois, S., 1970*b*, The *lac* repressor–operator interaction. II. Effect of galactosides and other ligands, *J. Mol. Biol.* **51**:303.

Riggs, A. D., Bourgeois, S., and Cohn, M., 1970*c*, The *lac* repressor–operator interaction. III. Kinetic studies, *J. Mol. Biol.* **53**:401.

Roberts, J., 1969, Termination factor for RNA synthesis, *Nature* **224**:1168.

Roberts, J., 1970, The ρ factor: Termination and antitermination in lambda, *Cold Spring Harbor Symp. Quant. Biol.* **35**:121.

Roberts, J., 1975, Transcription termination and late control in phage lambda, *Proc. Natl. Acad. Sci. U.S.A.* **72**:3300.

Roberts, J. W., 1976, Transcription termination and its control in *E. coli,* in: *RNA Polymerase* (R. Losick and M. Chamberlin, eds.), pp. 247–271, Cold Spring Harbor Lab., Cold Spring Harbor, New York.

Roberts, J. W., and Roberts, C. W., 1975, Proteolytic cleavage of bacteriophage lambda repressor in induction, *Proc. Natl. Acad. Sci. U.S.A.* **72**:147.

Rose, J. K., and Yanofsky, C., 1974, Interaction of the operator of the tryptophan operon with repressor, *Proc. Natl. Acad. Sci. U.S.A.* **71**:3134.

Rose, J. K., Squires, C. L., Yanofsky, C., Yang, H.-L., and Zubay, G., 1973, Regulation of *in vitro* transcription of the tryptophan operon by purified RNA polymerase in the presence of partially purified repressor and tryptophan, *Nature New Biol.* **245**:133.

Rosenberg, M., and Kramer, R. A., 1977, Nucleotide sequence surrounding a ribonuclease III processing site in bacteriophage T7 RNA, *Proc. Natl. Acad. Sci. U.S.A.* **74**:984.

Rosenberg, M., Weissman, S., and de Crombrugghe, B., 1975, Termination of transcription in bacteriophage λ. Heterogeneous 3′-terminal oligo-adenylate additions and the effects of ρ, *J. Biol. Chem.* **250**:4755.

Salstrom, J. S., and Szybalski, W., 1976, Phage lambda *nut*L mutants unable to utilize *N* product for leftward transcription, *Fed. Proc.* **35**:1538.

Sanger, F., and Coulson, A. R., 1975, A rapid method for determining sequences in DNA by primed synthesis with DNA polymerase, *J. Mol. Biol.* **94**:441.

Sanger, F., Air, G. M., Barrell, B. G., Brown, N. L., Coulson, A. R., Fiddes, J. C., Hutchison, C. A., III, Slocombe, P. M., and Smith, M., 1977, Nucleotide sequence of bacteriophage ϕX174 DNA, *Nature* **265**:687.

Saucier, J.-M., and Wang, J., 1972, Angular alteration of the DNA helix by *E. coli* RNA polymerase, *Nature* **239**:167.

Schaller, H., Gray, C., and Herrmann, K., 1975, Nucleotide sequence of an RNA polymerase binding site from the DNA of bacteriophage fd, *Proc. Natl. Acad. Sci. U.S.A.* **72**:737.

Seeburg, P. H., Nüsslein, C., and Schaller, H., 1977, Interaction of RNA polymerase with promoters from bacteriophage fd, *Eur. J. Biochem.* **74**:107.

Seeman, N. C., Rosenberg, J. M., and Rich, A., 1976, Sequence-specific recognition of double helical nucleic acids by proteins, *Proc. Natl. Acad. Sci. U.S.A.* **73**:804.

Sekiya, T., Takeya, T., Contreras, R., Küpper, H., Khorana, H. G., and Landy, A., 1976, Nucleotide sequences at the two ends of the *E. coli* tyrosine tRNA genes and studies on the promoter, in: *RNA Polymerase* (R. Losick and M. Chamberlin, eds.), pp. 455–472, Cold Spring Harbor Lab., Cold Spring Harbor, New York.

Shimizu, N., and Hayashi, M., 1974, *In vitro* transcription of the tryptophan operon integrated into a transducing phage genome, *J. Mol. Biol.* **84**:315.

So, A. G., and Downey, K. M., 1970, Studies on the mechanism of ribonucleic acid synthesis. II. Stabilization of the deoxyribonucleic acid–ribonucleic acid–polymerase complex by the formation of a single phosphodiester bond, *Biochemistry* **9**:4788.

Spiegelman, G. B., and Whiteley, H. R., 1974, *In vivo* and *in vitro* transcription by RNA polymerase from SP82-infected *Bacillus subtilis, J. Biol. Chem.* **249**:1843.

Squires, C., Lee, F., and Yanofsky, C., 1975, Interaction of the *trp* repressor and RNA polymerase with the *trp* operon, *J. Mol. Biol.* **92**:93.

Squires, C., Lee, F., Bertrand, K., Squires, C. L., Bronson, M. J., and Yanofsky, C., 1976, Nucleotide sequence of the 5′ end of tryptophan messenger RNA of *Escherichia coli, J. Mol. Biol.* **103**:351.

Stahl, S. J., and Chamberlin, M. J., 1977, An expanded map of T7 bacteriophage: Reading of minor T7 promoter sites *in vitro* by *E. coli* RNA polymerase, *J. Mol. Biol.* **112**:577.

Sugimoto, K., Okamoto, T., Sugisaki, H., and Takanami, M., 1975, The nucleotide sequence of an RNA polymerase binding site on bacteriophage fd DNA, *Nature* **253**:410.

Sugimoto, K., Sugisaki, T., Okamoto, T., and Takanami, M., 1977, Studies on bacteriophage fd DNA. IV. The sequence of messenger RNA for the major coat protein gene, *J. Mol. Biol.* **110**:487.

Szybalski, W., 1976*a,* A network of developmental controls in coliphage lambda, in: *Cell Differentiation in Microorganisms, Plants and Animals* (L. Nover and K. Mothes, eds.), VEB Fisher Verlag, Jena and Elsevier, Amsterdam.

Szybakski, W., 1976*b,* Genetic and molecular map of *Escherichia coli* bacteriophage lambda (λ), in: *Handbook of Biochemistry and Molecular Biology,* 3rd ed. *Nucleic Acids,* Vol. II (G. D. Fasman, ed.), pp. 677–685, CRC Press, Cleveland, Ohio.

Takanami, M., Sugimoto, K., Sugisaki, H., and Okamoto, T., 1976, Sequence of promoter for coat protein gene of bacteriophage fd, *Nature* **260**:297.

Travers, A., and Burgess, R., 1969, Cyclic reuse of RNA polymerase sigma factor, *Nature* **222**:537.

von Hippel, P. H., and McGhee, J. D., 1972, DNA–protein interactions, *Annu. Rev. Biochem.* **41**:231.

Walter, G., Zillig, W., Palm, P., and Fuchs, E., 1967, Initiation of DNA-dependent RNA synthesis and the effect of heparin on RNA polymerase, *Eur. J. Biochem.* **3**:194.

Walz, A., and Pirrotta, V., 1975, Sequence of the P_R promoter of phage λ, *Nature* **254**:118.

Wang, J., 1974, Interaction between twisted DNAs and enzymes: The effect of superhelical turns, *J. Mol. Biol.* **87**:797.

Wang, J., Barkley, M., and Bourgeois, S., 1974, Measurements of unwinding of *lac* operator by repressor, *Nature* **251**:247.

Whitely, H. R., Spiegelman, G. B., Lawrie, J. M., and Hiatt, W. R., 1976, The *in vitro* transcriptional specificity of RNA polymerase isolated from SP82-infected *Bacillus subtilis,* in: *RNA Polymerase* (R. Losick and M. Chamberlin, eds.), pp. 587–600, Cold Spring Harbor Lab., Cold Spring Harbor, New York.

Williams, R. C., and Chamberlin, M. J., 1977, Electron microscopic studies of transient complexes found between *E. coli* RNA polymerase holoenzyme and T7 DNA, *Proc. Natl. Acad. Sci. U.S.A.* **74**:3740.

Willson, C., Perrin, D., Cohn, M., Jacob, F., and Monod, J., 1964, Non-inducible mutants of the regulator gene in the "lactose" system of *Escherichia coli, J. Mol. Biol.* **8**:582.

Wu, F. Y.-H., Nath, K., and Wu, C. -W., 1974, Conformational transition of cyclic adenosine monophosphate receptor protein of *Escherichia coli.* A fluorescent probe study, *Biochemistry* **13**:2567.

Wulff, D. L., 1976, Lambda *cin-1,* a new mutation which enhances lysogenization by bacteriophage lambda, and the genetic structure of the lambda *cy* region, *Genetics* **82**:401.

Yanofsky, C., 1976, Regulation of transcription initiation and termination in the control of expression of the tryptophan operon of *E. coli,* in: *Molecular Mechanisms in the Control of Gene Expression* (D. P. Nierlich and W. J. Rutter, eds.), pp. 75–87, Academic Press, New York.

Yanofsky, C., and Soll, L., 1977, Mutations affecting $tRNA^{Trp}$ and its charging and their effect on regulation of transcription termination at the attenuator of the tryptophan operon, *J. Mol. Biol.* **113**:663.

Yarus, M., 1969, Recognition of nucleic acid sequences, *Annu. Rev. Biochem.* **38**:841.

Zillig, W., Fuchs, E., Palm, P., Rabussay, D., and Zechel, K., 1970*a,* On the different subunits of DNA-dependent RNA polymerase from *E. coli* and their role in the complex function of the enzyme, in: *Lepetit Colloquium on RNA Polymerase,* Vol. I (L. Silvestri, ed.), pp. 151–157, North-Holland, Amsterdam.

Zillig, W., Zechel, K., Rabussay, D., Schachner, M., Sethi, V., Palm, P., Heil, A., and Seifert, W., 1970*b,* On the role of different subunits of DNA-dependent RNA polymerase from *E. coli* in the transcription process, *Cold Spring Harbor Symp. Quant. Biol.* **35**:47.

Zillig, W., Palm, P., and Heil, A., 1976, Function and reassembly of subunits of DNA-dependent RNA polymerase, in: *RNA Polymerase* (R. Losick and M. Chamberlin, eds.), pp. 101–125, Cold Spring Harbor Lab., Cold Spring Harbor, New York.

Zubay, G., Schwartz, D., and Beckwith, J., 1970, Mechanism of activation of catabolite-sensitive genes: A positive control system, *Proc. Natl. Acad. Sci. U.S.A.* **66**:104.

Zubay, G., Morse, D. E., Schrenk, W. J., and Miller, J. H., 1972, Detection and isolation of the repressor protein for the tryptophan operon of *Escherichia coli, Proc. Natl. Acad. Sci. U.S.A.* **69**:1100.

8

On the Molecular Bases of the Specificity of Interaction of Transcriptional Proteins with Genome DNA

PETER H. VON HIPPEL

1 Introduction

In order to understand the transcriptional control of genome function, we must develop a detailed knowledge of the molecular interactions whereby transcriptional proteins recognize and interact with their specific target base pair sequences among the millions of sequences present in the superficially monotonic double-helical DNA of the genome. Most of the interactions that control cellular function and development in both prokaryotes and eukaryotes involve an orderly and progressive succession of DNA–protein interactions. The fraction of these interactions that control transcription are programmed to produce the required messenger RNA (mRNA) molecules, in proper amounts and with appropriate stabilities, in response to triggering signals about which we still know almost nothing (except in a few special cases).

Most of what we currently do understand about transcriptional control is based on the properties and behavior of a relatively few bacterial and viral systems. However, it seems likely that many of the principles derived from studies of these organisms may serve, in appropriately modified form, as models for some aspects

---

PETER H. VON HIPPEL • Institute of Molecular Biology and Department of Chemistry, University of Oregon, Eugene, Oregon

of the more complex control systems of eukaryotes. In any case, this is the current belief and hope which motivates many detailed studies of prokaryote control systems and also, in part, the writing of this chapter.

In a typical prokaryote transcription system (operon), the elements of DNA–protein recognition specificity include at least the following:

1. Recognition by DNA-dependent RNA polymerase of the promoter (the base pair sequence specific to the beginning of the relevant gene or polycistronically transcribed gene cluster). This recognition is followed by "melting in" of the polymerase into the double-helical DNA structure; at this step DNA strand separation and articulation of the polymerase with the messenger template strand of the DNA occur, as well as other rearrangements of the conformation of the polymerase and the DNA template involved in initiation of transcription. Different promoters appear to be recognized (utilized) with different efficiencies, since we know that promoters may be "set" high (as in the constitutive synthesis of the mRNA for β-galactosidase in the *lac* operon), or set low (as in the synthesis of the mRNA for *lac* repressor itself).
2. In prokaryote operons concerned with sugar utilization and processing (such as the *lac* operon), the initial functional interaction of the promoter with RNA polymerase is regulated in part by a *general* protein–small molecule effector system that couples the level of transcription of each operon to the level of transcription of the other operons of the system. This general "sugar–operon" modulating system consists of catabolite gene activating protein (CAP), which, when complexed to 2′,3′,5′-cyclic AMP (cAMP), binds to specific DNA sites near the promoter and facilitates mRNA initiation. It is not clear whether other groups of operons with related functions are also partially regulated by analogous "coupling" systems, though it is known that the production of functionally interdependent molecules (such as the ribosomal components) is somehow integrated by the cell.
3. Many prokaryote operons are also regulated by an *operon-specific* control system; that is, they contain base pair sequences that comprise (unique) binding loci for control proteins that specifically repress or activate the operon. These repressors or activators also generally function in collaboration with a small molecule effector, which is often an early product of, or an intermediate in, the enzymatic pathway subject to regulation.

Thus, in typical regulated prokaryote operons at least three proteins recognize specific target sequences within a relatively restricted length of DNA and, via control interactions with small molecule effectors that alter the affinity of these proteins for the target sequences, appropriately "set up" the RNA polymerase for its transcriptional function. In eukaryotes the situation is doubtless more complex, possibly involving elements of control via chromatin structure (nucleosomes) superimposed on specific operon-like interactions modulated by nonhistone proteins.

Our understanding of the facts of transcriptional control in eukaryotes is not yet sufficiently advanced to warrant detailed molecular discussion on most points, and the *facts* of prokaryote transcription and transcriptional control have been reviewed in detail by many authors over the past few years (see, for example, von Hippel and McGhee, 1972; Chamberlin, 1974; Bourgeois and Pfahl, 1976; Losick and Chamberlin, 1976). In this chapter, therefore, I propose to build on the background provided by previous reviews to organize the facts of prokaryotic transcriptional control in a semiquantitative framework, at least as they relate to the best understood system, the lactose operon. To do this, I propose to discuss first the molecular and statistical aspects of the interactions by means of which a specific sequence in a double-helical DNA genome can be recognized by a protein, discriminated from the myriad other sequences available in the chromosome, and interacted with functionally.* Then I will undertake a detailed analysis of the lactose operon, which will become less detailed and more qualitative as we move from repressor–DNA interactions, to RNA polymerase–DNA interactions, and to CAP–cAMP–DNA interactions. In conclusion, and on a very qualitative basis, I will indicate how these approaches might be applied to certain eukaryotic systems, notably the steroid receptor–steroid–DNA interaction which bears a striking superficial resemblance to the CAP–cAMP–DNA system.†

2 *Molecular Bases of Protein–Nucleic Acid Interactions*

The interactions of proteins with nucleic acids involve the same intermolecular forces (hydrogen bonds, charges, dipoles, and solvent-driven interactions) and the same relatively simple functional groups (amino acids, bases, sugars, and phosphates) that are involved in the interactions of these macromolecules with small molecules. The interactions of such simple groups can develop enormous overall specificity because both the proteins and the nucleic acids are folded into highly specific conformations, and thus the potentially interacting functional groups form multisite interaction domains of great positional and directional specificity. Figure 1 illustrates this obvious fact by showing how much more specific even bifunctional charge–charge or hydrogen-bonding interactions can be than their monofunctional equivalents, *when the relative positions of the two functional groups of the interacting molecules are fixed.* This specificity, of course, becomes even greater when both interactants are *poly*functional, adding three-dimensional elements of directionality, and so on.

The conformational basis of the specificity of protein–nucleic acid interactions also accounts for the facility with which small changes in the conformation of either component can alter the total stability of the complex (for example, a change in protein conformation evoked by the binding of a small molecule effector). Clearly, a minor mispositioning of some of the functional groups of a

*Obviously, these principles will also apply to protein–nucleic acid interactions other than those involved in control of transcription.

†I have approached this chapter in the spirit suggested by the editors; it is in no way intended as a comprehensive review, and, in fact, a good portion represents work from our laboratory which has not been previously published. Papers have been selected for discussion simply because they illustrate a point I am attempting to make, and not necessarily because they are either the best or the only articles on the subject. The fact that this chapter is intended in part as a personal view of the problem may also have led to excessive discussion of the work of our own group. I apologize to my colleagues elsewhere who may feel that this approach has resulted in a less than sufficient emphasis on, or an apparently inadequate appreciation of, their own major contributions to the field.

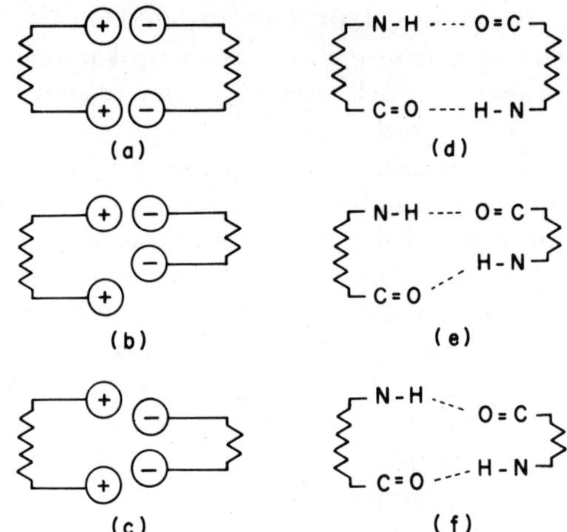

Figure 1. Specificity of bifunctional complementary intermolecular interactions: (a) A pair of charge–charge interactions properly spaced and oriented for maximal stability. (b, c) Arrangements that are less stable due to suboptimal interfunctional group spacing on the complementary molecules. (d) An optimally positioned bifunctional hydrogen-bond donor and acceptor complex. (e, f) Less favorable spacings and orientations.

component will result in a large decrease in the strength and specificity of the overall interaction.

2.1 *Protein Structures and Functional Groups*

Generalizing from the relatively few systems about which we have any detailed molecular information, it appears that most proteins involved in genome control are of the globular (as opposed to the fibrous) type, and carry one or more nucleic acid-binding sites. Many of these proteins bind as dimers or oligomers, which may consist of identical subunits for those proteins charged with fairly simple functions (repressors, for example) and nonidentical subunits for those with more complex functional responsibilities (RNA polymerase, for example). Many carry a net negative charge, but nevertheless have a strong favorable electrostatic component in their free energy of interaction with negatively charged nucleic acids, suggesting that the nucleic acid-binding site may contain a local excess of positive residues. Since many of these proteins must alter their affinity for the nucleic acid substrate as a consequence of interaction with other macromolecular constituents or small molecule allosteric effectors, the binding conformation must be in equilibrium with one or more protein conformational states of comparable total free energy.*

The functional groups available in proteins to react with nucleic acids include those discussed in the following subsections.

2.1.1 *Positively Charged Residues*

Positively charged residues have been reliably implicated in the interaction sites of several nucleic acid-binding proteins, and are thought to operate primarily

*Some of these generalizations may not apply to the histones, which are the major class of proteins interacting with eukaryotic DNA. However, histones play a major *structural* (as well as a presumably functional) role in organizing the eukaryotic chromosome.

via charge–charge interaction with the backbone phosphates of the DNA. These charge–charge interactions can provide some binding specificity, since if we view the binding surface of the protein as a "grid" of positive charges positioned for favorable interaction with the negative charges of the DNA backbone, the relative positions of the nucleic acid charges can be varied to some extent by altering the nucleic acid conformation (for example, the B-to-A form DNA transition, to name one; see Section 2.2.1).

2.1.2 *Hydrogen-Bonding Groups*

The primary source of interactional specificity between proteins and nucleic acids is doubtless hydrogen bonding, since hydrogen bonds exhibit strongly preferred lengths and dihedral angles, and a significant distortion of hydrogen-bond lengths or angles can be energetically costly. Almost all amino acid residues carry one or more hydrogen-bond donors or acceptors. As indicated earlier, the hydrogen-bond donors and acceptors of the protein site can be considered as a specifically positioned multisite network appropriately set up to interact in a complementary fashion with their proper nucleic acid counterparts (see Section 2.2.2). In general, hydrogen-bonding groups on the protein surface are involved in interactions either with properly positioned donors and acceptors on the surfaces of other macromolecules, or with solvent constituents. Thus, proper hydrogen bonding probably does not greatly stabilize macromolecular complexes. However, *improperly* positioned hydrogen-bond donors or acceptors which are not able to find properly oriented acceptors and donors in an otherwise correctly structured complex (and are therefore buried away from solvent without a proper partner) will markedly *decrease* the stability of the protein–nucleic acid interaction. Thus, just as for the stability of the Watson–Crick canonical base pairs relative to the many other hydrogen-bonded base pairs available, it is not so much that the correct structures are stable as that the incorrect structures are unstable. (This point is developed further in subsequent sections of this chapter.)

2.1.3 *Hydrophobic Groups*

Hydrophobic bonding (in the classical Kauzmann–Tanford sense: Kauzmann, 1959; Tanford, 1962) probably does not play a major role in the actual surface-to-surface protein–nucleic acid interaction, since most of the functional groups of the nucleic acid component (phosphates, sugars, and most parts of the bases) are quite hydrophilic in nature. The methyl and methylene groups of the aliphatic residues of the protein could interact favorably with the methyl group of thymine, and might, under certain special circumstances, find the anhydrous environment of one of the grooves of the double-helical structure preferable to the fully solvated state. Hydrophobic residues could also be involved in interactions with postreplicationally (or posttranscriptionally) modified nucleic acid units, since the modifying groups often have a hydrophobic aspect.

2.1.4 *Stacking Interactions*

Stacking interactions are the primary source of the structural stability of nucleic acid conformations, and some of the amino residues in proteins (Tyr, Trp,

His, Phe) have appropriate planar configurations and could be involved in stacking interactions (intercalation) with nucleic acid bases. Such interactions have been implicated both in complexes of nucleic acids and oligopeptides (for example, see Toulmé and Hélène, 1977) and in complexes of nucleic acids and proteins (for example, see Coleman *et al.*, 1976). These stacked structures doubtless stabilize the complex, but it is not clear whether they contribute only to stability or also to specificity by discriminating one nucleic acid base from another.

2.1.5 *Dipolar Interactions*

Proper alignment of dipoles is doubtless crucial in establishing stable protein and nucleic acid structures, as well as for the formation of stable protein–nucleic acid complexes. Most constituents of proteins (and nucleic acids, see later) that can be involved in dipolar interactions are also hydrogen-bond donors or acceptors, and thus probably contribute to the stability and specificity of protein–nucleic acid complexes primarily via hydrogen bonding.

2.2 *Nucleic Acid Conformations and Functional Groups*

Nucleic acids are characterized by a greater variety of potentially interactive functional groups than are proteins, even though they contain a smaller number of residue types (bases) than do proteins (amino acids).

2.2.1 *Polynucleotide Backbones*

Figure 2 shows the sugar–phosphate backbone of a single nucleic acid chain, indicating the potentially rotatable single bonds and the various atoms involved in one repeating unit. Clearly, most of the backbone groups are hydrophilic and most of the oxygens carry appreciable partial negative charge, making them excellent hydrogen-bond acceptors. The 2′-OH group of RNA is the only hydrogen-bond donor on the sugar–phosphate backbone; there are no hydrogen-bond donors on the DNA backbone. The phosphates represent the only ionized groups in polynucleotides at neutral pH; they carry a single negative charge at all pH values greater than about 1 (terminal phosphates carry a second negative charge with a pK of ~6.5). These negative phosphate groups are the primary loci of charge–charge interaction with the basic amino acid residues (Arg, Lys, His) of proteins, and contribute a major part of the interaction energy of many protein–nucleic acid complexes.

Despite the existence of seven single bonds (one within the pentose ring) between adjacent nucleotide bases along a nucleic acid chain, single-stranded polynucleotides show relatively little rotational freedom about these bonds primarily because most attainable positions tend to push negatively charged oxygen atoms into one another. As a result of this high degree of rotational hindrance the single-stranded backbone is quite stiff (even in the absence of base stacking), and is characterized by relatively few acceptable configurations. This finding, based both on model building and solution studies (for a review see Sundaralingam, 1975), led to the development of the "rigid nucleotide concept," which postulates that the position of most functional groups of the backbone is fixed and tightly coupled to the "pucker" of the ribose (or deoxyribose) ring. Figure 3 shows the two preferred (rigid) nucleotide backbone conformations characteristic of most known nucleic acid conformations: the C(3′)-*endo* form characteristic of the A form of double-stranded DNA and RNA is illustrated in the upper part of the figure, while the lower part shows the C(2′)-*endo* form, close to that found in the B form of double-stranded DNA. Clearly, the rise along the helix axis (per phosphate group) is appreciably greater in the C(2′)-*endo* form. Also, it should be stated explicitly here that, based on the rigid nucleotide concept, it is now generally accepted that the differences between the solution conformations of DNA and RNA arise primarily from the fact that the addition of the 2′-OH to the deoxyribose ring shifts the preferred ring conformation from the C(2′)-*endo* [or the closely related C(3′)-*exo*] to the C(3′)-*endo* form. This difference in preferred ring conformation results in the rearrangement of the surrounding groups from the conformation in the lower to that in the upper part of Fig. 3, and is primarily responsible for triggering the DNA A to B form interconversion as the solvent environment is changed.

In addition to sugar ring conformations, there are favored dihedral rotational angles around the glycosyl (χ) bond and the phosphodiester (ω and ω') bonds. The

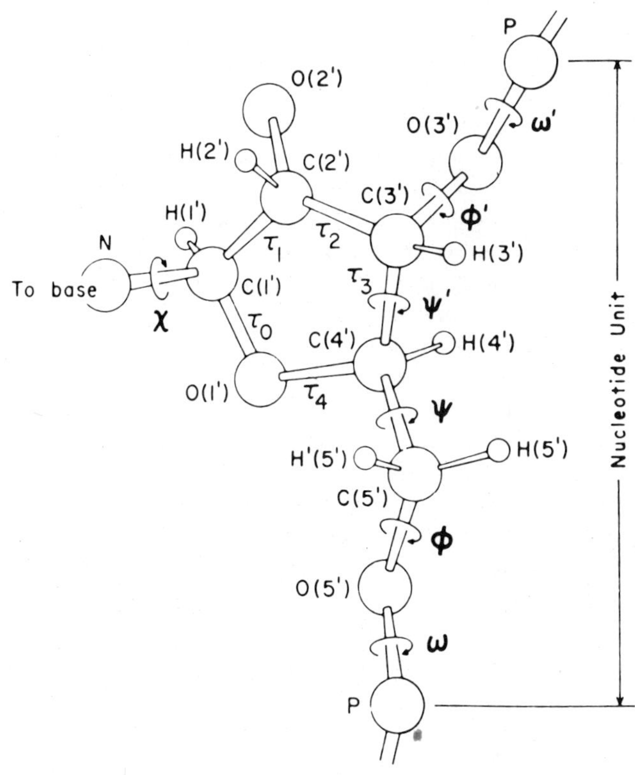

Figure 2. A nucleotide backbone unit, showing the six sugar–phosphate backbone dihedral angles (ω, ϕ, ψ, ψ', ϕ', and ω'), the torsional angles within the pentose ring (τ_0, τ_1, τ_2, τ_3, and τ_4), and the glycosyl dihedral angle (χ) between the sugar ring and the base. (Reproduced with permission from Sundaralingam, 1975.)

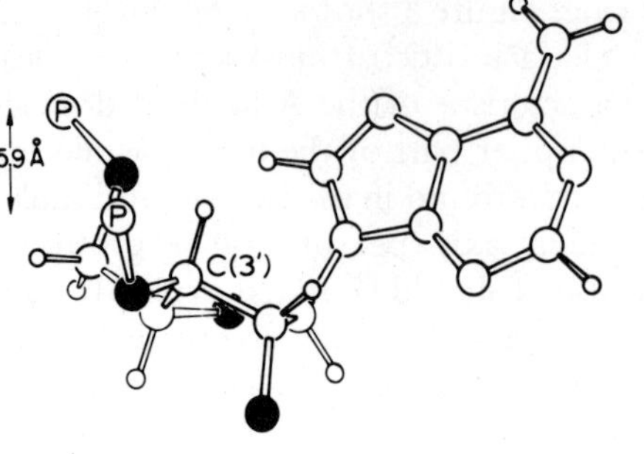

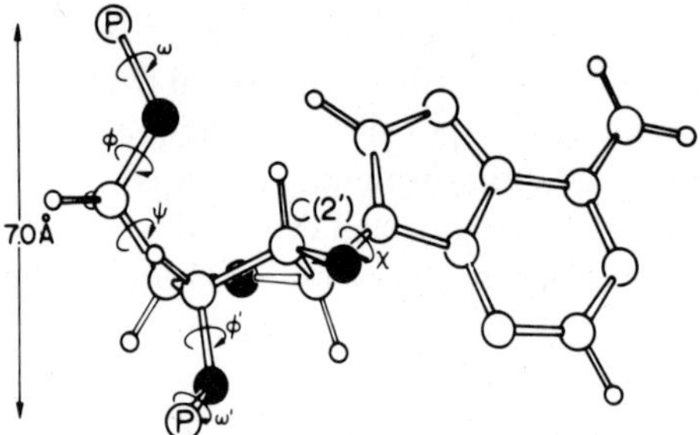

Figure 3. The two most preferred (stable) sugar–phosphate backbone units. Upper part, the C(3′)-*endo* form; lower part, the C(2′)-*endo* form. (Reproduced with permission from Sundaralingam, 1975.)

preferred rotational positions for these bonds are those shown in Fig. 3 (for further details see Sundaralingam, 1975).

From the point of view we are developing here, the major outcome of these studies of polynucleotide backbone geometry is that there exist a very few definite, preferred chain conformations that present a specific, multifaceted array of charges, hydrogen-bonding acceptors, and dipoles to a potentially interacting protein surface. These interactions can be very different for the different chain configurations, as illustrated in Fig. 4, which shows some of the changes in interfunctional group angles and distances resulting (ultimately) from these basic ring–pucker interconversions. A protein surface fitting better to one chain configuration than to the other can induce a conformational change in the polynucleotide chain if the free energy change of the nucleic acid conformation is more than offset by the increased free energy of interaction of the protein–nucleic acid complex. Alternatively, of two protein conformations in equilibrium, that which more closely approximates the interacting protein surface geometry thermodynamically preferred by a particular polynucleotide backbone conformation will be favored, and the protein conformational equilibrium will be perturbed accordingly.

2.2.2 *Nucleic Acid Bases*

The major information-containing variable which discriminates one segment of natural nucleic acid from another is, of course, base or base pair sequence. Figure 5 shows detailed views of the canonical base pairs of DNA and RNA (uracil is generated from thymine by replacing the 5-methyl with a hydrogen atom), indicating the various bond distances and angles involved in these entities. (The

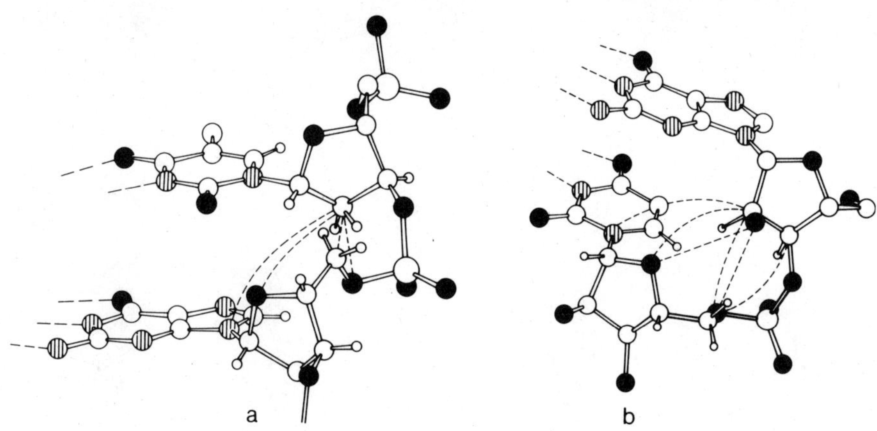

Figure 4. (a) A single-stranded section of B-form DNA showing some of the short-range sugar–sugar and sugar–phosphate intramolecular interactions between sequential nucleotide residues. (b) A similar view of a single-stranded section of the double-helical RNA conformation, showing how distances and angles between functional groups of sequential nucleotide residues are changed in this conformation. (Reproduced with permission from Sundaralingam, 1975.)

atoms of the individual bases are all coplanar, and the two hydrogen-bonded bases are almost coplanar in all double-helical conformations; see Section 2.2.3.)

It is clear from Fig. 5 that there are a number of loci available for discriminating bases in single-stranded DNA. These bases may be viewed as groups of hydrogen-bond donors and acceptors with their relative orientations fixed in space; specific complex formation with protein components probably involves two or more specific hydrogen-bonding interactions per base, just as it does for base–base interactions.* Recognition of single-stranded nucleic acid sequences can involve patterns of hydrogen-bonding interactions with the donors and acceptors normally involved in Watson–Crick base pairing, as well as the functional groups at the "top" or the "bottom" of the bases, as drawn in Fig. 5. In double-helical DNA and RNA, the Watson–Crick hydrogen-bonding groups are, of course, not available for protein–nucleic acid recognition reactions. Nevertheless, most proteins which show a high degree of sequence specificity appear to recognize double-helical sequences about 15 or more base pairs long. Though different base pair sequences can, in principle, have specific effects on backbone geometry, local conformational motility, and so on, the primary basis of recognition by protein functional groups must be the base-pair-specific groups that protrude into the major groove (groups at the top of the base pairs as drawn in Fig. 5) and the minor groove (groups at the bottom of the base pairs of Fig. 5) of the double helix. As pointed out by Yarus (1969) and von Hippel and McGhee (1972), and recently elaborated by Seeman *et al.* (1976), there is ample functional group and positional

*Interbase hydrogen bonding demonstrates particularly clearly the power of multisite functional group interactions; although the interactions are not very strong, they are very specific because the *relative* positioning of even two hydrogen-bond donor–acceptor groups (coupled with restraints on base positions due to backbone connectivity and conformation) is sufficiently definitive to exclude almost all incorrect pairings. To my knowledge no positionally sensitive specific biological complex ever depends on a single hydrogen bond or other noncovalent interaction, no matter how stable it might be.

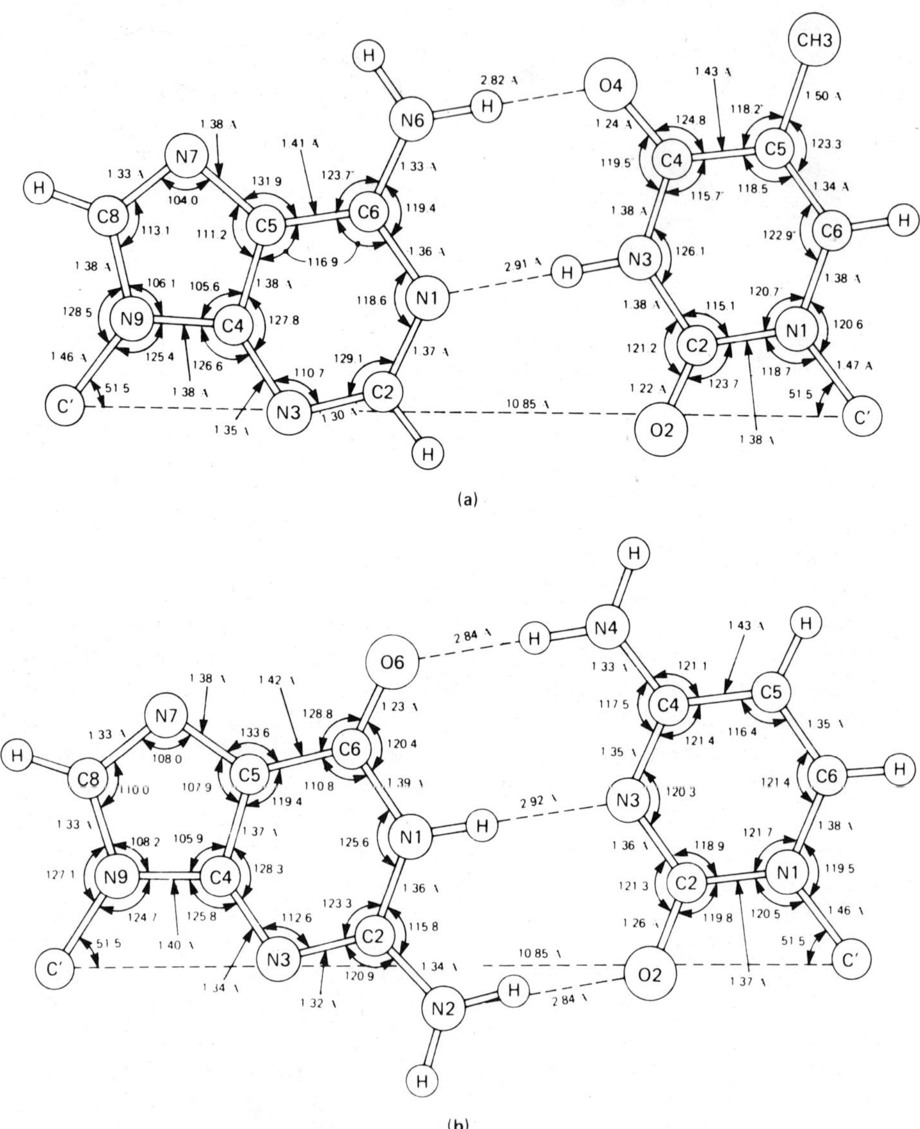

Figure 5. Functional groups, disatances, and angles for (a) the A·T and (b) G·C base pairs of DNA. The groups lying in the major groove of the DNA structure are at the "top" of each base pair as shown, while those accessible through the minor groove are at the "bottom." (Reproduced with permission from Arnott *et al.,* 1969.)

information available in both grooves to differentiate the four types of DNA (or RNA) base pairs (there are four base pair types because nucleic acid backbones are polar, and as a consequence an A·T pair can, and must, be discriminated from a T·A pair, and a G·C pair from a C·G pair).

Figure 6, taken from the work of Seeman *et al.* (1976), defines a nomenclature for the potential interactions in the major (W, for wide) and the minor (S, for small) grooves of double-helical A·U (=A·T) and G·C base pairs. Clearly, there are three potentially discriminatable loci (hydrogen-bond donors or acceptors) in the major groove, and two in the minor groove for base pairs of both types (oriented as either A·U or U·A, and so on). Seeman *et al.* (1976) then examined these potentially discriminating functional groups in turn, and attempted to see whether a single protein hydrogen-bonding donor or acceptor (presumably appropriately positioned over the correct part of the appropriate groove by other protein–nucleic acid interactions) could unambiguously discriminate a particular base pair from the other three canonical forms. They concluded that in most cases this is not possible, though all base pairs can be discriminated via either groove if any *two* of the functional groups in the groove are recognized by the protein simultaneously.

Figure 7 provides examples of particular amino acid residues that may serve as bifunctional recognition "templates" for discriminating individual base pairs. It is important to add that the necessary bifunctional recognition of a single base pair need not involve a single amino acid residue. Since particular protein conformations tend to lock hydrogen-bonding groups into matrices of fixed relative position, two properly spaced amino acid residues could also serve. Similarly, we should appreciate that a pair of functional hydrogen-bonding groups, whether located on a single amino acid residue or on two residues of fixed relative position, can also recognize base pair functional groups located on two adjacent base pairs. Since the interfunctional group distances or angles of such groups can be varied depending on local double-helix geometry (see later), this provides an additional variable for discriminating the same base pair sequence within two different double-helical geometries.

2.2.3 *Nucleic Acid Conformations*

Building on the components outlined in the preceding two sections, in Fig. 8 we show the overall geometries of the nucleic acid base pairs within two of the major double-helical backbone geometries that have been defined by fiber X-ray diffraction. Clearly the A form of DNA (and the closely related RNA structure) differs greatly from the B form of DNA with respect to base pair positioning relative to the sugar–phosphate backbones (due primarily to differing sugar ring pucker). In addition to A and B conformations, X-ray studies of synthetic polynucleotides have demonstrated the existence of various other sequence-dependent conformations of the DNA double helix. For example, the dimensional parameters of DNA sequences with the purine bases all in one chain and the pyrimidine bases all in the other, differ significantly from those for double helices in which purines and pyrimidines alternate in each chain, etc. (for a recent summary, see Arnott *et al.,* 1975). Obviously, a multisite protein recognition matrix or template which is keyed on adjacent base pairs (as well as perhaps on backbone charges or

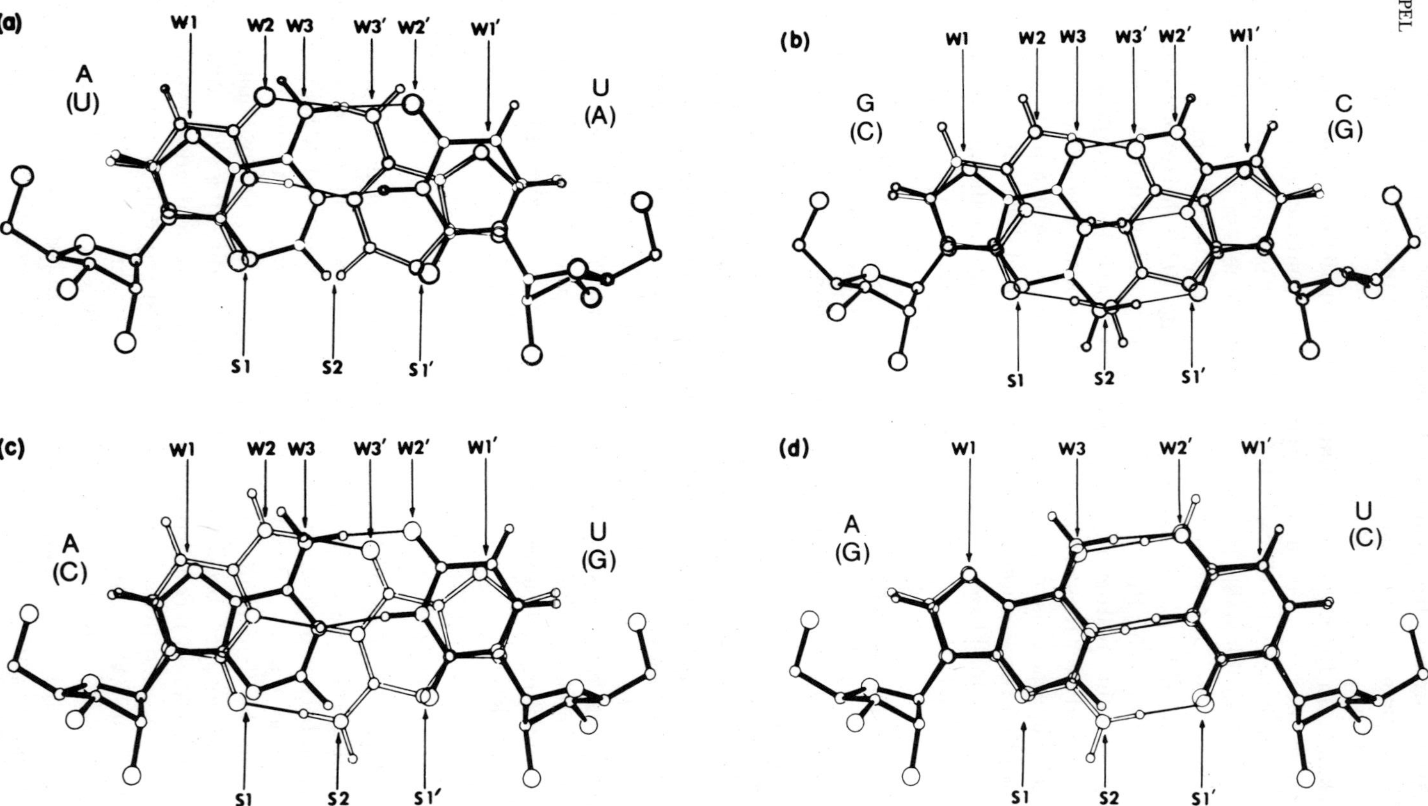

Figure 6. The overall stereochemistry of double-helical A·U and G·C base pairs. The pairs are superimposed on each other with one base drawn with solid bonds, and the other with outlined bonds. The upper letter at the side refers to the solid bases, and the lower letter in parentheses refers to the outlined bases. Both bases are drawn as attached to the same ribose residues in the antiparallel double-helical conformation. The sites designated W refer to hydrogen-bond donor or acceptor loci in the major (wide) groove and the sites designated S refer to hydrogen-bond donor or acceptor loci in the minor (small) groove. The dyad axis between the two antiparallel ribose residues is vertical in the plane of the paper. Diagrams (a) through (d) represent all of the possible base pair comparisons. (Reproduced with permission from Seeman *et al.*, 1976.)

hydrogen-bonding groups) can discriminate different chain conformations. This provides a major additional dimension for protein–nucleic acid recognition mechanisms.

2.2.4 *Other Structural Features Potentially Useful in Protein–Nucleic Acid Recognition Interactions*

Using the multisite functional group approach just outlined, there are a number of additional elements of nucleic acid conformation that could play a role in the recognition of specific nucleic acid sequences by proteins. Here they are simply listed, with one or two examples of where they have been (or might be) applied. Most will be taken up further in connection with the more detailed discussions of protein–nucleic acid interactions involved in transcription in subsequent sections.

2.2.4a Superhelix Formation. The addition of superhelical turns to circular, double-helical DNA can vary the relative positions of adjacent nucleic acid functional groups sufficiently to change both the strength and the specificity of protein–nucleic acid interactions. Examples include the binding of RNA polymerase to promoter (Wang, 1974), repressor to operator (Wang *et al.,* 1974), and so on.

2.2.4b Conformational Motility. Base sequences of varying composition show appreciable differences in conformational motility, or extent to which the base pairs are "open" (unstacked and/or not hydrogen bonded). Proteins that bind preferentially to single-strand regions (melting proteins, for example) interact with different sequences depending on the extent to which the sequences are open; AT-rich regions tend to react with such proteins in significant preference to GC-rich regions (for a discussion of sequence discrimination by melting proteins and melting protein models, see von Hippel *et al.,* 1977). Examples demonstrating such specificity include micrococcal nuclease attacking primarily double-helical DNA (Wingert and von Hippel, 1968) as well as the behavior, in part, of biologi-

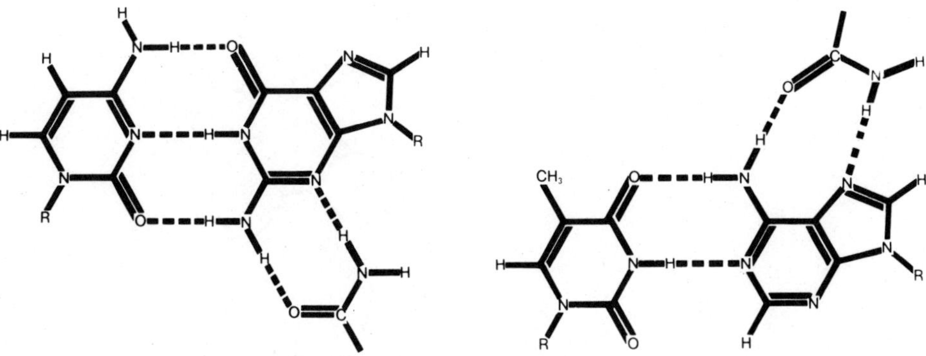

Figure 7. A G·C base pair (on the left) showing a possible bifunctional interaction with a glutamine (or asparagine) amino acid residue via the minor groove of the DNA double helix. An A·T base pair (on the right) showing a possible bifunctional interaction with a glutamine (or asparagine) amino acid residue via the major groove. (Reproduced with permission from Hélène, 1977.)

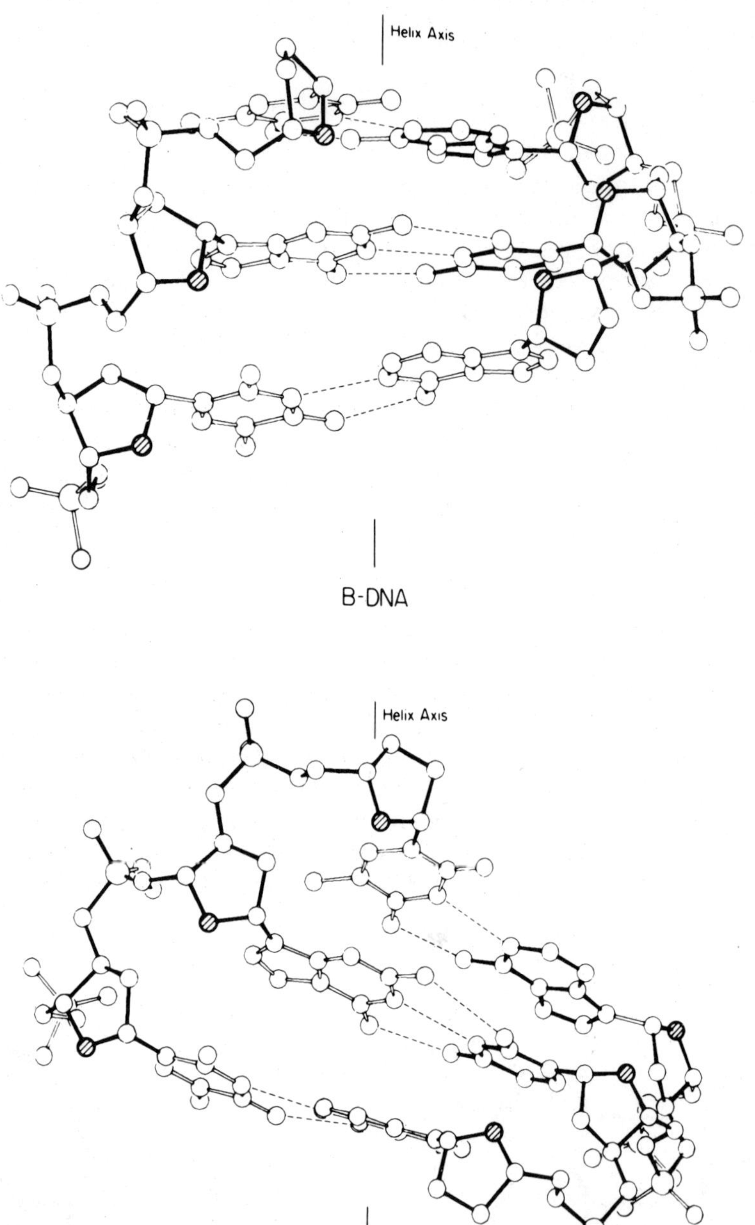

Figure 8. The overall geometry of (top) B-form DNA and (bottom) A-form DNA. Note the many changes in interfunctional group orientations and spacings in going from one of these conformations to the other. (Reproduced with permission from Sundaralingam, 1975.)

cally important proteins involving a DNA motility-dependent component, such as RNA polymerase (Chamberlin, 1974).*

2.2.4c Structural Rigidity. Certain double-helical DNA (or RNA) base pair sequences may be more rigid—that is, less subject to deformation as a consequence of complex formation with proteins—than are others. For example, poly-L-lysine binds preferentially to AT-rich DNA even though lysine monomers show no intrinsic AT-binding preference. Since polylysine binding (as well as the binding of some lysine-rich histones) results in deformation of the DNA backbone, as monitored by circular dichroism, hydrogen exchange, and other techniques, it may well be that the optimal fit of the double-stranded polynucleotide backbone conformation to polylysine involves a structure somewhat different from the equilibrium free solution B form, and that AT-rich sequences of DNA are more easily distorted into this structure than are GC-rich sequences (for a review of, and references to, these studies, see von Hippel and McGhee, 1972).

2.2.4d Palindromic Sequences. As more and more double-helical base pair sequences that bind proteins are isolated and identified, it is becoming ever more striking that many of these sequences are partially, or entirely, palindromic; that is, the base pair sequence encountered in reading from one end of the palindrome in the $5' \rightarrow 3'$ direction along one chain is identical to that found in reading from the other end of the palindrome in the $5' \rightarrow 3'$ direction along the other chain (for examples of such palindromic sequences, see Table I). Such sequences can potentially serve a number of functions in protein–nucleic acid interactions. For example, they may reflect the involvement of two identical protein subunits in recognition or functional interactions. Alternatively, such sequences make possible the formation of hairpin (or "doodad") structures (Gierer, 1966), which, though they are always less stable than the "straight" B-form double helix, will always exist in thermodynamic equilibrium with the straight double helix and under special circumstances may have sufficient stability to be recognized. Sobell (1973*a,b*) has written extensively on the possible significance of palindromic sequences in recognition mechanisms (see also Chapter 5 of this volume).

3 The Problem of the Other Sites

In the preceding sections we have discussed the molecular mechanisms by which a protein recognizes a specific sequence of nucleotide bases or base pairs. The obverse of that problem is the one we consider here, namely, the mechanisms by which a protein discriminates the "right" site from the overwhelming number of very similar "wrong" sites.

The dimensions of the problem become clear when we consider *Escherichia coli,* which, of course, has a much smaller genome than most eukaryotic cells. The *E. coli* cell contains, on the average, about 10^7 DNA base pairs. The number of

*Recently, potentially useful small molecule additives that bind preferentially to A·T (or G·C) base pairs, and thus can change the relative stability of base pairs of different composition, have been discovered (see, for example, Melchior and von Hippel, 1973). Such perturbants should make possible experiments to probe directly the involvement of conformational motility in the specificity of protein–nucleic acid recognition interactions.

copies per cell of any particular site-specific genome-controlling protein is much smaller than this (for example, $\sim 10^1$ for *lac* repressor, $\sim 10^3$ for RNA polymerase, and so on). Therefore, ignoring end effects, we can state that the *E. coli* genome contains approximately 10^7 potential protein-binding sites, since (regardless of the length of the site) every base pair can be considered to comprise the start of a potential (overlapping) binding site. Thus, the problem for the *lac* repressor, for example, becomes that of distinguishing the one (or two or three, depending on the stage of genome replication) operator site(s) from all the other potential binding loci. This problem has a molecular aspect, which has been discussed in general terms in Section 2. It also has informational (statistical) and energetic (thermodynamic) aspects. These components of the problem are considered in this section.

3.1 *Informational Aspects of Regulation*

Let us assume, to make the problem manageable, that the individual elements (informational "bytes") by which site-specific binding proteins recognize nucleic acids are the individual nucleotide bases (for proteins that bind single-stranded DNA) and individual base pairs (for proteins that prefer double-stranded DNA).*

3.1.1 *Probability of Particular Sequences*

We may then ask the following: How many bases or base pairs in a row (n) are required to define a unique protein-binding site (base or base pair sequence) within a genome N bases or base pairs in length? Neglecting end effects (or assuming the genome is circular), this means we must make n sufficiently large so that the expected mean number of random occurrences per genome of the same sequence is well below one.†

As indicated earlier, there are not only bases of four discriminatable types in single-stranded DNA (A, T, G, and C) but also, because of backbone polarity, there are base pairs of four types (A·T, T·A, G·C, and C·G). We will write single-strand sequences with the 5′ end on the left, and number the lattice positions from the left (5′) end of the sequence. For example,

$$\begin{array}{ccccccccccc} & 1 & 2 & 3 & 4 & 5 & 6 & 7 & 8 & \\ (5'-) & \mathrm{A} & \mathrm{T} & \mathrm{G} & \mathrm{C} & \mathrm{G} & \mathrm{T} & \mathrm{T} & \mathrm{C} & (-3') \end{array} \qquad \text{(I)}$$

*This is certainly true to a first approximation, in that all recognition must ultimately depend on base or base pair sequences. However, recognition of a base pair need not be totally nearest neighbor independent; to the extent that it involves recognition of subtle changes in backbone geometry or conformational motility, it will depend on the sequence and composition of larger clusters of base pairs, and these longer range sequence-dependent interactions must be taken into account explicitly (see, for example, Wartell, 1977). Also, as outlined in Section 2.2.2, multisite hydrogen-bonding recognition by an amino acid residue could involve functional groups on more than one base or base pair.

†Thomas (1966) has taken a similar approach to the distribution of base sequences in DNA in connection with the problem of genetic recombination.

Similarly, double-stranded sequences will be written with the 5′ end of the *top* strand on the left, and numbered in the same direction. For example,

	1	2	3	4	5	6	7	8		
(5′−)	A	T	G	C	G	T	T	C	(−3′)	(II)
(3′−)	T	A	C	G	C	A	A	G	(−5′)	

(Note that the two strands of all natural double-stranded nucleic acids run antiparallel.)

The probability that base A will occur in any particular lattice position (defined by position index i) in a single-stranded DNA sequence (or in the top chain in that position of a double-stranded DNA) will be $P_{i,A}$. Similarly, $P_{i,T}$, $P_{i,G}$, and $P_{i,C}$ will be the probabilities of the other bases appearing at position i. (The same nomenclature can be applied to single- and double-stranded RNA by replacing T by U.) If all base sequences are *chemically* random, then $P_{i,A}$ is the mole fraction of A in the nucleic acid, which we write P_A, and so on.
Several situations can exist:

1. For single-stranded DNA, the mole fractions of all four bases can be (and generally are) different: $P_A \neq P_T \neq P_G \neq P_C$.
2. For double-stranded DNA, $P_A = P_T \neq P_G = P_C$.
3. For double-stranded DNA containing 0.5 (mole fraction) AT (A·T + T·A) base pairs and 0.5 GC (G·C + C·G) base pairs, $P_A = P_T = P_G = P_C$.
4. In some sequences, certain positions (i) may be specified only as containing pyrimidines or purines. The probability that a purine or pyrimidine occurs in position i will be $P_{i,pu}$ and $P_{i,py}$, respectively. For double-stranded DNA, $P_{pu} = P_{py} = 0.5$; for single-stranded DNA, in general, $P_{pu} \neq P_{py}$.
5. In some sequences, certain positions may be specified as containing *any* base or base pair. These positions will be called blanks, and the probability that a blank in position i contains *some* base or base pair is $(P_{i,bl}) = 1.0$.

For a particular sequence n, within a large genome N (where $N >> n$), the probability of occurrence of a particular sequence can be shown by simple conditional probabilities to be the product of the individual probabilities of the occurrence of each base (or base pair) in its particular lattice position, i.

To illustrate, for a particular single-stranded sequence with $n = 11$:

1	2	3	4	5	6	7	8	9	10	11	
A	T	Pu	G	C	bl	G	Py	T	T	C	(III)
$P_{1,A}$	$P_{2,T}$	$P_{3,pu}$	$P_{4,G}$	$P_{5,C}$	$P_{6,bl}$	$P_{7,G}$	$P_{8,py}$	$P_{9,T}$	$P_{10,T}$	$P_{11,C}$	

where the position indices are indicated above the sequence and the individual positional probabilities below the sequence. The overall conditional probability of sequence (III) is therefore

$$P_{11} = (P_A)^1 (P_T)^3 (P_G)^2 (P_C)^2 (P_{pu})^1 (P_{py})^1 (P_{bl})^1$$

where P_n is the overall probability of a particular sequence n long; in this case $n = 11$.

If (III) represents (the top line of) a double-stranded sequence, the overall conditional probability of this sequence is

$$P_{11} = (P_{\mathrm{A}})^4\,(P_{\mathrm{G}})^4\,(0.5)^2\,(1)$$

since, for any double-stranded sequence, $P_{\mathrm{A}} = P_{\mathrm{T}}$, $P_{\mathrm{G}} = P_{\mathrm{C}}$, $P_{\mathrm{pu}} = P_{\mathrm{py}}$, and (as previously) $P_{\mathrm{bl}} = 1$.

If, in addition, the mole fractions of all four bases in the double-stranded (or single-stranded) genome are the same (that is, 0.25), we can simplify further to

$$P_{11} = (0.25)^8\,(0.5)^2\,(1) = 3.8 \times 10^{-6}$$

From this point on, since our focus is on transcriptional proteins which recognize primarily double-helical (base-paired) nucleic acid-binding sites, we will carry out calculations only for base-paired sequences (that is, where $P_{\mathrm{A}} = P_{\mathrm{T}}$, $P_{\mathrm{G}} = P_{\mathrm{C}}$, and $P_{\mathrm{pu}} = P_{\mathrm{py}}$). To simplify further, illustrative calculations will generally be carried out for double-stranded DNA for which the overall mole fractions of A, T, G, and C are equal, as is, for example, approximately true for bacteriophage λ or T7 DNA. Thus, assuming *chemical* base pair sequence randomness, in most of our model calculations $P_{\mathrm{A}} = P_{\mathrm{T}} = P_{\mathrm{G}} = P_{\mathrm{C}} = 0.25$. Furthermore, in writing double-stranded sequences we will write out only the sequence of the *top* [(5′-) . . . . . . (-3′)] chain. Based on the calculation rules outlined earlier, extensions to other situations are obvious.

What we really want to know is the mean frequency *per genome* that a given putative recognition sequence will reoccur at random. This predicted frequency of occurrence per genome is $P_n \cdot 2N$, where P_n is the intrinsic probability of the sequence, as defined earlier, and N is the number of base pairs in the genome and also the number of sequences of length N in the overall (circular or very long) genome lattice. The factor 2 enters because a sequence n base pairs in length arranged in a circular DNA lattice N base pairs long can be read in either direction, and will be different if read in the two directions except if the sequence is palindromic (see later). For a linear lattice this factor becomes $2(N - n - 1)$, but if $N >> n$, the additional terms can be (and generally here will be) ignored.

Some model calculations for the *E. coli* system (using $N = 10^7$ potential sites) are plotted as $\log (P_n \cdot 2N)$ versus n in Fig. 9 for a DNA with $P_{\mathrm{A}} = P_{\mathrm{T}} = P_{\mathrm{G}} = P_{\mathrm{C}} = 0.25$. The results show that for a genome of the size of *E. coli,* a sequence of 12 base pairs must be specified, since any shorter sequence has a predicted frequency greater than unity ($P_n \cdot 2N > 1$) of reoccurring somewhere in the genome by chance.

Some additional comments on these calculations may be useful.

3.1.1a Effects of "Blank" Loci. Any number of blanks can be inserted *in specified places* in any sequence without changing its total conditional probability. For example, the probability of the sequence

1	2	3	4	5	6	7	8	9	10	11	12	
A	A	T	G	T	C	C	T	A	G	C	A	(IV)

is $P_{12} = (0.25)^{12} = 6.0 \times 10^{-8}$; and that of the sequence

1	2	3	4	5	6	7	8	9	10	11	12	13	14	15	16	
A	A	T	bl	bl	G	T	C	bl	C	T	bl	A	G	C	A	(V)

is $P_{16} = (0.25)^{12}(1)^4 = 6.0 \times 10^{-8}$; that is, the probabilities are the same.

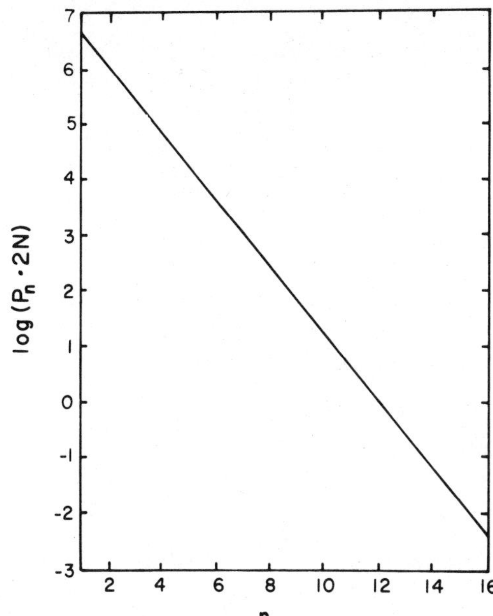

Figure 9. Plot (as a function of n) of the logarithm of the mean frequency of random occurrence of a specific base pair sequence of length n within a genome of length $N = 10^7$ sites, for a DNA in which the individual probabilities of A·T, T·A, G·C, and C·G base pairs are all equal ($P_A = P_T = P_G = P_C$).

3.1.1b Effects of Specifying Loci Only as Purines or Pyrimidines. For double-stranded DNA, it requires *two* loci specified only as containing a purine or pyrimidine (that is, a Pu·Py or a Py·Pu base pair, with the first base representing that in the top strand) to reduce the overall probability of a sequence to the same extent as *one* completely specified base pair locus (for DNA with $P_A = P_T = P_G = P_C$). Thus, n must be approximately equal to 24 base pairs to reduce to less than unity the mean frequency of random reoccurrence of a sequence specified in each position only as having either a purine or a pyrimidine in the top strand.

3.1.1c Effects of Repeated (Tandem) Sequences. Tandem sequences are also possible. Such repeated sequences have the same probability of occurrence as a nonrepeated sequence of the same length. For example, the sequence

$$\begin{matrix} 1 & 2 & 3 & 4 & 5 & 6 & 7 & 8 & 9 & 10 & 11 & 12 \\ A & T & G & C & A & T & G & C & A & T & G & C \end{matrix} \qquad \text{(VI)}$$

(for $P_A = P_T = P_G = P_C = 0.25$) has the same probability of occurrence [$P_{12} = (0.25)^{12} = 6.0 \times 10^{-8}$] as a nonrepeating sequence (IV) with $n = 12$.

3.1.1d Effects of Base Composition. The probability (P_n) of a particular sequence also depends, of course, on the actual base composition of the genome, or of the (large) section of the genome under consideration. Thus, for double-helical DNA in which $P_A = P_T \neq P_G = P_C$, we can generate Fig. 10, where we plot log $(P_n \cdot 2N)$ versus P_A for various sequences with $n = 12$ lattice units (each specified as *specific* base pairs). The different lines represent different numbers of (A + T) base pairs in the sequence (order unspecified); we define $n = l + m$, where l is the number of (A + T) residues in the upper strand, and m is the number of (G + C) residues. We note that in a GC-rich genome, the mean frequency of occurrence of a given sequence is smaller if more AT-containing base pairs, rather than GC-containing base pairs, are used, and so on. For example, in an extreme case [$l = 11$ at P_A ($=P_T$) $= 0.1$], the expected frequency of a sequence 12 base pairs long of that composition is reduced by more than four orders of magnitude from the value at $P_A = 0.25$ (the crossover point of the graph).

The equation used to calculate the predicted frequency per genome of the random occurrence of a particular sequence n base pairs long, containing l (A·T + T·A) base pairs and m (G·C + C·G) base pairs in double-helical DNA with $P_A = P_T \neq P_G = P_C$, is

$$(P_n \cdot 2N) = (P_A)^l (P_G)^m (2N) \tag{1}$$

Clearly, for DNA of base composition $P_A = P_T = P_G = P_C = 0.25$ (as in the case shown in Fig. 9), Eq. (1) reduces to

$$(P_n \cdot 2N) = 0.25)^n (2N)$$

3.1.2 Underlying Assumptions

Before going further, the assumptions underlying these calculations must be examined. We may ask: (i) how accurate are the probabilities calculated here likely to be on purely statistical grounds; and (ii) how close to *chemically* random are the base sequences of real DNA?

3.1.2a Variance of the Probabilities. For large genomes ($N >> n$), it can be shown to follow from Chebyshev's law of large numbers (see, for example, Gnedenko and Khinchin, 1962) that the variance of the probabilities calculated

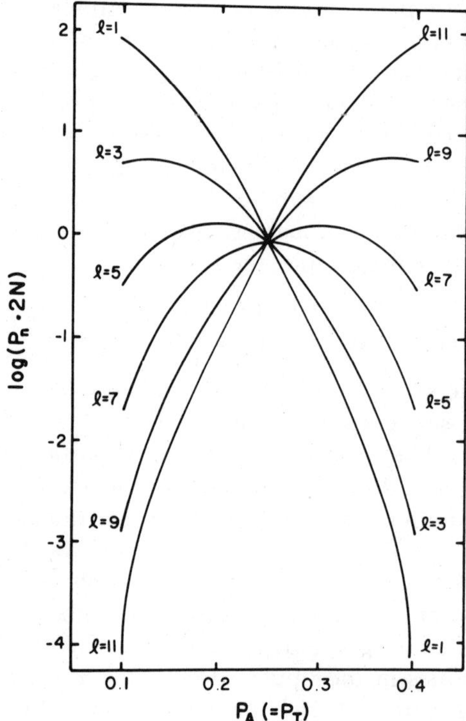

Figure 10. Plot (as a function of P_A) of the logarithm of the mean frequency of random occurrence of a specific base pair sequence of length $n = 12$ with a genome of length $N = 10^7$ sites, for genomes containing *different* mole fractions of A and G ($P_A = P_T \neq P_G = P_C$). l is the number of (A + T) residues in the "upper strand" of the specified sequence (see text).

here will be vanishingly small. Thus, for $N \simeq 10^7$ (the *E. coli* genome), the variance of the P_n should be much less than 1% of the calculated values.

3.1.2b Are DNA Sequences Chemically Random? Until recently, sufficient sequence data did not exist to test the assumption that DNA base pair sequences can be considered to be approximately *chemically* random. Of course, sequences are not *genetically* random, which, in itself, means that they cannot be *totally* chemically random. And other isolated facts—for example, one end of the λ genome contains approximately 0.48 (mole fraction) AT bases, and the other end contains approximately 0.52 AT bases—also mitigate against this assumption being *exactly* correct.

The recent availability of purified restriction enzymes, together with a knowledge of the double-stranded sequences for which each is specific, makes it possible to test the randomness assumption. In Table I we list a number of enzymes, their sequence specificities, and the number of "cuts" each has been observed to make in three different phage genomes. These data are taken from the very extensive recent compilation of Roberts (1976*b*).

The sequences recognized are short enough (four to six base pairs in length) to make it probable that they recur several times even in these relatively small genomes. It is clear from the results quoted that several cuts per genome are indeed observed. In the last column we calculate the number of cuts ($P_n \cdot N$) *expected* in each case.* These numbers are calculated using Eq. (1), with n, l, and m as appropriate for the sequence and using the values of P_A and P_G [mole fractions of A (=T) and G (=C)] as tabulated by Fasman (1976) for these phage genomes.

Clearly, the agreement between the number of cuts observed and the number calculated is reasonably good, given the likelihood of substantial errors in much of the experimental data (generally, estimates of the number of cuts per genome in data for systems with 30 to 50 or more cuts are likely to be inaccurate because of the difficult in resolving overlapping bands after electrophoresis of the DNA fragments in gels). It appears that the calculated frequencies of occurrence per genome are, on the average, somewhat *larger* than those actually observed. This suggests that there may have been some genetic selection *against* restriction-specific sequences. However, on the whole, the results of Table I provide good support for the assumption that the base pair sequences of these genomes may be considered to be essentially *chemically* random.†

3.1.3 Significance of "Blank" Loci in the Design of Protein "Recognition Sites"

The statistical effect of insertion of unspecified base pairs ("blanks") into specific lattice sequences warrants further discussion. As pointed out earlier, the overall probability of occurrence of a given sequence is not changed by inserting blanks *in defined places;* the overall sequence simply becomes longer, with no

*Here we use ($P_n \cdot N$) rather than ($P_n \cdot 2N$) to calculate the number of cuts expected for these sequences because all are palindromic; that is, a rotation of 180° about the center of the sequence generates the *same* sequence. Thus, a double-stranded (circular) lattice N base pairs long can contain only N, rather than $2N$, distinguishable sequences of this type.

†This assumption is certainly *not* true for the highly repetitious parts of certain eukaryotic genomes. However, since the reannealing rates (Britten and Kohne, 1968) for viral, prokaryotic, and nonrepetitive eukaryotic DNAs are approximately the same (after taking into account the length of the genome), we take it as approximately true for all *nonrepetitive* DNAs.

TABLE I. Distribution of Restriction Enzyme Recognition Sequences[a]

Enzyme	Recognition sequence (5′ → 3′)	Number of cuts per genome: Lambda		Ad2		SV40	
		Observed	Calculated	Observed	Calculated	Observed	Calculated
*Eco*RI	G↓AATTC	5	12	5	10	1	1
*Hind*III	A↓AGCTT	6	12	11	10	6	1
*Hpa*I	GTT↓AAC	11	12	6	10	3	1
*Hind*II	GTPy↓PuAC	34	49	>20	39	5	5
*Eco*RII	↓CCTGG	>35	49	>35	39	16	5
*Hae*III	CG↓CC	>50	195	>50	156	17	20
*Hpa*II	C↓CGG	>50	195	>50	156	1	20
*Alu*I	AG↓CT	>50	195	>50	156	>30	20

[a]Experimental (observed) data taken from compilation of Roberts (1976*b*). The recognition sequences listed show only the top strand of the actual palindromic double-stranded sequence; cuts are made at the position indicated (in *both* the upper and the lower strand). The calculated number of cuts per genome are estimated by calculating $P^n \cdot N$ for each genome, using $N = 5 \times 10^3$ base pairs for SV40 (simian virus 40) DNA, $N = 4 \times 10^4$ base pairs for Ad2 (adenovirus 2) DNA, and $N = 5 \times 10^4$ base pairs for lambda phage DNA. To simplify the calculations, we have taken $P_A = P_T = P_G = P_C$, both for the various recognition sites *and* for the genomes. This assumption tends to make some of the calculated number of cuts somewhat too large (see Fig. 11), especially for restriction sequences such as those for *Hae*III and *Hpa*II, which contain only GC base pairs.

concomitant change in information content. On the other hand, the insertion of blank sequences can have an appreciable effect from the point of view of design of the complementary protein template.

Consider sequence (IV), with $n = 12$. Let us imagine a linearly arranged complementary sequence of protein residues, the first residue (or cluster of residues) of which specifically recognizes the first (A·T) base pair in sequence (IV) by forming an appropriate pair of hydrogen bonds with it, the second residue interacts with the next base pair (A·T) of the sequence, the third with the third base pair (T·A), and so forth. Twelve sets of interactions, spaced one base pair apart, are required. *The same set* of protein residues, arranged in the same order, is required to recognize sequence (V). However, even though sequence (V) contains the same series of base pairs arranged in the same order as sequence (IV), there are blanks at positions 4, 5, 9, and 12, and $n = 16$ rather than 12. For the protein, this requires only a simple change in the structure of the linear template surface; that is, moving the residues responsible for recognizing base pairs 3 and 4 in sequence (IV) apart appropriately (by inserting "inert" amino acid residues?) so that they now recognize residues 3 and 6, and so on.

We pointed out earlier that the probabilities (P_n) of base pair sequences (IV) and (V) occurring within the nucleic acid genome are identical. Thus, *specifically placed blanks* do not alter the information content of the nucleic acid sequence in its role in recognizing a specific protein. On the other hand, the existence of blanks in the nucleic acid sequence can provide the system much more latitude in protein design. The particular ordering of residues that recognizes sequence (IV) can only recognize one sequence of $n = 12$. But for $n = 16$, with four blanks, that array of residues can recognize $(16!)/(4!)(12!) = 1820$ nucleic acid sequences with the *same* 12 base pairs in the same order as in sequences (IV) and (V), allowing simply that the blanks be placed in all possible positions in the sequence in turn. So here, by merely varying the positions of the blanks, we have introduced 1819 additional nucleic acid sequences that can be recognized by the same linear array of protein residues without changing their order and without providing *any additional recognition interactions* except for making the protein residue array longer. (Of course, a *particular* protein will still recognize only one nucleic acid sequence.)

In general, there are $[n!/k!(n-k)!]$ ways to arrange *k indistinguishable* objects (blanks) in n boxes (base pair lattice positions). Clearly, as n increases, with $(n-k)$ (the *number and sequence* of specified base pairs) fixed, the number of identical nucleic acid sequences (except for location of blanks) that can be recognized by the same protein residue array (with insertion of variable inert protein residues) increases very rapidly. Thus, in principle, a very few specific protein residue arrays, with differently located "inert" or "bridging" residues to serve as templates for the blank nucleic acid base pair loci, can recognize an enormous number of different nucleic acid sequences (for example, 1.26×10^5 for a single protein array with $n = 12$ and $k = 8$). Appreciation of this principle can greatly simplify the superficially vast problem of constructing adequate numbers of differentiable (by nucleic acid sequences) protein recognition arrays.

3.1.4 *Probabilities of Partially Correct Sequences*

We now ask the following: Given n base pairs arranged in a specified sequence and comprising a specific protein-binding site, what is the probability of occurrence of the same sequence containing 1, 2, . . ., $(n - 1)$ *incorrect* base pairs? This question can also be easily considered using the conditional probability approach.* For example, given sequence (IV), we may ask what is the probability of occurrence of a sequence that differs only in respect to an incorrect base pair at position 11? In other words, what is the probability that position 11 contains a G·C, an A·T, or a T·A base pair? (As before, we represent the top base of the sequence as the first base in the pair as written here.) For double-helical DNA with $P_A = P_T = P_G = P_C = 0.25$, the probability of this occurrence for sequence (IV) $(n = 12)$ is

$$P_{12,i=11} = (0.25)^{11}(0.75)^1 = 1.8 \times 10^{-7}$$

since, in any given locus for this base composition, there is a 25% probability that the base pair will be correct, and a 75% probability that it will be incorrect. The general designation, $P_{n,i=x}$, is used to represent the overall probability of a sequence of specified base pairs n long with an incorrect base pair in *specific position* x. The overall probability of a sequence n long with j incorrect base pairs in *unspecified* positions in the sequence will be written $P_{n,j}$ (see later).

We now ask the following: What is the probability $(P_{12,1})$ of any one locus in sequence (IV) $(n = 12)$ being incorrect? This is simply the foregoing probability times the number of ways (12) one incorrect base pair can be positioned within 12 sites, or

$$P_{12,1} = (0.25)^{11}(0.75)^1 \left[\frac{12!}{(1!)\,(11!)}\right] = 2.15 \times 10^{-6}$$

In general, for a specified base pair sequence n long (with $P_A = P_T = P_G = P_C = 0.25$), the probability of that sequence containing j incorrect base pairs, distributed into any of the specified loci, is

$$P_{n,j} = (0.25)^{n-j}(0.75)^j \left\{\frac{n!}{j!(n-j)!}\right\} \tag{2}$$

If the double-stranded DNA genome has an overall composition of $P_A = P_T \neq P_G = P_C$, then, by combining Eqs. (1) and (2), we may write

$$P_{n,j} = (P_A)^{[l-2P_{Aj}]}(P_G)^{[m-2P_{Gj}]}(0.75)^j \left\{\frac{n!}{j!(n-j)!}\right\} \tag{3}$$

where l and m are the mole fractions of A (=T) and G (=C) bases in the sequence, as for Eq. (2). [The intrinsic *average* probability of occurrence of an incorrect base pair in the various loci j (for example, P_{Aj}, P_{Gj}) is taken as equal to 0.75, as before.]

To illustrate the properties of Eq. (2), we calculate $\log(P_{n,j})$ as a function of j for $n = 10$, 12, and 14 (and $P_A = P_T = P_G = P_C = 0.25$), and plot the results in Fig. 11. First, by inspection of Fig. 11 we see that the *most probable* sequence of 12

*If a base pair different from that specified in the sequence occurs at site i, it is, by definition, an *incorrect* base pair. For purposes of this section, calculations are confined to sequences containing only fully specified base pairs (that is, no blanks or sites specified as Pu or Py). In addition, only DNA with $P_A = P_T = P_G = P_C = 0.25$ will be considered in most calculations. Extensions to more complex situations are explicitly indicated, or can easily be developed on the basis of the examples given.

base pairs contains nine incorrect and three correct base pairs (relative to an initial arbitrary fixed sequence defined as correct). That is, a 12-base-pair sequence with *all* base pairs incorrect is about tenfold less likely (for this case) than is one with only nine incorrect base pairs, and so on. In addition, we see that the probability of a sequence with *one* base pair incorrect (for $n = 12$) is 12 times that for the totally correct sequence, and so on. Thus, the probability of incorrect sequences increases rapidly with "degree of incorrectness." This will be important for the thermodynamic aspects of the recognition process considered in the following section; the consequences of these probabilities of incorrect sequences will also become apparent in our discussion of the *lac* system.

3.2 *Thermodynamic Aspects of Recognition*

In addition to defining the relative probabilities of the correct, and the various partially incorrect, nucleic acid sequences of potential binding sites for genome-controlling proteins, it is also necessary to consider the effects of various degrees of incorrectness on the binding free energy of the potential protein–nucleic acid recognition interaction. Because a meaningful discussion of thermodynamic effects involves knowing something about the molecular details of the system, we will take up most aspects of this question using the binding of *lac* repressor as an example, since most of the quantitative information we have about the interaction of genome-controlling proteins with specific (correct) and nonspecific (partially or totally incorrect) sequences has been obtained with the *lac* repressor–operator–DNA system. However, a few general comments may be useful.

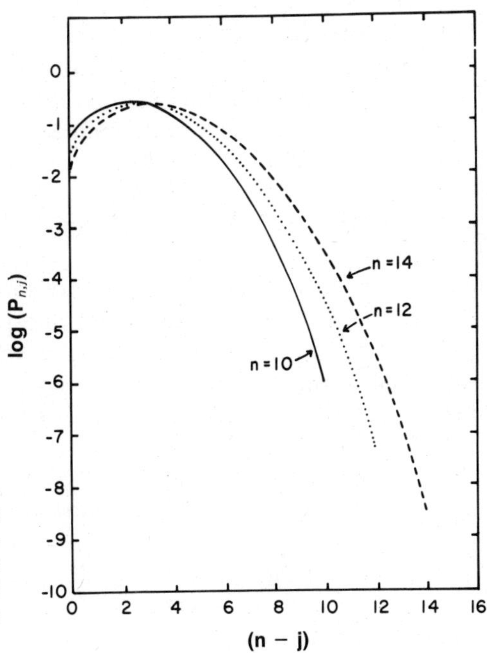

Figure 11. Plot (as a function of $n - j$) of the logarithm of the probability of random occurrence of a base pair sequence of length n (for $n = 10$, 12, and 14), when n contains j *randomly placed incorrect* base pairs and $(n - j)$ *correct* base pairs (see text). ($P_A = P_T = P_G = P_C$.)

3.2.1 *Distribution of Binding Free Energies among Specific and Nonspecific Interactions*

The free energy of interaction of a protein with its specific nucleic acid-binding site can be assumed to be distributed in a number of ways. The simplest model is to assume that the binding free energy is "equipartitioned" over all the *specified* lattice sites. This means that all the base pairs specified in the sequence as A·T, T·A, G·C, C·G, Pu·Py, and Py·Pu (but not those specified as blanks) are assigned *equal* contributions to the binding free energy. This means

$$\Delta G_{\text{int,total}} = n(\Delta G_{\text{int,bp}}) \tag{4}$$

where $\Delta G_{\text{int,total}}$ is the total protein–nucleic acid interaction free energy, and $\Delta G_{\text{int,bp}}$ is the free energy due to the interaction of one base pair with the relevant complementary amino acid residue(s) of the protein-binding site.

There are a number of reasons to suspect that this formulation is not totally correct. First, as written, Eq. (4) suggests that the interaction free energy must derive mostly or entirely from hydrogen-bonding contacts. The enormous dependence on salt concentration of the apparent binding constants for most protein–nucleic acid interactions suggests that a large charge–charge interaction component must also be involved. Also, taken at face value, Eq. (4) suggests that a broad range of apparent protein–nucleic acid binding constants should be observed for a specific protein binding to nucleic acid sequences of varying degrees of "correctness." Figure 11 has shown that significant populations of potential binding sites should exist (and *must* exist, if indeed the overall DNA sequence is approximately chemically random) that contain from 15 to 45% correctly specified lattice sites. Therefore, the simple interaction free energy equipartition model [Eq. (4)] predicts a significant *range* of nonspecific binding constants, rather than the fairly monodisperse nonspecific binding behavior actually observed (see the description of the *lac* system in Section 4.1.5). On the other hand, for the one system in which the effects of a limited series of alterations in specific base pairs on the binding of a genome-controlling protein to its specific target site *have* been tested [operator constitutive mutants of the *lac* operator (Gilbert *et al.*, 1975); see Section 4.1.3], the results are compatible with a limited equipartition of free energy over at least some of the specific binding loci.

On the basis of these and other considerations, a more plausible general binding model might be one that contains an appreciable component of binding free energy that is not base pair sequence specific. We presume that this interaction would have a significant electrostatic component, due to charge–charge interactions between backbone DNA phosphates and appropriate basic residues of the protein. However, interactions other than charge–charge interactions with the functional groups of the sugars could also contribute. Such interactions could serve to hold the protein–nucleic acid components together in a nonspecific way, and since (see Section 2.1.1) most nucleic acid-binding proteins seem to carry one or more areas, or "patches," of positive charge within a negatively charged molecule, they could also "position" the specific protein–nucleic acid surface hydrogen-bonding matrices appropriately for the specific interactions.

Proceeding on the postulate (developed briefly in Section 2.1.2) that it is the thermodynamic "unfavorableness" of incorrect interactions, rather than the "favorableness" of correct interactions, which is of primary importance in determining the stability of most intermolecular complexes, it seems reasonable to suggest that there is no interaction of the specific nucleic acid lattice sites (base pairs) with their specific protein counterparts until appreciably more than half of them are correctly aligned. The simplest *mechanistic* interpretation of this suggestion would be that the native binding protein can exist in (at least) two conformations of roughly comparable free energy; one in which only nonspecific interactions take place, and the other in which the specifically interacting protein and nucleic acid functional groups are brought into contact. This would account satisfactorily for the single nonspecific binding constant found for most genome-binding proteins, as well as for the fact that when all but one or two interactions are correctly made, the specific component of the free energy of interaction appears to be approximately equipartitioned among a number of correct base pair–protein residue contacts (see discussion of the *lac* repressor–operator interaction in Section 4.1.3).

Mathematically, in terms of this general model we can replace Eq. (4) with

$$\Delta G_{\text{int,total}} = p(\Delta G_{\text{int,bp}}) + q(\Delta G_{\text{int,nonelec}}) + m'(\Delta G_{\text{int,ch-ch}}) \tag{5}$$

where p ($\leqslant n$) represents the number of specific base pairs recognized, $\Delta G_{\text{int,bp}}$, as before, the interaction free energy per specific base pair, q the number of nonspecific and nonelectrostatic interactions between the protein and the DNA lattice, $\Delta G_{\text{int,nonelec}}$ the free energy contributed by each of these interactions, m' the number of charge–charge pairs formed (see Section 3.3) between protein and DNA, and $\Delta G_{\text{int,ch-ch}}$ the binding free energy contribution of each of these ion pairs. In Eq. (5) we have retained the concept of equipartition of free energy between individual interactions of like type, but, of course, in real systems this will apply only approximately.

3.2.1a Overspecification of Unique Sequences. In considering the energetics of specific protein–DNA interactions we must appreciate that Fig. 11 has important thermodynamic as well as statistical consequences. For example, in Section 3.1.4 it is shown that for a site-specific DNA-binding protein utilizing 12 recognition lattice sites (the minimum value of n for a unique *E. coli* protein-binding site), there are 12 times as many sequences with *one* incorrect base pair, and so on. If (as the operator constitutive mutation data suggest for *lac* repressor), the binding affinity of the protein for its target sequence increases approximately tenfold per correct base pair (that is, ~1.4 kcal per base pair in interaction free energy), then the aggregate binding of repressor to sequences with one base pair incorrect will exceed that to the correct site, the available protein will be largely utilized in binding to these incorrect sequences, and effective target-specific binding will not be observed. This situation, of course, can be rectified by "overspecifying" the target binding site; that is, by specifying base pairs (increasing n) above the point necessary to make $(P_n \cdot 2N) \leqslant 1$ (Fig. 9). Thus, for the situation described here, we must reduce $(P_n \cdot 2N)$ to about 0.01 to 0.05 to assure effective binding to the target site. The extent to which the sequence must be overspecified (n increased) will depend, of course, on the magnitude of the interaction free energy contributed by each correct base pair in the specified sequence.

3.2.2 Possible Roles of Nonspecific Binding

The foregoing overall model suggests that most genome-binding proteins will show an appreciable affinity for *all* DNA sites. This nonspecific binding may be functionally important (as well as interesting to study), not only because it can provide insight into, and a quantitative measure of, the nonspecific component of the specific binding interaction, but also because this (competitive with specific) binding to the overwhelming preponderance of nonspecific sites will control the *free concentration* (the chemical potential) and the *location in space* of the major fraction of each of the genome-controlling proteins (von Hippel *et al.,* 1974, 1975). The control of location of the protein in space afforded by nonspecific binding may be central to solving the very difficult problem of how a genome-controlling protein finds its specific target sequence in a reasonable time (as dictated by the biological time scale set by transcription, replication, recombination, and so on) in various organisms. These points are elaborated in the following with respect to the *lac* system.

3.3 Methods for Studying Specific and Nonspecific Binding of Proteins to Nucleic Acids and Definitions of Interaction Parameters

There are many approaches to the study of the specific or nonspecific interactions of proteins with nucleic acids; at least the same diversity of possibilities exists as is available for examination of the properties of any other macromolecular complex. In this section I propose only to outline the general types of parameters required to define a protein–nucleic acid interaction system thermodynamically. Details of the relevant measurements are described in the references cited.

The interactions of proteins with specific nucleic acid sequences have hitherto been difficult to study by conventional physicochemical techniques, not only because protein has usually been in very short supply (this problem has been fairly well solved for some of the better known prokaryotic genome-regulating proteins by utilizing physicochemical micromethods and developing overproducing strains of bacteria and viruses), but also because reasonable quantities of target sequences (at the micromolar level) have been virtually impossible to obtain. This means that most studies of genome control systems have been carried out by variants of the filter-binding method of Riggs *et al.* (1970*a*), which, while easy to apply, is often hard to make totally reproducible, quantitative, or sometimes even interpretable. This problem, thanks to recombinant DNA and cloning technology, is now on the way to solution, and we should soon be able to obtain quantities of operator or promoter DNA adequate for physicochemical study.

Meanwhile measurements on non-sequence-specific DNA–protein complexes in which protein is nonspecifically bound to DNA are becoming routine. In general, for any protein that binds to a single- or double-stranded nucleic acid lattice, one needs to determine only three parameters to have a complete (thermodynamic) description of the interaction. These parameters are the intrinsic association constant K, the binding site size n, defined as the number of bases or base pairs covered by the binding of a single protein, and ω, the (dimensionless) binding cooperativity parameter. These parameters are defined and illustrated schematically in Fig. 12 (for approaches to the measurement of these parameters in real protein–nucleic acid systems see, for example, McGhee and von Hippel, 1974; Schellman, 1974, 1975; McGhee, 1976; Jensen and von Hippel, 1977; Butler *et al.,* 1977; deHaseth *et al.,* 1977).

Since proteins bound to nucleic acids generally cover more than one nucleotide residue or base pair (nucleic acid lattice site), binding in such systems is of the "overlap" type; that is, the number of potential binding sites remaining on a DNA lattice is not a linear function of the number of protein ligands bound. This results in curved Scatchard type binding plots, even in the absence of protein–protein binding cooperativity, and somewhat modifies the way one extracts binding constants and site sizes from the binding data, compared to a system in which the binding sites are independent (see, for example, McGhee and von Hippel, 1974; Schellman, 1974).

Protein–nucleic acid interactions often involve an appreciable electrostatic (charge–charge interaction) component. As a consequence, binding parameters are generally measured as a function of ionic strength and pH. (The latter variable is often convenient for diagnosing the involvement of histidine residues in the binding reaction, which manifest themselves by a change in binding affinity with pH in the neutral range.)

In addition to a "bare bones" thermodynamic description of a protein–nucleic acid interaction, it is often possible to learn something about the protein-binding domain (and the complementary nucleic acid interaction surface), by a variety of other approaches. Thus, the number of bases or base pairs actually *interacting* with the protein (m), as opposed to merely being covered by it (n), can often be determined by measuring the apparent binding constants of a series of oligonucleotides to the site (see Fig. 13). When the length of the test oligonucleotide lattice (l) exceeds m, further increases in lattice length should merely increase the apparent oligonucleotide-binding constant by a statistical factor. [At $l > 2n$, the magnitude of the cooperativity parameter, ω, can also be measured (Kelly *et al.*, 1976; Draper and von Hippel, 1978).] In addition, at $l = m$ the apparent protein–oligonucleotide binding constant should equal the intrinsic K obtained with long

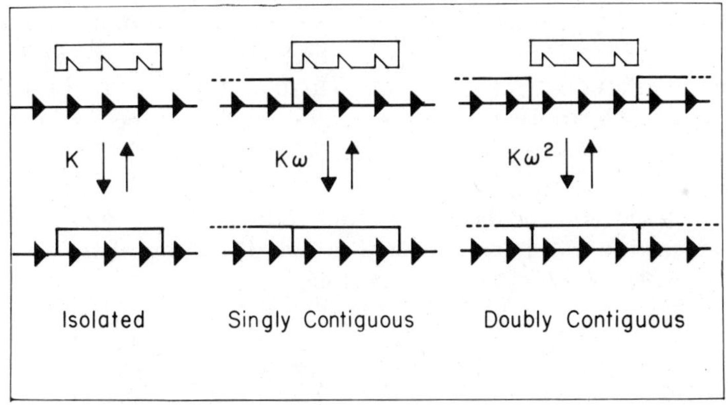

Figure 12. Definitions of the binding parameters of a binding protein to a nucleic acid lattice. The black arrowheads represent a lattice site (that is, a base or base pair) and the illustrated protein covers *three* such sites ($n = 3$). K (in M^{-1}) is the intrinsic binding constant for protein binding to the lattice at an isolated site, and ω (dimensionless) is the parameter representing the cooperativity of binding. If contiguous binding is favored, $\omega > 1$; if contiguous binding is disfavored, $\omega < 1$; and if binding is noncooperative, $\omega = 1$. Positive cooperativity can result either from direct interactions between contiguously bound proteins, or from lattice distortion induced by protein binding which favors the binding of a protein at a contiguous site (see McGhee and von Hippel, 1974).

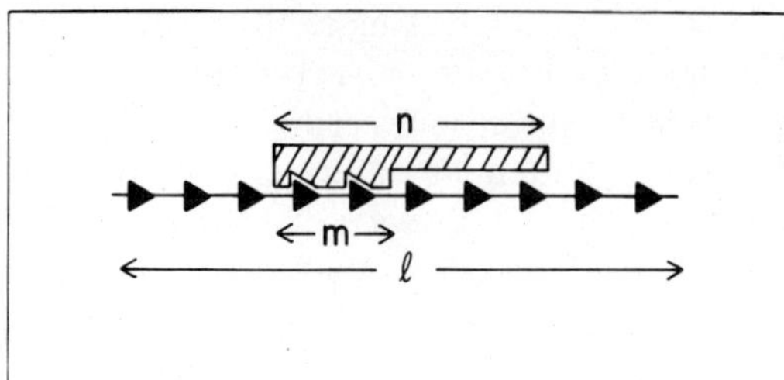

Figure 13. Definition of additional binding parameters. The illustrated protein binding to a lattice l bases or base pairs long, covers n lattice sites (here 5), but binds to only m lattice sites (here 2). More generally, m refers to the length (in lattice sites) of the nucleic acid lattice extending between the *farthest apart* residues of the protein that interact with the lattice (see text and Section 4.1.3).

DNA or polynucleotide lattices. Studies of the base and sugar dependence of the binding constants of oligonucleotides also provide information about binding specificities (see, for example, Draper *et al.*, 1977; Kelly *et al.*, 1976; Kelly and von Hippel, 1976).

Record *et al.* (1976), by an ingenious exploitation of the thermodynamics of the competitive binding of various mono-, oligo-, and polycations to nucleic acids, have been able to show that the change in the apparent protein-binding constant to the (single- or double-stranded) DNA lattice as a function of salt concentration provides a direct measure of m', defined as the number of ionic (DNA phosphate–basic amino acid residue) interactions involved in binding the protein to the nucleic acid.

Special features of the protein–nucleic acid interaction have also been exploited to map the protein-binding site. In favorable cases one can determine whether planar amino acid residues intercalate on binding to the nucleic acid partner (Coleman *et al.*, 1976; Toulmé and Hélène, 1977), whether tryptophans in the protein-binding site come in contact with quenching nucleotide phosphates (Kelly and von Hippel, 1976; Draper *et al.*, 1977), whether the DNA lattice is deformed in binding (Wang *et al.*, 1974; Anderson and Coleman, 1975; Butler *et al.*, 1977), and so on.

In addition, high-resolution electron microscopy is becoming quite useful in defining the number of specific (tight) binding sites for various proteins to well-defined genomes and genome segments (Hirsh and Schleif, 1976). In especially favorable cases, such studies can also be used to study nonspecific binding and to determine binding site size (n) (Williams and Chamberlin, 1977; Zingsheim *et al.*, 1977).

Specific applications of these and other techniques, and some of the results obtained, are considered in more detail in our discussion of the protein–DNA interactions of the lactose operon, which follows.

4 *The Lactose Operon of Escherichia coli*

The interactions controlling the metabolic processing of lactose by *E. coli* comprised the first genome-regulatory system to be extensively studied, and provided the prototypic results on which Jacob and Monod (1961) constructed their operon theory. The lactose operon is a classical negative control system. A set of enzymes (β-galactosidase, lactose permease, and transacetylase) involved in the transport and metabolism of a particular substrate (lactose) is coded on a set of contiguous genes and transcribed by DNA-dependent RNA polymerase as a polycistronic message from a single (DNA) promoter site. Control of *lac* enzyme concentrations is maintained primarily at the transcriptional level; in wild type *E. coli,* transcription from the *lac* promoter is largely prevented by the binding of a specific *lac* repressor protein to a vicinal operator (DNA) site. The repressor protein is coded on a neighboring cistron (the *I* gene) and transcribed at low levels (about ten repressor tetramers per cell) from a promoter that is not regulated by additional protein components.

Four identical repressor subunits, each specified by the *I* gene, combine to form active (tetrameric) repressor. Under noninducing conditions, the repressor is bound to the operator and the *lac* enzymes are synthesized at about 10^{-3} of the fully constitutive (or I^-) rate. A feedback mechanism operates in the presence of high concentrations of lactose. An early product of lactose metabolism (β-allolactose, produced from lactose by β-galactosidase) binds to specific sites on the *lac* repressor and triggers, or "traps" (see later), a conformational change in the protein that decreases its affinity for operator sufficiently to permit virtually complete dissociation of the repressor–operator complex *in vivo,* with consequent transcription of the *lac* genes at the unrestricted (constitutive) rate.

As pointed out in the Introduction, there is an additional, and more general, genome control system that also regulates the *lac* operon. This is the CAP–cAMP system which is involved in the regulation of most of the operons of *E. coli* concerned with processing simple sugars. This system couples the metabolism of the various sugars to the level of the preferred metabolite, glucose. When the glucose level in the cell is low, the intracellular concentration of 3′,5′-cyclic AMP (cAMP) is increased, and this small molecule effector binds to free catabolite activating protein (CAP). The CAP–cAMP complex binds specifically to a CAP-binding site on the DNA near the *lac* promoter, and increases the frequency with which *lac* mRNA is initiated. (Free CAP does not bind appreciably to the CAP-binding site.) This general CAP–cAMP control system operates in addition to the specific *lac* repressor–inducer–operator system, and serves to integrate the *lac* operon into the sugar metabolism of the entire *E. coli* cell. The isolation of *E. coli* with point mutations or deletions in either the CAP-binding site or the *lac* repressor or operator sequences has permitted the examination of these control systems independently of one another.

The *lac* system is outlined schematically in Fig. 14, and has been summarized and reviewed in detail many times (most recently by Bourgeois and Pfahl, 1976). My purpose here is not to add yet another epic in that vein; instead, I will focus on general principles since the physicochemical principles governing the *lac* system seem to be utilized (in various combinations) by many other prokaryotic (and perhaps some eukaryotic) genome control systems as well. The advantage of the *lac* system as a specific model of the application of these principles is that it is the only system on which sufficient quantitative work has been done to permit the articulation of these ideas in more than a descriptive manner.

4.1 *lac Repressor–Operator–Inducer–DNA Interactions*

4.1.1 *lac Repressor*

In its functional state, *lac* repressor consists of four identical subunits of molecular weight about 38,000. The complete amino acid sequence of the sub-

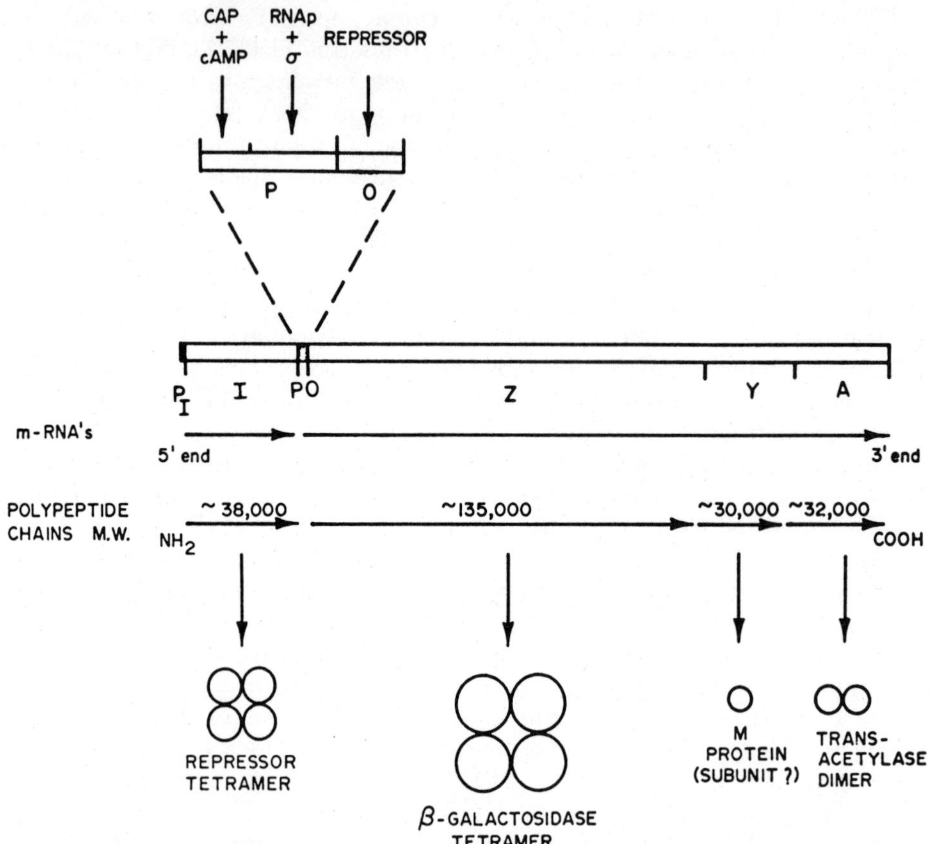

Figure 14. Schematic view of the lactose operon and its products. The genes and proteins are drawn roughly to scale. The promoter–operator region (P, O) for the *lac* enzyme genes is shown enlarged in the upper part of the figure to illustrate the relative locations of the binding sites for the cAMP-binding protein (CAP), the RNA polymerase protein (RNAp) with its σ factor, and the *lac* repressor. The (unregulated) *I* gene promoter which controls repressor synthesis is also shown (see text). (Redrawn and reproduced with permission from Bourgeois and Pfahl, 1976.)

units is known, and the protein contains no nucleotides (thus dashing an early hypothesis that internucleotide base pairing might still be somehow involved in protein–nucleic acid recognition specificity) or other prosthetic groups. Intersubunit binding is very tight, and appreciable subunit dissociation or interchange does not take place in the absence of denaturants.

As a consequence of the extensive genetic studies of Miller and his colleagues, of Müller-Hill and his colleagues, and of others (see Bourgeois and Pfahl, 1976), a great deal is known about the functional significance of various parts of the repressor sequence. These functional domains, together with key amino acid residues, are shown schematically in Fig. 15. A number of lines of evidence demonstrate that the residues of the N-terminal 58-residue peptide fragment of *lac* repressor comprise the DNA-binding part of the molecule (for a summary see Weber *et al.*, 1975). The remainder of the repressor subunit contains the inducer-binding domain, as well as the intersubunit contact sites involved in tetramer formation. "Core" tetramers (with each subunit genetically or enzymatically depleted of its N-terminal DNA-binding peptide) have approximately the same affinity for inducer as does the intact tetramer, but show no binding to either operator or nonoperator DNA. Primarily on the basis of electron microscopic evidence, the entire repressor tetramer appears to be quite compact in overall dimensions (~90 × 75 × 75 Å) (see, for example, Zingsheim *et al.*, 1977).

4.1.2 *Repressor–Inducer Interactions*

The *lac* repressor binds specifically to β-allolactose and related inducer molecules (Fig. 16). As indicated earlier, β-allolactose is the natural inducer of the *lac* operon, but other chemically related (gratuitous) inducers have been widely used in repressor research (in large measure because they are not further metabolized and thus their intracellular concentrations can be defined in *in vivo* experiments). Isopropyl-β-D-thiogalactoside (IPTG) is a particularly useful gratuitous inducer of this type, and binds repressor with approximately the same affinity as does β-allolactose. The binding constant (K_{RI}) for β-allolactose (and IPTG) to repressor subunits is approximately 10^6 M^{-1}; this parameter is relatively insensitive to salt concentration. Other inducers (and antiinducers, see further on) bind more weakly. Each repressor subunit contains one inducer-binding site, and thus the maximum inducer-binding stoichiometry is four per repressor tetramer. Inducer binding brings about a conformational change in the repressor subunit, decreasing the affinity of the tetramer for operator (but not for nonoperator DNA; see Section 4.1.5).

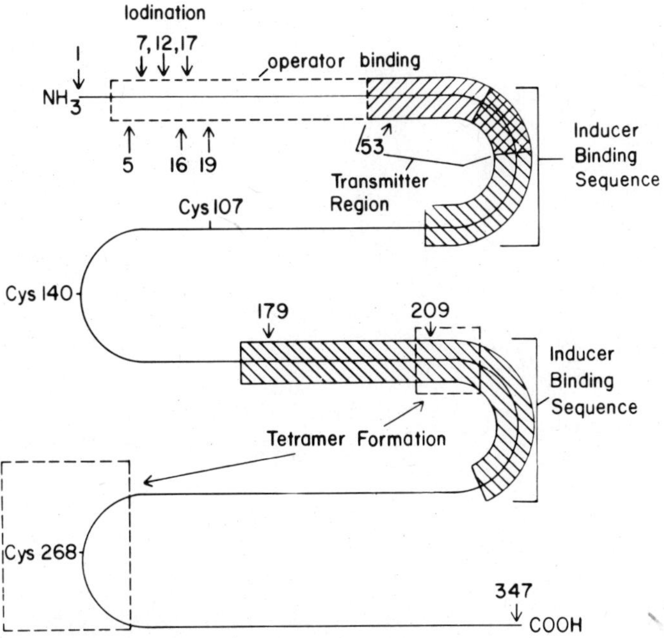

Figure 15. Functional binding domains of the *lac* repressor subunit polypeptide chain (see text). (Reproduced with permission from Butler, 1976.)

Other small molecule effectors, chemically similar to inducer, bind to the repressor subunits competitively with inducer, but bring about an *increase* in the affinity of the repressor for operator DNA. These effectors are called *antiinducers.* (Figure 16 shows the structure of *o*-nitrophenylfucoside, ONPF, a potent gratuitous antiinducer.) Glucose itself is a weak antiinducer, and thus can further inhibit, in a way complementary to the CAP–cAMP mechanism, the utilization of lactose in the presence of ample supplies of this preferred metabolite.

The conformational changes resulting from inducer (and antiinducer) binding have been studied by physicochemical techniques. These changes are obviously subtle, since the concomitant alterations in the circular dichroism and ultraviolet absorption spectra of the protein, as well as in the tetramer sedimentation velocity, are sufficiently small to verge on the undetectable (for a summary see Bourgeois and Pfahl, 1976). Intrinsic tryptophan protein fluorescence has been the most useful probe of this conformational change to date, and a combination of static, stopped-flow, and temperature-jump fluorescence measurements on wild type and genetically modified repressor has shown that the inducer binding perturbs both tryptophan residues in each subunit, bringing about a conformational change that results in burying try-209, particularly in a more hydrophobic environment (Laiken *et al.,* 1972; Sommer *et al.,* 1976; Wu *et al.,* 1976; Friedman *et al.,* 1977). These data, together with equilibrium dialysis measurements, have been used to show that the conformational changes resulting from inducer binding are noncooperative between subunits; that is, the inducer affinity of one subunit does not depend on whether other subunits of the same tetramer are also complexed with ligand. In addition, genetic studies have suggested that the amino acid residues designated as transmitter regions in Fig. 15 are particularly involved in facilitating the transmission of the conformational change from the inducer-

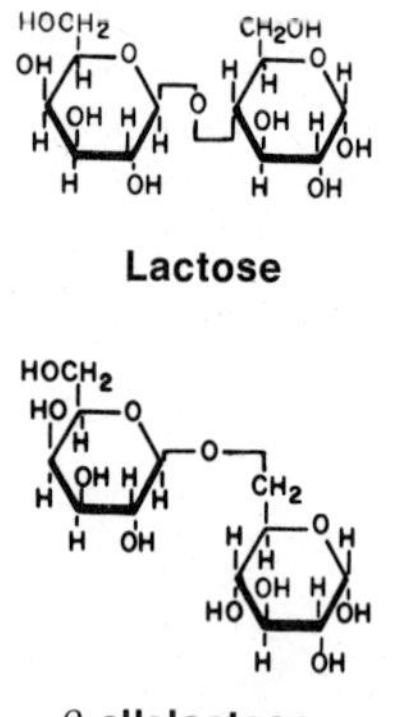

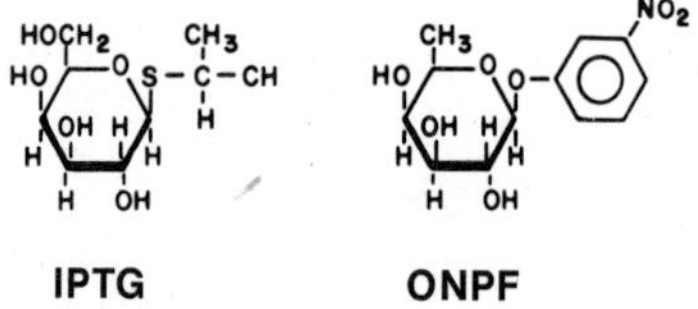

Figure 16. Structure of the natural "substrate" of the *lac* operon (lactose), the natural inducer (β-allolactose), a gratuitous inducer (isopropyl-β-D-thiogalactoside, IPTG), and a gratuitous antiinducer (*o*-nitrophenylfucoside, ONPF).

binding site to the operator-binding domain (for a summary of the relevant data see Müller-Hill, 1975; Miller *et al.*, 1975).

The results of the fluorescence studies just cited (see also Butler *et al.*, 1977) also show, in confirmation of allosteric theory, that the individual subunits of *lac* repressor can exist in at least two conformations which are in rapid equilibrium:

$$R_O \rightleftharpoons R_I$$

where R_O is the tight operator-binding form (and therefore, on the basis of considerations of microscopic reversibility, the loose inducer-binding form), and R_I is the form that binds to inducer tightly and to operator loosely. In the absence of inducer, R_O is the dominant species; when inducer is present, R_I is trapped and accumulated by a sequential reaction mechanism involving at least the following steps:

$$R_O \rightleftharpoons R_I + I \rightleftharpoons R_I I$$

In addition, a form of the repressor subunit with an even higher affinity for the operator must also be in equilibrium with R_O, and be subject to trapping by antiinducer.

4.1.3 Repressor–Operator Interactions

The *lac* repressor binds tightly and specifically to the particular sequence of double-helical DNA base pairs called the operator. Binding is very dependent on salt concentration, and typical binding constants (K_{RO}) for the wild type repressor–operator interaction, estimated from filter binding measurements at presumed physiological ionic strengths, range from 10^{12} to 10^{14} M^{-1}. Mutant repressors have been isolated that are characterized by both larger and smaller values of K_{RO}, and mutant operators (termed O^c, for operator constitutive) have also been isolated and found to have decreased affinity for repressor.

The wild type operator base pair sequence, as well as the changes in this sequence occurring for several single base pair O^c mutants, has been determined by a variety of techniques by Gilbert and co-workers (see Gilbert *et al.*, 1975). These sequences are shown in Fig. 17.

The operator sequence (Fig. 17) was initially defined as that sequence of base pairs to which repressor binds sufficiently tightly to protect it against digestion during a short exposure to deoxyribonuclease. Defined this way, the repressor-binding site size (*n;* see Section 3.3) has been shown to be 24 to 26 base pairs. However (as also discussed in Section 3.3), this does not necessarily mean that all these base pairs are directly involved in the binding of the repressor molecule, or even in contact with it. Thus, *m* (the number of base pairs actually articulating with repressor functional groups and contributing binding free energy to the interaction) may well be less than *n* (the number of base pairs *occluded* by the binding).

Various approaches have been taken to determining the number (*m*) and positions of base pair functional groups involved in the repressor–operator interaction. Gilbert *et al.* (1976*a*) have examined differences in the kinetics of methylation of the operator sequence by dimethyl sulfate in the presence and absence of repressor, respectively. The results are also shown in Fig. 17, and indicate that some base pair functional groups exposed in both the major and minor grooves of the DNA double helix are protected against methylation by repressor, while the reactivity of another set of functional groups is enhanced. It seems reasonable to assume that the protected groups are in close contact with repressor functional groups, and may well comprise at least some of the sites involved in both recognition specificity and binding affinity. (For reasons discussed in Section 2, both these roles may not be played by the same sets of functional groups.) The base pairs in the middle of the operator sequence, for which the reaction with dimethyl sulfate seems to be *enhanced* as a consequence of repressor binding, may be involved in hydrophobic contacts with repressor, the nonpolar protein surround perhaps serving to *concentrate* the dimethyl sulfate reagent (Gilbert *et al.*, 1976*a*). Gilbert *et al.* (1976*b;* see also Ogata and Gilbert, 1977) have also used chemical cross-linking techniques to determine which thymine residues of the operator sequence actually make contact with repressor. The thymine residues so identified are also indicated in Fig. 17. In aggregate, these results suggest that m for the R–O interaction is about 17 base pairs.

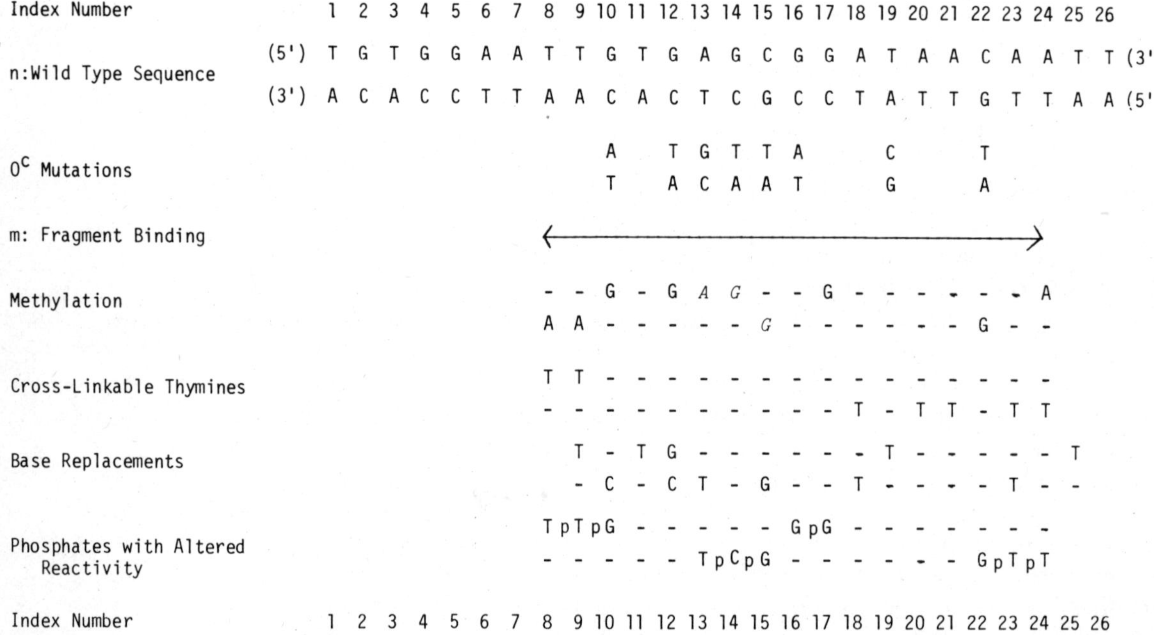

Figure 17. Lac operator modifications and their effects on repressor binding. The top sequence is that of the wild type *E. coli* operator (see Gilbert *et al.*, 1975). The second row indicates the positions (and base pair replacements) of O[c] mutations (Gilbert *et al.*, 1975). The row labeled "*m*: Fragment Binding" shows the minimum synthetic sequence necessary for a repressor-binding affinity approximately equal to that of the wild type (Bahl *et al.*, 1977). The row labeled "Methylation" indicates those bases of the operator sequence for which access of the dimethyl sulfate reagent is affected by the presence of repressor; methylation is at the N7 of G (major groove) and the N3 of A (minor groove). The italicized bases represent sites of *enhanced* methylation (over the operator control in the absence of repressor), and the remaining indicated bases represent sites of *reduced* methylation (Gilbert *et al.*, 1976*a*). The row labeled "Cross-Linkable Thymines" represents those sites in which BrU is incorporated in place of T cross-links to repressor on photoactivation (Gilbert *et al.*, 1976*b;* Ogata and Gilbert, 1977). The row marked "Base replacements" represents those operator loci at which a substituted base [BrU or U for T; BrC for C; H (hypoxathine) for G] changes (increases or decreases) the apparent free energy of binding to repressor by more than about 0.2 kcal/mol (Goeddel *et al.*, 1978). The "Phosphates with Altered Reactivity" are those whose rate of reaction with ethyl nitrosourea is altered by repressor binding (A. Maxam and W. Gilbert, private communication).

Another approach to identifying the interactions important in repressor binding to operator has been examination of the dependence of binding on ionic strength. Riggs *et al.* (1970*a*) have shown by filter binding techniques that decreasing the ionic strength enhances binding, suggesting that charge–charge interactions between DNA phosphates and basic protein residues are involved. Record and co-workers (1977) have analyzed the salt dependence of the repressor to operator-binding data of Riggs *et al.* (1970*a*), using the approach developed by Record *et al.* (1976) (see Section 3.3), and conclude that eight (± one) charge–charge interactions of this type are involved in the R–O interaction. Thus, in terms of the nomenclature set forth in Section 3.3, we conclude that $m' \sim 8$ ionic interactions for the formation of the R–O complex. Maxam and Gilbert (W. Gilbert, private communication) have compared the reactivity of the DNA phosphates of the operator to alkylation by ethyl nitrosourea in the presence and absence of bound repressor, and find that the alkylation rates of seven phosphates are modified (Fig. 17), in good accord with the value of m' calculated by Record *et al.* (1977).

Synthetic techniques offer another approach to determining m for this interaction. Bahl *et al.* (1977) and Goeddel *et al.* (1977) have synthesized various lengths of the double-stranded *lac* operator sequence *de novo,* and have measured repressor-binding affinity to these sequences. Although the binding affinity for all the short operator sequences appears to be smaller than that to the operator incorporated into a long DNA molecule, the results indicate that binding is relatively independent of sequence length as the sequence is shortened from 24 base pairs, until approximately the 17-base-pair sequence indicated in Fig. 17 is reached. Further removal of base pairs from this sequence reduces the apparent binding constant precipitously.

Goeddel *et al.* (1977, 1978) have taken the synthetic approach still further. By systematically replacing individual base pairs within the operator sequence with chemically modified pairs (for example, A·U and A·BrU for A·T; I·C for G·C; and so on), these authors have identified a number of loci within the operator sequence where such replacements cause an appreciable change in repressor-binding affinity and other loci where changes have little effect. Some of these extensive studies are also summarized in Fig. 17.

Smith and Sadler (1971) and Sadler and Smith (1971) isolated a number of mutants of *E. coli* in which repression of the *lac* enzymes seemed to be less effective than in the wild type strain. These were called operator constitutive (O^c) mutants, and were shown by genetic means to arise as a consequence of single base pair changes in the *lac* operator. A number of these mutant operators have been transferred to phages and the DNA isolated; R–O affinities were measured (by filter binding) for these DNAs *in vitro,* and the apparent binding constants (K_{RO}) have been shown to be decreased approximately tenfold per base pair substitution (Jobe *et al.,* 1974). (Some double mutants were also isolated and tested, and shown to have K_{RO} values approximately 100-fold smaller than those of the wild type.) The positions of several of these mutations were mapped biochemically by Gilbert *et al.* (1975), and were found to be located within the 13 central residues of the *lac* operator sequence (Fig. 17). No O^c mutations were found outside of this region, and no operator changes were identified that resulted in binding affinities *greater* than that of the wild type operator. Several of the O^c mutants isolated by Smith and Sadler (1971) were found by Gilbert and co-workers (1975) to be biochemically identical, suggesting that the identified loci might comprise a significant fraction of the mutable (O^c) sites available in the operator.

The fact that all the established O^c mutants fall within a 13-base-pair region in the middle of the 24- to 26-base-pair sequence identified as the *lac* operator by protection–digestion experiments strongly suggests that at least these seven base pair loci are specifically recognized by functional groups in the *lac* repressor-binding site. Furthermove, since each of the single base pair changes in the mutants seems to result in about a tenfold reduction in K_{RO}, in keeping with the suggestion of Section 3.2.1 that the *specific portion* of the binding free energy is approximately equipartitioned among the base pairs specifically recognized by the protein; in this system, at least, a specific binding free energy contribution of approximately −1.4 kcal/mol appears to be made by each specified base pair. Alternatively (Section 2.1.2), we can interpret this finding to mean that the binding interaction is *destabilized* by about +1.4 kcal/mol for every *incorrect* base pair replacement within the specific sequence. This conclusion is also in accord with the approximately thirty- to several hundred-fold weaker binding of repressor to the operator-like (pseudooperator) sequences that have been found in the *lacZ* (β-galactosidase) gene and in the *lacI* (repressor) gene (see Gilbert *et al.*, 1976*b*). Both differ from the wild type operator by two to three base pair changes.

It is interesting that no mutated operator sequence has been found that binds repressor more tightly than does the wild type operator. This might suggest that all the base pairs for which complementary functional group sequences are available in the protein-binding site are fully (optimally) utilized in the wild type operator. However, this possibility is rendered less likely when we consider the implications of the number of repressor subunits that are probably involved in operator binding, as well as perhaps by the existence of mutant *repressors* that bind operator more tightly than does the wild type protein.

Chamness and Wilson (1970) and Pfahl (1976) isolated and characterized one such tight-binding repressor mutant (X86). They found that the affinity of this repressor for operator seemed to be *enhanced* by low inducer concentrations, though at high levels of inducer significant derepression could be demonstrated. This behavior could reflect a change in the protein that generates an additional specific recognition of a previously undefined base pair in or near the operator, but since this mutant repressor also binds nonoperator DNA more tightly, the increase in K_{RO} may reflect an increase in the nonspecific binding component (Pfahl, 1976) (see Section 4.2.3f).

Several lines of evidence (discussed in Sections 4.1.4 and 4.1.5) suggest that *two* repressor subunits are in contact with the operator (or nonoperator) DNA in forming a specific or nonspecific repressor–DNA complex. Assuming symmetry of tetramer assembly, since repressor subunits have identical amino acid sequences and thus presumably also identical intersubunit contact sites, we would expect that each subunit would cover approximately 12 base pairs in the operator sequence and that at least those base pairs comprising the specific binding loci should exist as two identical sequences related by a central axis of symmetry. [Some of the topological aspects and consequences of varying repressor symmetries have been considered by Sobell (1973*a,b*) and are discussed by him in Chapter 5 of this volume.] Figure 17 shows that although some elements of centrosymmetry do exist, they are incomplete. Furthermore, O^c mutations can occur that either *increase* or *decrease* the apparent centrosymmetry. This suggests that one subunit must be making more specific contacts than the other, and that perhaps all available recognition loci are *not* used in the wild type complex. Complexes of maximal affinity (and complete symmetry?) are selected against, perhaps because they are too difficult to derepress (see Section 4.1.4).

4.1.4 *Interactions of Repressor–Inducer Complexes with Operator*

The affinity of inducer-saturated repressor tetramer for the operator sequence is approximately three orders of magnitude weaker than is the affinity of the unliganded repressor for operator (Barkley *et al.*, 1975). Thus, if K_{RO} for the wild type species is 10^{12} to 10^{14} M^{-1}, K_{RIO} (the constant for binding inducer-saturated repressor to operator) is 10^9 to 10^{11} M^{-1} under the same conditions. The various O^c mutants are also inducible (Jobe *et al.*, 1974). Although different inducers bind to repressor with differing affinities, at saturation all result in K_{RIO} values of about the same magnitude for any given repressor. Antiinducer-saturated repressor binds operator *more tightly* than unliganded repressor.

The binding affinity of inducer-saturated repressor for operator is still much higher than that of either R or RI complexes for nonoperator DNA, suggesting that not all the specific repressor–operator contacts are destroyed on derepression. Rather, in keeping with the discussion of the O^c mutations in the preceding section, we may conclude that the derepression process may be accounted for by disruption (by inducer) of as few as about three sets of specific base pair–repressor functional group contacts. (See Section 4.1.5 for a discussion of the role of nonspecific binding in derepression.) Yagil and Yagil (1971) have assembled the available physiological data relevant to determining the number of inducer molecules required for derepression. They conclude that about two are needed, in good accord with the various lines of evidence suggesting the involvement of two repressor subunits in the DNA-binding site (we recall that inducer binding is noncooperative; Section 4.1.2).

4.1.5 *Interactions of Repressor and Repressor–Inducer Complexes with Nonoperator DNA*

As indicated in Section 3.2, *lac* repressor, in common with other DNA-binding proteins highly specific for a particular target sequence of base pairs, also binds to other DNA sequences, but with lower affinity. For some particular sequences (such as the *lacZ* gene and *lacI* gene pseudooperators, or the O^c mutant operators; Section 4.1.3), the binding constant for repressor is only slightly decreased (one to three orders of magnitude in K_{RO}). However, for most nonoperator sequences, a single binding constant (K_{RD}) applies. K_{RD} is also very dependent on ionic strength and has been estimated to be 10^3 to 10^5 M^{-1} under physiological conditions (Kao-Huang *et al.*, 1977).

The repressor–nonoperator DNA-binding constant has been measured by direct physical chemical means under a variety of conditions. As indicated earlier, it is very dependent on salt concentration, and a value of m' of 11 ± 1 ionic interactions has been calculated for this interaction in two independent studies (Revzin and von Hippel, 1977; deHaseth *et al.*, 1977). We note that this number of phosphate–basic residue charge–charge interactions is somewhat larger than that calculated for the repressor–operator interaction. K_{RD} is apparently independent of base composition (Wang *et al.*, 1977), though poly[d(A–T)] and a few other synthetic double-stranded polynucleotides appear to bind repressor somewhat more tightly than does native nonoperator DNA. K_{RD} is also quite insensitive to changes in temperature, as long as the temperature remains below the T_m of the DNA and below the denaturation temperature of the repressor. The repressor–DNA binding constant also decreases with pH (between pH 7 and 8). On the basis of these latter results, two histidine residues (per tetramer) have been implicated in the binding interaction (deHaseth *et al.*, 1977; Revzin and von Hippel, 1977).

A number of lines of evidence suggest that *lac* repressor binds with approximately the same geometry to operator and nonoperator DNA (see summary in Butler *et al.*, 1977). The evidence includes the very similar ionic strength dependence of the two interactions, and particularly the fact that both interactions depend crucially on the 58-residue N-terminal sequence of amino acid residues of the repressor subunits; the removal of this sequence to produce "core repressor" abolishes both operator and nonoperator DNA binding.

The site size (n) for *lac* repressor binding to nonoperator DNA has been very carefully measured, and shown to be 12 ± 1 base pairs (Butler *et al.*, 1977). This appears incompatible with the nuclease-protected value of n = 24–26 base pairs determined for operator binding by Gilbert *et al.* (1975), and just marginally compatible with the value of m estimated from the positions of O^c mutations in the operator, the minimum length of synthetic operator required for tight binding, and so on (13–17 base pairs; see Section 4.1.3 and Fig. 17). A number of proposals have been made to rationalize these differences (see Butler *et al.*, 1977); perhaps the most straightforward is based on the assumption that *lac* repressor can bind nonspecifically to *two* sides of the double-helical DNA lattice. Since the site size of 12 ± 1 base pairs is calculated by assuming binding to *one* side of the lattice only, this modification makes n for nonoperator DNA equal to 24 ± 2 base pairs, in good agreement with the operator parameters. A recent electron microscopy study by Zingsheim *et al.* (1977) of repressor binding to nonoperator DNA shows that the dimensions of the complex are just what would be expected if binding did indeed take place on two sides of the lattice.

The binding of repressor to nonoperator DNA is also compatible with the involvement of two repressor subunits in the interaction. As indicated earlier, approximately 11 basic residues must contact DNA phosphates in complex formation. The N-terminal sequence of each repressor subunit contains only seven or eight basic residues; thus, at least two subunits must be involved. Similarly, the pH dependence of K_{RD} has been interpreted in terms of the titration of two histidine residues involved in the R–D interaction; yet each N-terminal repressor sequence contains only one. Finally, the site size for repressor binding to nonoperator DNA was determined, in part, by monitoring changes in the accessibility of repressor sulfhydryl groups to chemical probes as DNA binds (Butler *et al.*, 1977). On saturation with DNA, access to approximately one-half of the repressor cysteine residues seems to be limited; this finding is also compatible with two of the four repressor subunits coming in contact with the DNA as a consequence of complex formation.

Inducer binding weakens the affinity of repressor for operator; however, it does not affect the binding constant of repressor for nonoperator DNA (that is,

K_{RD} = K_{RID}) (Lin and Riggs, 1975*a;* von Hippel *et al.*, 1975; Revzin and von Hippel, 1977). This result is consistent with the model developed in Section 4.1.2, in which we suggested that the conformational changes involved in inducer binding result in less favorable contacts with two or three specific base pairs in the operator. The lack of change in the affinity for nonoperator DNA suggests that the nonspecific binding contacts between repressor and DNA remain largely unperturbed by these conformational changes.

The overall picture emerging from all these facts is that of a nonspecific interaction involving both favorable charge–charge interactions and nonelectrostatic (with functional groups of sugar residues?) contacts between repressor and DNA. These interactions provide approximately one-third of the free energy of the R–O interaction at physiological salt concentrations; the rest comes from favorable interactions (largely independent of salt concentration) between complementary proteins and base pair functional groups in the grooves of the double-helical DNA structure. If more than four or five base pairs are incorrectly placed relative to their complementary protein functional groups, the protein appears to undergo a conformational change resulting in the withdrawal of the remaining (favorable) specific contacts. This conclusion follows from the facts that K_{RD} = K_{RID} and that a single value of K_{RD} seems to be found with various nonoperator DNAs (Revzin and von Hippel, 1977). If, in binding to nonoperator DNA, the protein remains in contact with specific base pairs, a considerable *range* of values of K_{RD} would be expected, since the number of correct base pairs in nonoperator DNA sequences present in substantial concentrations should range at least from three to six (see Section 3.2.1 and Fig. 11).

4.1.6 *Summary: Thermodynamic Parameters and Molecular Properties of the lac Repressor–Operator–Inducer–DNA System*

Our thermodynamic and molecular conclusions concerning the direct interaction of the *lac* repressor with the other components of the *lac* control system are summarized schematically (and somewhat speculatively) in Figs. 18 and 19. In the following section we attempt to fit these parameters into a *quantitative* set of coupled equilibria to provide a thermodynamic rationale for the *in vivo* function of the lactose repressor system.

4.2 *Repression of the Lactose Operon as an Integrated Control System*

The fact that repressor and inducer-saturated repressor bind to nonoperator DNA and the direct observation that these interactions occur *in vivo* (Section 4.3), have important consequences for the regulation of the *lac* operon. We now know that repressor and inducer-saturated repressor do not simply exist free in solution when not bound to operator, but also bind to the myriad other nonspecific loci on the *E. coli* chromosome; thus, the whole system must be viewed as a set of coupled equilibria collaborating to control the concentrations of free repressor and repressor–inducer complexes, and, as a final consequence, the concentration of free operator as well.

Since many parameters and binding species are involved, we have carried out

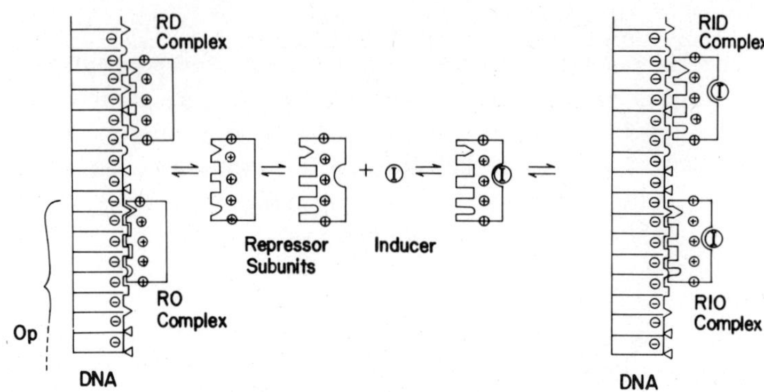

Figure 18. Schematic representation of the thermodynamics of binding of one (free or inducer-complexed) subunit of *lac* repressor to operator and nonoperator DNA. The "good fit" of the specific interactions of R with O, and the slightly less favorable specific interactions of the RI complex with O, are shown; we note that there is no specific interaction of either R or RI with D. The conformational equilibrium between a tight operator-binding and a tight inducer-binding subunit is also indicated; the latter is the form (R_I; see Section 4.1.2) that is "trapped" by I. Charges (+ and −) are shown to represent the nonspecific electrostatic interaction component.

a computer modeling study of the *in vivo* binding of *lac* repressor and repressor–inducer complexes to the *lac* operator and to nonoperator DNA. To keep the system quantitatively manageable, we focus here only on the control exercised by the binding of repressor to operator; that is, on the repressor-regulated portion of the *lac* transcriptional control system. Some details of this study have been reported elsewhere (von Hippel *et al.,* 1974, 1975). Here the previously reported results are summarized and the investigation is expanded into certain additional directions, in the hope that they might serve as a model for the extension of such quantitative approaches to other genome control systems as the relevant (*in vivo*) binding parameters become available. Some qualitative comments on the extension of these ideas to the interactions controlling the function of RNA polymerase, the cAMP–CAP system, and the steroid–steroid receptor complex system are made in subsequent sections.

The model presented is designed to be compatible with, and to explain, the following "facts" which summarize the relevant aspects of our present understanding of the *lac* control system. Some justification for these facts has been presented earlier; further support and documentation are available in the extensive *lac* literature; see, for example, Bourgeois and Pfahl (1976) for references.

4.2.1 *The Facts*

1. Repression of the *lac* operon of *E. coli* is a consequence of the binding of repressor tetramer (R) to operator (O), and induction (derepression) is due to inducer (I) binding to R, with the formation of an RI_n complex characterized by decreased affinity for O.
2. The constitutive rate of *lac* enzyme production in wild type *E. coli* cells is approximately the same in the presence of saturating inducer as it is in *lacI*$^-$ (defective R) mutant cells; that is, full derepression is possible with attainable inducer concentrations.

3. The constitutive rate of *lac* enzyme synthesis is about 10^3 times greater than the fully repressed (basal) rate in wild type cells.
4. An increase in intracellular R levels decreases the basal rate of *lac* enzyme synthesis in direct proportion.
5. Inducer concentrations necessary to achieve fully constitutive rates range from about 10^{-1} to about 10^{-3}M for inducers with repressor association constants (K_{RI}) ranging from 10^4 to $10^6 M^{-1}$; R mutants exhibiting decreased values of K_{RI} require increased concentrations of I to achieve full derepressions.
6. Cells carrying single (or double) base pair mutations in the operator (O^c) show an increase of 10- to 500-fold in basal rates; the isolated O^c-containing DNAs show a parallel 10- to 500-fold decrease from wild type levels in the measured R–O binding constants.
7. The affinity of the (I-saturated) RI_n complex for O is about 10^3 less than that of R for O.
8. R and RI_n bind to nonspecific DNA with approximately equal affinities.

4.2.2 *The Model*

The following parameters relevant to the wild type *E. coli* cell have been used to establish a "base set" of binding constants and concentration constraints to describe the coupled equilibria that apply *in vivo*. The internal volume of the cell is taken as 10^{-15} liters. We assume a total operator concentration $[O_T]$ of about 2×10^{-9} M (~1 per cell); a total repressor concentration $[R_T]$ of about 2×10^{-8} M (~10 per cell); a total nonspecific DNA site concentration $[D_T]$ of about 2×10^{-2} M (~10^7 per cell; the *E. coli* chromosome contains ~10^7 base pairs, and in principle every base pair represents the beginning of a separate nonspecific DNA-binding

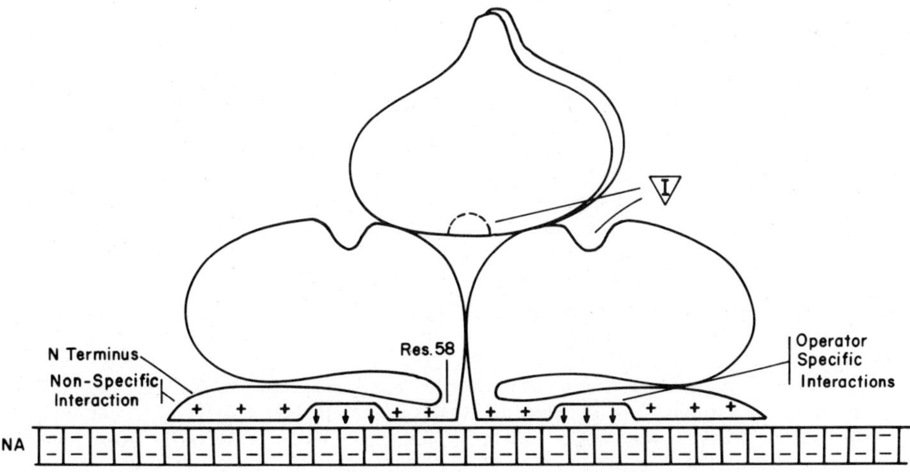

Figure 19. Schematic (and highly speculative) representation of *lac* repressor bound to nonoperator DNA. This sketch assembles the known or "most likely" molecular properties of the *lac* repressor, and is based on the evidence summarized in the text. Two subunits are shown binding to the DNA via their N-terminal residues by nonspecific (largely electrostatic) interactions (the operator-specific interactions are shown in the "withdrawn" conformation). The repressor subunit (and the inducer-binding site) are shown in the R_O (tight operator-binding) form.

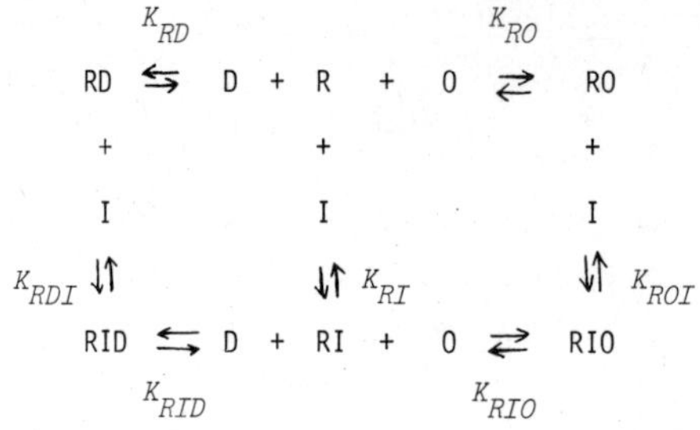

$$
\begin{array}{ccccccc}
 & K_{RD} & & & & K_{RO} & \\
RD & \leftrightarrows & D & + R & + O & \rightleftarrows & RO \\
+ & & & + & & & + \\
I & & & I & & & I \\
K_{RDI} \downarrow\uparrow & & & \downarrow\uparrow K_{RI} & & & \downarrow\uparrow K_{ROI} \\
RID & \leftrightarrows & D & + RI & + O & \rightleftarrows & RIO \\
 & K_{RID} & & & & K_{RIO} &
\end{array}
$$

Figure 20. Model of the repressor–operator–inducer–nonoperator–DNA system; R = repressor, O = operator, I = inducer, D = nonoperator DNA sites; RO, RD, RI, RIO, and RID are the various complexed species; and K_{RO}, K_{RD}, etc, represent the indicated association constants. (From von Hippel *et al.*, 1974.)

site); and a fully derepressing total inducer concentration $[I_T]$ of about 10^{-3} M (this applies to the gratuitous inducer, isopropylthiogalactoside, IPTG). The definitions of the necessary binding constants, together with the base set values used, are*

$$
\begin{aligned}
K_{RO} &= [RO]/[R]\cdot[O] \simeq 10^{14}\ M^{-1} \\
K_{RD} &= [RD]/[R]\cdot[D] \simeq 10^{5}\ M^{-1} \\
K_{RI} &= [RI]/[R]\cdot[I] \simeq 10^{6}\ M^{-1}\ (\text{for IPTG}) \\
K_{RIO} &= [RIO]/[RI]\cdot[O] \simeq 10^{-3}\ K_{RO} \\
K_{RID} &= [RID]/[RI]\cdot[D] \simeq K_{RD}
\end{aligned}
\tag{6}
$$

with conservation constraints:

$$
\begin{aligned}
[O_T] &= [O] + [RO] + [RIO] \\
[R_T] &= [R] + [RO] + [RD] + [RI] + [RIO] + [RID] \\
[D_T] &= [D] + [RD] + [RID] \\
[I_T] &= [I] + [RI] + [RIO] + [RID]
\end{aligned}
\tag{7}
$$

Figure 20 presents a model of the system, showing the relevant equilibrium constants and molecular species.† We have calculated the concentrations of all

*The exact internal ionic milieu of the *E. coli* cell is unknown. It is usually assumed that the cell is about 0.15 to 0.20 M in KCl, and several millimolar in Mg^{2+}, polyamines, and other polyvalent cations. However, these values are based on measurements of the total ionic species present, and since most of (at least) the di- and polyvalent cations doubtless exist primarily in fairly tight complexes with the various molecular and macromolecular constituents of the cell, they may not contribute appreciably to the actual ionic "activity" of the cytoplasm. The binding parameters used are a self-consistent set based on *in vitro* measurements under ionic conditions *assumed* to apply approximately within the cell (an *effective* cationic activity equivalent to ≈ 0.18 M Na^+). Direct measurement of the extent of repressor and repressor–inducer complex binding to the DNA genome of *E. coli in vivo* indicates that these parameters are physiologically reasonable (Kao-Huang *et al.*, 1977) (see Section 4.3).

†Note that here we consider only a single binding site for I to R in these equilibria. Of course, free *lac* repressor actually contains four sites to which inducer binds independently. It is very likely that more than one inducer must be bound per repressor to bring about complete derepression (Section 4.1.5), but since we define K_{RIO} (and K_{RID}) as the observed binding constants for the repressor–inducer complex at *saturating* inducer concentrations, for present purposes we can represent the situation in terms of a single inducer-binding interaction. By this definition of K_{RIO}, the concentration range of I over which the system is derepressed will be approximately independent of the actual value of n involved in the derepressing RI_n complex. On the other hand, the *shape* of the induction curve as a function of I concentration will depend on n. In Section 4.2.3g, we consider this aspect further.

We also list only a single mass action equation (and constant) for the RD interaction, since direct repressor-binding experiments (see earlier) show that the predominant class of sites can be characterized by a single binding constant, of about the order of magnitude indicated, which is essentially independent of DNA nucleotide composition and sequence. However, in principle, we can subdivide the D sites into i different types, each characterized by the mass action relation $K_{RD_i} = [RD_i]/[R]\cdot[D_i]$. The effects of subclasses of tighter binding sites on the repression system are considered in Section 4.2.3c.

species in the system using a computer program for solving the simultaneous equations (6) with constraints given by Eqs. (7). The results are generally expressed as fraction of free operator present in the cell $[O_{free}]/[O_T]$), which is assumed (see Sadler and Novick, 1965) to be directly proportional to the parameter actually measured *in vivo*—that is, the *ratio* of the repressed to constitutive intracellular β-galactosidase activity.

4.2.3 Conclusions

4.2.3a The Basal Level. Figure 21 shows the fraction of total cellular repressor bound to the *E. coli* genome as a function of either the repressor–nonoperator DNA-binding constant (for $[D_T] = 2 \times 10^{-2}$ M; upper abscissa and dashed curve) or of total cellular DNA (for $K_{RD} = 10^5$ M^{-1}; lower abscissa and solid curve). Clearly, for the binding constants and component concentrations assigned in Section 4.2.2, almost all the repressor is bound to the genome.

The role played by nonoperator DNA binding in controlling the magnitude

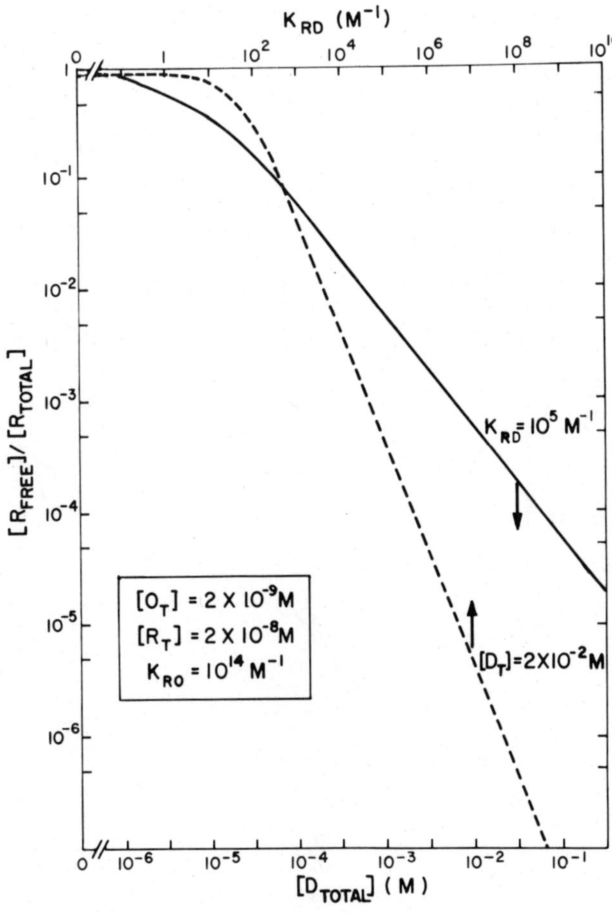

Figure 21. Fraction of repressor calculated to be free (uncomplexed with genome DNA) in the *E. coli* cell as a function of either K_{RD} (dashed curve and upper abscissa) or total D (solid curve and lower abscissa); see text.

of the observed repression of the *lac* operon ($[O_{free}]/[O_{total}]$) is illustrated in Fig. 22, in which the fraction of *free operator sites* is plotted as a function of either K_{RD} (for $[D_T] = 2 \times 10^{-2}$ M nonoperator sites; upper abscissa and dashed curve), or of $[D_T]$ (for $K_{RD} = 10^5\ M^{-1}$; lower abscissa and solid curve). No inducer is present, and only one class of nonoperator-binding sites is assumed. At low K_{RD} (or low $[D_T]$), there is no nonspecific binding and the system is repressed to a calculated basal level ($[O]/[O_T]) \simeq 5 \times 10^{-7}$ of the constitutive level! As we increase K_{RD} (or $[D_T]$), the extent of repression decreases, asymptotically approaching the totally unrepressed state (that is, constitutive levels of *lac* enzyme production) only at $K_{RD} > 10^9\ M^{-1}$, or $[D_T] > \sim 10$ M, under the conditions of the calculation. Note that the value of $[O]/[O_T]$ predicted for the levels of $[D_T]$ and K_{RD} that presumably apply in the *E. coli* cell is approximately equal to 10^{-3}, close to the basal level actually measured *in vivo* (fact 3 of Section 4.2.1). Thus, Fig. 22 shows that in the absence of R binding to nonspecific DNA sites, the basal rate of *lac* enzyme synthesis should be three to four orders of magnitude smaller than is actually observed, and that the observed basal level is primarily established by nonspecific binding of R to D sites, the latter acting as a "sink" for R, in competition with O.

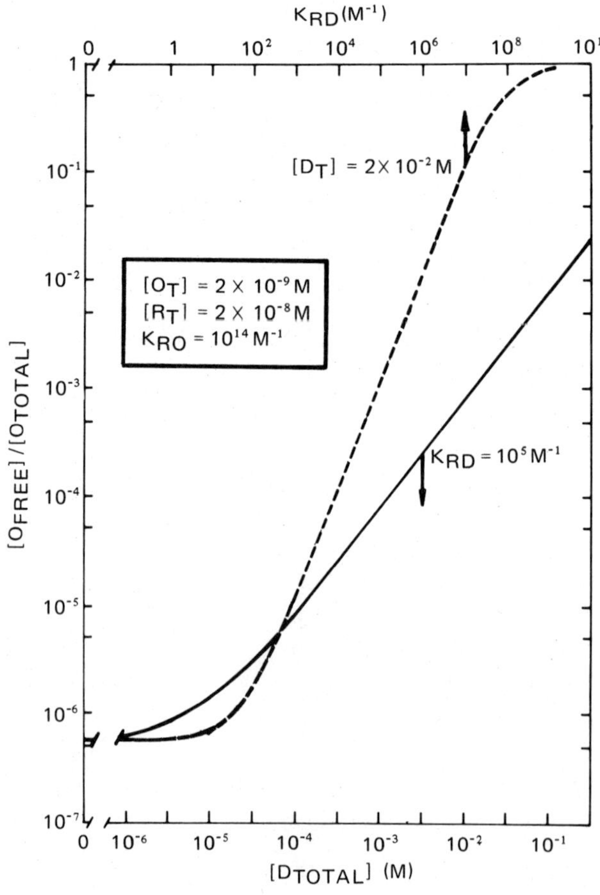

Figure 22. Fraction of *lac* operator calculated to be free (uncomplexed with repressor) in *E. coli* as a function of K_{RD} (dashed curve, upper abscissa) or $[D_T]$ (solid curve, lower abscissa); see text. (From von Hippel *et al.*, 1974.)

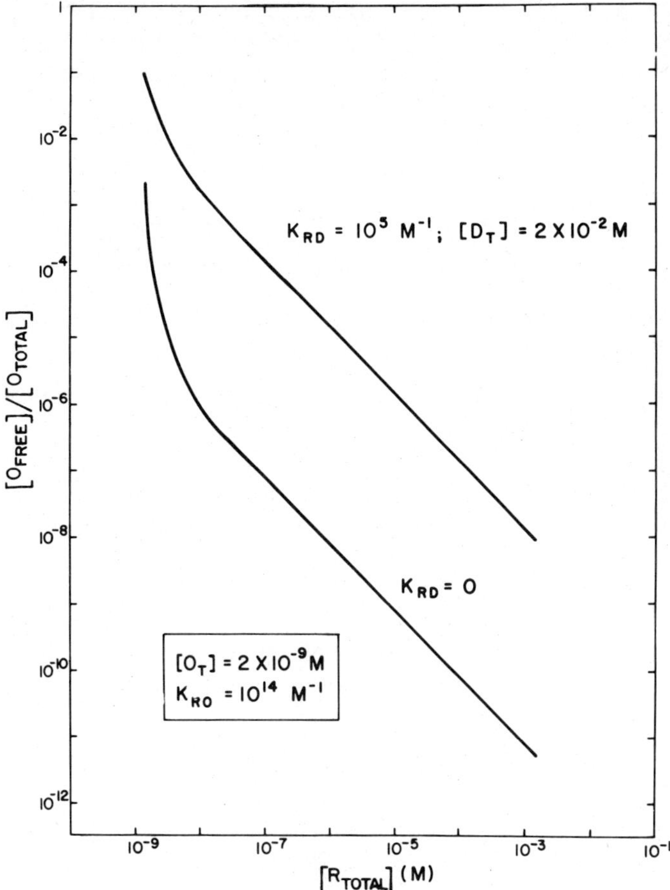

Figure 23. Fraction of *lac* operator calculated to be free as a function of increasing total cellular repressor.

4.2.3b Effects of Increasing $[R_T]$. In Fig. 23 the calculated basal level of *lac* enzyme synthesis is plotted as a function of $[R_T]$. These calculations show that in *E. coli* mutants overproducing repressor by 10- to 200-fold, the basal level is expected to decrease linearly with increasing $[R_T]$, from about 10^{-3} of the constitutive level for wild type cells, to about 3×10^{-5} for the most overproducing strains. These expectations are in excellent accord with the actual physiological measurements (fact 4 of Section 4.2.1) (Sadler and Novick, 1965; Smith and Sadler, 1971).

4.2.3c Effects of Additional Strong Repressor-Binding Sites. Figure 24 shows the calculated effects on $[O]/[O_T]$ of adding 100 stronger binding sites to the genome, in addition to the 10^7 sites per cell with $K_{RD} = 10^5\ M^{-1}$. From Fig. 24 it can be seen that to have a significant effect on the basal level, such sites must exhibit repressor affinities in excess of $10^{11}\ M^{-1}$ (compare curves 2–5, for $K_{RO} = 10^{14}\ M^{-1}$). Even a few (less than ten) sites having K_{RD} as large as $10^{13}\ M^{-1}$ have little effect on $[O]/[O_T]$.

4.2.3d Operator Constitutive (O^c) Mutations. Figure 24 also shows that decreasing K_{RO} by one or two orders of magnitude (O^c mutations) still does not bring $[O]/[O_T]$ to the observed basal level in the absence of competitive R binding

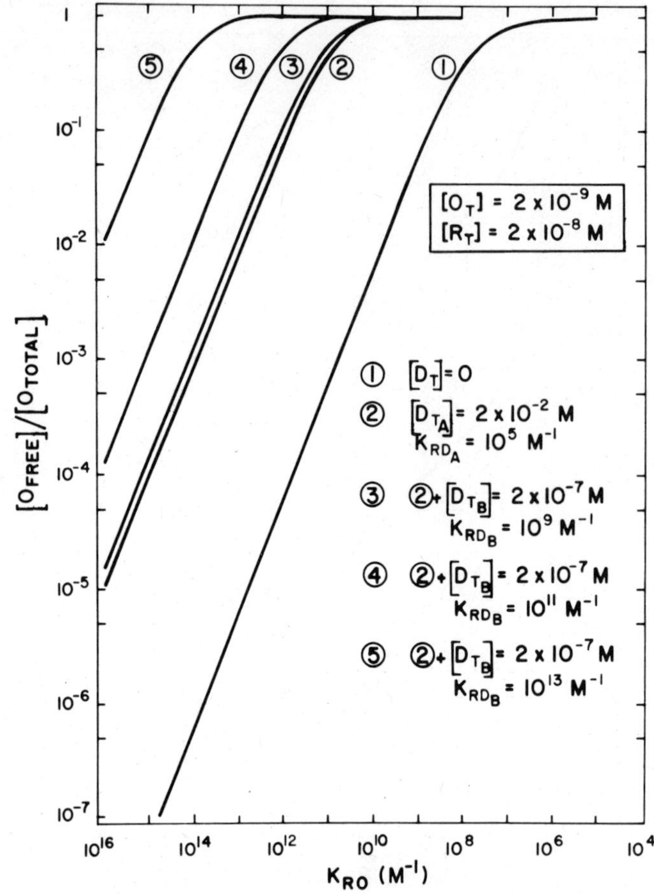

Figure 24. Effect of nonoperator DNA-binding sites on the calculated fraction of free operator for various values of the repressor–operator association constant (K_{RO}). The effects of a second class of nonoperator DNA sites, with elevated values of K_{RD}, are also shown (see text). (From von Hippel *et al.*, 1974.)

to nonspecific sites (curve 1), but does indeed show just the effects observed on the basal level with O^c mutants (fact 6 of Section 4.2.1) (Smith and Sadler, 1971; Jobe *et al.*, 1974) when nonspecific binding is included (curve 2).

4.2.3e Inducer-Binding Effects. In Fig. 25 are plotted the calculated effects of inducer binding on the fraction of *lac* operator that is free. The lower (solid) curve shows that, in the absence of nonspecific binding of R and RI complex, derepression of the *lac* operon, even at supersaturating concentrations of I, cannot take place. This follows because the RI complex binds to O with an affinity that is only about three orders of magnitude lower than that of R itself (Bourgeois and Jobe, 1970; Barkley *et al.*, 1975); thus, in the absence of the nonspecific sites as a "sink" for the RI complex, the RI complex itself binds to O and maintains repression at calculated values below the basal level. It is only if one makes the thermodynamically impossible assumption that I binding can reduce the affinity of the RI complex for O to values close to zero (Fig. 25, dashed curve), that total derepression in the absence of nonspecific binding can be achieved, and even then only at total inducer levels about 10^3 times greater than actually required *in vivo*.

The upper curves in Fig. 25 show the calculated effects of added I on the basal level at three different nonspecific binding affinities. [In constructing these curves we assume that $K_{RIO} = 10^{-3}K_{RO}$ (Section 4.2.1, fact 7) and that $K_{RID} = K_{RD}$ (Section 4.2.1, fact 8).] We note that only the curve for $K_{RD} = 10^5$ M^{-1} starts at the observed basal level at low $[I_T]$ and reaches approximately the constitutive level (within a factor of 2) at saturating concentrations of $[I_T]$. This observation provides an independent confirmation that the value of K_{RD} chosen for the base set of parameters is indeed a reasonable representation of the *in vivo* situation or, more accurately, that the *ratio* of K_{RO} to K_{RD} is approximately correct. Figure 25 also shows that derepression of the *lac* operon occurs at the I concentrations observed *in vivo* (Section 4.2.1, fact 5) when nonoperator DNA binding of R and RI is taken into account, but not otherwise.

4.2.3f Effects of Varying Ratios of K_{RIO}/K_{RO} and K_{RID}/K_{RD}. An additional mechanism for changing the effect of inducer on the level of repression of the *lac* operon, as controlled by the final value of $[O]/[O_T]$, is variation in the ratios of

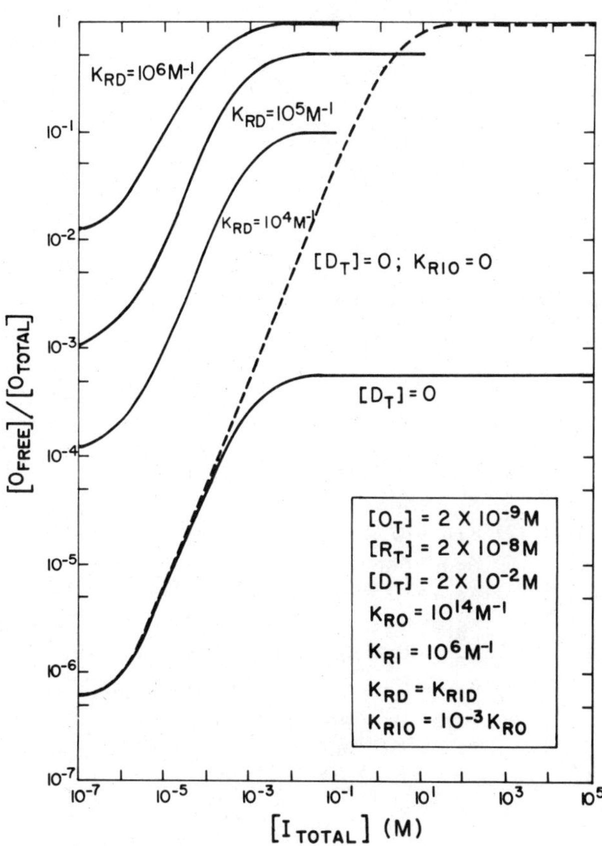

Figure 25. Fraction of free operator calculated as a function of total inducer concentration at various values of the association constants for repressor and repressor–inducer complex to nonoperator DNA sites (see text). (From von Hippel *et al.*, 1975.)

K_{RIO} to K_{RO}, and of K_{RID} to K_{RD}. That is, we suggest that certain types of mutant proteins may arise that, because of a relatively different positioning of functional groups in the induced and noninduced states, may show differences in K_{RD} and K_{RID}, or, alternatively, may demonstrate ratios of K_{RO}/K_{RIO} other than the canonical value of 10^3 used in the preceding calculations. Various possibilities are shown in the induction curves of Fig. 26, and show that certain ratios of K_{RIO} to K_{RO}, and of K_{RID} to K_{RD}, result in no change in $[O]/[O_T]$ as a function of I_T, or actually shows a *decrease* in $[O]/[O_T]$ (further *repression*) with increased inducer concentration. Such "computer-generated mutants" provide a plausible explanation of the phenotypic behavior of some members of the I^s class of repressor mutants (see Section 4.1.3), which actually do show tighter binding to operator with increased inducer concentration (Myers and Sadler, 1971; Jobe, Sadler, and Bourgeois, 1974). Pfahl (1976) has speculated qualitatively that the X86 tight-binding repressor phenotype may reflect an increase in repressor binding to nonoperator DNA; possible quantitative interpretations of this proposal may be derived from Fig. 26.

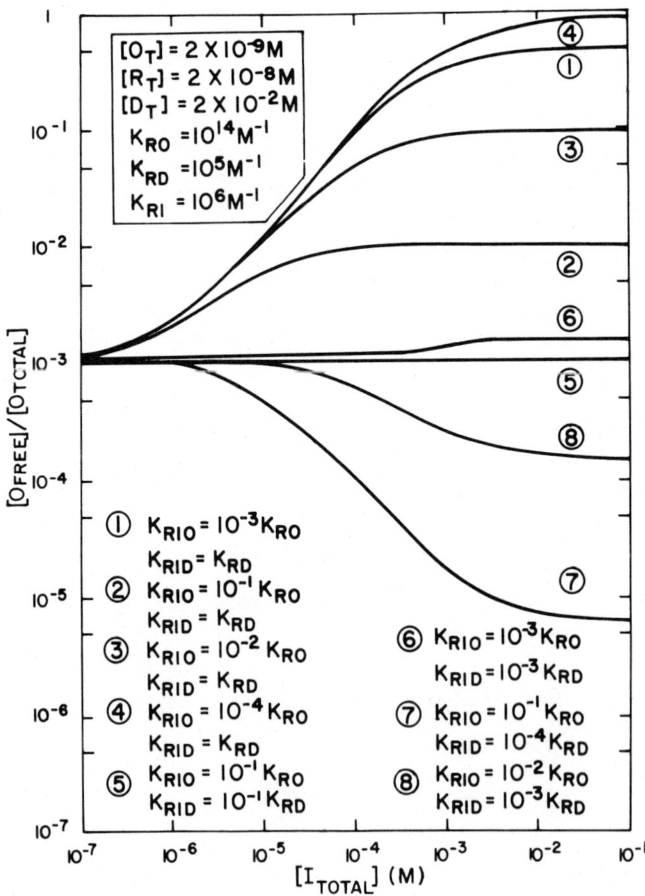

Figure 26. Effects on the fraction of free operator calculated as a function of varying ratios of K_{RIO} to K_{RO}, and of K_{RID} to K_{RD} (see text).

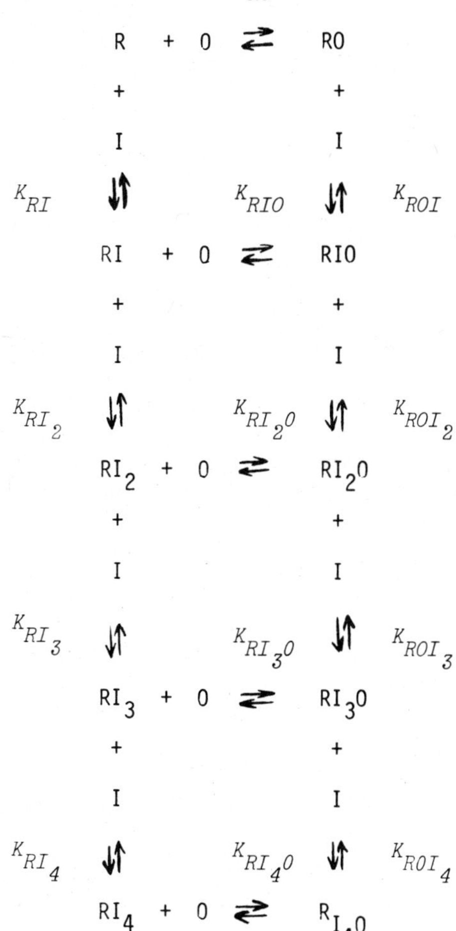

Figure 27. Model of the repressor–operator–inducer–nonoperator–DNA system including explicitly all the forms of the repressor–inducer complexes. For clarity, only the operator-binding interactions are shown; a more complete diagram would include a comparable matrix of interactions with nonoperator DNA.

4.2.3g Effects of Considering Partially Saturated Inducer–Repressor Complexes. Up to this point we have ignored the fact that *each* subunit of the repressor tetramer has an inducer-binding site, and that all can bind inducer, whether the repressor is free, bound to operator, or bound to nonoperator DNA (see Bourgeois and Pfahl, 1976; Butler *et al.*, 1977). This means that a more realistic model must consider not only K_{RI}, K_{RIO}, and K_{RID}, but also K_{RI_2}, . . . , K_{RI_4}, K_{RI_2O}, . . . , K_{RI_4O}, and K_{RI_2D}, . . . , K_{RI_4D}, and so on. A partial model of this expanded situation, defining the additional complexes and binding constants that must be considered, is shown in Fig. 27. (The equivalent set of equilibria involving binding to nonoperator DNA has been omitted, both for clarity and because all inducer–repressor complexes bind to nonoperator DNA with equal affinity; Section 4.2.1, fact 8.)

Perhaps the most reasonable model which takes these complexes into account, as well as the rest of the inferences about the interactions of repressor with operator, inducer, and nonoperator DNA summarized in Section 4.1.6, is the following:

1. All R subunits bind I independently and equally (the coefficients simply reflecting the statistical effects of binding four ligands to four independent sites; that is, there are four ways to bind the first inducer molecule, and so on):

$$K_{\mathrm{RI}} = 4K_{\mathrm{RI,int}}$$
$$K_{\mathrm{RI}_2} = 3/2\,(K_{\mathrm{RI,int}})$$
$$K_{\mathrm{RI}_3} = 2/3\,(K_{\mathrm{RI,int}})$$
$$K_{\mathrm{RI}_4} = 1/4\,(K_{\mathrm{RI,int}})$$

2. All complexes bind to nonoperator DNA sites equally:

$$K_{\mathrm{RD}} = K_{\mathrm{RID}} = K_{\mathrm{RI}_2\mathrm{D}} = K_{\mathrm{RI}_3\mathrm{D}} = K_{\mathrm{RI}_4\mathrm{D}}$$

3. Only two R subunits bind to operator, induction is noncooperative between subunits, and the free energy of induction is equipartitioned between the two subunits that do bind to the operator. (Note that this formulation requires that the nonoperator-bound repressor subunits bind inducer more tightly than do the operator-bound subunits.)

$$K_{\mathrm{RO}} = K_{\mathrm{RIO}} = K_{\mathrm{RI}_2\mathrm{O}} = 10^{1.5}K_{\mathrm{RI}_3\mathrm{O}} = 10^{3.0}K_{\mathrm{RI}_4\mathrm{O}}$$

Note also that after the model is set up as just outlined, the interrelationships of the model parameters (Fig. 27) fix the values of $K_{\mathrm{ROI}}, \ldots, K_{\mathrm{ROI}_4}$ in terms of the other binding constants.

Figure 28 compares the course of induction ($[\mathrm{O}]/[\mathrm{O_T}]$ versus $\mathrm{I_T}$) for the simple RI model with that for the specific RI_4 model just described (other plausible RI_4 models give similar results). We observe, as pointed out in Section 4.2.2, that these complications change the *shape* of the induction curve (make it more cooperative), but have little effect on the inducer concentration at which the operon is derepressed to any given extent.

4.3 *In Vivo Determination of the Thermodynamic Parameters of the lac System*

As indicated in Sections 4.1.3 and 4.1.5, the binding affinity of *lac* repressor for both operator and nonoperator DNA is extremely dependent on salt concentration. Since the effective ionic activity of the interior of the *E. coli* cell is not known, we also do not know the *absolute* values of $K_{\mathrm{RO}}, K_{\mathrm{RD}}, \ldots$, that apply to the *lac* system *in vivo*. The *relative* values of these parameters are known; they have been established *in vitro* under comparable environmental conditions (and knowledge of these relative values has permitted establishment of the system of coupled equilibria for controlling the level of repression of the lactose operon outlined in Section 4.2). Furthermore, since the relative values of the various equilibrium constants for the system are established by the interrelations given in Section 4.2, it is apparent that only one of these parameters needs to be measured *in vivo* to establish the *in vivo* values of all. In addition, a crucial feature of the model outlined herein is the existence of nonoperator DNA binding of repressor and inducer-saturated repressor. This depends, of course, on the assumption (Section

4.2.2) that the *in vivo* ionic activity of the *E. coli* cell is not so high that nonspecific DNA binding is prevented. In order to test the validity of the model outlined herein, and to establish intracellular values of the various binding parameters, we have attempted to measure the *in vivo* value of K_{RD} (and K_{RID}).

In principle, what one wants to know is the distribution of *lac* repressor tetramers between genome-bound species and molecules free in the cytoplasm (Fig. 21). Since the total concentration of repressor tetramers in the cell can be measured by standard extractive techniques, the key measurement is the determination of the cytoplasmic concentration of repressor. In theory, this could be established by puncturing the *E. coli* cell with a micropipet and withdrawing a bit of cytoplasm. In practice, of course, this cannot be done; but an equivalent measurement can be made using the "minicell" mutant of *E. coli,* which sheds, during cell division, small membrane-encapsulated minicells containing all the normal components of the cytoplasm, but no DNA (Adler *et al.,* 1967); see Fig. 29.

We (Kao-Huang *et al.,* 1977) have used this system to measure the total repressor concentration (per unit soluble protein) in the parent cells, and also in the minicells. The conclusion reached was that in both induced (with IPTG) and uninduced cells a *maximum* of 7% (±4%) of the total repressor tetramers of the cell are free in the cytoplasm. This is a *maximum* value because of unavoidable slight

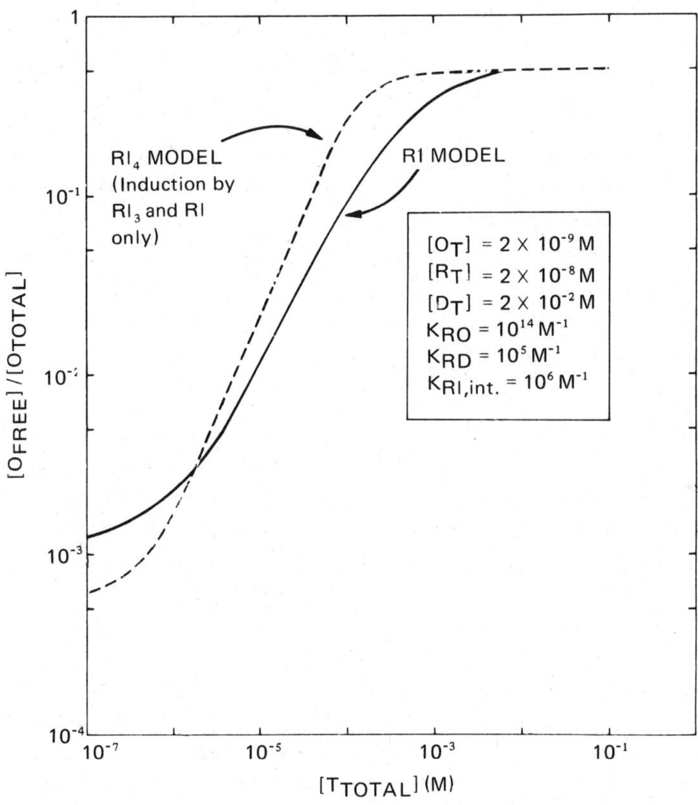

Figure 28. Effects of inducer binding on the fraction of free operator calculated as a function of total inducer concentration for the simple RI model, and for the specific RI_4 model described in the text.

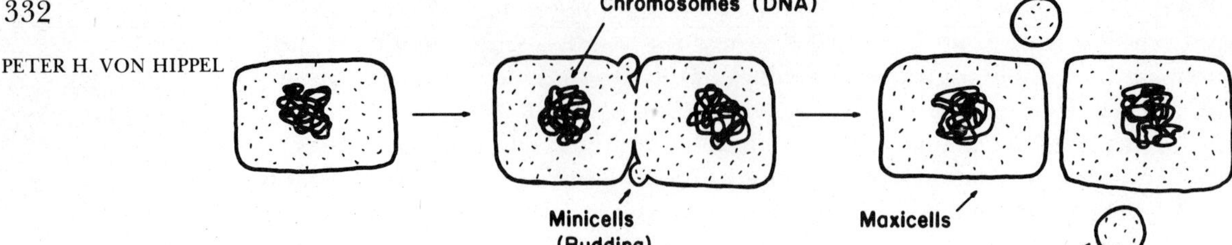

Figure 29. Schematic view of the production of minicells by the minicell mutant of *E. coli* described by Adler *et al.* (1967). Parent cells ("maxicells") and minicells are shown at various stages of cell division.

contamination of minicell preparations with parent cells; as detailed by Kao-Huang *et al.* (1977), the actual concentration of repressor in the cytoplasm could be as much as two orders of magnitude smaller.

This establishes the *in vivo* value of K_{RD} (and K_{RID}) at 10^3 to 10^5 M^{-1}, and (see Section 4.2) therefore also permits us to conclude that the *in vivo* values of K_{RO} and K_{RIO} are 10^{12} to 10^{14} M^{-1} and 10^9 to 10^{11} M^{-1}, respectively. As a serendipitous fringe benefit, knowledge of the steep *in vitro* dependence of K_{RD} on salt concentrations (Fig. 30) permits us to use the distribution of *lac* repressor between DNA and cytoplasm as an accurate *in vivo* "ionic strength meter" to provide an estimate of the internal cationic activity (chemical potential of effective cation; in Na^+ equivalents) within the *E. coli* cell. As Fig. 30 shows, this measure indicates that the effective net concentration of free intracellular cation ranges between 0.17 and 0.23 M in Na^+ equivalents. Since the *E. coli* cell contains not only monovalent cation (mostly K^+ at approximately this total concentration), but also Mg^{2+}, polyamines, and other multivalent species, this result indicates that most of the latter are complexed to various cellular constituents, such as nucleotide triphosphates, tRNA, and ribosomes, and thus not available to compete with repressor in DNA binding.

4.4 *Kinetics of Intracellular Repressor Transport*

The nonspecific binding of *lac* repressor, as well as of other genome-regulating proteins, to DNA sequences other than their specific genome target sites also means that the kinetics of the transport of these proteins to their respective DNA targets involves processes more complex than simple free diffusion.

The first inkling of this came when Riggs *et al.* (1970*b*) showed that the forward rate constant for the binding of *lac* repressor to operator seemed considerably larger than could be accounted for by a diffusion-controlled process. Forward rate constants (k_f) of about 10^{10} M^{-1} sec^{-1} were measured by these workers (using filter binding techniques with very dilute solutions to slow the reaction rates into the measurable range). These experimental values of k_f can be compared with theoretical values generated by assuming that the rate-limiting step is diffusion-controlled, and using the Debye–Smoluchowski equation:

$$k_f = 4\pi\kappa a_f (D_{\mathrm{R}} + D_{\mathrm{O}})\, N_0/1000 \tag{8}$$

where κ is a unitless steric interaction factor, a_f the interaction radius (in cm), D_R and D_O the free-volume diffusion constants for R and O, respectively (in cm²/sec), and N_0 Avogadro's number. (As written, the units of k_f are M^{-1} sec^{-1}.) Using reasonable estimates of D_O ($<10^{-8}$ cm²/sec), D_R ($\sim 2 \times 10^{-7}$ cm²/sec), a_f ($\sim$15 Å), and κ ($\sim$1/20), we estimate from Eq. (8) that k_f for the diffusion-controlled process should be approximately 10^7 M^{-1} sec^{-1} (that is, about three orders of magnitude less than the measured value).*

$$R + O \underset{k_b}{\overset{k_f}{\rightleftharpoons}} RO \tag{9}$$

Since it is manifestly impossible for a process to be faster than diffusion-controlled, this can only mean that the reaction is not properly represented by Eq. (9). The observed data can also be fit by writing

$$R + O' \underset{k_2}{\overset{k_1}{\rightleftharpoons}} RO' \underset{k_4}{\overset{k_3}{\rightleftharpoons}} RO \tag{10}$$

for which, given certain ratios of rate constants, the observed reaction will still appear bimolecular. A representation such as Eq. (10) means that the original diffusion–encounter complex is represented by O′ (that is, an "extended" operator) and that the RO′ complex then rearranges, via an *intramolecular* reaction, to form the final RO complex.

Proposals of at least two types have been put forward to account for the anomalously large experimental values measured for the apparent second-order forward rate constant. Both may be expressed in terms of the repressor–extended operator complex. One hypothesis, formulated qualitatively by Riggs *et al.* (1970*b*) and quantitatively by Richter and Eigen (1974) and Berg and Blomberg (1976), suggests that the extended operator is a larger piece of DNA encompassing approximately 500 Å on either side of the operator. In this model, a repressor molecule encountering DNA will be held to it by nonspecific binding interactions and will be able to "slide" (during the lifetime of the nonspecifically bound state, by a random walk process) to the operator before dissociating if the site of the initial binding lies within an appropriate distance from the operator. This model, represented schematically in Fig. 31, has been called the sliding model.

The other proposal, shown schematically in Fig. 32, was put forward by von Hippel *et al.* (1975). In this (intramolecular transfer) model it is explicitly taken into account that at the levels of dilution characteristic of the filter binding assay the operator-containing DNA molecules exist as separate stiff random coils, each occupying its own "domain," with most (>99.5%) of the solution volume being

*This estimate of κ is based on considering the operator as part of a cylinder that must be encountered within about one-fifth of all possible solid angles to permit correct complex formation, while the proper angle of approach to the repressor-binding site is taken as about one-fourth of the total (assumed spherical) interaction surface. The overall κ for the reaction is thus the product of these two parameters. Equation (8) could also include an electrostatic factor, which could increase or decrease k_f as much as tenfold. Since DNA is negatively charged, while the *active site* of repressor is positively charged and the overall molecule is negatively charged, we assume for present purposes that the latter two effects approximately cancel each other, meaning that the electrostatic part of k_f [as formulated in Eq. (8)] might be quite small. Recent approaches by Lohman and Record and by Berg (private communications) deal explicitly with electrostatic aspects of the RO interaction kinetics.

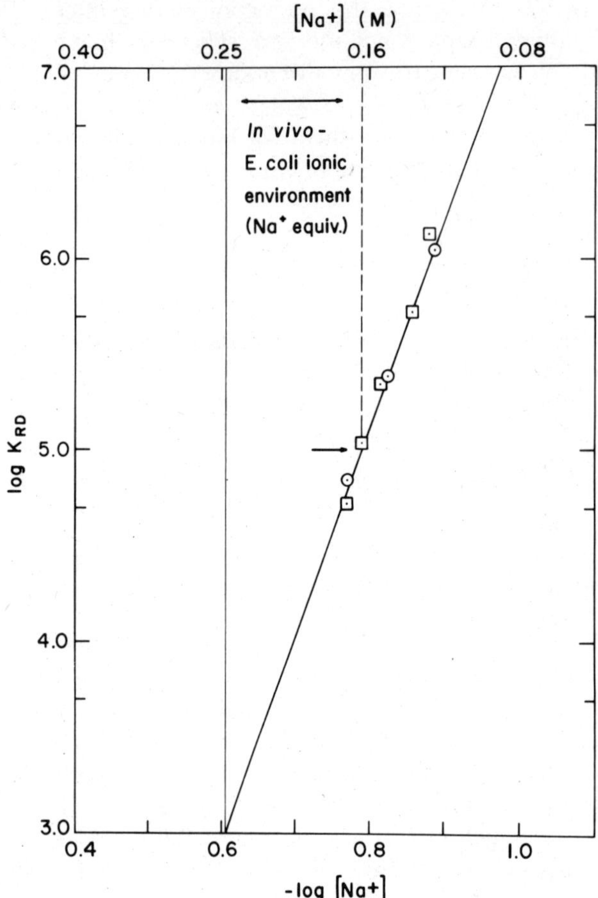

Figure 30. Log-log plot of the apparent *in vitro* association constant (K_{RD}) of *lac* repressor to nonoperator DNA, as a function of Na^+ concentration. Data (○) from Revzin and von Hippel (1977) and (□) from Record *et al.* (1977). The ionic concentration range indicated is the estimated effective intracellular ion activity in Na^+ equivalents (see text). (From Kao-Huang *et al.*, 1977.)

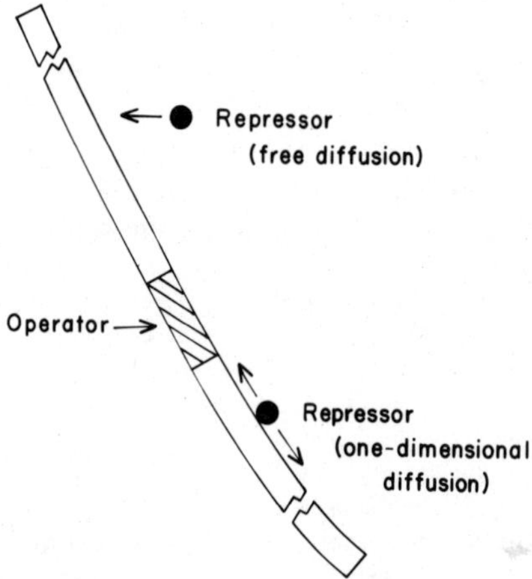

Figure 31. Sliding mechanism for transport of repressor to operator. Here the extended operator (O ′) is defined as the linear DNA segment contiguous to the actual operator site (see text).

empty. In this model the extended operator is the entire DNA molecule, the bimolecular step represents the diffusion-controlled encounter of a repressor molecule with the DNA domain, and rearrangement to the final RO complex involves intramolecular transfer of R to O via a number of intermediate RD complexes. For λp*lac* DNA (essentially a lambda DNA molecule into which the *lac* operon has been transferred), the domains will have characteristic radii (radii of gyration, r_g) of about 5000 Å under the conditions of the filter assay (~0.05 M Na^+). The intradomain concentration of nonspecific DNA-binding sites will be quite high (~2×10^4 M) and K_{RD} will be about 10^9 M^{-1} (see Wang *et al.*, 1977) under these ionic conditions. This suggests that a repressor molecule striking anywhere within the DNA domain will have a considerable probability of *remaining* bound within that domain. We visualize that the transfer of R within the domain proceeds rapidly by a direct DNA-to-DNA mechanism involving "ring closure" events* with the R molecule transiently bound between two DNA sites. (Our present model of the *lac* repressor, Fig. 19, with two subunits involved in each DNA-binding site, is consistent with such a double-bonded intermediate.) If both sites are of the nonoperator type, there should be an approximately 50% probability of intersite transfer. A series of such events (on the average involving transient binding to $1/e$ of the nonoperator sites of the DNA molecule) will eventually result in the transfer of R to O; this process may be very fast because the proposed mechanism circumvents the need to overcome the activation barrier for dissociation of nonspecifically bound repressor complexes into solution. Since $K_{RO} >> K_{RD}$, the probability that subsequent ring closure events will remove R from O is small.

In these terms the Debye–Smoluchowski equation [Eq. (8)] can be reinterpreted by inserting the *observed* value of k_f and solving for the κa_f product. In these terms, we find $\kappa a_f \simeq 660$ Å. If we then identify a_f with r_g (the average radius of gyration of the polymer domain) and κ with the probability that an initial collision will be fruitful (that is, will lead to the formation of an RO complex), we obtain $\kappa \simeq 0.15$, meaning that in terms of this model there is an approximately 85% probability of a repressor molecule that initially collides with the DNA being lost back into the "extradomain" solvent, and a 15% probability that an RO complex will be formed (via intermediate intradomain transfer events).

*Such a ring closure event can be visualized as follows. Consider a repressor tetramer bound (via a two-repressor subunit-binding site) to a particular nonoperator DNA sequence within the DNA domain. There is a finite probability that the relative (segmental) diffusion of other DNA segments will bring another nonoperator DNA site into contact with the uncomplexed subunits of the repressor, and that a transiently double-bonded repressor–DNA complex may form. We call this a ring closure event, since it involves the DNA double helix "looping back" on itself and, in essence, forming a closed ring via the doubly bonded repressor molecule. Diffusing apart of the DNA segments will break one of these contacts, and in the proposed model the repressor will have a 50% probability of being transferred, as a result, to a new site far removed (as measured along the actual extended DNA molecule) from the original nonoperator-binding locus. Bob Goldberger has pointed out to me that this process is very analogous to one invented by Madeleine L'Engle in her children's book, "A Wrinkle in Time" (L'Engle, 1962). L'Engle proposes that while time can be considered to be linear, like a piece of string (or DNA), an occasional loop can form between two segments of the time "string," and one may then simply step across a "wrinkle in time." Stepping across such looped structures in time is called "tessering" by L'Engle, and leads to interesting consequences.

We (R. Winter and P. H. von Hippel, in preparation) are currently conducting experiments to test and discriminate between these hypotheses. Clearly, the initial kinetic result [Eq. (10)] requires thinking in terms of transfer steps involving nonoperator DNA binding, and an understanding of these transfer pathways may well help to illuminate the intracellular mechanisms whereby genome-regulating proteins find their functional sites *in vivo*.

4.5 Other Components of the Lactose Operon

Up to this point this chapter has focused primarily on the *lac* repressor–inducer–operator–nonoperator DNA system. Of course this system does not function in isolation. As shown in Fig. 14, the operon also involves RNA polymerase; and it is the binding of this polymerase *to* the *lac* promoter (or the transcription of this polymerase *from* the *lac* promoter) that the *lac* repressor–inducer–operator–DNA system is designed to control. In addition, as pointed out earlier, the *lac* operon is further controlled (in a manner that is integrated with the control

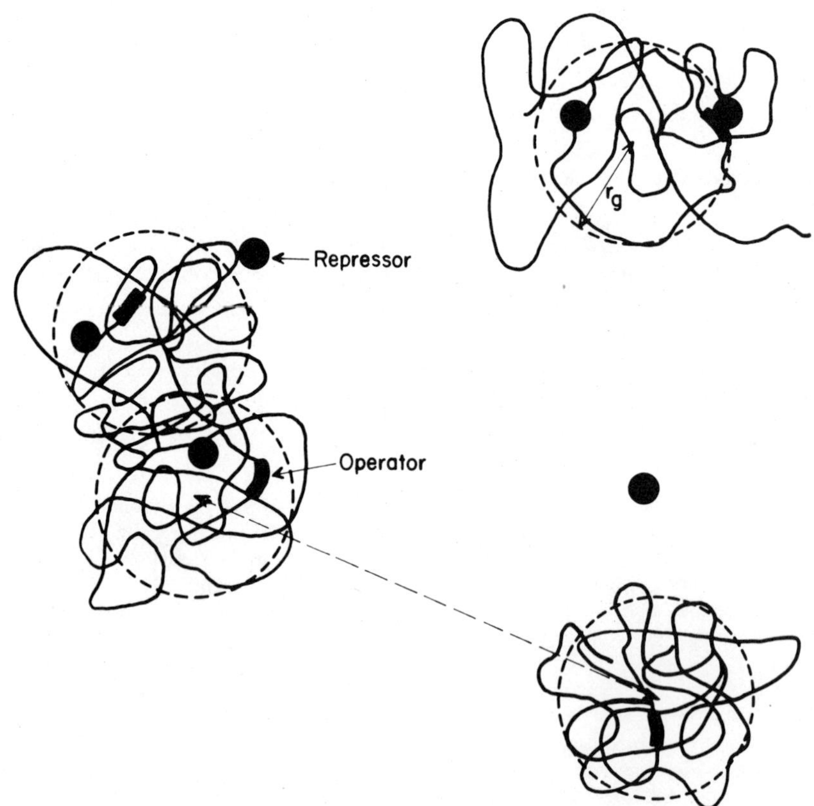

Figure 32. Intradomain transfer mechanism for transport of repressor to operator. Here the extended operator (O′) is defined as the entire random coil DNA molecule—that is, roughly that volume of the solution included within the radius of gyration (r_g) of the DNA molecule (see text).

of the other sugar-metabolizing operons of *E. coli*) via the catabolite activating protein (CAP)–cyclic AMP system. The properties of each of these components of the lactose operon and the integration of these components with the *lac* repressor system are considered in the following sections. The discussion of these components is much briefer than that of repressor, both because much less is currently known about them in a quantitative molecular sense and because we hope that the detailed and quantitative treatment of the repressor system outlined earlier will serve as a partial model for the inclusion of these components in a larger scale control scheme when sufficient quantitative measurements are at hand.

4.5.1 DNA-Dependent RNA Polymerase and the Polymerase–Promoter Interaction

As indicated earlier, RNA polymerase is the central molecule in any consideration of transcriptional control of genome function because it is the timing and quantitation of its activity (that is, mRNA synthesis) that transcriptional control systems are designed to regulate. A great deal has recently been written about the structure and properties of RNA polymerase (for a comprehensive set of reviews see the volume edited by Losick and Chamberlin, 1976). Recent reviews by Chamberlin (1974, 1976*a,b*) very effectively pull together the current state of knowledge of molecular aspects of the structure and function of this complex macromolecule. Here I propose merely to summarize the various molecular features of the function of this enzyme, and to indicate how it might be quantitatively integrated into the overall control scheme of the lactose operon.

4.5.1a Structure and Function of E. coli RNA Polymerase. The RNA polymerase of *E. coli* is a large macromolecular complex with an aggregate molecular weight of 400,000 to 500,000. The complex comprises a number of subunits, including α (mol. wt. ~40,000), β (mol. wt. ~150,000), β' (mol. wt. ~160,000), and σ (mol. wt ~86,000) as essential components. The so-called *core* enzyme, which is capable of discharging all the enzymatic functions of the polymerase, consists of two α, one β, and one β' subunit. The *holoenzyme,* which not only can function enzymatically but also is capable of specific recognition of promoter sequences, consists of the core enzyme complex plus a sigma subunit.

The sequence of steps involved in RNA polymerase function can be summarized as follows (for details, see Chamberlin, 1974, 1976*a,b;* von Hippel and McGhee, 1972):

1. *"Closed" RNA polymerase–promoter complex formation:* This step involves recognition of the promoter (DNA base pair) sequence by the holoenzyme and the formation of a specific protein–DNA complex. All the aspects and elements of recognition and specific complex formation considered previously for the *lac* repressor–operator interaction are presumed to be involved here, but much less is known about the molecular details for the polymerase–promoter interaction.

2. *"Open" polymerase–promoter complex formation:* This step involves an "opening" event, in which the base-paired promoter sequence is either actively forced open, or passively "trapped" (as a consequence of a local DNA "breathing" event) by the polymerase in a locally single-stranded state. The polymerase then articulates specifically with the proper template strand in preparation for the initiation of mRNA transcription of the proper polarity.

3. *mRNA chain initiation:* In this step a ternary initiation complex involving template, bound polymerase, and the initial base (in nucleotide triphosphate form) of the mRNA sequence is formed.

4. *mRNA chain elongation:* Once initiation has taken place, elongation involves the processive movement of the polymerase along the template strand, with the incorporation of precursor nucleoside triphosphates (as nucleoside monophosphates) into the mRNA strand in a sequence complementary to the template DNA strand. This step involves not only processive movement along the DNA template strand, but also DNA strand separation *ahead* of the active transcriptional complex, and presumably mRNA strand displacement and DNA double-helix reformation *behind* the complex. This and subsequent steps are not thought to require the sigma subunit—that is, they can be carried on by core enzyme alone. As a consequence, sigma subunits may be recycled between core enzyme molecules to permit more such molecules to participate in the formation of new initiation complexes.

5. *mRNA chain termination:* When the end of the operon is reached the nascent mRNA chain is terminated, the polymerase and the mRNA chain separate from one another and from the DNA template strand (the order of these events is unknown), and the DNA double helix recloses. Recognition of the terminal bases of operons (cistrons) appears to involve both direct recognition by polymerase of certain "terminator" base sequences and the participation of chain-terminating proteins such as rho factor (Roberts, 1976*a*).

Only the first two steps of mRNA synthesis just outlined are of further concern to us in this chapter.

As indicated earlier, the initial polymerase–promoter interaction must involve the recognition of a specific base pair sequence of the DNA by the polymerase, much as the operator sequence is recognized by repressor. Since specific recognition of the *lac* promoter by polymerase can occur in the *absence* of repressor in *in vitro* transcription systems, and, of course, also *in vivo* in I^- mutant *E. coli,* it is clear that recognition does not require simultaneous or previous complex formation between *lac* repressor and operator. A number of promoters have been sequenced; some parts of the promoter sequence have been shown to be common to many promotors, but appreciable differences (which may modulate the rate of mRNA initiation, perhaps by facilitating or slowing the "melting in" step involved in formation of the open polymerase–promoter complex) are also found. A detailed discussion of promoter sequences, and the consequences of these sequences for promoter function, is found in Chapter 7 of this volume.

We note that the *lac* promoter (and other promoters) has been defined primarily as the base pair sequence specifically protected, by polymerase (holoenzyme) binding, from DNase digestion. It has not been established whether this protective binding involves the closed or the open promoter–polymerase binding complex. Thus, since polymerase could well translocate on the DNA between the formation of the closed and the open complex (indeed some evidence exists that it does so, at least in some cases) (see Gilbert, 1976), the identity of the *initial* polymerase-binding (recognition) site is not yet unequivocally established. And this ambiguity, together with the difficulty involved in isolating closed and open complexes in separable and stable forms, has made it hard to establish definitive binding constants for these two promoter–polymerase complexes.

The melting-in event (that is, the transition from closed to open complex) is also not completely understood. Thus, we do not know whether recognition is total at the closed complex level, or whether there is some specific single-strand sequence recognition involved in the formation of the open complex. The nature of the closed $\rightleftharpoons$ open interconversion of the promoter DNA, and whether or not polymerase plays an active part in it, has also not been established. In its ultimate preference for the formation of a polymerase–single strand DNA complex, the polymerase (holoenzyme) molecule obviously functions as a DNA melting (or "helix destablizing") protein. Studies on simple DNA melting proteins and melting protein models have shown that they generally trap the open form generated as a consequence of a local, thermally driven breathing fluctuation, rather than prebinding to the double-helical form and forcing open the structure (for a general discussion of the specificity of melting protein–nucleic acid interactions see von Hippel *et al.,* 1977). However, because a specific prebinding event to double-helical promoter DNA seems to be required as a prelude to the formation of the open polymerase–promoter complex, polymerase may be an exception to this general dictum.

4.5.2 *Integration of RNA Polymerase Binding with Control of the Lactose Operon*

The complexities outlined earlier make it difficult to integrate quantitatively the binding of RNA polymerase with the thermodynamics of the *lac* repressor–inducer–operator–DNA system, though some preliminary attempts to do so have been made (see, for example, von Hippel *et al.,* 1974, 1975).

Figure 14 shows the *lac* operator and promoter side by side. Sequencing of the base pairs protected by binding of repressor and polymerase, respectively, suggests that these sequences may actually overlap. If the initial binding sites for these two proteins do overlap, then, of course, the system must be modeled as one involving competitive binding of repressor and polymerase from free solution (or from nonspecific binding sites) to the same (or partially the same) DNA-binding site. Then, if the *extent* of repression is to be thermodynamically controlled, we cannot consider the repressor–inducer–operator–DNA equilibria in isolation, but must incorporate the competitive binding of polymerase to the operator site (or at least to a part of it), as well. In principle, this feature can easily be added to the model described in Section 4.2 (see von Hippel *et al.,* 1974, 1975), but the necessary thermodynamic parameters to describe the binding of polymerase are not yet available. However, because the polymerase might be translocated from the "real" recognition sequence to the base pair sequence involved in the protected complex identified by nuclease digestion experiments, we do not currently know whether the formation of the initial repressor–operator and polymerase–promoter complexes should be considered competitive or noncompetitive events.*

*The binding of repressor to operator is reversible, whereas in functional terms the binding of polymerase to promoter is not; that is, during polymerase function, the bound and melted-in polymerase "leaves" the promoter not by dissociation but by traveling down the operon in the elongation step of mRNA synthesis. This consideration will greatly reduce the effective lifetime of the polymerase–promoter complex relative to that of the repressor–operator complex, and will make polymerase a less effective competitive binder (relative to repressor) than it would be in the absence of mRNA synthesis. (I am grateful to Mike Chamberlin for pointing this out to me.)

In either case, the binding parameters characterizing the polymerase–genome interaction must be determined. In an important series of studies, Hinkle, Chamberlin, and co-workers (for a review see Chamberlin, 1974) and most recently Williams and Chamberlin (1977) and deHaseth *et al.* (1978) have made a start on such measurements. These workers have shown that holoenzyme and core enzyme both bind, with varying affinities, to both specific and nonspecific (promoter and nonpromoter) DNA. In the earlier papers of the Chamberlin group it was shown that at least at some ionic strengths core polymerase binds *more tightly* to the nonpromoter sites than does holoenzyme, and that holoenzyme formation (from core) might be crucial to functional promoter–polymerase complex formation because of nonspecific binding competition from the rest of the genome for the core enzyme moiety. The parallels between this system and the coupled *lac* repressor–inducer–operator–DNA equilibria were sufficiently striking to induce von Hippel *et al.* (1974, 1975) to formulate a comparable control model for the RNA polymerase system, using the best estimates then available for the various necessary binding parameters. This suggested that control of polymerase binding to promoter sites might involve a role for sigma factor comparable to the role of inducer in the *lac* system. In the case of polymerase, the binding of sigma factor to core enzyme would "induce" the enzyme away from nonspecific sites to make it available, as holoenzyme, in appropriate quantities to permit effective binding to promoter. The recent measurements of deHaseth *et al.* (1978) on the salt concentration dependence of the binding of core and holoenzyme to nonpromoter DNA suggest that these modeling attempts may have been quantitatively premature, although the qualitative principles on which these models were based may still apply. In addition, a fully quantitative coupled equilibrium model must also take into account the competitive effects of the binding of both forms of polymerase in the various other mRNA synthesis complexes simultaneously present in the *E. coli* cell. The tight binding of the various forms of RNA polymerase to nonpromoter DNA suggests that transport mechanisms for polymerase to and about the genome may have many parallels to those described for the *lac* repressor in Section 4.4.

4.5.3 The CAP–cAMP System

Catabolite gene activating protein (Zubay and Chambers, 1971; Perlman and Pastan, 1971; Majors, 1975) is a recently isolated and purified component of the lactose operon (as well as of the other *E. coli* operons involved with the metabolism of sugars). It appears to exist in solution and to function as a dimeric protein, of overall molecular weight of about 50,000. The dimer consists of two identical subunits, and thus is presumed to carry two binding sites for cAMP. As indicated in Sections 1 and 4, it functions physiologically in response to the solution concentration of cAMP which, in turn, depends on the level of the preferred metabolite, glucose, in the cell. When glucose concentration is low, the level of cAMP rises; this effector then binds to CAP, presumably triggering a conformational change to (or trapping) a form of the protein that has an enhanced affinity for a specific CAP-binding site in the control region of the *lac* operon (and of certain other operons). The CAP-binding site has been located in the *lac* operon just to the left (Fig. 14) of the RNA polymerase-binding site by a combination of genetic (mutation) and chemical (protection from nuclease digestion and chemical methylation) experiments (for a review see Gilbert, 1976).

The binding of the CAP–cAMP complex serves as a positive control element in the function of these "secondary" (to glucose) sugar control operons, apparently by stimulating polymerase activity. The mechanism of this stimulation is not known; however, the proximity of the CAP- and the RNA polymerase-binding sites suggests that the binding of the CAP–cAMP complex may alter the conformation or stability of the vicinal RNA polymerase site so as to facilitate the melting-in step (that is, conversion from a closed to an open complex) in the establishment of the active mRNA initiation system.

Clearly, the formation and binding to DNA of the CAP–cAMP complex involves still another set of coupled equilibria in the overall lactose control system. Since neither the binding constants nor the relevant intracellular concentrations of the various components of these equilibria are known, this portion of the overall control system also cannot be formulated quantitatively at present. However, the elements that must be considered (and the parameters that must be measured) for a minimal model are shown in a highly speculative, schematic form in Fig. 33. When these parameters have been determined for the CAP–cAMP system (and for the RNA polymerase system), a complete thermodynamic description of the *lac* operon should be possible.

5 *Extension to Other Transcription Regulatory Systems*

5.1 *Prokaryotic Control Systems*

The extension of such physicochemical considerations to other bacterial and viral transcriptional control systems is obvious (though not easy). For several operons (phage lambda; the *E. coli* arabinose, galactose, and tryptophan systems; the *Salmonella typhimurium* histidine system; and so on), some or all of the small molecule, protein, and DNA participants have been identified and partially characterized, though for none of these systems is the catalog of interdependent component concentrations and binding parameters so complete as for the *lac* operon. One of my motives in writing this chapter is to stimulate others to

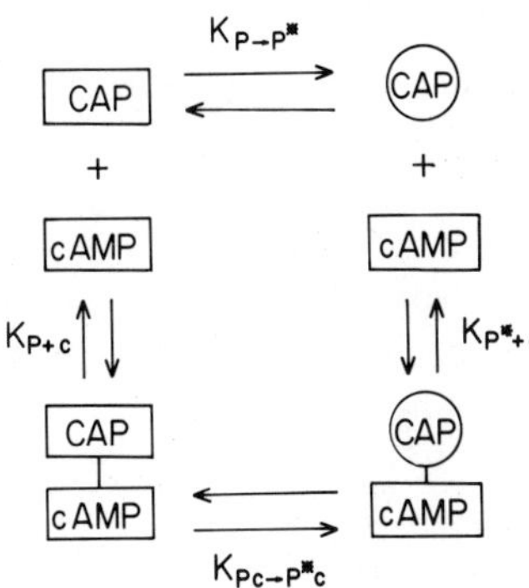

Figure 33. Speculative partial model for the cAMP–CAP interaction, showing a conformational equilibrium between two forms of free CAP, one which binds cAMP tightly (P*) and the other which binds cAMP weakly (P). The relevant equilibria and equilibrium constants are shown; a more complete model would also include the binding of all of these forms of both free and complexed CAP to the specific CAP-binding DNA base pair sequence, *and* to nonspecific DNA sequences.

participate actively in a systematic effort to determine the relevant parameters (measured under, or extrapolated to, physiological conditions) for enough of these operons to permit a physicochemical definition of several such control systems. Such quantitative definition is valuable not only for its own sake but because it can, in principle, reveal that our qualitative pictures of these systems are, in many cases, incomplete, or even internally incompatible. In addition, such thermodynamic formulations can lead to the elaboration of much more precise and definitive molecular questions than is possible at the descriptive (qualitative) level which characterizes our present understanding of most of these interlocking control systems.

Two final cautions should be added to these brave words:

1. We must not forget that life is, by definition, a dynamic process; therefore, in many cases elegantly described equilibria may not be attained. The measured dissociation rates of genome-regulatory proteins from their target DNA sites have been shown to be very dependent on ionic strength in many cases, and even at moderate ionic strengths one dissociation act may take longer than the entire generation time of the cell! Thus it is crucial that we determine and use *physiologically relevant* values of binding and rate constants, and that we be prepared to identify reactions that either do not go to completion within the biological time frame, or require special intracellular catalytic events such as special transport mechanisms (see Section 4.4) to permit equilibrium to be attained within the time available.

2. In addition, transcriptional control systems, as complex as they are, do not exist in isolation in the cell. Obviously, they do (and must) merge, and share control points, with systems regulating mRNA translation, as well as those regulating genome replication and repair. A complete physicochemical description of total cellular function, of the sort attempted here for a small portion of the *lac* operon, is not only beyond our grasp but is probably now even beyond our sight. However, we should not only hold such a description before us as an ultimate goal, but must also be prepared for the forced intrusion of other components of the cellular control apparatus into our present analyses. These may manifest themselves as an apparent inability to make a defined system behave in a thermodynamically closed manner (for example, the *lac* system in the absence of nonspecific DNA binding of repressor and repressor–inducer complexes), thus indicating the need to take into account previously unrecognized participants in the process at hand.

5.2 *Eukaryotic Control Systems*

Clearly, eukaryotic genome control systems are at present so incompletely understood that even to attempt a preliminary description in the terms outlined here would be, at best, premature and, at worst, misleading. Nevertheless, such an approach to eukaryotic systems will be both possible and useful when some of the fundamental molecular biological descriptions of these systems have achieved a more definitive status.

In closing, I would simply like to point out that at least one component of a putative eukaryotic genome-regulatory mechanism, namely, the steroid–receptor–DNA interaction, has many intriguing resemblances to the cAMP–CAP–DNA system. The steroid system features a receptor protein which, in the absence of hormonally active steroid, is largely free in the cytoplasm, but which migrates to the nucleus and shows a greatly increased affinity for chromosomal (DNA) binding sites when steroid hormone is bound. Various changes in the behavior of cellular metabolic pathways then follow. It is tempting to speculate that binding of the appropriate steroid "induces" ("traps"?) a protein conformation with increased affinity for specific DNA control sites. DNA-binding specificity of the receptor–steroid complex has not yet been demonstrated, but may also reflect the competitive participation of nonspecific DNA binding as a regulatory mechanism (see von Hippel *et al.,* 1974; Lin and Riggs, 1975*b;* Yamamoto and Alberts, 1975). The available facts about the steroid–receptor–DNA system, and how these interactions might regulate genetic expression, have recently been reviewed by Gorski and Gannon (1976). Also, Yamamoto and Alberts (1976) have surveyed some of the features of steroid receptors as elements of modulation of eukaryotic transcription, and have pointed out analogies with prokaryotic systems of gene control. While much remains to be done before we can know whether these resemblances are fundamental or merely superficial, the prospects, at least, are intriguing.

Acknowledgments

I am pleased to acknowledge that the perspectives on nucleic acid–protein interactions which underlie this chapter have been developed over a number of years, partly as an outcome of the experimental work of, and partly as a consequence of discussions with, my colleagues in our laboratory in the Institute of Molecular Biology. The magnitude and nature of their various contributions are indicated throughout the text in the form of references to the specific studies in which they have participated. I am also pleased to acknowledge that a major part of the support for the research work of our laboratory has been derived from USPHS Research Grant GM-15792.

References

Adler, H. I., Fisher, W. D., Cohen, A., and Hardigree, A. A., 1967, Miniature *Escherichia coli* cells deficient in DNA, *Proc. Natl. Acad. Sci. U.S.A.* **57**:321.

Anderson, R. A., and Coleman, J. E., 1975, Physicochemical properties of DNA binding proteins: Gene 32 protein of T4 and *Escherichia coli* unwinding protein, *Biochemistry* **14**:5485.

Arnott, S., Dover, S. D., and Wonacott, A. J., 1969, Least-squares refinement of the crystal and molecular structures of DNA and RNA from X-ray data and standard bond lengths and angles. *Acta Crystallogr.* **B25**:2192.

Arnott, S., Chandrasekaran, R., and Selsing, E., 1975, The variety of polynucleotide helices, in: *Structure and Conformation of Nucleic Acid and Protein–Nucleic Acid Interactions* (M. Sundaralingam and S. T. Rao, eds.), pp. 577–596, University Park Press, Baltimore, Maryland.

Bahl, C. P., Wu, R., Stawinsky, J., and Narang, S., 1977, Minimal length of the lactose operator sequence for the specific recognition by the lactose repressor, *Proc. Natl. Acad. Sci. U.S.A.* **74**:966.

Barkley, M. D., Riggs, A. D., Jobe, A., and Bourgeois, S., 1975, Interaction of effecting ligands with *lac* repressor and repressor–operator complex, *Biochemistry* **14**:1700.

Berg, O. G., and Blomberg, C., 1976, Association kinetics with coupled diffusional flows. Special application to the *lac* repressor–operator system, *Biophys. Chem.* **4**:367.

Burgeois, S., and Jobe, A., 1970, Superrepressors of the *lac* operon, in: *The Lactose Operon* (J. R. Beckwith and D. Zipsen, eds.), pp. 325–341, Cold Spring Harbor Laboratory, Cold Spring Harbor, New York.

Bourgeois, S., and Pfahl, M., 1976, Repressors, *Adv. Protein Chem.* **30**:1.

Britten, R. J., and Kohne, D. E., 1968, Repeated sequences in DNA, *Science* **161**:529.

Butler, A. P., 1976, Physical and Chemical Characterization of the Lactose Repressor, Ph.D. Thesis, University of Oregon, Eugene, Oregon.

Butler, A. P., Revzin, A., and von Hippel, P. H., 1977, Molecular parameters characterizing the interaction of *Escherichia coli lac* repressor with non-operator DNA and inducer, *Biochemistry* **16**:4757.

Chamberlin, M. J., 1974, The selectivity of transcription, *Annu. Rev. Biochem.* **43**:721.

Chamberlin, M. J., 1976*a,* RNA polymerase—An overview, in: *RNA Polymerase* (R. Losick and M. Chamberlin, eds.), pp. 17–67, Cold Spring Harbor Lab., Cold Spring Harbor, New York.

Chamberlin, M. J., 1976*b,* Interaction of RNA polymerase with the DNA template, in: *RNA Polymerase* (R. Losick and M. Chamberlin, eds.), pp. 159–191, Cold Spring Harbor Lab., Cold Spring Harbor, New York.

Chamness, G. C., and Willson, C. D., 1970, An unusual *lac* repressor mutant, *J. Mol. Biol.* **53**:561.

Coleman, J. E., Anderson, R. A., Ratcliffe, R. G., and Armitage, I. M., 1976, Structure of gene 5 protein-oligodeoxynucleotide complexes as determined by ^{1}H, ^{19}F, and ^{31}P nuclear magnetic resonance, *Biochemistry* **15**:5419.

deHaseth, P. L., Lohman, T. M., and Record, M. T., Jr., 1977, Nonspecific interaction of *lac* repressor with DNA: An association reaction driven by counterion release, *Biochemistry* **16**:4783.

deHaseth, P. L., Lohman, T. M., Record, M. T., Jr., and Burgess, R. R., 1978, Interactions of *E. coli* RNA polymerase with native and denatured DNA: Differences in the binding behavior of core and holoenzyme, *Biochemistry* **17**:1612.

Draper, D., Pratt, C., and von Hippel, P. H., 1977, *E. coli* ribosomal protein Sl has two polynucleotide binding sites, *Proc. Natl. Acad. Sci. U.S.A.* **74**:4786.

Draper, D., and von Hippel, P. H., 1978, Nucleic acid binding properties of *E. coli* ribosomal protein Sl. II. Cooperativity and specificity of binding site II, *J. Mol. Biol.* **122**:321.

Fasman, G. D., ed., 1976, *CRC Handbook of Biochemistry and Molecular Biology,* 3rd ed., Section B: *Nucleic Acids,* Vol. II, CRC Press, Cleveland, Ohio.

Friedman, B. E., Olson, J. S., and Matthews, K. S., 1977, Interaction of *lac* repressor with inducer. Kinetic and equilibrium measurements, *J. Mol. Biol.* **111**:27.

Gierer, A., 1966, Model for DNA and protein interactions and the function of the operator, *Nature* **212**:1480.

Gilbert, W., 1976, Starting and stopping sequences for the RNA polymerase, in: *RNA Polymerase* (R. Losick and M. Chamberlin, eds.), pp. 193–205, Cold Spring Lab., Cold Spring Harbor, New York.

Gilbert, W., Gralla, J., Majors, J., and Maxam, A., 1975, Lactose operator sequences and the action of *lac* repressors, in: *Protein–Ligand Interactions* (H. Sund and G. Blauer, eds.), pp. 193–210, de Gruyter, Berlin.

Gilbert, W., Maxam, A., and Mirzabekov, A., 1976*a,* Contacts between the *lac* repressor and DNA revealed by methylation, in: *Control of Ribosome Synthesis,* Alfred Benzon Symp. IX, pp. 139–148, Munksgaard, Copenhagen.

Gilbert, W., Majors, J., and Maxam, A. M., 1976*b,* How proteins recognize DNA sequences, in: *Organization and Expression of Chromosomes,* Life Sciences Research Report 4, pp. 167–178, Heyden and Son, London.

Gnedenko, B. V., and Khinchin, A. Ya., 1962, *An Elementary Introduction to the Theory of Probability* [translated from the 5th (Russian) edition by L. F. Boron], Dover, New York.

Goeddel, D. V., Yansura, D. G., and Caruthers, M. H. 1977, Binding of synthetic lactose operator DNAs to lactose repressors, *Proc. Natl. Acad. Sci. U.S.A.* **74**:3292.

Goeddel, D. V., Yansura, D. G., and Caruthers, M. H., 1978, How *lac* repressor recognizes *lac* operator, *Proc. Natl. Acad. Sci. U.S.A.* (in press).

Gorski, J., and Gannon, F., 1976, Current models of steroid hormone action: A critique, *Annu. Rev. Physiol.* **38**:425.

Hélène, C., 1977, La reconnaissance sélective des acides nucleiques par les proteins, *La Recherche* **8**: 122.

Hirsh, J., and Schleif, R., 1976, High resolution electron microscopic studies of genetic regulation, *J. Mol. Biol.* **108**:471.

Jacob, F., and Monod, J., 1961, Genetic regulatory mechanisms in the synthesis of proteins, *J. Mol. Biol.* **3**:318.

Jensen, D. E., and von Hippel, P. H., 1977, A boundary sedimentation velocity method for determining nonspecific nucleic acid–protein interaction binding parameters, *Anal. Biochem.* **80**:267.

Jobe, A., Sadler, J. R., and Bourgeois, S., 1974, *Lac* repressor–operator interaction. IX. The binding of *lac* repressor to operators containing O^c mutations. *J. Mol. Biol.* **85**:231.

Kao-Huang, Y., Revzin, A., Butler, A. P., O'Conner, P., Noble, D. W., and von Hippel, P. H., 1977, Nonspecific DNA binding of genome-regulating proteins as a biological control mechanism: Measurement of DNA-bound *Escherichia coli lac* repressor *in vivo, Proc. Natl. Acad. Sci. U.S.A.* **74**:4228.

Kauzmann, W., 1959, Some factors in the interpretation of protein denaturation, *Adv. Protein Chem.* **14**:1.

Kelly, R. C., and von Hippel, P. H., 1976, DNA "melting" proteins. III. Fluorescence "mapping" of the nucleic acid binding site of bacteriophage T4 gene 32-protein, *J. Biol. Chem.* **251**:7229.

Kelly, R. C., Jensen, D. E., and von Hippel, P. H., 1976, DNA "melting" proteins. IV. Fluorescence measurements of binding parameters for bacteriophage T4 gene 32-protein to mono-, oligo-, and polynucleotides, *J. Biol. Chem.* **251**:7240.

Laiken, S. L., Gross, C. A., and von Hippel, P. H., 1972, Equilibrium and kinetic studies of *Escherichia coli lac* repressor–inducer interactions, *J. Mol. Biol.* **66**:143.

L'Engle, M., 1962, *A Wrinkle in Time,* Dell, New York.

Lin, S.-Y., and Riggs, A. D., 1975*a,* A comparison of *lac* repressor binding to operator and to nonoperator DNA, *Biochem. Biophys. Res. Commun.* **62**:704.

Lin, S.-Y., and Riggs, A. D., 1975*b,* The general affinity of *lac* repressor for *E. coli* DNA: Implications for gene regulation in prokaryotes and eukaryotes, *Cell* **4**:107.

Losick, R., and Chamberlin, M., eds., 1976, *RNA Polymerase,* Cold Spring Harbor Lab., Cold Spring Harbor, New York.

Majors, J., 1975, Specific binding of CAP factor to *lac* promoter DNA, *Nature* **256**:672.

McGhee, J. D., 1976, Theoretical calculations of the helix-coil transition of DNA in the presence of large, cooperatively binding ligands, *Biopolymers* **15**:1345.

McGhee, J. D., and von Hippel, P. H. 1974, Theoretical aspects of DNA–protein interactions: Co-operative and non-co-operative binding of large ligands to a one-dimensional homogeneous lattice, *J. Mol. Biol.* **86**:469.

Melchior, W. B., Jr., and von Hippel, P. H., 1973, Alteration of the relative stability of dA·dT and dG·dC base pairs in DNA, *Proc. Natl. Acad. Sci. U.S.A.* **70**:298.

Miller, J. H., Coulondre, C., Schmeissner, U., Schmitz, A., and Lu, P., 1975. The use of suppressed nonsense mutations to generate altered lac repressor molecules, in: *Protein–Ligand Interactions* (H. Sund and G. Blauer, eds.), pp. 238–252, de Gruyter, Berlin.

Müller-Hill, B., 1975, Lac repressor and lac operator, *Prog. Biophys. Mol. Biol.* **30**:227.

Ogata, R., and Gilbert, W., 1977, Contacts between the *lac* repressor and thymines in the *lac* operator, *Proc. Natl. Acad. Sci. U.S.A.* **74**:4973.

Perlman, R. L., and Pastan, I., 1971, The role of cyclic AMP in bacteria, *Curr. Top. Cell. Regul.* **3**:117.

Pfahl, M., 1976, *Lac* repressor–operator interaction. Analysis of the X86 repressor mutant, *J. Mol. Biol.* **106**:857.

Record, M. T., Jr., Lohman, T. M., and deHaseth, P. L., 1976, Ion effects on ligand–nucleic acid interactions, *J. Mol. Biol.* **107**:145.

Record, M. T., Jr., deHaseth, P. L., and Lohman, T. M., 1977, Interpretation of monovalent and divalent cation effects on the *lac* repressor–operator interaction, *Biochemistry* **16**:4791.

Revzin, A., and von Hippel, P. H., 1977, Direct measurement of association constants for the binding of *Escherichia coli lac* repressor to non-operator DNA, *Biochemistry* **16**:4769.

Richter, P. H., and Eigen, M., 1974, Diffusion-controlled reaction rates in spheroidal geometry. Application to repressor–operator association and membrane bound enzymes, *Biophys. Chem.* **2**:255.

Riggs, A. D., Suzuki, H., and Bourgeois, S., 1970*a, Lac* repressor–operator interaction. I. Equilibrium studies, *J. Mol. Biol.* **48**:67.

Riggs, A. D., Bourgeois, S., and Cohn, M., 1970*b*, The *lac* repressor–operator interaction. III. Kinetic studies, *J. Mol. Biol.* **53**:401.

Roberts, J. W., 1976*a*, Transcription termination and its control in *E. coli*, in: *RNA Polymerase* (R. Losick and M. Chamberlin, eds.), pp. 247–271, Cold Spring Harbor Lab., Cold Spring Harbor, New York.

Roberts, R. J., 1976*b*, Restriction endonucleases, *CRC Crit. Rev. Biochem.* **4**:123.

Sadler, J. R., and Novick, A., 1965, The properties of repressor and the kinetics of its action, *J. Mol. Biol.* **12**:305.

Sadler, J. R., and Smith, T. F., 1971, Mapping of the lactose operator, *J. Mol. Biol.* **62**:139.

Schellman, J. A., 1974, Cooperative multisite binding to DNA, *Isr. J. Chem.* **12**:219.

Schellman, J. A., 1975, Macromolecular binding, *Biopolymers* **14**:999.

Seeman, N. C., Rosenberg, J. M., and Rich, A., 1976, Sequence-specific recognition of double helical nucleic acids by proteins, *Proc. Natl. Acad. Sci. U.S.A.* **73**:804.

Smith, T. F., and Sadler, J. R., 1971, The nature of lactose operator constitutive mutations, *J. Mol. Biol.* **59**:273.

Sobell, H. M., 1973*a*, Symmetry in protein–nucleic acid interaction and its genetic implications, *Adv. Genet.* **17**:411.

Sobell, H. M., 1973*b*, The stereochemistry of actinomycin binding to DNA and its implications in molecular biology, *Prog. Nucleic Acid Res. Mol. Biol.* **13**:153.

Sommer, H., Lu, P., and Miller, J. H., 1976, *Lac* repressor. Fluorescence of the two tryptophans, *J. Biol. Chem.* **251**:3774.

Sundaralingam, M., 1975, Principles governing nucleic acid and polynucleotide conformations, in: *Structure and Conformation of Nucleic Acids and Protein–Nucleic Acid Interactions* (M. Sundaralingam and S. T. Rao, eds.), pp. 487–524, University Park Press, Baltimore, Maryland.

Tanford, C., 1962, Contributions of hydrophobic interactions to the stability of the globular conformation of proteins, *J. Am. Chem. Soc.* **84**:4240.

Thomas, C. A., Jr., 1966, Recombination of DNA molecules, *Prog. Nucleic Acid Res. Mol. Biol.* **5**:315.

Toulmé, J.-J., and Hélène, C., 1977, Specific recognition of single-stranded nucleic acids, *J. Biol. Chem.* **252**:244.

von Hippel, P. H., and McGhee, J. D., 1972, DNA–protein interactions, *Annu. Rev. Biochem.* **41**:231.

von Hippel, P. H., Revzin, A., Gross, C. A., and Wang, A. C., 1974, Non-specific DNA binding of genome regulating proteins as a biological control mechanism: I. The *lac* operon: Equilibrium aspects, *Proc. Natl. Acad. Sci. U.S.A.* **71**:4808.

von Hippel, P. H., Revzin, A., Gross, C. A., and Wang, A. C., 1975, Interaction of lac repressor with non-specific DNA binding sites, in: *Protein–Ligand Interactions* (H. Sund and G. Blauer, eds.), pp. 270–288, de Gruyter, Berlin.

von Hippel, P. H., Jensen, D. E., Kelly, R. C., and McGhee, J. D., 1977, Molecular approaches to the interaction of nucleic acids with "melting" proteins, in: *Nucleic Acid–Protein Recognition* (H. J. Vogel, ed.), pp. 65–89, Academic Press, New York.

Wang, A. C., Revzin, A., Butler, A. P., and von Hippel, P. H., 1977, Binding of *E. coli lac* repressor to non-operator DNA, *Nucleic Acid Res. (Vinograd Memorial Issue)* **4**:1579.

Wang, J., 1974, Interactions between twisted DNAs and enzymes: The effects of superhelical turns, *J. Mol. Biol.* **87**:797.

Wang, J., Barkley, M. D., and Bourgeois, S., 1974, Measurements of unwinding of *lac* operator by repressor, *Nature* **251**:247.

Wartell, R. M., 1977, The transmission of stability or instability from site-specific protein–DNA complexes, *Nucleic Acid Res.* **4**:2779.

Weber, K., Files, J. G., Platt, T., Ganem, D., and Miller, J. H., 1975, Lac repressor, in: *Protein–Ligand Interactions* (H. Sund and G. Blauer, eds.), pp. 228–237, de Gruyter, Berlin.

Williams, R. C., and Chamberlin, M. J., 1977, Electron microscope studies of transient complexes formed between *Escherichia coli* RNA polymerase holoenzyme and T7 DNA, *Proc. Natl. Acad. Sci. U.S.A.* **74**:3740.

Wingert, L., and von Hippel, P. H., 1968, The conformation-dependent hydrolysis of DNA by micrococcal nuclease, *Biochim. Biophys. Acta* **157**:114.

Wu, F. Y.-H., Bandyopadhyay, P., and Wu, C.-W., 1976, Conformational transitions of the *lac* repressor from *Escherichia coli*, *J. Mol. Biol.* **100**:459.

Yagil, G., and Yagil, E., 1971, On the relation between effector concentration and the rate of induced enzyme synthesis, *Biophys. J.* **11**:11.

Yamamoto, K. R., and Alberts, B. M., 1975, The interaction of estradiol-receptor protein with the genome: An argument for the existence of undetected specific sites, *Cell* **4**:301.

Yamamoto, K. R., and Alberts, B. M., 1976, Steroid receptors: Elements for modulation of eukaryotic transcription, *Annu. Rev. Biochem.* **45**:721.

Yarus, M., 1969, Recognition of nucleotide sequences, *Annu. Rev. Biochem.* **38**:841.

Zingsheim, H. P., Geisler, N., Mayer, F., and Weber, K., 1977, Complexes of *E. coli lac* repressor with non-operator DNA revealed by electron microscopy: Two repressor molecules can share the same segment of DNA, *J. Mol. Biol.* **115**:565.

Zubay, G., and Chambers, D. A., 1971, Regulating the *lac* operon, in: *Metabolic Pathways,* Vol. 5 (H. J. Vogel, ed.), pp. 297–347, Academic Press, New York.

9

Genetic Signals and Nucleotide Sequences in Messenger RNA

JOAN ARGETSINGER STEITZ

1 Introduction

Today every schoolchild learns that DNA makes RNA and RNA makes protein. He is further taught that a messenger RNA molecule is like a string of beads consisting of A, C, G, and U residues. The genetic information is read out (like any written code) in a linear fashion, here in groups of three. Such imagery immediately engenders the notion that RNA functions purely as a one-dimensional structure and that all of its secrets are locked in its sequence of bases. Of course, the latter concept is, in essence, true. Just as the amino acid sequence of a polypeptide chain ultimately determines the three-dimensional folding of a protein, so does the sequence of nucleotides in an RNA molecule dictate its potential for forming internal Watson–Crick base pairs (RNA secondary structure) which may further interact specifically with each other (RNA tertiary structure). Indeed, it has long been accepted that tRNA molecules utilize both base pairing and intramolecular bonds of other types to assume a quite defined shape which allows their specific recognition by ribosomes, aminoacyl-tRNA synthetases, and so on. However, it is a more recent realization that messenger RNAs and ribosomal RNAs can form thermodynamically stable secondary and tertiary structures as well.

JOAN ARGETSINGER STEITZ • Department of Molecular Biophysics and Biochemistry, Yale University, New Haven, Connecticut

When we now consider exactly how a messenger RNA might interact with a multitude of other macromolecules during the translation process, we see that either its linear sequence of nucleotides or its preferred three-dimensional conformation could serve as a recognition element. Certainly, when interacting with another RNA molecule, an mRNA can most conveniently utilize the simple Watson–Crick base pairing rules. Here, the codon·anticodon interaction, by which the mRNA aligns successive tRNAs on the ribosome during polypeptide chain growth, is an obvious example. Similarly, very recent results (to be discussed later) strongly suggest that additional direct base pairing interactions are important to translation. By contrast, a protein which binds RNA can design its active site to select out a sequence of bases and/or a specific three-dimensional array of functional groups. Hence, what a protein recognizes in an mRNA molecule may not be obvious from knowing just the nucleotide sequence of the bound region. Ultimately, it must be remembered that the ribosome is a large and complex organelle containing both RNA and protein molecules. Its contacts with an mRNA therefore very likely occur at numerous sites, each of which will make a partial contribution to the overall binding energy required for any particular step of the translation process.

My objective here is to discuss what we currently know (or think we know) about the nature and functioning of genetic signals in messenger RNA molecules. Because our understanding is derived primarily from work with bacteria and phages, I shall concentrate on prokaryotic mRNA sequences and their interactions with various cellular components known to contribute to the specificity of translation. The mechanism and control of mRNA recognition by ribosomes during polypeptide chain initiation are considered in some detail. Likewise, recent ideas, which are derived from the complete sequence analysis of several bacteriophage genomes and concern the selectivity of codon usage during polypeptide chain elongation, are presented. The role of mRNA sequences in suppression and termination events is the subject of another chapter in this volume and is not discussed here. Finally, I review currently perceived differences in the structures and functioning of mRNAs in prokaryotes versus eukaryotes, where our knowledge is only beginning to penetrate to the level of specific molecular interactions.

2 *Ribosome Recognition of Initiation Signals*

2.1 *A Bit of History*

The course of our thinking about how ribosomes correctly recognize initiation signals in messenger RNA can be traced through rather discrete stages. Early work on the genetic code revealed that AUG, GUG, and perhaps several other triplets could direct the binding of *N*-formylmethionyl transfer RNA_f^{Met}, which initiates polypeptide chains in prokaryotes, to *Escherichia coli* ribosomes. Also in the middle 1960s (for a review of all this early material see Lucas-Lenard and Lipmann, 1971) it became clear that GTP, the 30 S ribosomal subunit, and at least three proteins found in the ribosomal wash—the initiation factors IF-1, IF-2, and IF-3—were required for the first steps of the initiation process on natural messenger RNAs. Synthetic polyribonucleotides having an AUG or GUG codon at their 5′ termini could likewise direct the synthesis of polypeptides initiated with *N*-formylmethionine. This led to the idea that ribosomes start protein synthesis directly at the 5′ end of an mRNA molecule with no special signal other than the presence of an initiator triplet.

However, the first nucleotide sequence data on the RNA bacteriophages and several plant viruses revealed that these naturally occurring mRNA molecules do not begin with AUG or GUG (see Dahlberg, 1968). Similarly, the possibility that ribosomes thread on to the message and slide along to the first AUG or GUG triplet was ruled out by experiments showing that *E. coli* ribosomes can initiate polypeptide chains very efficiently on a circular messenger molecule (Bretscher, 1969). Thus, ribosomes must be capable of selecting out and binding to true initiation signals internally on a messenger RNA.

Several groups (Steitz, 1969; Hindley and Staples, 1969; Gupta *et al.*, 1970) in the late 1960s therefore attempted the isolation and sequence analysis of ribosome-binding sites (initiator regions) from the single-stranded genomes of RNA bacteriophages. The hope here was that knowledge of the nucleotide sequence surrounding initiator triplets would reveal some obvious and unusual feature, which would say to us (as well as to the ribosome) "I am that special something which distinguishes true initiator triplets from the many in-phase and out-of-phase internal AUGs and GUGs in every messenger RNA." Procedurally, ribosomes were bound to ^{32}P-labeled phage RNA under conditions of polypeptide chain initiation *in vitro,* the resulting complexes treated with ribonuclease to trim away portions of the message not associated with the ribosome, and the protected RNA pieces sequenced utilizing the newly developed methods of Sanger (see Barrell, 1971). However exciting the first glimpses at initiation sites proved to be, their sequences did not quite live up to expectation. We did learn (1) that sites protected by ribosomes under such conditions are exclusively at the beginnings of genes; (2) that there exist intercistronic spaces in the RNA phage genomes; (3) that ribosomes can bind (even in the absence of translation) to internal initiator regions on a polycistronic mRNA; and (4) that previous assignments for the genetic code seemed to be correct. But the much sought-after special sequence in initiator regions was simply not apparent.

During the period (1969–1974) when sequences of RNA bacteriophage initiator regions and of many of the other ribosome-binding sites listed in Table I were being accumulated, it was generally believed that ribosome recognition of initiator regions was an example of a specific protein–nucleic acid interaction. One or several of the ribosomal proteins, or of the initiation factors, were presumed to bind selectively to the as yet unidentified important sequence, which could exist either in single-stranded form or folded into a particular RNA secondary structure. This dogma dominated thinking about the problem until 1974, when two Australian scientists, Shine and Dalgarno (1974), reported a correct sequence for the 3′ terminal dodecanucleotide of *E. coli* 16 S ribosomal RNA; these authors suggested that direct base pairing between this region and a sequence 5′ to an initiator triplet in the message might be instrumental in ribosome recognition of a true initiator region. Shine and Dalgarno's insight provided an impetus for renewed interest and experimentation. Our molecular understanding of the details of ribosome–mRNA interaction has therefore taken a giant step forward in the past few years.

2.2 A Look at Today's Catalog of Ribosome-Binding Sites

Table I presents a complete list of currently known sequences for ribosome-binding sites from coliphage and *E. coli* messenger RNAs. Each of these has been identified as a start site for protein synthesis either by ribosome protection after initiation complex formation *in vitro* or by comparison with the known N-terminal amino acid sequence of the protein specified. To my knowledge, so far a perfect correlation has been observed; all sites isolated from native mRNAs using *E. coli* ribosomes have proven to be initiators for translation. Additional sequences which are highly likely to be initiation sites (particularly in the case of the single-stranded DNA phage, ϕX174) but have not yet satisfied one or both of these criteria are not included. Sequences are given as RNA or DNA, according to the form in which they were first analyzed. Only in the case of the ϕX174 gene G initiator region has the single-stranded DNA been shown to function as a ribosome-binding site.

The initiator triplets, which appear near the middle of the approximately 30-nucleotide-long ribosome-binding sites, are indicated by boldface type in Table I. Of the possible initiator codons, AUG clearly predominates. GUG is utilized by wild type messengers only in the case of the MS2 A protein and the *E. coli* lactose repressor protein *(lacI)*.

Most of the protected regions recovered from pancreatic ribonuclease-treated initiation complexes contain five or six triplets 5′ to the initiator codon and extend about three triplets beyond the initiator into the coding region on the 3′ side. Although variations in mRNA sequence and in digestion conditions can obviously alter such results, these values probably represent the length of mRNA which makes close contact with the ribosome. Thus, it is interesting that six triplets is also the shortest observed distance between the 5′ terminus of an mRNA and a known initiator triplet (the λp_R mRNA, Table I). The possible exception here is a short form of the λc_I mRNA, which is considered later.

Normally, even the first cistron of a polycistronic mRNA is located some distance from the 5′ end of the mRNA. An extreme example is the A protein gene in the R17 and MS2 bacteriophage genomes, which starts at nucleotide 129; 26 or more residues precede the initiator triplet in all *E. coli* mRNAs examined so far. The length of intercistronic regions (the spaces between genes on a polycistronic mRNA) is also highly variable. It can be as long as several hundred nucleotides, or so short that the initiator triplet and the terminator codon for the previous cistron overlap. Functions for lengthy extracistronic regions in mRNAs and the phenomenon of gene overlap are discussed later.

One revealing feature of ribosome-binding sites is that about a third of them contain more than one AUG or GUG sequence. Since the additional triplet(s) can appear either 5′ or 3′ to the real initiator codon, it seems highly unlikely that the ribosome first recognizes some distant site on the mRNA and then slides (forward or backward) to the next AUG. However, within the binding site regions there also appears no strictly conserved sequence homology to which the ribosome might directly respond.

One early hypothesis was that all initiation sites might be capable of folding into a common secondary structure which signals ribosome recognition. Hairpin loops with slightly negative free energy of formation can indeed be formed by several of the sequences listed in Table I (see Steitz, 1975), but this now turns out to be the exception rather than the rule. Moreover, even when loops can be drawn, they are predicted to be in an open form a significant fraction of the time under physiological conditions (Gralla and Crothers, 1973).* Our current understanding suggests that there is good reason for the initiator triplet and the sequence preceding it to remain extended and available; as we will see, these must be free for interaction with other components during the initiation event on the ribosome.

The single feature common to all initiation sequences, as first pointed out by Shine and Dalgarno (1974), is indicated by underlining in Table I. This is a region complementary to some portion of the 3′ end sequence of 16 S ribosomal RNA (also given in Table I). The length of the complementarity varies between three and nine nucleotides in the examples shown in Table I, the average being about 4.5. The number of nucleotides separating the complementary region from the initiator triplet also varies; here the average is about ten nucleotides to the middle of the potentially base-paired stretch. All of the underlined mRNA regions are rich in purine residues; at least two (and usually three or four) residues from the CCUCC sequence near the 3′ end of the 16 S molecule participate in the predicted base pairing. It is relevant to mention here that adjacent G·C base pairs vastly increase the stability of a double helical region relative to the contribution of A·U/A·U or mixed G·C/A·U stacks (Gralla and Crothers, 1973). Finally, it should be pointed out that in every case in which two or more initiator triplets appear in a single initiator region, the one preceded by the strongest and most appropriately positioned complementarity (based on the preceding average value) is the one that functions in polypeptide chain initiation.

2.3 *mRNA and rRNA Pair during Initiation*

In addition to the appearance of a sequence complementary to 16 S rRNA in each known ribosome-binding site, other lines of evidence made the idea of mRNA·rRNA base pairing during initiation seem attractive. First, chemical cross-linking studies (Kenner, 1973; Hawley *et al.,* 1974; Bollen *et al.,* 1975; Czernilofsky *et al.,* 1975; van Duin *et al.,* 1975, 1976; Heimark *et al.,* 1976) had located the 3′ end of 16 S rRNA, the binding sites for initiation factors, and certain ribosomal proteins implicated in initiation (Held *et al.,* 1974; van Duin and van Knippenberg, 1974; van Dieijen *et al.,* 1975; Szer and Leffler, 1974; Dahlberg, 1974; Dahlberg and Dahlberg, 1975; Fiser *et al.,* 1975) in the same neighborhood on the 30 S ribosome. Recent data place the 16 S 3′ terminus on the interface between the two ribosomal subunits, exactly where the mRNA is presumed to bind (Chapman and Noller, 1977; Santer and Shane, 1977). Second, several antibiotic inhibitors of initiation also seem to act near the 16 S 3′ end (Dahlberg *et al.,* 1973; Helser *et al.,* 1971). Third, random copolymers that are rich in A and G had been found to be the best competitive inhibitors of initiation on natural mRNAs (Revel and Greenshpan, 1970), underscoring the importance of polypurines in ribosome

*Throughout this work the rules developed by Gralla and Crothers (1973) have been utilized to assess the stability of RNA secondary structures, since T_m values calculated in this way have subsequently proven to agree better with experimentally determined values than have T_m values calculated by other published methods.

Table I. Initiation Sequences Recognized by E. coli Ribosomes[a]

mRNA	Ribosome-binding site	References
R17 A	GAU UCC UAG GAG GUU UGA CCU **AUG** CGA GCU UUU AGU G	Steitz, 1969
MS2 A	GAU UCC UAG GAG GUU UGA CCU **GUG** CGA GCU UUU AGU G	Fiers *et al.*, 1975
Qβ A	UCA CUG AGU AUA AGA GGA CAU **AUG** CCU AAA UUA CCG CGU	Staples *et al.*, 1971
R17 coat	CC UCA ACC GGG GUU UGA AGC **AUG** GCU UCU AAC UUU	Steitz, 1969
Qβ coat	AAA CUU UGG GUC AAU UUG AUC **AUG** GCA AAA UUA GAG ACU	Hindley and Staples, 1969; Steitz, 1972
f2, MS2 coat	CC UCA ACC GAG GUU UGA AGC **AUG** GCU UCC AAC UUU ACU	Gupta *et al.*, 1970[c]; Min Jou *et al.*, 1972
R17, MS2 replicase	AA ACA UGA GGA UUA CCC **AUG** UCG AAG ACA ACA AAG	Steitz, 1969; Min Jou *et al.*, 1972
Qβ replicase	AG UAA CUA AGG AUG AAA UGC **AUG** UCU AAG ACA G	Staples and Hindley, 1971; Steitz, 1972
f1 coat	UUU AAU GGA AAC UUC CUC **AUG** AAA AAG UCU UU	Pieczenik *et al.*, 1974
f1 gene 5	A AGG UAA UUC ACA **AUG** AUU AAA GUU GAA AU	Pieczenik *et al.*, 1974
fl gene ?	A AAA AAG GUA AUU CAA **AUG** AAA UU	Pieczenik *et al.*, 1974
T3 *in vitro*[b]	AAC AUG AGG UAA CAC CAA **AUG** AUU UUC ACU AAA GAG	Arrand and Hindley, 1973
T7 gene *0.3* site a	AAC UGC ACG AGG UAA CAC AAG **AUG** GCU AUG UCU AAC AUG	Steitz and Bryan, 1977
T7 gene *0.3* site b	GUA CGA GGA GGA UGA AGA GUA **AUG** U	Steitz and Bryan, 1977[c]
λp_R	ppp AUG UAC UAA GGA GGU UGU **AUG** GAA CAA CGC	Steege, 1977*a*
λc_I	TTG CGG TGA TAG ATT TAA CGT **ATG** AGC ACA AAA AAG	Ptashne *et al.*, 1976; Walz *et al.*, 1976

ϕX174 G	TTT CTG CTT AGG AGT TTA ATC **ATG** TTT CAG ACT TTT ATT	Robertson *et al.*, 1973
ϕX174 F	CCT ACT TGA GGA UAA AUU **AUG** UCU AAU AUU CAA ACU	Ravetch *et al.*, 1977; Sanger *et al.*, 1977
ϕX174 D	ACC ACT AAT AGG TAA GAA ATC **ATG** AGT CAA GTT ACT	Barrell *et al.*, 1976
ϕX174 H	ACT TAA GTG AGG TGA TTT **ATG** TTT GGT GCT ATT	Sanger *et al.*, 1977
ϕX174 B	UAA AGG UCC AGG AGC UAA AGA **AUG** GAA CAA CUC ACU	Ravetch *et al.*, 1977
ϕX174 J	ACG TGC GGA AGG AGT GAT GTA **ATG** TCT AAA GGT AAA	Barrell *et al.*, 1976
araB	UUU UUU GGA UGG·AGU GAA ACG **AUG** GCG AUU GCA AUU	Lee and Carbon, 1977
trp leader	CAC GUA AAA AGG GUA UCG ACA **AUG** AAA GCA AUU UUC GUG	Bertrand *et al.*, 1975
trpE	GAA CAA AAU UAG AGA AUA ACA **AUG** CAA AGA CAA AAA CCG	Bertrand *et al.*, 1975
trpA	GAA AGC ACG AGG GGA AAU CUG **AUG** GAA CGC UAC GAA UCU	Platt and Yanofsky, 1975
lacZ	AAU UUC ACA CAG GAA ACA GCU **AUG** ACC AUG AUU ACG GAU	Maizels, 1974
lacI	AGU CAA UUC AGG GUG GUG AAU **GUG** AAA CCA GUA ACG	Steege, 1977*b*[c]
galE	AUA AGC CUA AUG GAG CGA AUU **AUG** AGA GUU CUG GUU ACC	Musso *et al.*, 1974
galT	TAT CCC GAT TAA GGA ACG ACC **ATG** ACG CAA TTT AAT CCC	N. Grindley, personal communication
	16 S RNA 3′ end $_{HO}$A U U C C U C C A C U A $G_{5'}$	Shine and Dalgarno, 1974; Noller and Herr, 1974; Ehresmann *et al.*, 1974; Sprague and Steitz, 1975

[a]Underlining indicates contiguous bases complementary to the 3′-oligonucleotide of *E. coli* 16 S rRNA. Dots indicate G· U base pairs. Although in several cases base pairing of an additional noncontiguous region might be predicted to increase the stability of the mRNA· rRNA interaction (Gralla and Crothers, 1973), these are not indicated because of uncertainty about the accuracy of predictions for structures including internal and bulge loops. Initiator triplets are indicated by boldface type.

[b]What was originally thought to be bacteriophage T7 has turned out to be T3 (F. W. Studier, personal communication).

[c]In each of these cases unordered nucleotides, which have been subsequently assigned, appeared in the original RNA sequence (see references listed).

binding to initiator regions. However, direct experimental evidence was required to demonstrate that the mRNA and rRNA do interact during the initiation of protein biosynthesis.

2.3.1 *Demonstration of an mRNA·rRNA Complex*

A critical test of the Shine and Dalgarno hypothesis was first performed by myself and K. Jakes in 1975 (Steitz and Jakes, 1975). Our idea was first to preform protein synthesis initiation complexes with a ^{32}P-labeled messenger fragment of known sequence; the A protein initiator region of R17 bacteriophage RNA was chosen because it possessed the longest complementarity to rRNA known at that time (seven base pairs). Next, the initiation complexes were treated with colicin E3, which cleaves the 16 S RNA at a specific site about 50 nucleotides from its 3′ terminus (Senior and Holland, 1975; Bowman *et al.*, 1971). The 70 S ribosome was then disassembled by treatment with 1% sodium dodecyl sulfate (SDS) and its components fractionated on a polyacrylamide gel. The gel revealed a novel entity containing the ^{32}P-labeled mRNA fragment, which had properties expected for an mRNA·rRNA hybrid consisting of the 30-nucleotide-long initiator region and the colicin fragment.

Additional experiments (Steitz and Jakes, 1975) provided evidence that the appearance of the initiator region·colicin fragment complex was dependent on involvement of the labeled mRNA in a 70 S initiation complex before ribosome disassembly. As expected, the complex was found to contain one mole of initiator region per mole of colicin fragment. Moreover, it appeared to be a pure RNA·RNA hybrid (not requiring the presence of residual ribosomal proteins for maintenance), which could be dissociated by mild heat treatment.

2.3.2 *Structure of the Complex*

The presumed structure of the mRNA·rRNA complex recovered from the initiating ribosome is depicted in Fig. 1. The figure suggests that a specific secondary structure is assumed by the 3′-terminal region of the 16 S rRNA and that upon mRNA binding some of the *intra*molecular base pairs may be exchanged for *inter*molecular hydrogen bonds. What evidence do we have to support this proposal?

The secondary structure of the isolated colicin fragment from the 3′ end of *E. coli* 16 S rRNA has been investigated using temperature-jump relaxation methods (Yuan *et al.*, 1976) and nuclear magnetic resonance techniques (Baan *et al.*, 1977). Data of both types strongly suggest the existence of a quite stable helical region (T_m = 80°C by temperature jump) containing the nine base pairs pictured above the bulge loop in Fig. 1. Temperature-jump studies also reveal a second transition of 21°C that could correspond to the melting of additional base pairs below the bulge loop. Although it seems likely that only the upper, more stable portion of the hairpin is formed under physiological conditions, the existence of the lower transition is relevant to studies probing the role of ribosomal protein S1 in initiation (see Section 2.4.3).

Several lines of evidence also support the placement of the illustrated intermolecular hydrogen bonds in the isolated complex containing the colicin fragment and the R17 A protein initiator region (Steitz and Steege, 1977). For instance, the G residues involved in the postulated Watson–Crick base pairs are found to be more resistant to digestion by T_1 ribonuclease than are other G residues in the mRNA fragment. Likewise, thermal melting studies of this complex (with seven base pairs) and a comparable mRNA·rRNA hybrid containing the λp_R ribosome-binding site (nine base pairs, see Table I) reveal a 5°C higher T_m for the latter complex, in accord with the predicted values (Gralla and Crothers, 1973), assuming the structure shown. The absolute T_m values are about 32 and 37°C, respectively, in the same buffer as that used for initiation complex formation *in vitro.*

2.3.3 *Questions Concerning mRNA·rRNA Interaction*

The studies just described provide strong circumstantial evidence for a base-pairing interaction between mRNA and rRNA during the initiation of protein synthesis in *E. coli.* What has been demonstrated is that in a 70 S initiation complex the initiator region is positioned such that upon gentle dissociation of the ribosome it can be recovered noncovalently complexed with a sequence near the 3′ terminus of 16 S rRNA. However, since protein synthesis can begin *in vivo* and *in vitro* at temperatures far exceeding the T_m values of the most stable isolated mRNA·rRNA hybrids (see earlier), we can conclude that the mRNA·rRNA interaction must be supplemented by other mRNA·ribosome contacts. Presumably the requirements for such supplementary contacts will be very great in the case of most of the initiation sites shown in Table I, which exhibit much less mRNA·rRNA complementarity than does the R17 A site or the λp_R initiator region.

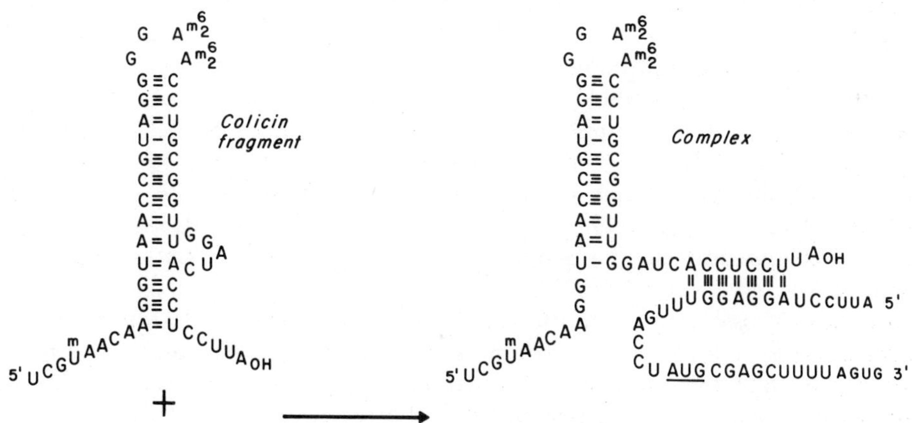

Figure 1. Postulated hydrogen bonding between the colicin fragment from the 3′ end of *E. coli* 16 S rRNA (Ehresmann *et al.,* 1975) and the bacteriophage R17 initiator region (Steitz, 1969). Evidence for the illustrated secondary structure of the colicin fragment and for the position of the intermolecular base pairs is discussed in the text. The isolated A protein initiator fragment is predicted not to assume a stable secondary structure under physiological conditions (Gralla and Crothers, 1973).

The necessity of supplementary contacts between mRNA and ribosome, plus the fact that both the length of the complementary region and its distance from the initiator triplet vary substantially, raises many questions concerning the role of the mRNA·rRNA interaction in initiation. How many of the potentially interacting nucleotides actually do base-pair during formation of the initiation complex? When during the first steps of protein synthesis is the interaction broken? Is the geometry of the ribosome really such that the variable distance between the mRNA initiator triplet and the region complementary to 16 S rRNA can easily be accommodated? Does this variability affect the efficiency of initiation?

Perhaps most important are several more general questions. First, is the presence of an initiator triplet preceded by a sequence complementary to 16 S rRNA sufficient to define a true ribosome-binding site in mRNA? Second, is the calculated strength of the mRNA·rRNA interaction correlated in any direct way with the efficiency of initiation at a particular site? Third, what are the other mRNA–ribosome interactions that contribute so vastly to the stability of the initiation complex? Partial answers to several of these questions are provided in the succeeding sections. The remainder are reconsidered at the end of the section on initiation.

2.4 *Proteins as Determinants in Initiation*

In addition to mRNA·rRNA base pairing, at least one other well-documented RNA·RNA interaction, between the $tRNA_f^{Met}$ anticodon and the mRNA initiator triplet, is of prime importance to messenger selection and binding by ribosomes. An initiator trinucleotide not only can independently direct the specific binding of initiator tRNA to the ribosome (Clark and Marcker, 1966) but also is instrumental in the correct phasing of translation when present in a synthetic polynucleotide lacking sequences complementary to 16 S rRNA (Stanley *et al.*, 1966). In the case of either initiator trinucleotides or synthetic messengers, the level of functioning is low (5–10% relative to natural mRNAs), but initiation does occur.

Several ribosome-associated proteins have also been implicated in messenger binding during polypeptide chain initiation. For instance, since its discovery, initiation factor IF-3 has been widely believed to be endowed with specific messenger-selecting abilities (see Lucas-Lenard and Lipmann, 1971; Steitz, 1975). Likewise, S1, the largest protein of the 30S subunit, was first identified as an "interference factor" possessing a high affinity for RNA (Groner *et al.*, 1972); it was only later shown to be a positive translational element (van Duin and van Knippenberg, 1974) present in molar amounts on active ribosomes. Although my arguments will surely represent an oversimplification of the actual situation, I should like to suggest in the three following sections that these and other "mRNA-selecting proteins" do not operate by direct binding to specific recognition sites on mRNAs. Rather, their effects can just as easily be explained by the idea that they facilitate the two RNA·RNA interactions described earlier.

2.4.1 *Differential Requirements for Factors and S1 at Different Initiator Regions*

It was observed quite some time ago that ribosomes washed in high salt [which removes the initiation factors (Stanley *et al.*, 1966) and S1 (Wahba *et al.*,

1974; Inouye *et al.*, 1974; Szer *et al.*, 1975)] retain their ability, albeit at a very low level, to recognize correctly the three initiator regions in R17 RNA (Steitz, 1973*a*). Readdition of the crude ribosomal wash, however, does not stimulate initiation at the three sites equally. Instead, binding to the A protein initiator region (with a seven-base-pair mRNA·rRNA complementarity) increases only about twofold; the replicase initiation site (four base pairs plus a terminal G·U pair) is five- to eightfold stimulated; and recognition of the coat protein initiator region (five base pairs including an internal G·U pair) rises over tenfold. Thus, relative requirements for components present in the ribosomal wash seem to be inversely correlated with the degree of mRNA·rRNA complementarity exhibited by the regions at the beginnings of the three R17 cistrons. A much higher requirement for IF-3 has also been observed for translation of ϕ80*trp* mRNA than for phage T7 mRNA (Benne and Pouwels, 1975).

Recent experiments (Steitz *et al.*, 1977), using purified initiation factors, S1, and ribosomal subunits either containing or lacking S1, further revealed that S1 as well as the factors contribute to the differential stimulation of ribosomal binding. Can our current knowledge of the mechanism of action of initiation factors and of S1 be reconciled with these results?

2.4.2 *Initiation Factor Function*

Despite varying opinions concerning the detailed functioning of the *E. coli* initiation factors, it is generally agreed that their primary task is to position the initiator tRNA in the P site* on the ribosome. Here IF-2, which specifically binds fMet-tRNA$_f^{Met}$, acts most directly (see Miller and Wahba, 1973). IF-1 has been shown to function catalytically, recycling IF-2 on and off the ribosome (Benne *et al.*, 1972). Early findings that IF-3 stimulates translation of natural mRNAs (Iwasaki *et al.*, 1968) were interpreted to indicate a direct interaction with the messenger molecule; however, more recent work has shown that IF-3 can promote translation of synthetic polynucleotides and even stimulate AUG trinucleotide-directed ribosome binding of initiator tRNA (Wahba *et al.*, 1969; Grunberg-Manago *et al.*, 1971; Miller and Wahba, 1974; Schiff *et al.*, 1974; Dondon *et al.*, 1974; Sobura *et al.*, 1977). Therefore, IF-3 may also directly facilitate tRNA placement. Presumably, it accomplishes this by inducing a conformational change in the 30 S ribosome (see Pon and Gualerzi, 1974; Ewald *et al.*, 1976; Michalski *et al.*, 1976), which could also explain the capacity of IF-3 to cause ribosomal subunit dissociation (Noll and Noll, 1972; Kaempfer, 1972). Since IF-3 has a high affinity for RNA, it seems likely that its major contact with the 30 S subunit is in the 16 S rRNA molecule (Pon and Gualerzi, 1976).

If we now consider the mRNA·rRNA interaction and the fMet-tRNA·codon interaction to be two primary contributors to mRNA recognition by ribosomes, it is reasonable that initiation at the three R17 cistrons might be differentially dependent on factors. The high degree of complementarity between the A site and the 3′ end of 16 S RNA may provide sufficient binding energy and therefore substitute for the usual factor requirements. Conversely, the coat site, with its weak complementarity, might be expected to require the presence of initiator tRNA in

*The two tRNA-binding sites on the ribosome are called the P site, which holds the peptidyl-tRNA (and fMet-tRNA$_f^{Met}$), and the A site, which accepts the incoming aminoacyl-tRNA during peptide chain elongation (see Lucas-Lenard and Lipmann, 1971).

order for stable interaction with the ribosome to occur. Interesting in this regard are experiments suggesting that fMet-tRNA can be positioned on the 30 S subunit before intact R17 RNA (in which the coat protein initiator region is the primary available site; see Section 2.6.3) is bound and phased correctly (Jay and Kaempfer, 1975*a*). However, since other studies indicate that the mRNA and the tRNA can be bound in either order (Gualerzi *et al.*, 1977), this finding should probably not be used to make a generalization for all initiator regions; indeed, the preferred sequence may vary just as individual requirements for initiation factors vary.

2.4.3 S1 Acts at the 16 S 3′ End

In contrast to IF-3, ribosomal protein S1 appears to be required only when mRNAs longer than a trinucleotide are used for initiation. More specifically, ribosomes lacking S1 can bind fMet-tRNA in the presence of AUG trinucleotide, while intact phage RNA and various polynucleotides are bound and translated only by ribosomes that contain S1 (van Duin and van Knippenberg, 1974; van Dieijen *et al.*, 1975; Dahlberg, 1974; Szer and Leffler, 1974; Hermoso and Szer, 1974; Sobura *et al.*, 1977). Data from chemical crosslinking experiments have located S1 in the mRNA-binding site of the 30 S ribosome (Fiser *et al.*, 1975), directly adjacent to the 3′ end of 16 S rRNA (Kenner, 1973; Czernilofsky *et al.*, 1975). The work of Dahlberg and Dahlberg (1975) further suggests that this protein may interact specifically with the pyrimidine-rich terminal dodecanucleotide of the rRNA molecule. Physical (Bear *et al.*, 1976; Draper and von Hippel, 1976; Szer *et al.*, 1976) and other studies (Tal *et al.*, 1972; Miller *et al.*, 1974; Miller and Wahba, 1974; Carmichael, 1975; Jay and Kaempfer, 1975*b*; Senear and Steitz, 1976) indicate that S1 is a polynucleotide-binding protein with high affinity for pyrimidine-rich single-stranded regions in RNA.

Why then should one natural initiator region have a higher requirement for S1 than another? A most attractive hypothesis is that S1 may function, as first suggested by Dahlberg and Dahlberg (1975), by correctly positioning the 3′ end of 16 S rRNA for subsequent base pairing with the complementary region in the mRNA. This idea is now supported by physical evidence from temperature-jump melting studies (Yuan *et al.*, 1978), which show that addition of 1 mol of S1 protein to 1 mol of colicin fragment disrupts the base-pairing interactions below the bulge in the proposed secondary structure shown in Fig. 1. Thus, the requirement for S1 would be expected to be highest for ribosome binding to initiator regions with weak potential for hydrogen bonding to 16 S rRNA. Conversely, if the mRNA·rRNA interaction is strong (that is, thermodynamically stable under physiological conditions), a hydrogen-bonded complex may well be able to form without the assistance of a "helper protein" such as S1.

On the other hand, alternative possibilities for S1 action should not be disregarded at this time. For instance, S1 could utilize its ability to bind single-stranded RNA to disrupt mRNA secondary structures adjacent to initiator triplets (van Dieijen *et al.*, 1976) or its preference for certain pyrimidine-rich sequences to select out particular mRNA regions for ribosome binding. In the case of the initiator region for the coat protein cistron of bacteriophage Qβ (which exhibits rather weak mRNA·rRNA complementarity), stable interaction with the ribosome may be facilitated by specific S1 recognition of a sequence appearing about 35 nucleotides 5′ to the initiator AUG (Goelz and Steitz, 1977). The possibility of additional direct S1 interactions with the mRNA is made plausible by recent observations that this protein may possess several independent RNA-binding sites (Draper and von Hippel, 1976). In general, it would not be surprising to find that the protein requirements for ribosome recognition of each naturally occurring initiator region differ simply because different nucleotide sequences provide varying potential for the establishment of intermolecular contacts.

2.5 *RNA versus Proteins in Species Specificity*

Shortly after analyzing the 12 residues at the 3′ end of *E. coli* 16 S rRNA, Shine and Dalgarno (1975) examined comparable regions from several other bacterial species. These they found invariably to be pyrimidine-rich, but with slightly differing nucleotide sequences. This result was most intriguing because it seemed to provide a molecular explanation for widely reported differences in the capacity of protein-synthesizing systems from various prokaryotes to translate each others' messenger RNAs (see Stallcup *et al.,* 1975).

Only one interspecies system had been characterized thoroughly enough to provide a critical test of the importance of mRNA·rRNA complementarity in determining the ribosome's translational specificity. In 1969 Lodish (1969*a*) reported that at temperatures above 50°C extracts from the thermophilic bacterium *Bacillus stearothermophilus* selectively initiate and translate only the A protein cistron of bacteriophage f2. Discrimination against the other two cistrons was traced to the 30 S ribosomal subunit, not (as expected at that time) to the initiation factors (Lodish, 1970*a*). In experiments designed to show that ribosome binding and protection was also limited to the A site in R17 and Q*β* RNAs, I obtained an unpredicted result (Steitz, 1973*a*). Whereas thermophilic ribosomes did bind, at temperatures above 60°C, to the A protein initiator region of R17 (a close relative of f2), they ignored all three normal initiation sites of Q*β* and instead tenaciously bound to two other regions in the RNA molecule (see Table II). One of these was recovered as a simple purine-rich nonanucleotide and the other as an appropriately sized (35 nucleotide) ribosome-binding site. The latter contains an 11-nucleotide polypurine tract, but totally lacks an initiator triplet; consequently, it does not direct the synthesis of any formylmethionyl dipeptide when used in the dipeptide assay for translational initiation (Lodish, 1969*b*). Also at lower temperatures, where ribosomes from *B. stearothermophilus* and *E. coli* can be directly compared, marked differences in initiation specificity are observed. Relative to *E. coli* ribosomes (which can bind all three initiators in both phage RNAs), the coat and replicase sites in Q*β* and the replicase initiator region in R17 RNA are recognized only weakly by thermophilic ribosomes at 49°C (Goldberg and Steitz, 1974); attachment to the beginning of the R17 coat cistron has never been observed with *B. stearothermophilus* ribosomes.

Table II shows that the polypurine tracts of the two noninitiator Q*β* regions do indeed exhibit substantial complementarity to the extreme 3′-terminal sequence of *B. stearothermophilus* 16 S rRNA. Thus, it is reasonable that these sequences should be efficiently bound by thermophilic ribosomes at high temperatures, as is the A site of R17. Their selection provides additional support for the general conclusion that the 3′ termini of prokaryotic 16 S rRNAs are positioned so as to be active determinants in mRNA–ribosome recognition events.

On the other hand, analysis of a more extensive region at the 3′ end of 16 S rRNA from *B. stearothermophilus,* which we hoped would explain the translational specificity of these ribosomes at 49°C, surprisingly revealed a sequence identical to that of *E. coli,* except that the terminal adenosine of the latter is replaced by $UCUA_{OH}$ in the thermophile (Sprague *et al.,* 1977). Ribosomes from the two species therefore possess the same potential for mRNA·rRNA interaction, yet exhibit different recognition patterns for the R17 and Qβ phage initiator regions! Thus, some other component of the 30 S ribosome must be responsible.

In fact, several observations had already suggested important roles for certain proteins in determining differences in initiation specificity between *E. coli* and *B. stearothermophilus* ribosomes. Reconstitution studies monitoring the capacity of hybrid ribosomes from the two species to recognize the R17 and Qβ initiator regions had suggested that the 30 S protein fraction (rather than the 16 S RNA) is primarily responsible for determining specificity (Goldberg and Steitz, 1974). One of the contributors to this difference in specificity is apparently protein S12 (Held *et al.,* 1974), which may exert its effect by altering the association of initiation factors with the ribosome (Hawley *et al.,* 1974; Bollen *et al.,* 1975; Heimark *et al.,* 1976). Also, protein S1 has been found to be either functionally distinct in or absent from *B. stearothermophilus* ribosomes (Isono and Isono, 1976). Again, these proteins may be viewed as exerting their specificity via the RNA·RNA interactions

Table II. RNA Phage Sites Bound by B. stearothermophilus Ribosomes

Binding temperature[a] (°C)	Site	Sequence[b]	Theoretical T_m of mRNA·rRNA complex[c] (°C)
65	R17 A	CCUAGGAGGUUUGACCU**AUG**[d]	38
	Qβ noninitiator *a*	CUGAAAGGGGAGAUUACUCG[e]	29
	Qβ noninitiator *b*	GGAAGGAGC[f]	29
49	R17 A	CCUAGGAGGUUUGACCU**AUG**[d]	38
	R17 replicase	AUGAGGAUUACCC**AUG**[d]	−7
	Qβ replicase	AACUAAGGAUGAA**AUG**[d]	−7
	Qβ coat	AAACUUUGGGUCAAUUUGAUC**AUG**[d]	−25
Not bound	R17 coat	CAACCGGGGUUUGAAGC**AUG**[d]	−46
B. stearothermophilus	16 S 3′ end	$_{HO}$*AUCU*UUCCUCCACUAG$_{5'}$[g]	

[a]Ribosome-binding experiments are reported by Steitz (1973*a*) and Goldberg and Steitz (1974). Note that only the R17 A site is bound strongly by *B. stearothermophilus* ribosomes at 49°C; the other regions are recognized much more weakly by thermophilic than by *E. coli* ribosomes at 49°C. Lack of binding of Qβ noninitiators and the R17 A site at 49°C is probably due to interfering secondary structures in the phage RNAs (see Sprague *et al.,* 1977).

[b]Underlining indicates phage RNA regions which are complementary to the 3′-end sequence of *B. stearothermophilus* 16 S rRNA aligned below. Dots represent G·U base pairs. Initiator AUGs are in boldface type. Italics indicate that portion of the 16 S terminal sequence which differs in *E. coli* and *B. stearothermophilus.*

[c]T_m values calculated according to Gralla and Crothers (1973). Note correlation between the T_m of the mRNA·rRNA complex and the temperature at which each site binds stably to *B. stearothermophilus* ribosomes.

[d]See Table I for sequence references.

[e]From Steitz (1973*a*) and Sprague *et al.* (1977).

[f]From Sprague *et al.* (1977).

[g]From Shine and Dalgarno (1975), Woese *et al.* (1976), and Sprague *et al.* (1977).

occurring during initiation: Since both S1 and properly functioning initiation factors are stringently required for *E. coli* ribosomes to recognize those R17 initiator regions that have weak complementarity to 16 S rRNA (Steitz *et al.*, 1977), it is not surprising that *B. stearothermophilus* ribosomes encounter difficulty interacting with sites other than the highly complementary R17 A protein initiator region. Support for this interpretation comes from the work of Isono and Isono (1975), who were able to achieve novel translation of the f2 coat and replicase cistrons simply by adding *E. coli* S1 to *B. stearothermophilus* ribosomes! We conclude that both proteins and rRNA must be regarded as potential contributors to initiation specificity in prokaryotic systems.

2.6 *mRNA Structure and Initiation*

So far in our discussion of the nature and functioning of initiation signals, we have focused only on limited regions of the mRNA adjacent to authentic initiator codons. However, recognition by ribosomes is also likely to be influenced by intramolecular interactions within an mRNA molecule. For instance, if either the initiator triplet, the polypurine tract, or both of these elements were to be made unavailable for ribosome attachment (for example, by the assumption of a stable RNA secondary or tertiary structure), a potential initiation signal could be rendered inactive. On the other hand, the possibility that RNA conformation might contribute in a positive way to ribosome recognition of initiation sites should not be ignored.

2.6.1 *Evidence for mRNA Secondary and Tertiary Structure*

Certainly the best characterized mRNAs from a structural standpoint are the RNA bacteriophage genomes. Here not only are extensive nucleotide sequences known, but also, because of the availability of large amounts of material, physical–chemical studies of these polycistronic mRNA molecules have been conducted.

The complete nucleotide sequence of the MS2 phage genome (Min Jou *et al.*, 1972; Fiers *et al.*, 1975, 1976) suggests that this mRNA folds into a series of well-defined hairpin loops involving at least 65% of its 3569 residues in Watson–Crick base pairs (see Fig. 2). The existence of the proposed RNA secondary structures is supported both by the results of partial ribonuclease digestion experiments, which yield a characteristic and highly reproducible pattern of oligonucleotide products, and by the observed specificity of chemical mutagenesis performed on whole phage particles (Min Jou *et al.*, 1972). According to the theoretical calculations of Gralla and Crothers (1973), the maintenance of most of these loop structures is also favored in solution under physiological conditions—that is, inside a phage-infected cell or in an *in vitro* protein-synthesizing system. In fact, direct measurements using temperature-jump relaxation (Gralla *et al.*, 1974) and nuclear magnetic resonance methods (Hilbers *et al.*, 1974) of the thermal melting behavior of one specific phage RNA fragment (a 59-nucleotide segment of the R17 genome) yielded T_m values within 5°C of predicted values (see Fig. 3 and Section 2.7.1a). A possible danger in extrapolating from detailed knowledge of the secondary structure of a particular isolated region is that its environment in the intact phage

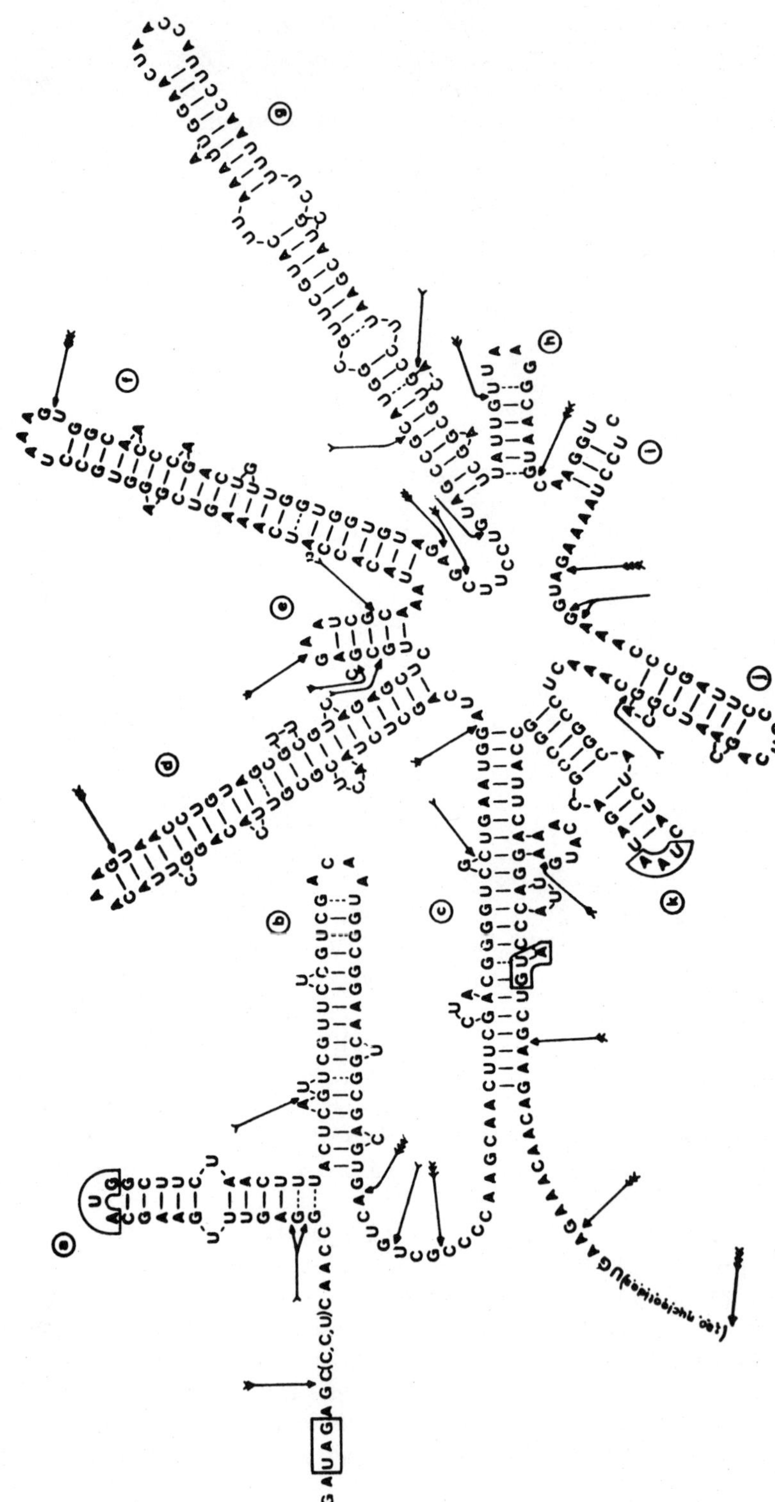

Figure 2. Nucleotide sequence and predicted secondary structure ("flower model") for the coat protein cistron and adjacent regions of bacteriophage MS2 RNA. Terminator triplets (at the ends of the A and coat cistrons) and initiator codons (for the coat and replicase genes) are blocked in. Arrows indicate sites of cleavage of the MS2 RNA molecule by T_1 ribonuclease under conditions of partial digestion; the number of feathers provides a measure of the relative susceptibility of each site, where four feathers indicate bonds always split. (From Min Jou *et al.*, 1972, with permission of the authors and publishers.)

mRNA may provide an opportunity for the formation of alternative conformations that are even more stable. However, since the likelihood of interaction with an adjacent sequence is always higher than with a more distant site, in most cases the MS2 RNA secondary structures that have been proposed are probably correct (Min Jou *et al.*, 1972; Fiers *et al.*, 1975, 1976).

Assuming that the phage RNA does undergo extensive local helix formation, we can further ask whether such structures associate in any preferred way to yield a defined tertiary conformation. Evidence here is much less direct, but nonetheless persuasive. Even the earliest physical–chemical measurements on phage RNA molecules indicated not only a high degree of helicity (60–80%) but an unusually compact overall structure (for a review see Boedtker and Gesteland, 1975). The observation that various ribonucleases first cleave the RNA phage genome within a limited region located about 40% from the 5′ end (again see Boedtker and Gesteland, 1975) is most reasonably explained by a folding pattern which renders this site particularly accessible. Moreover, even at very low salt concentrations, MS2 RNA molecules spread on an electron microscope grid reproducibly display interactions involving distant portions of the genome (Jacobson, 1976). Whether or not these structural features are a consequence of the requirement for the phage RNA to fold into a minimum-sized sphere capable of being encapsulated by a protein coat, the concept of higher order organization in these RNAs is most helpful in explaining many of the translational phenomena discussed later.

In the case of messenger RNAs other than the RNA phage genomes, direct evidence for the formation of preferred three-dimensional structures is sparse. The best data are from experiments with mRNAs isolated from phage T4-infected cells (Ricard and Salser, 1974, 1975); melting studies and partial nuclease digestion experiments suggest a less compact tertiary folding than in R17 RNA. Nonetheless, the extent of T4 mRNA secondary structure appears comparable to

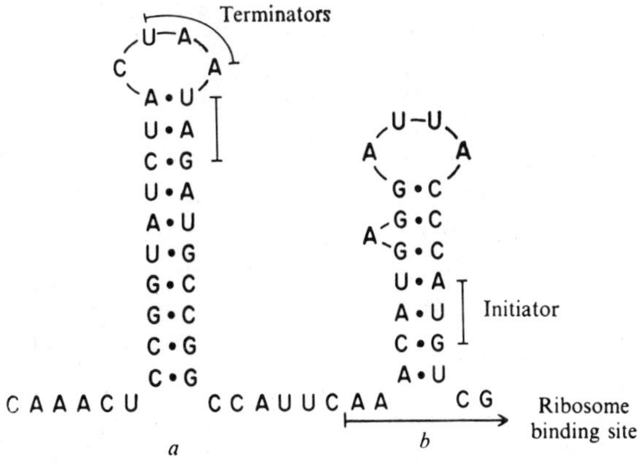

Figure 3. The segment of bacteriophage R17 RNA which is protected from RNase digestion by coat protein (Bernardi and Spahr, 1972; Gralla *et al.*, 1974). It includes an entire intercistronic region and portions of the coat and replicase genes, as indicated. Temperature-jump relaxation studies (Gralla *et al.*, 1974) yielded T_m values of 83 and 61°C for hairpin loops *a* and *b*, respectively. Binding of coat protein to the fragment alters the melting behavior of helix *b* only. (Reproduced from Gralla *et al.*, 1974, by permission of the authors and publisher.)

rRNA, from which reproducible oligonucleotide fragments indicative of the existence of many defined hairpin loops have been obtained during sequence analysis (Ehresmann *et al.*, 1975). The limited work to date on bacterial mRNAs suggests that they may undergo significantly less internal base pairing, which could be related to their short half-lives *in vivo.* However, since even totally random RNA sequences are predicted to form secondary structures involving about 50% of their residues in base pairs (Fresco *et al.*, 1960; Gralla and Delisi, 1974), it seems unlikely that bacterial mRNAs have evolved to resist any loop formation under all physiological conditions. Again, perhaps the best argument for the existence of secondary structures in non-RNA phage messengers is that mRNA folding can provide plausible molecular explanations for otherwise puzzling behavior; several examples are discussed in the sections on translational control and reinitiation.

2.6.2 *Secondary Structure Prevents False Starts in RNA Phage Messengers*

Since the length of the region complementary to 16 S rRNA in authentic messenger initiation sites can be as short as three nucleotides (Table I), it is obvious that the potential for base pairing plus the presence of an initiator triplet is not sufficient to describe a true mRNA initiator region. Indeed, a quick scan of nucleotide sequences in the MS2 RNA phage genome (Min Jou *et al.*, 1972; Fiers *et al.*, 1975, 1976) reveals a number of internal and out-of-phase AUG and GUG triplets that are preceded by appropriately situated polypurine tracts. Yet, only the three regions listed in Table I are recognized by *E. coli* ribosomes as functional start signals in the intact RNA molecule. If secondary structure does play an active negative role in limiting initiation to the proper regions, we would expect that sequences surrounding each of the additional potential initiators should be at least partially sequestered by RNA secondary structure. In fact, examination of Fig. 2 and the MS2 A (Fiers *et al.*, 1975) and replicase (Fiers *et al.*, 1976) cistrons shows this to be largely true.

Substantial experimental evidence can also be marshalled to support the notion that the RNA phages utilize secondary structure to restrict initiation. Most compelling is the finding that partial unfolding of the MS2 or f2 RNA molecule with formaldehyde (which has been shown to disrupt base pairing) (Feldman, 1973) results in the synthesis of a discrete set of novel fMet-dipeptides (Lodish, 1970*b*) that can be elongated to produce a heterogeneous population of polypeptide products. Although no sequence analysis has yet been undertaken on these "counterfeit" initiator regions, translation presumably originates at internal and out-of-phase initiation triplets that are normally blocked. Interestingly, utilization of these sites is independent both of the presence of S1 protein on the ribosome (van Dieijen *et al.*, 1976) and of initiation factor IF-3 (Berissi *et al.*, 1971), suggesting that their complementarity to 16 S rRNA may be rather strong. Evidence that RNA secondary, rather than tertiary, structure is responsible for inactivating these internal initiators comes from the observation that fragmentation of the phage RNA does not yield new fMet-dipeptides until the fragments approach 4S in size (less than 100 nucleotides) (Staples and Hindley, 1971; Steitz, 1973*b*; Voorma *et al.*, 1971). Furthermore, statistical analysis of the differences in nucleotide sequence among the group I RNA phages (MS2, f2, R17) suggests that there has been selective pressure to preserve the RNA secondary structures (Min

Jou and Fiers, 1976). Whether messenger RNAs other than the RNA phage genomes have evolved a similar mechanism for limiting translational initiation is not yet known.

2.6.3 *Structural Effects at Authentic RNA Phage Initiation Sites*

Extensive *in vitro* manipulation of the RNA phage genome has also provided excellent evidence that three-dimensional RNA structure effects a significant decrease in the rate at which *E. coli* ribosomes attach to two of the three true phage initiator regions (for reviews see Lodish and Robertson, 1969; Steitz, 1975; Lodish, 1975). With intact RNA from either the group I phages or Qβ, protein synthesis is initiated efficiently at only one of the three ribosome-binding sites, that of the coat cistron (for a review see Kozak and Nathans, 1972). Replicase initiations occur at about one-third this rate and are dependent on prior translation of a portion of the coat protein cistron (the so-called polarity effect; see later). A protein is synthesized at only about one-twentieth the rate of coat protein. At least three treatments, all of which are predicted to disrupt the native secondary/tertiary conformation of the RNA molecule, stimulate both replicase and A protein synthesis dramatically. These are mild fragmentation (for references see Steitz, 1973*b*), heat treatments under various ionic conditions (Fukami and Imahori, 1971), and formaldehyde unfolding (Lodish, 1970*b*, 1971). Since the newly exposed authentic initiator regions are bound by either *E. coli* or *B. stearothermophilus* ribosomes in accord with the documented species-specificity pattern (see Section 2.5), apparently only the availability of the sites for ribosome attachment is being altered. Accordingly, it has also been observed that nascent RNA molecules are significantly better messengers for both f2 (Robertson and Lodish, 1970) and Qβ (Staples *et al.*, 1971) A protein synthesis (the A cistron is the first gene) than are mature RNA molecules. This finding is consistent with the idea that A protein initiations are severely limited in the intact phage mRNA. In fact, Fiers *et al.* (1975) have recently noted that a nucleotide sequence appearing two-thirds of the way along the A cistron in MS2 RNA can base-pair with approximately 15 residues just preceding the A protein initiator triplet; an interaction between these two regions could be that which restricts initiation.

Experiments which examined the ability of isolated RNA phage initiator regions to rebind to ribosomes further suggest a positive contribution of RNA structure to initiation at some sites. Whereas the R17 A protein initiator region becomes an exceedingly efficient initiation site when it is released from the remainder of the molecule (Steitz, 1973*b*), surprisingly the region at the beginning of the R17 coat protein cistron decreases in initiation potential upon release (Adams *et al.*, 1972; Steitz, 1973*b*). Two molecular explanations are possible: Either the coat protein site folds into a stable secondary structure that precludes ribosome binding when it is an isolated fragment but not when it is in the intact molecule, or the overall tertiary structure of the phage RNA may be designed such that this region is normally so exposed that it cannot be ignored by a passing ribosome. Long-range positive effects of RNA structure also seem to operate at the Qβ coat protein initiator; here it has been found that fragments containing this region do not rebind detectably to ribosomes unless they are at least 100 nucleotides long (Porter and Hindley, 1973).

The experiments just discussed suggest that each individual initiator region in mRNA may have slightly different requirements for the various components normally considered essential for initiation, such as the initiation factors and S1. Thus, general control over translational specificity could theoretically be achieved by manipulation of the availability of these components. More specifically, it is quite conceivable that the changes in a cell's physiology which occur after bacteriophage infection or upon slowing from logarithmic growth to stationary phase could decrease the levels of initiation factors or S1 protein. Preferential utilization of those initiator regions with relatively low requirements for these proteins would result. Indeed, numerous reports of alterations in ribosome-associated proteins have appeared in the literature (see Ochoa and Mazumder, 1974; Lodish, 1976; Scheps *et al.,* 1971), but mostly these have proven difficult to substantiate, especially with respect to their *in vivo* relevance.

Hence, our discussion of translational control is confined to those situations in which we have both a molecular understanding of the mechanism and a reasonable assurance that its functioning is physiologically important. It is interesting that all well-documented examples stem from studies on bacteriophage systems; often, the availability of mutants with altered translational expression has proven invaluable in elucidating the regulatory mechanism. The systems to be discussed have been divided into two groups on the basis of whether a protein or an RNA structure appears to be the direct controlling element. However, this division is somewhat artificial, and the interplay between these two entities cannot escape notice.

2.7.1 *Proteins as Translational Repressors*

Negative control over gene expression is classically exemplified by the *E. coli* lactose repressor protein, which binds to a specific recognition site on DNA (the lactose operator) and thereby prevents initiation of transcription of *lac* mRNA (for a review see Bourgeois and Pfahl, 1976). Parallels between this system and the translational repressors discussed later are striking. First, in each case, ribosome attachment appears to be blocked directly by specific protein binding to an initiator region in mRNA, exactly as the *lac* repressor–DNA interaction precludes the attachment of RNA polymerase to the *lac* promoter. Second, in all instances at least a portion of the nucleic acid-binding site is intercistronic, its structure presumably having evolved especially to allow recognition by the controlling protein. However, unlike the *lac* repressor, each of the translational repressors primarily fulfills another function in phage physiology and apparently only secondarily plays a regulatory role. It should also be noted that examples of proteins recognizing single-stranded regions and specific RNA secondary structures in mRNA are represented in the following discussion.

2.7.1a RNA Phage Coat Protein Turns off Replicase Synthesis. Late during the intracellular growth of an RNA bacteriophage, replicase synthesis is shut down as the rate of coat protein production peaks to provide enough (approximately 180 per particle) coat monomers for phage assembly. Mutants carrying missense or nonsense codons in the coat cistron are defective in this repressor activity and overproduce replicase late in infection (see Kozak and Nathans, 1972).

Elucidation of the molecular mechanism of replicase turnoff began with early observations that (1) addition of coat protein to an f2-directed *in vitro* translation system specifically decreases replicase synthesis, and (2) a few molecules of coat protein bind directly to the phage RNA (for references see Capecchi and Webster, 1975). Bernardi and Spahr (1972) next utilized R17 coat protein to trap on a Millipore filter a specific fragment of R17 RNA; this contained 59 nucleotides, including the entire coat protein/replicase intercistronic region plus a few residues from each gene (Fig. 3). Direct physical studies by both temperature-jump relaxation (Gralla *et al.,* 1974) and nuclear magnetic resonance (Hilbers *et al.,* 1974) techniques revealed that this segment of R17 RNA assumes the predicted double hairpin structure pictured in Fig. 3. Furthermore, the data (Gralla *et al.,* 1974) argue that the coat protein recognizes and binds to helix b only when this region of the RNA is folded into the hairpin loop illustrated. Thus, the R17 coat protein appears to function as a translational repressor by stabilizing a preformed RNA secondary structure that sequesters both the replicase initiator AUG and its polypurine tract, thereby rendering this initiation site unrecognizable by ribosomes.

Although the comparable Qβ system has not been studied in such detail, an identical mechanism seems to operate. The Qβ coat protein has been shown to bind to a site preceding and including the Qβ replicase initiator codon (Weber, 1976); the sequence of this region is again capable of assuming a secondary structure which would block ribosome attachment (see Steitz, 1975; Weber, 1976).

2.7.1b Qβ Replicase Uses a Host Polypeptide to Repress Coat Protein Synthesis. The virus-specified subunit of Qβ replicase orchestrates the assembly of three host polypeptides with itself to produce an enzyme complex capable of replicating the Qβ genome (Kamen, 1970; Kondo *et al.,* 1970). The *E. coli* proteins are the translational elongation factors, EFTu and EFTs (Blumenthal *et al.,* 1972), and ribosomal protein S1 (Wahba *et al.,* 1974; Inouye *et al.,* 1974). Apparently disregarding the fact that it must initiate RNA synthesis at the 3′ end of the Qβ genome, the replicase complex binds most tightly to a site in the middle of Qβ RNA; ribonuclease protection experiments yield a piece containing about 100 residues which precede (and just include) the initiator AUG for the coat protein gene (Weber *et al.,* 1972). A biological rationale (Kolakofsky and Weissmann, 1971*a*) for this unexpected behavior of the replicase is that the entry of ribosomes at the major translational initiation site would thereby be prevented and the RNA cleared for collision-free replication. *In vitro* experiments demonstrating that replicase binding indeed inhibits the initiation of Qβ coat protein synthesis (Kolakofsky and Weissmann, 1971*b*) and that ribosomes translating the RNA halt the progress of replicase along its Qβ template (Kolakofsky *et al.,* 1973) substantiate this model.

Somewhat surprisingly, host protein S1, rather than the phage-coded polypeptide, seems to direct the specific binding of replicase to the region preceding the Qβ coat cistron. Replicase lacking S1 cannot recognize this site (Kamen, 1975); and S1 alone has been shown to bind selectively to a pyrimidine-rich oligonucleotide 21 residues long derived from this region (Goelz and Steitz, 1977). Apparently, it is the primary structure of the RNA that is recognized both in this case and in the other S1-binding sites which have been identified (see Table III). This conclusion is based on the finding that the isolated oligonucleotides of three of these sites (none of which is predicted to form intramolecular base pairs) can

Table III. *Known S1-Binding Sites*

Oligonucleotide	Sequence[a]
16 S rRNA (3′ end)[b]	AUC A CCU CCUUA$_{OH}$
Qβ (near 3′ end)[c]	AAUAAAUUAUCACAAUCA C U C UUACG
Qβ (region preceding coat cistron)[d]	UAUCUUUUUAUUAACCCAACG
23 S rRNA[e]	AAAU ACC CUUUACAAUG

[a]Sequences are aligned to show maximum homology.
[b]Sequence data from references in Table I; S1 binding by Dahlberg and Dahlberg (1975).
[c]Sequence from Weissmann *et al.* (1973); S1 binding by Senear and Steitz (1976).
[d]Sequence from Weber *et al.* (1972); S1 binding by Goelz and Steitz (1977).
[e]Sequence and S1 binding by A. Krol, L. Visentin, and C. Branlant (personal communication).

efficiently rebind S1 protein (Dahlberg and Dahlberg, 1975; Senear and Steitz, 1976; Goelz and Steitz, 1977).

2.7.1c Phage T4 Gene 32 Protein Represses Its Own Translation. In a cell infected by bacteriophage T4, the gene 32 protein plays an essential role in genetic recombination, DNA synthesis, and repair by virtue of its strong cooperative binding to single-stranded DNA (Alberts and Frey, 1970). Curiously, amber and missense mutants in gene 32 vastly overproduce the altered gene 32 polypeptide (Krisch *et al.,* 1974; Gold *et al.,* 1976); and during infections by other mutants, gene 32 expression appears to be correlated directly with the quantity of single-stranded DNA present (Gold *et al.,* 1976). A model has therefore been postulated in which gene 32 protein binds preferentially to all available single-stranded DNA and then, presumably as a buffer against the lethal consequences of high cellular concentrations of a potent DNA-binding protein, represses its own synthesis.

Control here is apparently exerted at the translational level (Russel *et al.,* 1976), since the gene 32 mRNA (even in repressed cells) is exceedingly stable and derepression of gene 32 protein synthesis can occur in the presence of the drug rifamycin, which blocks transcriptional initiations. Although more detailed molecular information is only now being gathered, a plausible picture is that the gene 32 protein binds to its own mRNA and directly competes for ribosome attachment (Russel *et al.,* 1976). Gene 32 protein does in fact bind cooperatively to single-stranded RNA (Delius *et al.,* 1973; Kelly and von Hippel, 1976; Anderson and Coleman, 1975; Kelly *et al.,* 1976), as well as to DNA. Here it will be interesting to learn whether its specific interaction with its own message depends on an as yet undetected sequence preference of the protein or on specialized construction of the gene 32 mRNA (Russel *et al.,* 1976) relative to all other T4 messages (such that it alone contains a single-stranded site for protein binding). Whatever the answer, this system is significant in that it represents the first example of a translational control mechanism which could have been exerted at the level of transcription.

2.7.2 *Regulation via RNA Secondary Structure*

In the preceding discussion of RNA structure as a determinant in initiation, I summarized evidence from *in vitro* experiments which established that manipulation of mRNA conformation not only can alter the ribosome's ability to discriminate true initiators from internal AUGs and GUGs but also can change the relative utilization of authentic initiator regions. However, one thoroughly investigated and two less well-documented examples of translational control falling into the latter category were not considered. Instead, they are presented here because awareness of their existence originated with *in vivo* experiments that showed that the level of protein synthesis could be altered by changing conditions inside a living cell. Thus, these systems, plus phenomena discussed later in the section on translational restarts, constitute excellent evidence that changes in mRNA structure *in vivo* can mediate temporal regulation of translational expression.

2.7.2a Molecular Basis of the "Polarity Effect" in the RNA Bacteriophages. Amber mutations early in the coat protein gene of both the group I (MS2, f2, R17) and Qβ RNA bacteriophages have long been known to exert a strong polar effect on expression of the replicase gene (see Horiuchi, 1975; Kozak and Nathans, 1972). For example, position 6 amber mutants in the group I coat proteins produce almost no replicase, whereas coat nonsense mutations at amino acids 50, 54, and 70 are nonpleiotropic. A molecular explanation for this phenomenon was proposed by Min Jou *et al.* (1972) on the basis of the complete nucleotide sequence of the MS2 coat protein cistron and adjacent regions (Fig. 2). In the "flower model" of these authors, the entire replicase initiation site is hydrogen-bonded to an early segment of the coat cistron (encoding amino acids 24–32). Consequently, only when ribosomes translate beyond the sixth codon of the coat cistron and disrupt this structure (helix c of Fig. 2) would the replicase initiator region become accessible for ribosome attachment. Thus, the existence of two classes of coat mutants is logically explained.

The existence of the controlling structure shown in Fig. 2 has been experimentally substantiated by the finding that the region surrounding the replicase initiator and the interacting segment of the coat gene are isolated together from partial nuclease digests of MS2 RNA (Min Jou *et al.*, 1972). Note that this conformation at the replicase initiator is an alternative to the hairpin structure that is recognized by the coat protein (Fig. 3). The realization that at least two identifiable conformations can be assumed by a single region in the phage RNA suggests that this and presumably other predetermined structural transitions will occur during the course of intracellular phage growth. Certainly the progress of ribosomes and of replicase in turn along the RNA molecule cannot help but unravel certain interactions, thereby allowing others to form, with functional consequences.

2.7.2b RNase III Processing Is Essential for Translation of a Phage T7 mRNA. When bacteriophage T7 infects *E. coli,* the leftmost 20% of its genome is transcribed by the host RNA polymerase to yield a precursor RNA which is cut into relatively stable, monocistronic messengers by a host endonuclease, RNase III (Dunn and Studier, 1973*a, b*). Whereas the translation of most of these early T7 mRNAs is not significantly affected by the cleavage process, synthesis of the gene *0.3* protein is drastically reduced in RNase III$^-$ host cells or in *in vitro* systems lacking the enzyme (Dunn and Studier, 1975). Recent analysis of T7 ribosome-binding sites (Steitz and Bryan, 1977) has located the putative initiator AUG for this protein (T7 gene *0.3* site a, Table I) just 35 nucleotides beyond the site of RNA processing at the left end of the *0.3* mRNA (J. J. Dunn, personal communication). Although the sequence (and hence the secondary structure) for the RNA surrounding this particular RNase III cleavage point is not yet known, our knowledge of several other T7 regions in which this nuclease acts strongly argues that initiation of the gene *0.3* protein is negatively controlled by RNA secondary structure.

The natural substrates of RNase III include not only bacteriophage T7 messengers but also primary transcripts of the *E. coli* ribosomal RNA cistrons (Dunn and Studier, 1973*b*; Nikolaev *et al.*, 1973), which contain the 16 S, 23 S, and 5 S rRNA sequences (Ginsburg and Steitz, 1975). In each of these precursor RNAs cleavage is reproducibly executed at specific sites (Rosenberg *et al.*, 1974; Ginsburg and Steitz, 1975), but the recognition element appears to involve RNA secondary structure rather than a particular surrounding sequence (Crouch, 1974; Robertson and Dunn, 1975). Figure 4 illustrates the RNA structure at one of two extensively analyzed RNase III cleavage regions (Rosenberg and Kramer, 1977), that at the right end of the T7 *0.3* gene; a second more limited sequence (Robertson *et al.*, 1977) available at the T7 gene *1.1*/*1.3* junction reveals ten base pairs identical to those appearing in the upper portion of the hairpin loop structure shown. The effect of T7 deletions destroying the lower helical region in Fig. 4 suggests that both double-stranded stretches are required for recognition by RNase III (Rosenberg and Kramer, 1977), which interestingly is a dimeric enzyme (Dunn, 1976).

Returning to the ribosome-binding site near the RNase III cleavage point at

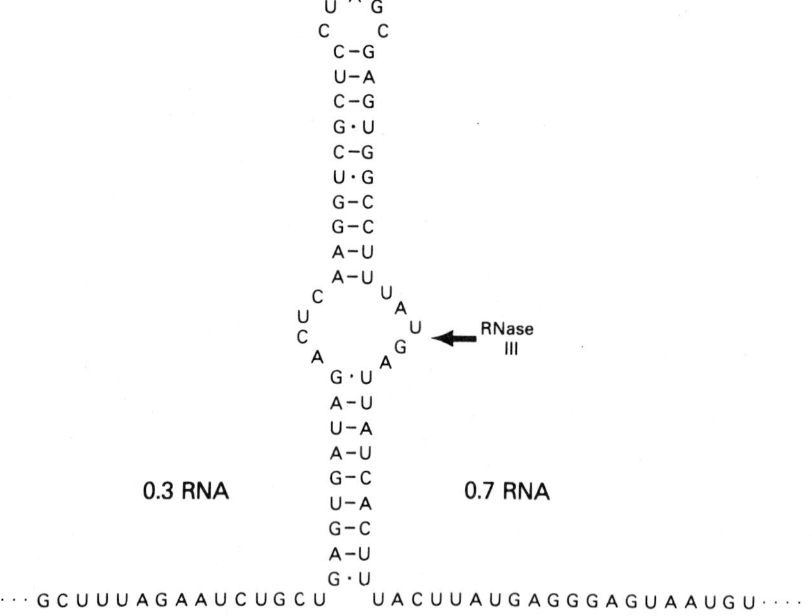

Figure 4. Sequence of the region surrounding a ribonuclease III processing site between genes *0.3* and *0.7* in the early mRNA precursor of bacteriophage T7 (Rosenberg and Kramer, 1977). The secondary strucutre illustrated is supported by the results of partial ribonuclease digestion (Rosenberg and Kramer, 1977). A sequence which can form a hairpin loop containing in its stem ten base pairs identical to those appearing in the upper portion of this structure is found adjacent to another RNase III cleavage site between T7 genes *1.1* and *1.3* (Robertson *et al.*, 1977). (Reproduced by permission of the authors and publisher.)

the left end of the T7 *0.3* mRNA, it is easy to imagine that a similar secondary structure might prevent ribosomes from attaching to this region in the precursor RNA. Surely, activation of an initiator triplet by mRNA scission at a site 35 nucleotides distant is consistent with the existence of an interfering nuclease-sensitive RNA secondary or tertiary structure. It would not be surprising to discover that RNase III cleavage of the ribosomal precursor RNA similarly activates binding sites for ribosomal proteins or other enzymes involved in ribosome maturation.

2.7.2c Active and Inactive mRNAs for the ϕX174 A Protein. The A cistron of bacteriophage ϕX174 is included in two types of messenger RNAs which can be isolated from infected cells: one is unstable and active; the other, stable but inactive (Hayashi *et al.,* 1976). Although the A protein ribosome-binding site has not been identified by either of the criteria required for inclusion in Table I, it is highly likely that polypeptide chain initiation occurs at the AUG triplet indicated in Fig. 5. The complete DNA sequence of the ϕX174 genome (Sanger *et al.,* 1977) predicts that the preceding mRNA region (including the appropriately positioned GGAGG complementarity to 16 S rRNA) would be sequestered in the hairpin loop structure shown in Fig. 5. However, analysis of *in vitro* ϕX transcripts (Smith and Sinsheimer, 1976; Grohmann *et al.,* 1975; Axelrod, 1976) has identified the AAAT sequence appearing near the turn in the loop as the 5′ terminus of a ϕX174 mRNA. Hence, assuming that some A protein mRNA begins at this site whereas other molecules are initiated at an earlier promotor (and read through the potential termination signal indicated), the existence of the observed active and inactive mRNAs with different lifetimes would be explained (Hayashi *et al.,* 1976; Sanger *et al.,* 1977). Another indication that altered mRNA termini can affect the rate of messenger degradation comes from studies of the P_L transcript of λ phage, which is apparently processed *in vivo* by RNase III (Lozeron *et al.,* 1977).

2.8 The Why and Wherefore of Translational Restarts

One of the most intriguing examples of ribosome recognition of novel initiator regions is the phenomenon of "reinitiation" which occurs *in vivo* after a polypeptide chain has been terminated at the site of a nonsense mutation. Although reinitiation has been demonstrated in several *E. coli* cistrons, the best characterized system is the lactose repressor gene (*lacI* gene). Here, Weber and

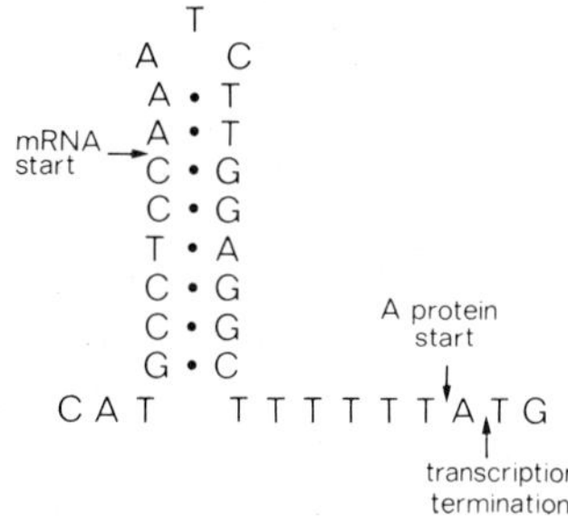

Figure 5. Predicted secondary structure of mRNA preceding the presumptive initiator triplet for the ϕX174 A protein. Note that only in mRNA molecules initiated at promoters 5′ to the A promoter (indicated by "mRNA start") would the GGAGG sequence, which is complementary to 16 S rRNA, be sequestered in a stable double-helical region. (From Sanger *et al.,* 1977, with permission of the authors and publisher.)

Miller and their colleagues (Platt *et al.*, 1972; Ganem *et al.*, 1973; Files *et al.*, 1974, 1975) have identified three discrete restart sites which give rise to C-terminal repressor fragments; the codons at valine 23, methionine 42, and leucine 62 of the repressor amino acid sequence can each be used to reinitiate translation beginning with *N*-formylmethionine. Restart polypeptides are not detected in wild type *E. coli,* but accumulate in suitable *lacI* nonsense mutant strains at approximately 10% of the repressor concentration found in wild type cells. It is interesting that a single nonsense mutation can activate more than one reinitiation site.

To examine RNA sequences and structures in the vicinities of these restart signals, D. Steege (1977*b*) recently determined the sequence of 214 nucleotides at the 5′ terminus of the *lacI* mRNA. Table IV shows that the wild type lactose repressor protein is initiated at a GUG codon preceded by a sequence exhibiting substantial complementarity to 16 S rRNA; note again that of the three in-phase GUG triplets clustered in this region, the real initiator is that codon which is most appropriately situated relative to the polypurine stretch. The sequence data confirm earlier predictions that *N*-formylmethionine is encoded by GUG and AUG at the first two restarts, respectively; the UUG assignment at the third site was deduced previously from genetic analysis (Files *et al.*, 1975). Relative to the true *lacI* initiator region, mRNA complementarity to rRNA is weak at the restart sites. This situation presumably contributes both to the tenfold lower level of restart polypeptides found relative to the wild type repressor protein *in vivo* and to the observation that only the wild type initiator region binds detectably to ribosomes *in vitro* (Steege, 1977*b*).

A most significant feature of the *lacI* mRNA sequence is the presence of several additional in-phase GUG codons which have *not* been observed to serve as reinitiation signals in the nonsense mutant strains examined so far. Thus, it is clear that ribosomes do not restart following termination simply by moving to the next available initiator codon.

Table IV. Sequences of lacI mRNA and the Repressor Protein[a]

pppG GAA GAG AGU CAA UUC AGG **GUG** (**GUG** AAU) **GUG** AAA CCA GUA ACG UUA UAC GAU GUC GCA GAG UAU
Met – Lys – Pro – Val – Thr – Leu – Tyr – Asp – Val – Ala – Glu – Tyr
1 10

GCC GGU GUC UCU UAU CAG ACC GUU UCC CGC **GUG GUG** AAC CAG CCC AGC CAC GUU UCU
Ala – Gly – Val – Ser – Tyr – Gln – Thr – Val – Ser – Arg – Val – Val – Asn – Gly – Ala – Ser – His – Val – Ser
20 23 restart 30

GCG AAA ACG CGG GAA AAA **GUG** GAA GCG GCG **AUG** GCG GAG CUG AAU UAC AUU CCC AAC CGC
Ala – Lys – Thr – Arg – Glu – Lys – Val – Glu – Ala – Ala – Met – Ala – Glu – Leu – Asn – Tyr – Ile – Pro Asn – Arg
40 42 restart 50

GUC GC(A CAA C)AA CUG GCG GGC AAA CAG UCG (**UUG**)
Val – Ala – Gln – Gln – Leu – Ala – Gly – Lys – Gln – Ser – Leu
60 62 restart

[a]Sequence of the *lacI* mRNA (Steege, 1977*b*) and the repressor protein (Beyreuther *et al.,* 1973), arranged so that the initiator codons for the three restart proteins are aligned under that for the wild type repressor. All in-phase initiator triplets are shown in boldface type. Sequences which are complementary to the 3′ end of 16 S rRNA and precede the utilized initiators are indicated by underlining (with G·U pairs denoted by dots). The ribosome-protected region (Steege, 1977*b*) is indicated by a dashed line.

A plausible explanation for the selective utilization of *lacI* restart signals can be formulated by examining secondary structures that might be assumed by the mRNA beyond the site of early *I* gene nonsense codons. Figure 6 shows the position of a predicted hairpin loop (closed by a very stable helical stem) relative to several potential restart sites. This structure effectively buries the GGGA sequence complementary to rRNA that appears four nucleotides prior to a GUG at valine 38; it can thereby account for the observed lack of reinitiation at this site. Also, the curious observation that reinitiation at the valine 23 site always occurs concomitantly with the production of the methionine 42 restart protein (Files *et al.*, 1974) can be explained: Since ribosome binding at the accessible methionine 42 site should destabilize the helical region (it falls within the six triplets 5′ to the initiator that are normally protected by the ribosome), the sequence preceding the valine 23 GUG would be exposed only as a consequence of reinitiation at position 42. It is hoped that future analyses of restart peptides synthesized by strains carrying *lacI* mutations located in the double-helical stem shown in Fig. 6 will solidify evidence for the *in vivo* formation and functioning of this structure.

2.9 *Mutations in Ribosome-Binding Sites*

In contrast to promoter and operator mutants, which have been most helpful in identifying nucleotide sequences crucial to transcriptional initiation and control, mutants affecting translational initiation have been slow to appear. Part of the problem certainly stems from the fact that even a "noninitiator" AUG (for example, in a synthetic mRNA or at a restart site) can, under the appropriate conditions, direct initiation at a 5 or 10% level (see earlier). Thus, it is more difficult to devise selection schemes for mutants altered in the initiation of translation than of transcription, where the level of functioning may change 100- to 1000-fold.

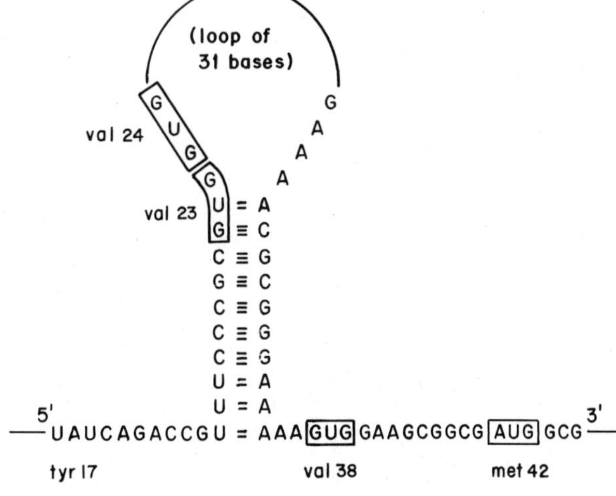

Figure 6. A secondary structure which may be assumed by a portion of the *lacI* mRNA beyond the sites of early *I* gene amber mutations. The stem region illustrated has a predicted T_m of 97°C (Gralla and Crothers, 1973). Potential initiator triplets are blocked in. Restart repressor peptides initiating only at Val 23 and Met 42 have been observed. (From Steege, 1977*a,* with permission of the author and publisher.)

Nonetheless, both to ensure that our current ideas about the important elements in initiator regions are correct and to refine further our understanding of their functioning, characterization of ribosome-binding-site mutants will be required. At present, analysis of the first such mutants is only beginning. Hence, the discussion here is confined to the types of mutations we might expect and how the preliminary data fit our predictions.

One class of ribosome-binding-site mutants should be those affecting the triplet interaction between the initiator codon and the fMet-tRNA anticodon sequence. Here Taniguchi and Weissmann (1978) have obtained such mutants in the $Q\beta$ coat protein cistron. In these mutants the third position of the initiator AUG and the next residue (the first position of the succeeding alanine codon) are altered. Potential base pairing between the initiator tRNA anticodon region and the relevant $Q\beta$ RNA sequences is presented in Table V, along with the relative efficiencies of *in vitro* ribosome attachment to the same four initiator regions. Since both of the mutant RNAs that lack the AUG triplet retain considerable ribosome-binding activity, we can conclude (as with the UUG restart in the *lacI* gene) that other ribosome–mRNA interactions (presumably including mRNA·rRNA base pairing) suffice for the formation of a correctly situated initiation complex. However, it is most striking that a mutation in the first position of the alanine triplet significantly increases ribosome binding! A tantalizing idea here (to be discussed in more detail shortly) is that the nucleotides flanking the tRNA anticodon have the potential for forming additional base pairs with the mRNA and can thereby further stabilize the ribosome–mRNA interaction (Taniguchi and Weissmann, 1978).

Ribosome-binding-site mutations of a second class should alter the extent of mRNA·rRNA complementarity. Analysis of such mutants would not only allow an assessment of the importance of this RNA·RNA interaction to the initiation process, but also provide a measure of the exact length of intermolecular helix formed on the ribosome. Naturally occurring mutants of this type actually exist: the coat protein ribosome-binding sites of R17 versus MS2 and f2 phages (Table I) differ only in a G→ A transition, which slightly increases the predicted stability of the mRNA·rRNA interaction in the latter cases. However, the relative initiation capacities of these initiator regions have never been directly compared. Luckily, another such mutant has recently been obtained and studied (J. J. Dunn, E. Pollert, and F. W. Studier, personal communication). It affects the T7 gene *0.3*

Table V. mRNA·tRNA Interaction at $Q\beta$ Coat Protein Initiator Region[a]

	mRNA sequences	Relative ribosome binding
Wild type	C**AUG**G	1
Mutants	C**AUG**A	2.8
	C**AUA**G	<0.1
	C**AUA**A	0.33
Anticodon loop sequence of $tRNA_f^{Met}$	$_{3'}$AUACU$_{5'}$	

[a]Mutant $Q\beta$ RNAs and ribosome-binding data from Taniguchi and Weissmann (1978). Underlining indicates possible base pairing between the mRNA and the anticodon loop sequence of $tRNA_f^{Met}$ (Dube *et al.*, 1968). The "initiator triplet" appears in boldface type.

ribosome-binding-site *a* (Table I) and transforms its five-base match with 16 S rRNA (GAGGU) into a less complementary sequence (GAAGU). Concomitantly, a drop of about tenfold in the efficiency of synthesis of the gene *0.3* protein in T7-infected cells is observed. This important observation provides direct evidence that mRNA·rRNA base pairing makes a significant contribution to ribosome recognition of true initiation signals in mRNA *in vivo*.

Finally, if RNA secondary structure is indeed a potent negative regulator of initiation, both "up" and "down" mutations that alter the ability of the mRNA region surrounding an initiator triplet to assume a stable secondary conformation should be discovered. A most intriguing possibility (Singer and Gold, 1976; Belin and Epstein, 1977) is HD263, a recessive temperature-sensitive mutation which affects the synthesis of the bacteriophage T4 rIIB protein. At 25°C, the amount of mutant protein produced is 10 to 20% of the wild type level; at 37°C and above, no detectable synthesis occurs. Reduced transcription, messenger instability, and differences in protein sequence or degradation have been ruled out as possible explanations. Rather, the existence of a "melting curve" for rIIB expression in HD263 suggests that RNA conformation plays a role in translational initiation at this site. Presumably, sequence analysis of the region surrounding the beginning of the rIIB cistron will establish whether this intuition is correct.

2.10 *Perspectives and Problems*

If now asked to summarize how a ribosome recognizes a true initiation signal in messenger RNA, we can come up with the following molecular picture. The mRNA primary structure participates in at least two RNA·RNA interactions during the positioning of the ribosome to form the initiation complex: specific portions of the message base pair with the initiator tRNA on the one hand and with the 3′ end of the 16 S rRNA on the other. Also, during the ribosome-binding process, mRNA secondary structure functions as a controlling element, covering unwanted internal initiators and regulating the availability of authentic initiation sites. However, simple accessibility of an mRNA region including an initiator triplet and a sequence complementary to 16 S rRNA is not sufficient to define an initiator region; we have seen that *B. stearothermophilus* ribosomes fail to recognize several RNA phage sites that are bound by *E. coli* ribosomes under identical conditions. Thus, proteins, several of which can be viewed as acting through the two RNA·RNA interactions, enter the picture as additional contributors to mRNA recognition by ribosomes. Not bad for just the ribosome and the mRNA!

Considering what defines an active initiator region inside a living cell brings up numerous complications. In addition to variables such as the level of transcription and messenger lifetime, there exists the possibility of competition between mRNAs. Under circumstances in which ribosomes are limiting, presumably those initiator regions whose individual interactions sum to give the highest ribosome affinity will win out over weaker sites (for an analytical treatment see Lodish, 1974). Further, we have seen that some initiator regions are designed to be sensitive to the intracellular concentration of a regulatory protein, whose binding effectively eliminates its target site from the competing pool. Certain temporal factors must also be important *in vivo*. For instance, all buried internal codons must be exposed at least momentarily both during transcription and as a translating ribosome successively unfolds mRNA secondary structures; yet only under special circumstances (for example, the *lacI* restarts, which occur exclusively after translational termination) are certain internal sites utilized. This leads to the suspicion that there may exist a minimum time during which a potential initiator region must be uncovered and available to participate in the assembly of an initiation complex in order for the process to reach completion.

Although much more work will be required to understand the *in vivo* interplay of all these variables, we can still measure our progress by posing a simple and important question. Namely, is it possible to predict the relative translational efficiencies of known mRNA initiator regions from the strength of the mRNA–ribosome interactions we can currently identify?

We have already discussed three situations in which ribosome binding proceeds in the absence of an AUG or GUG initiator triplet. These are (1) reinitiation at the leucine 62 (UUG) codon in the *lacI* mRNA; (2) ribosome recognition of the mutant Qβ coat protein initiator region with the AUG → AUA transition; and (3) selection of Qβ sites lacking initiator triplets by *B. stearothermophilus* ribosomes. Thus, we know that other contacts between mRNA and the ribosome can substitute for a proper codon·anticodon match both *in vivo* and *in vitro.* Yet, the low efficiency with which the Qβ mutant sequence binds to ribosomes *in vitro* (about 10% relative to the wild type site) indicates that mRNA·tRNA base pairing can contribute significantly to the energy required for formation of the initiation complex.

Is there evidence that translational efficiency is likewise correlated with the degree of mRNA·rRNA complementarity? Certainly, it is not valid to compare initiation sites with totally different sequences (as in Table I) both because of the negative effects of mRNA structure and the variable possibility of additional interactions between the messenger and the ribosome. Rather, this question must be posed with initiator regions which are identical except in their degree of complementarity to rRNA. So far, only the T7 gene *0.3* ribosome-binding-site *a* mutant (with the GAGGU → GAAGU transition) discussed previously fulfills this qualification; its lower expression *in vivo* constitutes strong evidence for a direct relationship between mRNA·rRNA complementarity and translational efficiency. Another more extreme situation involves the λ repressor (c_I) protein, which can be synthesized from either of two mRNAs originating at different promoters (Walz *et al.,* 1976; Ptashne *et al.,* 1976); one message possesses the complete ribosome-binding-site sequence shown in Table I, whereas the other begins at the initiator triplet with pppAUG. . . . Again, the 10- to 20-fold lower level of protein synthesized in the latter case (Ptashne *et al.,* 1976) argues for the importance of mRNA·rRNA base pairing to initiation efficiency.

Thus, through analysis of mutants, we have begun to dissect the initiation process and delineate what contribution each of the RNA·RNA interactions makes to ribosome binding. However, there are still several problem areas in our molecular picture of mRNA–ribosome interaction which can be tackled in other ways.

First, since the distance between the mRNA region complementary to rRNA and the initiator triplet is not constant (see Table I), we would like to know how much variability can be accommodated without introducing strain into the initiation complex. The involvement of an apparently flexible 3′ terminus on the 16 S rRNA (van Duin *et al.,* 1976; Chapman and Noller, 1977; Santer and Shane, 1977) is unquestionably helpful. Further, paper and pencil model building with all the sites listed in Table I suggests that fixed positions for both the mRNA initiator triplet and the upper portion of the 16 S rRNA hairpin loop region (Fig. 1) can be found that do allow interaction of the complementary regions from the two RNA molecules. An ideal test of this parameter would require chemical synthesis of a series of polynucleotides containing a complementary region separated from an initiator triplet by a homopolymer sequence varying in length from 2 to 12 residues (suggested by O. Uhlenbeck).

Second, since we do not know how or when the mRNA·rRNA bonds are broken, it is conceivable that weaker messenger complementarity to rRNA might facilitate more rapid elongation of the polypeptide chain. Thus, we may eventually discover that initiation efficiency is only indirectly related to the strength of the mRNA·rRNA interaction. Here, it is relevant that the measured T_m values for the most stable isolated mRNA·rRNA hybrids are no higher than physiological temperatures (see earlier), indicating that a large energy input is not required in most cases for disrupting these interactions. Also, the finding that ribosomes whose 16 S rRNA has suffered cleavage by colicin E3 (the terminal fragment remains on the ribosome) can nevertheless correctly form initiation complexes (Ravetch and Jakes, 1976) and synthesize fMet-dipeptides (Baan *et al.,* 1976) suggests that the mRNA·rRNA base pairs are not severed until, or after, the first translocation event. However, even subsequent elongation steps could be rate limiting to translation. Basically, what we require are data comparing ribosome binding with *in vitro* translation efficiency for cistrons in mRNAs other than the RNA phages, where a reasonably good correlation of these two parameters is observed (see Steitz, 1975).

Perhaps most important, we must continue to gather data on RNA secondary and tertiary structure and to incorporate any new principles which might emerge into our formulation of what constitutes an initiation signal. For instance, the recently documented hydrogen bonding of 2′-OH groups to phosphate residues in the backbone of phenylalanine tRNA (Ladner *et al.,* 1975; Quigley and Rich, 1976) suggests quite different possibilities for the tertiary folding of RNA relative to DNA molecules, even though these two polynucleotides admittedly have the same potential for forming Watson–Crick base pairs. Such conformational differences might even explain the previously puzzling observation that the ØX174 gene G initiator region in the single-stranded phage DNA molecule is bound almost exclusively by ribosomes (Robertson *et al.,* 1973), whereas in mRNA this site is weak relative to other true initiators (Ravetch *et al.,* 1977). Clearly, we have a long way to go in the area of three-dimensional RNA structure. It cannot be considered anything but an exceedingly challenging molecular problem.

3 Sequences Directing Elongation of Polypeptide Chains

The genetic signals appearing in the interior of an mRNA cistron are the triplet code words. Their major function is, of course, to position successive aminoacyl-tRNAs as the ribosome moves along the messenger molecule during polypeptide chain elongation. Here we presume that only the linear structure of the RNA is important; once protein synthesis is initiated, the ribosome can unravel even very stable mRNA secondary structures.

The genetic code, as we know it today, was established by the mid-1960s. Therefore, mRNA sequence analysis has simply confirmed the validity of earlier assignments rather than discovering new genetic signals. However, the degeneracy of the code (use of more than one codon to specify a given amino acid) has long provided fertile ground for speculations concerning the functional implications of flexibility in codon usage. Indeed, as more and more mRNA sequences have accumulated, distinctive examples of selective code word utilization and of overlapping function have begun to emerge. The nature and possible molecular origins of these phenomena are considered next.

3.1 *Selective Codon Usage in Bacteriophage Messengers*

At present, nucleotide sequences from only two genomes are extensive enough to allow meaningful statistical analysis of primary structure. These are the single-stranded RNA molecule of bacteriophage MS2 and the circular single-stranded DNA genome of phage ϕX174.

3.1.1 *MS2*

The entire sequence of 3569 residues in the MS2 bacteriophage genome has recently been completed by Fiers and his co-workers, using RNA sequencing methods (Min Jou *et al.*, 1972; Fiers *et al.*, 1975, 1976). This monumental feat allows examination of a total of 1068 code words comprising the three bacteriophage cistrons. Table VI, which compares data for the entire genome (A) with that for the 132-codon coat cistron (B), shows that the choice of triplets in MS2 is certainly not random.

In the case of four amino acids, MS2 codon utilization patterns are particularly distinctive (Fiers *et al.*, 1976). For glycine, GGU is definitely preferred (in the entire genome) over the other three glycine codons; however, a third position U preference is not observed for other amino acids. Discrimination against certain isoleucine, tyrosine, and arginine triplets hints that these may function as modulating codons which slow the speed of translation. Namely, UAU for tyrosine, CGPu and AGPu for arginine, and AUA for isoleucine are totally absent from the coat cistron (which requires efficient translation), whereas they do appear (albeit with relatively low frequency) in the A protein and replicase genes. The observation that the isoleucine tRNA species which responds to AUA is present in only very low amounts in *E. coli* (Harada and Nishimura, 1974) supports the modulation hypothesis (Ames and Martin, 1964).

One early explanation for nonrandom codon usage in MS2 RNA was that the phage RNA secondary structure might be maximized by a proper choice of degenerate triplets. Although in the coat cistron third codon letters do most often appear base-paired to first and second position residues (Min Jou *et al.*, 1972), this pattern has not been substantiated by subsequent data from the other two genes

A. Entire genome[a]											
Phe	UUU	19	Ser	UCU	15	Tyr	UAU	9	Cys	UGU	6
	UUC	29		UCC	20		UAC	32		UGC	6
Leu	UUA	17		UCA	16	Ter	UAA	1	Ter	UGA	0
	UUG	11		UCG	22	Ter	UAG	2	Trp	UGG	23
Leu	CUU	15	Pro	CCU	17	His	CAU	6	Arg	CGU	21
	CUC	26		CCC	10		CAC	9		CGC	20
	CUA	15		CCA	9	Gln	CAA	17		CGA	10
	CUG	9		CCG	13		CAG	22		CGG	11
Ile	AUU	12	Thr	ACU	19	Asn	AAU	17	Ser	AGU	8
	AUC	25		ACC	21		AAC	28		AGC	16
	AUA	19		ACA	13	Lys	AAA	19	Arg	AGA	7
Met	AUG	20		ACG	14		AAG	26		AGG	6
Val	GUU	21	Ala	GCU	26	Asp	GAU	28	Gly	GGU	37
	GUC	21		GCC	21		GAC	22		GGC	16
	GUA	16		GCA	21	Glu	GAA	16		GGA	12
	GUG	18 + 1[b]		GCG	23		GAG	28		GGG	16
B. Coat protein cistron[c]											
Phe	UUU	1	Ser	UCU	3	Tyr	UAU	0	Cys	UGU	1
	UUC	3		UCC	2		UAC	4		UGC	1
Leu	UUA	1		UCA	2	Ter	UAA	1	Ter	UGA	0
	UUG	0		UCG	2	Ter	UAG	1	Trp	UGG	2
Leu	CUU	2	Pro	CCU	2	His	CAU	0	Arg	CGU	3
	CUC	2		CCC	1		CAC	0		CGC	1
	CUA	2		CCA	2	Gln	CAA	1		CGA	0
	CUG	0		CCG	1		CAG	5		CGG	0
Ile	AUU	4	Thr	ACU	4	Asn	AAU	4	Ser	AGU	0
	AUC	4		ACC	4		AAC	6		AGC	4
	AUA	0		ACA	0	Lys	AAA	5	Arg	AGA	0
Met	AUG	3		ACG	1		AAG	1		AGG	0
Val	GUU	4	Ala	GCU	5	Asp	GAU	1	Gly	GGU	3
	GUC	4		GCC	2		GAC	3		GGC	3
	GUA	3		GCA	6	Glu	GAA	2		GGA	2
	GUG	3		GCG	1		GAG	3		GGG	1

[a] Data for 1068 codons from Fiers *et al.* (1976).
[b] One GUG triplet codes for fMet.
[c] Data for 132 codons from Min Jou *et al.* (1972).

(Fiers *et al.,* 1976). Also, the number of nonsense triplets appearing in the two nonreading frames is no higher than expected from a random distribution (Fiers *et al.,* 1975, 1976) and therefore cannot explain selective codon usage. On the other hand, it is not unreasonable to expect that requirements for forming specific RNA tertiary structures or for preventing interaction of the phage minus strand with cellular components such as ribosomes will take advantage of third position flexibility in the code. At this time we have no way of evaluating the extent of such constraints on code word utilization.

A most intriguing correlation has recently been established between the frequency of appearance of degenerate codons in MS2 RNA and the predicted lifetimes of the corresponding codon·anticodon interactions (H. Grosjean, B. Cedergren, D. Sankoff, W. Min Jou, and W. Fiers, personal communication). Previously, both the anticodon base composition and the presence (or absence) of certain modified nucleotides in the anticodon loop region of a tRNA molecule had been shown to influence greatly the stability of *in vitro* complexes formed between two tRNAs with complementary anticodons (Grosjean *et al.*, 1976, 1978). Assuming comparable effects of these variables on base pairing between an mRNA triplet and the decoding tRNA on the ribosome, the codon usage pattern in MS2 RNA hints that there might exist an optimal lifetime for efficient codon · anticodon interaction. A simplified example of what has been observed is that if the first two letters of a codon produce G · C base pairs, the third pair tends to be G · U; and conversely, when the first two are A · U, the third tends to be G · C. Thus, intermediate strength codon · anticodon interactions are apparently preferred over either very strong or very weak base pairing. These findings support current coding theories based on kinetic parameters of the codon · tRNA association (Ninio, 1974; Hopfield, 1974).

3.1.2. ϕX174

Sanger and his collaborators (Sanger *et al.*, 1977) have utilized elegant DNA sequencing methods to elucidate the complete nucleotide sequence of the ϕX174 genome. It contains some 5375 residues, which code for nine known proteins. The data for 1346 ϕX codons (from fully confirmed portions of the sequence) are presented in Table VII. Most striking is the observation that over 40% of the code words terminate in T (Sanger *et al.*, 1977), a peculiarity which is reflected in the high (31%) T content of ϕX174 DNA. Another nonrandom feature of codon usage is the very low occurrence of triplets starting with AG.

Statistical analysis of portions of the ϕX174 (and MS2) sequence has led Pieczenik (1972) to an exciting and profound realization concerning selective codon utilization in this phage genome. His studies revealed that the probability of being able to form four or five base pairs between the messenger RNA and the anticodon loop sequence of each decoding tRNA is significantly higher in the reading frame than in the other two frames. Currently, we can only speculate on

Table VII. Codons Used in ϕX174[a]

Phe	TTT	39	Ser	TCT	35	Tyr	TAT	36	Cys	TGT	12
	TTC	26		TCC	9		TAC	15		TGC	10
Leu	TTA	19		TCA	16	Ter	TAA	3	Ter	TGA	5
	TTG	26		TCG	14		TAG	0	Trp	TGG	16
Leu	CTT	36	Pro	CCT	34	His	CAT	16	Arg	CGT	40
	CTC	15		CCC	6		CAC	7		CGC	29
	CTA	3		CCA	6	Gln	CAA	27		CGA	4
	CTG	24		CCG	21		CAG	34		CGG	8
Ile	ATT	45	Thr	ACT	40	Asn	AAT	37	Ser	AGT	9
	ATC	12		ACC	18		AAC	25		AGC	5
	ATA	2		ACA	13	Lys	AAA	47	Arg	AGA	6
Met	ATG	42		ACG	19		AAG	31		AGG	1
Val	GTT	53	Ala	GCT	64	Asp	GAT	44	Gly	GGT	38
	GTC	14		GCC	17		GAC	35		GGC	28
	GTA	10		GCA	12	Glu	GAA	27		GGA	13
	GTG	11		GCG	12		GAG	34		GGG	3

[a] Data for 1346 codons from Sanger *et al.* (1977).

whether this is a vestigial pattern reflecting the evolutionary development of a functional genetic code (Crick *et al.*, 1976) and whether the potential formation of base pairs flanking the codon·anticodon interaction is utilized by today's ribosome. *In vivo* tests of the latter possibility could now be undertaken in several systems in which a particular (eukaryotic) cell type has been shown to produce elevated amounts of certain tRNAs to facilitate translation of a specialized protein whose mRNA sequence is known. For example, in the fibroin messenger RNA from the silk gland of *Bombyx mori,* sequences encoding the reiterated amino acid sequence Gly–Ser–Gly–Ala are distinctively nonrandom (Suzuki and Brown, 1972): Only UCA and GCU are utilized for serine and alanine, respectively; either GGU or GGA can be used for glycine when it is followed by alanine, but only GGU appears 5′ to serine codons (K. U. Sprague, unpublished observations). Sequence analysis of the correspondingly abundant alanine, glycine, and serine tRNA species from the silk gland (Garel *et al.*, 1971) should establish whether more than a three-base-pair mRNA·tRNA match has evolved (or been maintained) in this system. Note that if some degree of codon overlap is in fact used by the present translational machinery, an intrinsic self-buffering capacity of the genetic material will have been discovered, rendering it more resistant to outside manipulation than previously believed.

3.2 *Overlapping Genes and Signals in Messenger RNA*

It was once fashionable to believe that coding regions in a messenger RNA were monofunctional. That is, it seemed quite enough to expect that these internal sequences should direct the synthesis of a protein and, at the same time, possess sufficient ribonuclease-resistant features to avoid immediate obliteration by cellular enzymes. In fact, the first mRNA sequences obtained fortified this myth. In the RNA bacteriophage genome, signals specifying ribosome binding or recognition by components of the RNA replication machinery are almost exclusively confined to rather lengthy extracistronic regions whose sequences interestingly turn out to be more highly conserved among the closely related phages than the coding regions (Min Jou and Fiers, 1976)! However, the subsequent elucidation of sequences from other phages and *E. coli* mRNAs has devastated this tidy model; the frequent discovery of new overlapping genes and genetic signals now emphasizes the multifunctional potential of RNA molecules and the ability of an organism to utilize its genetic material to the fullest.

The ultimate example of multiply functioning coding sequences became apparent during the recent DNA sequence work on bacteriophage ØX174 (Sanger *et al.*, 1977). Here, two cases of overlapping cistrons which are translated in different reading frames have been documented! Mutant analysis places the B protein coding sequence (360 nucleotides) completely within the A gene (1536 nucleotides) (Sanger *et al.*, 1977). Likewise, amber mutants in gene E and the amino acid sequence of the D protein demonstrate that the E coding sequence (273 nucleotides) lies entirely within the D cistron (456 nucleotides) (Barrell *et al.*, 1976). Such extensive codon overlap must place severe restrictions on the amino acid sequences of these protein pairs. Moreover, it is obvious that sequences signaling the initiation (and termination) of the included polypeptide chains must simultaneously specify the amino acid sequences of the two longer proteins. Nevertheless, both the ϕX174 gene B ribosome-binding site (Table I) and the proposed E initiator region (Barrell *et al.,* 1976) exhibit the usual appropriately positioned complementarity to 16 S rRNA.

Wild type ribosome-binding sites which completely or partially overlap other structural genes are actually nothing new. Several examples had previously been identified in both *E. coli* and bacteriophage messengers. The AUG initiator codon for the *trp A* protein (the sixth and last known polypeptide encoded by the polycistronic tryptophan mRNA of *E. coli*) overlaps in its first position the UGA termination triplet of the preceding *trp B* gene (Table I). Again in ϕX174, the A protein has long been known to appear in two forms, called A and A*. The shorter A* protein is basically a restart peptide (Linney and Hayashi, 1974), presumably initiated at an internal in-phase methionine codon of the A gene, but it is synthesized concomitantly with the full-length A protein by wild type phage. Also, the ϕX174 D gene terminator codon has been found to overlap the initiation triplet for the J protein (*TAATG,* Table I).

The list of various other genetic signals that either appear within structural genes or overlap ribosome recognition sites is also rapidly growing. For instance, the origin of ØX174 viral strand DNA synthesis has been located within the gene for the A protein (Baas *et al.,* 1976), which functions in DNA replication by introducing a specific nick (Henry and Knippers, 1974) into its own coding region in the replicative form DNA. In the double-stranded DNA genome of bacteriophage lambda, there exists a region about ten nucleotides long which (1) can be bound by the lambda repressor protein, (2) is recognized by RNA polymerase as it initiates the short version of the λc_{I} mRNA (see earlier), and (3) also possesses the sequence complementary to 16 S rRNA which precedes the c_{I} coding region (Table I) (Walz *et al.,* 1976; Ptashne *et al.,* 1976). In the lactose operon of *E. coli,* not only does the major repressor-recognized DNA sequence overlap the 5′ end of the ribosome-binding site for the *lacZ* gene (Maizels, 1974), but a second (weaker) repressor-binding site is located in the middle of the *Z* cistron (Reznikoff *et al.,* 1974). Even in the RNA bacteriophages, binding sites for proteins involved in replication are now being discovered within coding regions: A second replicase-binding site (see earlier) in Qβ RNA is probably located in the replicase gene (Weber *et al.,* 1974; Meyer *et al.,* 1975) as is one of two regions bound by host factor (Senear and Steitz, 1976), an *E. coli* protein necessary for the initiation of Qβ minus strand synthesis using plus strand template. Taken together, these examples of gene–signal overlap erase any lingering doubt about the potential multifunctionality of mRNA sequences and structures.

4 *RNA·RNA Interactions in Ribosome Function*

The acquisition of vast amounts of RNA sequence information has been instrumental in refocusing our attention on the central role of RNA molecules in protein biosynthesis. Assuredly, our molecular understanding of translation began with these polynucleotides: The discovery of messenger RNA and the realization that tRNAs serve as adaptors in the decoding process nicely complemented the earlier finding that ribosomes contain RNA. Then, for some years we seemed to have lost sight of the fact that about two-thirds of the mass of each ribosomal subunit is RNA. Enumerating and delineating the functions of at least 56 ribosomal proteins, three initiation factors, three elongation factors, and two termination factors was so all-consuming that proteins were assumed to carry out most of the molecular recognition events occurring on the ribosome. Certainly the essential role of each of these proteins in a particular stage of translation cannot be disputed. However, recent developments indicate that the three ribosomal RNA molecules (16 S, 23 S, and 5 S) are much more than just convenient racks on which to hang ribosomal proteins and factors (see also Kurland, 1974).

In fact, intermolecular RNA·RNA interactions may provide the basis for almost every reaction which takes place on the *E. coli* ribosome. A list, beginning with the three interactions just discussed, can be compiled as follows:

Accepted:

1. The anticodon of the initiator tRNA base-pairs with the mRNA initiator triplet during polypeptide chain initiation.
2. The anticodons of elongator tRNAs base-pair with mRNA codons during polypeptide chain elongation.

Reasonably solid:

3. The 3′ end of 16 S rRNA base-pairs with a purine-rich region in the mRNA during polypeptide chain initiation.
4. The TψC region of elongator tRNAs base-pairs with 5 S RNA during polypeptide chain elongation. Here the work of Erdmann and his colleagues (Erdmann *et al.*, 1973; Richter *et al.*, 1973) provides strong evidence that GTψC interacts with a complementary GAAC sequence in 5 S rRNA during positioning of an incoming amino acyl-tRNA on the ribosome. Specifically, the tetranucleotide, TψCG, which binds stoichiometrically to the 50 S subunit (Erdmann, 1976), prevents tRNA binding to the ribosomal A site* but not initiator tRNA binding to the P site* (Richter *et al.*, 1973; Erdmann, 1976). The postulated intermolecular base pairing would require an opening of the TψC loop region from its structure in $tRNA^{Phe}$, as deduced by X-ray crystallographic analysis (see Rich and Raj Bhandary, 1976); such a conformational change in the tRNA molecule could be triggered by codon·anticodon interaction (Schwarz *et al.*, 1976). Finally, the tetranucleotide, TψCG, is sufficient to induce stringent factor-mediated production of ppGpp by the ribosome (Richter *et al.*, 1974), suggesting that this curious reaction is also dependent on direct 5 S·tRNA base pairing.

Still speculative:

5. Bases in the D (dihydrouracil) loop of the initiator tRNA may pair with 23 S rRNA during joining of the 50 S ribosomal subunit to the 30 S initiation complex. Substantial complementarity between the D loop and stem region of $tRNA_f^{Met}$ and a sequence appearing somewhere in 23 S RNA has recently been detected by E. Lund and J. Dahlberg (personal communication). Although it is perhaps unlikely that a tRNA stem unwinds on the ribosome, utilization of unpaired nucleotides in the D loop for interaction with an RNA in the 50 S subunit seems most reasonable.

*See the footnote on page 359.

6. 16 S and 23 S rRNAs may interact in the 70 S ribosome. The finding that initiation factor IF-3 can be cross-linked to the 3′ end of 23 S RNA as well as to 16 S RNA has led Van Duin *et al.* (1976) to suggest that the 3′ termini of these two RNA molecules base-pair upon formation of the 70 S ribosome. Their proposal is supported both by reasonable complementarity between the two sequences (assuming a certain 16 S RNA secondary structure) and by observed differences in the nuclease susceptibility of the 16 S molecule in the 30 S versus the 70 S ribosomal subunit (Chapman and Noller, 1977; Santer and Shane, 1977). Moreover, two other remarkably extensive base-pairing possibilities between 16 S and 23 S rRNAs have recently been discussed (Branlant *et al.*, 1976). Whether or not these exact interactions occur, the general notion of 23 S·16 S base pairing during ribosomal subunit association is appealing.
7. 23 S and 5 S rRNAs may base-pair during some phase of ribosome function. Herr and Noller (1975) have pointed out a 12-base complementarity between 5 S RNA and a sequence in 23 S RNA which lies adjacent to an exposed G residue in the 50 S ribosomal subunit. This particular region (nucleotides 72–83) in the 5 S molecule has been implicated in 5 S·23 S interaction mediated by proteins (Gray *et al.,* 1973) and is also highly conserved among prokaryotic organisms (Fox and Woese, 1975). Although no direct evidence for intermolecular base pairing has yet been obtained, lively speculations concerning the functional implications of such an interaction have appeared (Weidner *et al.*, 1977).
8. Initiator and elongator tRNAs may base-pair with 5 S RNA in the P site on the ribosome. Nuclease digestion experiments, which reveal that the TψC loop and the anticodon region are uniquely protected in $tRNA_f^{Met}$ when bound to the 70 S ribosome, have led Dube (1973) to suggest that a GTψC·5 S interaction also occurs in the P site on the ribosome. Despite the seemingly contrary results mentioned in item 4, he argues that since prokaryotic 5 S RNA does have two GAAC sequences, similar RNA·RNA interactions could occur in both tRNA-binding sites.
9. The 3′ end of 16 S rRNA may participate in polypeptide chain termination. Because *E. coli* 16 S rRNA terminates with UUA_{OH}, which (allowing wobble) is complementary to all three translational terminator triplets, Shine and Dalgarno (1974) have discussed the possibility that the 3′ end of 16 S rRNA recognizes mRNA signals for termination as well as for initiation. One attractive aspect of this proposed interaction is that it could explain why discrimination between individual terminator trinucleotides by isolated protein release factors has not been observed (Capecchi and Klein, 1969). However, since the extreme 3′-terminal sequences of 16 S rRNA from all bacterial species are not identical (Shine and Dalgarno, 1975), it is perhaps more reasonable to search for another rRANA region which might base-pair with terminator triplets during the termination step of protein biosynthesis.

These speculations lead to two intriguing questions. First, how many more undiscovered, but functionally significant RNA·RNA interactions take place on the prokaryotic ribosome? Second, are these interactions trying to tell us something about the evolution of the translational apparatus? To dispose of the unanswerable question first, Crick (1968), Orgel (1968), and Woese (1967) have all ventured the opinion that the primitive translational apparatus may have consisted entirely of RNA molecules. Knowing what we now know about the importance of substrate positioning in enzymatic catalysis, it is perhaps not so farfetched to suggest that at an early time rRNA molecules may have functioned as rudimentary enzymes; later this particular role was taken over by the more precise and efficient proteins. With respect to the experimentally vulnerable former question, only time will tell how many more RNA·RNA interactions we can identify in ribosome function. My guess is at least several.

5 *Are Eukaryotic Messengers Different?*

At first glance, the nature and functioning of genetic signals in eukaryotic messenger RNAs appear very similar to those in prokaryotic cells. The genetic code (with its matching set of tRNAs) has been preserved; even a special methionyl tRNA is utilized for polypeptide chain initiation, although in this case the charged $tRNA_f^{Met}$ does not become formylated. The eukaryotic translational apparatus consists of the same general array of large and small subunit ribosomal proteins, ribosomal RNAs, and initiation, elongation, and termination factors. Only the numbers, sizes, and shapes of these molecules differ somewhat (for reviews see Wool and Stöffler, 1974; Weissbach and Ochoa, 1976).

On the other hand, the nuclear membrane of a eukaryotic cell necessitates certain modifications in the mode of messenger utilization. In eukaryotic organisms, translation cannot be initiated as soon as a portion of a cistron has been transcribed, as it can be in prokaryotic organisms, because the mRNA must first be transported from its site of synthesis on the nuclear chromatin to the ribosomes in the cytoplasm. During this process, it seems that a large amount of eukaryotic RNA is discarded in the nucleus (see Perry, 1976). Although it is not clear whether certain sequences are destined for total destruction or whether a fraction of all RNA transcripts turn over, the existence of longer nuclear precursors to certain specific cytoplasmic mRNAs has been well documented (Ross, 1976; Curtis and Weissmann, 1976).

The messenger RNA molecules which ultimately appear in the eukaryotic cytoplasm also differ from their prokaryotic counterparts in a number of structural features. First, for reasons we do not understand, eukaryotic mRNAs always seem to be monocistronic, in the sense that they contain only one functional ribosome binding site (for a discussion see Lodish, 1976). Here, we know that eukaryotic ribosomes can initiate and translate internal genes on a polycistronic bacterial mRNA and that certain viral transcripts contain internal initiation sequences. Yet, in no case is there convincing evidence that ribosomes begin protein synthesis without prior RNA processing which moves the internal initiator region closer to the 5′ end of the eukaryotic messenger molecule.

Second, eukaryotic mRNAs acquire certain unique covalent modifications. Specifically, the 5′ termini of most (but not all) messengers possess a peculiar blocking group of the general structure $m^7G(5')ppp(5')ppp(5')Nmp$ . . . (Adams and Cory, 1975; also see Lodish, 1976, for further references). When present, this methylated "cap" moiety is required for ribosome binding and translation (Both *et al.,* 1975*a, b;* Muthukrishnan *et al.,* 1975). At the 3′ end, most eukaryotic messenger molecules carry a poly-A tail of some 50 to 200 residues, which has been added posttranscriptionally (Darnell *et al.,* 1973; Brawerman, 1974). This modification is obligatory neither for transport out of the nucleus nor for translation, since it is not present on all functional mRNAs (for example, histone messenger) (Adesnik and Darnell, 1972; Greenberg and Perry, 1972). Finally, a limited number of other methyl groups are introduced into apparently specific sites on eukaryotic mRNAs (for references see Perry, 1976); their function is also unknown.

Another distinguishing feature of eukaryotic mRNAs is that they interact quite specifically with different sets of proteins at various stages in their lifetime. In addition to the enzymes involved in mRNA processing and modification, both nuclear and cytoplasmic RNA-binding proteins have been characterized (for references see Brawerman, 1974). Certain of these are present in rather large amounts in cells. Some may move with the RNA from the nucleus to the cytoplasm. Others may specifically recognize particular mRNAs and regulate their utilization in the cytoplasm. Finally, during translation, every mRNA interacts with a ribosome that is richer in protein than is the prokaryotic organelle (Wool and Stöffler, 1974). Also, the initiation factors are at least six in number, consist of up to ten subunits, and catalyze more steps than are required for assembly of a prokaryotic initiation complex (see Weissbach and Ochoa, 1976).

What have RNA sequences revealed concerning the nature of genetic signals in eukaryotic messengers? The first data on internal regions of mRNA confirmed code word assignments and suggested that eukaryotic mRNAs might be capable of assuming relatively stable secondary conformations (see, for example, Marotta *et al.,* 1976; Salser *et al.,* 1976). Substantial extracistronic regions have been identified at both the 5′ and 3′ termini of every eukaryotic mRNA examined. Analysis of 3′ untranslated regions from eight different vertebrate mRNAs has revealed a conserved sequence AAUAAA (Proudfoot and Brownlee, 1976; M. Billeter and W. Fiers, personal communication) which appears about 20 residues 5′ to the poly-A tail in each molecule. Amusingly, it is exactly this sequence which is recognized in RNA bacteriophage mRNAs by the *E. coli* protein (of unknown cellular function) called host factor (see Section 3.2) (Senear and Steitz, 1976).

How do ribosomes recognize eukaryotic initiator regions? Analysis of ribosome-bound sites (Dasgupta *et al.,* 1975; Kozak, 1977; Lazarowitz and Robertson, 1977; Kozak and Shatkin, 1977) and of 5′-terminal regions from both cellular (Baralle, 1977*a, b*) and viral (Shine *et al.,* 1977; Haseltine *et al.,* 1977; W. Fiers and K. Richards, personal communication) mRNAs currently provides us with a list of nearly ten (probable) eukaryotic initiator regions. Here, segments protected by the 40 S ribosome are significantly longer at their 5′ ends than are those protected by the 80 S ribosome (Kozak and Shatkin, 1976; Legon, 1976; Lazarowitz and Robertson, 1977), in contrast to the nearly equal-sized initiator regions recovered from *E. coli* 30 S and 70 S ribosomes (my unpublished observations; Legon *et al.,* 1977). The sequences directly preceding each eukaryotic initiator AUG have, of course, been closely scrutinized for complementarity to the 3′ end of 18 S ribosomal RNA, whose eight terminal nucleotides have suggestively been conserved throughout the eukaryotic kingdom (Dalgarno and Shine, 1973; Eladari and Galibert, 1975; Sprague *et al.,* 1975), excepting higher plants (Oakden and Lane, 1975). Alas, no regularly positioned mRNA sequence which could pair with the terminus of the 18 S rRNA is evident in these regions. However, results hinting at a specific 18 S·globin mRNA interaction have been obtained in several laboratories (Kabat, 1975; B. Erni and T. Staehelin, personal communication). Thus, the possibility that a sequence complementary to mRNA initiator regions resides somewhat further from the terminus of the 18 S molecule is now being investigated.

Overall, it seems quite reasonable to expect that recognition of genetic signals in eukaryotic mRNAs will involve molecular interactions of the same types as those that have been documented in prokaryotic systems. For instance, during eukaryotic (like prokaryotic) initiation complex formation, contributions are apparently made by multiple individual contacts between the ribosome and the mRNA, any one of which may be omitted if replaced by other compensating interactions [for example, poliovirus mRNAs, which lack the 5′ cap (Nomoto *et al.,* 1976; Hewlett *et al.,* 1976; Fernandez-Munoz and Darnell, 1976), can still function]. Likewise, it is probable that tRNA binding to the ribosome involves direct base pairing with 5.8 S rRNA in the P site and with 5 S rRNA in the A site (Erdmann, 1976) (the 60 S ribosomal subunit contains two small RNAs), suggesting that the eukaryotic ribosome has retained at least some RNA·RNA interactions in its functional repertoire. We can also expect that certain eukaryotic proteins which bind mRNA will recognize nucleotide sequences whereas others will select RNA secondary structures. Certainly, some of the most intriguing questions concern the exact roles of the many proteins implicated in eukaryotic mRNA metabolism, for they presumably determine which nuclear RNA sequences are transformed into cytoplasmic messengers and how mRNA utilization is regulated.

Acknowledgments

I am grateful to my Göttingen colleagues, Anthony Bretscher, David Henderson, and Norbert Geisler for their cogent comments, Mary Osborn and Klaus Weber for the hospitality of their laboratory, and Lorraine Henderson for her patient typing. I also thank all my collaborators over the past 7 years at Yale, without whom this chapter would not have been written. I was supported in part by the Josiah Macy Jr. Foundation and by Grant Number AI10243 from the National Institutes of Health.

References

Adams, J. M., and Cory, S., 1975, Modified nucleosides and bizarre 5′-termini in mouse myeloma mRNA, *Nature* **255**:28.

Adams, J. M., Cory, S., and Spahr, P. F., 1972, Nucleotide sequences of fragments of R17 bacteriophage RNA from the region immediately preceding the coat-protein cistron, *Eur. J. Biochem.* **29**:469.

Adesnik, M., and Darnell, J. E., 1972, Biogenesis and characterization of histone messenger RNA in the HeLa cell, *J. Mol. Biol.* **67**:397.

Alberts, B. M., and Frey, L., 1970, T4 bacteriophage gene 32: A structural protein in the replication and recombination of DNA, *Nature* **227**:1313.

Ames, B. N., and Martin, R. G., 1964, Biochemical aspects of genetics: The operon, *Annu. Rev. Biochem.* **33**:235.

Anderson, R. A., and Coleman, J. E., 1975, Physicochemical properties of DNA binding proteins: Gene 32 protein of T4 and *Escherichia coli* unwinding protein, *Biochemistry* **14**:5485.

Arrand, J. R., and Hindley, J., 1973, Nucleotide sequence of a ribosome binding site on RNA synthesized *in vitro* from coliphage T7, *Nature New Biol.* **244**:10.

Axelrod, N., 1976, Transcription of bacteriophage ϕX174 *in vitro:* Selective initiation with oligonucleotides, *J. Mol. Biol.* **108**:753.

Baan, R. A., Duijfjes, J. J., Van Leerdam, E., Van Knippenberg, P. H., and Bosch, L., 1976, Specific *in situ* cleavage of 16S ribosomal RNA of *Escherichia coli* interferes with the function of initiation factor IF-1, *Proc. Natl. Acad. Sci. U.S.A.* **73**:702.

Baan, R. A., Hilbers, C. W., Van Chardorp, R., Van Leerdam, E., Van Knippenberg, P. H., and Bosch, L., 1977, A high resolution proton magnetic resonance study of the secondary structure of the 3′ terminal 49 nucleotide fragment of 16S rRNA from *Escherichia coli, Proc. Natl. Acad. Sci. U.S.A.* **74**:1028.

Baas, P. D., Jansz, H. S., and Sinsheimer, R. L., 1976, Bacteriophage ϕX174 DNA synthesis in a replication-deficient host: Determination of the origin of ϕX DNA replication, *J. Mol. Biol.* **102**:633.

Baralle, F. E., 1977*a,* Complete nucleotide sequences of the 5′ noncoding region of rabbit β-globin in RNA, *Nature* **267**:279.

Baralle, F. E., 1977*b,* Structure–function relationship of 5′ noncoding sequences of rabbit α and β-globin mRNA, *Nature* **267**:279.

Barrell, B. G., 1971, Fractionation and sequence analysis of radioactive nucleotides, in: *Procedures in Nucleic Acid Research,* Vol. 2 (G. L. Cantoni and D. R. Davies, eds.), pp. 751–779, Harper, New York.

Barrell, B. G., Air, G. M., and Hutchinson, C. A., III, 1976, Overlapping genes in bacteriophage ϕX174, *Nature* **264**:34.

Bear, D. G., Ng, R., Van Derveer, D., Johnson, N. P., Thomas, G., Schleich, T., and Noller, H., 1976, Alteration of polynucleotide secondary structure by ribosomal protein S1, *Proc. Natl. Acad. Sci. U.S.A.* **73**:1824.

Belin, D., and Epstein, R. H., 1977, A temperature-sensitive rIIB mutation which affects the synthesis of bacteriophage T4 rIIB protein, *Virology* **78**:537.

Benne, R., and Pouwels, P. H., 1975, The role of IF-3 in the translation of T7- and ϕ80*trp* messenger RNA, *Mol. Gen. Genet.* **139**:311.

Benne, R., Arentzen, R., and Voorma, H.O., 1972, The mechanism of action of initiation factor F1 from *Escherichia coli, Biochim. Biophys. Acta* **269**:304.

Berissi, H., Groner, Y., and Revel, M., 1971, Effect of a purified initiation factor IF3 (B) on the selection of ribosomal binding sites on phage MS2 RNA, *Nature New Biol.* **234**:44.

Bernardi, A., and Spahr, P., 1972, Nucleotide sequence at the binding site for coat protein on the RNA of bacteriophage R17, *Proc. Natl. Acad. Sci. U.S.A.* **69**:3033.

Bertrand, K., Korn, L., Lee, F., Platt, T., Squires, C. L., Squires, C., and Yanofsky, C., 1975, New features of the regulation of the tryptophan operon, *Science* **189**:22.

Beyreuther, K., Adler, K., Geisler, N., and Klemm, A., 1973, The amino acid sequence of *lac* repressor, *Proc. Natl. Acad. Sci. U.S.A.* **70**:3576.

Blumenthal, T., Landers, T. A., and Weber, K., 1972, Bacteriophage Qβ replicase contains the protein biosynthesis elongation factors EF Tu and EF Ts, *Proc. Natl. Acad. Sci. U.S.A.* **69**:1313.

Boedtker, H., and Gesteland, R. F., 1975, Physical properties of RNA bacteriophages and their RNA, in: *RNA Phages* (N. D. Zinder, ed.), pp. 1–28, Cold Spring Harbor Lab., Cold Spring Harbor, New York.

Bollen, A., Heimark, R. L., Cozzone, A., Traut, R. R., Hershey, J. W. B., and Kahan, L., 1975, Cross-linking of initiation factor IF-2 to *Escherichia coli* 30 S ribosomal proteins with dimethylsuberimidate, *J. Biol. Chem.* **250**:4310.

Both, G. W., Furiuchi, Y., Muthukrishnan, S., and Shatkin, A. J., 1975*a,* Ribosome binding to reovirus RNA in protein synthesis requires 5′ terminal 7-methylguanosine, *Cell* **6**:185.

Both, G. W., Bannerjee, A. K., and Shatkin, A. J., 1975*b,* Methylation-dependent translation of viral messenger RNAs *in vitro, Proc. Natl. Acad. Sci. U.S.A.* **72**:1189.

Bourgeois, S., and Pfahl, M., 1976, Repressors, *Adv. Protein Chem.* **30**:1.

Bowman, C. M., Dahlberg, J. E., Ikemura, T., Konisky, J., and Nomura, M., 1971, Specific inactivation of 16S ribosomal RNA induced by colicin E3 *in vivo*, *Proc. Natl. Acad. Sci. U.S.A.* **68**:964.

Branlant, C., Sri Widada, J., Krol, A., and Ebel, J. P., 1976, Extensions of the known sequences at the 3′ and 5′ ends of 23S ribosomal RNA from *Escherichia coli,* possible base pairing between these 23S regions and 16S ribosomal RNA, *Nucleic Acids Res.* **3**:1671.

Brawerman, G., 1974, Eukaryotic messenger RNA, *Annu. Rev. Biochem.* **43**:621.

Bretscher, M. S., 1969, Direct translation of bacteriophage fd DNA in the absence of neomycin B, *J. Mol. Biol.* **42**:595.

Capecchi, M. R., and Klein, H. A., 1969, Characterization of three proteins involved in polypeptide chain termination, *Cold Spring Harbor Symp. Quant. Biol.* **34**:469.

Capecchi, M. R., and Webster, R. E., 1975, Bacteriophage RNA as template for *in vitro* protein synthesis, in: *RNA Phages* (N. D. Zinder, ed.), pp. 279–299, Cold Spring Harbor Lab., Cold Spring Harbor, New York.

Carmichael, G. G., 1975, Isolation of bacterial and phage proteins by homopolymer RNA-cellulose chromatography, *J. Biol. Chem.* **250**:6160.

Chapman, N. M., and Noller, H. F., 1977, Differential accessibility of specific sites in 16S RNA in 30S and 70S ribosomes, *J. Mol. Biol.* **109**:131.

Clark, B. F. C., and Marcker, K. A., 1966, The role of *N*-formylmethionyl-sRNA in protein biosynthesis, *J. Mol. Biol.* **17**:394.

Crick, F. H. C., 1968, The origin of the genetic code, *J. Mol. Biol.* **38**:367.

Crick, F. H. C., Brenner, S., Klug, A., and Pieczenik, G., 1976, A speculation on the origin of protein synthesis, *Origins of Life* **7**:389.

Crouch, R. J., 1974, Ribonuclease III does not degrade deoxyribonucleic acid–ribonucleic acid hybrids, *J. Biol. Chem.* **249**:1314.

Curtis, P. J., and Weissmann, C., 1976, Purification of globin messenger RNA from dimethylsulfoxide-induced Friend cells and detection of a putative globin messenger RNA precursor, *J. Mol. Biol.* **106**:1061.

Czernilofsky, A. P., Kurland, C. G., and Stöffler, G., 1975, 30S ribosomal proteins associated with the 3′-terminus of 16S RNA, *FEBS Lett.* **58**:281.

Dahlberg, A. E., 1974, Two forms of the 30S ribosomal subunit of *Escherichia coli, J. Biol. Chem.* **249**:7673.

Dahlberg, A. E., and Dahlberg, J. E., 1975, Binding of ribosomal protein S1 of *Escherichia coli* to the 3′ end of 16S rRNA, *Proc. Natl. Acad. Sci. U.S.A.* **72**:2940.

Dahlberg, A. E., Lund, E., Kjeldgaard, N. O., Bowman, C. M., and Nomura, M., 1973, Colicin E3-induced cleavage of 16S ribosomal ribonucleic acid; blocking effects of certain antibiotics, *Biochemistry* **12**:948.

Dahlberg, J. E., 1968, Terminal sequences of bacteriophage RNAs, *Nature* **220**:548.

Dalgarno, L., and Shine, J., 1973, Conserved terminal sequence in 18S rRNA may represent terminator anticodons, *Nature New Biol.* **245**:261.

Darnell, J. E., Jelinek, W. R., and Molloy, G. R., 1973, Biogenesis of mRNA: Genetic regulation in mammalian cells, *Science* **181**:1215.

Dasgupta, R., Shih, D. S., Saris, C., and Kaesberg, P., 1975, Nucleotide sequence of a viral RNA fragment that binds to eukaryotic ribosomes, *Nature* **256**:624.

Delius, H., Westphal, H., and Axelrod, N., 1973, Length measurements of RNA synthesized *in vitro* by *Escherichia coli* RNA polymerase, *J. Mol. Biol.* **74**:677.

Dondon, J., Godfrey-Colburn, T., Graffe, M., and Grunberg-Manago, M., 1974, IF-3 requirements for initiation complex formation with synthetic messengers in *E. coli* system, *FEBS Lett.* **45**:82.

Draper, D. E., and von Hippel, P. H., 1976, Gene expression: Polynucleotide binding properties of *E. coli* ribosomal protein S1, *ICN-UCLA Symp. Mol. Cell. Biol.* **5**:421.

Dube, S. K., 1973, Recognition of tRNA by the ribosome: A possible role of 5S RNA, *FEBS Lett.* **36**: 39.

Dube, S. K., Marcker, K. A., Clark, B. F. C., and Cory, S., 1968, Nucleotide sequence of *N*-formylmethionyl-transfer RNA, *Nature* **218**:232.

Dunn, J. J., 1976, RNase III cleavage of single-stranded RNA: Effect of ionic strength on the fidelity of cleavage, *J. Biol. Chem.* **251**:3807.

Dunn, J. J., and Studier, F. W., 1973*a,* T7 early RNAs are generated by site-specific cleavages, *Proc. Natl. Acad. Sci. U.S.A.* **70**:1559.

Dunn, J. J., and Studier, F. W., 1973*b*, T7 early RNAs and *Escherichia coli* ribosomal RNAs are cut from large precursor RNAs *in vitro* by ribonuclease III, *Proc. Natl. Acad. Sci. U.S.A.* **70**:3296.

Dunn, J. J., and Studier, F. W., 1975, Effect of RNAase III cleavage on translation of bacteriophage T7 messenger RNAs, *J. Mol. Biol.* **99**:487.

Ehresmann, C., Stiegler, P., and Ebel, J.-P., 1974, Sequence analysis of the 3′-T_1 oligonucleotide of 16S ribosomal RNA from *Escherichia coli, FEBS Lett.* **49**:47.

Ehresmann, C., Stiegler, P., Mackie, G. A., Zimmermann, R. A., Ebel, J.-P., and Fellner, P., 1975, Primary sequence of the 16S ribosomal RNA of *Escherichia coli, Nucleic Acids Res.* **2**:265.

Eladari, M.-E., and Galibert, R., 1975, Sequence determination of 5′-terminal and 3′-terminal oligonucleotides of 18S ribosomal RNA of a mouse cell line (L 5178 Y), *Eur. J. Biochem.* **55**:247.

Erdmann, V. A., 1976, Structure and function of 5S and 5.8S RNA, *Prog. Nucleic Acid Res. Mol. Biol.* **18**:45.

Erdmann, V. A., Sprinzl, M., and Pongs, O., 1973, The involvement of 5S RNA in the binding of tRNA to ribosomes, *Biochem. Biophys. Res. Commun.* **54**:942.

Ewald, R., Pon, C., and Gualerzi, C., 1976, Reactivity of ribosomal sulfhydryl groups in 30S ribosomal subunits of *Escherichia coli* and 30S-IF-3 complexes, *Biochemistry* **15**:4786.

Feldman, M. Y., 1973, Reactions of nucleic acids and nucleoproteins with formaldehyde, *Prog. Nucleic Acid Res. Mol. Biol.* **13**:1.

Fernandez-Munoz, R., and Darnell, J. E., 1976, Structural difference between the 5′ termini of viral and cellular mRNA in poliovirus-infected cells: Possible basis for the inhibition of host protein synthesis, *J. Virol.* **18**:719.

Fiers, W., Contreras, R., Duerinck, F., Haegeman, G., Merregaert, J., Min Jou, W., Raeymaekers, A., Volckaert, G., Ysebaert, M., Van de Kerckhove, J., Nolf, F., and Van Montagu, M., 1975, A-protein gene of bacteriophage MS2, *Nature* **256**:273.

Fiers, W., Contreras, R., Duerinck, F., Haegeman, G., Iserentant, D., Merregaert, J., Min Jou, W., Molemans, F., Raeymaekers, A., Van den Berghe, A., Volckaert, G., and Ysebaert, M., 1976, Complete nucleotide sequence of bacteriophage MS2 RNA: Primary and secondary structure of the replicase gene, *Nature* **260**:500.

Files, J. G., Weber, K., and Miller, J. H., 1974, Translational reinitiation: Reinitiation of *lac* repressor fragments at three internal sites early in the *lac i* gene of *Escherichia coli, Proc. Natl. Acad. Sci. U.S.A.* **71**:667.

Files, J. G., Weber, K., Coulondre, C., and Miller, J. H., 1975, Identification of the UUG codon as a translational initiation codon *in vivo, J. Mol. Biol.* **95**:327.

Fiser, I., Margaritella, P., and Kuechler, E., 1975, Photoaffinity reaction between polyuridylic acid and protein S1 on the *Escherichia coli* ribosome, *FEBS Lett.* **52**:281.

Fox, G. E., and Woese, C. R., 1975, 5S RNA secondary structure, *Nature* **256**:505.

Fresco, J. R., Alberts, B. M., and Doty, P., 1960, Some molecular details of the secondary structure of ribonucleic acid, *Nature* **188**:98.

Fukami, H., and Imahori, K., 1971, Control of translation by the conformation of messenger RNA, *Proc. Natl. Acad. Sci. U.S.A.* **68**:570.

Ganem, D., Miller, J. H., Files, J. G., Platt, T., and Weber, K., 1973, Reinitiation of a *lac* repressor fragment at a codon other than AUG, *Proc. Natl. Acad. Sci. U.S.A.* **70**:3165.

Garel, J. P., Mandel, P., Chavancy, G., and Daillie, J., 1971, Functional adaptation of tRNAs to protein biosynthesis in a highly differentiated cell system, III. Indication of isoacceptor tRNAs during the secretion of fibroin in the silk gland of *Bombyx mori* L, *FEBS Lett.* **12**:249.

Ginsburg, D., and Steitz, J. A., 1975, The 30S ribosomal precursor RNA from *Escherichia coli:* A primary transcript containing 23S, 16S, and 5S sequences, *J. Biol. Chem.* **250**:5647.

Goelz, S., and Steitz, J. A., 1977, *E. coli* ribosomal protein S1 recognizes two sites in bacteriophage $Q\beta$ RNA, *J. Biol. Chem.* **252**:5177.

Gold, L., O'Farrell, P. Z., and Russel, M., 1976, Regulation of gene 32 expression during bacteriophage T4 infection of *Escherichia coli, J. Biol. Chem.* **251**:7251.

Goldberg, M. L., and Steitz, J. A., 1974, Cistron specificity of 30S ribosomes heterologously reconstituted with components from *Escherichia coli* and *Bacillus stearothermophilus, Biochemistry* **13**:2123.

Gralla, J., and Crothers, D. M., 1973, The free energy of imperfect nucleic acid helices. II. Small hairpin loops, *J. Mol. Biol.* **73**:497.

Gralla, J., and DeLisi, C., 1974, mRNA is expected to form stable secondary structures, *Nature* **248**:330.

Gralla, J., Steitz, J. A., and Crothers, D. M., 1974, Direct physical evidence for secondary structure in an isolated fragment of R17 bacteriophage mRNA, *Nature* **248**:204.

Gray, P. N., Bellemare, G., Monier, R., Garrett, R. A., and Stöffler, G., 1973, Identification of the nucleotide sequences involved in the interaction between *Escherichia coli* 5S RNA and specific 50S subunit proteins, *J. Mol. Biol.* **77**:133.

Greenberg, J. R., and Perry, R. P., 1972, Relative occurrence of polyadenylic acid sequences in messenger and heterogeneous nuclear RNA of L cells as determined by poly(U)-hydroxylapatite chromatography, *J. Mol. Biol.* **72**:91.

Grohmann, K., Smith, L. H., and Sinsheimer, R. L., 1975, New method for isolation and sequence determination of 5′-terminal regions of bacteriophage ØX174 *in vitro* mRNAs, *Biochemistry* **14**:1951.

Groner, Y., Pollack, Y., Berissi, H., and Revel, M., 1972, Cistron-specific translation control protein in *Escherichia coli, Nature New Biol.* **239**:16.

Grosjean, H., Söll, D. G., and Crothers, D. M., 1976, Studies of the complex between transfer RNAs with complementary anticodons. I. Origins of enhanced affinity between complementary triplets, *J. Mol. Biol.* **103**:499.

Grosjean, H. J., de Henau, S., and Crothers, D. M., 1978, On the physical basis for ambiguity in genetic coding, *Proc. Natl. Acad. Sci. U.S.A.* **75**:610.

Grunberg-Manago, M., Rabinowitz, J. C., Dondon, J., Lelong, J. C., and Gros, F., 1971, Different classes of initiation factors F3 and their dissociation activity, *FEBS Lett.* **19**:193.

Gualerzi, C., Risuleo, G., and Pon, C. L., 1977, Initial rate kinetic analysis of the mechanism of initiation complex formation and the role of initiation factor IF-3, *Biochemistry* **16**:1684.

Gupta, S. L., Chen, J., Schaefer, L., Lengyel, P., and Weissman, S. M., 1970, Nucleotide sequence of a ribosome attachment site of bacteriophage f2 RNA, *Biochem. Biophys. Res. Commun.* **39**:883.

Harada, F., and Nishimura, S., 1974, Purification and characterization of AUA specific isoleucine transfer ribonucleic acid from *Escherichia coli* B, *Biochemistry* **13**:300.

Haseltine, W. A., Maxam, A. M., and Gilbert, W., 1977, The Rous sarcoma virus genome is terminally redundant: The 5′ sequence, *Proc. Natl. Acad. Sci. U.S.A.* **74**:989.

Hawley, D. A., Slobin, L. I., and Wahba, A. J., 1974, The mechanism of action of initiation factor 3 in protein synthesis, II. Association of the 30S ribosomal protein S12 with IF-3, *Biochem. Biophys. Res. Commun.* **61**:544.

Hayashi, M., Fujimura, F. K., and Hayashi, M., 1976, Mapping of *in vivo* messenger RNAs for bacteriophage ϕX-174, *Proc. Natl. Acad. Sci. U.S.A.* **73**:3519.

Heimark, R. L., Kahan, L., Johnston, K., Hershey, J. W. B., and Traut, R. R., 1976, Cross-linking of initiation factor IF 3 to proteins of the *Escherichia coli* 30S ribosomal subunit, *J. Mol. Biol.* **105**:219.

Held, W. A., Gette, W. R., and Nomura, M., 1974, Role of 16S ribosomal ribonucleic acid and the 30S ribosomal protein S12 in the initiation of natural messenger ribonucleic acid translation, *Biochemistry* **13**:2115.

Helser, T. L., Davies, J. E., and Dahlberg, J. E., 1971, Change in methylation of 16S ribosomal RNA associated with mutation to kasugamycin resistance in *Escherichia coli, Nature New Biol.* **233**:12.

Henry, T. J., and Knippers, R., 1974, Isolation and function of the gene *A* initiator of bacteriophage ØX174, a highly specific DNA endonuclease, *Proc. Natl. Acad. Sci. U.S.A.* **71**:1549.

Hermoso, J. M., and Szer, W., 1974, Replacement of ribosomal protein S1 by interference factor iα in ribosomal binding of phage MS2 RNA, *Proc. Natl. Acad. Sci. U.S.A.* **71**:4708.

Herr, W., and Noller, H. F., 1975, A fragment of 23S RNA containing a nucleotide sequence complementary to a region of 5S RNA, *FEBS Lett.* **53**:248.

Hewlett, M. J., Rose, J. K., and Baltimore, D., 1976, 5′ terminal structure of polio virus polyribosomal RNA is pUp, *Proc. Natl. Acad. Sci. U.S.A.* **73**:327.

Hilbers, C. W., Shulman, R. G., Yamane, T., and Steitz, J. A., 1974, High resolution proton NMR study of an isolated fragment of R17 bacteriophage mRNA, *Nature* **248**:225.

Hindley, J., and Staples, D. H., 1969, Sequence of a ribosome binding site in bacteriophage Qβ-RNA, *Nature* **224**:964.

Hopfield, J. J., 1974, Kinetic proofreading: A new mechanism for reducing errors in biosynthetic processes requiring high specificity, *Proc. Natl. Acad. Sci. U.S.A.* **72**:4135.

Horiuchi, K., 1975, Genetic studies of RNA phages, in: *RNA Phages* (N. D. Zinder, ed.) pp. 29–50, Cold Spring Harbor Lab., Cold Spring Harbor, New York.

Inouye, H., Pollack, Y., and Petre, J., 1974, Physical and functional homology between ribosomal protein S1 and interference factor i, *Eur. J. Biochem.* **45**:109.

Isono, K., and Isono, S., 1976, Lack of ribosomal protein S1 in *Bacillus stearothermophilus, Proc. Natl. Acad. Sci. U.S.A.* **73**:767.

Isono, S., and Isono, K., 1975, Role of ribosomal protein S1 in protein synthesis: Effects of its addition to *Bacillus stearothermophilus* cell-free system, *Eur. J. Biochem.* **56**:15.

Iwasaki, K., Sabol, S., Wahba, A. J., and Ochoa, S., 1968, Translation of the genetic message. VII. Role of initiation factors in formation of the chain initiation complex with *Escherichia coli* ribosomes, *Arch. Biochem. Biophys.* **125**:542.

Jacobson, A. B., 1976, Studies on secondary structure of single-stranded RNA from bacteriophage MS2 by electron microscopy, *Proc. Natl. Acad. Sci. U.S.A.* **73**:307.

Jay, G., and Kaempfer, R., 1975*a,* Initiation of protein synthesis; binding of messenger RNA, *J. Biol. Chem.* **250**:5742.

Jay, G., and Kaempfer, R., 1975*b,* Translational repression of a viral messenger RNA by a host protein, *J. Biol. Chem.* **250**:5749.

Kabat, D., 1975, Potentiation of hemoglobin messenger ribonucleic acid, *J. Biol. Chem.* **250**:6085.

Kaempfer, R., 1972, Initiation factor IF-3: A specific inhibitor of ribosomal subunit association, *J. Mol. Biol.* **71**:583.

Kamen, R., 1970, Characterization of the subunits of Qβ replicase, *Nature* **228**:527.

Kamen, R. I., 1975, Structure and function of the Qβ RNA replicase, in: *RNA Phages* (N. D. Zinder, ed.), pp. 203–234, Cold Spring Harbor Lab., Cold Spring Harbor, New York.

Kelly, R. C., and von Hippel, P. H., 1976, DNA "melting" proteins. III. Fluorescence "mapping" of the nucleic acid binding site of bacteriophage T4 gene 32-protein, *J. Biol. Chem.* **251**:7229.

Kelly, R. C., Jensen, D. E., and von Hippel, P. H., 1976, DNA "melting" proteins. IV. Fluorescence measurements of binding parameters for bacteriophage T4 gene 32-protein to mono-, oligo-, and polynucleotides, *J. Biol. Chem.* **251**:7240.

Kenner, R. A., 1973, A protein–nucleic acid crosslink in 30S ribosomes, *Biochem. Biophys. Res. Commun.* **51**:932.

Kolakofsky, D., and Weissmann, C., 1971*a,* Possible mechanism for transition of viral RNA from polysome to replication complex, *Nature New Biol.* **231**:42.

Kolakofsky, D., and Weissmann, C., 1971*b*, Qβ replicase as repressor of Qβ RNA-directed protein synthesis, *Biochim. Biophys. Acta* **246**:596.

Kolakofsky, D., Billeter, M. A., Weber, H., and Weissmann, C., 1973, Resynchronization of RNA synthesis by coliphage Qβ replicase at an internal site of the RNA template, *J. Mol. Biol.* **76**:271.

Kondo, M., Gallerani, R., and Weissmann, C., 1970, Subunit structure of Qβ replicase, *Nature* **228**:525.

Kozak, M., 1977, Nucleotide sequences of 5′-terminal ribosome-protected initiation regions from two reovirus messages, *Nature* **269**:390.

Kozak, M., and Nathans, D., 1972, Translation of the genome of a ribonucleic acid bacteriophage, *Bacteriol. Rev.* **36**:109.

Kozak, M., and Shatkin, A. J., 1976, Characterization of ribosome-protected fragments from reovirus messenger RNA, *J. Biol. Chem.* **251**:4259.

Kozak, M., and Shatkin, A. J., 1977, Sequences of two 5′-terminal ribosome-protected fragments from reovirus messenger RNAs, *J. Mol. Biol.* **112**:75.

Krisch, H. M., Bolle, A., and Epstein, R. H., 1974, Regulation of the synthesis of bacteriophage T4 gene *32* protein, *J. Mol. Biol.* **88**:89.

Kurland, C. G., 1974, Functional organization of the 30S ribosomal subunit, in *Ribosomes* (M. Nomura, A. Tissières, and P. Lengyel, eds.), pp. 309–331, Cold Spring Harbor Lab., Cold Spring Harbor, New York.

Ladner, J. E., Jack, A., Robertus, J. D., Brown, R. S., Rhodes, D., Clark, B. F. C., and Klug, A., 1975, Structure of yeast phenylalanine transfer RNA at 2.5 Å resolution, *Proc. Natl. Acad. Sci. U.S.A.* **72**:4414.

Lazarowitz, S. G., and Robertson, H. D., 1977, Ribosome-protected regions of reovirus *s* mRNA, *J. Biol. Chem* **252**:7842.

Lee, N., and Carbon, J., 1977, Nucleotide sequence of the 5′ end of *araBAD* operon messenger RNA in *Escherichia coli* B/r, *Proc. Natl. Acad. Sci. U.S.A.* **74**:49.

Legon, S., 1976, Characterization of the ribosome-protected regions of ^{125}I-labelled rabbit globin messenger RNA, *J. Mol. Biol.* **106**:37.

Legon, S., Model, P., and Robertson, H. D., 1977, The interaction of rabbit reticulocyte ribosomes with bacteriophage f1 messenger RNA and *Escherichia coli* ribosomes with rabbit globin messenger RNA, *Proc. Natl. Acad. Sci. U.S.A.* **74**:2692.

Linney, E., and Hayashi, M., 1974, Intragenic regulation of the synthesis of ØX174 gene A proteins, *Nature* **249**:345.

Lodish, H. F., 1969*a*, Species specificity of polypeptide chain initiation, *Nature* **224**:867.

Lodish, H. F., 1969*b*, Independent initiation of translation of two bacteriophage f2 proteins, *Biochem. Biophys. Res. Commun.* **37**:127.

Lodish, H. F., 1970*a*, Specificity in bacterial protein synthesis: Role of initiation factors and ribosomal subunits, *Nature* **226**:705.

Lodish, H. F., 1970*b*, Secondary structure of bacteriophage f2 ribonucleic acid and the initiation of *in vitro* protein biosynthesis, *J. Mol. Biol.* **50**:689.

Lodish, H. F., 1971, Thermal melting of bacteriophage f2 RNA and initiation of synthesis of the maturation protein, *J. Mol. Biol.* **56**:627.

Lodish, H. F., 1974, Model for the regulation of mRNA translation applied to haemoglobin synthesis, *Nature* **251**:385.

Lodish, H. F., 1975, Regulation of *in vitro* protein synthesis by bacteriophage RNA by RNA tertiary structure, in: *RNA Phages* (N. D. Zinder, ed.), pp. 301–318, Cold Spring Harbor Lab., Cold Spring Harbor, New York.

Lodish, H. F., 1976, Translational control of protein synthesis, *Annu. Rev. Biochem.* **45**:39.

Lodish, H. F., and Robertson, H. D., 1969, Regulation of *in vitro* translation of bacteriophage f2 RNA, *Cold Spring Harbor Symp. Quant. Biol.* **34**:655.

Lozeron, H. A., Anevski, P. J., and Apirion, D., 1977, Antitermination and absence of processing of the leftward transcript of coliphage lambda in the RNAase III-deficient host, *J. Mol. Biol.* **109**:359.

Lucas-Lenard, J., and Lipmann, F., 1971, Protein biosynthesis, *Annu. Rev. Biochem.* **40**:409.

Maizels, N., 1974, *E. coli* lactose operon ribosome binding site, *Nature* **249**:647.

Marotta, C. A., Forget, B. G., Cohen-Solal, M., and Weissman, S. M., 1976, Nucleotide sequence analysis of coding and noncoding regions of human β-globin mRNA, *Prog. Nucleic Acid Res. Mol. Biol.* **19**:165.

Meyer, R., Weber, H., Vollenweider, H. J., and Weissmann, C., 1975, The binding sites of Qβ replicase on Qβ RNA, *Experientia* **31**:743.

Michalski, C. J., Sells, B. H., and Wahba, A. J., 1976, Molecular morphology of ribosomes: Effect of chain initiation factor 3 on 30S subunit conformation, *FEBS Lett.* **71**:347.

Miller, M. J., and Wahba, A. J., 1973, Chain initiation factor 2: Purification and properties of two species from *Escherichia coli* MRE 600, *J. Biol. Chem.* **248**:1084.

Miller, M. J., and Wahba, A. J., 1974, Inhibition of synthetic and natural messenger translation. II. Specificity and mechanism of action of a protein isolated from *Escherichia coli* MRE 600 ribosomes, *J. Biol. Chem.* **249**:3808.

Miller, M. J., Niveleau, A., and Wahba, A. J., 1974, Inhibition of synthetic and natural messenger translation. I. Purification and properties of a protein isolated from *Escherichia coli* MRE 600 ribosomes, *J. Biol. Chem.* **249**:3803.

Min Jou, W., and Fiers, W., 1976, Studies on the bacteriophage MS2: XXXIII. Comparison of the nucleotide sequences in related bacteriophage RNAs, *J. Mol. Biol.* **106**:1047.

Min Jou, W., Haegeman, G., Ysebaert, M., and Fiers, W., 1972, Nucleotide sequence of the gene coding for the bacteriophage MS2 coat protein, *Nature* **237**:82.

Musso, R., de Crombrugghe, B., Pastan, I., Sklar, J., Yot, P., and Weissman, S., 1974, The 5′-terminal nucleotide sequence of galactose messenger ribonucleic acid of *Escherichia coli, Proc. Natl. Acad. Sci. U.S.A.* **71**:4940.

Muthukrishnan, S., Both, G. W., Furiuchi, Y., and Shatkin, A. J., 1975, 5′ terminal 7-methylguanosine in eukaryotic mRNA is required for translation, *Nature* **255**:33.

Nikolaev, N., Silengo, L., and Schlessinger, D., 1973, Synthesis of a large precursor to ribosomal RNA in a mutant of *Escherichia coli, Proc. Natl. Acad. Sci. U.S.A.* **70**:3361.

Ninio, J., 1974, A semi-quantitative treatment of missense and nonsense suppression of *str*A and *ram* ribosomal mutants of *Escherichia coli:* Evaluation of some molecular parameters of translation *in vivo, J. Mol. Biol.* **84**:297.

Noll, M., and Noll, H., 1972, Mechanism and control of initiation in the translation of R17 RNA, *Nature New Biol.* **238**:225.

Noller, H. F., and Herr, W., 1974, Nucleotide sequence of the 3′ terminus of *E. coli* 16 S ribosomal RNA, *Mol. Biol. Rep.* **1**:437.

Nomoto, A., Lee, Y., and Wimmer, E., 1976, The 5′ end of poliovirus mRNA is not capped with $m^7G(5')ppp(5')Np$, *Proc. Natl. Acad. Sci. U.S.A.* **73**:375.

Oakden, K. M., and Lane, B. G., 1975, Wheat embryo ribonucleates. VI. Comparison of the 3′-hydroxyl termini in "rapidly labelled" RNA from metabolizing wheat embryos with the corresponding termini in ribosomal RNA from differentiating embryos of wheat, barley, corn and pea, *Can. J. Biochem.* **54**:261.

Ochoa, S., and Mazumder, R., 1974, Polypeptide chain initiation, in: *The Enzymes,* Vol. X (P. D. Boyer, ed.), pp. 1–51, Academic Press, New York.

Orgel, L. F., 1968, Evolution of the genetic apparatus, *J. Mol. Biol.* **38**:381.

Perry, R. P., 1976, Processing of RNA, *Annu. Rev. Biochem.* **45**:605.

Pieczenik, G., 1972, Ph.D. Thesis, New York University.

Pieczenik, G., Model, P., and Robertson, H. D., 1974, Sequence and symmetry in ribosome binding sites of bacteriophage f1 RNA, *J. Mol. Biol.* **90**:191.

Platt, T., and Yanofsky, C., 1975, An intercistronic region and ribosome-binding site in bacterial messenger RNA, *Proc. Natl. Acad. Sci. U.S.A.* **72**:2399.

Platt, T., Weber, K., Ganem, D., and Miller, J. H., 1972, Translational restarts: AUG reinitiation of a *lac* repressor fragment, *Proc. Natl. Acad. Sci. U.S.A.* **69**:897.

Pon, C. L., and Gualerzi, C., 1974, Effect of initiation factor 3 binding on the 30S ribosomal subunits of *Escherichia coli, Proc. Natl. Acad. Sci. U.S.A.* **71**:4950.

Pon, C. L., and Gualerzi, C., 1976, The role of 16S rRNA in ribosomal binding of IF-3, *Biochemistry* **15**:804.

Porter, A. G., and Hindley, J., 1973, The binding of Qβ initiator fragments to *E. coli* ribosomes, *FEBS Lett.* **33**:339.

Proudfoot, N. J., and Brownlee, G. G., 1976, 3′ noncoding region sequences in eukaryotic messenger RNA, *Nature* **263**:211.

Ptashne, M., Backman, K., Humayun, M. Z., Jeffrey, A., Maurer, R., Meyer, B., and Sauer, T., 1976, Autoregulation and function of a repressor in bacteriophage lambda, *Science* **194**:156.

Quigley, G. J., and Rich, A., 1976, Structural domains in transfer RNA molecules, *Science* **194**:796.

Ravetch, J. V., and Jakes, K., 1976, Intact 3′ end of 16S rRNA is not required for specific mRNA binding, *Nature* **262**:150.

Ravetch, J. V., Model, P., and Robertson, H. D., 1977, Isolation and characterization of the ØX174 ribosome, *Nature* **265**:698.

Revel, M., and Greenshpan, H., 1970, Specificity in the binding of *Escherichia coli* ribosomes to natural messenger RNA, *Eur. J. Biochem.* **16**:117.

Reznikoff, W. S., Winter, R. B., and Katovich Hurley, C., 1974, The location of the repressor binding sites in the *lac* operon, *Proc. Natl. Acad. Sci. U.S.A.* **71**:2314.

Ricard, B., and Salser, W., 1974, Size and folding of the messenger for phage T4 lysozyme, *Nature* **248**:314.

Ricard, B., and Salser, W., 1975, Secondary structures formed by random RNA sequences, *Biochem. Biophys. Res. Commun.* **63**:548.

Rich, A., and Raj Bhandary, U. L., 1976, Transfer RNA: Molecular structure, sequence, and properties, *Annu. Rev. Biochem.* **45**:805.

Richter, D., Erdmann, V. A., and Sprinzl, M., 1973, Specific recognition of GTψC loop (loop IV) of tRNA by 50S ribosomal subunits from *E. coli, Nature New Biol.* **246**:132.

Richter, D., Erdmann, V. A., and Sprinzl, M., 1974, A new transfer RNA fragment reaction: Tpψp-CpGp bound to a ribosome–messenger RNA complex induces the synthesis of guanosine tetra- and pentaphosphates, *Proc. Natl. Acad. Sci. U.S.A.* **71**:3226.

Robertson, H. D., and Dunn, J. J., 1975, Ribonucleic acid processing activity of *Escherichia coli* ribonuclease III, *J. Biol. Chem.* **250**:3050.

Robertson, H. D., and Lodish, H. F., 1970, Messenger characteristics of nascent bacteriophage RNA, *Proc. Natl. Acad. Sci. U.S.A.* **67**:710.

Robertson, H. D., Barrell, B. G., Weith, H. L., and Donelson, J. E., 1973, Isolation and sequence analysis of a ribosome-protected fragment from bacteriophage ϕX174 DNA, *Nature New Biol.* **241**:38.

Robertson, H. D., Dickson, E., and Dunn, J. J., 1977, A nucleotide sequence from a ribonuclease III processing site in bacteriophage T7 RNA, *Proc. Natl. Acad. Sci. U.S.A.* **74**:822.

Rosenberg, M., and Kramer, R., 1977, Nucleotide sequence surrounding a ribonuclease III processing site in bacteriophage T7 RNA, *Proc. Natl. Acad. Sci. U.S.A.* **74**:984.

Rosenberg, M., Kramer, R. A., and Steitz, J. A., 1974, T7 early messenger RNAs are the direct products of ribonuclease III cleavage, *J. Mol. Biol.* **89**:777.

Ross, J., 1976, A precursor of globin messenger RNA, *J. Mol. Biol.* **106**:403.

Russel, M., Gold, L., Morrissett, H., and O'Farrell, P. Z., 1976, Translational, autogenous regulation of gene 32 expression during bacteriophage T4 infection, *J. Biol. Chem.* **251**:7263.

Salser, W., Browne, J., Clarke, P., Heindell, H., Higuchi, R., Paddock, G., Roberts, J., Studnicka, G., and Zakar, P., 1976, Determination of globin mRNA sequences and their insertion into bacterial plasmids, *Prog. Nucleic Acid Res. Mol. Biol.* **19**:177.

Sanger, F., Air. G. M., Barrell, B. G., Brown, N. L., Coulson, A. R., Fiddes, J. C., Hutchinson, C. A., III, Slocombe, P. M., and Smith, M., 1977, Nucleotide sequence of bacteriophage ØX174 DNA, *Nature* **265**:687.

Santer, M., and Shane, S., 1977, The area of 16S RNA at or near the interface between 30S and 50S ribosomes of *E. coli, J. Bacteriol.* **130**:900.

Scheps, R., Wax, R., and Revel, M., 1971, Reactivation *in vitro* of inactive ribosomes from stationary phase *Escherichia coli, Biochim. Biophys. Acta* **232**:140.

Schiff, N., Miller, M. J., and Wahba, A. J., 1974, Purification and properties of chain initiation factor 3 from T4-infected and uninfected *Escherichia coli* MRE 600: Stimulation of translation of synthetic and natural messengers, *J. Biol. Chem.* **249**:3797.

Schwarz, U., Menzel, H. M., and Gassen, H. G., 1976, Codon-dependent rearrangement of the three-dimensional structure of phenylalanine tRNA, exposing the T-ψ-C-G sequence for binding to the 50S ribosomal subunit, *Biochemistry* **15**:2484.

Senear, A. W., and Steitz, J. A., 1976, Site-specific interaction of $Q\beta$ host factor and ribosomal protein S1 with $Q\beta$ and R17 bacteriophage RNAs, *J. Biol. Chem.* **251**:1902.

Senior, B. W., and Holland, I. B., 1971, Effect of colicin E3 upon the 30S ribosomal subunit of *Escherichia coli, Proc. Natl. Acad. Sci. U.S.A.* **68**:959.

Shine, J., and Dalgarno, L., 1974, The 3′-terminal sequence of *Escherichia coli* 16S ribosomal RNA: Complementarity to nonsense triplets and ribosome binding sites, *Proc. Natl. Acad. Sci. U.S.A.* **71**:1342.

Shine, J., and Dalgarno, L., 1975, Determinant of cistron specificity in bacterial ribosomes, *Nature* **254**:34.

Shine, J., Czernilofsky, A. P., Friedrich, R., Bishop, J. M., and Goodman, H. M., 1977, Nucleotide sequence at the 5′ terminus of the avian Rous sarcoma virus genome, *Proc. Natl. Acad. Sci. U.S.A.* **74**:1473.

Singer, B. S., and Gold, L., 1976, A mutation that confers temperature sensitivity on the translation of *rIIB* in bacteriophage T4, *J. Mol. Biol.* **103**:627.

Smith, L. H., and Sinsheimer, R. L., 1976, The *in vitro* transcription units of bacteriophage ØX174. II. *In vitro* initiation sites of ØX174 transcription, *J. Mol. Biol.* **103**:699.

Sobura, J. E., Chowdhury, M. R., Hawley, D. A., and Wahba, A. J., 1977, Requirement of chain initiation factor 3 and ribosomal protein S1 in translation of synthetic and natural messenger RNA, *Nucleic Acids Res.* **4**:17.

Sprague, K. U., and Steitz, J. A., 1975, The 3′ terminal oligonucleotide of *E. coli* 16S ribosomal RNA: The sequence in both wild-type and RNase III$^-$ cells is complementary to the polypurine tracts common to mRNA initiator regions, *Nucleic Acids Res.* **2**:787.

Sprague, K. U., Kramer, R. A., and Jackson, M. B., 1975, The terminal sequences of *Bombyx mori* 18S ribosomal RNA, *Nucleic Acids Res.* **2**:2111.

Sprague, K. U., Steitz, J. A., Grenley, R. M., and Stocking, C. E., 1977, 3′ terminal sequences of 16S rRNA do not explain translational specificity differences between E. coli and *B. stearothermophilus* ribosomes, *Nature* **267**:462.

Stallcup, M. R., Sharrock, W. J., and Rabinowitz, J. C., 1975, Specificity of bacterial ribosomes and messenger ribonucleic acids in protein synthesis reactions *in vitro, J. Biol. Chem.* **251**:2499.

Stanley, W. M., Jr., Salas, M., Wahba, A. J., and Ochoa, S., 1966, Translation of the genetic message: Factors involved in the initiation of protein synthesis, *Proc. Natl. Acad. Sci. U.S.A.* **56**:290.

Staples, D. H., and Hindley, J., 1971, Ribosome binding site of $Q\beta$ RNA polymerase cistron, *Nature New Biol.* **234**:211.

Staples, D. H., Hindley, J., Billeter, M. A., and Weissmann, C., 1971, Localization of $Q\beta$ maturation cistron ribosome binding site, *Nature New Biol.* **234**:202.

Steege, D. A., 1977*a,* A ribosome binding site from the P_R RNA of bacteriophage λ, *J. Mol. Biol.* **114**:559.

Steege, D. A., 1977*b,* The 5′ terminal nucleotide sequence of the *E. coli* lactose repressor messenger RNA: Features of translational initiation and reinitiation sties, *Proc. Natl. Acad. Sci. U.S.A.* **74**:4163.

Steitz, J. A., 1969, Polypeptide chain initiation: Nucleotide sequences of the three ribosomal binding sites in bacteriophage R17 RNA, *Nature* **224**:957.

Steitz, J. A., 1972, Oligonucleotide sequence of the replicase initiation site in *Qβ* RNA, *Nature New Biol.* **236**:71.

Steitz, J. A., 1973*a*, Specific recognition of non-initiator regions in RNA bacteriophage messengers by ribosomes of *B. stearothermophilus, J. Mol. Biol.* **73**:1.

Steitz, J. A., 1973*b*, Discriminatory ribosome rebinding of isolated regions of protein synthesis initiation from the ribonucleic acid of bacteriophage R17, *Proc. Natl. Acad. Sci. U.S.A.* **70**:2605.

Steitz, J. A., 1975, Ribosome recognition of initiator regions in the RNA bacteriophage genome, in: *RNA Phages* (N. D. Zinder, ed.), pp. 319–352, Cold Spring Harbor Lab. Cold Spring Harbor, New York.

Steitz, J. A., and Bryan, R. A., 1977, Two ribosome binding sites from the gene *0.3* mRNA of bacteriophage T7, *J. Mol. Biol.* **114**:527.

Steitz, J. A., and Jakes, K., 1975, How ribosomes select initiator regions in mRNA: Base pair formation between the 3′ terminus of 16S rRNA and the mRNA during initiation of protein synthesis in *Escherichia coli, Proc. Natl. Acad. Sci. U.S.A.* **72**:4734.

Steitz, J. A., and Steege, D. A., 1977, Characterization of two mRNA·rRNA complexes implicated in the initiation of protein biosynthesis, *J. Mol. Biol.* **114**:545.

Steitz, J. A., Wahba, A. J., Laughrea, M., and Moore, P. B., 1977, Differential requirements for polypeptide chain initiation complex formation at the three bacteriophage R17 initiator regions, *Nucleic Acids Res.* **4**:1.

Suzuki, Y., and Brown, D. D., 1972, Isolation and identification of the messenger RNA for silk fibroin from *Bombyx mori, J. Mol. Biol.* **63**:409.

Szer, W., and Leffler, S., 1974, Interaction of *Escherichia coli* 30S ribosomal subunits with MS2 phage RNA in the absence of initiation factors, *Proc. Natl. Acad. Sci. U.S.A.* **71**:3611.

Szer, W., Hermoso, J. M., and Leffler, S., 1975, Ribosomal protein S1 and polypeptide chain initiation in bacteria, *Proc. Natl. Acad. Sci. U.S.A.* **72**:2325.

Szer, W., Hermoso, J. M., and Boublik, M., 1976, Destabilization of the secondary structure of RNA by ribosomal protein S1 from *Escherichia coli, Biochem. Biophys. Res. Commun.* **70**:957.

Tal, M., Aviram, M., Kanarek, A., and Weiss, A., 1972, Polyuridylic acid binding and translating by *Escherichia coli* ribosomes: Stimulation by protein I, inhibition by aurintricarboxylic acid, *Biochim. Biophys. Acta* **281**:381.

Taniguchi, T., and Weissmann, C., 1978, *J. Mol. Biol.* **118**:533.

Van Dieijen, G., Van der Laken, C. J., Van Knippenberg, P. H., and Van Duin, J., 1975, Function of *Escherichia coli* ribosomal protein S1 in translation of natural and synthetic messenger RNA, *J. Mol. Biol.* **93**:351.

Van Dieijen, G., Van Knippenberg, P. H., and Van Duin, J., 1976, The specific role of ribosomal protein S1 in the recognition of native phage RNA, *Eur. J. Biochem.* **64**:511.

Van Duin, J., and Van Knippenberg, P. H., 1974, Functional heterogeneity of the 30S ribosomal subunit of *Escherichia coli.* III. Requirement of protein S1 for translation, *J. Mol. Biol.* **84**: 185.

Van Duin, J., Kurland, C. G., Dondon, J., and Grunberg-Manago, M., 1975, Near neighbors of IF3 bound to 30S ribosomal subunits, *FEBS Lett.* **59**:287.

Van Duin, J., Kurland, C. G., Dondon, J., Grunberg-Manago, M., Branlant, C., and Ebel, J. P., 1976, New aspects of the IF3-ribosome interaction, *FEBS Lett.* **62**:111.

Voorma, H. O., Benne, R., den Hertog, T. J. A., 1971, Binding of aminoacyl-tRNA to ribosomes programmed with bacteriophage MS2 RNA, *Eur. J. Biochem.* **18**:451.

Wahba, A. J., Iwasaki, K., Miller, M. J., Sabol, S., Sillero, M. A. G., and Vasquez, C., 1969, Initiation of protein synthesis in *Escherichia coli.* II. Role of the initiation factors in polypeptide synthesis, *Cold Spring Harbor Symp. Quant. Biol.* **34**:291.

Wahba, A. J., Miller, M. J., Niveleau, A., Landers, T. A., Carmichael, G. G., Weber, K., Hawley, D. A., and Slobin, L. I., 1974, Subunit I of *Qβ* replicase and 30S ribosomal protein S1 of *Escherichia coli:* Evidence for the identity of the two proteins, *J. Biol. Chem.* **249**:3314.

Walz, A., Pirrotta, V., and Ineichen, K., 1976, λ repressor regulates the switch between P_R and P_{rm} promoters, *Nature* **262**:665.

Weber, H., 1976, The binding site for coat protein on bacteriophage *Qβ* RNA, *Biochim. Biophys. Acta* **418**:175.

Weber, H., Billeter, M. A., Kahane, S., Weissman, C., Hindley, J., and Porter, A., 1972, Molecular basis for the repressor activity of *Qβ* replicase, *Nature New Biol.* **237**:166.

Weber, H., Kamen, R., Meyer, F., and Weissmann, C., 1974, Interactions between Qβ replicase and Qβ RNA, *Experientia* **30**:711.

Weidner, H., Yuan, R., and Crothers, D. M., 1977, Does 5S RNA function by a switch between two secondary structures? *Nature* **266**:193.

Weissbach, H., and Ochoa, S., 1976, Soluble factors required for eukaryotic protein synthesis, *Annu. Rev. Biochem.* **45**:191.

Weissmann, C., Billeter, M. A., Goodman, H. M., Hindley, J., and Weber, H., 1973, Structure and function of phage RNA, *Annu. Rev. Biochem.* **42**:303.

Woese, C. R., 1967, *The Genetic Code,* Harper, New York.

Woese, C., Sogin, M., Stahl, D., Lewis, B. J., and Bowen, L., 1976, A comparison of the 16S ribosomal RNAs from mesophilic and thermophilic bacilli: Some modifications in the Sanger method for RNA sequencing, *J. Mol. Evol.* **7**:197.

Wool, I. G., and Stöffler, G., 1974, Structure and function of eukaryotic ribosomes, in: *Ribosomes* (M. Nomura, A. Tissières, and P. Lengyel, eds.), pp. 417–460, Cold Spring Harbor Lab., Cold Spring Harbor, New York.

Yuan, R. C., Steitz, J. A., and Crothers, D. M., 1976, Direct evidence for secondary structure in the colicin E 3 released 3′-terminal 16S RNA fragment, *Fed. Proc.* **35**:1351.

Yuan, R. C., Steitz, J. A., Crothers, D. M., and Moore, P. B., 1978, The secondary structure of the 3′ terminal region of *E. coli* 16S rRNA and its complex with ribosomal protein S1, *J. Mol. Biol.* (in press).

10

The Role of tRNA in Regulation

RICCARDO CORTESE

1 *Introduction*

tRNA molecules were discovered about 20 years ago and the subsequent clarification of their role in protein biosynthesis developed from two different approaches. One approach, stemming from thermodynamic considerations, led Lipmann (1941) to postulate that amino acids had to be activated before they could be polymerized. Subsequently, Hoagland *et al.* (1957) discovered the formation of high-energy anhydride bonds between ATP and the carboxyl groups of amino acids. These authors reported that a soluble protein fraction from rat liver catalyzed the exchange of ^{32}P with ATP, a reaction enhanced severalfold by the addition of pure amino acids. This protein fraction was shown to contain an RNA species of low molecular weight, called soluble RNA or sRNA, which binds amino acids and these bound amino acids could be transferred to proteins. Second, from another view, Crick (1958) pointed out that nucleic acids could not form highly specific templates for binding the side chains of the amino acids. Nucleic acids lack both the charges required to bind amino acids and the hydrophobic cavities necessary for interaction with the aliphatic amino acids. Moreover, Crick noted that a particular sequence of bases can provide a highly specific pattern of sites for hydrogen bonding and suggested that each amino acid is combined with a special *adaptor,* which is in turn capable of forming a definite pattern of hydrogen

RICCARDO CORTESE • Institute of Biological Chemistry, Faculty of Medicine and Surgery, University of Naples, Naples, Italy. *Present address:* MRC Laboratory of Molecular Biology, Cambridge, England

bonding with a nucleic acid template. It was soon realized that the RNA discovered by Hoagland *et al.* (1957) could perform just such a function. Subsequent research has shown that sRNA, now known as *transfer* RNA or tRNA, is the adaptor for amino acids in protein synthesis and that tRNA plays a central role in the transfer of information from DNA to proteins.

From this background the details of the involvement of tRNA in protein biosynthesis have been elucidated: its interaction with aminoacyl-tRNA ligases, the binding to elongation factor Tu and ribosomal sites, and the translocation reaction (for a review see Haselkorn and Rothman-Denes, 1973; Kim, 1976; Rich and Rajbhandary, 1976). The first complete tRNA sequence, published 12 years ago (Holley *et al.,* 1965), made it possible to investigate the common features of tRNA. Every tRNA sequence can be arranged in the same secondary structure, the *cloverleaf.* This common structure gives reason for the lack of discrimination among tRNA species in many reactions of protein synthesis. On the other hand, sequence differences are the basis for the high specificity of the aminoacylation reaction. This research culminated in 1973, when the three-dimensional structure of $tRNA^{Phe}$ was determined by X-ray crystallography, as the result of the independent effort of two groups (Kim *et al.,* 1973; Robertus *et al.,* 1974).

Simultaneously, the genetic studies of tRNA progressed rapidly with the isolation of suppressor mutants, most of which have alterations in genes specifying tRNA (Smith, 1972). tRNA genes have been identified and mapped both in bacteria and yeast (Smith, 1972). More recently, utilizing *in vitro* recombinant DNA techniques and gene amplification in plasmid vectors, it has been possible to isolate pure tRNA genes from eukaryotes (Beckman *et al.,* 1977).

tRNA has been implicated in several cellular processes other than protein sythesis. For example, tRNA plays a role in transcriptional regulation of certain operons in bacteria (Brenner and Ames, 1971; Brenchley and Williams, 1975), in cell wall synthesis (Roberts, 1972), in nonribosomal transfer of amino acids (Nesbitt and Lennarz, 1968; Deutch *et al.,* 1977), and in DNA synthesis as a primer for reverse transcriptase (Dahlberg *et al.,* 1974). In some of these processes the involvement of tRNA has been called *regulatory* as opposed to the *structural* role as amino acid adaptor in protein synthesis. In this chapter I will discuss exclusively some of the cellular processes wherein tRNA has been assigned a regulatory role. After a brief description of tRNA biosynthesis, I will discuss the involvement of tRNA in the control of transcription (the stringent response and the operon-specific controls), examine the involvement of tRNA in the modulation of translation, and finally consider the participation of tRNA as a primer for reverse transcriptase.

1.1 Biosynthesis of tRNA

Most research concerned with the biosynthesis of tRNA has been done in bacterial and phage systems. Genetic and biochemical approaches have shown that tRNA genes are scattered along the *Escherichia coli* chromosome, often in small clusters of identical or different tRNA species (Smith, 1972; Carbon *et al.,* 1974). Similarly, tRNA genes appear to be dispersed in the genomes of some eukaryotes, such as yeast (Beckman *et al.,* 1977) and Drosophila (G. M. Tener, personal communication). The immediate transcriptional product of tRNA genes is a polynucleotide chain longer than tRNA and, in some cases, it contains the sequences for several tRNA molecules (Schedl and Primakoff, 1973; Schedl *et al.*, 1974; Carbon *et al.*, 1974).

The precursor molecules undergo *maturation* processes composed of two classes of reactions: (1) *size reduction,* in which long precursor polynucleotide chain is processed to the length of a tRNA molecule, and (2) *modification,* such as methylation, thiolation, and pseudouridylation, to name a few.

The isolation of mutants affecting different steps in the maturation pathway has provided a deeper insight into the sequence of these various reactions, their interdependence, and the enzymes involved (Schedl and Primakoff, 1974; Sakano and Shimura, 1975; Ilgen *et al.*, 1976; Bossi and Cortese, 1977; Bossi *et al.*, 1978). The maturation process involves specific endonucleases, the best characterized of which is the RNase P (Altman *et al.*, 1974). This enzyme cleaves the tRNA precursor, removing a segment from the 5′ end, and thereby generates the mature 5′ terminus. Evidence for the participation of other specific nucleases has been presented (Schedl and Primakoff, 1974; Sakano and Shimura, 1975), but at the present time these enzymes are less well characterized.

In some cases, an interdependence has been found between steps of the maturation process. For instance, the trimming of the tRNA's 3′ end and the addition of the terminal CCA must precede the RNase P reaction (Seidman and McClain, 1975). Similarly, the biosynthesis of the modified nucleoside 2-*O*-methylguanosine occurs only after the tRNA precursor has been cleaved to its mature size (Sakano *et al.*, 1974*a*). On the other hand, several modification reactions clearly occur on the uncleaved precursor molecule (Sakano *et al.*, 1974*a,b;* McClain and Seidman, 1975).

The half-life of the maturation intermediates is very short and their steady state concentration is scarcely detectable. It is not known whether some of these molecules have any function other than being the precursor to tRNA (later I will examine this possibility in the discussion of *hisT* mutants). As yet the basic logic of the tRNA maturation process still escapes our understanding. Further studies are necessary to complete the description of *how* maturation occurs, and should eventually provide a richer background of information for speculation as to *why* it occurs.

1.2 tRNA in Protein Synthesis

Mature tRNA molecules are aminoacylated by a set of specific aminoacyl-tRNA ligases. Aminoacyl-tRNAs (with the exception of formylmethionyl-$tRNA_f^{Met}$, which follows another route because of its unique role as initiator) are specifically recognized by the elongation factor Tu (EFTu). In a manner not yet completely understood, the aminoacyl-tRNA–EFTu complex interacts with the 70 S ribosome particle at the A site. If the mRNA is exposing a codon complementary to this particular tRNA, a sequence of reactions follows which includes the synthesis of a peptide bond, translocation of tRNA from the A to the P site, and the intervention of the G factor and the hydrolysis of GTP. The tRNA is then released from the P site and made ready for another synthetic cycle. This process is repeated for every amino acid in the elongation of the protein chain. Finally, nonsense codons act as termination signals, interrupting the mechanism and leading to the release of a protein.

2 *tRNA as a Regulatory Molecule*

The various reactions of the cell can be grouped into a *structural* class, represented by the fundamental reactions which constitute the basic network of metabolic activities, and a *regulatory* class, represented by reactions whose primary role is the coordination and the synchronization of the rate and extent of the structural reactions. Thus, one can theoretically distinguish between structural and regulatory functions that are analogous to the structural and regulatory genes of the lactose operon, for example. In practice, however, the interaction of the various molecules and the interdependence of many reactions in the cell are intricately related; what is regulatory from one point of view may appear structural in the light of additional information or with respect to a different area of metabolism.

Perhaps it will be useful to specify here a clear-cut use of the word *regulatory* when referring to biochemical reactions involving tRNA. In general terms, given a process A → B, consisting of one or more reactions which transform molecule A into molecule B (a metabolic conversion or any change of state), its rate can change for a variety of reasons. In particular, if a molecule X interacts with the A → B transformation process, modifying its rate so as to coordinate it with some other event in the cell, X can be considered a *regulatory molecule.* X can be the *signal* for regulation and exert its effect on the structural reaction A → B, with A and B being the *structural molecules.* The specific component of the A → B system interacting with the regulatory molecule X is the *target molecule.* Here B is a structural molecule whose intracellular concentration is regulated by X; however, it is possible that B acts as a regulatory molecule for the control of other reactions in different areas of metabolism.

In the last decade studies have indicated that tRNA, in addition to its structural role as amino acid adaptor in protein synthesis, also plays a regulatory role in the cell. These studies can be divided into two broad categories: (1) studies that have revealed that tRNA is a *regulatory molecule,* a *signal* for regulation, in the sense just specified; and (2) studies showing that tRNA is a *target* for regulation.

Only in the first case does tRNA perform a regulatory role; in the second case tRNA is only a component of the structural system upon which the regulatory effect is exerted. In the following pages I will first consider tRNA as a *regulatory molecule* and discuss its function as a signal for the synthesis of guanosine triphosphate (ppGpp) in the stringent control and as a signal for regulation (repression) of amino acid operons. I shall then examine tRNA as a *target* of regulation in prokaryotic and eukaryotic systems. Finally, I will briefly mention the function of tRNA as primer for the reverse transcriptase as an example of a category of reactions in which tRNA seems to be involved as a structural molecule in a reaction unrelated to protein synthesis.

2.1.1 Uncharged tRNA Is the Signal for ppGpp Synthesis

The term *stringent control* originally referred to the abrupt arrest of RNA synthesis following the cessation of protein synthesis (Stent and Brenner, 1961). Subsequently, it has become increasingly clear that stringent control is a more complex phenomenon involving several metabolic functions, such as membrane transport, lipopolysaccharide synthesis, and the activation and inhibition of transcription (for a recent review see Gallant and Lazzarini, 1976). Stringent control can be regarded as a major readjustment of the general metabolic strategy when the cell is confronted with environmental changes.

Stringent control is usually studied under extreme conditions, such as in a state of amino acid starvation. In this case, one observes exaggerated properties of the same control mechanism which normally operates under less dramatic conditions. During normal growth, the relative amounts of macromolecules are precisely equilibrated, guaranteeing an optimal growth rate for a given environment. To maintain this equilibrium there are several regulatory mechanisms, some of which consist of repression or derepression of gene sets. Stringent control is one such regulatory mechanism and involves the coordinate expression of a large number of genes, having a *negative effect* on the transcription of the genes for rRNA, tRNA and ribosomal proteins, and a *positive effect* on some biosynthetic operons (histidine, tryptophan, isoleucine–valine) and degradative operons (lactose) (Stephens *et al.*, 1975; Yang *et al.*, 1974).

The current view (but see also Cozzone and Donini, 1973; Donini *et al.*, 1978; Gallant *et al.*, 1977) is that the pleiotropic effect of stringent control is mediated by a signal molecule, ppGpp. The hypothesis that ppGpp is an allosteric effector for RNA polymerase, influencing its transcriptional selectivity, has received some support from *in vitro* studies (Debenham and Travers, 1977). The biosynthesis of ppGpp has received considerable attention in recent years and there is strong evidence that tRNA plays an essential role in this process.

Mature tRNA molecules exist into two forms: charged and uncharged. In logarithmically growing cells the charged species constitutes 65 to 90% of the total depending on the culture medium. As mentioned before, it is charged tRNA that participates in protein synthesis; however, uncharged tRNA can also interact with the protein-synthesizing machinery, binding to the ribosomal A site. In this case, a different reaction takes place: the biosynthesis of ppGpp.

During protein synthesis, after the addition of the nth amino acid and following translocation, the whole system is reset for the addition of the $(n + 1)$th amino acid. Charged and uncharged tRNA molecules interact continuously with the ribosome, but if the anticoden is not complementary to the codon present in the appropriate site, the molecules are released. When a tRNA molecule with an anticodon complementary to the $(n + 1)$th codon enters the A site, there are two possibilities: If it is charged, there is, so to speak, another *click* in the elongation of the growing polypeptide chain, with a repetition of the previous process. If it is uncharged, there is still an interaction (mediated by the so-called *stringent factor,* a product of the *relA* gene) and the system responds with an alternative program: the biosynthesis of ppGpp.

The biosynthesis of ppGpp has been well characterized *in vitro* (Haseltine *et al.*, 1972; Haseltine and Block, 1973) and the requirements of its *in vitro* synthesis correspond authentically to what is known to occur *in vivo*. It has been shown that the mRNA–ribosome complex binds uncharged tRNA both at the A site and at the P site, but in the presence of purified stringent factor the binding to the A site is favored. This process is cyclic, tRNA being released and repeatedly bound to the A site, whereas the stringent factor stays bound to the 70 S ribosome (Richter, 1976). In order for ppGpp synthesis to occur, the uncharged tRNA must have the anticodon complementary to the mRNA codon that is being presented. In addition, it has been established that the 3′-OH terminus of the tRNA must be intact, since removal or oxidation of the terminal adenosine eliminates the ability to promote ppGpp synthesis (Sprinzl and Richter, 1976). The process is not, however, dependent on an intact terminal *adenosine* because substitution of adenosine with a cytosine residue does not affect ppGpp synthesis (G. Chinali and E. J. Ofengand, unpublished observations). Actually, Chinali and Ofengand (unpublished observations) have established, by modifying tRNA *in vitro* with nucleotidyltransferase, that if the CCA stem is shortened, there is no ppGpp synthesis, whereas extending the CCA stem by one residue does not affect ppGpp synthesis. These results emphasize the importance of the state of 3′-OH terminus of the tRNA. It is interesting that periodate-treated tRNA is not an inhibitor of ppGpp synthesis, suggesting that the 3′-OH end is the first feature of the tRNA molecule to be recognized by the ribosome. Notably, the oligonucleotide, TΨCG, can substitute for uncharged tRNA in ppGpp synthesis (Richter *et al.*, 1974). In this case, there cannot be a codon–anticodon base pairing, nor is there a 3′-OH end to be recognized.

These results help to dissect the interaction of uncharged tRNA with the ribosome. Apparently, the 3′-OH of the uncharged tRNA first interacts with the ribosome (oxidized tRNA fails this first binding and is immediately expelled); next, the anticodon and codon interact; and finally the right positioning of the GTψC arm triggers ppGpp synthesis. The isolated TψGC oligonucleotide is able to bypass the 3′-OH terminus scrutiny and the codon–anticodon interaction, directly binding at the right place and promoting ppGpp synthesis.

In summary, the ribosomal machinery has two alternative programs governed by the state of tRNA. Thus, tRNA functions as a *regulatory* molecule, serving as *signal* for the program choice. The cell continuously monitors the amount of charged and uncharged tRNA. An increase in the concentration of the latter is interpreted by the cell as a sign of decreased supply of nutrients from the environment and the response is stringent control. This mechanism is responsible for maintaining an optimal internal chemical balance between the basic cellular machinery (that is, rRNA, tRNA, ribosomal proteins) and the products of this machinery. Specific mechanisms that regulate the amount of energy used for *machinery* and for *products* are very important. In fact, *rel* mutants, which lack stringent control, encounter severe growth difficulties when there are sudden changes in the environment (Gallant and Lazzarini, 1976).

2.1.2 Are Aminoacyl-tRNA Ligases under Stringent Control?

Since the function of the aminoacyl-tRNA ligases requires the participation of the amino acids, it seems logical to consider these enzymes as components of the protein-synthesizing machinery. This view is supported by the finding that in bacteria and in some eukaryotes (Neidhart *et al.,* 1975; Johnson *et al.,* 1977) the concentrations of most aminoacyl-tRNA ligases are proportional to the growth rate. In *E. coli* their concentration increases by a factor proportional to the square of the growth rate, as does ribosome synthesis. This mode of regulation, named *metabolic regulation* (Neidhart *et al.,* 1975), is most likely a manifestation of stringent control. In eukaryotes, similar results have been obtained (Johnson *et al.,* 1977) and, in general, it is commonly observed that the activities of the ligases are highest in those cells and in those tissues in which protein synthesis is most active.

In contrast with these results, there are numerous reports indicating that the biosynthesis of some aminoacyl-tRNA ligases is coupled to the cognate amino acid biosynthetic pathway (for a review see Brenchley and Williams, 1975). Quite often, the effects observed are minor and transient, yet they constitute a strong indication that aminoacyl-tRNA ligases and biosynthetic pathway regulation share some elements.

Recent studies on the sensitivity of several ligases to stringent control induced specifically by amino acid starvation show that there are heterogeneous responses, each ligase behaving in a different manner (Blumehthal *et al.,* 1976). This heterogeneity of response is to be contrasted with the uniformity with which all ligases respond to changes in steady state growth or in a metabolic shift-up, and is indicative of a complex and specialized regulation. It may be due to the existence of two control mechanisms, the general metabolic control (stringent control) on the one hand, and a variety of operon-specific controls, quantitiatively and/or qualitatively different for each ligase, on the other.

2.1.3 *Direct Effect of tRNA$_f^{Met}$ on RNA Polymerase Activity*

In stringent control, uncharged tRNA is *indirectly* involved as the starting signal for a chain of reactions which starts with the synthesis of *ppGpp* and leads to the switching on and off of several genes. The exact mechanisms of the positive and negative effects of ppGpp are not known, but a hypothesis that has received some experimental support is that ppGpp binds to RNA polymerase and thereby affects promoter choice (Travers, 1974, 1976).

There is one case, however, in which the involvement of tRNA in regulation of gene expression is more direct. It has been shown (Pongs and Ulbrich, 1976) that formylmethionyl-tRNA$_f^{Met}$ binds specifically to RNA polymerase. Furthermore, *in vitro* transcription studies have revealed that RNA polymerase bound to formylmethionyl-tRNA$_f^{Met}$ preferentially transcribes *lac* operon genes as opposed to stable RNA genes. In a subsequent study it has been shown that the protein synthesis initiation factor IF-2 also binds to RNA polymerase, but with the opposite effect of favoring the transcription of rRNA genes. These results can be explained by postulating the existence of two systems that compete for binding formylmethionyl-tRNA$_f^{Met}$: (1) RNA polymerase, and (2) the mRNA–30 S ribosome complex. If the total production of mRNA decreases, then there is more formylmethionyl-tRNA$_f^{Met}$ available for binding to RNA polymerase, with the effect of restoring higher production of mRNA. The reciprocal argument can be made for IF-2 which could preferentially bind to RNA polymerase when the 30 S ribosomal subunit is engaged with mRNA.

The picture emerging from these studies is that RNA polymerase can exist in different forms, each capable of preferentially transcribing a specific set of promoters. There are many factors in the cell that may induce the RNA polymerase to shift from one form to the other; IF-2 would favor the form that preferentially transcribes stable rRNA genes (but also the genes for ribosomal proteins). ppGpp and formylmethionyl-tRNA$_f^{Met}$ would favor the form that preferentially transcribes another set of genes, including those of the *lac* operon. Other molecules are also known to affect the activity of RNA polymerase, for instance, EFTuTs and EF-G (Travers, 1976; P. Debenham, A. Travers, and O. Pongs, personal communication). In this manner, the fine balance of all these molecules and the ratios of their free versus bound forms would characterize the cell's physiological state and would determine, quantitiatively and qualitatively, the transcription of the genome.

2.1.4 *Analogous Mechanisms in Eukaryotes?*

The maintenance of an internal chemical balance compatible with growth under various environmental conditions seems to be the primary role of stringent control, and it is achieved by coordinating the expression of several genes. Obviously, coordinate gene expression must be at the very heart of the process of cellular differentiation in eukaryotes, and it is likely that many aspects of cellular differentiation are related the regulation of cell growth. It is not known whether or not there is a system in eukaryotes that is identical to stringent control; but the general principles underlying this control mechanism undoubtedly apply to regulation in higher organisms.

Mammalian cells growing in culture can be shifted from a proliferative state to a quiescent one by a variety of suboptimal conditions (Pardee, 1974; Holley and Kiernan, 1974). Entry into the quiescent state leads to a set of metabolic reactions that have been called the *pleiotypic response* (Hershko *et al.*, 1971). Pardee (1974) has studied the induction of the quiescent state and the kinetics of reentry from quiescence into the S phase under various conditions. His experiments have shown that all cells presented with a variety of suboptimal conditions arrest growth at the same point in their growth cycle, early in the G-1 phase. Pardee calls this point the *restriction point* and suggests that it is the time at which the cell's subsequent growth is determined. Furthermore, he proposes that normal animal cells have evolved the ability to shift from a proliferative to a quiescent state as a survival mechanism for dealing with varying nutritional conditions. Entering the quiescent state implies an adjustment of the metabolic activities to maintain the capacity for growth when the adverse external conditions cease. It is, like stringent control, a strategy aimed at maintaining an internal chemical balance for optimizing energy utilization.

The pleiotypic response, that of the cell entering the quiescent state at the restriction point, is similar to the stringent response in bacteria. Hershko *et al.* (1971) have found that inhibition of stable RNA synthesis, glucose transport, and uptake of nucleic acid precursors and stimulation of protein degradation are common to both phenomena. The analogy also extends to the peculiar effects of certain antibiotics. In *E. coli* the arrest of protein synthesis induced by chloramphenicol treatment is not followed by a stringent response; the cell continues RNA synthesis, mimicking a relaxed phenotype (Gallant and Lazzarini, 1976). In eukaryotes, serum-deprived 3T3 cells treated with cycloheximide show a stimulation of RNA synthesis, in contrast to the usual inhibition observed in the pleiotypic response. Since, in bacteria, the stringent response is triggered by deacylation of tRNA, one understands the effect of chloramphenicol that leads to an accumulation rather than depletion of aminoacyl-tRNA. One could explain the effects of cycloheximide and puromycin in a similar manner. The former compound, an inhibitor of the transfer reaction in protein synthesis, stimulated RNA synthesis in cells experiencing an amino acid shift-down. The latter enhanced the inhibtion of RNA synthesis, as expected, since it inhibits protein synthesis after the utilization of aminoacyl-tRNA. Similar differential effects of puromycin and cycloheximide on rRNA synthesis have been reported (Soeiro *et al.,* 1968).

In a recent study, Grummt and Grummt (1976) have used histidinol, a competitive inhibitor of the aminoacylation of $tRNA^{His}$, as a tool to induce high levels of uncharged $tRNA^{His}$ in mammalian cells without lowering the pool of free histidine. The results of these authors are compatible with the hypothesis that uncharged tRNA is the signal for the onset of the pleiotypic response. Grummt and Grummt (1976) have also shown that ribosomes from mammalian cells *in vitro* degrade GTP to guanine when uncharged tRNA is present at the ribosomal A site. They have proposed that also *in vivo* the depletion of the GTP pool is important for the onset of the pleiotypic response (so far ppGpp has not been found in mammalian cells).

A restriction point in the G-1 phase of the cell cycle has been described in the yeast *Saccaromyces cerevisiae* (Hartwell, 1974). Starvation of wild type yeast for any of the essential elements (C, N, P, or S) prevents reinitiation of a new cell cycle and unbudded cells accumulate. Wolfner *et al.* (1975) have isolated a temperature-sensitive mutant strain, TRA-3, which shows a block in G-1 and also shows a pleiotropic derepression of the genes for the biosynthesis of histidine, arginine, lysine, and tryptophan. These authors propose that the TRA-3 gene is a special sensor gene whose product coordinates the biosynthesis of amino acids with the cell cycle at the G-1 restriction point. They further suggest, in analogy with the bacterial systems, that tRNA might be involved as the signal.

One aspect of the phenotype of the TRA-3 mutant, simultaneous derepression of several amino acid biosynthetic pathways, is commonly observed in neurospora and yeast. This is believed to be induced by the deacylation of tRNA because it occurs in aminoacyl-tRNA ligase mutants (McLaughlin *et al.,* 1969; Nazazio *et al.,* 1971) (see also later in this chapter). It is therefore possible that the TRA-3 gene codes for an aminoacyl-tRNA ligase. The general effect of the arrest in G-1 could then be induced by a condition of amino acid starvation, and might therefore be due to the same signal as that responsible for derepression of amino acid biosynthetic pathways. Such a signal could be identified with some form of charged or uncharged tRNA.

tRNA is implicated in the arrest of cell growth at the G-1 phase in *S. cerevisiae* by the recent results of Unger and Hartwell (1976). These authors report that when cultures of wild types strains are shifted from minimal medium to minimal medium without sulfur, the cells are arrested at the usual point, early in G-1. Apparently, a decrease in concentration of sulfate or of some intermediate in the sulfate assimilation pathway is the signal for specific inhibition of an early event in G-1. In an effort to identify such a signal, it was found that all the metabolic intermediates from sulfate to methionine could not be the control signal because auxotrophic mutants altered in various steps of methionine biosynthesis normally induced the arrest in G-1. Experiments performed on a temperature-sensitive mutant in the gene for the methionyl-tRNA$_f^{Met}$ ligase also showed that starvation induced by the reduced availability of methionyl-tRNA$_f^{Met}$ caused the cell to stop in G-1. The authors concluded that if a unique signal exists for impending sulfate starvation, it follows the step catalyzed by methionyl-tRNA$_f^{Met}$ ligase. Thus, the involvement of uncharged tRNA^{Met} as signal for sulfate starvation, though not conclusively proven, is highly likely.

2.2 *Operon-Specific Control*

In the preceding section on stringent control and analogous mechanisms in eukaryotes, I have emphasized that the cell can monitor the amount of charged and uncharged tRNA, and accordingly adjust its metabolic activities. On the basis of what is known about the participation of uncharged tRNA in the synthesis of ppGpp, it is reasonable to conclude that all uncharged tRNA species participate in an equivalent manner. In contrast, bacteria show additional specific effects induced by individual tRNA species in regulating the biosynthesis of the cognate amino acids. I shall examine separately three examples of such tRNA-dependent control.

2.2.1 *tRNA^{His} Is Involved in Regulation of the Histidine Operon*

The histidine operon consists of nine adjacent genes that are expressed coordinately. The basal level of the histidine biosynthetic enzymes in cells growing in minimal medium is enhanced 10- to 12-fold (derepression) under conditions of histidine starvation or in histidine constitutive mutants. Conversely, there is a repression in minimal medium supplemented with all amino acids (or, which is essentially the same, in nutrient broth), to approximately one-fifth of the basal level in minimal medium. General reviews of the physiological and genetic studies on the histidine operon have been published elsewhere (Brenner and Ames, 1971; Goldberger and Kovach, 1972).

Schlesinger and Magasanik (1964) showed that derepression of the histidine operon is induced by growing the cells in the presense of α-methylhistidine, an inhibitor of the histidyl-tRNA^{His} ligase. This was taken as evidence that the pool of free histidine is not the signal for regulation and as an indication that tRNA^{His} could be such a signal. Moreover, there is now strong genetic evidence for the involvement of tRNA^{His} in regulation of the histidine operon. Mutations that have been mapped in six different loci were found to result in constitutive derepression of the histidine operon:

hisO (mutants mapping in a region adjacent to the histidine operon): For some time they were all considered to be classical operator mutants, but more recent studies have revealed additional properties that favor an operator–attenuator model (discussed later).

hisS (the structural gene for the histidyl-tRNA ligase): Mutants with defective ligase show a decrease in the ratio of charged versus uncharged tRNA, thereby causing derepression of the operon (Roth and Ames, 1966).

hisR: In these mutants the intracellular concentration of $tRNA^{His}$ is only about 50% of that found in the wild type organism. Since the concentrations of other species of tRNA tested appear to be normal, it is believed that *hisR* is the structural gene for $tRNA^{His}$ (Brenner and Ames, 1971; Lewis and Ames, 1972). This view is supported by the observation that cells harboring an episome covering the *hisR* region have higher than normal levels of $tRNA^{His}$ (Brenner and Ames, 1971). However, direct proof that *hisR* codes for $tRNA^{His}$ has not yet been provided.

hisU: Similarly, in these mutants the level of $tRNA^{His}$ is lower than that found in wild type, but the levels of all other tRNA species tested are decreased as well. A biochemical analysis of these mutants (Bossi and Cortese, 1977; Bossi *et al.*, 1978) distinguishes two classes that are probably not allelic. Mutants of class I are altered in a nuclease, probably RNase P, which is involved in the maturation of all tRNA species, with the consequence of an accumulation of tRNA precursors and a decrease of the mature tRNA levels. Mutants of class II are also altered in the maturation of tRNA, but in a manner different from those of class I, not showing accumulation of high-molecular-weight tRNA precursors.

hisW: These mutants also show a lower concentration of several tRNA species, but differ phenotypically from the *hisU* mutants because they do not accumulate tRNA precursors. The nature of the *hisW* mutants is poorly understood. One such mutant, *hisW3333,* is a cold-sensitive lethal (Brenchley and Ingraham, 1973), indicating that the *hisW* gene codes for an essential function.

hisT: Mutants in this gene have normal levels of $tRNA^{His}$ (as well as of all other tRNAs), but the tRNA lacks the pseudouridine modification in the anticodon region (Singer *et al.,* 1972). The gene product of *hisT* has been identified as one of the two pseudouridylate synthetases present in *S. typhimurium* and in *E. coli* (Cortese *et al.,* 1974*a*). The enzyme has been purified and partially characterized (Ciampi *et al.,* 1977). Thus, *hisT* mutants are unable to modify a specific uridylic acid residue in the anticodon loop of $tRNA^{His}$ (as well as of about 50% of all other species of tRNA). This unmodified $tRNA^{His}$, lacking the normally occurring pseudouridine, is perfectly normal in the aminoacylation reaction and in protein synthesis (Brenner *et al.,* 1972), yet is unable to participate in regulation of the histidine operon.

The *hisT* mutants are especially interesting from a regulatory point of view because they show that the structural features necessary for the normal performance of the structural role of tRNA in protein synthesis are independent and separate from those necessary for the regulatory role.

The *hisT* mutants were originally isolated on the basis of the fact that they were derepressed for the histidine operon (Brenner and Ames, 1971). Subsequent studies (Cortese *et al.,* 1974*b*) revealed that other operons are also derepressed in these mutants. The *hisT* mutants are therefore pleiotropic, and reveal an interesting point with respect to the possible function of pseudouridine in the anitcodon region of tRNA. Since the absence of pseudouridine does not impair the participation of tRNA in protein synthesis (this is true both *in vivo* and *in vitro*), one wonders whether its existence in tRNA is not exclusively for regulatory purposes. If that is so, then it would be a likely candidate for a signal of the pleiotropic control of amino acid biosynthesis (this hypothesis had been briefly discussed previously by Cortese *et al.,* 1974*b*). Regulation of amino acid biosynthesis by pseudouridine, in its pleiotropy, would be similar to the simultaneous derepression of several amino acid biosynthetic pathways in yeast and neurospora (see later). This could be a mechanism specifically aimed at derepressing the amino acid operons, even when the corresponding tRNAs are normally aminoacylated. One can imagine that such a mechanism could be very useful to the cell when high concentrations of an amino acid are required for functions other than protein synthesis. For instance, histidine could be utilized as a source of nitrogen via its degradative pathway (Meiss *et al.,* 1969), could be decarboxylated to histamine (Meister, 1965), and so on. In view of this hypothesis, studies on the pseudouridine in the tRNA may well further our understanding of the regulatory role of tRNA.

Another question raised by the studies of *hisT* mutants is the following: By what mechanism does the lack of pseudouridine induce derepression of the histidine and other operons? Since the mechanism of action of tRNA in derepression is unknown, the possibility of understanding the details of the function played by pseudouridine is even more remote. What we know is that changes in expression of several amino acid biosynthetic operons are correlated with changes in the amounts of the corresponding charged tRNA (or of its reciprocal, the uncharged tRNA; see later). Now, there is evidence that tRNA undergoes a conformational change upon aminacylation (Kan *et al.,* 1976). In the case of *hisT* mutants, charged and uncharged tRNAs differ not only in the presence or absence of the amino acid esterified at the 3′ end, but also in another structural aspect. The pseudouridine modification in the anticodon region might be required for an aminoacylation-dependent conformational change. One can imagine that tRNA, even when it is fully aminoacylated but without pseudouridine, is *frozen* in a conformation which has some features of the uncharged tRNA. Furthermore, one has to assume that the protein-synthesizing machinery accepts these molecules as *bona fide* charged tRNA, wherease the controlling elements of the histidine and other operons are more sensitive to the alteration and detect the uncharged appearance of a fully aminoacylated but nonpseudouridylated tRNA.

In the framework of this hypothesis, one might expect *hisT* mutants to be altered in the biosynthesis of ppGpp, because this process is dependent on the presence of uncharged tRNA on the ribosome. The prediction would be that *hisT* mutants might be "superstringent." Preliminary results (P. Donini and R. Cortese, unpublished observations) suggest that the kinetics of synthesis of ppGpp in *hisT* mutants is indistinguishable from those of the wild type organism. The results are consistent with the fact that most of the molecules involved in ppGpp synthesis are also involved in protein synthesis and since they accept *hisT* tRNA as normal for the latter process, it is not really surprising that they do the same for the former.

Lewis and Ames (1972) have provided evidence indicating that regulation of the histidine operon depends on the absolute concentration of charged $tRNA^{His}$, the system being insensitive to the absolute concentration of the uncharged species or to the ratio of the two forms. This interpretation of their data rests on two main points: (1) The degree of derepression of the *his* operon is correlated with the absolute concentration of histidyl-$tRNA^{His}$ in a variety of cases in which the ratios between charged and uncharged tRNA differ widely. However, this finding is also compatible with the possibility that uncharged $tRNA^{His}$ is the signal for derepression of the operon. (2) The results with *hisR, hisU,* and *hisW* mutants are compatible with the possibility that histidyl-tRNA^{His} functions as a corepressor and apparently rules out the possibility that uncharged tRNA^{His} functions as an inducer. These conclusions rest on the underlying assumption that in *hisR, hisU,* and *hisW* mutants there is, *in vivo,* a diminution of both charged and uncharged tRNA^{His}. It is possible, however, that the total pool of tRNA^{His} sequences present in the cytoplasm of these mutants is the same as that of the wild type organism but that only a part of it is capable of aminoacylation. This altered fraction could still exert an effect on control of the histidine operon. These considerations revive the hypothesis that uncharged tRNA^{His} is a positive effector of the histidine operon. In fact, we do not really know what structural features tRNA^{His} molecules must possess in order to participate in regulation of the histidine operon. We do know, however, from the *hisT* mutant data, that the lack of pseudouridine impairs the regulatory role of tRNA without interfering with its capacity to participate in the aminoacylation reaction. The exact reciprocal situation might be true of *hisR, hisU,* and hisW mutants. In these mutants the transcriptional product of the tRNA^{His} gene might be altered so as to be unsuitable for the aminoacylation reaction, but perfectly able to perform its regulatory role.

Some *hisU* mutants have an altered nuclease responsible for the maturation of tRNA precursors (Bossi and Cortese, 1977; Bossi *et al.,* 1978). In these *hisU* mutants the tRNA^{His} gene is transcribed but a precursor accumulates in the cell which cannot be charged with histidine. This tRNA precursor differs from the mature form only in a few extra nucleotides at the 5′ terminus (R. Cortese, unpublished observations; Vogeli et al , 1977). There is abundant evidence that tRNA precursors, especially the small ones, have a tRNA-like conformation (McClain and Seidman, 1975). Must we rule out the possibility that a molecule, identical to mature tRNA^{His} save for a few extra nucleotides at the 5′ end, can participate effectively in regulation?

Assuming that *hisR* is the structural gene for tRNA^{His}, there are several ways to explain the phenotype of *hisR* mutants (Brenner and Ames, 1971; Singer, 1971). For instance, in the two-gene hypothesis, the *hisR* gene is present as a tandem duplication; *hisR* mutants could be altered in only one of the two genes, hence the concentration of tRNA^{His} in the mutant would be 50% of that of the wild type organism. The transcriptional product of the altered gene is not a good substrate for the aminoacylation reaction but may still work effectively in regulation. Alternatively, there may be only a single *hisR* gene; *hisR* mutants could carry a mutation in this gene that alters the transcriptional product so as to slow down its maturation, resulting in an accumulation of a tRNA^{His} precursor, for which the same arguments presented for *hisU* mutants are valid.

An objection to these arguments is that the idea that uncharged tRNA^{His} acts as a positive effector necessarily implies that *hisR* mutants should be dominant with respect to the wild type allele. In the one case investigated, *hisR1223,* the mutant was found to be recessive to the wild type allele with respect to regulation of the histidine operon (Fink and Roth, 1968). However, nothing else is known about the nature of the *hisR1223* mutation. Perhaps the mutation alters the structure of the tRNA^{His} gene transcriptional product so as to slow down its maturation, as discussed before. Mutants of this type have indeed been found in the tRNA^{Tyr} gene (Smith, 1976). If this were true, then the recessivity of *hisR1883* could be an artifact resulting from the use of an episome, F′14, which also carries the *hisU* gene. Since the *hisU* gene codes for a nuclease responsible for the maturation of tRNA, one can imagine that an increase in the concentration of this nuclease in the strain harboring the episome would accelerate the maturation of the altered $tRNA^{His}$ precursor, and consequently repress the histidine operon. These considerations on *hisR* and *hisU* mutants support the hypothesis that uncharged tRNA is a positive effector. Admittedly the case for charged tRNA as a negative effector is stronger because fewer *ad hoc* assumptions are needed. I think, however, that more experiments are required in order to settle the question. Mutants altered in the transcription of the $tRNA^{His}$ gene, such as promoter mutants, might be very informative. I will return to these points in the discussion of the isoleucine–valine and leucine operons, where the hypothesis that uncharged tRNA is a positive effector seems to fit best the available data.

Clearly, we do not know the molecular mechanisms by which tRNA is involved in regulation of transcription. However, it may be useful to review briefly what is known about the other aspects of histidine operon regulation so as to put the role of $tRNA^{His}$ in prespective.

Goldberger and co-workers observed anomalous kinetics in the repression and derepression of the histidine operon in some strains carrying a mutation in the *histG* gene (the gene that codes for the first enzyme of the histidine biosynthetic pathway) (Goldberger and Kovach, 1972). In these strains the catalytic properties of the *hisG* enzyme were unaffected but the enzyme was no longer sensitive to "feedback" inhibition by histidine (feedback-resistant enzyme). The possibility that the *hisG* enzyme is directly responsible, as a regulatory molecule, for controlling the rate of transcription of the *his* operon has been suggested (Goldberger and Kovach, 1972). In favor of such a proposal, it was found that purified *hisG* enzyme binds specifically to purified histidine operon DNA (Meyers *et al.,* 1975) and that $tRNA^{His}$ binds specifically to the *hisG* enzyme (Blasi *et al.,* 1971). Moreover, the aminoacylated form binds with a higher affinity than the uncharged species (Deeley *et al.,* 1975). No differences in binding affinities were found between normal tRNA and tRNA lacking pseudouridine. Recently (Kleeman and Parsons, 1977), the binding of histidyl-$tRNA^{His}$ to the *hisG* enzyme was studied in more detail and it was found that the binding was inhibited in the presence of histidine and ppGpp. This ppGpp-dependent pheonomenon is particularly interesting in view of the well-established role for ppGpp as a positive effector for expression of the *his* operon (see later) (Stephens *et al.,* 1975). Blasi and co-workers, using an *in vitro* DNA–RNA polymerase system, discovered that addition of *hisG* enzyme specifically inhibited transcription of the histidine operon (Di Nocera *et al.,* 1975). One important observation prevents us from accepting the conclusion that the *hisG* enzyme plays an essential role in regulation of the histidine operon. Mutant strains, carrying a deletion of the *hisG* gene, are apparently normal in regulation of the histidine operon (Scott *et al.,* 1975). The possibility of more than one mechanism of regulation, acting in a complementary or even alternative way, must be borne in mind.

At any rate, there are two reasons for believing that regulation of the histidine operon is highly complex and involves multiple control points: (1) *ppGpp-dependent transcription. In vitro,* the extent of transcription of the *his* operon is proportional to the ppGpp concentration, in the range between 0 and 10 mM (Stephens *et al.,* 1975). Stephens and co-workers suggest that *in vivo* changes in the level of histidine operon expression in cells shifted from minimal medium to nutrient broth are entirely due to changes in ppGpp concentration which are associated with the two different media. The analysis of the repression–derepression behavior in isogenic *rel*$^+$ and *rel*$^-$ strains (these latter are incapable of synthesizing ppGpp following amino acid starvation) supports their contention. Stephens *et al.* (1975), observing that ppGpp acts as positive effector in other systems (Yang *et al.,* 1974), propose that ppGpp is a general signal for amino acid deficiency. The ppGpp-dependent regulation appears to be a pleiotropic mechanism of control aimed at readjusting cellular metabolism to prevent an imbalance of amino acid supply. ppGpp is considered to be an example of a special kind of signal molecule, for which Ames used the term *alarmone.* Alarmones, such as ppGpp (cAMP could be another example), serve to reorient the cell's energy investments in response to stress in a particular metabolic area. The specific effect of ppGpp on the histidine operon serves to adjust the synthesis of histidine biosynthetic enzymes with respect to the need for histidine and the relative supply of all amino acids. (2) *Operon-specific regulation.* This is an additional mechanism, probably acting in conjunction with the ppGpp pleiotropic control, but specifically responsive to the cell's histidine requirement. It is believed that tRNA exerts its effect at this level. Kasai (1974) on the basis of evidence from his study of the *in vitro* transcription of the histidine operon, and fine physiological and genetic analysis of *hisO* mutants (Ely, 1974; Ely *et al.,* 1974), has proposed a novel type of regulation. He postulates the existence of a particular structural feature in the *hisO* gene, called the *attenuator,* which constitutes a barrier to transcription by RNA polymerase. Artz and Broach (1975), using an *in vitro* coupled transcription–translation system, have found support for the attenuator hypothesis and have extended it to a more comprehensive model for the *his* operon regulation. This model is based on the existence of a positive activator. Under conditions of histidine deprivation, derepression is accomplished by a positive factor which interacts with the *hisO* region, allowing the RNA polymerase to bypass the attentuator and transcribe histidine structural genes. In this model, repression is maintained by inactivation of the positive factor. Artz and Broach also consider the possibility that the *hisG* enzyme exerts a role (albeit a subsidiary one) at the level of termination of transcription at the attenuator site, with the possibility that the binding to histidyl-tRNA^{His} will affect its action.

2.2.2 The Role of tRNA in Regulation of the Isoleucine–Valine and Leucine Operons

Comparatively less information is available on the involvement of tRNA in regulation of the isoleucine–valine and leucine operons. The effect of the *hisT* mutation on expression of these operons is similar to the effect on the histidine operon (Allaudeen *et al.,* 1972; Cortese *et al.,* 1974*b;* Rizzino *et al.,* 1974). It is interesting to analyze the effect on the *ilv* and *leu* operons further because there are some details that indicate that uncharged tRNA could be the signal molecule for regulation, acting as positive effector. The isoleucine–valine and leucine operons have minimal levels of expression in a medium supplemented with all the branched-chain amino acids. In the absence of these amino acids there is a four- to five-fold derepression in wild type strains, but if a complete starvation can be induced, for instance in auxotrophs, then there is a 20- to 30-fold derepression. In the *hisT* mutant the enzyme levels are at a value four to five times higher than in the wild type and there is no further repression (Cortese *et al.*, 1974*b;* Rizzino *et al.*, 1974), the system being insensitive to the presence of the branched-chain amino acids. Notably, this level rises very little during starvation, reaching about tenfold elevation, whereas the wild type strains reach a value of 20- to 30-fold. This special behavior of *hisT* mutants was observed both by Cortese *et al.* (1974*b*) and by Rizzino *et al.* (1974), but only the latter authors emphasized the possible importance of it (Bresalier *et al.*, 1975). In the case of the isoleucine–valine and leucine operons, the lack of pseudouridine in tRNA does not simply lead to derepression, as in the case of histidine. Rather, the effect is one of reducing the range of the levels within which the operon can be regulated. One might say that the lack of pseudouridine confers rigidity on the system.

Elaborating the hypothesis that when tRNA lacks pseudouridine it is "frozen" in an uncharged-like conformation, one can postulate that tRNA is frozen in a conformation intermediate between uncharged and charged. One can interpret the incapacity of the unmodified tRNA of *hisT* mutants to induce full derepression or full repression as the consequence of its intermediate conformation, which imposes a fixed intermediate level of operon expression. A logical consequence of this idea is that normal uncharged tRNA is a positive effector. Though indirect, the argument is compelling.

A further indication that uncharged tRNA might be a positive effector comes from studies of *hisU* mutants. In these strains, tRNA maturation is altered and, in particular $tRNA^{Leu}$ (at least some isoaccepting species) accumulates as a very long multimeric precursor whose structure is sufficiently different from tRNA to escape recognition by most enzymes (R. Cortese, unpublished observations). In these mutants the isoleucine–valine and the leucine operons are *hyperrepressed* (R. Cortese, unpublished observations; Davidson *et al.*, 1977), having a value of 1 in minimal medium supplemented with the amino acids, compared to the wild type value of 4 to 5. This could be easily explained if uncharged tRNA species were positive effectors and their concentrations were lower in *hisU* mutants. Furthermore, this suggests that the $tRNA^{Leu}$ multimeric precursor cannot participate in regulation. (The argument differs for the histidine operon, but there we were dealing with a precursor that is only very slightly larger than mature tRNA).

2.2.3 The Role of tRNA^{Trp} in Regulation of the Tryptophan Operon

The role of $tRNA^{Trp}$ in regulation of the tryptophan operon has long been controversial. Recently, however, more penetrating studies (Morse and Morse, 1976; Yanofsky and Soll, 1977) have indicated a definite role for $tRNA^{Trp}$. The tryptophan operon is subject to multiple levels of regulation. A classical repressor–operator mechanism exists, with the repressor molecule (the product of the *trpR* gene) activated by free tryptophan, a classical corepressor. This repressor-mediated control is the dominant one and usually masks another mechanism which can be detected only in *trpR*$^-$ strains. In the absence of a normal repressor protein there is still a residual derepression of the tryptophan operon induced by tryptophan starvation. Morse and Morse (1976) showed that the tryptophan analog 7-azatryptophan can effectively substitute for tryptophan in this residual regulation, whereas 5-methyltryptophan cannot (exactly the opposite is true for the *trpR*-mediated regulation). An important difference between these two analogs is that the former is a substrate for the tryptophanyl-tRNA^{Trp} ligase and is charged on tRNA^{Trp} *in vivo,* whereas the latter is not. Morse and Morse (1976) have also found that in a *trpR*$^-$ *trpS*$^-$ double mutant, which produces a temperature-sensitive tryptophanyl-tRNA^{Trp} ligase that does not use 7-azatryptophan as a substrate, the *trpR*-independent regulation of the tryptophan operon no longer responds to 7-azatryptophan. These data were taken as evidence that tRNA^{Trp} (or, more cautiously, a product of the aminoacylation reaction) is the signal for regulation. In this same study the authors used a set of mutants carrying deletions covering various segments of the *trp* operator–promoter region to show that tRNA^{Trp} acts at the level of the previously characterized *attenuator* site (Bertrand *et al.,* 1975). (tRNA^{His} is believed to act at the analogous site in the histidine operon). The sequence of the 5′ region of the trp mRNA (Squires *et al.,* 1976) reveals some short regions of homology with the tRNA^{Trp} sequence (Hirsh, 1970), but no data are available on a possible direct interaction.

Direct evidence for the participation of tRNA^{Trp} in the regulation of the *trp* operon has been presented by Yanofsky and Soll (1977). A mutant altered in the *trpT* gene, the structural gene for tRNA^{Trp}, and carrying a trpR mutation which eliminates the tryptophan aporepressor-mediated regulation, shows a sevenfold derepression of *trp* enzymes. This mutant is analogous to the *hisR* mutants in the *his* operon system; however, in this case studies on merodiploid strains have shown that the *trpT*$^-$ allele is partially dominant to the wild type allele, indicating a possible role for uncharged tRNA^{Trp} in the *trp* operon regulation. This is in accordance with and strengthens the arguments presented in Sections 2.2.1 and 2.2.2 concerning the possible role of uncharged tRNA in the regulation of *his, ilv,* and *leu* operons. Yanofsky and Soll have also shown that *trpT* mutants are not derepressed in strains carrying deletions of the attenuator region. These authors conclude that tRNA^{Trp} exerts its effects at the level of the attenuator-mediated termination of *trp* mRNA transcription.

2.2.4 Does tRNA Control the Expression of Genes in Eukaryotes?

The possible involvement of tRNA in regulation of amino acid biosynthesis in yeast and neurospora has been mentioned in Section 2.1.4. In neurospora, histidine starvation results in derepression of the synthesis of tryptophan and arginine biosynthetic enzymes as well as of the enzymes of the histidine pathway, and the reciprocal is also true (Carsiotis *et al.,* 1974; Carsiotis and Jones, 1974). Analogous results were obtained in yeast (Schurch *et al.,* 1971; Messenguy and Delforge, 1976). The involvement of tRNA in these derepressions is suggested by work on isoleucyl-tRNA^{Ile} ligase (McLaughin *et al.,* 1969) and tryptophanyl-tRNA^{Trp} ligase mutants (Nazario *et al.,* 1971). More recently, Spurgeon and Matchett (1977) have shown that in neurospora conditions that induce *in vivo* inhibition of the activity of histidyl-tRNA^{His} ligase or tryptophanyl-tRNA^{Trp} ligase (causing a lower concentration of charged tRNA^{His} or tRNA^{Trp}, respectively) are associated with derepression of the histidine, tryptophan, and arginine pathways. It is clear that in both yeast and neurospora the biosynthesis of various amino acids is jointly regulated by a common mechanism, so that starvation for one amino acid induces a multiple derepression in several amino acid biosynthetic operons.

There is no obvious reason for the interdependent regulation of the biosynthesis of arginine, lysine, histidine, and tryptophan (Spurgeon and Mitchell, 1977; Wolfner *et al.,* 1975), unless one considers that the biosynthesis of all amino acids is under the same pleiotropic control. This situation would then be similar to the ppGpp-mediated regulation of amino acid biosynthesis as proposed by Stephens *et al.* (1975). It seems probable that in bacteria there are two levels of regulation of the biosynthesis of amino acids: a ppGpp-mediated pleiotropic control which ensures a well-balanced endogenous production of all amino acids relative to each other, and a (tRNA-mediated?) operon-specific control which ensures the continuous supply of specifically required amino acids. In yeast and neurospora there seems to be only one mechanism of regulation, one which almost looks like a hybrid of the two bacterial mechanisms in that it is pleiotropic but tRNA-mediated. In this respect I may mention that ppGpp has been recently found in yeast, but has not yet been associated with any particular regulatory function (Pao *et al.,* 1977).

The preceding data on the pleiotropic controls of amino acid biosynthesis in yeast and neurospora do not necessarily rule out the existence of amino acid-specific controls. It may well be possible that such specific mechanisms are active only in special circumstances. Alternatively, the general control may be epistatic to the specific ones, masking them in ordinary experiments, very much like the *attenuator tRNA*^{Trp}-dependent control of the tryptophan operon, which reveals itself only when the epistatic tryptophan repressor control is rendered inoperative by mutation. The existence of multiple levels of control, which can partially or totally substitute for each other, is emerging as a general feature in many different systems, suggesting that redundancy of regulatory mechanisms is the rule rather than the exception.

From what I have said, it appears that tRNA-dependent control of gene transcription is a widespread phenomenon (for a more comprehensive review of all cases of operon control in which tRNA has been claimed to play a regulatory role, see Brenchley and Williams, 1975). So far, most research has concentrated on establishing, beyond doubt, the involvement of tRNA. Future efforts should focus on *how,* not *if,* tRNA functions in operon control. There are several possibilities: (1) tRNA can interact directly with DNA by base pairing. Even though tRNA^{His} does not hybridize with purified *his* operon DNA (R. Cortese, unpublished observations), this does not rule out the possibility of a direct interaction *in vivo,* perhaps requiring a cofactor. Such direct interaction with DNA is necessary for the tRNA-dependent priming of reverse transcriptase (see later). (2) Since *rho* factor is involved in termination of transcription at the attenuator site in the *trp* operon (Korn and Yanofsky, 1976), tRNA could exert its action by interacting with *rho* factor. (3) The segment of RNA that starts at the initiation of transcription and ends at the attenuator site is called the *leader* (Bronson *et al.,* 1973). When repression of the operon occurs by termination at the attenuator site (for histidine see Artz and Broach, 1975; for tryptophan see Morse and Morse, 1976; Yanofsky, and Söll, 1977), there is always a full rate of synthesis of the leader RNA. If this leader RNA were translated, one could imagine that its translational product plays a role in the regulation of the corresponding operon. tRNA could be involved in the regulation of its translation as a special case of tRNA-dependent *modulation* of translation (discussed later). For instance, if the leader RNA of the histidine operon contained a high number of codons for histidine, then its rate of translation would be dependent on the concentration of histidyl-$tRNA^{His}$. This hypothesis was advanced many years ago by B. Ames and R. Martin (personal communication). The sequence of the leader RNA for the *trp operon* has been established (Bronson *et al.*, 1973). Several features of the sequence suggest that it might be translated. For example, preceding the attenuator site there is a ribosomal binding site, an AUG triplet and two adjacent *trp* codons in the same reading frame, all of which are followed by a termination triplet. Lee and Yanofsky (1977) have observed that the leader RNA could form two alternative secondary structures. They postulate that translation of the leader is coupled with its transcription and determines whether there will be transcriptional termination at the attenuator. This could be effected if the presence of ribosomes near the attenuator favors a secondary structure of the leader RNA which terminates transcription. In this manner active translation of the leader RNA is equivalent to *repression* of the *trp operon.* Accordingly, low level of tryptophanyl-$tRNA^{Trp}$ would impede ribosomal movement at the two adjacent *trp* codons. This "idling" of the ribosomes will favor a secondary structure of the leader RNA which will allow for the continuation of transcription beyond the attenuator. Thus low levels of tryptophanyl-$tRNA^{Trp}$ are equivalent to *derepression* of the *trp operon.* This model is attractive in both its simplicity and explanatory value. However there are a few objections to this hypothesis which are discussed elsewhere (Lee and Yanofsky, 1977). This model will be better evaluated once similar information about other operons is at hand. (4) tRNA could function as corepressor in conjunction with a yet unidentified repressor.

Why is tRNA, and not the free amino acid, the regulatory signal? Since this is a prevalent phenomenon, one should look for a general explanation. Perhaps the physicochemical properties of tRNA confer an intrinsic capacity for interaction with the transcriptional apparatus. Alternatively, the reason may be historical: One can speculate that tRNA originally had the capacity to interact with the corresponding operon, for some function relating to a primitive translational mechanism (Woese, 1967; Crick *et al.*, 1976). This capacity to interact with the operon may have been converted into a regulatory function through evolutionary selection.

3 *tRNA as a Target for Regulation*

3.1 *tRNA-Dependent Modulation of Translation: An Evolutionary Equilibrium*

tRNA is a structural component of the protein-synthesizing machinery and its availability at the ribosmoal A site is probably important in determining the rate of translation (Anderson, 1969). Consequently, a possible factor in the control of translation is the correspondence between the relative frequencies of the various amino acids in the proteins and the relative abundances of the various tRNA species. The idea of a tRNA-dependent translational control is an old one: Ames and Hartman (1963), in an attempt to explain polarity in the synthesis of histidine operon enzymes in *S. typhimurium,* suggested that in its travel along the mRNA, the ribosome slows down when encountering codons whose corresponding tRNA is present in low concentrations in the cell. At these points a ribosome would have a higher probability of falling off the mRNA with consequent termination of protein synthesis. Ames and Hartman (1963) called this phenomenon *modulation* of the rate of translation by the distribution of codons whose tRNA is present in limiting concentrations. The ribosome would not move along the mRNA at a constant rate but rather proceed in a stuttering fashion (evidence for this has been provided by Talkad *et al.,* 1976).

Since the code is degenerate for all amino acids one has to extend the modulation hypothesis to include the possibility of a control exerted on the choice of which codon, and therefore which isoaccepting tRNA species, is used in any particular case. It is a common observation that the relative concentrations of isoaccepting species are very different, so that it is usually possible to distinguish major and minor isoaccepting species. If the distribution of the various synonymous codons in the various mRNAs is not random, then the rate of translation of different mRNAs might not be the same. In support of this possibility, it has been shown that, using characterized eukaryotic isoaccepting tRNA species in *in vitro* translation of globin mRNAs, there is a preferential use of particular codons for valine (Claker and Hilse, 1974), glutamine (Hilse and Rudtoff, 1975), and lysine (Woodward and Herbertt, 1972). More recently, the nucleotide sequences of several genes have revealed a nonrandom distribution of synonymous codons (Fiers *et al.,* 1976; Sanger *et al.,* 1977; Efstradiatis *et al.,* 1977). Thus, it is conceivable that there are both quantitative and qualitative (discriminative) tRNA-dependent controls on the rate of translation.

In the light of these considerations it is interesting, but not surprising, that in various organisms, including *E. coli,* one finds a good correlation between the relative abundances of tRNA species and the frequencies of amino acids in proteins (Garel, 1974). This could be the result of an evolutionary equilibrium, aimed at the optimization of the rate of translation and at saving energy by not synthesizing more tRNA than is needed. Thus, if in any independent genome there is such an optimization, one would expect mRNA translation using heterologous tRNA to be suboptimal. Some *in vivo* and *in vitro* studies confirm this expectation.

In vivo, phage T4 codes for eight species of tRNA that are normally transcribed, matured, and used in protein synthesis during infection. It is not clear why the phage carries information for this limited set of tRNA molecules. There is some indication that they serve the purpose of supplementing the cellular tRNA pool with anticodons complementary to codons more frequent in the T4 mRNA (Sherberg and Weiss, 1972). A more direct experiment showing the importance of the phage-coded tRNA has been reported by Wilson (1973). He found a mutant T4 strain, Δ*33* which has a deletion for all tRNA genes and is therefore totally dependent on cellular tRNA for translation of its messages. The Δ*33* strain is nevertheless able to go through a normal infective cycle, showing that the phage-coded tRNAs are not essential. However, the burst size is significantly reduced (about 50% of that for wild type phage). Analysis of the proteins synthesized during Δ*33* infection shows an altered pattern of the late proteins, with at least one species greately reduced. Moreover, the DNA synthesis pattern is different from the wild type pattern, being similar to the prolonged synthesis observed in mutants lacking the late proteins necessary for encapsulation (Fario *et al.,* 1977). A simple interpretation of these data is that one or more cellular tRNA species is present in a limiting concentration, so that the synthesis of some phage proteins is optimal only with the addition of the phage-coded tRNAs. The rate of synthesis of these proteins is accordingly reduced in the Δ*33* strain.

By using cell-free protein-synthesizing systems it is possible to combine components derived from different tissues or organisms, and thereby test the influence of different tRNA populations on mRNA translation. Early work of Anderson and Gilbert (1969) showed that deletion of a tRNA fraction in the translation of globin mRNA differentially reduces the synthesis of one of the globin chains. More recently, Sharma *et al.* (1976) have found that ascites tRNA and rabbit liver tRNA could promote efficient *in vitro* translation of globin mRNA, oviduct mRNA, and encephalomyocarditis viral RNA. In constrast, reticulocyte tRNA participates efficiently in the translation of only globin mRNA, causing early termination of translation of other mRNAs. Le Meur *et al.* (1976) have also studied the *in vitro* translation of hen oviduct and rabbit reticulocyte mRNA. Their results show that homologous tRNA populations are always more efficient, both in the rate and the extent of protein synthesis. These authors were, however, unable to ascribe this effect to a specific subset of tRNA species and concluded that the homologous system works better because of the precise relative abundance of the various tRNA species.

Individual tRNA species were, on the contrary, identified as responsible for the translational block observed in interferon-treated cells (Zilbertstein *et al.,* 1976). Extracts of mouse L cells which had previously been exposed to interferon lost their ability to translate mRNA *in vitro.* It had been observed (Revel *et al.,* 1975) that the addition of tRNA to the extract promoted translation, overcoming the block. Moreover, the effect could be obtained using a subset of tRNA species. Extensive purification identified minor $tRNA^{Leu}$ species which are capable of restoring the ability to translate specific messages. $tRNA^{Leu}$ minor species A, capable of recognizing the codon CUG, is able to restore hemoglobin mRNA translation. Similarly, $tRNA^{Leu}$ minor species B, recognizing codon UUG/UUA, overcomes the block of translation of Mengo virus mRNA. This study provides additional evidence for the importance of tRNA concentration in translation, at the same time furnishing a clear example of *codon-dependent translational selectivity.* It also raises a new problem: The $tRNA^{Leu}$ species A, uniquely capable of restoring hemoglobin mRNA translation in the mouse cell *in vitro* system, recognizes the codon CUG, which is precisely the same codon recognized by the major $tRNA^{Leu}$ species. However, the major $tRNA^{Leu}$ species is incapable of restoring translational activity. It seems, therefore, that the codon–anticodon recognition process is a complicated phenomenon. One has to postulate that mRNA-dependent restrictions are imposed on the indiscriminate utilization of two different tRNA species which carry the same anticodon in translation. If this were shown to be the case, then posttranscriptional modifications of the same tRNA species, such as methylations and pseudouridylations, might play a role by providing structural features which are specifically required for recognizing specific codons in some mRNAs but not in others.

From the studies mentioned previously one sees that the particular composition of tRNA populations is probably the result of selective pressure to reach an optimal equilibrium. In addition to this evloutionary steady state it is now clear that cells (both of microorganisms and of multicellular organisms) have the capacity to modify the relative concentration of tRNA species in response to changes in physiological conditions and during differentiation and development. Littauer and Inouye (1973) reviewed the numerous reports on a variety of physiological conditions that affect the distribution of tRNA species. These studies commonly utilize chromatographic fractionations and thereby compare the elution profiles of the various tRNA populations. The majority of these studies do not distinguish between differences in tRNA populations owing to posttranscriptional modifications or to newly expressed genes. Nor do they provide any correlation between the tRNA changes observed and the specific physiological situation compared.

Changes in tRNA populations are best analyzed when the change in the relative abundance of some tRNA species corresponds to a parallel change in the frequency with which the cognate amino acid appears in proteins. A good example of this phenomenon occurs in the reticulocyte. This cell is well suited for such studies because it synthesizes essentially only one protein, hemoglobin, and can easily be obtained free of other contaminating cells (Smith, 1975). A comparison of the tRNA population in rabbit reticulocytes and rabbit liver shows differences that could have been predicted on the basis of the average amino acid composition of the proteins of the two cell types. Isoleucine is more abundant in the liver and histidine is more abundant in hemoglobin; correspondingly, there is twofold less $tRNA^{Ile}$ and twofold more $tRNA^{His}$ in reticulocytes than in liver cells (Smith and McNamara, 1971; Smith, 1975). Even though tRNA differences in various organs of the same organism appear to be the exception rather than the rule, these studies on reticulocytes and liver cells show that the same genome can produce varying amounts of tRNA in two separate differentiated cells. Another example of developmental adaptation of tRNA populations occurs in the reticulocytes of sheep. Anemic sheep synthesize a hemoglobin with β- chain of abnormal amino acid composition (Litt and Kabat, 1972). The tRNA extracted from these reticulocytes is different from that extracted from normal reticulocytes and the difference corresponds to the changed amino acid composition of the abnormal hemoglobin.

Yet another system showing a highly developed specialization in protein biosynthesis is the posterior silk gland of the silkworm *Bombyx mori.* During its larval stage it produces only a single protein, silk fibroin. This protein has an unusual amino acid composition, containing primarily alanine, glycine, serine, and tyrosine, with glycine comprising 45% and alanine 30% of the total (Tashiro *et al.,* 1968). The abundances of $tRNA^{Ala}$, $tRNA^{Gly}$, $tRNA^{Tyr}$, and $tRNA^{Ser}$ mirror the frequencies of the corresponding amino acids in silk fibroin (Garel *et al.,* 1970). In the development of the posterior gland, several phases can be distinguished: In the fifth instar, at stage V-1, there is active growth but very little fibroin synthesis, whereas at V-8 there is maximal fibroin synthesis. Enrichment of $tRNA^{Gly}$ and $tRNA^{Ala}$ begins during the early period of V-1 and from V-1 to V-8 there is an increase of about 15-fold. [The other tRNA species also increase, but only five- to sixfold (Delaney and Siddiqui 1975).] This increase appears even more dramatic if one considers only the amount of tRNA found aminoacylated *in vivo*. In this case $tRNA^{Ala}$ and $tRNA^{Gly}$ increase about 30-fold, and all the other tRNAs only six- to eightfold. Relative changes within the three $tRNA^{Gly}$ isoaccepting species are also observed; the minor one is only 6% of the total in V-1 but becomes 17% in V-8. More interesting are recent experiments (Meza *et al.*, 1977; Sprague *et al.*, 1977) showing that among the isoaccepting $tRNA^{Ala}$ present in the posterior gland there is one that is present in this organ alone. This seems to be a clear case (and the only one known so far) of a tRNA species uniquely transcribed in one organ; in the other cases posttranscriptional modification differences could not be ruled out.

Changes in posttranscriptional modifications have been demonstrated in *Drosophila* (White *et al.*, 1973*a*). These authors discovered that marked changes in the chromatrographic elution profile of four tRNA species occur during the development from the larva to adult. For three of these species ($tRNA^{His}$, $tRNA^{Asp}$, $tRNA^{Tyr}$) the larval stage shows only one peak in the elution profile, indicative of one major isoaccepting species. In contrast, the elution profile of tRNA from the adult stage shows an additional peak. Similarly, for $tRNA^{Asn}$ there are four peaks in the elution profile of tRNA of the larval stage which become six in the adult tRNA. The uniform pattern of these changes suggests that a posttranscriptional modification is involved. White *et al.* (1973*b*) discovered that the adult tRNA species lack a modified base present in the larval tRNA species. This modified base shows the same chromatographic properties as those of the well-characterized nucleoside *Q* in *E. coli*. (It is interesting that the nucleoside *Q* is found in *E. coli* exclusively in the anticodon region *of the same* four tRNA species.) Recently, an enzyme has been described in rabbit reticulocytes (Farkas and Singh, 1973) that is able to replace the *Q* nucleoside with guanine, which is known to be the presursor of *Q* (Okada *et al.*, 1976). Furthermore, it has been shown that only the tRNA species containing nucleoside *Q* are substrates for this enzyme (Okada *et al.*, 1976). A comparison of the activity of this enzyme in the larva and adult *Drosophila* could shed light on the mechanism by which posttranscriptional modifications of tRNA species are developmentally regulated.

Similarly, differences in the distribution of tRNA species probably due to changes in the activity of modifying enzymes, are observed in comparisons of tumor and normal cells (Borek and Kerr, 1972; Littauer and Inouye, 1973). More recently, Grumberger *et al.* (1975) have found that in Novikoff hepatoma cells changes in the $tRNA^{Phe}$ elution profile are due to the undermodification of the nucleoside Y in the anticodon loop. Salomon *et al.* (1976), using a Sepharose-bound antinucleoside Y antibody, have shown that 90% of $tRNA^{Phe}$ from a variety of normal tissues contains the Y base, whereas only 10 to 15% of $tRNA^{Phe}$ from a variety of tumors contains the Y base (the other 80–85% being unmodified).

In these and similar but less defined observations in tumor tissues, it is not possible to distinguish between a tumor-specific phenomenon and an epiphenomenon of disordered growth. It is possible that the tRNA changes in tumor cells are the result of an internal chemical imbalance, such as an imbalance between the synthesis of tRNA and modifying enzymes. A parallel with this situation is found in bacteria, in which an internal chemical imbalance between nucleic acids and protein synthesis can be artificially induced. When protein synthesis is arrested by chloramphenicol, which prevents the normal stringent response responsible for maintaining the balance between nucleic acids and protein synthesis, many changes in the chromatographic behavior of tRNA are observed. In general, it has been shown that they are due to posttranscriptional alterations (Kitchingman and Fournier, 1974). Of course, there is no proof that tumor cells are similarly affected, but I think that this possibility is a very likely one.

From this discussion it is evident that changes in the relative amounts of tRNA species are a widespread phenomenon. In some cases, it has been shown that these changes reflect parallel changes in the amino acid compositions of proteins. However, in view of the studies with bacteria and yeast (see Section 2.2), specific tRNA molecules may well be involved in other functions, such as control of gene transcription.

The tRNA population present in a cell is adapted to the specific protein synthesis repertoire of that cell (Garel, 1974, 1976). How is this accomplished? In addition to the evolutionary selection for translational optimization, the tRNA population is evidently subject to physiologically or developmentally determined changes. This could happen in at least two different ways: (1) There could be a general mechanism, present in every cell, which monitors continuously the rate of utilization of all tRNA species in protein synthesis. This common mechanism would confer an unlimited capacity to adjust the relative abundances of the various tRNA species to optimize translation of the various mRNAs during development, differentiation, or pathological events. (2) tRNA genes could be regulated as is any other gene, as an integral part of a developmental program. tRNA changes could be the expression of the same regulatory mechanisms responsible for all other characteristics of the differentiated cell. In this case, the capacity of the cell to adapt its tRNA population to the special need of protein synthesis would be a *rigid* one, as opposed to the *plastic* one of the previously mentioned more general mechanism. Of course, both mechanisms could function in different or even in the same cell.

In conclusion, considerable effort has been directed toward detecting tRNA changes in various physiological or pathological conditions, and indeed, interesting changes have been found. However, very little has been done toward understanding how these tRNA changes are accomplished.

4 *tRNA Has Other Functions*

In the preceding two sections I have summarized some aspects of the current knowledge of the involvement of tRNA in regulation. In all the cases described, tRNA participates in regulation by a mechanism that is related to a structural function in protein synthesis. The capacity to exist in two forms, charged and uncharged, makes a regulatory role for tRNA possible. One could speculate that the particular way in which tRNA performs its structural role as adaptor of amino acids is responsible for it being the molecule "chosen" by natural selection to fulfill additional regulatory tasks. Its participation in protein synthesis provides the opportunity for the emergence of other capacities as a regulatory molecule. It is probably for this reason that tRNA is used as aminoacyl donor in reactions not involving ribosomes, such as the biosynthesis of aminoacylphosphatidylglycerol (Nesbitt and Lennarz, 1968) and glycyllipopolysaccharide (Gentner and Berg, 1971), the terminal addition of amino acid residues to proteins (Deutch *et al.*, 1977), and cell wall biosynthesis (Roberts, 1972). These additional reactions, which I have not discussed, have been reviewed by Littauer and Inouye (1973).

On the contrary, the function of tRNA as a primer for reverse transcriptase does not seem to be related to protein synthesis.

It has been shown that tRNA, frequently found in virion particles, binds specifically to reverse transcriptase (Panet *et al.*, 1975) and can function *in vitro* as a primer for reverse transcriptase (Dahlberg *et al.*, 1974; Faras *et al.*, 1974). In a detailed study, it has been shown that $tRNA^{Trp}$, found in the Rous sarcoma virion, works as a primer because it contains a sequence of 16 nucleotides conplementary to a region of the Rous sarcoma virus RNA located nearly 100 bases from the 5′ terminus (Haseltine *et al.*, 1977).

It has been shown that $tRNA^{Trp}$ can form base pairs elsewhere along the viral RNA sequence (Haseltine *et al.*, 1977), suggesting that there are other possible interactions and that $tRNA^{Trp}$ may participate in the formation of Rous sarcoma virus RNA dimers which are the usual form in the virion. This implies that $tRNA^{Trp}$ might have a *structural* role in the proper packaging of the viral RNA into the virion. Furthermore, since it has been noted that $tRNA^{Trp}$ can form base pairs in a region near a putative site for ribosomal binding, the possibility of its involvement in regulation of translation has also been considered (Haseltine *et al.*, 1977).

To this long list of possible functions of $tRNA^{Trp}$ in the life cycle of Rous sarcoma virus, I might add another. We know that charged tRNA can hybridize to DNA, losing its tertiary folding, *but still maintaining the amino acid esterified at its 3′-OH end* (see Clarkson and Kurer, 1976). Perhaps both charged and uncharged $tRNA^{Trp}$ bind to Rous sarcoma virus RNA. But of course only the uncharged $tRNA^{Trp}$ can function as a primer, because the amino acid of charged RNA acts as a kind of "primer plug," blocking the 3′-OH end, and thereby preventing elongation by reverse transcriptase. If this is the case, then a very profound physiological meaning can be attached to the ratio of charged to uncharged tRNA, as we have noted before in quite different situations. In this case, it would determine the extent of complementary DNA synthesis.

The involvement of $tRNA^{Trp}$ as a primer for reverse transcriptase recalls another situation: the tRNA-like sequences present at the 3′-OH end (and elsewhere) in certain viral RNAs (Littauer and Inouye, 1973). The expression "tRNA-like structure" is entirely justified because these viral RNAs can be specifically aminoacylated and, at least in some cases (Haenni *et al.*, 1973; Prochiantz and Haenni, 1973), specifically cleaved by tRNA maturation enzymes. The function of these sequences is still unknown.

Could there be a general principle connecting the involvement of $tRNA^{Trp}$ in reverse transcription and the presence of tRNA-like structures in viral RNA? Perhaps they represent different aspects of the same strategy aimed at RNA replication. RNA viruses face a unique problem, because their survival depends on the availability of an RNA-replicating machinery which, as far as we know, is of no use to the host cell. Thus, viral RNA does not find a ready-to-use or "experienced" machinery designed for its own replication and must therefore create one. In general, parasitic organisms exploit the available material of the host; in this case RNA viruses may use the host RNA-binding proteins as building blocks for their own replicase. Thus, there would be a selective advantage in evolving RNA sequences or conformations mimicking those of the host. Transfer RNA-like structures are an example of this. Perhaps many different hostlike structures have evolved in viral RNA, but our present technology has detected only one subset, the tRNA-like structures. A variation of this theme might be the Rous sarcoma virus and related viruses, which developed the capacity to bind specifically to tRNA.

A general consequence of this hypothesis is that viral RNA replicases should contain host-coded subunits which normally have some function in host RNA metabolism. The case of Qβ RNA virus could be taken as evidence supporting this view. Three essential Qβ replicase subunits are the elongation factor Tu, the elongation factor Ts, and the ribsomal protein Sl, all of which bind specifically to RNA when exerting their cellular role (Blumenthal *et al.,* 1972).

5 *Concluding Remarks*

The conceptual framework for this chapter is that tRNA has a central and primary function as the amino acid adaptor and that its other functions are secondary to, or derivations of, this primary one. For example, such view is clearly appropriate in considering tRNA as the target of regulation (Section 3). Moreover, the capacity of tRNA to function as a truly regulatory molecule in stringent control (Section 2.1) or in the control of transcription (Section 2.2) can be viewed as a consequence of its involvement in protein synthesis. The continuous and reversible transformation from the charged to the uncharged form makes tRNA an easily detectable and precise index of the state of the amino acid pools available for protein synthesis. The available data indicate that both prokaryotes and eukaryotes have mechanisms to measure the amount of tRNA in the two forms and consequently to adjust their metabolism. Similarly, cell wall biosynthesis and nonribosomal amino acid transfer are just variants of the same theme, both using tRNA as an amino acid donor. For the function as primer of reverse transcriptase (Section 4), I have argued that the use of tRNA could be considered a particular application of a general strategy used by RNA viruses to solve the problem posed by their replication. The tRNA-like structures present in plant and animal viruses are another particular case of the same strategy. In these last two cases there would be selective pressure to use tRNA, or to evolve tRNA-like structures, in order to exploit for viral purposes the proteins with which tRNA normally interacts in performing its role as adaptor of amino acids.

At the present time the participation of tRNA in a variety of reactions is a well-documented phenomenon. With few exceptions, however, we ignore the molecular details of the multiple functions of tRNA. It is to be hoped that future research will concentrate not on enriching the catalog of the many "involvements" of tRNA in metabolism but on an understanding of how tRNA performs its multiple roles.

I am very thankful to D. Melton for helping in the writing of this chapter and to R. Goldberger for reading the manuscript critically.

References

Allaudeen, M. S., Yang, S. K., and Söll, D., 1972, Leucine $tRNA_1$ from *hisT* mutant of *Salmonella typhimurium* lacks two pseudouridines, *FEBS Lett.* **28**:205.

Altman, S., Bothwell, A. L. M., and Stark, B. C., 1974, Processing of *Escherichia coli* $tRNA^{Tyr}$ precursor RNA *in vitro, Brookhaven Symp. Biol.* **26**:12.

Ames, B. N., and Hartman, P. E., 1963, The histidine operon, *Cold Spring Harbor Symp. Quant. Biol.* **28**:349.

Anderson, W. F., 1969, The effect of tRNA concentration on the rate of protein synthesis, *Proc. Natl. Acad. Sci. U.S.A.* **62**:566.

Anderson, W. F., and Gilbert, Y. M., 1969, tRNA-dependent translational control of *in vitro* haemoglobin synthesis, *Biochem. Biophys. Res. Commun.* **36**:456.

Artz, S. W., and Broach, J. R., 1975, Histidine regulation in *Salmonella typhimurium:* An activator–attenuator model of gene regulation, *Proc. Natl. Acad. Sci. U.S.A..* **72**:3453.

Beckman, J. S., Johnson, P. F., and Abelson, J., 1977, Cloning of yeast tRNA genes in *E. coli, Science* **196**:205.

Bertrand, K., Korn, L., Lee, F., Platt, T., Squires, C. C., Squires, C., and Yanofsky, C., 1975, New features of the regulation of the *tryptophan operon, Science* **189**:22.

Blasi, F., Barton, R. W., Kovach, J. S., and Goldberger, R. F., 1971, Interaction between the first enzyme for histidine biosynthesis and histidyl-$tRNA^{His}$, *J. Bacteriol.* **106**:508.

Blumenthal, T., Saunders, T. A., and Weber, K., 1972, Bacteriophage *QB* replicase contains the protein biosynthesis elongation factors EFTu and EFTs, *Proc. Natl. Acad. Sci. U.S.A..* **69**:1313.

Blumenthal, R. M., Lemaux, P. G., Neidhardt, F. C., and Dennis, P. P., 1976, The effects of the rel A gene on the synthesis of aminoacyl-tRNA synthetases and other transcription and translation proteins in *Escherichia coli* B, *Mol. Gen. Genet.* **149**:291.

Borek, E., and Kerr, S., 1972, Atypical transfer RNA's and their origin in neoplastic cells, *Adv. Cancer Res.* **15**:163.

Bossi, L., and Cortese, R., 1977, Biosynthesis of tRNA in histidine regulatory mutants of *Salmonella typhimurium, Nucleic Acids Res.* **4**:1945.

Bossi L., Ciampi M.S., and Cortese R., 1978, Characterization of an *hisU* mutant of *Salmonella typhimurium* defective in tRNA precursor processing, *J. Bacteriol.* **134**:612–620.

Brenchley, J. E., and Ingraham, J. L., 1973, Characterization of a cold-sensitive *hisW* mutant of *Salmonella typhimurium, J. Bacteriol.* **114**:528.

Brenchley, J. E., and Williams, L. S., 1975, Transfer RNA involvement in the regulation of enzymes synthesis, *Annu. Rev. Microbiol.* **29**:251.

Brenner, M., and Ames, B. N., 1971, The histidine operon and its regulation, in: *Metabolic Pathways,* Vol. 5 (H. J. Vogel, ed.), pp. 349–387, Academic Press, New York.

Brenner, M., Lewis, J. A., Strauss, D. S., De Lorenzo, F., and Ames, B. N., 1972, Histidine regulation in *Salmonella typhimurium.* XIV: Interaction of the histidyl tRNA synthetase with $tRNA^{His}$, *J. Biol. Chem.* **247**:4333.

Bresalier, R. S., Rizzino, A. A., and Freundlich, M., 1975, Reduced maximal levels of derepression of isoleucine–valine and leucine enzymes in *hisT* mutants of *Salmonella typhimurium, Nature* **253**:279.

Bronson, M. Y., Squires, C., and Yanofsky, C., 1973, Nucleotide sequences from tryptophan mRNA of *E. coli:* The sequence corresponding to the amino terminal region of the first polypeptide specified by the operon, *Proc. Natl. Acad. Sci. U.S.A..* **70**:2335.

Calker, D., and Hilse, K., 1974, Properties of isoaccepting $tRNA^{Val}$ from rabbit recticulocytes: Fractionation and codon recognition, *FEBS Lett.* **39**:56.

Carbon, J., Chang, S., and Kirk, L. L., 1974, Clustered tRNA genes in *Escherichia coli,* transcription and processing, *Brookhaven Symp. Biol.* **26**:26.

Carsiotis, M., and Jones, R. F., 1974, Cross-pathway regulation: Tryptophan-mediated control of histidine and arginine biosynthetic enzymes in *Neurospora crassa, J. Bacteriol.* **119**:889.

Carsiotis, M., Jones, R. F., and Wesseling, A. C.,1974, Cross-pathway regulations: Histidine-mediated control of histidine, tryptophan and arginine biosynthetic enzymes in *Neurospora crassa, J. Bacteriol.* **119**:893.

Ciampi, M. S., Arena, F., Cortese, R., and Daniel, V., 1977, Biosynthesis of pseudouridine in the *in vitro* transcribed $tRNA^{Tyr}$ precursor, *FEBS Lett.* **77**:75.

Clarkson, S. G., and Kurer, V., 1976, Isolation and some properties of DNA coding for $tRNA^{Met}$ from *Xenopus laevis, Cell* **8**:183.

Cortese, R., Kammen, H. O., Spengler, S. H., and Ames, B. N., 1974*a,* Biosynthesis of pseudouridine in tRNA, *J. Biol. Chem.* **249:**1103.

Cortese, R., Landsberg, R. M., Van der Haar, R. A., Umbarger, H. E., and Ames, B. N., 1974*b,* Pleiotropy of *hisT* mutants blocked in pseudouridine synthesis in tRNA: Leucine and isoleucine–valine operons, *Proc. Natl. Acad. Sci. U.S.A.* **71**:1857.

Cozzone, A., and Donini, P., 1973, Turnover of polysomes in amino-acid starved *Escherichia coli, J. Mol. Biol.* **76**:149.

Crick, F. H. C., 1958, On protein synthesis, *Symp. Soc. Exp. Biol.* **12**:138.

Crick, F. H. C., Brenner, S., Klug, A., and Pieczenik, G., 1976, A speculation on the origin of protein synthesis, *Origins of Life* **7**:389.

Dahlberg, J. E., Sawyer, R. C., Taylor, S. M., Faras, A. J., Levinson, W. E., Goodman, H. M., and Bishop, J. M., 1974, Transcription of DNA from the 70S RNA of Rous sarcoma virus. Identification of a specific 4S RNA which serves as primer, *J. Virol.* **13**:1126.

Davidson, J. P., Davis, L., and Williams, L. S., 1977, Control of isoleucine–valine biosynthesis in tRNA ribonucleic acid mutants of *Salmonella typhimurium, Fed. Proc.* **36**:659 (abstract).

Debenham, P., and Travers, A., 1977, Selective inhibition of $tRNA^{Tyr}$ transcription by guanosine 3′-diphosphate-5′-diphosphate, *Eur. J. Biochem.* **72**:515.

Deeley, R. G., Goldberger, R. F., Kovach, J., Meyers, M., and Mullinix, K., 1975, Interaction between phosphoribosyltransferase and purified histidine tRNA from wild type *Salmonella typhimurium* and a derepressed *hisT* mutant strain, *Nucleic Acids Res.* **2**:545.

Delaney, P., and Siddiqui, M. A. Q., 1976, Changes in *in vivo* levels of charged tRNA species during development of the posterior silk-gland of *Bombix mori, Dev. Biol.* **44**:54.

Deutch, C. E., Scarpulla, R. C., Sonnenblick, E. B., and Soffer, R. L., 1977, Pleiotropic phenotype of an *E. coli* mutant lacking leucyl-, phenylalanyl-transfer ribonucleic acid-protein transferase, *J. Bacteriol.* **129**:544.

Di Nocera, P. P., Avitabile, A., and Blasi, F., 1975, *In vitro* transcription of the *Escherichia coli his operon* primed by dinucleotides. Effect of the first histidine biosynthetic enzyme, *J. Biol. Chem.* **250**:8376.

Donini, P., Santonastaso, V., Roche, J., and Cozzone, A. J., 1978, The relationship between guanosine tetraphosphate, polysomes and RNA synthesis in amino acid-starved *Escherichia coli, Molec. Biol. Rep.* **4**:15–19.

Efstradiatis A., Kafatos, F. C., and Maniatis, T., 1977, The primary structure of rabbit β-globin mRNA as determined from cloned DNA, *Cell* **10**:571.

Ely, B., 1974, Physiological studies of *Salmonella* histidine operator–promoter mutants, *Genetics* **78**:593.

Ely, B., Fankhauser, B. D., and Hartman, P. E., 1974, A fine structure map of the *Salmonella typhimurium* histidine operator–promoter, *Genetics* **78**:607.

Faras, A. J., Dahlberg, J. E., Sawyer, R. C., Harada, F., Taylor, J. M., Levinson, W. E., Bishop, J. M., and Goodman, H. M., 1974, Transcription of DNA from the 70S RNA of Rous sarcoma virus, II: Structure of a 4S RNA primer, *J. Virol.* **13**:1133.

Fario, M., Cascino, A., and Cortese, R., 1977, Regulation of the intracellular concentration of T4 induced tRNA, *Mol. Gen. Genet.* **155**:61.

Farkas, W. R., and Singh, R., 1973, Guanylation of tRNA by cell-free lysate of rabbit reticulocytes, *J. Biol. Chem.* **248**:7780.

Fiers, W., Contreras, R., Duerinck, F., Haegeman, G., Isertant, D., Meviegaert, J., Min-Jon, W., Molemans, F., Rawymackers, A., Van den Berghe, A., Volckaert, G., and Ysebaert, M., 1976, Complete nucleotide sequence of bacteriophage MS2 RNA: Primary and secondary structure of the replicase gene, *Nature* **260**:500.

Fink, G. R., and Roth, J. R., 1968, Histidine regulatory mutants in *Salmonella typhimurium,* VI. Dominance studies, *J. Mol. Biol.* **33**:547.

Gallant, J., and Lazzarini, R. A., 1976, The strigent control, in: *Protein Synthesis. A Series of Advances,* Vol. 2 (E. H. McConkey, ed.), pp. 309–359, Marcel Dekker, New York.

Gallant, J., Palmer, C., and Pao, C. C., 1977, Anomalous synthesis of ppGpp in growing cells, *Cell* **11**:181.

Garel, J. P., 1974, Functional adaptation of tRNA populations, *J. Theor. Biol.* **43**:211.

Garel, J. P., 1976, Quantitative adaptation of isoacceptor tRNA's to mRNA codons of alanine, glycine and serine, *Nature* **60**:805.

Garel, J. P., Mandel, P., Chavancy, G., and Dallie, J., 1970, Functional adaptation of tRNA's to fibroin biosynthesis in the silk-gland of *Bombix mori, FEBS Lett.* **7**:327.

Gentner, H., and Berg, P., 1971, Occurrence of a glycyl-lipopolysaccharide structure in *Escherichia coli* and its enzymatic formation form glycyl-tRNA, *Fed. Proc.* **30**:1218 (abstract).

Goldberger, R. F., and Kovach, J. S., 1972, Regulation of histidine biosynthesis in *Salmonella typhimurium, Curr. Top. Cell. Regul.* **5**:285.

Grumberger, D., Weinstein, F. B., and Mushinsky, J. F., 1975, Deficiency of the Y base in a hepatoma phenylalanine tRNA, *Nature* **253**:66.

Grummt, F., and Grummt, I., 1976, Studies on the role of uncharged tRNA in the pleiotypic responses of animal cells, *Eur. J. Biochem.* **64**:307.

Haenni, A. L., Prochiantz, A., Bernard, O., and Chapeville, F., 1973, TYMV valyl-RNA as an amino acid donor in protein biosynthesis, *Nature New Biol.* **241**:166.

Hartwell, L. H., 1974, *Saccharomyces cerevisiae* cell cycle, *Bacteriol. Rev.* **38**:164.

Haselkorn, R., and Rothman-Denes, L. B., 1973, Protein synthesis, *Annu. Rev. Biochem.* **42**:397.

Haseltine, W. A., and Block, R., 1973, Synthesis of guanosine tetra- and pentaphosphate requires the presence of codon-specific, uncharged tRNA in the acceptor site of ribosome, *Proc. Natl. Acad. Sci. U.S.A.* **70**:1564.

Haseltine, W. A., Block, R., Gilbert, W., and Weber, K., 1972, MSI and MSII made on ribosomes in idling step of protein synthesis, *Nature* **238**:381.

Haseltine, W. A., Maxam, A. M., and Gilbert, W., 1977, Rous sarcoma virus genome is terminally redundant. The 5′ sequence, *Proc. Natl. Acad. Sci. U.S.A.* **74**:989.

Hershko, A., Mamont, P., Shields, R., and Tomkins, G. M., 1971, Pleiotypic response, *Nature New Biol.* **232**:206.

Hilse, K., and Rudloff, E., 1975, Glutamine cognate codons in rabbit haemoglobin mRNA's, *FEBS Lett.* **60**:380.

Hirsh, D., 1970, Tryptophan tRNA of *E. coli, Nature* **228**:57.

Hoagland, M. B., Zamecnick, P. C., and Stephenson, M. L., 1957, Intermediate reactions in protein biosynthesis, *Biochim. Biophys. Acta* **24**:215.

Holley, R. W., and Kiernan, J. A., 1974, Control of the initiation of DNA synthesis in 3T3 cells: Low molecular weight nutrients, *Proc. Natl. Acad. Sci. U.S.A.* **71**:2942.

Holley, R. W., Apgar, H., Everett, G. A., Madison, J. T., Marquisec, M., Merrill, S. H., Penswick, J. R., and Zamir, A., 1965, Structure of a ribonucleic acid, *Science* **147**:1462.

Ilgen, C., Kirk, L. L., and Carbon, J., 1976, Isolation and characterization of large tRNA precursors from *E. coli, J. Biol. Chem.* **251**:922.

Johnson, R. C., Vanatta, P. R., and Fresco, J., 1977, Metabolic regulation of aminoacyl-tRNA synthetase biosynthesis in baker's yeast, *J. Biol. Chem.* **252**:878.

Kan, L. S., Ts'o, P. O. P., Sprinzl, M., Van der Haar, F., and Cramer, F., 1976, PMR studies on the NH-H hydrogen-bonded enol methyl, methylene proton resonances of $tRNA^{Phe}$ and phe-$tRNA^{Phe}$, *Biophys. J.* **16**:11a.

Kasai, T., 1974, Regulation of the expression of the histidine operon in *Salmonella typhimurium, Nature* **249**:523.

Kim, S. H., 1976, Tridimensional structure of transfer RNA, *Prog. Nucleic Acid Res. Mol. Biol.* **17**:182.

Kim, S. H., Quigley, G. J., Suddath, F. L., McPherson, A., Sneden, D., Kim, J. J., Weinzierl, J., and Rich, A., 1973, Three-dimensional structure of yeast $tRNA^{Phe}$: Folding of the polynucleotide chain, *Science* **179**:285.

Kitchingman, G. R., and Fournier, M. J., 1974, Inhibition of posttranscriptional modification in *E. coli, Brookhaven Symp. Biol.* **26**:44.

Kleeman, J. E., and Parsons, S. M., 1977, Inhibition of histidyl-tRNA-adenosine triphosphate phosphoribosyltransferase complex formation by histidine and by guanosine tetraphosphate, *Proc. Natl. Acad. Sci. U.S.A.* **74**:1535.

Korn, L. J., and Yanofsky, C., 1976, Polarity suppressors increase expression of the wild-type tryptophan operon in *E. Coli, J. Mol. Biol.* **103**:395.

Lee F., and Yanofsky C., 1977, Transcription termination at the *trp* operon attenuators of *Escherichia coli* and *Salmonella typhimurium*: RNA secondary structure and regulation of termination, *Proc. Nat. Acad. Sci. USA.* **74**:4365–4369.

Le Meur, M. A., Gerlinger, P., and Ebel, J. P., 1976, Messenger RNA translation in the presence of homologous and heterologous tRNA, *Eur. J. Biochem.* **67**:519.

Lewis, J. A., and Ames, B. N., 1972, Histidine regulation in *Salmonella typhimurium.* XI. The percentage of $tRNA^{His}$ charged *in vivo* and its relation to the repression of the histidine operon, *J. Mol. Biol.* **66**:131.

Lipmann, F., 1941, The metabolic generation and utilization of phosphate bond energy, *Adv. Enzymol.* **1**:99.

Litt, M., and Kabat, D., 1972, Studies of tRNA's and haemoglobin synthesis in sheep reticulocytes, *J. Biol. Chem.* **247**:6659.

Littauer, U. Z., and Inouye, H., 1973, The regulation of tRNA, *Annu. Rev. Biochem.* **42**:439.

McClain, W. H., and Seidman, J. G., 1975, Genetic perturbations that reveal tertiary conformation of tRNA precursor molecules, *Nature* **257**:106.

McLaughlin, C. S., Magee, P. T., and Hartwell, L. H., 1969, Role of isoleucyl-transfer ribonucleic acid synthetase in ribonucleic acid synthesis and enzyme repression in yeast, *J. Bacteriol.* **100**:579.

Meiss, H. K., Brill, W. J., and Magasanik, B., 1969, Genetic control of histidine degradation in *Salmonella typhimurium* strain LT-2, *J. Biol. Chem.* **244**:5382.

Meister, A., 1965, *Biochemistry of the Amino Acids,* Academic Press, New York.

Messenguy, F., and Delforge, J., 1976, Role of tRNA in the regulation of several biosynthesis in *Saccharomyces cerevisiae, Eur. J. Biochem.* **67**:335.

Meyers, M., Blasi, F., Bruni, C. B., Deeley, R. G., Kovach, J. S., Levinthal, M., Mullinix, K. P., Vogel, T., and Goldberger, R. F., 1975, Specific binding of the first enzyme for histidine biosynthesis to the DNA of the histidine operon, *Nucleic Acids Res.* **2**:2021.

Meza, L., Araya, A., Leon, G., Kranskoff, M., Siddiqui, M. A. Q., and Garel, J. P., 1977, Specific alanine-tRNA species associated with fibroin biosynthesis in the posterior silk-gland of *Bombyx mori, FEBS Lett.* **77**:255.

Morse, D. E., and Morse, N. C. A., 1976, Dual control of the *trp* operon is mediated by both tryptophanyl-tRNA synthetase and the repressor, *J. Mol. Biol.* **103**:209.

Nazario, M., Kinsey, J. A., and Ahmad, M., 1971, *Neurospora* mutant deficient in the tryptophanyl-tRNA synthetase activity, *J. Bacteriol.* **105**:121.

Neidhart, F. C., Parker, J., and McKeever, W. G., 1975, Function and regulation of aminoacyl-tRNA synthetases in prokaryotic and eukaryotic cells, *Annu. Rev. Microbiol.* **29**:215.

Nesbitt, J. A., and Lennarz, W. J., 1968, Participation of aminoacyl tRNA in aminoacyl phosphatidylglycerol synthesis, *J. Biol. Chem.* **243**:3088.

Okada, H., Horada, F., and Nishimura, S., 1976, Specific replacement of Q base in the anticodon of tRNA by guanine catalysed by a cell-free extract of rabbit reticulocytes, *Nucleic Acids Res.* **3**:2593.

Panet, A., Haseltine, W. A., Baltimore, D., Peters, G., Horada, F., and Dahlberg, J. E., 1975, Specific binding of tryptophan tRNA to avian myeloblastosis virus RNA-dependent DNA polymerase (reverse transcriptase), *Proc. Natl. Acad. Sci. U.S.A.* **72**:2535.

Pao, C. C., Paietta, J., and Gallant, J., 1977, Synthesis of guanosine tetraphosphate (magic spot I) in *Saccharomyces cerevisiae, Biochem. Biophys. Res. Commun.* **74**:314.

Pardee, A. B. , 1974, A restriction point for control of normal animal cell proliferation, *Proc. Natl. Acad. Sci. U.S.A.* **71**:1286.

Pongs, O., and Ulbrich, M., 1976, Specific binding of formylated initiator-tRNA to *Escherichia coli* RNA polymerase, *Proc, Natl. Acad. Sci. U.S.A.* **73**:3064.

Prochiantz, A., and Haenni, A. L., 1973, TYMV RNA as a substrate of tRNA maturation endonuclease, *Nature New Biol.* **241:**168.

Revel, M., Content, J., Zilberstein, A., Nudel, U., Berissi, H., and Dudock, B., 1975, Control of mRNA translation by specific tRNA's in extracts from interferon treated mouse cells, *Colloq. Inst. Natl. Santé Rech. Med.* **47**:397.

Rich A., and Rajbhandary, U. L., 1976, Transfer RNA: Molecular structure, sequence and properties, *Annu. Rev. Biochem.* **45**:805.

Richter, D., 1976, Stringent factor from *Escherichia coli* directs ribosomal binding and release of uncharged tRNA, *Proc. Natl. Acad. Sci. U.S.A.* **73**:707.

Richter, D., Erdman, V. A., and Sprinzl, M., 1974, A new transfer RNA fragment reaction: TpψpCpGp bound to a ribosome mRNA complex induces the synthesis of guanosine tetra- and pentaphosphate, *Proc. Natl. Acad. Sci. U.S.A.* **71**:3226.

Rizzino, A. A., Bresalier, R. S., and Freundlich, M., 1974, Derepressed levels of the isoleucine–valine and leucine enzymes in *hisT1504,* a strain of *Salmonella typhimurium* with altered leucine tRNA, *J. Bacteriol.* **117**:449.

Roberts, R. J., 1972, Structures of two glycyl tRNA's from *Staphyloccus epidermidis, Nature New Biol.* **237**:44.

Robertus, J. D., Ladner, J. E., Finch, J. T., Rhodes, D., Brown, R. S., Clark, B. F. C. , and Klug, A., 1974, Structure of yeast phenylalanine tRNA at 3 Å resolution, *Nature* **250**:546.

Roth, J. R., and Ames, B. H., 1966, Histidine regulatory mutants in *Salmonella typhimurium.* II. Histidine regulatory mutants having altered histidyl-tRNA synthetase, *J. Mol. Biol.* **22**:325.

Sakano, H., and Shimura, Y., 1975, Sequential processing of precursor tRNA molecules in *Escherichia coli, Proc. Natl. Acad. Sci. U.S.A.* **72**:3369.

Sakano, H., Shimura, Y., and Ozeky, H., 1974*a,* Selective modification of nucleosides of tRNA precursors accumulated in a temperature-sensitive mutant of *E. Coli, FEBS Lett.* **48**:118.

Sakano, H., Yamada, S., Ikemura, T., Shimura, Y., and Ozeky, M., 1974*b,* Temperature-sensitive mutants of *Escherichia coli* for tRNA synthesis, *Nucleic Acids Res.* **1**:355.

Salomon, P., Giveon, D., Kimhi, Y., and Littauer, U. Z., 1976, Abundance of $tRNA^{Phe}$ lacking the peroxy-Y base in mouse neuroblastoma, *Biochemistry* **15**:5258.

Sanger, F., Air, G. M., Barrell, B. G., Brown, H. L., Coulson, A. R., Fiddles, J. C., Hutchinson, C. A., Slocombe, P. M., and Smith, M., 1977, Nucleotide sequence of bacteriophage ϕX174 DNA, *Nature* **265**:687.

Schedl, P., and Primakoff, P., 1973, Mutants of *E. coli* thermosensitive for the synthesis of tRNA, *Proc. Natl. Acad. Sci. U.S.A.* **70**:2091.

Schedl, P., Primakoff, P., and Roberts, J., 1974, Processing of *E. coli* tRNA precursors, *Brookhaven Symp. Biol.* **26**:53.

Schlesinger, S., and Magasanik, B., 1964, Effect of α-methylhistidine on the control of histidine synthesis, *J. Mol. Biol.* **9**:670.

Schurch A., Miozzari J., and Hutter R., 1974, Regulation of tryptophan biosynthesis in *Saccaromyces cerevisiae:* Mode of action of 5-methyl-tryptophan and 5-methyl-tryptophan sensitive mutants, *J. Bacteriol.* **117**:1131–1140.

Scott, J. F., Roth, J. R., and Artz, S. W., 1975, Regulation of histidine operon does not require hisG enzyme, *Proc. Natl. Acad. Sci. U.S.A.* **72**:5021.

Seidman, J. G., and McClain, W. H., 1975, Three steps in conversion of large precursor RNA into serine and proline tRNA's *Proc. Natl. Acad. Sci. U.S.A.* **72**:1491.

Sharma, O. K., Beezley, D. N., and Roberts, W. K., 1976, Limitation of reticulocyte tRNA in the translation of heterologous mRNA's, *Biochemistry* **15**:4313.

Sherberg, H. H., and Weiss, S. B., 1972, T4 transfer RNA's: Codon recognition and translational properties, *Proc. Natl. Acad. Sci. U.S.A.* **69**:1114.

Singer, C. E., 1972, Ph.D. Thesis, University of California, Berkeley.

Singer, C. E., Smith, G. R., Cortese, R., and Ames, B. H., 1972, Mutant $tRNA^{His}$ ineffective in repression and lacking two pseudouridine modifications, *Nature New Biol.* **238**:72.

Smith, D. W. E., 1975, Reticulocyte transfer RNA and haemoglobin synthesis, *Science* **190**:529.

Smith, D. W. E., and McNamara, A L., 1971, Specialization of rabbit reticulocyte tRNA content for haemoglobin synthesis, *Science* **171**:577.

Smith, J. D., 1972, Genetics of transfer RNA, *Annu. Rev. Genet.* **6**:235.

Smith, J. D., 1976, Transcription and processing of transfer RNA precursors, *Prog. Nucleic Acid Res. Mol. Biol.* **16**:25.

Soeiro, R., Vaughan, M. H., and Darnell, J. E., 1968, The effect of puromycin on intranuclear steps in ribosome biosynthesis, *J. Cell Biol.* **36**:91.

Sprague, U. K., Hagenbuckle, O., and Zuniza, M. C., 1977, The nucleotide sequence of two silk gland alanine tRNA's: Implications for fibroin synthesis and for initiator tRNA structure, *Cell* **11**:561–570.

Sprinzl, M., and Richter, D., 1976, Free 3′-OH group of the terminal adenosine of tRNA molecules is essential for the synthesis *in vitro* of guanosine tetra-phosphate and penta-phosphate in a ribosomal system from *Escherichia coli, Eur. J. Biochem.* **71**:171.

Spurgeon, S. L., and Matchett, W. H., 1977, Inhibition of aminoacyltransfer ribonucleic acid synthetases and the regulation of amino-acid biosynthetic enzymes in *Neurospora crassa, J. Bacteriol.* **129**:1303.

Squires, G., Lee, F., Bertrand, K., Squires, C., Bronson, J. M., and Yanofsky, C., 1976, Nucleotide sequence at the 5′ end of tryptophan mRNA of *E. coli, J. Mol. Biol.* **103**:351.

Stent, G. S., and Brenner, S., 1961, A genetic locus for the regulation of RNA synthesis, *Proc. Natl. Acad. Sci. U.S.A.* **47**:2005.

Stephens, J. C., Artz, S. W., and Ames, B. N., 1975, Guanosine-5′-diphosphate-3′ diphosphate (ppGpp). Positive effector for histidine operon transcription and general signal for amino acid deficiency. *Proc. Natl. Acad. Sci. U.S.A.* **72**:4389.

Talkad, V., Schneider, E., and Kennell, D., 1976, Evidence for variable rates of ribosome movement in *Escherichia coli, J. Mol. Biol.* **104**:299.

Tashiro, Y., Morimoto, T., Matsura, S., and Hayata, S., 1968, Studies on the posterior silk gland cells and biosynthesis of fibroin during the Vth larval instar, *J. Cell Biol.* **38**:574.

Travers, A., 1974, RNA polymerase promoter interaction. Some general principles, *Cell* **3**:93.

Travers, A., 1976, Modulation of RNA polymerase specificity by ppGpp, *Mol. Gen. Genet.* **147**:225.

Unger, M. W., and Hartwell, L. H., 1976, Control of cell division in *Saccharomyces cerevisiae* by methionyl-tRNA, *Proc. Natl. Acad. Sci. U.S.A.* **73**:1664.

Vogeli, G., Stewart, T. S., McCutchan, T., and Soll, D., 1977, Isolation of *Escherichia coli* precursor tRNA's containing modified nucleoside Q, *J. Biol. Chem.* **252**:2311.

White, B. H., Tener, G. M., Holden, J., and Suzuki, D. T., 1973*a,* Analysis of tRNA's during the development of *Drosophilia, Dev. Biol.* **33**:185.

White, B. H., Tener, G. M., Holden, J., and Suzuki, D. T., 1973*b,* Activity of a tRNA modifying enzyme during the development of *Drosophila* and its relationship to the Su(s) locus, *J. Mol. Biol.* **74**:635.

Wilson, J. H., 1973, Function of the bacteriophage T4 transfer RNA's *J. Mol. Biol.* **74**:753.

Woese, C. R., 1967, *The Genetic Code,* Harper, New York.

Wolfner, M., Yep, D., Messenguy, F., and Fink, G. R., 1975, Integration of amino acid biosynthesis into the cell cycle of *Saccharmyces cerevisiae, J. Mol. Biol.* **96**:273.

Woodward, W. R., and Herbert, E., 1972, Coding properties of reticulocyte lysine tRNA's in haemoglobin synthesis, *Science* **177**:1197.

Yang, H. L., Zubay, G., Urm, E., Reiness, G., and Cashel, M., 1974, Effects of guanosine tetraphosphate, guanosine pentaphosphate and β-γ-methylenyl-guanosine pentaphosphate on gene expression of *E. coli in vitro, Proc. Natl. Acad. Sci. U.S.A.* **71**:63.

Yanofsky, C., and Soll, L., 1977, Mutations affecting $tRNA^{Trp}$ and its charging and their effect on regulation on transcription termination at the attenuator of the tryptophan operon, *J. Mol. Biol.* **113**:663.

Zilbertstein, A., Dudock, B., Berissi, H., and Revel, M., 1976, Control of messenger RNA translation by minor species of leucyl-tRNA in extracts from interferon-treated cells, *J. Mol. Biol.* **108**:43.

Note Added in Proof. Important contributions have recently been made toward the clarification of the role of tRNA in the operon-specific control (see sec. 2.2). The model of Lee and Yanofsky (1977), discussed on page 419, has received strong support from the data of Barnes (personal communication) and P. P. Di Nocera *et al.* (personal communication). These authors have sequenced the leader region of the *his* operon in *Salmonella* and *E. coli* respectively and found in both cases *seven adjacent histidine codons.*

Also, for the role of pseudouridine in the anticodon region of tRNA, there are now evidences (L. Bossi and J. R. Roth, personal communication) that the lack of pseudouridine is not innocuous in protein synthesis, as formerly believed. These authors have observed that su_2 suppression, in $hisT^-$ background, is variably efficient depending on the position of the mutation on the mRNA. Bossi and Roth suggest that pseudouridine in the anticodon region is important for tRNA–tRNA interactions on the ribosome. This hypothesis explains both the context-dependent effects of *hisT* mutation on suppression and, in the framework of the Lee and Yanofsky model, also the role of pseudouridine in *his operon* regulation.

11

Suppression

DEBORAH A. STEEGE and DIETER G. SÖLL

1 Introduction

Suppression is the genetic term used to describe the effect of a general class of secondary mutations that restore a wild or pseudo-wild type phenotype to a mutant organism in which the primary mutation is still maintained. The isolation of a suppressor serves to identify a compensating mutation in the second of a pair of genes which may encode macromolecules interacting at some point in viral or cell metabolism. Since the molecular mechanism by which a wild type phenotype appears usually is not prescribed by the genetic selection imposed, one would predict that biochemical alterations of many different types could correct a given deficiency. In fact, evidence accumulated since the first report of suppression in 1927 (Bonnier, 1927) clearly shows that correction mechanisms operate both during the transcription and translation steps of gene expression, and also at the level of the final gene products. The many types of genetic suppression were enumerated several years ago in the comprehensive article of Hartman and Roth (1973). The experimental evidence which has identified the molecular basis for nonsense suppression (Garen, 1968; Körner *et al.,* 1978), missense suppression (Hill, 1975), frameshift suppression (Roth, 1974), and ribosomal suppression (Gorini, 1970, 1974) has been discussed. Hawthorne and Leupold (1974) have recently summarized our knowledge of genetic suppression in yeast. These remain excellent discussions of the basic principles for the class of suppressor mutations, termed informational suppressors, that affects the macromolecular components of the transcription and translation processes.

DEBORAH A. STEEGE • Department of Biochemistry, Duke University Medical Center, Durham, North Carolina DIETER G. SÖLL • Department of Molecular Biophysics and Biochemistry, Yale University, New Haven, Connecticut

Our aim here is not to undertake the formidable task of updating progress for all aspects of genetic suppression; the subject, indeed, is represented by experimental efforts sufficiently diverse that exciting breakthroughs in specific areas will be summarized more appropriately as they emerge. Rather, we wish to focus on two subjects: elucidation of the biochemical basis of a suppression event and use of suppressors as genetic tools. These topics are discussed in terms of the contributions they have made to our understanding of (1) the arrangement, structure, and expression of genes in bacterial and bacteriophage genomes; and (2) the macromolecules that transmit genetic information from a DNA sequence into the amino acid sequence of a protein. Although the basic outlines of many cellular processes can be sketched, the individual steps of those processes may not yet be fully understood. For example, in protein biosynthesis neither the basis of codon·anticodon selectivity nor the precise role of ribosomal RNA and proteins in the correct positioning of mRNA and tRNA on the ribosome during the translation process is clear. Moreover, the molecular interactions triggered by the appearance of a termination triplet in the messenger RNA have not yet been identified. We would like to reexamine certain conclusions drawn previously on the basis of information available from a few cases of suppression, with the hope that further work with an expanded set of well-characterized genetic suppression systems will either support current ideas or allow more accurate generalizations to be formulated.

A major reason for presenting yet another review on suppression at this time stems from the fact that the study of suppression in eukaryotes, aided by recent developments in technology, is gaining momentum. Methods for the isolation of specific gene sequences from any organism by recombinant DNA techniques, rapid DNA and RNA sequencing techniques (Maxam and Gilbert, 1977; Sanger *et al.*, 1977*b;* Donis-Keller *et al.*, 1977; Simoncsits *et al.*, 1977), and improved *in vitro* protein-synthesizing systems (see, for example, Capecchi *et al.*, 1977; Gesteland *et al.*, 1976) permit the exploration of many aspects of the organization, functioning, and regulation of eukaryotic genomes. In addition, after many futile attempts, work with several eukaryotic systems is now yielding some information on suppression at the molecular level (see Section 5). Some of the questions about the similarities and possible differences in the cellular processes of prokaryotic and eukaryotic cells may be approached by appropriate use of informational suppressors. Careful comparisons of tRNA anticodon sequences (Sprinzl, 1978), patterns of modified bases in tRNA (McCloskey and Nishimura, 1977), and selectivity of codon usage among organisms representing different stages of evolutionary development may offer insights into the evolution of the genetic code. Finally, the availability of suppressible mutations as a class of conditional lethal mutants of mammalian cells in culture (Capecchi *et al.*, 1977) will most likely become an important criterion for identifying genes encoding proteins.

Thus, after a brief description of the major types of informational suppressors, we shall consider their impact on our knowledge of (1) protein biosynthesis, (2) the ribosome's contribution to faithful translational processes, and (3) tRNA genetics, biosynthesis, and structure. We shall also describe the way in which polarity suppressors have revealed the basis for coupling of transcription to translation in the bacterial cell. These sections will deal exclusively with information obtained from studies of bacteria and their viruses. We will conclude with a discussion of the recent developments in eukaryotic nonsense and frameshift suppression.

2 *A Short Synopsis of Suppression*

2.1 *How It Started*

In 1958, in an attempt to "run the genetic map of the rII region of bacteriophage T4 into the ground" (Benzer, 1966), Benzer examined a number of rII mutants isolated by Tessman in connection with his studies on mutagenesis. To his consternation, some of the rII mutants exhibited an "ambivalent" phenotype—that is, the mutant phenotype did not appear when the phage was grown on certain bacterial hosts. Benzer correctly recognized that the ambivalence could be due to heritable differences in the decoding properties among several bacterial strains. The differences were seen to constitute a class of suppressor type mutations (Benzer and Champe, 1962). These observations were made in the context of intensive efforts on the part of many laboratories to elucidate the nature of the genetic code. Subsequent analysis of suppression events in several laboratories played two important roles in that elucidation.

First, the properties of two classes of rII mutants induced by the acridine dyes suggested that these classes resulted from nucleotide insertions (+1) or deletions (−1) in the coding region of a gene. This shifted the reading frame during translation such that the protein specified by that gene had an altered amino acid sequence beyond the position of the insertion/deletion. However, the effect of (+) frameshift mutations, as these are called, could be suppressed by mutations of the opposite (−) type. The fact that three mutations of the same class taken together could also yield a functional protein, containing only a limited region of altered amino acid sequence, suggested strongly that coding units were read sequentially in nonoverlapping units of three nucleotides, and that these units were not punctuated by spacer nucleotides (Crick *et al.,* 1961). This means of restoring a wild type phenotype through compensatory mutations within a cistron is referred to as an *intragenic suppression* event. It is distinct from a true reversion event in that reversion involves a base change at the site of the original mutation which yields the wild type or synonymous coding unit. *Extragenic suppression* of frameshift mutations, originating from a suppressor mutation in a second cistron, are discussed in Sections 2.5 and 3.1.

Second, the characterization of Benzer's "ambivalent" mutants led eventually to the conclusion that only 3 of the 64 possible triplet codons—UAG, UAA, and UGA—do not specify any of the 20 amino acids. These nonsense triplets normally signal polypeptide chain termination at the end of a cistron. When introduced via mutation into the reading phase of a messenger RNA sequence encoding a polypeptide, a nonsense mutation triggers a premature termination event that releases the truncated polypeptide product. Suppressors of these mutations are the major topics of discussion in this chapter. To provide a starting point for discussion, we will outline the basic facts about informational suppression in bacterial cells. For detailed background information the reader is referred to the articles previously listed.

2.2 The Problem of Nomenclature

The nomenclature describing wild type and suppressor alleles unfortunately has not yet been standardized despite the existence of uniform format rules. In general, *su* or *sup* is used to denote a suppressor allele. For the specific case of frameshift suppressors, the abbreviation *suf* is used. In *Drosophila* and yeast the symbol + represents the wild type form, which does not contain a suppressor activity. In bacterial systems, however, the symbol su^+ (= sup^-) often applies to strains containing the suppressor activity, while wild type strains are defined as su^- (= sup^+). Since our purpose is to point out the general usefulness of suppressors for work in many areas of research, we have conformed to the practices of individual authors in most cases.

2.3 Nonsense Suppression

As already described, a mutation that causes the conversion of an amino acid-specifying codon into one of the three terminator codons, UAG, UAA, and UGA, in a messenger RNA leads to premature termination of translation. The effect of such a mutation can be reversed by secondary extragenic suppressor mutations, *nonsense suppressors.* These are classified according to the nonsense codons on which they act: *amber* suppressors read UAG, *ochre* suppressors read both UAA and UAG, and *opal* suppressors read UGA. (The term UGA suppressor is now used more commonly than *opal* suppressor.) Each suppressor inserts a substitute amino acid into a polypeptide chain in response to the appropriate nonsense codon. In *Escherichia coli,* for example, the amber suppressor su^+1 inserts serine; su^+2, glutamine; su^+3, tyrosine; su^+6, leucine; and su^+7, glutamine [and low levels of tryptophan (Celis *et al.,* 1976)]. The ochre suppressors su^+4 and su^+5 insert tyrosine and lysine, respectively. A UGA suppressor inserts tryptophan. So long as the nature of the substituted amino acid allows the resulting protein to function, a suppression event restores a wild type phenotype.

Many cases of nonsense suppression have been shown to be mediated by tRNA. For some of these, the locus of the suppressor mutation has been traced directly to the structural gene for a tRNA. As would follow from the observation that tRNAs responding to nonsense codons are not normally present in wild type cells, tRNA sequence analysis has revealed that nonsense suppression generally results from a mutation in a "sense" anticodon that confers on a tRNA species the capacity to recognize a nonsense codon. Amber suppressor tRNAs have the anticodon sequence CUA complementary to the UAG codon. Ochre suppressor tRNAs have the anticodon UUA or U*UA [U* is a modified uridine (Altman, 1976)] which can recognize both UAA and UAG codons. The UGA suppressor arginine tRNA from bacteriophage T4 contains the anticodon U*CA which can base-pair with the UGA codon (Kao and McClain, 1977). These suppressor tRNAs no longer respond to the codons read by their wild type counterparts. For this reason, the isolation of viable bacterial strains carrying suppressor tRNAs with altered anticodons is, in general, restricted to cases in which there are "redundant" tRNA species (Söll *et al.,* 1966)—that is, isoaccepting tRNA species which respond to the same codon. Suppressor tRNA species are often found to be minor members of *E. coli* isoacceptor families, and hence are present only in small amounts in the cell (Söll, 1968).

A mutation in the anticodon is not the only way to generate a nonsense suppressor tRNA. The *E. coli* UGA-suppressing tryptophan tRNA is derived from the wild type tRNA by a base change in the stem of the D-loop (Hirsh, 1971). As described in more detail later, this base change leads to efficient recognition of the UGA codon as well as the tryptophan codon UGG.

2.4 *Missense Suppression*

A missense mutation is a base change in DNA that alters an mRNA codon such that a substitute amino acid is inserted in a polypeptide chain at the position of the altered codon. If the presence of that amino acid interferes with the activity of the gene product, function can be restored by secondary extragenic suppressor mutations that lead to the insertion of an acceptable amino acid in response to the mutant mRNA codon. Best studied are suppressors of missense mutations corresponding to glycine-requiring sites in the *E. coli* tryptophan synthetase α protein. The suppressors, $tRNA^{Gly}$ species with altered anticodon sequences, insert glycine in response to codons for other amino acids (for discussion see Hill, 1975). Recent efforts to extend the range of missense codons in the gene encoding the tryptophan synthetase α protein which are amenable to genetic selections for missense suppressors may yield novel families of mutant tRNAs (Murgola and Yanofsky, 1974; E. Murgola, personal communication).

2.5 *Frameshift Suppression*

Bacterial extragenic suppressors can correct frameshift mutations by restoring the proper reading frame of translation. In *Salmonella typhimurium,* six unlinked genetic loci have been found to give rise to frameshift-specific suppressors which act on (+1) frameshift mutations (Yourno and Tanemura, 1970; Riddle and Roth, 1972*a*). *SufA, SufB,* and *SufC* suppress one series of frameshift mutations in the histidine operon; *SufD, SufE,* and *SufF* correct a different set of *his* mutations. Suppressors of the first class appear to restore the reading frame at the site of mutations resulting from the insertion of a C residue into a sequence of several C residues; the second class responds similarly to an additional G in a run of G residues. The *SufD* suppressor has been shown to be a $tRNA^{Gly}$ species (Riddle and Carbon, 1973) with an anticodon loop of eight nucleotides [all other known tRNAs contain seven nucleotides in the anticodon loop (Sprinzl, 1978)]. Its tetranucleotide anticodon CCCC may read a four-base GGGG codon and thereby bring translation beyond an insertion back to the wild type reading frame.

2.6 *Ribosomal Suppression*

Suppression of nonsense and frameshift mutations in the absence of suppressor tRNAs can also be achieved by alterations in the components of the ribosome. Ribosomes with altered proteins allow a range of codon misreading to give a weak suppression. In addition, structural changes in ribosomal proteins modulate the efficiencies of tRNA suppressors. For example, mutations in two genes, *strA* and *ram,* define the S12 and S4 proteins of the small ribosomal subunit of *E. coli.* Work in the laboratory of Gorini (Gorini, 1970; Biswas and Gorini, 1972) has demonstrated that *strA* mutations restrict the efficiency of suppression of nonsense and missense mutations, whereas *ram* mutants restore the efficiency of nonsense suppressors (Rosset and Gorini, 1969).

2.7 *Polarity Suppression*

In a group of genes organized as a single transcriptional unit (operon), the presence of a nonsense mutation in one gene frequently not only inactivates that gene product but also results in reduced expression of genes which, relative to the origin of the transcriptional unit, are located distal to the nonsense mutation. This pleiotropic effect is called *polarity.* Some frameshift mutations and larger DNA insertions, which presumably introduce a nonsense triplet into the reading frame, are also polar. The degree to which distal gene expression is limited has been shown to be a function of the distance between the nonsense triplet and the next polypeptide chain initiation site (Newton *et al.,* 1965).

Polarity can be relieved by several types of intragenic suppressor events which do not act directly on the polar mutation to restore the translation of a functional gene product. For example, second site mutations that delete either the polar mutation or some of the DNA intervening before the next gene boundary, or those that create new promoters, can restore distal gene function. In addition, the effect of a polar mutation in a cistron can be neutralized at the level of translation by the creation of new signals for the reinitiation of polypeptide synthesis just beyond the polar mutation. Examples of each of these cases have been described (Hartman and Roth, 1973). In a few cases, second-site mutations that allow translational reinitiation have been shown actually to return function to the gene product in which the primary polar mutation occurs (Sarabhai and Brenner, 1967; Michels and Zipser, 1969), most likely by producing a carboxy-terminal polypeptide fragment that can still interact with the N-terminal nonsense fragment.

More relevant to the focus of this discussion is the finding that extragenic suppressors can also eliminate the polar effects of a nonsense mutation without restoring the mutant function (Beckwith, 1963). These are not codon specific, and presumably can relieve the polarity caused by nonsense mutations in any operon. These have now been shown to act at the level of transcription termination, a finding that underscores dramatically the surprise element inherent in the characterization of genetic suppressors (Section 3.6).

3 Some Topics in Molecular Biology Influenced by Analysis of Genetic Suppression

Genetic, biochemical, and biophysical studies over the past 15 years have shown that many of the component structural elements of a cell's translational apparatus contribute to the faithful transmission of the information contained in messenger RNA into the amino acid sequences of proteins. Although at first glance this statement might appear to be an obvious consequence of the fact that a large number of macromolecules cooperate in carrying out this process, the steps of protein biosynthesis themselves do not necessarily reveal the types of molecular interactions that underlie their specificities. In this section, we would like to draw together the major conclusions derived from the analysis of informational suppressors which have led to a more detailed description of several steps in protein biosynthesis.

3.1 Phasing of Messenger RNA

The phasing of mRNA to be read three bases at a time could be attributed to those components of the ribosome that carry out translocation—that is, one might argue that reading three bases at a time results from a strict ribosomal movement of a certain distance as aminoacyl-tRNA is translocated from the A site to the P site of the ribosome. Alternatively, one might propose that once the ribosome has selected the appropriate mRNA reading frame defined by a genuine initiator region containing an AUG or GUG triplet, the interaction between tRNA and mRNA thereafter contributes strongly to correct phasing of the codons in the mRNA. Studies of frameshift suppression in *S. typhimurium* (for a review see Roth, 1974) have provided some support for the latter view.

Suppression of frameshift insertions by dominant extragenic mutations has been shown to be mediated by tRNAs with altered anticodons. As mentioned earlier, the existence of a CCCC tetranucleotide anticodon of the $\text{tRNA}^{\text{Gly}}_{sufD}$ suppressor (Riddle and Carbon, 1973) raises the possibility that the larger anticodon loop reads four bases at the site of an inserted base in a message to bring subsequent elongation steps back into the wild type phase. Other dominant suppressors (*sufA* and *sufB*) are probably proline tRNA species with anticodons GGGG. Although no proline suppressor tRNAs have been sequenced, the *sufB* suppressor has been shown to insert proline in response to $\text{CCC}^{\text{U}}_{\text{C}}$ (Yourno and Tanemura, 1970).

The frameshift suppressors analyzed thus far appear to act only at positions at which G and C are repeated in the DNA. The absence of suppressors derived from tRNA^{Lys} and tRNA^{Phe}, which presumably would act on insertions into consecutive sequences of A and T residues, respectively, may simply be due to the specificity of the acridine mutagen (ICR 191) used in the original experiments to obtain most of the frameshift mutants in *Salmonella*. Alternatively, dispensable tRNA species with the required codon specificities may not exist for tRNA^{Lys} or tRNA^{Phe}. Given the known multiplicity of *E. coli* tRNA species for most amino acids, this is an unlikely possibility. Attempts are being made to generate new frameshift mutations with proflavin, which in bacteriophage T4 has been shown to cause frameshift mutations in a variety of sequences (Roth, 1974). Revertants derived from these mutants should include novel frameshift suppressor tRNAs; these would help us to understand which features of the anticodon loop are important for strict mRNA phasing during the translocation process.

So far no clear-cut evidence has been obtained for suppressors of frameshift deletion mutations that result in the deletion of one nucleotide from the anticodon to give a six-base anticodon loop. It is not clear whether this implies that tRNAs with a six-base anticodon loop do not survive processing to yield mature tRNA species, that two base pairs are insufficient for forming a stable codon·anticodon complex, or that the additional stacking interactions required to stabilize codon·anticodon binding (Grosjean *et al.,* 1976) are not favored by the six-base anticodon loop structure. Certainly, the actual pattern of codon use has suggested to some investigators that often only two bases appear to specify an amino acid (Jukes, 1977; Lagerkvist, 1978). It is now possible to construct tRNAs with six nucleotides in the anticodon loop *in vitro,* as shown by Kaufmann and Littauer (1974), by joining appropriate tRNA fragments with RNA ligase (Last and Anderson, 1976). Preparation of such tRNAs should allow critical tests both of their ability to recognize two-base codons and also of the structural properties of six-base anticodon loops.

One of the suppressors acting on insertions into stretches of G residues behaves as a recessive mutation, *sufF* (Riddle and Roth, 1972*a*). A simple model for the origin of a recessive suppressor is that a defect in a tRNA-modifying enzyme normally acting on a glycine tRNA species produces an undermodified glycine tRNA with the capacity to recognize GGGG. Although *sufF* mutations do lead to alterations in the chromatographic mobility of $tRNA_2^{Gly}$ (Riddle and Roth, 1972*b*), it is not certain which tRNA species in *sufF* strains serves as the frameshift suppressor. Until this is identified, the possibility that other tRNA species with C-rich anticodon loops may mediate *supF* suppression should also be considered. An *E. coli* tRNA that satisfies this criterion is $tRNA^{Trp}$, which has the anticodon loop sequence $C_mUCCAms^2i^6AA$. It is tempting, therefore, to speculate that an undermodified $tRNA^{Trp}$ species might utilize the CUCC for hydrogen binding with a GGGG tetranucleotide. We currently have no information on whether codon·anticodon interaction ever involves pairing with other parts of the anticodon loop.

No evidence is currently available to suggest a mechanism whereby an anticodon loop of normal size (seven bases) containing one unmodified residue could recognize a tetranucleotide codon. One might propose either that improper nucleoside modification reduces the strength of codon·anticodon base pairing such that the precision of the translocation step is impaired, or that the absence of base modification relaxes the conformation of the anticodon loop such that four bases rather than three are capable of hydrogen bonding in certain mRNA contexts. Improper methylation of the 5′ anticodon base is proposed to explain the frameshift-suppressing activity exhibited by strains carrying the recessive *supK* mutation. Riyasaty and Atkins (1968) found that a frameshift mutation in the tryptophan operon of *S. typhimurium* is suppressed, albeit inefficiently, by an extragenic suppressor which they later showed (Atkins and Ryce, 1974) to be identical to the recessive UGA suppressor, *supK* (Reeves and Roth, 1971). *supK* strains contain a defective tRNA methylase; as a consequence, the modified nucleoside V, uridine-5-oxyacetic acid methyl ester (McCloskey and Nishimura, 1977), lacks its methyl group (Pope and Reeves, 1978*a,b*). The undermethylated V is designated V*. The nucleotide V occurs in the 5′ position of the anticodon in four *E. coli* tRNA species of known sequence, $tRNA^{Val}$, $tRNA^{Ala}$, $tRNA^{Ser}$, and $tRNA^{Pro}$ (Sprinzl, 1978); it also is found in four other unidentified species (Dudock *et al.*, 1978). The two predominant undermethylated species in the *supK* strain are a major $tRNA^{Ala}$ species and a minor $tRNA^{Ser}$ species. Based on the expected anticodons for these tRNAs, one would predict that the undermodified $tRNA^{Ser}$ (anticodon V*GA) may be able to suppress the UGA codon via a one-base mismatch in the second position of the anticodon, while the $tRNA^{Ala}$ species (anticodon V*GC) may serve as frameshift suppressor. More detailed characterization of these tRNAs is needed to confirm these assignments and to explain how the undermodification changes their coding capacities.

Thus, there may be several means by which frameshift suppressors can be generated. Since only one frameshift-suppressing tRNA has been analyzed in detail, we cannot even conclude at the present time that all observed suppressors are due to altered tRNA anticodons. As described earlier in the short summary of nonsense suppression, UGA suppression mediated by an altered $tRNA^{Trp}$ in *E. coli* is brought about by a base change in the dihydrouridine stem of $tRNA^{Trp}$ (Hirsh, 1971). The suppressor mutation may lead to less stable tertiary interactions in the molecule and possibly to a more flexible anticodon loop, with the end result that the mutant tRNA reads both UGA and the tryptophan codon, UGG. It is noteworthy that even wild type tryptophan tRNA translates UGA codons as tryptophan (UGG) at a surprisingly high rate (Hirsh and Gold, 1971). The anticodon conformations of su^+ $tRNA^{Trp}$ and su^- $tRNA^{Trp}$ in solution appear to be the same, as judged by their ability to form complexes with the complementary anticodon of proline tRNA (Buckingham, 1976). This would suggest that components of the translation machinery, presumably the ribosome (Kurland *et al.*, 1975; Kurland, 1977), must facilitate recognition of the UGA codon. If indirect effects transmitted by a change in tRNA structure outside the anticodon triplet are important to proper codon·anticodon recognition, it becomes more plausible that tRNA-modifying enzymes, causing under- or overmodification of tRNA, might also act as recessive suppressors.

3.2 *UAG, UAA, and UGA in Polypeptide Chain Termination*

In this section the problem of polypeptide chain termination is considered. In the context of nonsense suppression, questions from three points of view can be posed. First, does current evidence support the widely held view that chain termination is a unique example of a protein·mRNA interaction? Second, does nonsense suppression tell us anything about mRNA nucleotide sequence patterns in the vicinity of natural termination signals? And finally, does the wide fluctuation in measurable suppression efficiencies imply that different mRNA contexts provide a spectrum of mRNA·tRNA binding strengths?

3.2.1 The Role of Release Factors in the Termination Process

The final step in the synthesis of a polypeptide is triggered by the appearance of a UAG, UAA, or UGA mRNA codon in the A site of the ribosome. Despite substantial effort, our knowledge of the molecular details of the termination reaction is still sketchy. Basically, what occurs is this: first, the hydrolysis of the peptidyl-tRNA ester linkage to release the completed polypeptide from the ribosome and, subsequently, the disassembly of the 70 S ribosome·mRNA·tRNA complex to allow recycling of these components for new rounds of protein synthesis. The aspect of termination of primary interest in the context of this discussion is terminator codon recognition. Other topics are dealt with elsewhere (Tate and Caskey, 1974; Martin and Webster, 1975), and are not considered in detail here.

It was first thought that a bacterial cell might contain tRNAs that recognize terminator triplets but that cannot be charged with any one of the 20 amino acids. Such chain-terminating tRNAs, however, are not natural components of the cell's tRNA population. The possibility that chain termination signals function simply in a passive way, as unreadable codons, was ruled out by the demonstration that in a cell-free amino acid-incorporating system depleted of all the tRNAs for a given amino acid codon, chain elongation stops at that codon, but the chain is not released (Bretscher, 1968; Fox and Ganoza, 1968). In other words, termination requires active recognition of the specific mRNA triplets UAG, UAA, and UGA. Three soluble release factors (RF), in concert with ribosomal functions, are required for peptide chain termination in *E. coli.* Two of these factors, RF1 and RF2, are protein molecules that bind to ribosomes in response to specific codons. The activities of these purified proteins are insensitive to treatment with T1 or pancreatic ribonuclease, and both appear to be free of RNA (Capecchi and Klein, 1969; Smrt *et al.,* 1970). A recent probe of the molecular structures of RF1 (mol. wt. 49,000) and RF2 (mol. wt. 50,000), using antibodies raised against each purified factor separately, indicates that the two proteins share antigenic determinants (Ratliff and Caskey, 1977). This is consistent with the fact that they interact with similar arrays of ribosomal proteins (Tate *et al.,* 1973, 1975; Brot *et al.,* 1974) (see later) and have identical activities *vis-à-vis* the ribosomal peptidyltransferase function. However, RF1 is specific for UAG and UAA codons, whereas RF2 responds only to UAA and UGA (Scolnick *et al.,* 1968; Scolnick and Caskey, 1969). The mechanism of selective terminator codon recognition is thus of considerable significance, since all other codons are recognized by tRNAs.

The role of release factors in termination has been deduced from their behavior *in vitro.* A simple model assay for the termination reaction is achieved by incubating washed *E. coli* ribosomes with AUG triplets and fMet·$tRNA_f^{Met}$. Under these conditions, an $tRNA_f^{Met}$·AUG·ribosome complex is formed. Release of formylmethionine from this complex in the presence of terminator trinucleotides is then used as a measure of termination. Formylmethionine release is promoted by RF1 in the presence of UAG or UAA, and by RF2 in the presence of UAA or UGA. The rate of release for both factors is enhanced by GTP (or GDP) and by a third release factor protein, RF3 (Beaudet and Caskey, 1972). The precise function of guanine nucleotides in the prokaryotic termination system is not yet clearly understood. RF3 may act to stimulate binding of RF1 and RF2 to the ribosome and/or to promote dissociation of the release factors from the RF·terminator codon·ribosome intermediate. In any case, once they are on the ribosome, the release factors interact functionally with the "catalytic center," peptidyltransferase, to facilitate the hydrolysis of peptidyl-tRNA.

Information regarding the ribosomal components required for the formation of release factor·termination codon recognition complexes is now emerging. Taken together, the effects of antibodies directed against specific ribosomal proteins and of antibiotic inhibitors on RF interaction with the ribosome identify L7/L12, S9, S11, and S3 as elements of the RF-binding domain (Caskey and Beaudet, 1971; Tate *et al.,* 1975). Recently, Caskey *et al.* (1977) have presented evidence which indicates that the 3′ terminus of 16 S rRNA must be intact for successful RF binding to the ribosome. These authors find that pretreatment of ribosomes with cloacin DF13, a bacteriocin which specifically cleaves 49 nucleotides from the 3′-OH terminus of 16 S rRNA in the ribosome, significantly impairs RF2-mediated release of fMet from $tRNA_f^{Met}$· AUG·ribosome complexes as well as the formation of RF·[^{3}H]UAA·ribosome complexes. The capacity of treated ribosomes for aminoacyl-tRNA binding in the A site is similarly reduced (Baan *et al.,* 1976). By contrast, fMet release from cloacin-treated ribosome in reactions containing 20% ethanol, a protocol used routinely to measure codon-independent peptide release (Tompkins *et al.,* 1970), is unaffected. These results suggest that the 3′-terminal nucleotides of 16 S rRNA play some role in the terminator codon-directed association of release factors with the ribosome. Moreover, they add to the growing body of evidence that underscores the integral involvement of this region of 16 S rRNA in an important functional site on the interface between the 30 S and 50 S ribosomal subunits (see Kurland, 1977, and Chapter 9 in this volume).

If we now consider how the release factors RF1 and RF2 read termination signals, the obvious possibilities are either that they specifically recognize the nucleotide sequences or that they promote an interaction between a terminator codon and some ribosomal component that subsequently promotes polypeptide release. The results of several experiments have been used as evidence that there is direct recognition of termination codons by release factors. First, the difference in codon specificity exhibited by RF1 and RF2 suggests that they may have the ability to distinguish among nucleotide sequences. The specificity was suggested by the finding (Scolnick and Caskey, 1969) that the binding of radioactive trinucleotide UA[^{3}H]A or UA[^{3}H]G to ribosomes in the presence of RF1 is subject to competition by nonradioactive UAA orUAG, but not by UGA. Second, Capecchi and Klein (1969) have found that the factors themselves demonstrate a limited degree of specificity in equilibrium dialysis binding to tetranucleotides. However, the differences between terminator and nonterminator sequence binding observed in the experiments of these authors were only 2- to 2.5-fold. These data thus may indicate alternatively that at least part of the RF molecule has the general capacity to interact with RNA. Third, the competition of nonsense suppressor tRNAs and release factors for the translation of terminator codons has also been viewed as support for RF codon recognition. Using a cell-free protein synthesis system programmed by mRNA templates containing a nonsense codon, both

Beaudet and Caskey (1970) and Ganoza and Tomkins (1970) have shown that the ratio of suppressor tRNA to release factors determines the ratio of readthrough translation to chain termination. These results indicate that aminoacylated suppressor tRNAs and release factors cannot be accommodated simultaneously on the ribosome, but they do not demonstrate specifically that both molecules compete for terminator codon binding *per se.*

The first data which raised the possibility that RNA·RNA rather than protein·mRNA codon interaction may mediate release factor function were provided by experiments designed to explore the relationship between the structure of the terminator codons and their activity as mRNA templates for RF1 and RF2. Smrt *et al.* (1970) measured fMet release from RF·AUG·$tRNA_f^{Met}$·ribosome complexes, directed by a series of UAA and UAG derivatives containing substituted uridine moieties. The pattern of response obtained suggested that the N_3 proton and C_4 carbonyl moieties of the pyrimidine ring are essential for the recognition of terminator codons, whereas substituents in the C_5 position do not impair function. It is striking that N_3 and C_4 are the positions required for hydrogen bonding of Watson–Crick base pairs. Although a protein·RNA interaction might utilize these same hydrogen-bonding atoms, it is not clear that such an interaction should necessarily be restricted to the same atoms of the pyrimidine ring as those required for a base-paired RNA·RNA interaction.

Dalgarno and Shine (1973) have suggested that in the 3′-terminal hexanucleotide $UCAUUA_{OH}$ of the eukaryotic 18 S rRNA, UCA might pair with the UGA codon during the termination process, while the UUA sequence could interact with both UAA and UAG codons. By analogy with the eukaryotic case, they point out (Shine and Dalgarno, 1974) that the 3′-terminal UUA_{OH} of *E. coli* 16 S rRNA could recognize all three terminator codons via standard or wobble (U·G base pairs) complementarity. While the ribosomal RNA sequences focus attention on the possibility that release factors "recognize" terminator codons by promoting specific mRNA·rRNA interactions which then lead to peptidyl-tRNA hydrolysis on the ribosome, more recent sequence information has revealed considerable variation in the 3′-terminal nucleotides of prokaryotic 16 S rRNAs. For example, *Bacillus subtilis* 16 S rRNA terminates with $CUUCU_{OH}$; *Caulobacter crescentus,* with $CCUUUCU_{OH}$ (Shine and Dalgarno, 1975*a,b*); and *B. stearothermophilus,* with $UUUCUA_{OH}$ (Sprague *et al.,* 1977). Whereas in most eukaryotic organisms examined, the sequence of the 18 S rRNA 3′ terminus is highly conserved, a 3′-terminal UUG_{OH} is found in plants (Oakden and Lane, 1976; Hagenbüchle *et al.,* 1978). Thus, this particular region of ribosomal RNA, which does not provide a consistent potential for base pairing with terminator codons, cannot be involved in termination by a mechanism in which such pairing occurs. Moreover, when the terminator codon recognition properties of RF1 and RF2 from *E. coli* and *B. subtilis* are examined on their homologous and heterologous ribosomes, no differences in ribosomal specificity are observed (Caskey *et al.,* 1977). In light of the codon recognition studies emphasizing the resemblance of RF·codon interactions to Watson–Crick base pairing (Smrt *et al.,* 1970) and of more recent work implicating specific RNA·RNA interactions in many ribosomal functions (see Chapter 9 in this volume), other regions of rRNA could reasonably be examined for potential base pairing with terminator codons during polypeptide chain termination.

The characterization of the properties of nonsense mutations within genes clearly established the capacity of UAG, UAA, and UGA to terminate polypeptide chains prematurely. The question then is, which of these codons serve as signals for normal chain termination? An indirect approach to this question is to examine the physiological effects of nonsense suppressors on general cell function. One would presume that in many circumstances, extension of a wild type polypeptide via suppressor tRNA action at the normal termination site would yield a nonfunctional protein and hence adversely affect cell growth. Circumstantial evidence based on this assumption was initially marshalled to suggest that termination of many *E. coli* proteins may be directed by UAA (Brenner and Beckwith, 1965). Whereas it appeared that suppression of UAG nonsense mutations could be as high as 76% efficient without having any measurable pleiotropic effect, and UGA suppressors seemed similarly to be tolerated well by bacterial strains, ochre (UAA) suppressors were, by contrast, always of low efficiency and inhibitory to cell growth. Several groups found that strains carrying ochre suppressors were slow growers, whereas the isogenic parents carrying the allelic amber suppressors grew normally (Ohlsson *et al.*, 1968; Person and Osborn, 1968). Ohlsson *et al.* (1968) noted, in addition, that if the efficiency of an ochre suppressor was reduced by the introduction into the ochre-suppressing strains of a ribosomal mutation restricting the cell's translation (streptomycin resistance; see Section 3.5), growth impairment was coordinately reduced. Collectively, these data and the fact that both prokaryotic release factors recognize UAA seemed to emphasize the functional importance of UAA as a natural termination signal and to argue against common physiological use of UAG and UGA.

RNA sequence analysis of viral genome fragments provided direct information on the nature of the terminator codons used. Among the first termination sequences determined were those for the coat proteins of the two closely related RNA bacteriophages, R17 and f2. In both instances, the pair of nonsense codons UAAUAG was found (Nichols, 1970; Nichols and Robertson, 1971). The use of tandem termination codons seemed to provide a convenient way for the cell to ensure that readthrough, seen frequently as the "leakiness" of nonsense mutations within genes, could not occur at the ends of cistrons. Lu and Rich (1971) followed up this discovery with experiments designed to estimate the frequency of tandem termination codons in *E. coli* genes. These authors measured the increase in the proportion of C-terminal tyrosine in the total protein extracted from *E. coli* strains carrying tyrosine-inserting UAA and UAG suppressors; from their data, they suggested that approximately 13% of termination signals might contain a UAG or UAA in tandem with a second nonsense triplet.

How do these initial predictions conform to the pattern of termination codons actually used? Table I summarizes currently known nucleotide sequences in the vicinity of those natural polypeptide chain termination sites for which carboxy-terminal protein sequence information is available. Sequences from bacterial and animal viruses and from prokaryotic and eukaryotic organisms are included. One can clearly see that UAG, UAA, and UGA are able to function singly as termination signals. In the limited number of available examples derived from *E. coli* cistrons (three cases: tryptophan synthetase α and β, and *lac* repressor), it is not

Table I. *Natural Polypeptide Chain Termination Sequences*[a]

Cistron	Termination sequence	References
A. *UAG*		
MS2A	Ser Arg UCU AGA **UAG** AGC CCU	Contreras *et al.*, 1973; Remaut and Fiers, 1972
MS2 replicase	Pro Arg CCU CGG **UAG** CUG ACC	Fiers *et al.*, 1976
Human chorionic somatomammotropin	Gly Phe GGC UUC **UAG** GUG CCC	Seeburg *et al.*, 1977
B. *UAA*		
λCro repressor	Thr Ala ACA GCA **UAA**	Hsiang *et al.*, 1977; Roberts *et al.*, 1977
φX174 D	Val Met GUG AUG **UAA** UGU CUA	Barrell *et al.*, 1976
φX174 J	Gln Phe CAA UUU **UAA** UUG CAG	Freymeyer *et al.*, 1977; Sanger *et al.*, 1977*a*
E. coli tryptophan synthetase α	Arg Ser CGC AGU **UAA** UCC CAC	A. Wu and T. Platt, personal communication
Human α-globin Constant Spring	Phe Glu UUU GAA **UAA** AGU CUG	Proudfoot, 1977; Wilson *et al.*, 1977
Human α-globin	Tyr Arg UAC CGU **UAA** GCU GGA	Clegg *et al.*, 1974; Proudfoot and Longley, 1976
Rabbit α-globin	Tyr Arg UAU CGU **UAA** GCU GGA	Proudfoot *et al.*, 1977
Human β-globin	Tyr His UAU CAC **UAA** GCU CGC	Proudfoot, 1977
Tandem termination triplets		
MS2 coat	Ile Tyr AUC UAC **UAA UAG** ACG CCG	Min Jou *et al.*, 1972
R17 coat	Ile Tyr AUC UAC **UAA UAG** AUG CCG	Nichols, 1970
f2 coat	Ile Tyr AUC UAC **UAA UAG** ACG CCG	Nichols and Robertson, 1971
C. *UGA*		
Qβ coat	Ala Tyr GCX UAPy **UGA** ACX UUPu CUX	Weiner and Weber, 1973
φX174 G	Leu Lys CUU AAG **UGA** GGU GAU	Air *et al.*, 1976
λC_I repressor	Phe Gly UUU GGC **UGA** UCG GCA	Humayun, 1977
E. coli lac repressor	Gly Gln GGG CAG **UGA** GCG CAA	Dickson *et al.*, 1975
E. coli tryptophan synthetase β	Glu Ile GAA AUC **UGA** UGG AAC	Platt and Yanofsky, 1975
SV40 VP1	Met Gln AUG CAG **UGA**	Pan *et al.*, 1977
Rabbit β-globin	Tyr His UAC CAC **UGA** GAU CUU	Proudfoot, 1977; Efstratiadis *et al.*, 1977
Tandem termination triplets		
φX174 F	Thr Ser ACU UCG **UGA UAA** AAG AUU	Fiddes, 1976

[a]mRNA sequences and the RNA equivalents deduced from DNA sequences are shown. Terminator codons are given in boldface type. Only those sequences for which the corresponding carboxy-terminal protein sequences have been determined are included.

obvious at this time which of the codons is used predominantly. If one looks at the termination triplets for the bacteriophage and *E. coli* genes which are all translated by *E. coli* ribosomes (the UAAUAG for the closely related MS2, R17, and f2 coat protein cistrons are counted as one case), UAG is used twice, UAA, five times, and UGA, six times. Tandem triplets are found only at the ends of the RNA phage coat protein genes and the cistron encoding the ϕX174 F protein. Although the frequency of the tandem codons, 15%, is in the same range as the estimate made by Lu and Rich (1971), this may be purely fortuitous, based on a small sample size of 13 cases.

Since both UAG and UGA, in addition to UAA, serve as termination signals for the *E. coli* termination process, the pleiotropic effects of ochre suppressors remain an unsolved problem. In the absence of more explicit evidence to the contrary, it may be reasonable to maintain the view that readthrough of a very limited number of UAA termination signals yields extended polypeptides whose malfunction impairs cellular processes.

Finally, one might ask why UAG, UAA, and UGA in particular have evolved as terminator codons and not as sense codons specifying amino acids? The only suggestive evidence on this point comes from studies of bimolecular complexes formed with tRNAs containing complementary anticodon sequences. Grosjean *et al.* (1978) find that the lifetime for the complex between $tRNA^{Leu}$ (anticodon UAG) and the $su^{+}3$ amber suppressor $tRNA^{Tyr}$ (anticodon CUA) is anomalously low, compared to all other examples of fully complementary interactions tested. This is regarded as support for the speculation that the UAG sequence may have evolved as a termination codon because of its limited capacity for stable codon·anticodon pairing. It remains to be seen whether complex lifetimes in the same range are obtained for the appropriate anticodon·anticodon pairs prepared for UAA and UGA.

3.2.3 *mRNA Reading Context: Termination Signals and the Efficiency of Suppression*

In contrast to the situation at the beginning of a cistron, where additional recognition elements are required for ribosomal discrimination of true initiator codons from the many noninitiator AUG and GUG triplets encoded in a messenger RNA, there is no compelling reason to expect that a UAG, UAA, or UGA codon may comprise only part of a more complex termination signal. Wherever these triplets appear in a coding frame, they do lead to polypeptide chain termination. In fact, the only common feature among the sequences listed in Table I appears to be the presence of at least one terminator codon. Note that all four bases are represented in the nucleotides flanking the terminator triplets. Likewise, sequences farther downstream from the termination point (data not shown) reveal no obvious pattern of conserved residues.

It is of interest to ask whether the nature of the nucleotides in the immediate vicinity of a termination triplet, as well as the sequence of the terminator codon itself, may govern the efficiency of polypeptide release at a given site. For the case of natural termination signals, in general, this problem has not been addressed. While the discrete carboxyl termini of proteins attest to the high efficiency of the process, chain termination in bacterial strains lacking nonsense suppressors is not always 100% efficient. For example, two interesting readthrough proteins, each resulting from sense translation of a UGA codon, are discussed under the subject of ribosomal suppression (see Section 3.5). The particular features of these cistron ends that lead to inefficient termination have not been identified. We have more detailed information, however, concerning the role of mRNA context in suppressor tRNA· terminator codon interaction. In numerous instances, systematic variations in suppression efficiencies have been cited as evidence that neighboring nucleotides influence the recognition of terminator codons (Salser, 1969; Salser *et al.,* 1969; Yahata *et al.,* 1970; Comer *et al.,* 1974, 1975; Colby *et al.,* 1976; Akaboshi *et al.,* 1976; Fluck *et al.,* 1977; Feinstein and Altman, 1977, 1978); the level of suppression occurring at the site of a terminator codon then does indirectly affect the overall efficiency of a termination event.

Suppression efficiency, at best, is a difficult parameter to quantitate. In order to represent a measure of the affinity of a terminator codon for an anticodon sequence, several variables must be carefully controlled. These include the genetic alleles for numerous ribosomal proteins and the aminoacylation capacities of suppressor tRNAs for both the correct and possibly incorrect amino acids. If one looks at the suppression of a series of amber mutations within a single gene by one suppressor tRNA species, nonuniform responses usually stem from the variable degree to which the amino acid residue inserted by the suppressor is tolerated at various positions in the protein product. Similarly, at a given nonsense site, the efficiencies of several suppressor tRNA species of different nucleotide sequence may reflect other aspects of ribosome·tRNA interaction superimposed on the codon·anticodon interaction. However, if one compares the UAA/UAG suppression ratios obtained for pairs of allelic UAA and UAG mutations (homotopic pairs) by a single ochre suppressor, the contributions of mRNA reading context can most clearly be seen.

A combination of amino acid sequence and direct nucleotide sequence data provides a series of homotopic pairs in the T4 rIIB, T4 lysozyme, *E. coli* β-galactosidase, and two *E. coli trp* genes for study. The coding properties of three ochre suppressors have now been characterized in this fashion (Yahata *et al.,* 1970; Comer *et al.,* 1975; Feinstein and Altman, 1977, 1978). The nucleotide sequences for two of the suppressors, namely the *E. coli* $su^{+}3_{oc}$ $tRNA^{Tyr}$ (Altman *et al.,* 1971) and a glutamine-inserting tRNA specified by the T4 phage genome (Comer *et al.,* 1974; Seidman *et al.,* 1974), have been determined. The finding that UAA/UAG suppression ratios obtained for the series of homotopic nonsense sites vary over a tenfold range confirms that some aspect of the mRNA reading context must play a role in nonsense codon recognition. At most sites, response to the amber codon is favored, but at a few, UAA is read more efficiently that UAG. Different ratios for the same homotopic site are obtained, as expected, when ochre suppressors with nonidentical anticodon loop sequences and patterns of nucleoside modification are tested. More intriguingly, secondary mutations that alter the tRNA sequence outside the anticodon region also change the ratios obtained for a given suppressor tRNA, suggesting that other parts of the tRNA molecule influence the final mRNA· tRNA anticodon interaction.

From available mRNA sequence information at the homotopic sites, one can begin to see characteristic effects of certain residues on UAA and UAG suppression by the ochre suppressors. However, one contribution to the observed fluctuation in suppression efficiencies from one homotopic pair to another may result from differences in the particular translational kinetics of the codons flanking each pair. Therefore, it should be possible to extend present data by systematic manipulation of the nucleotide sequences surrounding a single nonsense site in a messenger RNA. This could be done either via fine structure genetic dissection of cistrons such as the *E. coli lacI* gene (see Section 4) or by appropriate application of localized mutagenesis techniques (Weissmann *et al.,* 1977; C. A. Hutchison III, personal communication). At a single homotopic site, the effects on suppression efficiency of base changes adjacent to the 5′ and 3′ codon positions may more clearly reveal the direct role of context on codon·anticodon interaction.

3.3 *Effect of tRNA Modification on Codon Specificity*

The function of modified nucleotides in controlling the specificity of the codon·anticodon interaction during protein synthesis is not well understood. Crick's wobble hypothesis (Crick, 1966) provides a physical rationale for the fact that one tRNA species can recognize several codons. It states that the nucleotide in the first (5′) position of the anticodon may recognize certain nucleotides in addition to its strict complement if slight deviations in the position of the C_1' atoms at the glycosidic bond are allowed. The five pairings proposed by Crick are listed in Table II. It is noteworthy that with nearly 100 tRNA sequences available, adenosine has not been found in the first position of any anticodon. Thus, there is no experimental evidence for the use of A·U wobble base pair. At the time the hypothesis was formulated, inosine was the only modified nucleoside considered. In the intervening 12 years, nucleotide sequence analysis of many tRNA species has revealed the presence of other modified nucleosides in the first and second positions of the anticodon (Sprinzl, 1978; McCloskey and Nishimura, 1977; Agris and Söll, 1977). Some of these modified bases have pairing properties different from those of their unmodified counterparts. For instance, derivatives of 2-thiouridine are found in the first position of the anticodon of several tRNAs (for example, $tRNA^{Lys}$, $tRNA^{Glu}$, and $tRNA^{Gln}$). Trinucleotide-binding studies with a $tRNA^{Glu}$ species that should recognize the codons GAA and GAG have shown that multiple base-pairing ability is restricted; only GAA is recognized efficiently (Yoshida *et al.,* 1970, 1971). This suggests that if 2-thiouridine replaces uridine in the first position of a tRNA anticodon, that tRNA species should only respond to codons ending in A.

Another example of an anticodon modification that affects the coding prop-

Table II. Nonstandard Base Pairs Allowed by the Wobble Hypothesis (Crick, 1966) for mRNA Codon · tRNA Anticodon Interaction

5′-Anticodon base	3′-Codon base
G	U or C
C	G
A	U
U	A or G
I	A, U, or C

erties of the corresponding tRNAs is the nucleoside V, uridine-5-oxyacetic acid methyl ester. In the first position of the anticodon, it leads to recognition of codons ending in U, A, or G (McCloskey and Nishimura, 1977). As discussed previously in the context of frameshift suppression, the unmethylated V in a $tRNA^{Ser}$ anticodon, V*GA, is proposed to confer on the suppressor tRNA the ability to read UGA via a one-base mismatch in the second anticodon position (Pope and Reeves, 1978*b*).

Several other lines of evidence suggest that modified nucleosides also affect the strength of the codon·anticodon interaction. Gefter and Russell (1969) have shown that the efficiency of the *E. coli* $su^{+}3$ amber suppressor $tRNA^{Tyr}$ in protein synthesis is greatly reduced if the A residue in position 38, adjacent to the 3′ terminus of the anticodon (see Fig. 4), is incompletely modified to methylthioisopentenyladenosine. When the suppressor tRNA is used to direct the *in vitro* translation of bacteriophage f2 *sus*4A (a coat protein amber mutant) RNA, those forms containing only adenosine or isopentenyladenosine adjacent to the 3′ end of the anticodon support protein synthesis poorly compared to the fully modified tRNA. However, other *in vivo* and *in vitro* experiments with tRNAs lacking isopentenyladenosine and its derivatives indicate that the presence of this modification in tRNA is not an absolute prerequisite for protein synthesis (Litwack and Peterkofsky, 1971; Kimball and Söll, 1974). Recent experiments may provide a physical explanation for the role of modified bases adjacent to the 3′ terminus of the anticodon. Temperature-jump relaxation studies of $tRNA^{Phe}$·$tRNA^{Glu}$ complexes, a tRNA pair with complementary anticodons, have shown that the modification of the nucleotide in this position significantly enhances the stability of the anticodon·anticodon complex (Grosjean *et al.*, 1976). These results are consistent with the finding that the fully modified $su^{+}3$ tRNA functions optimally in protein synthesis.

Other cases are known in which base modification of the nucleotide adjacent to the 3′-terminal base of the anticodon may affect codon·anticodon interaction. During studies of missense suppression in *E. coli*, glycine tRNAs with base changes in the third anticodon position were identified. When the mutation in this position gives a C → A change, the adjacent adenosine (originally unmodified) becomes isopentenyladenosine (Carbon and Fleck, 1974); when the mutation is a C → U change, the adjacent base becomes threonylcarbamoyladenosine (Roberts and Carbon, 1974). Based on the results of studies with the anticodon·anticodon tRNA complexes mentioned previously (Grosjean *et al.*, 1976), it is plausible to suppose that the introduction of these modifications may change the capacity of each species for mRNA codon binding. Alternatively, these nucleotide modifications may occur without regard to suppressor function whenever a particular base occupies the third anticodon position of a tRNA. This is in keeping with the regular occurrence of these two hypermodified nucleosides in various tRNA species (McCloskey and Nishimura, 1977; Agris and Söll, 1977). In any case, systematic characterization of the function of these tRNAs in protein synthesis should prove quite interesting.

Recently, an *E. coli* mutant was isolated (Colby *et al.*, 1976) which is deficient in the formation of 2-thiouridine derivatives in the 5′ position of the anticodon of several tRNAs (Sprinzl, 1978). In this strain, the efficiencies of T4-encoded ochre-suppressing glutamine and serine tRNAs, which normally contain this modified

U, are significantly reduced. In particular, reading of both ochre and amber codons by the glutamine-inserting ochre suppressor is severely restricted. By contrast, the allelic amber suppressor tRNA, which normally possesses an unmodified C in the wobble position and should not be affected by the cellular mutation, functions normally. Analysis of the ochre-suppressing glutamine tRNA confirms that the U modification is completely absent in the mutant cells. It is interesting that when the efficiency of the undermodified ochre suppressor is measured at five nonsense sites derived from wild type glutamine codons in the T4 lysozyme gene, the range of values found is much larger than that obtained for those same sites tested with ochre suppressor tRNA contained in wild type cells. To the extent that the phage plaque assays used for this measurement actually quantitate lysozyme production and therefore measure suppression efficiency, the data suggest that the undermodified ochre suppressor is sensitized to fluctuations in the mRNA reading context. The results are puzzling, since by the rules of the wobble hypothesis (Crick, 1966), the undermodified tRNA should be able to recognize the ochre codon. One possible function for the wobble base anticodon modification raised by these and other studies (Ghosh and Ghosh, 1970) would be to increase the flexibility of the anticodon loop in the tRNAs such that it responds more uniformly to codons in a variety of contexts. This aspect of anticodon loop structure might also explain the differences generally seen in the suppression efficiencies exhibited by allelic amber and ochre suppressor tRNAs.

Other cellular mutations affecting tRNA modifications have been shown to alter suppressor tRNA function. For example, in mutants unable to form some 7-methylguanosine residues in tRNA, the efficiency of the *E. coli* $su^{+}2$ (glutamine) suppressor on some amber mutations is greatly changed (Marinus *et al.,* 1975). In this case, however, a relationship between undermodification and the effect on suppression has not yet been established.

3.4 How Specific Is Codon·Anticodon Interaction?

If one looks for heterogeneity in the amino acid sequences of purified proteins, it is clear that the fidelity of the translation process is high. Loftfield and Vanderjagt (1972) have estimated from the number of valine substitutions for isoleucine at certain positions in chicken ovalbumin and rabbit hemoglobin that protein synthesis errors occur with a frequency of about 3×10^{-4} per codon. Since valine and isoleucine have very similar structures, one might argue that rather than a decoding error, this value represents a measure of incorrect aminoacylation of $tRNA^{Ile}$. However, an error frequency in the same range has recently been obtained by Edelmann and Gallant (1977) from a separate set of experiments. These authors measured trace quantities of [^{35}S]cysteine incorporated into highly purified preparations of the flagellin protomer of *E. coli,* a protein that contains no cysteine residues. Their data provide an upper limit estimate of the mistranslation probability per codon in wild type *E. coli* strains of 10^{-4}.

To give this level of accuracy, the first requirement would appear to be highly selective matching of codon triplets with appropriate tRNA anticodons during the elongation step of protein synthesis. However, several lines of evidence now suggest that coding specificity at the level of hydrogen bonding between the component nucleotide triplets may be more relaxed than was previously anticipated. Grosjean *et al.* (1976, 1978) have formed dimeric tRNA complexes with pairs of purified tRNAs exhibiting various degrees of complementarity in their anticodon sequences. The surprising result of these experiments is that even when pairs are selected such that deliberate Watson–Crick pairing mismatches are generated in various positions of the anticodons, the complex lifetimes observed indicate that anticodon · anticodon interaction does occur.

These tRNA complexes may not be an adequate model for codon · anticodon pairing on the ribosome, where other molecular interactions may contribute significantly to binding stability and specificity. Two recent *in vitro* translation studies, however, also point to a relaxation in coding specificity. Both use *E. coli* cell-free systems that include only one species of a family of tRNA isoacceptors ($tRNA^{Val}$ or $tRNA^{Leu}$) to direct the synthesis of bacteriophage MS2 polypeptides. Analysis of the coat protein product reveals that *E. coli* $tRNA_1^{Val}$, which is thought to be specific for GUA, GUG, and GUU (McCloskey and Nishimura, 1977), also recognizes GUC. Similarly, $tRNA_2^{Val}$, which should be specific for GUUand GUC (Sprinzl, 1978), also recognizes GUA and GUG. Thus, each of the two valine tRNA anticodons, VAC and GAC, was able to respond to all four valine codons (Mitra *et al.,* 1977). A similar degeneracy in anticodon specificity has been observed for *E. coli* leucine tRNAs (Holmes *et al.,* 1977). In addition, a report by Weissenbach *et al.* (1977) makes it likely that a yeast $tRNA^{Leu}$ species with the anticodon UAG may respond to all six leucine codons.

In contrast to the implications of these results, there is evidence that in at least some circumstances, the pairing of a tRNA anticodon with an mRNA codon may be more stringently controlled than suggested by the wobble hypothesis. When exogenous mRNAs are translated in extracts prepared from interferon-treated eukaryotic cells (Zilberstein *et al.,* 1976), the translation of certain CUG codons can be accomplished only by a minor leucine tRNA species. The major CUG-specific leucine tRNA is inactive at these sites. In the absence of sequence information for the eukaryotic leucine tRNAs, the reason for this restriction is not clear.

Crick's wobble hypothesis was the first rationalization for the occurrence of some nonstandard base pairs which could fit the "mold" of the ribosome because of their close approximation to the atomic dimensions of standard base pairs. The identification of modified nucleosides in tRNAs has extended the list of base pairs used *in vivo.* A more radical approach to thinking about the codon · anticodon interaction may be required, however, to accommodate the examples of degeneracy cited above if, indeed, additional cases confirm the validity of this pattern. Recently, Lagerkvist (1978) has used steric and thermodynamic considerations to explain how the response of a single aminoacyl-tRNA to U, C, A, and G in the third codon position does not lead to misreading in cases where XY_G^A and XY_C^U specify two different amino acids. He proposes that the probability of misreading is governed by base-pairing strength. Strong interactions involving either G · C base pairs in the first two positions of the codon · anticodon complex or one G · C base pair coupled with a purine in the central position of the anticodon (Orgel, 1972) are suggested to give a high misreading probability, while weak interactions dominated by A · U pairs would minimize misreading. Lagerkvist points out with reference to the table of codons (Fig. 1) that those codons which are capable of interacting strongly with tRNA anticodons are confined to families in which the first two nucleotides of the codon suffice to specify an amino acid. By contrast, codon families in which coding errors would result if only two out of three codons are read (for example, Asn-AA$^{U}_{C}$; Lys-AA$^{A}_{G}$) are made up of triplets capable of relatively weak interactions with the anticodon. From this vantage, the code appears to have evolved so as to minimize the probability of coding errors.

An alternative explanation for the evidence demonstrating relaxed anticodon specificity can be envisioned if the interaction between mRNA and tRNA extends beyond three nucleotide pairs. Pieczenik (1972) has described a provocative idea based on his statistical analysis of some regions of the bacteriophage ϕX174 DNA sequence. He found that the probability of a four- or five-base-pair mRNA·tRNA interaction, which utilizes nucleotides flanking the codon·anticodon complex, is significantly higher for the mRNA reading frame actually used *in vivo* than for the other two frames. Whether adjacent interactions contribute positively (or, on the other hand, negatively) to the *in vivo* decoding process is not known at present. Continued progress in the determination of DNA, mRNA, and tRNA sequences should provide data for a critical evaluation of this possibility. Whether such "expanded" mRNA·tRNA interaction is related to the finding thus far that codon usage *in vivo* is not random (Fiers *et al.*, 1976) remains to be seen.

Thinking about codon·anticodon interaction, as exemplified by the proposals of Crick, Lagerkvist, and Pieczenik, is based in large part on a concern with

CODON					
1st letter	2nd letter				3rd letter
	U	C	A	G	
U	PHE	SER	TYR	CYS	U
	PHE	SER	TYR	CYS	C
	LEU	SER	C.T.	C.T.	A
	LEU	SER	C.T.	TRP	G
C	LEU	PRO	HIS	ARG	U
	LEU	PRO	HIS	ARG	C
	LEU	PRO	GLN	ARG	A
	LEU	PRO	GLN	ARG	G
A	ILE	THR	ASN	SER	U
	ILE	THR	ASN	SER	C
	ILE	THR	LYS	ARG	A
	MET	THR	LYS	ARG	G
G	VAL	ALA	ASP	GLY	U
	VAL	ALA	ASP	GLY	C
	VAL	ALA	GLU	GLY	A
	VAL	ALA	GLU	GLY	G

Figure 1. The Genetic Code. C.T. denotes polypeptide chain terminator.

hydrogen-bonding possibilities and with steric fit of codon and anticodon on the ribosome. An entirely different view of this interaction as it is manifest in translation errors focuses on the kinetics of peptide bond formation (Ninio, 1974; Kurland *et al.,* 1975; Kurland, 1977). One of the major features of any mechanism whereby the mRNA·ribosome complex selects a correct aminoacyl-tRNA is that rapid release of incorrect tRNA species must occur. This feature alone would suggest that an increase in the protein synthesis elongation rate would enhance the probability of an error by "fixing" an incorrect amino acid in a polypeptide chain before the tRNA esterified to that amino acid is rejected from codon association in the ribosomal A site. The frequency of translation errors, then, would be dependent not primarily on the steric "fit" of codon and anticodon, but rather on the time constants both of the mRNA·tRNA interaction on the ribosome and also of the peptidyltransferase reaction leading to peptide bond formation.

This concept provides a consistent context for considering both the effects of ribosomal mutations on translational fidelity and the problem of varied nonsense suppressor tRNA efficiency. For example, Ninio (1974) characterizes the affinities of several nonsense suppressor tRNAs for mRNA codons on the ribosome in terms of their "sticking times." Further, he suggests that mutant ribosomes can be described by a parameter defining the rates of peptide bond formation they permit. The *ram* mutation (ribosomal ambiguity) (Rosset and Gorini, 1969), which alters the small ribosomal subunit protein S4 (Zimmermann *et al.,* 1971), is said to give an increased rate of peptide bond formation, while the *strA* mutation (streptomycin-resistant phenotype), which affects protein S12 (Ozaki *et al.,* 1969), leads to a reduced rate relative to that for wild type ribosomes. Given these parameters, Ninio finds that he can describe accurately the increase in nonsense suppressor efficiency produced by *ram* mutations and the restrictive effects on nonsense suppression brought about by *str* mutations. In fact, this may be a good representation of the *in vivo* situation, since recent experiments have confirmed that ribosomes from *strA* strains do exhibit a lower rate of peptide bond formation than do ribosomes from wild type strains (Galas and Branscomb, 1976; Zengel *et al.,* 1977).

3.5 *Other Errors in Translation*

During the course of this discussion, we have considered how changes in tRNA structure or patterns of nucleoside modification are thought to modulate the fidelity of the codon·anticodon interaction in the elongation steps of protein synthesis. In describing the properties of nonsense, missense, and frameshift suppressors, however, we have repeatedly alluded to the active contribution made by ribosomal components to the specificity and efficiency of tRNA-mediated suppression. We are thus led to ask what current evidence tells us about the level of coding ambiguity inherent in wild type ribosomes, and how phenotypic and genotypic alterations of ribosome structure affect translational errors.

The effects of streptomycin on translational fidelity have been reviewed by Gorini (1974), and the properties of well-characterized *strA* and *ramA* ribosomal protein mutations were summarized briefly in the preceding section. Gorini and his co-workers (Breckenridge and Gorini, 1970) have distinguished four classes of *strA* strains, all of which exhibit altered S12 proteins, on the basis of their restriction of nonsense and missense suppression efficiencies. In addition, a group of neamine-resistant strains *(neaA)* represents mutations in S17 and shares the restrictive properties of *strA* mutants (Topisirovic *et al.*, 1977). S4 mutations (*ramA*), S5 mutations (*ramC*) that relieve a streptomycin-dependent phenotype (Piepersberg *et al.*, 1975), and the presence of streptomycin have the opposite effect; they increase misreading and the efficiency of suppression. Indeed, in the recent study of Edelmann and Gallant (1977), the presence of streptomycin (2 μg/ml) in cultures of wild type *E. coli* strains increased the frequency of [^{35}S]cysteine incorporation into flagellin protomers by about sixfold. Gorini's observations on misreading levels in *strA* and *ramA* mutants led him to conclude that an appreciable amount of ambiguity is intrinsic to codon·anticodon pairing on the ribosome. Further, he postulated that the 30 S ribosomal proteins altered by these mutations function as part of a "recognition screen" (Gorini, 1971) to discriminate among incoming tRNAs on the basis of the codon·anticodon pairing strengths they provide. The effectiveness of this screen would be increased or decreased by ribosomal mutations.

Evidence from the laboratory of Nomura confirms Gorini's view that wild type ribosomes exhibit a higher level of misreading than do ribosomes from *strA* (streptomycin-resistant) strains. Yates *et al.* (1977) have developed an *in vitro* system in which they demonstrate directly that ribosomes from *strA* strains restrict the readthrough of chain terminator codons and the efficiency of amber suppression. Using a DNA-dependent translation system including wild type ribosomes to direct the synthesis of bacteriophage λ proteins, these authors find that two products, O and O′, are specified by the *O* gene. Only the smaller polypeptide is produced in reactions containing ribosomes from a strain representing Gorini's most restrictive class of *strA* mutants. If DNA from an *O* gene amber mutant is used with both types of ribosomes, neither O nor O′ is made, but rather, a short amber fragment. From these and other data, Yates and co-workers deduce that O′ is synthesized by misreading of the normal *O* gene polypeptide chain terminator codon(s). Further, they show that addition of UGA suppressor tRNA to the system with wild type ribosomes specifically causes a substantial increase in the yield of the readthrough protein. This implies that the natural terminator codon in the *O* gene messenger RNA is UGA. The presence of UGA suppressor tRNA does not allow detectable synthesis of O′ by the *strA* ribosomes. Finally, data from a parallel series of experiments clearly demonstrate the markedly reduced capacity of ribosomes from several *strA* strains to mediate suppression of a λ *O* gene amber codon in the presence of amber suppressor tRNA.

Although the role of the *O* gene readthrough protein in the λ life cycle is unknown, there is circumstantial evidence to suggest that it might carry out an essential function. The growth of λ in *strA* strains is generally observed to be poor relative to that obtained in streptomycin-sensitive strains. Moreover, Friedman and Yarmolinsky (1972) have noticed that some *strA* alleles interfere with the lethal effects on bacterial cells usually produced by the induction of certain λ prophages. These authors present several lines of evidence that suggest that nonlethality is due to diminished expression of the λ *O* gene. By contrast, for the case of a bacteriophage *Qβ* protein produced by extension of the coat protein chain beyond a UGA codon (Weiner and Weber, 1973), there is direct evidence that the readthrough product is essential for phage infectivity (Hofstetter *et al.*, 1974). Qβ fails to grow on a *strA* mutant (Engelberg-Kulka *et al.*, 1977), and *strA* ribosomes do not allow readthrough of the coat protein terminator codon *in vitro*, either in the presence or absence of UGA suppressor tRNA (Yates *et al.*, 1977).

Taken together, these results suggest that readthrough of polypeptide chain termination signals may be required for the production of other as yet unidentified proteins, and that a certain intrinsic level of misreading by the mRNA·tRNA·ribosome complex may be optimal for cell function. As discussed in the context of theories on codon·anticodon selectivity, the *strA* mutation appears to reduce the rate of peptide chain elongation (Galas and Branscomb, 1976; Zengel *et al.*, 1977). Current evidence thus provides a fresh perspective for considering a number of problems related to translational specificity. To give one example of a puzzling phenomenon, the introduction of *strA* mutations into male strains of *E. coli* frequently makes them permissive for the growth of the female-specific bacteriophage, T7. Intriguingly, some rifampicin-resistant derivatives of permissive *strA* mutants, which presumably now possess altered RNA polymerases, no longer support T7 growth (Chakrabarti and Gorini, 1975). Although the primary mechanism by which T7 infection in males is restricted is not known, we would like to suggest that reexamination of this problem in light of the data discussed may prove productive.

Another class of ribosomal mutations, isolated as extragenic suppressors of *E. coli* mutations which give thermolabile alanyl- and valyl-tRNA synthetases (Buckel *et al.*, 1976; Wittmann *et al.*, 1974), is proposed to function by lowering the rate of translation. Among derivatives of alanyl-tRNA synthetase mutants which are able to grow at the previously nonpermissive temperatures, strains with an altered ribosomal protein S5 or S20 are found; for valyl-tRNA synthetase mutants, isolates with an altered S8 or S20 are observed. In both cases, the thermolabile aminoacyl-tRNA synthetases persist in the suppressed strains, and lead to very low levels of the relevant aminoacylated tRNA species. Buckel *et al.* (1976) suggest from the slow growth of these strains that the mutant ribosomes suppress the effects of synthetase mutations by reducing the rate of polypeptide synthesis so that amino acid incorporation is kept in step with the limited aminoacylation capacity of the defective synthetases.

The examples of ribosomal protein mutations we have discussed fall into two general genetic categories on the basis of their specificity. The suppressors of aminoacyl-tRNA synthetase mutations are cistron specific—that is, an alteration in the 30 S subunit protein S8 restores a wild type phenotype to strains carrying mutations at one or a limited number of genetic loci. Many other ribosomal suppressors share this property (for a review see Nomura *et al.*, 1977). For example, out of 120 streptomycin-independent strains derived from one streptomycin-dependent mutant characterized by an altered protein S8, 13 show changes in 30 S proteins and 15 show changes in 50 S proteins (Dabbs and Wittmann, 1976). Further efforts to isolate suppressors that compensate for structural and functional changes in the translational machinery should contribute useful information on the molecular architecture of the ribosome.

In contrast to the cistron-specific ribosomal suppressors, the *strA, ramA,* and *neaA* mutations influence the accuracy with which many genes are translated. Appropriate use of nonsense suppressors should facilitate a search for additional mutants of this type that alter other ribosomal proteins or the polypeptide chain release factors. As discussed earlier, Colby *et al.* (1976) have isolated a mutant defective in tRNA modification by a strategy that identifies strains unable to synthesize a functional T4-coded ochre suppressor tRNA. Elseviers and Gorini (1975) have employed a similar selection scheme to look for novel mutations that restrict translational ambiguity. Although the molecular alteration in the restrictive mutant they obtained has not been identified or clearly established as ribosomal, genetic mapping has placed the mutation near the isoleucine–valine operon, in the vicinity of several rRNA operons (Nomura *et al.*, 1977). In view of recent work suggesting that specific RNA · RNA interactions may mediate several functions on the ribosome, it may be possible to produce mutants in ribosomal RNA that change its structure to allow increased or decreased translational fidelity.

3.6 Polarity and the Coupling of Transcription to Translation in Bacteria

The polar effects of nonsense mutations on distal gene expression within a transcriptional unit have received much attention because polarity appears in some way to reflect an obligate coupling of transcription to translation in prokaryotic cells. During the last 2 years, experimental evidence from investigations of polarity, of the properties of extragenic polarity suppressors, and of several unrelated areas has begun to yield a clearer description of this phenomenon (for recent summary articles, see Adhya *et al.,* 1976; Franklin and Yanofsky, 1976; Galluppi *et al.,* 1976; Ratner, 1976*b*).

A major feature of polarity which must be considered is whether reduced yields of the distal polypeptides of an operon are a direct result of altered controls at the level of translation or simply a consequence of the fact that distal mRNA templates are not present for translation. Clear evidence for the *E. coli* tryptophan operon indicates that the limitation in gene expression distal to a polar mutation derives from the markedly reduced levels of the relevant messenger RNA sequences (Morse and Yanofsky, 1969; Hiraga and Yanofsky, 1972; Segawa and Imamoto, 1974). The results of hybridizations of *trp* mRNA to discrete segments of the *trp* operon incorporated into transducing phage genomes suggest that mRNA beyond the position of a nonsense mutation is present but hyperlabile up to the next gene boundary; beyond that, little mRNA can be detected. The loss of mRNA sequences as a result of premature translation termination at polar sites was at first interpreted as evidence of either mRNA degradation (Morse and Yanofsky, 1969; Morse and Primakoff, 1970; Morse and Guertin, 1972) or obligate coupling of transcription and translation (Segawa and Imamoto, 1974). It was argued that the *suA* mutation, the original polarity suppressor (Beckwith, 1963), restored downstream gene function either by promoting continued ribosome movement on the polycistronic messenger RNA in the absence of protein synthesis, or by removing from the cell an endonuclease responsible for mRNA cleavage and subsequent degradation.

A series of studies designed to determine the means by which the N protein of bacteriophage λ positively regulates early λ transcription has recently converged on the problem of mutational polarity and offers an alternate proposal for its mechanism. Roberts (1969) first showed that an *E. coli* protein factor, rho, which did not act as a ribonuclease, caused specific termination of λ transcription *in vitro*. Subsequent evidence has supported his suggestion that the λ *N* gene product, in contrast to rho, acts as an antiterminator to extend early λ transcription beyond the *N* and *cro* genes, most likely by altering the capacity of RNA polymerase to stop at termination sites (for a review see Franklin and Yanofsky, 1976). The intriguing observation that the *N* gene product additionally functions to relieve the polar effects of nonsense mutations and of some DNA insertions in the *trp* and *gal* operons, when transcription of these units is initiated at early λ promoters (Adhya *et al.,* 1974; Franklin, 1974), suggests that transcription termination at the end of a gene, rather than mRNA degradation, is the cause of the reduction in mRNA distal to a polar mutation. *E. coli* mutants of at least one class that affects *N* function or bypasses the need for its activity (Inoko and Imai, 1976; Inoko *et al.,* 1977; Brunel and Davison, 1975) have been shown to possess altered rho termination factor activity (Inoko and Imai, 1976; Inoko *et al.,* 1977). Another class represents mutations in the *β* subunit of *E. coli* RNA polymerase (Georgopoulos, 1971; Ghysen and Pironio, 1972). The properties of these mutants are consistent with the suggestion that it is the absence of transcription termination which maintains distal mRNA levels beyond the position of a polar mutation.

Richardson *et al.* (1975) have presented evidence that the *suA* mutation initially characterized as a polarity suppressor leads to altered rho factor activity. They find that although the rho protein purified from one *suA* strain appears in normal amounts, it does not terminate transcription at sites recognized by wild type rho; moreover, it exhibits markedly reduced RNA-dependent ATPase activity, which has been shown to be required for the termination of RNA synthesis by rho (Galluppi *et al.,* 1976). The *suA* allele, as well as several more recently isolated polarity suppressors shown to affect rho (Korn and Yanofsky, 1976*a,b;* Das *et al.,* 1976), has been located at minute 83 on the map of the *E. coli* genome (Bachmann *et al.,* 1976) near the isoleucine–valine (*ilv*) operon. Independent genetic evidence obtained by exploiting the distinct electrophoretic mobilities of rho factor proteins from *E. coli* strains K12 and B has placed the *rho* gene in this same region (Ratner, 1976*a*). The temperature sensitivity and pleiotropic effects on cellular functions exhibited by the polarity suppressors of Das *et al.* (1976), one of which has also been identified as a *rho* mutation, suggest that rho has an essential function in *E. coli.*

Polarity suppressors have some capacity to obviate the requirement for λ *N* gene function. Conversely, other *E. coli* mutations, obtained by requiring expression of *N*-dependent functions in the absence of *N* protein, have map positions tightly linked to the locus of polarity suppressors and produce rho proteins with reduced stability and activity (Inoko and Imai, 1976; Inoko *et al.,* 1977). However, not all *ilv*-linked bacterial mutations selected to alter or bypass λ *N* gene function act as polarity suppressors (Brunel and Davison, 1975). Present evidence does not rigorously exclude the possibility that the various mutations discussed define several distinct cistrons in the *ilv* region of the *E. coli* genome. However, it seems reasonable to proceed at this time with the hypothesis that genetic selections applied to the *rho* gene may indeed yield separate classes of altered proteins, each representing a limited subset of rho's essential functions in the cell.

Polarity suppressors have been shown to affect transcription termination in an additional context. Korn and Yanofsky (1976*b*) have observed that some of the suppressors selected to relieve mutational polarity in both the *trp* and *lac* operons of *E. coli* simultaneously increase wild type *trp* mRNA synthesis. The region of the *trp* operon important for the action of the polarity suppressors has been identified as the *trp* attenuator, a segment of the untranslated *trp* mRNA leader sequence which previous work (see Chapter 7 of this volume) has shown to function as a strong natural transcription termination site. These findings indicate that, to some degree, rho acts to promote transcription termination at the *trp* attenuator *in vivo.* Regulation of this termination event has been implicated as an important means of controlling *trp* operon enzyme production and is apparently mediated by $tRNA^{Trp}$ (Yanofsky and Soll, 1977).

Collectively, present evidence strongly suggests that mutational polarity is brought about by transcription termination provoked by premature polypeptide chain termination. Just why the absence of translation affects the continuity of transcription is not understood. The mechanism of rho-mediated transcription termination itself is not yet clear. Likewise, although common sequence patterns and projected secondary structures have been noted at the 3′-OH termini of several primary transcripts, the specific requirements for rho-dependent termination have not been identified. However, the availability of both well-characterized termination sequences and a variety of mutationally altered rho proteins should greatly accelerate progress in solving these problems.

The occurrence of polarity as it is described here should be restricted to prokaryotic organisms, in which translation of a messenger RNA accompanies its transcription. In eukaryotes, cellular mRNAs are translated in the cytoplasm as monocistronic units after transport out of the nucleus. Although "polar" mutations have been described, the mechanisms operating to reduce distal expression are most likely not equivalent (Bigelis *et al.*, 1977); one would therefore not expect to find extragenic polarity suppressors of the bacterial type in eukaryotic cells.

3.7 Genetics of tRNA

The genome of *E. coli* includes about 60 tRNA genes (Brenner *et al.*, 1970), about twice the minimum number needed to translate the 61 mRNA sense codons if wobble base pairing (Crick, 1966) is permitted. Genetic analysis of *E. coli* tRNAs initially proceeded at a slower pace than studies of structural genes encoding polypeptides. This stems partly from the fact that tRNAs comprise a class of macromolecules with essential roles in protein biosynthesis and possibly other cellular functions. Mutations in some tRNA structural genes, therefore, are presumably lethal. In addition, mutations in other tRNA genes may not result in an observable phenotype if the decoding capacity they exhibit is duplicated in the cell. The phenotype brought about by a suppressor mutation, which changes the codon response of a tRNA species, thus serves as a crucial basis for genetic selections, and has permitted straightforward analysis of several *E. coli* and bacteriophage T4 tRNA genes.

The genes for six *E. coli* tRNA species were first detected as nonsense suppressors. The map locations of the genes for species of $tRNA^{Ser}$ ($su^{+}1$), $tRNA^{Gln}$ ($su^{+}2$), $tRNA^{Tyr}$ ($su^{+}3$), and $tRNA^{Trp}$ (the UGA and $su^{+}7$ suppressors) have been identified on this basis. It is likely that $su^{+}5$ and $su^{+}6$ specify altered forms of $tRNA^{Lys}$ (Garen, 1968) and $tRNA^{Leu}$, respectively (Gopinathan and Garen, 1970; Hayashi and Söll, 1971). The isolation of missense suppressors has generated valuable information about the four genetic loci that specify the family of glycine tRNA isoacceptors (Hill, 1975). All of the other genetic assignments have come from fortuitous discoveries of tRNA genes tightly linked either to suppressor loci or to the genes encoding ribosomal RNAs. For example, the $tRNA_3^{Thr}$ gene (*thrT*) is carried on a λ transducing phage selected to include the *sup*A36 missense suppressor (Squires *et al.,* 1973). Using rRNA transducing phages and plasmids, several groups have recently shown that the genes for $tRNA_2^{Glu}$, $tRNA_1^{Ile}$, and $tRNA_{1B}^{Ala}$ are located in the spacer regions between the cistrons for 16 S and 23 S rRNAs (Wu and Davidson, 1975; Lund *et al.,* 1976; Lund and Dahlberg, 1977; Morgan *et al.,* 1977). Finally, a gene for $tRNA_1^{Asp}$ has been found near the distal ends of some rRNA operons (Morgan *et al.,* 1977).

Taken together, information from all of these approaches provides map assignments for approximately one-third of the tRNA genes estimated for *E. coli* (Fig. 2). We need to acquire much more data for a complete description of the arrangement of tRNA genes in the bacterial genome. The isolation of novel suppressors would certainly make an important contribution to this effort. On the

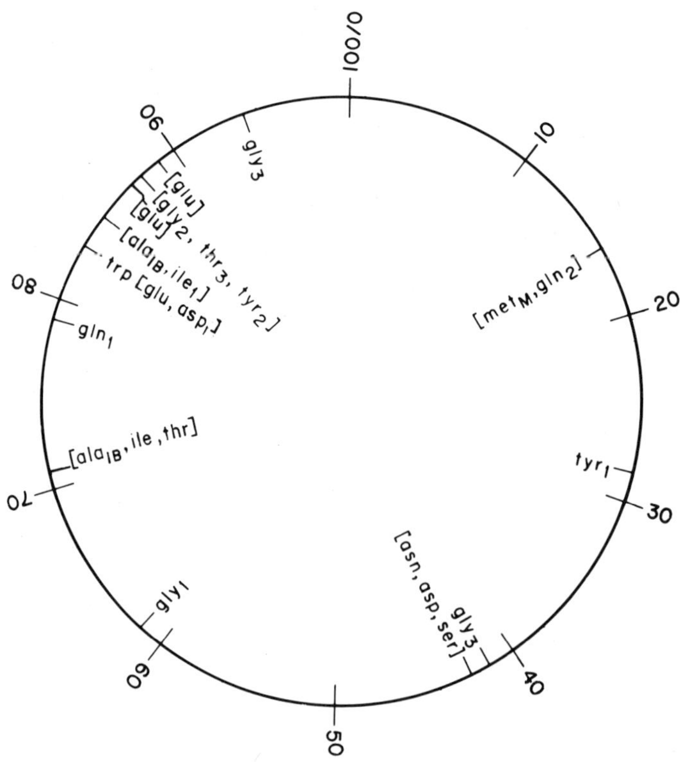

Figure 2. Map locations of tRNA genes identified in *E. coli.* Numbers and letters in subscripts refer to tRNA isoacceptor assignments. Bracketed assignments denote linkages to non-tRNA markers established by other techniques (hybridization to DNA of restriction fragments or of transducing phages).

basis of present evidence, however, we can rule out the possibility that tight spatial regulation serves to regulate coordinately the expression of all tRNA species or even a subset specific for a single amino acid. Nevertheless, the finding of several tRNA genes within the short segments of the bacterial genome included in transducing bacteriophages and the isolation of multimeric tRNA precursors (see later) suggest that some tRNA gene clustering does occur.

The products of all eight bacteriophage T4 genes which code for tRNA species have been identified (McClain *et al.,* 1972). These include tRNAs specific for the amino acids arginine, glutamine, glycine, isoleucine, leucine, proline, serine, and threonine. Single base changes in the anticodons of four of these tRNAs, namely $tRNA^{Arg}$, $tRNA^{Gln}$, $tRNA^{Leu}$, and $tRNA^{Ser}$, have generated nonsense suppressors for amber, ochre, and UGA codons (see Kao and McClain, 1977). In contrast to the situation in *E. coli,* the T4 genes are tightly linked; it is reasonable to suggest, on the basis of currently available data, that these genes may be transcribed as a single unit (McClain, 1977).

A wide variety of problems have been explored via the genetic analysis of suppressors (Smith, 1972) in bacteriophage T4, *E. coli,* and, as is discussed in a later section, in yeast. Mutations in tRNAs that alter the specificity of their interactions with aminoacyl-tRNA synthetases and lead to charging of a tRNA with a novel amino acid provide some insight into the basis for one important example of protein · nucleic acid recognition. Both intra- and extragenic mutations selected to affect the expression of the suppressor phenotype have been indispensable aids in efforts to define the steps of tRNA biosynthesis (Smith, 1976; McClain, 1977; Altman, 1978). The availability of temperature-sensitive nonsense suppressors derived from the *E. coli* $su^{+}1$ gene (Nagata and Horiuchi, 1973; Oeschger and Woods, 1976) enables one to manipulate the levels of individual proteins in intact cells. Finally, the recent use of point mutations within the gene for the T4-specified glutamine tRNA for fine structure genetic mapping is of special interest in this context. Comer (1977) has determined the recombination frequencies between pairs of the following mutations: the primary ochre suppressing mutation and a secondary anticodon change that converts the ochre suppressor to an amber suppressor, and three mutations that inactivate the tRNA. The nucleotide alterations responsible for these mutations have all been localized in the known $tRNA^{Gln}$ sequence. The genetic map intervals obtained for these point mutations, therefore, can be related directly to the actual distance in nucleotides between mutations, and yield an average recombination frequency of 0.014 per nucleotide. The clustered T4 tRNA genes provide an excellent system for further examination of genetic recombination events in regions of known sequence.

3.8 Biosynthesis of tRNAs

Initial transcription of tRNA genes to give an RNA product is followed by a series of enzymatic steps (Fig. 3) in which the RNA transcript is trimmed by endo- and exonucleases to a chain length characteristic of the mature tRNA species, and specific nucleosides in the RNA sequence become modified (Smith, 1976; McClain, 1977; Altman, 1978). The RNA precursors for some bacterial tRNAs contain the sequences for single tRNAs, while other RNA transcripts include the sequences corresponding to several tRNAs (Ilgen *et al.,* 1976; Sakano and Shimura, 1975). Present evidence suggests that approximately half of the *E. coli* tRNA species originate from multimeric precursors (Y. Shimura, personal communication).

Although the basic outlines of tRNA biosynthesis are clear at this time, a detailed description of the enzymes involved and the precise nature of the intermediates cannot yet be formulated. Elucidation of the individual steps in the biosynthetic pathway, however, depends heavily on the availability of the tRNA precursors. This is normally a serious experimental problem, since the half-lives of bacterial tRNA precursors are so short that purification of individual species in amounts sufficient for their characterization is difficult. Altman and Smith (1971) exploited the phenotype of the $su^{+}3$ nonsense suppressor in two ways in order to increase the transient accumulation of a precursor for the $su^{+}3$ $tRNA^{Tyr}$ sequence, and thereby achieved the first isolation of a specific bacterial tRNA precursor. First, the authors employed second-site suppressor mutants which exhibited reduced suppressor activity because less mature tRNA was synthesized; the presence of mutations with the prescursor sequence appeared to interfere with the processing steps. Second, they made use of a $\phi 80$ transducing bacteriophage selected to carry the $su^{+}3$ gene (Smith *et al.,* 1966) to amplify the yield of the tyrosine tRNA precursor from phage-infected cells. The monomeric $su^{+}3$ $tRNA^{Tyr}$ precursor thus obtained has been instrumental in subsequent efforts to explore the way in which tRNA mutations alter the efficiency of the processing steps, and as a substrate in the purification and characterization of *E. coli* (Bothwell *et al.,* 1976; Schmidt *et al.,* 1976) and eukaryotic (Koski *et al.,* 1976) enzymes involved in tRNA processing.

Another important approach to the problem of tRNA biosynthesis has also taken advantage of the $\phi 80psu^{+}3$ tranducing phage. Schedl and Primakoff (1973)

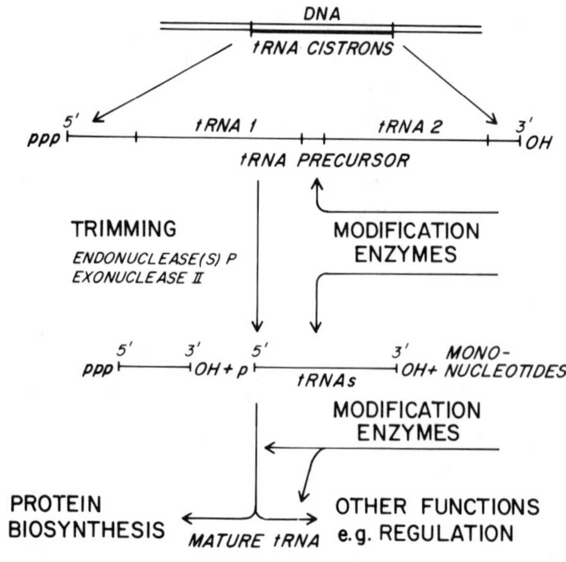

Figure 3. Scheme of the steps involved in the biosynthesis of tRNA.

devised a selection method for isolating bacterial mutants which, when infected with ϕ80p*su*$^+$3, do not express the amber suppressor at 42°C. By retaining only those mutants which were capable of both suppressing a nonsense mutation in the gene for β-galactosidase at low temperature and synthesizing active wild type β-galactosidase encoded by the wild type ϕ80p*lac*, Schedl and Primakoff selected specifically for mutants blocked in the synthesis of functional *su*$^+$3 tRNA^{Tyr} at the restrictive temperature. Among the many mutants obtained by this and an analogous strategy (Sakano *et al.*, 1974), some have been shown to be defective in the steps of tRNA transcription and processing. Other uncharacterized mutants may be strains unable to form some of the modified nucleosides normally contained in the suppressor tRNA. Apart from extensively studied mutations that lead to temperature sensitivity in RNase P, the enzyme that removes the extra segment from the 5′ end of the *su*$^+$3 tRNA precursor (Schedl and Primakoff, 1973; Sakano *et al.*, 1974; Ikemura *et al.*, 1975), no rigorous genetic or biochemical characterization of any of the other mutants (Sakano and Shimura, 1975) has been carried out.

In the case of bacteriophage T4, the availability of nucleotide sequence information and nonsense suppressing derivatives of several tRNAs has led to the development of a beautiful experimental system for elucidating the details of tRNA biosynthesis (for a review see McClain, 1977). Following T4 infection, phage DNA is transcribed to yield several precursors which are then processed by host enzymes. The shutoff of *E. coli* DNA transcription in T4-infected cells facilitates selective ^{32}P labeling of the tRNAs and their precursors. From the nature of the intermediates accumulated in the *E. coli* mutant strains discussed previously, McClain and his co-workers have been able to deduce the order of steps which complete the synthesis of mature tRNA^{Pro} and tRNA^{Ser} from a dimeric Pro–Ser precursor. Further, in the course of studying the consequences on various stages of processing of second-site mutations in the tRNA genes that result in the loss of suppressor function, these authors have observed that the major reductions in nucleotide modifications of precursor RNAs are confined to the tRNA sequence bearing the mutant nucleotide (McClain and Siedman, 1975). In earlier work of others (Kuchino *et al.*, 1971; Shershneva *et al.*, 1971), the effects of tRNA fragmentation of the specificity of methylation had demonstrated that the conformation of mature tRNA is of crucial importance for proper nucleoside modification. These and other studies led McClain and Seidman (1975) to conclude from the pattern of nucleoside modifications exhibited by the mutant Pro–Ser tRNA precursors that each tRNA component serves independently as a substrate for modification. This implies that during the processing steps, the precursor exhibits a tertiary conformation dominated by interactions between residues of the individual tRNA sequences, which may closely resemble their final "native" structures.

3.9 tRNA Structure–Function Relationships: Mischarging Suppressor tRNAs

Both the ease with which secondary mutations within the *su*$^+$3 gene can be isolated from mutagenized stocks of ϕ80p*su*$^+$3 and the relative simplicity of obtaining radiochemically pure preparations of the altered *su*3 tRNAs from ϕ80p*su*3-infected cells encouraged Smith, Brenner, and their colleagues to undertake the first extensive biochemical and genetic study of a tRNA. The essential question they have attempted to address is the following: What nucleotide substitutions in the mature tRNA sequence lead to loss of charging ability or to reduced functioning of mutant tRNAs in protein synthesis? With this general aim, these authors have selected mutants in which *su*3 suppressor activity is reduced or abolished. Subsequent nucleotide sequence analysis of the mutant tRNAs has identified those base changes which lead to a decrease in the ability of *su*3 tRNA to be aminoacylated, to a loss of its stable tertiary structure, or to inefficient interaction with the components of the translational apparatus (for a review see Smith, 1972).

An extension of this early work has contributed importantly to our present understanding of the high degree of specificity displayed in aminoacyl-tRNA synthetase recognition of appropriate tRNA species (for discussion see Söll and Schimmel, 1974; Rich and Schimmel, 1977). Smith and his colleagues (Hooper *et al.*, 1972) and Ozeki and his colleagues (Shimura *et al.*, 1972) have manipulated nonsense suppressor tRNAs in an ingenious way to identify base changes in the mature tRNA sequence that result in the aminoacylation of these mutant tRNAs with novel amino acids. These authors have made use of the fact that the phenotype produced by nonsense mutations corresponding to certain amino acid positions in a polypeptide can be reversed only by suppressors that insert amino acids compatible with the function of that gene product. Nonsense mutations in several *E. coli* cistrons, for example, can be suppressed by su^+2 (glutamine), but not by su^+3 (tyrosine). Thus, in cells carrying only the su^+3 suppressor, by requiring wild type phenotypes at several such nonsense mutant loci simultaneously, one has in effect a powerful genetic selection scheme for obtaining variants of the su^+3 tRNA which accept other amino acids.

In this way, several mutations in the amino acid acceptor stem of the su^+3 tRNA that cause this molecule to be mischarged with glutamine have been identified (Hooper *et al.*, 1972; Shimura *et al.*, 1972). The locations of these mutations in the *E. coli* su^+3 tRNA sequence are indicated in Fig. 4. The *E. coli* $tRNA^{Gln}$ primary structure is shown also, in order to point out that the two sequences differ significantly. Some of the su^+3 mischarging tRNAs have hybrid specificity, and are aminoacylated *in vivo* and *in vitro* with both glutamine and tyrosine (Celis *et al.*, 1973).

Since all the mutants characterized thus far have been shown to cause mischarging with glutamine, it is conceivable that the consistent isolation of glutamine-inserting suppressors may have the trivial explanation that there is an absolute requirement for that specific amino acid at the nonsense mutant sites employed in genetic selection schemes. Alternatively, it may reflect the possibility that there are only a few aminoacyl-tRNA synthetases for which the correct recognition properties can be generated by a single base change in $tRNA^{Tyr}$. Recently, several λ transducing phages carrying the su^+1 and su^+2 genes have been isolated (Steege and Low, 1975; Inokuchi *et al.*, 1975). Selection of mischarging derivatives of these two tRNA species may yield mutants with novel amino acid specificities. There is some evidence from work with suppressors of *E. coli* tryptophan synthetase *A* gene missense mutations that mischarging mutants originating from the su^+1 gene may insert glycine or alanine (E. Murgola, personal communication). In any case, characterization of several distinct mischarged tRNAs, combined with the investigation of the solution structure of these molecules, should identify the nucleotide residues and structural features involved in the specific interaction of aminoacyl-tRNA synthetases with their cognate tRNAs. Moreover, if the selection strategy used to obtain mischarging tRNA mutants is modified slightly so that it can be used to screen bacterial mutants, one would anticipate finding extragenic suppressor mutations in aminoacyl-tRNA synthetases that also lead to altered amino acid specificity. Mutations of this type might be viable only in diploid strains, since in bacteria there is one synthetase (specified by a single gene) for each amino acid (Söll and Schimmel, 1974).

Another very intriguing mischarging mutant has appeared unexpectedly from characterization of the recessive lethal *E. coli* glutamine-inserting $su^{+}7$ amber suppressor (Soll and Berg, 1969; Soll, 1974). The suppressor has been shown to be an altered tryptophan tRNA, with a CUA anticodon sequence comple-

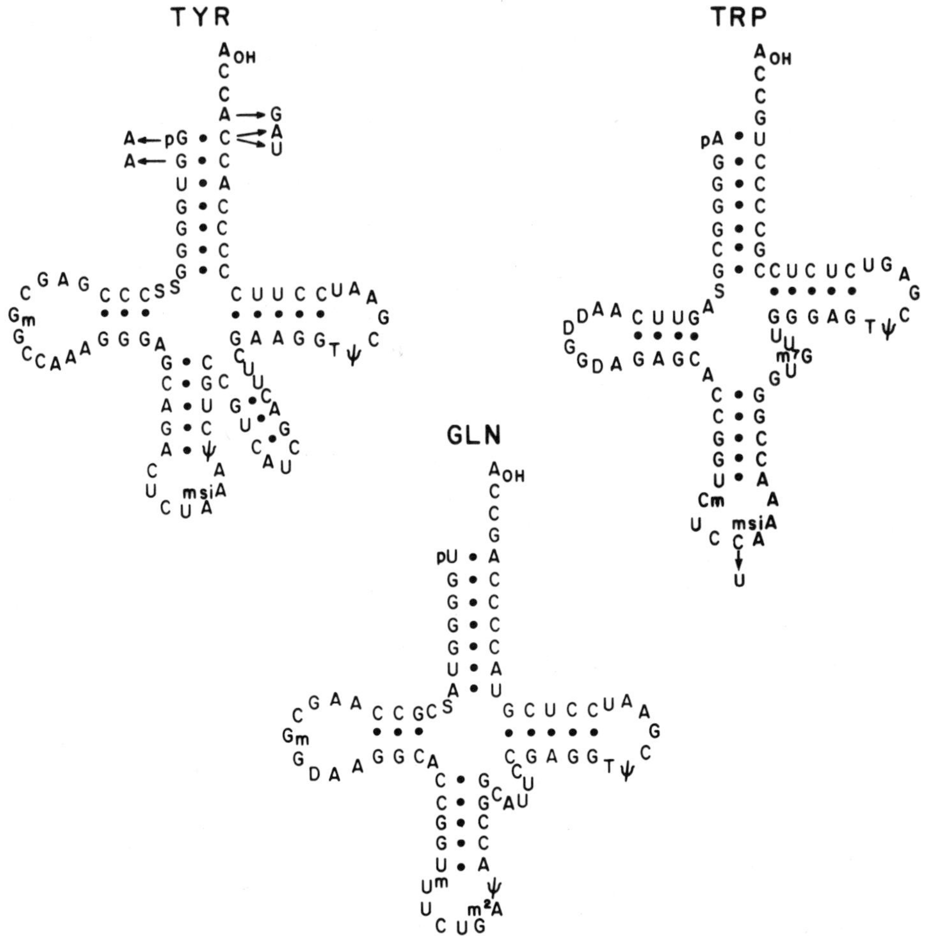

Figure 4. Summary of tRNA mischarging mutants. The nucleotide sequences of *E. coli* $tRNA^{Tyr}$, $tRNA^{Trp}$, and $tRNA^{Gln}$ are shown. The arrows indicate the nucleotide substitutions in certain positions of $tRNA^{Tyr}$ and $tRNA^{Trp}$ that allow these tRNAs to be charged with glutamine. (Data taken from Smith and Celis, 1973; Shimura *et al.*, 1972; Yaniv *et al.*, 1974.)

mentary to the UAG codon (Yaniv *et al.,* 1974). Thus, the single nucleotide change in the cell's unique structural gene for $tRNA^{Trp}$ leads simultaneously to the ability of the mutant tRNA to serve as an amber suppressor and as a substrate for glutaminyl-tRNA synthetase. From the specific patterns of suppression observed for su^+7 at several amber sites in the *E. coli lacI* gene and from fingerprint analysis of suppressed T4 phage head proteins, Celis *et al.* (1976) have shown that su^+7 inserts glutamine and tryptophan in a ratio of 9:1. These results demonstrate that tryptophanyl-tRNA synthetase can still recognize su^+7 $tRNA^{Trp}$, albeit with a lowered efficiency. The properties of the su^+7 suppressor indicate that the anticodon of this tRNA molecule serves as one recognition element for glutaminyl-tRNA synthetase, and that the tRNA structure presents recognition sites for both synthetase enzymes. Since there is no obvious similarity between the $tRNA^{Trp}$ and $tRNA^{Gln}$ sequences (Fig. 4), analysis of additional mutants is needed to define this recognition system further.

4 *Nonsense Mutations in the Escherichia coli lacI Gene*

The lactose repressor protein, encoded by the *lacI* gene, exerts negative control over the expression of the lactose operon of *E. coli.* In the early 1960s, when the model of Jacob and Monod for the regulation of operon expression was being tested, a view generally held was that repressors would prove to be RNA molecules. The isolation of *I* gene mutations leading to temperature-sensitive repressor function (Sadler and Novick, 1965) and, more convincingly, the characterization of nonsense mutations in the *lacI* gene (Bourgeois *et al.*, 1965; Müller-Hill, 1966) provided strong support for the idea that the *lac* repressor is a protein.

Extensive genetic analysis of the *I* gene and biochemical characterization of the repressor protein have led to a basic understanding of the molecular mechanism of repression (for a review see Müller-Hill, 1975). With the amino acid sequence of the repressor protein (Beyreuther *et al.*, 1973; Platt *et al.*, 1973) and the DNA sequence of the *lac* operator (Gilbert and Maxam, 1973) available, current work is directed toward a detailed description of the repressor–operator interaction and eventual elucidation of the three-dimensional crystal structure of the repressor protein.

The sites of 90 nonsense mutations in the *lacI* gene have been characterized with respect to the base changes required to generate amber, ochre, and UGA codons (Miller *et al.,* 1977; Coulondre and Miller, 1977*b*). Many of the intervals in the genetic map have been correlated with known segments of the repressor protein sequence, by mapping a large set of point mutations against more than 400 deletions representing an average endpoint separation of less than three nucleotides (Schmeissner *et al.,* 1977). The sites of distinct nonsense mutations within the same deletion interval have been differentiated by their characteristic patterns of response to a number of nonsense suppressors. Thus, virtually all of the nonsense mutations have been correlated with wild type codons for specific amino acid residues of the lactose repressor protein (Fig. 5). The *lacI* gene nucleotide sequence (Steege, 1977; P. Farabaugh, 1978) has verified most of these assignments.

Although nonsense mutants represent only one of several types of *I* gene mutants that have been useful for studies of the *lac* repressor, they are of special value for certain problems. For example, the possibility of using the *I* gene to obtain information concerning the specificity of translation initiation was raised by the discovery that nonsense mutations early in the gene activate reinitiation of

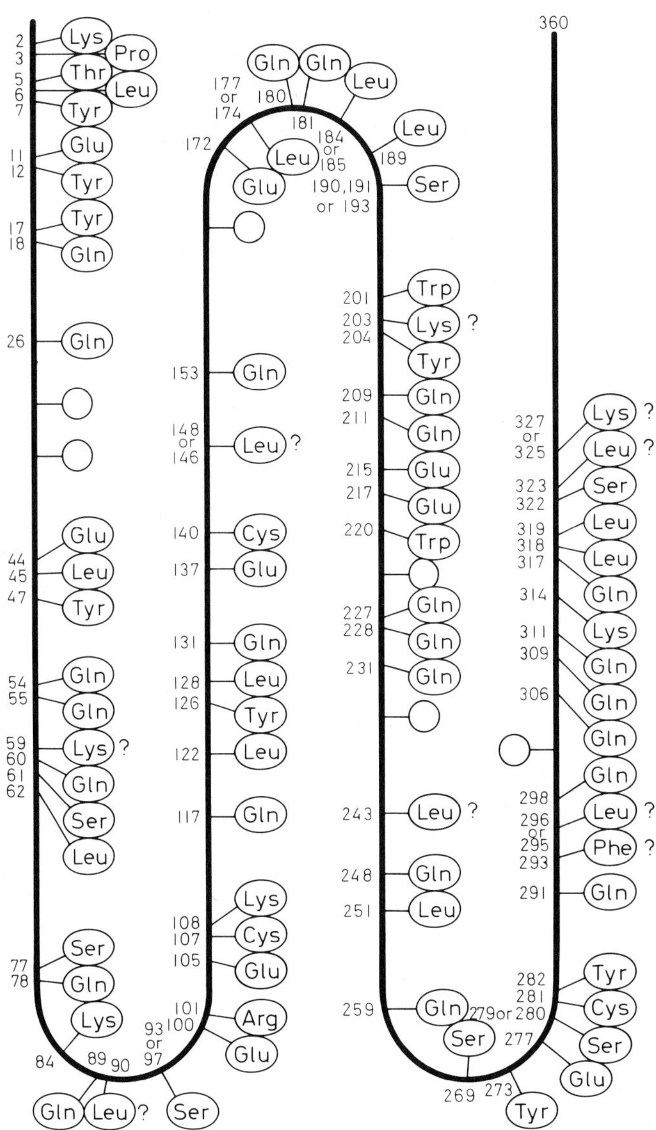

Figure 5. Schematic diagram of the sites of 90 nonsense mutations within the *E. coli lacI* gene. Virtually all of the mutations have been correlated with wild type codons for specific residues in the lactose repressor protein by genetic methods (Miller *et al.*, 1977; Coulondre and Miller, 1977*a*). The nucleotide sequence of the *lacI* gene (Steege, 1977; P. J. Farabaugh, 1978) has verified most of these assignments. Numbers indicate amino acid positions in the wild type repressor protein (Beyreuther *et al.*, 1973; Platt *et al.*, 1973). (Courtesy of J. H. Miller.)

repressor synthesis beyond the position of the chain-terminating codon. At least three sites within that portion of the mRNA corresponding to the first 62 amino acids of the wild type repressor protein direct the synthesis of carboxy-terminal restart polypeptides (Files *et al.,* 1974). To examine the nucleotide sequences that signal these reinitiation events, the *I* gene mRNA sequence in this region was recently determined (Steege, 1977) (for a discussion, see Chapter 9 of this volume). Genetic information indicating whether an amber or ochre codon is generated at a given amino acid position and data obtained from isolating nonsense mutations with mutagens of known specificity were particularly useful in the initial purification of *I* gene mRNA fragments; they predicted several unique oligonucleotide sequences for the region encoding the first 20 amino acids of the repressor protein (Miller *et al.,* 1977; Coulondre and Miller, 1977*a,b*).

The relative ease with which the sites of nonsense mutations can now be correlated with the base changes that generate them has enabled Coulondre and Miller (1977*b*) to examine the results of more than 4000 forward mutational events at many different sites in the *lacI* gene. These authors have compared the relative frequencies of mutations representing known base changes at each of many sites. This capability is not generally provided by other genetic loci used to test the specificity of chemical mutagens, since with these systems reversion events at a relatively small number of mutant sites are tabulated. Coulondre and Miller (1977*b*) have determined the mutagenic specificity of *N′*-methyl-*N′*-nitro-*N*-nitrosoguanidine, ethyl methanesulfonate, 4-nitroquinoline-1-oxide, 2-aminopurine, and ultraviolet light. In addition, they have analyzed the spectrum of nonsense codons produced by spontaneous mutations, and have established the preferred sites, or "hot spots," for several of these mutagens. Since the nucleotide sequences in the vicinity of the hot spots are known, the reasons for a high mutation rate at certain sites may soon emerge.

4.1 *Suppression of Nonsense Mutations Generates Altered lac Repressor Molecules*

A direct approach toward identifying the features of the *lac* repressor protein structure responsible for its self-association to form tetramers and for its specific interaction with operator DNA and with sugar derivatives that serve as inducers of the *lac* operon is to examine the effects of amino acid changes on repressor functions. Suppression permits the introduction of a defined amino acid into the repressor polypeptide at the site of a nonsense codon, and has enabled Miller and his colleagues to generate more than 300 altered repressors of known sequence. In this way, they have shown that, for example, a substitution for tyrosine at some positions with a different amino acid does not perturb repressor activity, whereas at other sites, the presence of tyrosine is specifically required (Miller *et al.,* 1975). Substitution of tyrosine 269 (residue 282 in the revised amino acid sequence shown in Fig. 5) (Coulondre and Miller, 1977*a*) with serine, via $su^{+}1$ suppression of the appropriate amber mutant, yields a repressor unable to form its normal tetrameric structure (Schmitz *et al.,* 1976). Sommer *et al.* (1976) have prepared two altered *lac* repressor proteins by growing mutants carrying amber mutations at the codons specifying tryptophans 190 and 209 (residues 201 and 220 in the revised amino acid sequence, Fig. 5) in su^+3 (tyrosine) suppressor strains. From studies of the fluorescence emission properties exhibited by the wild type and altered repressors, Sommer and co-workers propose that the tryptophan residue 209 does not make direct contact with inducer molecules, but may be involved in a conformational change of the protein when inducer is bound. Finally, Lu *et al.* (1976) have isolated functional *lac* repressor with tyrosine residues substituted with the analog 3-fluorotyrosine, and have examined the fluorine-19 nuclear magnetic resonance spectrum of the purified protein. In the spectrum, they see resonance peaks corresponding to the eight tyrosine residues of the repressor protein sequence. Because a nonsense mutation corresponding to each position has been identified, isolation of *lac* repressor from appropriate suppressed mutants will permit the straightforward assignment of resonance peaks to specific residues. This provides a powerful approach for probing the solution structure and dynamics of the *lac* repressor molecule.

5 Current Developments in Eukaryotic Suppression

5.1 tRNA-Mediated Suppression in Yeast

One would anticipate that of the 340 to 400 tRNA genes thought to be contained in the genome of the yeast *Saccharomyces cerevisiae* (Schweizer *et al.*, 1968), many specifying redundant isoaccepting species could be converted, via mutation, into nonsense, frameshift, or missense suppressors. For example, there are at least eight genes for tyrosine tRNA (Olson *et al.*, 1977) which generate mature molecules of identical nucleotide sequence. In fact, amber and ochre suppressing alleles for each of these genes have been selected (Hawthorne, 1976). The genetic analysis of yeast suppression is well underway (for a comprehensive review see Hawthorne and Leupold, 1974). The two species studied most extensively, *Saccharomyces cerevisiae* and *Schizosaccharomyces pombe*, can be manipulated in both the haploid and diploid states and thus are particularly amenable to fine structure genetic analysis. The availability of nonsense mutations at numerous *S. cerevisiae* loci, for example *cyc*1 (iso-1-cytochrome *c*), *his*4 (phosphoribosyl-AMP cyclohydrolase, phosphoribosyl-AMP pyrophosphorylase, and histidinol dehydrogenase), *arg*4 (argininosuccinate lyase), and *trp*1 [*N*-(5′-phosphoribosyl) anthranilate isomerase], has facilitated the selection of extragenic suppressors.

In general, many of the properties of nonsense suppression in yeast parallel those found for bacterial cells. In *S. cerevisiae,* the isolation of suppressors that simultaneously reverse the phenotypes of mutations in as many as five different loci first raised the possibility that nonsense suppression occurs in yeast. UAG, UAA, and UGA have since been identified as nonsense codons. Eight independent dominant suppressors (*SUP*2, *SUP*3, *SUP*4, *SUP*5, *SUP*6, *SUP*7, *SUP*8, and *SUP*11) have been shown to insert tyrosine at positions of the iso-1-cytochrome *c* protein sequence corresponding to sites of appropriate nonsense mutations (Gilmore *et al.*, 1971; Sherman *et al.*, 1973; Liebman *et al.*, 1976). *SUPQ*5 defines the gene for one serine-inserting suppressor which has been examined in both its amber (Capecchi *et al.*, 1975) and ochre allele (Liebman *et al.*, 1975). A different serine-inserting amber suppressor causes lethality in haploid strains (Brandriss *et al.*, 1976). Finally, the *SUP*52 amber suppressor has been shown to insert leucine (Liebman *et al.*, 1977).

In *S. pombe,* four dominant nonsense suppressors (*sup3, sup8, sup9,* and *sup10*) have been isolated (Hawthorne and Leupold, 1974). *sup3* and *sup8* each exist in two allelic forms that define two classes of suppressors. The efficient form (for example, *sup3*-e) suppresses one series of nonsense mutations, whereas the inefficient one (for example, *sup3*-i) responds weakly to a different series of nonsense mutations. On the basis of the characteristic patterns of suppression, *sup3* and *sup9* are thought to insert the same amino acid. Both *sup8* and *sup10,* on the other hand, insert a different amino acid. The amino acid accepted by the purified *sup3* suppressor tRNA species is serine, and for the *sup8* suppressor species, leucine (J. Kohli and F. Altruda, personal communication). The codon specificity of *S. pombe* suppressors is presently not clearly established. Contrary to earlier predictions based on genetic data (Hawthorne and Leupold, 1974), the recent biochemical demonstration that tRNAs mediate *sup3*-e and *sup8*-e suppression has shown that these suppressors respond to the UGA codon *in vitro* (Kohli *et al.,* 1978). These results imply that the inefficient alleles of *sup3* and *sup8* may be UAA suppressors, since genetic studies suggest that the inefficient and efficient suppressor alleles are related by single base changes (Hawthorne and Leupold, 1974). It is possible that the inefficient suppressors, like bacterial ochre suppressors, recognize both UAA and UAG codons, since an extensive search has not revealed a third distinct class of *S. pombe* suppressors.

Certain differences between bacterial and *S. cerevisiae* nonsense suppressors are apparent from the outset. As in bacteria, amber suppressors respond to UAG codons; ochre suppressors, however, recognize UAA uniquely. Moreover, it is the amber, not ochre, suppressors which frequently have deleterious effects on yeast cell growth (Liebman and Sherman, 1976). Finally, a function [PSI^+], encoded by a non-Mendelian genetic element localized in the cytoplasm, enhances the efficiencies of many ochre and frameshift suppressors encoded by the nuclear genome (Cox, 1965). In combination with some suppressors, this increase in efficiency leads to lethality. In [PSI^+] strains lacking any known suppressors, however, weak ochre suppression is observed. The molecular identities of [PSI] function and the genetic element encoding it are at present still mysteries (McCready *et al.,* 1977). The potential effects on suppressor function of interactions between nuclear and nonnuclear genomes point out one feature of genetic suppression which may be unique to eukaryotic cells.

The development of efficient eukaryotic cell-free protein-synthesizing systems in which amino acid incorportion is directed by exogenously added mRNA templates (Capecchi *et al.,* 1975; Gesteland *et al.,* 1976; Pelham and Jackson, 1976) has led recently to convincing demonstrations that suppression of UAG, UAA, and UGA codons in *S. cerevisiae* and *S. pombe* is mediated in several cases by tRNA. With a complex mammalian cell-free system programmed with Qβ bacteriophage RNA containing an amber mutation in either the synthetase or coat protein gene, one can assay added tRNA samples for amber suppressor activity (Capecchi *et al.,* 1975; Gesteland *et al.,* 1976). In addition, the capacity of these cell-free systems to extend polypeptide chains beyond a wild type termination signal in the presence of appropriate suppressor tRNAs has made it possible to employ a variety of natural mRNA templates for *in vitro* translation. Readthrough of the QB synthetase cistron is enhanced by tRNA from ochre suppressing strains (Gesteland *et al.*, 1976), and extension of rabbit globin chains is effected by tRNA from both ochre and UGA suppressing strains (Kohli *et al.*, 1978) (see also Table I). Murine leukemia virus RNAs similarly serve as a test for amber suppressor tRNAs (Philipson *et al.*, 1978).

With the assurance that tRNA-mediated suppression occurs in yeast cells, efforts to identify the suppressing species have quickly ensued. From differences in the oligonucleotides obtained from the unfractionated tyrosine tRNAs of wild type and *SUP*5-a amber suppressor *S. cerevisiae* strains, Piper *et al.* (1976) have deduced that a single base change in the GψA anticodon of the wild type tRNA is responsible for generating the CψA anticodon of the suppressor. To date this is the only yeast suppressor for which a tRNA primary structure has been established by RNA sequencing methods. Sequence information for a second yeast suppressor has appeared very recently, however, as part of intensive efforts to identify many of the tRNA genes in yeast. Among many yeast DNA fragments cloned in *E. coli* plasmids, Goodman *et al.* (1977) have identified four encoding the tyrosine ochre suppressor tRNA *SUP*4-o, the corresponding wild type tRNA, and two other $tRNA^{Tyr}$ species. When the sequences of these DNA fragments were determined, two important findings emerged. First, it is now clear that the wobble base in the ochre suppressor $tRNA^{Tyr}$ anticodon must be occupied by a U or a U derivative because the corresponding deoxyribonucleotide is T. The presence of a U provides strong support for the suggestion that a 2-thiouridine derivative in this position of the anticodon, which should lead to specific recognition of A (Yoshida *et al.*, 1970, 1971), gives the ochre suppressor coded for by this tyrosine tRNA gene its specificity for UAA. It remains to be seen whether, as previously proposed, other yeast ochre suppressors may utilize inosine in this position as an alternative means of generating the specificity for UAA codon (Hawthorne and Leupold, 1974). A second feature of the yeast tyrosine tRNA genes, which was entirely unexpected, is that all four genes studied contain a 14-nucleotide insert sequence adjacent to the base corresponding to the 3′ terminus of the anticodon. This insert is not present in the mature tRNA. Similar insert sequences, in the same location within the tRNA gene, have been found in the genes for $tRNA^{Phe}$ from *S. cerevisiae* (Valenzuela *et al.*, 1978). The relationship of these insert sequences to the much longer insertions in mammalian genes (Breathnach *et al.*, 1977; Jeffreys and Flavell, 1977) is not known.

Suppressors of frameshift mutations in *S. cerevisiae* have recently been reported (Culbertson *et al.*, 1977) and appear to have properties analogous to the bacterial frameshift suppressors. Two distinct classes of dominant suppressors act on ICR-170-induced frameshift mutations. Glycyl-tRNA from strains representing one class shows altered chromatographic mobility on Sepharose 4B. This group of suppressors, therefore, may represent mutations in structural genes for glycyl-tRNAs.

Genetic analysis of yeast suppressor strains has produced a number of mutants in which reduced levels of suppression are observed. The secondary mutations located within the suppressor genes are of great interest for studies of tRNA structure–function relationships analogous to those that have been carried out in bacteria. In addition, however, the level of genetic fine structure mapping available for a few suppressor genes makes these loci excellent substrates for probing the problem of "hot spots" in genetic recombination (Hawthorne and Leupold, 1974). Biochemical analysis of the many extragenic mutations that affect suppression levels (Thuriaux *et al.,* 1975) has just begun. It is quite probable that some mutants that interfere with suppressor tRNA expression may possess altered tRNA processing enzymes, tRNA modifying enzymes, or RNA polymerases specific for the transcription of tRNA genes. The absence of isopentenyladenosine in the tRNA of two mutant strains, the *S. cerevisiae mod5*-1 (J. W. Gorman, J. D. Young, H. M. Laten, F. H. Webb, and R. M. Bock, personal communication) and the *S. pombe sin1* (G. Vögeli and F. Hubschmid-Janner, personal communication), provides suggestive evidence that these strains may be defective in tRNA isopentenyl transferase (EC2.5.1.8), the enzyme responsible for isopentenyladenosine formation in tRNA (Fittler *et al.,* 1968; Bartz and Söll, 1972; Rosenbaum and Gefter, 1972).

5.2 The Search for Nonsense Mutations and Their Suppressors in Drosophila and Mammalian Cells

The large number of tRNA genes present in higher eukaryotic cells present certain problems for the isolation of suppressor mutations. Using the same argument we applied to yeast, one would presume that of the approximately 750 tRNA genes estimated for the *Drosophila* genome (Ritossa *et al.,* 1966), suppressor tRNAs could easily be generated from the many "spare" tRNA genes. Both the design of *in vivo* genetic selection systems and the development of highly efficient *in vitro* suppression assays may require stringent optimization, however, if we are to detect the result of a suppressor mutation in one among many reiterated genes specifying a given isoacceptor. It is not surprising, then, that while the phenomenon of genetic suppression is well known in *Drosophila,* and many suppressor-like mutations have been described (Lindsley and Grell, 1968), there is no biochemical evidence for the occurrence of tRNA-mediated suppression. Moreover, genetic and biochemical approaches to nonsense suppression in *Drosophila* are severely limited by the fact that no nonsense mutations have been characterized despite general acknowledgment of their usefulness. One straightforward but laborious method for obtaining nonsense mutations is the screening of many altered protein products for incomplete polypeptide chains resulting from mutations within a gene specifying a nonessential function. Such a study was undertaken for acid phosphatase-1 mutants (Bell and MacIntyre, 1973), but no definite conclusions were reached regarding the nature of these mutants.

The exciting proposal put forth several years ago that a species of *Drosophila* $tRNA^{Tyr}$ regulates the activity of the enzyme tryptophan pyrrolase in the *vermilion* mutant and is itself controlled by an extragenic suppressor is not supported by current evidence. Interest in this species of $tRNA^{Tyr}$ was aroused when Twardzik *et al.* (1971) found differences in the chromatographic profiles of tyrosyl-tRNAs from wild type *Drosophila* and strains carrying the recessive suppressor of *sable,* *su(s)*[2]. Of the two separable $tRNA^{Tyr}$ species in wild type strains, one appeared to inhibit the activity of tryptophan pyrrolase, an enzyme that is inactive in flies carrying the *vermilion* eye color mutation. Jacobson (1971) found that tryptophan pyrrolase extracts from *vermilion* flies could be restored to normal activity by treatment with T1 ribonuclease. The correlation between the restoration of tryptophan pyrrolase activity in suppressible *vermilion* mutants carrying the *sus(s)*^{2} allele and the absence of the inhibiting tRNA species suggested a mechanism for *su(s)*^{2} suppression. These data could not be reproduced, however, and subsequent experiments have shown that the reappearance of tryptophan pyrrolase activity was not a result of ribonuclease digestion but rather due to the use of EDTA in T1 RNase treatments (Mischke *et al.*, 1975). Further, the chromatographic separation of the various tRNA^{Tyr} species has been found to stem solely from differences in the levels of the modified nucleoside Q (White *et al.*, 1973), which occurs in the first anticodon position of several tRNAs (McCloskey and Nishimura, 1977). The reduced levels of the nucleoside Q in several tRNAs from *su(s)*^{2} strains, and the fact that *su(s)*^{2} acts simultaneously on mutations at several loci in *Drosophila* have raised the encouraging possibility that the undermodified tRNAs may function as suppressors during development of *Drosophila* (White *et al.*, 1973). New information casts doubt on even this notion, however, since the levels of the nucleoside Q found in *Drosophila* tRNAs are markedly dependent on growth conditions (Wosnick and White, 1977). Thus, at this time there is no positive evidence that implicates a direct or indirect role for tRNA in suppressor function.

Despite the lack of clear-cut nonsense or frameshift mutations in *Drosophila,* the availability of *in vitro* protein-synthesizing systems has led to systematic testing of tRNA samples prepared from selected suppressor strains. Several among the many *Drosophila* suppressor mutations are prime candidates for nonsense suppressors because they simultaneously reverse mutations at a variety of loci. These include the suppressors of *sable* [*su(s)*^{2}], *forked* [*Su(f)*], *hairy wing* [*su(Hw)*], *spineless* [*Su(ss)*^{2}], *tan* [*su(t)*], and *star* [*Su(S)*]. Kiger (1974) has shown that, whereas *Drosophila* tRNA supports the synthesis of the wild type bacteriophage f2 coat protein when wild type f2 RNA is used as a template in an *E. coli* cell-free system, tRNA from *Drosophila* strains carrying *su(Hw)* does not increase amino acid incorporation above control levels when RNA from an f2 amber mutant is used as a template. The results of *in vitro* suppression assays using two mammalian cell-free systems to test for response to amber (Gesteland *et al.*, 1976), ochre, and UGA codons (Kohli *et al.*, 1978) by tRNA from a number of *Drosophila* suppressor strains have been negative (E. Kubli, personal communication). While this does not represent a conclusive argument for the absence of suppressor tRNAs in *Drosophila,* biochemical experiments to date provide no evidence for the occurrence of tRNA-mediated nonsense suppression. It is quite possible that use of more refined detection methods (Capecchi *et al.*, 1977) and partially purified tRNAs will yield a clear-cut answer.

In contrast to the outlook for suppression in *Drosophila,* recent progress in mammalian systems is encouraging. Several groups have characterized shortened polypeptides that result from potential nonsense mutations in viral and host genes. Adetugbo *et al.* (1977) have described a mutant myeloma cell line which produces a truncated immunoglobulin H chain, and Summers *et al.* (1975) have reported mutants of Herpes simplex virus (I) which contain discrete fragments of thymidine kinase. Clear-cut identification of nonsense mutations, however, is restricted to two cases. First, Capecchi *et al.* (1977) have shown that a mouse L cell line deficient in the activity of hypoxanthine-guanine phosphoribosyltransferase (HGPRT) does contain enzyme molecules altered in the carboxy-terminal peptide. When ochre suppressor tRNA from *S. cerevisiae* is introduced into these cells by microinjection techniques (Schlegel and Rechsteiner, 1975), the cells regain HGPRT activity. Amber suppressor and wild type tRNA do not restore enzyme activity. Thus, the structural gene for HGPRT contains an ochre mutation which can be suppressed by exogenously supplied tRNA. Because a very simple genetic selection scheme can be used to obtain revertants of this cell line in which HGPRT activity is restored, the isolation of this HGPRT ochre mutation will now facilitate the direct selection for nonsense suppressors. Recently, Gesteland *et al.* (1977) have characterized one ochre and two amber mutants within a series of host-range mutants in an adeno(2)-SV40 hybrid virus which have lost the ability to grow on monkey cells. The specific polypeptide fragments altered in these mutants can be restored to wild type size by the addition of yeast suppressor tRNA to an *in vitro* translation reaction directed by SV40-specific mRNAs from the host-range mutants.

6 *Outlook*

Contrary to the feeling in the late 1960s that most of the useful information had been obtained from the analysis of suppression in prokaryotes, continued interest in this phenomenon has led to exciting advances in a number of areas. As we have tried to point out in this chapter, our current understanding at the molecular level of the transcription and translation processes could not have been assembled without the analytical power provided by the availability of well-characterized suppressors and suppressible mutations in genes. Since we have by no means arrived at a final description of these processes, there is clearly room for further work. It is our guess that continued exploitation of suppressors, combined with intensive efforts to determine the structure of the ribosome and to refine the three-dimensional structure of tRNA, will provide answers to many of the questions that have been raised in our discussion.

Recent demonstration of the direct involvement of tRNA in yeast nonsense suppression has given fresh momentum to the biochemical and genetic investigation of suppression in this lower eukaryotic organism. Likewise, the unexpected finding of nucleotide insertion sequences in yeast tRNA genes encourages our hope that a detailed study of suppression in yeast will uncover distinctly eukaryotic processes and help in their characterization.

The recent identification of nonsense mutations in a few mammalian genes, as well as the existence of sophisticated systems to screen for additional mutants (Schlegel and Rechsteiner, 1975; Capecchi *et al.*, 1977) indicates that more of these useful mutations will be isolated in other genes. Although the large number of tRNA genes may complicate the search for nonsense suppressors in higher eukaryotic cells, the availability of sensitive *in vitro* assay systems for eukaryotic nonsense suppressor tRNAs (Philipson *et al.*, 1978; Capecchi *et al.*, 1975; Kohli *et al.*, 1978) will prove invaluable.

We thank T. Platt, P. Modrich, and the members of our groups for their perceptive comments on an earlier version of this chapter, and A. Leigh-Brown, J. Carey, and D. LaMarche for their patient contributions to the preparation of the manuscript. Studies in the laboratory of D.G.S. were supported by grants from the National Institutes of Health (GM 22854 and HD 09167); D.A.S. was supported in part by funds from the National Cancer Institute (1PO1-CA-14236-05).

References

Adetugbo, K., Milstein, C., and Secher, D. S., 1977, Molecular analysis of spontaneous somatic mutants, *Nature* **265**:299–304.

Adhya, S., Gottesman, M., and de Crombrugghe, B., 1974, Release of polarity in *Escherichia coli* by gene *N* of phage λ: Termination and antitermination of transcription, *Proc. Natl. Acad. Sci. U.S.A.*, **71**:2534–2538.

Adhya, S., Gottesman, M., de Crombrugghe, B., and Court, D., 1976, Transcription termination regulates gene expression, in: *RNA Polymerase* (R. Losick and M. Chamberlin, eds.), pp. 719–730, Cold Spring Harbor Lab., Cold Spring Harbor, New York.

Agris, P., and Söll, D., 1977, The modified nucleosides in transfer RNA, in: *Nucleic Acid–Protein Recognition* (H. Vogel, ed.), pp. 321–344, Academic Press, New York.

Air, G. M., Sanger, F., and Coulson, A. R., 1976, Nucleotide and amino acid sequences of gene G of φX174, *J. Mol. Biol.* **108**:519–533.

Akaboshi, E., Inouye, M., and Tsugita, A., 1976, Effect of neighboring nucleotide sequences on suppression efficiency in amber mutants of T4 phage lysozyme, *Mol. Gen. Genet.* **149**:1–4.

Altman, S., 1976, A modified uridine in the anticodon of *E. coli* $tRNA_I^{Tyr}$ su_{oc}^+, *Nucleic Acids Res.* **3**:441–448.

Altman, S., 1978, tRNA biosynthesis, in: *Transfer RNA* (S. Altman, ed.), pp. 48–77, MIT Press, Cambridge, Massachusetts.

Altman, S., and Smith, J. D., 1971, Tyrosine tRNA precursor molecule polynucleotide sequence, *Nature New Biol.* **233**:35–39.

Altman, S., Brenner, S., and Smith, J. D., 1971, Identification of an ochre-suppressing anticodon, *J. Mol. Biol.* **56**:195–197.

Atkins, J. F., and Ryce, S., 1974, UGA and non-triplet suppressor reading of the genetic code, *Nature* **249**:527–530.

Baan, R. A., Duijfjes, J. J., van Leerdam, E., van Knippenberg, P. H., and Bosch, L., 1976, Specific *in situ* cleavage of 16S ribosomal RNA of *Escherichia coli* interferes with the function of initiation factor IF-1, *Proc. Natl. Acad. Sci. U.S.A.* **73**:702–706.

Bachmann, B. J., Low, K. B., and Taylor, A. L., 1976, Recalibrated linkage map of *Escherichia coli* K-12, *Bacteriol. Rev.* **40**:116–167.

Barrell, B. G., Air, G. M., and Hutchison, C. A., III, 1976, Overlapping genes in bacteriophage φX174, *Nature* **264**:34–40.

Bartz, J., and Söll, D., 1972, N^6-(Δ^2-Isopentenyl)adenosine: Biosynthesis *in vitro* in transfer RNA by an enzyme purified from *Escherichia coli, Biochimie* **54**:31–39.

Beaudet, A. L., and Caskey, C. T., 1970, Release factor translation of RNA phage terminator codons, *Nature* **227**:38–40.

Beaudet, A. L., and Caskey, C. T., 1972, Polypeptide chain termination, in: *The Mechanism of Protein Synthesis and Its Regulation* (L. Bosch, ed.), pp. 133–172, American Elsevier, New York.

Beckwith, J., 1963, Restoration of operon activity by suppressors, *Biochim. Biophys. Acta* **76**:162–164.

Bell, J., and MacIntyre, R., 1973, Characterization of acid phosphatase-1 null activity mutants of *Drosophila melanogaster, Biochem. Genet.* **10**:39–55.

Benzer, S., 1966, Adventures in the rII region, in: *Phage and the Origins of Molecular Biology* (J. Cairns, G. S. Stent, and J. D. Watson, eds.), pp. 157–165, Cold Spring Harbor Lab., Cold Spring Harbor, New York.

Benzer, S., and Champe, S. P., 1962, A change from nonsense to sense in the genetic code, *Proc. Natl. Acad. Sci. U.S.A.* **48**:1114–1121.

Beyreuther, K., Adler, K., Geisler, N., and Klemm, A., 1973, The amino acid sequence of *lac* repressor, *Proc. Natl. Acad. Sci. U.S.A.* **70**:3576–3580.

Bigelis, R., Keesey, J., and Fink, G. R., 1977, The *his-4* fungal gene cluster is not polycistronic, in: *Eukaryotic Genetic Systems,* (G. Wilcox, ed.), ICN-UCLA Symp. Vol. VIII, pp. 179–187, Academic Press, New York.

Biswas, D. K., and Gorini, L., 1972, Restriction, de-restriction and mistranslation in missense suppression. Ribosomal discrimination of transfer RNA's, *J. Mol. Biol.* **64**:119–134.

Bonnier, G., 1927, Note on the so-called vermilion-duplication, *Hereditas* **7**:229–232.

Bothwell, A. L. M., Stark, S. C., and Altman, S., 1976, Ribonuclease P substrate specificity: Cleavage of a bacteriophage ϕ80-induced RNA, *Proc. Natl. Acad. Sci. U.S.A.* **73**:1912–1916.

Bourgeois, S., Cohn, M., and Orgel, L. E., 1965, Suppression of and complementation among mutants of the regulatory gene of the lactose operon of *Escherichia coli, J. Mol. Biol.* **14**:300–302.

Brandriss, M. C., Stewart, J. W., Sherman, F., and Botstein, D., 1976, Substitution of serine caused by a recessive lethal suppressor in yeast, *J. Mol. Biol.* **102**:467–476.

Breathnach, R., Mandel, J. L., and Chambon, P., 1977, Ovalbumin gene is split in chicken DNA, *Nature* **270**:314–319.

Breckenridge, L., and Gorini, L., 1970, Genetic analysis of streptomycin resistance in *Escherichia coli, Genetics* **65**:9–25.

Brenner, D. J., Fournier, M. J., and Doctor, B. P., 1970, Isolation and partial characterization of the transfer ribonucleic acid cistrons from *Escherichia coli, Nature* **227**:448–451.

Brenner, S., and Beckwith, J. R., 1965, Ochre mutants, a new class of suppressible nonsense mutants, *J. Mol. Biol.* **13**:629–637.

Bretscher, M. S., 1968, Polypeptide chain termination: An active process, *J. Mol. Biol.* **34**:131–136.

Brot, N., Tate, W. P., Caskey, C. T., and Weissbach, H., 1974, The requirement for ribosomal proteins L7 and L12 in peptide chain termination, *Proc. Natl. Acad. Sci. U.S.A.* **71**:89–92.

Brunel, F., and Davison, J., 1975, Bacterial mutants able to partly suppress the effect of *N* mutations in bacteriophage λ, *Mol. Gen. Genet.* **136**:167–180.

Buckel, P., Piepersberg, W., and Böck, A., 1976, Suppression of temperature-sensitive aminoacyl-tRNA synthetase mutations by ribosomal mutations: A possible mechanism, *Mol. Gen. Genet.* **149**:51–61.

Buckingham, R. H., 1976, Anticodon conformation and accessibility in wild type and suppressor tryptophan tRNA from *E. coli, Nucleic Acids Res.* **3**:965–975.

Capecchi, M. R., and Klein, H. A., 1969, Characterization of three proteins involved in polypeptide chain termination, *Cold Spring Harbor Symp. Quant. Biol.* **34**:469–477.

Capecchi, M. R., Hughes, S. H., and Wahl, G. M., 1975, Yeast super-suppressors are altered tRNAs capable of translating a nonsense codon *in vitro, Cell* **6**:269–277.

Capecchi, M. R., von der Haar, R. A., Capecchi, N. E., and Sveda, M. M., 1977, The isolation of a suppressible nonsense mutant in mammalian cells, *Cell* **12**:371–381.

Carbon, J., and Fleck, E. W., 1974, Genetic alteration of structure and function in glycine transfer RNA of *Escherichia coli:* Mechanism of suppression of the tryptophan synthetase A78, *J. Mol. Biol.* **85**:371–391.

Caskey, C. T., and Beaudet, A. L., 1971, *Proceedings of the Symposium on Molecular Mechanisms of Antibiotic Action on Protein Biosynthesis and Membranes* (E. Munoz, F. Garvia-Fernandiz, and D. Vazquez, eds.), pp. 326–336, Granada, Spain.

Caskey, C. T., Bosch, L., and Konecki, D. S., 1977, Release factor binding to ribosome requires an intact 16S rRNA 3′ terminus, *J. Biol. Chem.* **252**:4435–4437.

Celis, J. E., Hooper, M. L., and Smith, J. D., 1973, Amino acid acceptor stem of *E. coli* suppressor $tRNA^{Tyr}$ is a site of synthetase recognition, *Nature New Biol.* **244**:261–264.

Celis, J. E., Coulondre, C., and Miller, J. H., 1976, Suppressor $su^{+}7$ inserts tryptophan in addition to glutamine, *J. Mol. Biol.* **104**:729–734.

Chakrabarti, S. L., and Gorini, L., 1975, A link between streptomycin and rifampicin mutation, *Proc. Natl. Acad. Sci. U.S.A.* **72**:2084–2087.

Clegg, J. B., Weatherall, D. J., Contopolou-Griva, I., Caroutsos, K., Poungouras, P., and Tsevrenis, H.,

1974, Haemoglobin Icaria, a new chain-termination mutant which causes α-thalassaemia, *Nature* **251**:245–247.

Colby, D. S., Schedl, P., and Guthrie, C., 1976, A functional requirement for modification of the wobble nucleotide in the anticodon of a T2 suppressor tRNA, *Cell* **9**:449–463.

Comer, M. M., 1977, Correlation between genetic and nucleotide distances in a bacteriophage T4 transfer RNA gene, *J. Mol. Biol.* **113**:267–271.

Comer, M. M., Guthrie, C., and McClain, W. H., 1974, An ochre suppressor of bacteriophage T4 that is associated with a transfer RNA, *J. Mol. Biol.* **90**:665–676.

Comer, M. M., Foss, K., and McClain, W. H., 1975, A mutation of the wobble nucleotide of a bacteriophage T4 transfer RNA, *J. Mol. Biol.* **99**:283–293.

Contreras, R., Ysebaert, M., Min Jou, W., and Fiers, W., 1973, Bacteriophage MS2 RNA: Nucleotide sequence of the end of the A protein gene and the intercistronic region, *Nature New Biol.* **241**:99–102.

Coulondre, C., and Miller, J. H., 1977*a*, Genetic studies of the *lac* repressor: III. Additional correlation of mutational sites with specific amino acid residues, *J. Mol. Biol.* **117**:525–575.

Coulondre, C., and Miller, J. H., 1977*b*, Genetic studies of the *lac* repressor: IV. Mutagenic specificity in the *lacI* gene of *E. coli*, *J. Mol. Biol.* **117**:577–606.

Cox, B. S., 1965, *ψ*, a cytoplasmic suppressor of super-suppressor in yeast, *Heredity* **20**:505–521.

Crick, F. H. C., 1966, Codon–anticodon pairing: The wobble hypothesis, *J. Mol. Biol.* **19**:548–555.

Crick, F. H. C., Barnett, L., Brenner, S., and Watts-Tobin, R. J., 1961, Triplet nature of the code, *Nature* **192**:1227–1232.

Culbertson, M. R., Charnes, L., Johnson, M. T., and Fink, G. R., 1977, Frameshifts and frameshift suppressors in *Saccharomyces cerevisiae*, *Genetics* **86**:745–764.

Dabbs, E. R., and Wittmann, H. G., 1976, A strain of *Escherichia coli* which gives rise to mutations in a large number of ribosomal proteins, *Mol. Gen. Genet.* **149**:303–309.

Dalgarno, L., and Shine, J., 1973, Conserved terminal sequence in 18S rRNA may represent terminator anticodons, *Nature New Biol.* **245**:261–262.

Das, A., Court, D., and Adhya, S., 1976, Isolation and characterization of conditional lethal mutants of *Escherichia coli* defective in transcription termination factor rho, *Proc. Natl. Acad. Sci. U.S.A.* **73**:1959–1963.

Dickson, R. C., Abelson, J., Barnes, W. M., and Reznikoff, W. S., 1975, Genetic regulation: The *lac* control region, *Science* **187**:27–35.

Donis-Keller, H., Maxam, A. M., and Gilbert, W., 1977, Mapping adenines, guanines, and pyrimidines in RNA. *Nucleic Acids Res.* **4**:2527–2538.

Dudock, B., Lesiewicz, J., Greenberg, R., and Wu, E., 1978, A new class of modification reactions in *E. coli* and rabbit liver mitochondrial tRNA: The formation of methyl esters (submitted for publication).

Edelmann, P., and Gallant, J., 1977, Mistranslation in *E. coli*, *Cell* **10**:131–137.

Efstratiadis, A., Kafatos, F. C., and Maniatis, T., 1977, The primary structure of rabbit β-globin mRNA as determined from cloned DNA, *Cell* **10**:571–585.

Elseviers, D., and Gorini, L., 1975, Direct selection of mutants restricting efficiency of suppression and misreading levels in *E. coli B*, *Mol. Gen. Genet.* **137**:277–287.

Engelberg-Kulka, H., Dekel, L., and Israeli-Reches, M., 1977, Streptomycin-resistant *Escherichia coli* mutant temperature sensitive for the production of QB-infective particles, *J. Virol.* **21**:1–6.

Farabaugh, P. J., 1978, Sequence of the *lacI* gene, Nature **274**:765–769.

Feinstein, S. I., and Altman, S., 1977, Coding properties of an ochre-suppressing derivative of *Escherichia coli* $tRNA_{I}^{Tyr}$, *J. Mol. Biol.* **112**:453–470.

Feinstein, S. I., and Altman, S., 1978, Context effects on nonsense codon suppression in *Escherichia coli*, *Genetics* **88**:201–219.

Fiddes, J. C., 1976, Nucleotide sequence of the intercistronic region between genes G and F in bacteriophage φX174 DNA, *J. Mol. Biol.* **107**:1–24.

Fiers, W., Contreras, R., Duerinck, F., Haegeman, G., Iserentant, D., Merregaert, J., Min Jou, W., Molemans, F., Raeymaekers, A., Van den Berghe, A., Volckaert, G., and Ysebaert, M., 1976, Complete nucleotide sequence of bacteriophage MS2 RNA: Primary and secondary structure of the replicase gene, *Nature* **260**:500–507.

Files, J. G., Weber, K., and Miller, J. H., 1974, Translational reinitiation: Reinitiation of *lac* repressor fragments at the three internal sites early in the *lac i* gene of *Escherichia coli*, *Proc. Natl. Acad. Sci. U.S.A.* **71**:667–670.

Fittler, F., Kline, L. K., and Hall, R. H., 1968, Biosynthesis of N^6-(Δ^2-isopentenyl)adenosine. The precursor relationship of acetate and mevalonate to the Δ^2-isopentenyl group of the transfer ribonucleic acid of microorganisms, *Biochemistry* **7**:940–944.

Fluck, M. M., Salser, W., and Epstein, R. H., 1977, The influence of the reading context upon the suppression of nonsense codons, *Mol. Gen. Genet.* **151**:137–149.

Fox, I. L., and Ganoza, M. C., 1968, Chain termination *in vitro.* Studies on the specificity of amber and ochre triplets, *Biochem. Biophys. Res. Commun.* **32**:1064–1070.

Franklin, N. C., 1974, Altered reading of genetic signals fused to the *N* operon of bacteriophage λ: Genetic evidence for modification of polymerase by the protein product of the *N* gene, *J. Mol. Biol.* **89**:33–48.

Franklin, N. C., and Yanofsky, C., 1976, The *N* protein of λ: Evidence bearing on transcription termination, polarity and the alteration of *E. coli* RNA polymerase, in: *RNA Polymerase* (R. Losick and M. Chamberlin, eds.), pp. 693–706, Cold Spring Harbor Lab., Cold Spring Harbor, New York.

Freymeyer, D. K., II, Shank, P. R., Edgell, M. H., Hutchison, C. A., III, and Vanaman, T. C., 1977, Amino acid sequence of the small core protein from bacteriophage ϕX174, *Biochemistry* **16**:4550–4556.

Friedman, D. I., and Yarmolinsky, M., 1972, Prevention of the lethality of induced λ prophage by an isogenic λ plasmid, *Virology* **50**:472–481.

Galas, D. J., and Branscomb, E. W., 1976, Ribosome slowed by mutation to streptomycin resistance, *Nature* **262**:617–619.

Galluppi, G., Lowery, C., and Richardson, J. P., 1976, Nucleoside triphosphate requirement for termination of RNA synthesis by rho factor, in: *RNA Polymerase* (R. Losick and M. Chamberlin, eds.), pp. 657–665, Cold Spring Harbor Lab., Cold Spring Harbor, New York.

Ganoza, M. C., and Tomkins, J. K. N., 1970, Polypeptide chain termination *in vitro:* Competition for nonsense codons between a purified release factor and suppressor tRNA, *Biochem. Biophys. Res. Commun.* **40**:1455–1463.

Garen, A., 1968, Sense and nonsense in the genetic code, *Science* **160**:149–159.

Gefter, M. L., and Russell, R. L., 1969, Role of modifications in tyrosine transfer RNA: A modified base affecting ribosome binding, *J. Mol. Biol.* **39**:145–157.

Georgopoulos, C. P., 1971, Bacterial mutants in which the gene *N* function of bacteriophage lambda is blocked have an altered RNA polymerase, *Proc. Natl. Acad. Sci. U.S.A.* **68**:2977–2981.

Gesteland, R. F., Wolfner, M., Grisafi, P., Fink, G., Botstein, D., and Roth, J. R., 1976, Yeast suppressors of UAA and UAG nonsense codons work efficiently *in vitro* via tRNA, *Cell* **7**:381–390.

Gesteland, R. F., Wills, N., Lewis, J. B., and Grodzicker, T., 1978, Identification of amber and ochre mutants of the human virus Ad2$^+$ND1, *Proc. Natl. Acad. Sci. U.S.A.* **74**:4567–4571.

Ghosh, K., and Ghosh, H. P., 1970, Role of modified nucleoside adjacent to 3′-end of anticodon in codon–anticodon interaction. *Biochem. Biophys. Res. Commun.* **40**:135–143.

Ghysen, A., and Pironio, M., 1972, Relationship between the *N* function of bacteriophage λ and host RNA polymerase, *J. Mol. Biol.* **65**:259–272.

Gilbert, W., and Maxam, A., 1973, The nucleotide sequence of the *lac* operator. *Proc. Natl. Acad. Sci. U.S.A.* **70**:3581–3584.

Gilmore, R. A., Stewart, J. W., and Sherman, F., 1971, Amino acid replacements resulting from super-suppression of nonsense mutants of iso-l-cytochrome *c* from yeast, *J. Mol. Biol.* **61**:157–173.

Goodman, H. M., Olson, M. V., and Hall, B. D., 1977, Nucleotide sequence of a mutant eukaryotic gene: The yeast tyrosine-inserting ochre suppressor SUP4-O, *Proc. Natl. Acad. Sci. U.S.A.* **74**:5453–5457.

Gopinathan, K. P., and Garen, A., 1970, A leucyl-transfer RNA by the amber suppressor gene *Su6*$^+$, *J. Mol. Biol.* **47**:393–401.

Gorini, L., 1970, Informational suppression, *Annu. Rev. Genet.* **4**:107–134.

Gorini, L., 1971, Ribosomal discrimination of tRNAs, *Nature New Biol.* **234**:261–264.

Gorini, L., 1974, Streptomycin and misreading of the genetic code, in: *Ribosomes* (M. Nomura, A. Tissières, and P. Lengyel, eds.), pp. 791–803, Cold Spring Harbor Lab., Cold Spring Harbor, New York.

Grosjean, H., Söll, D. G., and Crothers, D. M., 1976, Studies of the complex between transfer RNAs with complementary anticodons. I. Origins of enhanced affinity between complementary triplets, *J. Mol. Biol.* **103**:499–519.

Grosjean, H. J., de Henau, S., and Crothers, D. M., 1978, On the physical basis for ambiguity in genetic coding interactions, *Proc. Natl. Acad. Sci. U.S.A.* **75**:610–614.
Hagenbüchle, O., Santer, M., Steitz, J. A., and Mans, R. J., 1978, Conservation of the primary structure at the 3′-end of 18S rRNA from eucaryotic cells, *Cell* **13**:551–563.
Hartman, P. E., and Roth, J. R., 1973, Mechanisms of suppression, *Adv. Genet.* **17**:1–105.
Hawthorne, D. C., 1976, UGA mutations and UGA suppressors in yeast. *Biochimie* **58**:179–182.
Hawthorne, D. C., and Leupold, U., 1974, Suppressors in yeast, *Curr. Top. Microbiol. Immunol.* **64**:1–47.
Hayashi, H., and Söll, D., 1971, Purification of an *E. coli* leucine suppressor transfer RNA and its aminoacylation by the homologous leucyl-tRNA synthetase, *J. Biol. Chem.* **246**:4951–4954.
Hill, C. W., 1975, Informational suppression of missense mutations, *Cell* **6**:419–427.
Hiraga, S., and Yanofsky, C., 1972, Hyperlabile messenger RNA in polar mutants of the tryptophan operon of *E. coli, J. Mol. Biol.* **72**:103–110.
Hirsh, D., 1971, Tryptophan transfer RNA as the UGA suppressor, *J. Mol. Biol.* **58**:439–458.
Hirsh, D., and Gold, L., 1971, Translation of the UGA triplet *in vitro* by tryptophan transfer RNAs, *J. Mol. Biol.* **58**:459–468.
Hofstetter, H., Monstein, H. J., and Weissmann, C., 1974, The readthrough protein A_1 is required for *in vitro* reconstitution of infectious QB particles, *Experimentia* **30**:687.
Holmes, W. M., Goldman, E., Miner, T. A., and Hatfield, G. W., 1977, Differential utilization of leucyl-tRNAs by *Escherichia coli, Proc. Natl. Acad. Sci. U.S.A.* **74**:1393–1397.
Hooper, M. L., Russell, R. L., and Smith, J. D., 1972, Mischarging in mutant tyrosine transfer RNAs, *FEBS Lett.* **22**:149–155.
Hsiang, M. W., Cole, R. D., Takeda, Y., and Echols, H., 1977, Amino acid sequence of Cro regulatory protein of bacteriophage lambda, *Nature* **270**:275–277.
Humayun, Z., 1977, DNA sequence at the end of the C_I gene in bacteriophage λ, *Nucleic Acids Res.* **4**:2137–2144.
Ikemura, T., Shimura, Y., Sakano, H., and Ozeki, H., 1975, Precursor molecules of *Escherichia coli* transfer RNAs accumulated in a temperature-sensitive mutant, *J. Mol. Biol.* **96**:69–86.
Ilgen, C., Kirk, L. L., and Carbon, J., 1976, Isolation and characterization of large transfer ribonucleic acid precursors from *Escherichia coli, J. Biol. Chem.* **251**:922–929.
Inoko, H., and Imai, M., 1976, Isolation and genetic characterization of the *nitA* mutants of *Escherichia coli* affecting the termination factor rho, *Mol. Gen. Genet.* **143**:211–221.
Inoko, H., Shigesada, K., and Imai, M., 1977, Isolation and characterization of conditional-lethal rho mutants of *Escherichia coli, Proc. Natl. Acad. Sci. U.S.A.* **74**:1162–1166.
Inokuchi, H., Yamao, F., Yamagishi, H., Sakano, H., and Ozeki, H., 1975, Isolation of mischarging mutants of suppressor tRNA by using the transducing phages carrying su_1^+ and su_2^+ in *E. coli, Genet. Soc. (Can.) Bull.* **6**:37.
Jacobson, K. B., 1971, Role of an isoacceptor transfer ribonucleic acid as an enzyme inhibitor: Effect on tryptophan pyrrolase of *Drosophila, Nature New Biol.* **231**:17–19.
Jeffreys, A. J., and Flavell, R. A., 1977, The rabbit *β*-globin gene contains a large insert in the coding sequence, *Cell* **12**: 1097–1108.
Jukes, T. H., 1977, How many anticodons? *Science* **198**:319–320.
Kao, S.-H., and McClain, W. H., 1977, UGA suppressor of bacteriophage T4 associated with arginine transfer RNA, *J. Biol. Chem.* **252**:8254–8257.
Kaufmann, G., and Littauer, U. Z., 1974, Covalent joining of phenylalanine transfer ribonucleic acid half-molecules by T4 RNA ligase, *Proc. Natl. Acad. Sci. U.S.A.* **71**:3741–3745.
Kiger, J. A., Jr., 1974, Participation of *Drosophila* tRNA in protein synthesis in an *E. coli* protein synthesizing system, *Nucleic Acids Res.* **1**:1269–1277.
Kimball, M. E., and Söll, D. G., 1974, The phenylalanine tRNA from Mycoplasma sp. (Kid): A tRNA lacking hypermodified nucleosides functional in protein synthesis, *Nucleic Acids Res.* **1**:1713–1720.
Kohli, J., Kwong, T. C., Altruda, F., Söll, D., and Wahl, G., 1978, Nonsense suppression in *S. pombe:* The efficient suppressors recognize UGA, *J. Biol. Chem.* (in press).
Korn, L. J., and Yanofsky, C., 1976*a,* Polarity suppressors increase expression of the wild type tryptophan operon of *Escherichia coli, J. Mol. Biol.* **103**:395–409.
Korn, L. J., and Yanofsky, C., 1976*b,* Polarity suppressors defective in transcription termination at the attenuator of the tryptophan operon of *Escherichia coli* have altered rho factors, *J. Mol. Biol.* **106**:231–241.

Körner, A., Feinstein, S. I., and Altman, S., 1978, tRNA-mediated suppression, in: *Transfer RNA* (S. Altman, ed.), pp. 105–135. MIT Press, Cambridge, Massachusetts.

Koski, R. A., Bothwell, A. L. M., and Altman, S., 1976, Identification of a ribonuclease P-like activity from human KB cells, *Cell* **9**:101–116.

Kuchino, Y., Seno, T., and Nishimura, S., 1971, Fragmented *E. coli* methionine $tRNA_f$ as methyl acceptor for rat liver tRNA methylase: Alteration of the site of methylation by the conformational change of tRNA structure resulting from fragmentation, *Biochem. Biophys. Res. Commun.* **43**:476–483.

Kurland, C. G., 1977, Structure and function of the bacterial ribosome, *Annu. Rev. Biochem.* **46**:173–200.

Kurland, C. G., Rigler, R., Ehrenberg, M., and Blomberg, C., 1975, Allosteric mechanism for codon-dependent tRNA selection on ribosomes, *Proc. Natl. Acad. Sci. U.S.A.* **72**:4248–4251.

Lagerkvist, U., 1978, Two out of three: An alternative method for codon reading, *Proc. Natl. Acad. Sci. U.S.A.* **75**: 1759–1762.

Last, J. A., and Anderson, W. F., 1976, Purification and properties of bacteriophage T4-induced RNA ligase, *Arch. Biochem. Biophys.* **174**:167–176.

Liebman, S. W., and Sherman, F., 1976, Inhibition and growth by amber suppressors in yeast, *Genetics* **82**:233–249.

Liebman, S. W., Stewart, J. W., and Sherman, F., 1975, Serine substitution caused by an ochre suppressor in yeast, *J. Mol. Biol.* **94**:595–610.

Liebman, S. W., Sherman, F., and Stewart, J. W., 1976, Isolation and characterization of amber suppressors in yeast, *Genetics* **82**:251–272.

Liebman, S. W., Stewart, J. W., Parker, J. H., and Sherman, F., 1977, Leucine insertion caused by a yeast amber suppressor, *J. Mol. Biol.* **109**:13–22.

Lindsley, D. L., and Grell, E. H., 1968, Genetic variations of *Drosophila melanogaster, Carnegie Institute Washington Publication* No. 627.

Litwack, M., and Peterkofsky, A., 1971, Transfer ribonucleic acid deficient in N^6-(Δ^2-isopentenyl)adenosine due to mevalonic acid limitation, *Biochemistry* **10**:994–1001.

Loftfield, R. B., and Vanderjagt, D., 1972, The frequency of errors in protein biosynthesis, *Biochem. J.* **128**:1353–1356.

Lu, P., and Rich, A., 1971, The nature of the polypeptide chain termination signal, *J. Mol. Biol.* **58**:513–531.

Lu, P., Jarema, M., Mosser, K., and Daniel, W. E., Jr., 1976, *lac* repressor: 3-Fluorotyrosine substitution for nuclear magnetic resonance studies, *Proc. Natl. Acad. Sci. U.S.A.* **73**:3471–3475.

Lund, E., and Dahlberg, J. E., 1977, Spacer transfer RNAs in ribosomal RNA transcripts of *E. coli:* Processing of 30S ribosomal RNA *in vitro, Cell* **11**:247–262.

Lund, E., Dahlberg, J. E., Lindahl, L., Jaskunas, S. R., Dennis, P. P., and Nomura, M., 1976, Transfer RNA genes between 16S and 23S rRNA genes in rRNA transcription units of *E. coli, Cell* **7**:165–177.

Marinus, M. G., Morris, N. R., Söll, D., and Kwong, T. C., 1975, Isolation and partial characterization of three *Escherichia coli* mutants with altered transfer ribonucleic acid methylases, *J. Bacteriol.* **122**:257–265.

Martin, J., and Webster, R. E., 1975, The *in vitro* translation of a terminating signal by a single *Escherichia coli* ribosome, *J. Biol. Chem.* **250**:8132–8139.

Maxam, A. M., and Gilbert, W., 1977, A new method for sequencing DNA, *Proc. Natl. Acad. Sci. U.S.A.* **74**:560–564.

McClain, W. H., 1977, Seven terminal steps in a biosynthetic pathway leading from DNA to transfer RNA, *Acc. Chem. Res.* **10**:418–425.

McClain, W. H., and Seidman, J. G., 1975, Genetic perturbations that reveal tertiary conformation of tRNA precursor molecules, *Nature* **257**:106–110.

McClain, W. H., Guthrie, C., and Barrell, B. G., 1972, Eight transfer RNAs induced by infection of *Escherichia coli* with bacteriophage T4, *Proc. Natl. Acad. Sci. U.S.A.* **69**:3703–3707.

McCloskey, J. A., and Nishimura, S., 1977, Modified nucleosides in tRNA, *Acc. Chem. Res.* **10**:403–410.

McCready, S. J., Cox, B. S., and McLaughlin, C. S., 1977, The extrachromosomal control of nonsense suppression in yeast: An analysis of the elimination of [psi^+] in the presence of a nuclear gene PNM^-, *Mol. Gen. Genet.* **150**:265–270.

Michels, C. A., and Zipser, D., 1969, Mapping of polypeptide reinitiation sites within the β-galactosidase structural gene, *J. Mol. Biol.* **41**:341–347.

Miller, J. H., Coulondre, C., Schmeissner, U., Schmitz, A., and Lu, P., 1975, The use of suppressed nonsense mutations to generate altered *lac* repressor molecules, in: *Protein–Ligand Interactions* (H. Sund and G. Blauer, eds.), pp. 238–252, de Gruyter, Berlin.

Miller, J. H., Ganem, D., Lu, P., and Schmitz, A., 1977, Genetic studies of the *lac* repressor. I. Correlation of mutational sites with specific amino acid residues: Construction of a colinear gene–protein map, *J. Mol. Biol.* **109**:275–301.

Min Jou, W., Haegeman, G., Ysebaert, M., and Fiers, W., 1972, Nucleotide sequences of the gene coding for the bacteriophage MS2 coat protein, *Nature* **237**:82–88.

Mischke, D., Kloetzel, P., and Schwochau, M., 1975, Tryptophan pyrrolase activity regulation in *Drosophila:* Role of an isoacceptor tRNA unsettled, *Nature* **255**:79–80.

Mitra, S. K., Lustig, F., Åkesson, B., Lagerkvist, U., and Strid, L., 1977, Codon–anticodon recognition in the valine codon family, *J. Biol. Chem.* **252**:471–478.

Morgan, E. A., Ikemura, T., and Nomura, M., 1977, Identification of spacer tRNA genes in individual ribosomal RNA transcription units of *Escherichia coli, Proc. Natl. Acad. Sci. U.S.A.* **74**:2710–2714.

Morse, D. E., and Guertin, M., 1972, Amber *suA* mutations which relieve polarity, *J. Mol. Biol.* **63**:605–608.

Morse, D. E., and Primakoff, P., 1970, Relief of polarity in *E. coli* by *"suA," Nature* **226**:28–31.

Morse, D. E., and Yanofsky, C., 1969, Polarity and the degradation of mRNA, *Nature* **224**:329–331.

Müller-Hill, B., 1966, Suppressible regulator constitutive mutants of the lactose system in *Escherichia coli, J. Mol. Biol.* **15**:374–376.

Müller-Hill, B., 1975, Lac repressor and lac operator, *Prog. Biophys. Mol. Biol.* **30**:227–252.

Murgola, E. J., and Yanofsky, C., 1974, Suppression of glutamic acid codons by mutant glycine transfer ribonucleic acid, *J. Bacteriol.* **117**:439–443.

Nagata, T., and Horiuchi, T., 1973, Isolation and characterization of a temperature-sensitive amber suppressor mutant of *Escherichia coli* K12, *Mol. Gen. Genet.* **123**:77–88.

Newton, W. A., Beckwith, J. P., Zipser, D., and Brenner, S., 1965, Nonsense mutants and polarity in the *lac* operon of *Escherichia coli, J. Mol. Biol.* **14**:290–296.

Nichols, J. L., 1970, Nucleotide sequence from the polypeptide chain termination region of the coat protein cistron in bacteriophage R17 RNA, *Nature* **225**:147–151.

Nichols, J. L., and Robertson, H. D., 1971, Sequences of RNA fragments from the bacteriophage f2 coat protein cistron which differ from their R17 counterparts, *Biochim. Biophys. Acta* **228**:676–681.

Ninio, J., 1974, A semi-quantitative treatment of missense and nonsense suppression in the *str*A and *ram* ribosomal mutants of *Escherichia coli:* Evaluation of some molecular parameters of translation *in vivo, J. Mol. Biol.* **84**:297–313.

Nomura, M., Morgan, E. A., and Jaskunas, S. R., 1977, Genetics of bacterial ribosomes, *Annu. Rev. Genet.* **11**:297–347.

Oakden, K. M., and Lane, B. G., 1976, Wheat embryo ribonucleates. VI. Comparison of the 3′-hydroxyl termini in "rapidly labelled" RNA from metabolizing wheat embryos with the corresponding termini in ribosomal RNA from differentiating embryos of wheat, barley, corn, and pea, *Can. J. Biochem.* **54**:261–271.

Oeschger, M. P., and Woods, S. L., 1976, A temperature-sensitive suppressor enabling the manipulation of the level of individual proteins in intact cells, *Cell* **7**:205–212.

Ohlsson, B. M., Strigini, P. F., and Beckwith, J. R., 1968, Allelic amber and ochre suppressors, *J. Mol. Biol.* **36**:209–218.

Olson, M. V., Montgomery, D. L., Hopper, A. K., Page, G. S., Horodyski, F., and Hall, B. D., 1977, Molecular characterization of the tyrosine tRNA genes of yeast, *Nature* **267**:639–641.

Orgel, L. E., 1972, Possible step in the origin of the genetic code, *Isr. J. Chem.* **10**:287–292.

Ozaki, M., Mizushima, S., and Nomura, M., 1969, Identification and functional characterization of the protein controlled by the streptomycin-resistant locus in *E. coli, Nature* **222**:333–339.

Pan, J., Reddy, V. B., Thimmappaya, B., and Weissman, S. M., 1977, Nucleotide sequence of the gene for the major structural protein of SV40 virus, *Nucleic Acids Res.* **4**:2539–2548.

Pelham, R. B., and Jackson, R. J., 1976, An efficient mRNA-dependent translation system from reticulocyte lysates, *Eur. J. Biochem.* **67**:247–256.

Person, S., and Osborn, M., 1968, The conversion of amber suppressors to ochre suppressors, *Proc. Natl. Acad. Sci. U.S.A.* **60**:1030–1037.

Philipson, L., Andersson, P., Olshevsky, U., Weinberg, R., Baltimore, D., and Gesteland, R., 1978, Translation of MuLV and MSV RNAs in nuclease-treated reticulocyte extracts: Enhancement of the gag-pol polypeptide with yeast suppressor tRNA, *Cell* **13**:189–199.

Pieczenik, G., 1972, Ph.D. Thesis, New York University, New York City.

Piepersberg, W., Böck, A., and Wittmann, H. G., 1975, Effect of different mutations in ribosomal protein S5 of *Escherichia coli* on translational fidelity, *Mol. Gen. Genet.* **140**:91–100.

Piper, P. W., Wasserstein, M., Engbaek, F., Kaltoft, K., Celis, J. E., Zeuthen, J., Liebman, S., and Sherman, F., 1976, Nonsense suppressors of *Saccharomyces cerevisiae* can be generated by mutation of the tyrosine tRNA anticodon, *Nature* **262**:757–761.

Platt, T., and Yanofsky, C., 1975, An intercistronic region and ribosome-binding site in bacterial messenger RNA, *Proc. Natl. Acad. Sci. U.S.A.* **72**:2399–2403.

Platt, T., Files, J. G., and Weber, K., 1973, *lac* repressor. Specific proteolytic destruction of the NH_2-terminal region and loss of the deoxyribonucleic acid-binding activity, *J. Biol. Chem.* **248**:110–121.

Pope, W. T., and Reeves, R. H., 1978*a*, Purification of the *supK* methylase, *J. Bacteriol.* (in press).

Pope, W. T., Brown, A., and Reeves, R. H., 1978*b*, The identification of the tRNA substrates for the *supK* tRNA methylase, *Nucleic Acids Res.* **5**:1041–1057.

Proudfoot, N. J., 1977, Complete 3′ noncoding region sequences of rabbit and human β-globin messenger RNAs, *Cell* **10:**559–570.

Proudfoot, N. J., and Longley, J. I., 1976, The 3′ terminal sequences of human α and β globin messenger RNAs: Comparison with rabbit globin messenger RNA, *Cell* **9**:733–746.

Proudfoot, N. J., Gillam, S., Smith, M., and Longley, J., 1977, Nucleotide sequence of the 3′ terminal third of rabbit α-globin messenger RNA: Comparison with human α-globin messenger RNA, *Cell* **11**:807–818.

Ratliff, J. C., and Caskey, C. T., 1977, Immunologic evidence for structural homology between the release factors of *Escherichia coli, Arch. Biochem. Biophys.* **181**:671–677.

Ratner, D., 1976*a*, Evidence that mutations in the *suA* polarity suppressing gene directly affect termination factor rho, *Nature* **259**:151–153.

Ratner, D., 1976*b*, The rho gene of *E. coli* maps at *suA*, in: *RNA Polymerase* (R. Losick and M. Chamberlin, eds.), pp. 645–655, Cold Spring Harbor Lab., Cold Spring Harbor, New York.

Reeves, R., and Roth, J. R., 1971, A recessive UGA suppressor, *J. Mol. Biol.* **56**:523–533.

Remaut, E., and Fiers, W., 1972, Studies on the bacteriophage MS2 XVI. The termination signal of the A protein cistron, *J. Mol. Biol.* **71**:243–261.

Rich, A., and Schimmel, P. R., 1977, Structural organization of complexes of transfer RNAs with aminoacyl transfer RNA synthetase, *Nucleic Acids Res.* **5**:1649–1665.

Richardson, J. P., Grimley, C., and Lowery, C., 1975, Transcription termination factor rho activity is altered in *Escherichia coli* with *suA* gene mutations, *Proc. Natl. Acad. Sci. U.S.A.* **72**:1725–1733.

Riddle, D., and Carbon, J., 1973, A nucleotide addition in the anticodon of a glycine transfer RNA, *Nature New Biol.* **242**:230–234.

Riddle, D., and Roth, J., 1972*a*, Frameshift suppressors. II. Genetic mapping and dominance studies, *J. Mol. Biol.* **66**:483–493.

Riddle, D., and Roth, J., 1972*b*, Frameshift suppressors. III. Effects of suppressor mutations on transfer RNA, *J. Mol. Biol.* **66**:495–506.

Ritossa, F. M., Atwood, K. C., and Spiegelman, S., 1966, On the redundancy of DNA complementary to amino acid transfer RNA and its absence from the nucleolar organizer region of *Drosophila melanogaster, Genetics* **54**:663–676.

Riyasaty, S., and Atkins, J., 1968, External suppression of a frameshift mutant in *Salmonella, J. Mol. Biol.* **34**:541–557.

Roberts, J. W., 1969, Termination factor for RNA synthesis, *Nature* **224**:1168–1174.

Roberts, J. W., and Carbon, J., 1974, Molecular mechanism for missense suppression in *E. coli, Nature* **250**:412–414.

Roberts, T. R., Shimatake, H., Brady, C., and Rosenberg, M., 1977, Sequence of *cro* gene of bacteriophage lambda, *Nature* **270**:274–275.

Rosenbaum, N., and Gefter, M. L., 1972, Δ^2-Isopentenylpyrophosphate:transfer ribonucleic acid Δ^2-isopentenyltransferase from *Escherichia coli, J. Biol. Chem.* **247**:5675-5680.

Rosset, R., and Gorini, L., 1969, A ribosomal ambiguity mutation, *J. Mol. Biol.* **39**:95–112.

Roth, J. R., 1974, Frameshift mutations, *Annu. Rev. Genet.* **8**:319–346.

Sadler, J., and Novick, A., 1965, The properties of repressor and the kinetics of its action, *J. Mol. Biol.* **12**:305–327.

Sakano, H., and Shimura, Y., 1975, Sequential processing of precursor tRNA molecules in *Escherichia coli, Proc. Natl. Acad. Sci. U.S.A.* **72**:3369–3373.

Sakano, H., Yamada, S., Ikemura, T., Shimura, Y., and Ozeki, H., 1974, Temperature sensitive mutants of *Escherichia coli* for tRNA synthesis, *Nucleic Acids Res.* **1**:355–371.

Salser, W., 1969, The influence of the reading context upon the suppression of nonsense codons, *Mol. Gen. Genet.* **105**:125–130.

Salser, W., Fluck, M., and Epstein, R., 1969, The influence of the reading context upon the suppression of nonsense codons. III. *Cold Spring Harbor Symp. Quant. Biol.* **34**:513–520.

Sanger, F., Air, G. M., Barrell, B. G., Brown, N. L., Coulson, A. R., Fiddes, J. C., Hutchison, C. A., III, Slocombe, P. M., and Smith, M., 1977*a*. Nucleotide sequence of bacteriophage ϕX174 DNA, *Nature* **265**:687–695.

Sanger, F., Nicklen, S., and Coulson, A. R., 1977*b*, DNA sequencing with chain terminating inhibitors, *Proc. Natl. Acad. Sci. U.S.A.* **74**:5463–5467.

Sarabhai, A., and Brenner, S., 1967, A mutant which reinitiates the polypeptide chain after chain termination, *J. Mol. Biol.* **27**:145–162.

Schedl, P., and Primakoff, P., 1973, Mutants of *Escherichia coli* thermosensitive for the synthesis of transfer RNA, *Proc. Natl. Acad. Sci. U.S.A.* **70**:2091–2095.

Schlegel, R., and Rechsteiner, M. C., 1975, Microinjection of thymidine kinase and bovine serum albumin into mammalian cells by fusion with red blood cells, *Cell* **5**:371–379.

Schmeissner, U., Ganem, D., and Miller, J. H., 1977, Genetic studies of the *lac* repressor. II. Fine structure deletion map of the *lac I* gene, and its correlation with the physical map, *J. Mol. Biol.* **109**:303–326.

Schmidt, F. J., Seidman, J. G., and Bock, R. M., 1976, Transfer ribonucleic acid biosynthesis. Substrate specificity of ribonuclease P, *J. Biol. Chem.* **251**:2440–2445.

Schmitz, A., Schmeissner, U., Miller, J. H., and Lu, P., 1976, Mutations affecting the quaternary structure of the *lac* repressor, *J. Biol. Chem.* **251**:3359–3366.

Schweizer, E., Mackechnie, C., and Halvorson, H. O., 1968, The redundancy of ribosomal and transfer RNA genes in *Saccharomyces cerevisiae, J. Mol. Biol.* **40**:261–277.

Scolnick, E. M., and Caskey, C. T., 1969, Peptide chain termination. V. The role of release factors in mRNA terminator codon recognition, *Proc. Natl. Acad. Sci. U.S.A.* **64**:1235–1241.

Scolnick, E., Tompkins, R., Caskey, T., and Nirenberg, M., 1968, Release factors differing in specificity for terminator codons, *Proc. Natl. Acad. Sci. U.S.A.* **61**:768–774.

Seeburg, P. H., Shine, J., Martial, J. A., Ullrich, A., Baxter, J. D., and Goodman, H. M., 1977, Nucleotide sequence of part of the gene for human chorionic somatomammotropin: Purification of DNA complementary to predominant mRNA species, *Cell* **12**:157–165.

Segawa, T., and Imamoto, F., 1974, Diversity of regulation of genetic transcription. II. Specific relaxation of polarity in read-through transcription of the translocated *trp* operon in bacteriophage lambda *trp, J. Mol. Biol.* **87**:741–754.

Seidman, J. G., Comer, M. M., and McClain, W. H., 1974, Nucleotide alterations in the bacteriophage T4 glutamine transfer RNA that affect ochre suppressor activity, *J. Mol. Biol.* **90**:677–689.

Sherman, F., Liebman, S. W., Stewart, J. W., and Jackson, M., 1973, Tyrosine substitutions resulting from suppression of amber mutants of iso-l-cytochrome *c* in yeast, *J. Mol. Biol.* **78**:157–168.

Shershneva, L. D., Venkstern, T. V., and Bayev, A. A., 1971, A study of tRNA methylases by the dissected molecule method, *FEBS Lett.* **14**:297–298.

Shimura, Y., Aono, H., Ozeki, H., Sarabhai, A., Lamfrom, H., and Abelson, J., 1972, Mutant tyrosine tRNA of altered amino acid specificity, *FEBS Lett.* **22**:144–148.

Shine, J., and Dalgarno, L., 1974, The 3′-terminal sequence of *Escherichia coli* 16S ribosomal RNA: Complementarity to nonsense triplets and ribosome binding sites, *Proc. Natl. Acad. Sci. U.S.A.* **71**:1342–1346.

Shine, J., and Dalgarno, L., 1975*a*, Determinant of cistron specificity in bacterial ribosomes, *Nature* **254**:34–38.

Shine, J., and Dalgarno, L., 1975*b*, Terminal sequence analysis of bacterial ribosomal RNA. Correlation between the 3′-terminal polypyrimidine sequence of 16S RNA and translational specificity of the ribosome, *Eur. J. Biochem.* **57**:221–230.

Simoncsits, A., Brownlee, G. G., Brown, R. S., Rubin, J. R., and Guilley, H., 1977, New rapid gel sequencing method for RNA, *Nature* **269**:833–836.

Smith, J. D., 1972, Genetics of transfer RNA, *Annu. Rev. Genet.* **6**:235–256.

Smith, J. D., 1976, Transcription and processing of transfer RNA precursors, *Prog. Nucleic Acid Res. Mol. Biol.* **16**:25–73.

Smith, J. D., and Celis, J. E., 1973, Mutant tyrosine transfer RNA that can be charged with glutamine, *Nature New Biol.* **243**:66–71.

Smith, J. D., Abelson, J. N., Clark, B. F. C., Goodman, H. M., and Brenner, S., 1966, Studies on amber suppressor tRNA, *Cold Spring Harbor Symp. Quant. Biol.* **31**:479–485.

Smrt, J., Kemper, W., Caskey, T., and Nirenberg, M., 1970, Template activity of modified terminator codons, *J. Biol. Chem.* **245**:2753–2757.

Söll, D., 1968, Studies on polynucleotides. LXXXV. Partial purification of an amber suppressor tRNA and studies on *in vitro* suppression, *J. Mol. Biol.* **34**:175–187.

Söll, D., and Schimmel, P. R., 1974, Aminoacyl-tRNA synthetases, in: *The Enzymes,* Vol. X (P. Boyer, ed.), pp. 489–538, Academic Press, New York.

Söll, D., Cherayil, J., Jones, D. S., Faulkner, R. D., Hampel, A., Bock, R. M., and Khorana, H. G., 1966, sRNA specificity for codon recognition as studied by the ribosomal binding technique, *Cold Spring Harbor Symp. Quant. Biol.* **31**:51–61.

Soll, L., 1974, Mutational alterations of tryptophan-specific transfer RNA that generate translation suppressors of the UAA, UAG and UGA nonsense codons, *J. Mol. Biol.* **86**:233–243.

Soll, L., and Berg, P., 1969, Recessive lethal nonsense suppressor in *Escherichia coli* which inserts glutamine, *Nature* **223**:1340–1342.

Sommer, H., Lu, P., and Miller, J. H., 1976, *Lac* repressor. Fluorescence of the two tryptophans, *J. Biol. Chem.* **251**:3774–3779.

Sprague, K. U., Steitz, J. A., Grenley, R. M., and Stocking, C. E., 1977, 3′ terminal sequences of 16S rRNA do not explain translational specificity differences between *E. coli* and *B. stearothermophilus* ribosomes, *Nature* **267**:462–465.

Sprinzl, M., Grüter, F., and Gauss, D. H., 1978, Collection of published tRNA sequences, *Nucleic Acids Res.* **5**:r15–r27.

Squires, C., Konrad, B., Kirschbaum, J., and Carbon, J., 1973, Three adjacent transfer RNA genes in *Escherichia coli, Proc. Natl. Acad. Sci. U.S.A.* **70**:438–441.

Steege, D. A., 1977, 5′ terminal nucleotide sequence of *Escherichia coli* lactose repressor mRNA: Features of translational initiation and reinitiation sites, *Proc. Natl. Acad. Sci. U.S.A.* **74**:4163–4167.

Steege, D. A., and Low, B., 1975, Isolation and characterization of lambda transducing bacteriophages for the *su1*$^+$ (supD$^-$) amber suppressor of *Escherichia coli, J. Bacteriol.* **122**:120–128.

Summers, W. P., Wagner, M., and Summers, W. C., 1975, Possible peptide chain termination mutants in thymidine kinase gene of a mammalian virus, herpes simplex virus, *Proc. Natl. Acad. Sci. U.S.A.* **72**:4081–4084.

Tate, W. P., and Caskey, C. T., 1974, The mechanism of peptide chain termination, *Mol. Cell. Biochem.* **5**:115–126.

Tate, W. P., Beaudet, A. L., and Caskey, C. T., 1973, Influence of guanine nucleotides and elongation factors on interaction of release factors with the ribosome, *Proc. Natl. Acad. Sci. U.S.A.* **70**:2350–2355.

Tate, W. P., Caskey, C. T., and Stöffler, G., 1975, Inhibition of peptide chain termination by antibodies specific for ribosomal proteins, *J. Mol. Biol.* **93**:375–389.

Thuriaux, P., Minet, M., Hofer, F., and Leupold, U., 1975, Genetic analysis of antisuppressor mutants in the fission yeast *Schizosaccharomyces pombe, Mol. Gen. Genet.* **142**:251–261.

Tompkins, R. K., Scolnick, E. M., and Caskey, C. T., 1970, Peptide chain termination. VII. The ribosomal and release factor requirements for peptide release, *Proc. Natl. Acad. Sci. U.S.A.* **65**:702–708.

Topisirovic, L., Villarroel, R., DeWilde, M., Herzog, A., Cabezón, T., and Bollen, A., 1977, Translational fidelity in *Escherichia coli:* Contrasting role of *nea*A and *ram*A gene products in the ribosome functioning, *Mol. Gen. Genet.* **151**:89–94.

Twardzik, D. R., Grell, E. H., and Jacobson, K. B., 1971, Mechanism of suppression in *Drosophila:* A change in tyrosine transfer RNA, *J. Mol. Biol.* **57**:231–245.

Valenzuela, P., Venegas, A., Weinberg, F., Bishop, R., and Rutter, W. J., 1978, Structure of yeast phenylalanine-tRNA genes. An intervening DNA segment within the region coding for the tRNA, *Proc. Natl. Acad. Sci. U.S.A.* **75**:190–194.

Weiner, A. M., and Weber, K., 1973, A single UGA codon functions as a natural termination signal in the coliphage QB coat protein cistron, *J. Mol. Biol.* **80**:837–855.

Weissmann, C., Taniguchi, T., Domingo, E., Sabo, D., and Flavell, R. A., 1977, Site-directed mutagenesis as a tool in genetics, in: *14th Miami Winter Symposium* (J. Schultz and Z. Brada, eds.), pp. 11–36, Academic Press, New York.

Weissenbach, J., Dirheimer, G., Falcoff, R., Sanceau, J., and Falcoff, E., 1977, Yeast $tRNA^{Leu}$ (anticodon U-A-G) translates all six leucine codons in extracts from interferon treated cells, *FEBS Lett.* **82**:71–76.

White, B. N., Tener, G. M., Holden, J., and Suzuki, D. T., 1973, Activity of a transfer RNA modifying enzyme during the development of *Drosophila* and its relationship to the su(s) locus, *J. Mol. Biol.* **74**:635–651.

Wilson, J. T., de Riel, J. K., Forget, B. G., Marotta, C. A., and Weissman, S. M., 1977, Nucleotide sequence of 3′ untranslated portion of human alpha globin mRNA, *Nucleic Acids Res.* **4**:2353–2368.

Wittmann, H. G., Stöffler, G., Piepersberg, W., Buckel, P., Ruffler, D., and Böck, A., 1974, Altered S5 and S20 ribosomal proteins in revertants of an alanyl-tRNA synthetase mutant of *Escherichia coli*, *Mol. Gen. Genet.* **134**:225–236.

Wosnick, M. A., and White, B. N., 1977, A doubtful relationship between tyrosine tRNA and suppression of the vermilion mutant in *Drosophila*, *Nucleic Acids Res.* **4**:3919–3930.

Wu, M., and Davidson, N., 1975, Use of gene 32 protein staining of single-strand polynucleotides for gene mapping by electron microscopy: Application to the $\phi 80d_3 ilvsu^+7$ system, *Proc. Natl. Acad. Sci. U.S.A.* **72**:4506–4510.

Yahata, H., Ocada, Y., and Tsugita, A., 1970, Adjacent effect on suppression efficiency. II. Study of ochre and amber mutants of T4 phage lysozyme, *Mol. Gen. Genet.* **106**:208–212.

Yaniv, M., Folk, W. R., Berg, P., and Soll, L., 1974, A single modification of a tryptophan-specific transfer RNA permits aminoacylation by glutamine and translation of the codon UAG, *J. Mol. Biol.* **86**:245–260.

Yanofsky, C., and Soll, L., 1977, Mutations affecting $tRNA^{Trp}$ and its charging and their effect on regulation of transcription termination at the attenuator of the tryptophan operon, *J. Mol. Biol.* **113**:663–677.

Yates, J. L., Gette, W. R., Furth, M. E., and Nomura, M., 1977, Effects of ribosomal mutations on the read-through of a chain termination signal: Studies on the synthesis of bacteriophage λ *O* gene protein *in vitro*, *Proc. Natl. Acad. Sci. U.S.A.* **74**:689–693.

Yoshida, M., Takeishi, K., and Ukita, T., 1970, Anticodon structure of a GAA-specific glutamic acid tRNA from yeast, *Biochem. Biophys. Res. Commun.* **39**:852–857.

Yoshida, M., Takeishi, K., and Ukita, T., 1971, Structural studies on a yeast glutamic acid tRNA specific to GAA codon, *Biochim. Biophys. Acta* **228**:153–166.

Yourno, J., and Tanemura, S., 1970, Restoration of in-phase translation by an unlinked suppressor of a frameshift mutation in *Salmonella typhimurium*, *Nature* **225**:422–426.

Zengel, J. M., Young, R., Dennis, P. P., and Nomura, M., 1977, Role of ribosomal protein S12 in peptide chain elongation: Analysis of pleiotropic, streptomycin-resistant mutants of *Escherichia coli*, *J. Bacteriol.* **129**:1320–1329.

Zilberstein, A., Dudock, B., Berissi, H., and Revel, M., 1976, Control of messenger RNA translation by minor species of leucyl-transfer RNA in extracts from interferon-treated L cells, *J. Mol. Biol.* **108**:43–54.

Zimmermann, R. A., Garvin, R. T., and Gorini, L., 1971, Alteration of a 30S ribosomal protein accompanying the *ram* mutation in *Escherichia coli*, *Proc. Natl. Acad. Sci. U.S.A.* **68**:2263–2267.

12

Regulation of the Protein-Synthesizing Machinery—Ribosomes, tRNA, Factors, and So On

O. MAALØE

1 *Introduction*

The long title of this chapter was suggested by the editor. I find it excellent because, right away, it points to the complexity and vagueness of the subject. This in turn forces the author to clarify his position. First of all I shall not deal with eukaryotic cells; partly because I lack the necessary specialized knowledge, but also because, as a rule, these cells contain more than one protein-synthesizing system (PSS), making them particularly difficult to analyze. So, this chapter is about the PSS in bacteria with the usual bias toward *Escherichia coli.*

The boundaries of this system and the concepts necessary to describe it have no commonly accepted definitions. In the most general sense the PSS of *E. coli* is the cell itself, since protein synthesis consumes some 80% of the carbon and energy that flow through the system (Lehninger, 1965). Of course, a growing bacterium is in constant rapport with the surrounding medium, and our goal must be to understand how a host of individual control mechanisms and couplings, most of them poorly defined, cooperate to produce the relationships we can

O. MAALØE • Institute of Microbiology, University of Copenhagen, Copenhagen, Denmark observe between the growth conditions on the one hand and the size to which the PSS develops on the other.

Work toward this goal has some rather stringent requirements. First of all I want to emphasize the need for reducing the complexity of the problem as best we can. The most important single step in this direction is *never* to start an experiment until the cells have grown for a long time in an essentially unchanging environment—that is, until they approach the ideal steady state of growth (Maaløe and Kjeldgaard, 1966). A second rule is never to rely on comparisons between strains whose genetic relationship is doubtful.

Of course, the final goal will not be reached until we have fairly exhaustive descriptions of the behavior of a great many components and control functions in the bacterial cell. Our present, incomplete knowledge is reviewed herein.

2 *The Concepts and Elements*

This section is largely descriptive; it deals with the basic concepts of regulation by control of transcription and with the elements of the PSS. The somewhat elaborate accounts that follow are needed to define the system and to introduce the many and diverse parameters and concepts later used to analyze it. The word *control* will be used to refer to processes whose rates are governed by the concentrations in the cells of one or several effectors; the word *regulation* is used in a broader, less specific sense.

2.1 *Characteristics of Control at the Operon Level*

A small number of operons in *E. coli* have been analyzed in great detail (namely, the negatively controlled *lac, trp,* and *gal* operons and the positively controlled *ara* operon); and in a fair number of cases similar mechanisms seem to operate. I shall stretch the available evidence by assuming that the synthesis of a large fraction of *E. coli* proteins is subject to control. In the present context fine points such as promoter and operator sequences are of little use, and the different types of control of protein synthesis will be collected under one heading: They can all be described as mechanisms through which *access to transcription* of an operon is made dependent on the presence in the cell of a class of small molecules, the cogent and highly specific effector. Most likely the gate controlled by an effector will be either closed or open at any given time, and the effector concentration will determine the fraction of time that the gate is open.* Notice that the effector,

*It might be argued that valuable information is sacrificed by adopting this simple and general description. This is not so, and for two reasons: First, without the detailed biochemical and structural work on a few operons generalization would certainly not be warranted; and second, in this chapter we are concerned mainly with problems that can be analyzed without knowing precisely how individual operators work. To put it differently, the flow through a set of gates can be analyzed if one knows how the signals are set to which the individual gate locks respond; the detailed construction of the locks is not relevant.

while regulating access to an operon, probably does *not* in any direct way influence the frequency of initiation of transcription (see Section 3.4.1).

In addition to repressor-mediated control, other mechanisms exist, such as catabolite repression and the specifically controlled *attenuation* of transcription, demonstrated in the *trp* operon (Yanofsky, 1976) and suspected to exist in other systems (see, for example, Artz and Broach, 1975). All these mechanisms serve to control access to transcription at individual promoters or at groups of promoters.

2.2 *Transcription and the Regulation of the Protein-Synthesizing System*

The subject we now discuss differs in two important ways from that of Section 2.1. When a suitable inducer, or a single amino acid (*aa*), is added to a bacterial culture the effect on the synthesis of a few specific proteins hardly affects the PSS. However, when the PSS as a whole is the target for regulation the situation is quite different: First, the changes in cell composition from which we infer that the PSS is under some form of regulation affect a large number and a large fraction of the cell's proteins; and second, these changes are the results of extensive rearrangements affecting the entire system, which we observe as transients—that is, shifts up or down—that lead from one steady state of growth to another.

We must distinguish between three different levels at which control of, and within, the PSS can be envisaged:

(a) Whatever the mechanism(s) responsible, the fraction of total protein that is r-protein (α_r)* is strikingly and positively correlated with the steady state growth rate (μ is doublings per hour)† of the culture (Section 3.1; Fig. 2). Thus, the parameter α_r monitors the degree to which a major component of the PSS responds when the cell adjusts to growth in a particular medium.

(b) Inside the cell a host of controls serves to regulate the flow of amino acids into protein. The efficiency of a ribosome actively engaged in protein synthesis (r_{act} denotes active ribosome) can be measured in several ways (see Section 2.4.1) and expressed as the average polypeptide chain growth rate (cgr_p). This important parameter turns out to be independent of μ, except under abnormal stresses (see Sections 4.3 and 4.4). In other words, the assimilation and/or synthesis of the *aa* is regulated such that all 20 amino acid pools are always kept at levels that suffice to maintain the high and constant cgr_p. Arguments that the observed cgr_p corresponds to near saturation of the system with *aa* were presented by Maaløe and Kjeldgaard (1966).

(c) Within the PSS proper, the *core* of Fig. 1, a third set of couplings and/or controls must operate; the syntheses of r-proteins, from a number of separate transcriptional units, and of rRNA appear to be tightly coupled although the mechanisms involved are unknown. Similarly, the rate of synthesis of some of the protein factors directly involved in protein synthesis seems to be balanced against the rate of ribosome synthesis (Gordon, 1970; Pedersen *et al.*, 1978*a*.). This harmony within the PSS appears in some cases to be the result of transcriptional

*See Section 2.4.4. Here and elsewhere r stands for ribosome or ribosomal.

†μ ln 2 is the specific growth rate. In steady states of growth the rate of synthesis of component X is therefore $dX/dt = X\ \mu \ln 2$.

coupling (see, for example, Lindahl *et al.*, 1975; Yamamoto *et al.*, 1976; Nomura, 1976) but differential effects of guanosine tetraphosphate (guanosine, 3′-diphosphate, 5′-diphosphate: ppGpp) on the synthesis of mRNA of various classes and of stable RNA may be involved (Travers, 1976*a,b*) (see Section 4).

All the interactions just listed have been more or less intensively studied, but mostly in isolation. A certain degree of specialization is unavoidable, and up to a point the properties of, say, the *lac* and *ara* operons can and must be studied in their own rights. However, when the data from such studies are used to suggest mechanisms for the reglation of an entity so large as the PSS, rules imposed by work outside the domain of the specialist become relevant. The following two sections contain a formal description of the bacterial cell as a self-contained PSS (Section 2.3), and an annotated list of parameters that have been studied as functions of μ (Section 2.4).

2.3 *Relations between the Major Synthetic Activities*

Figure 1 is constructed as a visual aid to show how the main synthetic activities in the bacterial cell are involved in regulation of the PSS (the pattern is that of another visual aid, the mandala). The outer ring represents the genome, and I have grouped the genes according to function: Group I is composed of the genes that specify the proteins of the PSS proper, the *core* in Fig. 1; group II, the rRNA and tRNA genes; group III, the genes involved in catabolism; and group IV, the genes relating to biosynthesis. Amphibolism, the functional overlap between groups III and IV, is discussed in Section 3.3.

Consider first the *core:* it is composed of ribosomes, tRNA, the DNA-dependent RNA polymerase (RNA-P), the aminoacyl-tRNA synthetases, plus all factors,

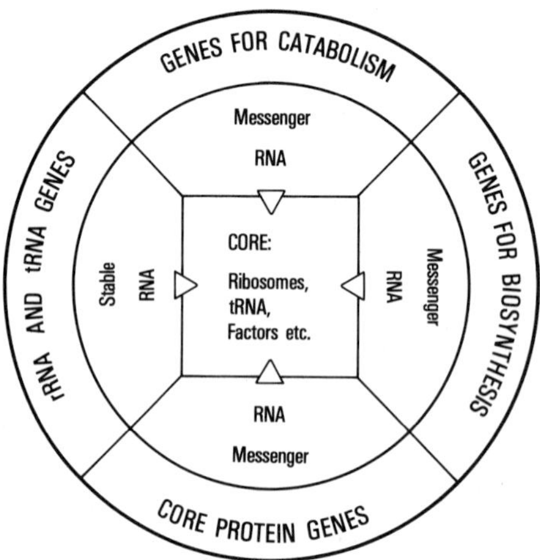

Figure 1. The PSS and the groups of genes upon which it depends. The outer ring, the genome, is divided into four parts. The core protein genes are described in the text as group I; the other groups are numbered II, III, and IV.

known and unknown, needed to ensure correct initiation, termination, and processing of the products of transcription as well as of translation. Among "factors" I include the proteins involved in the metabolism of ppGpp. An imaginary system containing these and no other elements would synthesize RNA and proteins if offered suitable DNA templates and adequate supplies of *aa* and ribonucleoside triphosphates. In principle, this *in vitro* system could be optimized with respect to the absolute and relative concentrations of all its component parts, including *aa,* triphosphates, and salts. The first criterion to be applied would be that the cgr_p should be brought up to the value characteristic of *in vivo* protein synthesis. A second criterion might be that small, but significant concentrations of ppGpp and pppGpp should be maintained as indication of an adequate but not excessive supply of *aa* (see Section 2.4.4). Such a balanced "core system" would be the ideal "S-30" for studying protein synthesis *in vitro.*

The purpose of this exercise is to single out a set of proteins, all of which (except, possibly, the stringent factor; see Section 4.3) are essential for RNA and protein synthesis to proceed as they do *in vivo* and then to examine the regulatory problems that emerge when the system is made dependent on active transcription from one or more of the four groups of genes of Fig. 1:

Group I, the core genes: The problem to consider first is very general: Is transcription organized so as to provide adequate and not excessive supplies of mRNA? Adequate amounts require a sufficient concentration of free, active RNA-P and enough open promoters (see Sections 2.4 and 3.4). *In vivo,* and at growth rates of one or more doublings per hour, we know that a high and constant fraction of all the ribosomes is active (Section 2.4.2). Under these conditions the mRNA supply therefore seems to be adequate, but transcription might still be in excess of demands. This brings up the important problem of coupling between initiation of transcription and of translation to be discussed in Section 3.4.2.

The core genes themselves must be regulated. Most of the r-proteins and some of the factors seem to be expressed coordinately, although they are known to be grouped into several transcription units (Gordon, 1970; Pedersen *et al.,* 1978a). But the synthesis of certain components of the core system, such as the ribosomal protein L 7/12 (Deusser, 1972), the α subunit of the RNA-P, and some of the synthetases (Reeh *et al.,* 1976), may be adjusted by secondary regulatory processes, although they clearly belong to the larger group of proteins that, collectively, increases with the growth rate more or less the way ribosomes do (see Figs. 2 and 7).

Group II, the stable RNA genes: With transcripts from the rRNA genes, complete ribosomes might be assembled in the core system, but the apparent coordination between the syntheses of rRNA and r-protein presents a serious problem since one round of transcription of the r-protein genes must be balanced by 20 to 30 rounds of transcription from one of the rRNA genes (Maaløe, 1969). This coupling could be achieved in several ways, but evidence for a specific mechanism is still lacking (Gausing, 1977).

The balance between rRNA and tRNA might be the result of coordinate transcription of all the corresponding genes, since in terms of nucleotides, the RNA in one ribosome corresponds roughly to one complete set of tRNA molecules. Assuming each species of tRNA to be represented by a single gene, and the rRNA by six or seven *rrn* genes per genome, the tRNA/rRNA ratio would not be far from that observed at medium and high growth rates (Section 2.4.2). However, at low growth rates this ratio increases, and, moreover, individual tRNA species seem to be synthesized in different amounts; this could reflect differences in promoter strength, but more sophisticated controls may be involved.

Group III, the genes for catabolic enzymes and their control proteins: Expression of these genes would provide the system with the capacity to generate its own triphosphates, and it would now be able to carry out most of the important syntheses of a cell growing in a medium containing all the amino acids, nucleosides, and vitamins. Only DNA replication and lipid synthesis, which, together, account for no more than 5 to 10% of the carbon and energy consumed during growth (Lehninger, 1965), are still left out. Group III displays some of the best known cases of transcriptional control, the *lac, gal,* and *ara* operons. Unfortunately, little is known about regulation of the synthesis of the large number of enzymes responsible for some of the indispensable catabolic and amphibolic pathways (see, for example, Henning *et al.,* 1968). The demand made on these enzyme systems under various conditions of growth is discussed in Section 3.

Group IV, the genes for biosynthetic enzymes and control proteins: When this group is included, the entire genetic potential of the cell is accounted for (including DNA replication and lipid biosynthesis). The overall system is now capable of all the syntheses needed for growth in a minimal salts medium containing one or another of a number of different carbon and energy sources. Not many of the biosynthetic operons have been thoroughly analyzed, and only the *trp* operon is understood in the same degree of detail as, say, the *lac* operon. Nevertheless, from the point of view of the overall balance of gene expression a very important principle has emerged: In the cases so far examined, the biosynthetic operons turn out to be under considerable *internal* repression when cells grow in a simple glucose minimal medium (Yanofsky and Ito, 1966). From this intermediate position, repression is known to increase if the appropriate effector is added to the medium. Often ignored, however, is the concept that *derepression* must be envisaged if the intracellular concentration of this effector were to decrease. It is therefore important to learn something about effector concentrations in cells growing under various conditions. Data are now becoming available for some of the amino acids, and their relevance to the general problem of gene expression is discussed in Section 3.3.

2.4 *Parameters Characterizing Steady States of Growth*

The various measurements reviewed in the following were carried out over a wide range of growth rates (μ). The maximum value of μ is around 2.8 doublings per hour, and in some cases glucose-limited chemostat cultures were used to realize a μ value as low as about 0.1. As discussed in detail in Section 3.1, most of the parameters studied bear simple relations to μ at medium and high growth rates—that is, they are invariant or they are linearly related to μ. At low growth rates ($\mu < 1.0$) systematic deviations from these simple rules are observed, as discussed in Section 3.2.

Within the accuracy of the measurements, polypeptide and RNA chain growth rates appear to be truly invariant with μ. The average cgr_p has been measured by several methods: from the RNA/protein ratio combined with estimates of the number of active ribosomes (see Section 2.3.2)*; from the kinetics of labeling of polypeptide chains of various lengths (Gausing, 1974); from the time of appearance of N-terminal label in free, finished β-galactosidase molecules (Engbaek *et al.,* 1973); and from the induction lag in β-galactosidase synthesis (Coffman *et al.,* 1971; Johnsen *et al.,* 1977) (see Fig. 4). In steady states of growth the average cgr_p is 17 to 18 amino acids per second at 37°C, with 16 to 20 as limits. A few estimates falling below this range have appeared in the literature, and, as far as the medium to high growth rates are concerned, they may be disregarded because they fail to account for the overall rate of protein synthesis. The reasons for asserting that the cgr_p remains high and constant even at the lowest growth rates are discussed in Section 3.2.1.

Recent work by Molin (1976) shows that the average chain growth rate of ribosomal RNA (cgr_{rRNA}) is invariant with μ. His technique is simple and the data obtained are correspondingly easy to interpret. It should be emphasized, however, that the total length of the rRNA transcript, before trimming, is not precisely known and that the average cgr_{rRNA} therefore must be stated in terms of a *minimal* number of about 60 ribonucleotide residues per second. Molin's data actually leave open the possibility that the average chain growth rate of messenger RNA (cgr_{mRNA}) may be significantly lower than the average cgr_{rRNA}.

2.4.2 *Partitioning of the Stable RNA*

Recent measurements by column chromatography indicate that the tRNA/total RNA ratio varies little, if at all, for μ values over about 0.6. In this range the best estimate of this ratio is around 0.12; however, at very low growth rates it gradually increases to about 0.2 to 0.25 (see Fig. 5). This increase could reflect breakdown of rRNA (see Section 4.1.1) without damage to the tRNA cotranscribed with it. Early experiments with *Salmonella typhimurium* suggested that the tRNA/total RNA ratio increased as the growth rate decreased (Maaløe and Kjeldgaard, 1966). This discrepancy is probably due to the lack of specificity of the gradient centrifugation technique used in the early studies (Kjeldgaard, 1967).

The fraction of rRNA that is in ribosomes actively engaged in protein synthesis—that is, in polysomes—has also been measured as a function of μ (Forchhammer and Lindahl, 1971). At medium to high growth rates, approximately 15% of all ribosomes are present as free 70 S, 50 S, and 30 S particles, and the remaining 85% are associated with mRNA in polysomes. Thus, the RNA in

*If N *aa* constitute a representative sample of a steady state system doubling μ times per hour, then $dN/dt = N\mu \ln 2/3600$ *aa* per second. The average cgr_p is obtained by dividing this rate by the number of active ribosomes corresponding to N. This number is k_1k_2N(RNA/protein), where k_1 is the average molecular weight of an amino acid residue divided by the molecular weight of the ribonucleotide residues in a ribosome, and k_2 is the fraction of all RNA found in polysomes (see Section 2.4.2). Figure 5 shows a set of measurements from which such calculations have been made.

active ribosomes, r_{act}, represents about 75% of the total RNA. At very low growth rates this fraction is greatly reduced, as discussed in Sections 3.2 and 4.3.

2.4.3 *Amounts and Stability of mRNA*

The messenger activity, per milligram of RNA, has been measured *in vitro* as a function of μ (Forchhammer and Lindahl, 1971). Again, at medium to high growth rates no variation was observed, and it was concluded that the messenger activity per ribosome did not vary significantly with μ. At the lowest growth rate examined, the activity per milligram of RNA seemed to be somewhat reduced.

Several attempts have been made to measure the mRNA/total RNA ratio. The techniques used so far are discussed by Summerton (1976), who designed a new and improved method. He estimates that 4.1% of the RNA in cells of *E. coli,* strain B, in a Tris–glucose–amino acid medium is unstable. Based on hybridization experiments, Norris and Koch (1972) estimated the mRNA content to be between 3 and 4.5%, and they found it to be independent of μ. The instability of mRNA is such that to maintain a mRNA/total RNA ratio of about 0.04, stable and unstable RNA must be synthesized at comparable rates. The instantaneous rate of total RNA synthesis can be measured from labeling kinetics; by hybridizing to DNA carrying the rRNA genes and by taking into account the known contribution of tRNA, the rate of mRNA synthesis can be calculated by subtraction from the figure for total RNA. Extensive measurements of this kind have been done by Gausing (1977), and her conclusion is that the calculated mRNA/total RNA ratio is about 0.04, and that it varies little, if at all, between μ = 2.8 and μ = 0.2 (K. Gausing, personal communication). Thus, the slight drop in messenger activity per milligram of RNA seen by Forchhammer and Lindahl (1971) at μ = 0.5 is probably not significant. The constant mRNA/total RNA ratio will be used later to estimate frequencies of initiation of transcription (Section 3.4).

Messenger stability has been evaluated in a number of ways: from the kinetics of labeling during the "run-out" of bulk protein synthesis after addition of rifampicin (Pato and v.Meyenburg, 1970); from the kinetics of induction and *de*induction (see, for example, Johnsen *et al.*, 1977); and from specific hybridization experiments (Forchhammer *et al.,* 1972). Most individual species of mRNA have half-lives between 1 and 2 min at 37°C, and little, if any, variation with μ has been observed. Pedersen *et al.* (1978*b*) report a mean value of 2.5 min at 30°C, corresponding to about 1.6 min at 37°C. The messengers for proteins that can be produced in very large amounts may have relatively long half-lives; thus the *tuf*A messenger (elongation factor TuA) has a half-life of 2.6 min at 37°C. The values registered by Pedersen *et al.* (1978*b*) range from about 30 s (r-protein S1, 37°C) to about 12–14 min (unidentified protein, 37°C). The run-out experiments just quoted show that messengers with exceptionally long half-lives code for no more than one or a few percent of the protein of *E. coli.*

2.4.4 *The Relative Rate of Synthesis of the Ribosomal Proteins, α_r*

The relative rate of synthesis of the ribosomal proteins, α_r, was first considered by Schleif (1967), and his technique has been refined by Gausing (1974). The characteristic and monotonic increase of α_r with μ is analyzed in detail in Section

3.1; here, it will suffice to recall that, in a steady state of growth, the relative rate of synthesis equals the relative amount. Thus, α_r becomes an estimate of the r-protein/total protein ratio.

2.4.5 *The Relative Rate of Synthesis of Proteins Other than the r-Proteins*

The powerful two-dimensional gel technique of O'Farrell (1975) has been used to study the synthesis of a large number of *E. coli* proteins as a function of μ (Pedersen *et al.*, 1978*a*). At the moment, only a few of the 140 proteins monitored have been identified, but the data obtained so far are highly relevant for this study. Some of the present uncertainties may disappear when more proteins have been identified (see Section 3.1).

2.4.6 *Levels of Guanosine Tetraphosphate*

Guanosine tetraphosphate, an effector that signals suboptimal charging of one or more species of tRNA, has been implicated in many control schemes. It is important to note that ppGpp concentrations may vary over a 1000-fold range under extreme conditions, whereas the steady state levels increase only three- to fourfold as μ decreases from about 2.8 to 0.6. The ppGpp level probably increases further at growth rates below $\mu \simeq 0.6$. The involvement of ppGpp in regulation of the PSS is discussed at the end of Section 3.1 and in Section 4.

2.4.7 *Relative Gene Numbers*

The pattern of DNA replication is such that the ratio between the numbers of copies of early and late replicating genes increases with μ (Yoshikawa and Sueoka, 1963; Chandler and Pritchard, 1975). The facts that replication is bidirectional and that the rRNA and most of the r-protein genes cluster on either side of the origin of replication need to be taken into account in discussing the possible mechanism(s) that govern the relationship between μ and the number of ribosomes in a cell (see Fig. 3).

3 *Patterns and Frequencies of Transcription*

In a steady state of growth all the operon controls must have assumed definite settings, or modes of oscillation, and averaging over the cell population, transcription will be initiated at a characteristic frequency at a given promoter. Let p_i be the promoter of the ith transcriptional unit from which the corresponding species of mRNA and of protein are derived. Two different but equally important parameters are needed to define the transcriptional activity at p_i: first, the probability that a promoter of this class is open, $p_i(\mathrm{o})$; and second, the frequency, F_i, with which the synthesis of a productive mRNA of class i will be initiated once p_i is open. Evidently, the average number of transcripts initiated at an i promoter per time unit is $p_i(\mathrm{o})F_i$. The sum of these products taken over all promoters of all classes in a culture of unit volume and density will be referred to as the transcriptional cross section of that culture. This sum represents the intensity of transcription required to sustain the overall rate of protein synthesis, dP/dt; and the individual terms of the sum define the pattern of transcription and of protein synthesis characteristic of the state of growth of the culture.

Note that $p_i(o)$ approaches unity with increasing degrees of induction or derepression, and that there must be an upper limit, $F_i(\max)$, for the frequency with which transcription can be initiated at p_i. Some promoters appear to have very low $F_i(\max)$ values, and we shall later examine how this affects the pattern of protein synthesis at different growth rates (see Section 3.4).

Several years ago, an attempt was made to analyze the regulation of the PSS of *E. coli* in terms of the pattern of transcription (Maaløe, 1969). The analysis was based on the obvious fact that, in a system with a finite capacity for protein synthesis, the share allotted to one species of protein cannot be changed independently of the pattern of syntheses as a whole. In particular, I explored the possibility of controlling the synthesis of the r-proteins indirectly, or passively, by increasing or decreasing access to mRNA transcription in other parts of the genome. Since that time, more has been learned about the general properties of the PSS, as detailed in Section 2.4, and some arguments against passive control of r-protein synthesis have been advanced. I shall therefore reexamine the situation, beginning with the range of medium and high growth rates, in which passive control may be an important factor (Section 3.1). At the low growth rates (Section 3.2) other, or additional, factors should probably be looked for. The complex mechanism of initiation of transcription must be considered throughout (see Section 3.4).

3.1 Protein Synthesis at Medium and High Growth Rates

Let us consider a quantity of protein with a total of N *aa*, constituting a representative sample from a steady state system with a growth rate of μ doublings per hour. The rate of increase of N, dN/dt, which equals $N\mu \ln 2$, can also be expressed as the product of the number of active ribosomes that corresponds to the quantity of protein represented by N, $r_{act}(N)$, and the polypeptide chain growth rate, cgr_p:

$$dN/dt = r_{act}(N) \times cgr_p = N\mu \ln 2$$

We have evidence that the cgr_p is independent of μ (Section 2.4.1) and that $r_{act}(N)$ is a constant fraction of all the ribosomal elements, r_{total}, except at low growth rates (Section 2.4.2). Like r_{act}, r_{total} is therefore proportional to μ at medium and high growth rates. This means that the efficiency of the ribosomes is constant. This important property of the PSS can also be deduced from direct measurements of $\alpha_r = dN_r/dN$. The extensive data obtained by Gausing (1974) are reproduced in Fig. 2. They show that α_r is proportional to μ for growth rates higher than about 1.0 doubling per hour. Since α_r is an estimate of the ratio N_r/N in steady states of growth, and since the pool of free r-proteins is very small (Gausing, 1974), the proportionality between r_{total} and μ is independently established by the α_r measurements.

It was shown by Gordon (1970) that elongation factors, G and Ts, are synthesized more or less coordinately with the ribosomes. This observation has been confirmed and extended by Pedersen *et al.* (1978*a*), who quantitated 140, largely nonribosomal, *E. coli* proteins. About 50 of these proteins, including those identified as core proteins, behaved like the r-proteins, as illustrated in Fig. 2. The authors draw the plausible conclusion that these 100 or so polypeptides comprise most of the known and unknown proteins directly involved in protein synthesis, namely, the core proteins of Fig. 1. The r-proteins represent about half of this class (by weight) and at maximum growth rate, this class accounts for some 50% of the total protein of the cells. It is the regulation of this class of essential proteins (which are indicated by the subscript ess throughout this text) with which we are now concerned. The fraction of total protein that makes up this class is the α_{ess} of Fig. 2.

The extreme opposite of passive control, as already described, would be a situation in which the core genes were specifically and increasingly repressed as μ decreases, while no change in degree of repression or induction occurred elsewhere in the genome. The two opposing views can be compared by considering the transcriptional cross section which was defined as the sum, taken over all promoters of all mRNA classes, of the product $p_i(\mathrm{o})F_1$. The first thing to observe is that only acts of transcription which are followed up by translation are relevant when protein synthesis is being measured, and that the rate at which *productive* mRNA molecules are produced therefore must be limited by the number of active

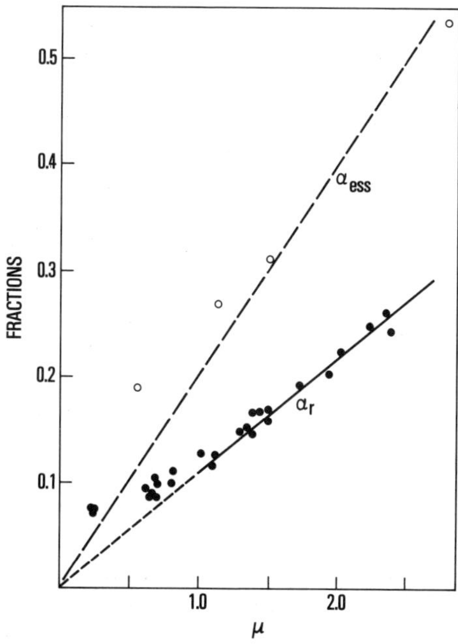

Figure 2. Relative rates of synthesis of the r-proteins, α_r, and of all the core proteins, α_{ess}. The α_r and α_{ess} values were measured by Gausing (1976) and by Pedersen *et al.* (1978*a*), respectively, and they are redrawn and reproduced with permission of the authors and publishers. These relative rates were measured on steady state cultures; therefore, they are equal to the relative abundances of the various proteins.

ribosomes in the system (see Section 3.4.1). Second, it is to be noted that the term $F_i(\max)$ expresses "promoter strength," or efficiency, and that differences between promoters in this respect affect the pattern of protein synthesis, especially at high growth rates when the overall rate of initiation of mRNA synthesis is high (see Sections 3.4.3 and 3.4.4).

For the moment, we shall disregard such differences and concentrate on the parameter $p_i(\text{o})$ which indicates to what extent the ith operon is available for transcription. If all promoters were of equal strength, the relative yield, α_i, obtained from this operon would be $p_i(\text{o})$, times the number (a_i) of i operons in the system, divided by the sum of such terms taken over all n transcriptional units from which mRNA is synthesized*:

$$\alpha_i = a_i p_i(\text{o}) \Big/ \sum_{i=1}^{i=n} a_i p_i(\text{o})$$

The denominator can be divided into the sums SI, SIII, and SIV, taken over the promoters of the core genes, the genes for catabolism, and the genes for biosynthesis, respectively (gene groups I, III, and IV of Fig. 1). In this notation we can write

$$\alpha_{\text{ess}} = \text{SI}/(\text{SI} + \text{SIII} + \text{SIV})$$

and as Fig. 2 shows, α_{ess} equals approximately $2\alpha_{\text{r}}$.

It is now possible to illustrate the basic difference between passive and active control of the synthesis of the core proteins in a simple manner. Passive control is based on the notion that the core proteins are synthesized constitutively; in other words, that the term SI is invariant with μ (when correction is made for the fact that the core genes tend to cluster on either side of the origin of replication). The observed decrease in α_{ess} with μ is thus thought to reflect derepression of many operons in sectors III and IV; this is illustrated in Fig. 3A. The opposite view, totally active control, implies the existence of a control system whose effector or effectors act on the promoters of the core sector causing SI to decrease sufficiently to account for the changes in α_{ess}. This extreme and somewhat unrealistic model is illustrated in Fig. 3B; it is presented as a boundary condition on the basis of which we may try to judge, in a qualitative way, the extent to which factors *other* than specific control of transcription from the core genes may contribute to the observed changes in their relative output.

Below the maximum growth rate ($\mu \approx 2.8$) the distances between the two curves in Fig. 3B probably underestimate (SIII + SIV), representing the noncore genes. Changing from rich to minimal medium, this cross section almost certainly expands as a result of derepression of many operons. First, and most obviously, it

*Note that the degree of repression exerted on the ith operon, $p_i(\text{o})$, does not define the relative yield of the corresponding protein(s), p_i, unambiguously. Imagine two states of a system that differ only in the sense that in one state all operons are fully open and in the other they are repressed to the same degree: In both states the overall rate of protein synthesis is set by the number of r_{act} in the system, and the relative yields of the individual proteins, which are defined by the *partitioning* of transcription, will also be the same. On the other hand, if *one* of the operons were insensitive to repression (of any kind), the *relative* yield of the protein(s) encoded in that operon would increase with the degree of repression exerted on all the other operons. Thus, a single, fully constitutive operon might act as a probe monitoring the general state of repression throughout the genome; this notion is developed further in Section 3.4.3.

expands in the biosynthetic segment of the genome (group IV in Fig. 1). It probably expands in the catabolic segment as well (group III), because in a minimal medium all the carbon needed for biosynthesis must be supplied by the catabolic pathways *(amphibolism)*.

The extent of expansion of the transcriptional cross section cannot be assessed now, but it may be recalled that many biosynthetic operons are under considerable internal repression during growth in a glucose minimal medium ($\mu \approx 1.5$), and that there is room not only for the increased repression observed in rich media, but also for relaxing repression when changing to media giving growth rates below 1.5 doublings per hour (Maaløe, 1969). This problem is discussed in Section 3.3, where we consider the actual concentrations of some of the *aa* as they relate to μ.

It must be recalled that, up to this point, the control of synthesis of the core proteins has been discussed on the assumption that all promoters are of equal "strength"—that is, one and the same $F_i(\max)$ for all of them. This is almost certainly not true and a realistic description of the system requires that the effects of differences between promoters in this respect be analyzed. This is done in Sections 3.4.3 and 3.4.4, where it is shown that sufficiently low $F_i(\max)$ values cause the relative rate of synthesis, α_i, of the corresponding protein(s) to decrease when μ increases. Phenotypically, this phenomenon cannot be distinguished from repression, and since it increases with μ, it contributes to the passive mode of control in the same way as does repression mediated by specific effectors.

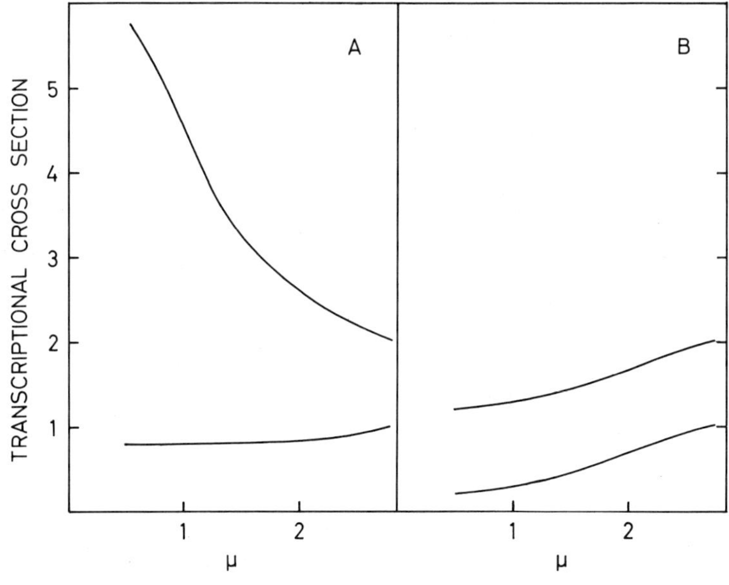

Figure 3. The transcriptional cross section for core and noncore mRNA. The upper curves represent the total cross sections, which can be divided into the sums SI, SIII, and SIV (see text). The lower curves show the core cross section, SI, by itself. (A) Passive control—that is, constitutive synthesis of all core proteins (SI is corrected for gene position, assuming the core genes duplicate during the first 20% of the replication cycle). (B) Totally active control. The graph is constructed such that, at all μ values, the ordinate on the lower curve (SI) divided by the ordinate on the upper curve (SI + SIII + SIV) equals α_{ess} (estimated from the α_r values of Fig. 2). SI has arbitrarily been assigned unit value at μ_{max}.

Finally, I want to make a point about frames A and B of Fig. 3 on the one hand, and the measurements by Pedersen *et al.* (1978*a*) of the relative yields of many of the *E. coli* proteins, at several growth rates, on the other. Their data are extremely valuable, and will increase in value as more proteins are identified; however, they do not help us much in distinguishing between the relative importance of active and passive regulation of the synthesis of the core proteins. This can be seen as follows: At maximum growth rate the total transcriptional cross section is equally divided between the core and noncore contributions, each of which has been assigned unit value in Fig. 3. The high, absolute rate of synthesis of the core proteins characteristic of the rich medium may require that at least some of the core operons are fully open.* In glucose minimal medium μ is reduced from 2.8 to 1.5 at 37°C, and the overall rate of protein synthesis, dP/dt, and the ratio of core to total protein, α_{ess}, are reduced by the same factor. Intuitively, one might think that this could be achieved by closing all core operons down in proportion to the change in growth rate. However, to achieve the desired effect on α_{ess}, the transcriptional cross section would have to remain constant, necessitating an expansion of the cross section representing the noncore operons. *Within* the latter, the individual operons could be readjusted to fit the partitioning of the noncore proteins observed in glucose minimal medium. What this example shows is that the degree of active control can be chosen freely as long as the availability of templates for transcription does not limit the growth rate and as long as the noncore cross section *can* expand adequately.

Note that this treatment of passive control as an obligatory, but quantitatively ill-defined, element in the regulation of the PSS does not take into account the synthesis of the stable RNA species. This reflects the author's belief that rRNA and tRNA synthesis is regulated by being coupled somehow to protein synthesis. A formally simple, positive control mediated by an r-protein has been suggested (Maaløe, 1969), but without additional, *ad hoc* assumptions it fails to account for the known cases in which rRNA and protein synthesis seem to have been uncoupled. Thus, when a *relA*$^-$ strain, growing in glucose minimal medium, is starved for a required *aa*, or when chloramphenicol is added to a similar culture of a *relA*$^+$ strain, the rate of rRNA and tRNA synthesis quickly increases beyond the steady state rate, even if protein synthesis is massively inhibited (Kurland and Maaløe, 1962). To a lesser degree, uncoupling is observed in steady states of growth at very low rates (Gausing, 1977). It is therefore most natural to envisage a mechanism involving *negative* control, and ppGpp is the one known effector that, by itself or in combination with unidentified cofactors, might control the rate of synthesis of rRNA and tRNA. This problem is discussed in Section 4.1.1.

3.2 *Protein Synthesis at Low Growth Rates*

It is common belief that bacteria, such as *E. coli,* compete in their natural habitat by growing fast and efficiently when conditions permit, and by surviving

*Throughout this general discussion the core genes are treated as a unit since they respond very similarly to changes of μ. Physical limitations on transcription frequencies (Section 3.4.3) and second-order control mechanisms seem to exist and to make adjustments *within* the core group of proteins (Reeh *et al.,* 1976).

well under adverse conditions (see, for example, Koch, 1971). We have just seen that when growth at medium or high rates is possible, the PSS operates with a high and constant efficiency. As μ falls below about one doubling per hour, however, the efficiency of the PSS decreases. This is evident from Fig. 2, where the deviation from proportionality between μ and α_r is seen to become more and more pronounced as μ decreases.

3.2.1 Polypeptide Chain Growth at Low Growth Rates

Formally, the loss of efficiency of the PSS at low growth rates could be due to one cause or to a combination of causes: (a) the cgr_p might go down; (b) protein turnover might increase; (c) initiation of translation might become less and less efficient; and/or (d) transcription might be reduced so that a decreasing fraction of the ribosomes could be accommodated on the available mRNA. Mechanisms (c) and (d) might both be significant, but the evidence for this cannot be presented until the regulatory role of ppGpp at low growth rates is examined in Section 4.4. Meanwhile, the first two possibilities are discussed.

(a) It has already been stated, quite categorically, that the cgr_p equals about 17 to 18 *aa* per second, and that it is invariant with μ (see Section 2.4.1). This property of the PSS is of the utmost importance for the present analysis, and it is therefore necessary to go into detail about experiments that seem to disagree either with our estimates of the rate itself or with its constancy. At medium and high growth rates, a high and constant fraction of all the ribosomal particles appears to be active in protein synthesis, and in this range a cgr_p of less than 17 to 18 *aa* per second is in fact insufficient to account for the overall rate of protein synthesis, dP/dt. This can be calculated on the basis of corresponding values of μ and of the RNA/protein ratio (see Section 2.4.1). A large number of chemical and radiochemical determinations of this ratio have been made, and at μ values above unity they agree with the several more direct measurements of the cgr_p referred to in Section 2.4.1. Thus, the low estimates of the cgr_p implied in the paper by Baker and Yanofsky (1968), and reported by Talkad *et al.* (1976), must be in error. The latter put the average cgr_p at 11 *aa* per second, and argued that it varies between 8 and 15. They studied β-galactosidase synthesis and their data can therefore be compared with the results of Engbaek *et al.* (1973), who found that N-terminal label appears in that enzyme after a lag of 60 to 70 s and thereafter accumulates at a constant rate. These experiments are sensitive to variations around the average cgr_p, and if the time of synthesis varied substantially from one β-galactosidase molecule to another the N-terminal label would, initially, accumulate at an increasing and not at a constant rate, as observed.

Below a μ value of about unity, estimates of the cgr_p based on the RNA/protein ratio keep going down with μ, whereas most direct measurements indicate that the rate remains unchanged. It has been argued, very tentatively, that the cgr_p might be somewhat reduced at low growth rates (Maaløe, 1969), and experiments by Forchhammer and Lindahl (1971) and by Dalbow and Young (1975) seem to support this view. However, the lowest μ values in their studies were between 0.50 and 0.65 and the interpretation of the data required several assumptions.

This uneasy situation was changed by the work of Coffman *et al.* (1971), which was continued and extended in our laboratory by Helle Jacobsen (unpublished thesis, 1974). In both studies, the time needed to complete the synthesis of a β-galactosidase molecule was used to estimate the cgr_p, and measurements were extended to very low values of μ. Figure 4 shows that the induction kinetics are identical in a glucose batch culture and a glucose-limited chemostat culture with μ values of 1.5 and 0.1, respectively. Here we need not be concerned with the precise time course of the process of induction or with the assembly of the four monomers of β-galactosidase; the important feature of the experiment is that no significant change that could be interpreted as a decrease of the cgr_p is seen at μ values way below the region discussed by Dalbow and Young (1975) and others. It is therefore unlikely that the trend toward lower values of the cgr_p they proposed is real, and it is certain that the cgr_p does not keep decreasing with μ.

Such a strong statement should not be made on the basis of the induction kinetics of a single enzyme, and I want to emphasize that this and other arguments about the cgr_p are based on a whole series of measurements by various methods and involve bulk protein synthesis as well as individual proteins. The data from all these experiments demonstrate that the cgr_p in steady states of growth at 37°C is 16 to 20 *aa* per second, most likely 17 to 18 (Gausing, 1972; Engbaek *et al.*, 1973; Coffman *et al.*, 1971; v.Meyenburg, 1971; H. Jacobsen, unpublished thesis; Johnsen *et al.*, 1977). Later, we shall discuss special cases of stress, which has the effect of reducing the cgr_p (see Sections 4.2 and 4.3).

(b) Protein turnover could seriously affect our estimates of the cgr_p. Nath and Koch (1970) examined the stability of the proteins of the B strain of *E. coli* over a wide range of growth rates. For $\mu > 0.6$, an unstable fraction of 3 to 4% and with a half-life of about 1 h was detected; at lower μ values this fraction decreased but so did the half-life. If their figures account for the total turnover of protein during normal growth, some 20 to 30% of the protein synthesized at μ values as

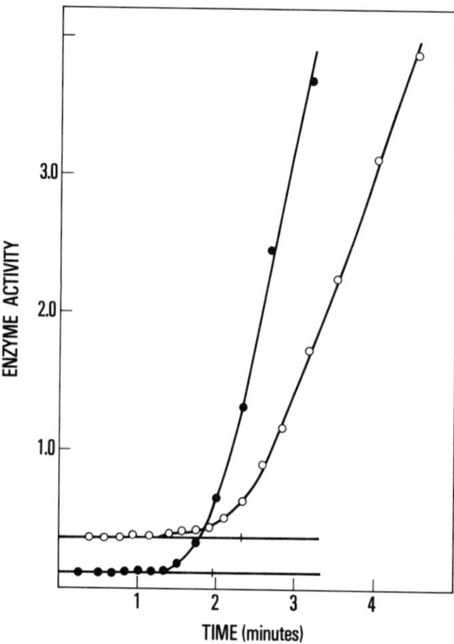

Figure 4. Induction lags at extreme growth rates. At t = O, IPTG was added to induce maximal synthesis of β-galactosidase. The ordinate shows enzyme activity in arbitrary units. •, μ = 1.47 in glucose minimal medium; ○ μ = 0.1 in glucose-limited chemostat. The data are reproduced from an unpublished thesis by Helle Jacobsen (1974), with the author's permission.

low as 0.02 would be unstable. However, at the very low growth rates, two-thirds or more of the protein would have to turn over to match the apparent excess of ribosomes (see Fig. 2).

The experiments of Nath and Koch (1970) were not designed to register the kind of rapid turnover elicited by fusidic acid, which causes premature termination of peptide chains and rapid degradation of proteins (Hansen *et al.*, 1973). Similar experiments were done during steady states of growth (K. Gausing, personal communication), and no trend toward greater instability at the lower growth rates was observed. It should also be noted that, according to Pine (1970), some 5% of the leucine incorporated during a 1-min pulse administered to a rapidly growing culture become acid soluble during a 1-min chase. A considerable fraction of this decay may represent removal of only the N-terminal segments of polypeptide chains and thus be independent of μ.

All told, protein turnover does not seem to contribute very significantly to the progressive loss of efficiency of the ribosome population at the low growth rates.

3.2.2 *The Energetics of Growth and of Protein Synthesis*

One of the basic observations in bacterial growth physiology is that, at a given temperature in the range from about 20 to about 40°C, the composition of the cells in terms of protein, RNA, and DNA bears a unique relation to the growth rate. This can be illustrated by comparing a slow-growing batch culture with, say, succinate as carbon source on the one hand, and a glucose-limited chemostat culture adjusted to give the same growth rate on the other. As Fig. 5 shows, the composition of the cells in two such cultures is the same. In the case of a transport mutant isolated and studied by v.Meyenburg (1971), μ can be varied by changing the concentrations of lactose in the medium; again the RNA/protein ratio observed at a given μ value is identical to the ratio found in a batch culture growing at the same rate but with a different carbon source.

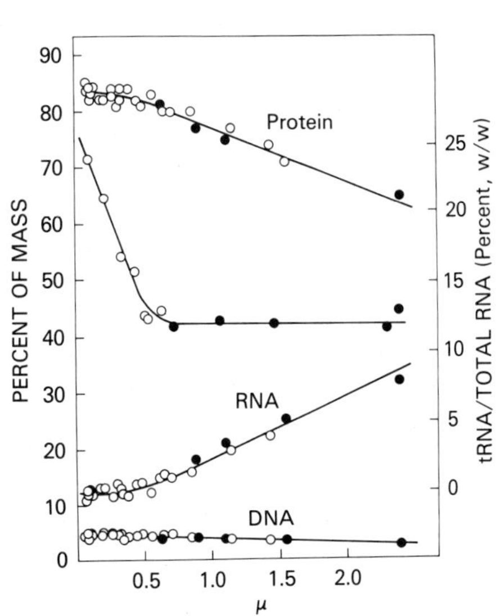

Figure 5. Cell composition at high, medium, and low growth rates. •, Batch cultures with various carbon sources and, for the highest μ values, with addition of *aa;* ○, growth in a glucose-limited chemostat. Before sampling, each culture had been observed to grow for several doublings at a constant μ. The data are from an unpublished thesis by Helle Jacobsen (1974), reproduced with the author's permission.

It is not obvious why a unique cell composition should establish itself at a given growth rate, irrespective of the chemical nature of the primary carbon source. However, recent work by Andersen (1974) in our laboratory may provide a lead. He measured oxygen consumption and CO_2 production during balanced, aerobic growth in batch and chemostat cultures representing growth rates and carbon sources similar to those of the experiments illustrated in Fig. 5. The results he obtained with *batch* cultures can be summarized in a surprisingly simple way—namely, the oxygen consumption per minute and per gram of cell material varies little, if at all, with μ.

This statement implies that in order to double a given quantity of protein the amount of energy consumed is inversely proportional to μ (disregarding minor corrections for maintenance energy). The greater part of the energy involved is used to activate *aa* and link them into peptide chains, and to support the necessary biosyntheses. The energy needed per step in polypeptide synthesis can be assumed to be independent of μ. We must therefore conclude that the energy requirements for the biosyntheses depend on the chemical nature of the carbon source, and that the growth rate decreases the higher this requirement is. Andersen has calculated that the excess energy needed to double the cell mass with, say, acetate as carbon source relative to the requirement with glucose, agrees reasonably well with available biochemical data. The main increase in energy consumption in a case like this seems to be due to the need to force many biosynthetic reactions to proceed "upstream," with the formation of a large number of carbon-to-carbon bonds not already present in the substrate molecules. As a working hypothesis, Andersen suggested that cells growing aerobically, with nonlimiting substrate concentrations, develop a capacity for oxidation that cannot be exceeded for physical reasons, and that the growth rate that results in a given medium depends on the efficiency with which biosynthesis proceeds.

This hypothesis is simple and attractive and leads to the following reasoning: Consider again the observations on pairs of cultures with identical growth rates and cell composition: one, a batch culture with a "poor" carbon source; the other, a glucose-limited chemostat culture. In the former culture much more substrate is oxidized per minute and per cell mass than in the latter; however, the two cultures, having the same protein and nucleic acid composition and growing at the same rate, *must consume energy at the same rate at the level of macromolecular synthesis.* The composition would therefore seem to be defined by the fraction of the total energy consumed that is available at that level. This seems reasonable, since the substrates for macromolecular syntheses, and such derivatives as cAMP and ppGpp, are themselves regulatory elements and probably play a large part in controlling the synthesis of ribosomes and other parts of the PSS.

3.3 The Amino Acids—Substrates and Effectors

The dual role of the *aa* and other substrates for macromolecular syntheses must be examined further. The fact that the cgr_p is always maintained near its maximal value demands that all the tRNA species capable of being charged with *aa* in fact are adequately charged at all times. This, in turn, sets a lower limit on the *aa* concentrations, or pool levels. The limit is not necessarily the same at all growth rates, but since the number of tRNA molecules to be charged per unit time *and* the specific activities of the synthetases are more or less proportional to μ, there is no apparent reason why the *aa* pool levels should vary in order to ensure adequate charging of the tRNA at the different growth rates.

The complementary aspect of the problem is the effector function of the *aa* and other substrate molecules at the same level. Again a very simple statement can be made: During steady state growth in a glucose minimal medium the *aa* pool levels must be high enough to produce the relatively strong internal repression observed in such cultures. When *aa* are added exogenously repressions increase as the intracellular *aa* pools swell. Up to this point, the reasoning is straightforward; it is based firmly on measurements of the $cgr_{\rm p}$ and of internal repression. The latter is readily demonstrated by comparing the yield of, say, the *trp* enzymes in an R^- strain to the much lower yield in the wild type strain (Yanofsky and Ito, 1966) (see Section 2.3).

At the highest growth rate attainable, the biosynthetic and amphibolic operons can be assumed to be under the maximal degree of repression; their contribution to the noncore sector of the cross section of transcription must be correspondingly small (see Fig. 3). The effect of these repressions could be compensated, however, if other sectors outside the core expanded simultaneously to make a *greater* contribution to the cross section of transcription. At least two specific functions should be considered candidates for such expansion, namely, (a) energy production and (b) transport.

(a) The arguments that can be brought up in this case are indirect, but they further illustrate an important property of the overall system. We have seen that the energy consumed in steady states of aerobic growth tends to be constant per unit mass and per minute. The enzymes involved may therefore also be a constant fraction of total protein. Let us assume that the specific activities of these enzymes are in fact constant and examine what this means in terms of the transcriptional cross section: If, for the reasons we have discussed, this whole section is reduced during growth in a rich as compared to a minimal medium, moderate *repression* of the operons coding for proteins such as the cytochromes is in fact required if the yield from them is to remain a constant fraction of total protein. In general, to maintain constant relative output of a given protein, the degree of repression exerted on the relevant operon must be such that the term by which this operon is represented in the transcriptional cross section is always a constant fraction of the total cross section.

(b) Not much need be said here about the transport proteins. The fact that cells respond *very* rapidly in terms of $\alpha_{\rm r}$ and the rate of rRNA and tRNA synthesis following a shift from glucose minimal medium to rich medium indicates that, at the time of the shift, the cells already possess the ability to take up and to concentrate *aa* added from outside.

All told, it seems evident that the noncore transcriptional cross section is smaller in rich than in minimal medium, and that, whatever else may be involved, this favors transcription from the core genes. I believe that, at medium to high growth rates, passive control of the synthesis of the PSS proper is a very significant factor.

In vitro studies of the charging of tRNA by purified synthetases have yielded

K_m values for various *aa* in the range between 10^{-6} and 10^{-8} M. These values are very low compared to the pool levels of 10^{-3} to 10^{-4} M estimated from the kinetics of uptake of labeled *aa* (see, for example, Schleif, 1967). This suggests that a wide gap exists between the pool levels needed to maintain high degrees of charging of tRNA, and the concentration range within which the *aa* express their effector properties. It appears reasonable that the substrate and the control functions should be separated in this way, but no very precise conclusions can be made by comparing data from *in vitro* and *in vivo* experiments.

It has been pointed out that the internal repression observed in cells growing in a glucose minimal medium leaves plenty of room for *de*repression of the noncore operons at lower growth rates and, consequently, room for expanding the transcriptional cross section beyond its size at a μ value of about 1.5 (glucose media). It should also be clear that further expansion of the noncore sector would mean a relative diminution of the core sector of the transcriptional cross section and a corresponding reduction of the relative output from it. This is a built-in property, or eigenbehavior, of the system, which of course does not exclude the simultaneous operation of some form of active control.

An indication that passive control plays a role even at relatively low growth rates has been obtained by measuring *aa* pool levels. Studies of the kinetics of assimilation of *aa* suggested that the pools of leucine and lysine are considerably reduced at μ values below 1.5 (K. Svensson, unpublished thesis, 1972). These experiments were, however, not easy to interpret, and additional and more direct estimates were therefore desirable. Table I shows the results obtained with a conventional bioassay adapted to measure very low *aa* concentrations.

These preliminary studies by two entirely different methods indicate that the pool levels of each of the few *aa* tested follow the growth rate and that some remain constant below a μ value of about unity. As far as they go, these data show that internal repression is less in the glycerol than in the glucose minimal medium, and hence that the noncore cross section of transcription expands somewhat. Passive control thus seems to contribute positively to the regulation of the PSS at medium and high growth rates. It is worth noting that an *increase* of the *aa* pool levels at low growth rates would have argued strongly for a dominating and active control.

TABLE I. Amino Acid Pool Levels

		μg amino acid/ml cell volume[a]			
Carbon source	μ	Met	Trp	Lys	Leu
Glucose	1.46	34 (~0.24 mM)	14 (~0.07 mM)	142 (~1.0 mM)	108 (~0.8 mM)
Glycerol	1.15	22	7.6	116	71
Succinate	0.56	25	9.6	69	75

[a] Samples from steady state cultures of a wild type strain were transferred into boiling medium and the *aa* extracts thus obtained were tested by means of a strain requiring two *aa*, the one under study and a second one that was supplied in radioactive form to the assay medium. A large inoculum of the requiring strain was used to ensure exhaustion of the extract within about 10 min. The concentration of the *aa* in the extract was estimated from linear standard curves obtained by plotting counts per minute against known concentration of the *aa* being measured. These experiments were done in 1976 in Copenhagen by Robert Spees, a premedical student at Stanford University holding a Rotary Scholarship.

The mechanism(s) responsible for the further reduction of α_r and α_{ess} below $\mu = 1$ remains unknown. But it must also be remembered that α_r is only moderately reduced in this range (Figs. 2 and 3), and that the overall efficiency of the ribosome falls off drastically (Section 4.4).

3.4 *Transcription and Translation Frequencies*

The initiation of transcription has already been singled out as the key process on which most controls act, and also as a particularly complex event. This complexity has become increasingly clear in recent years with reports on a number of proteins (see, for example, Ramakrishnan and Echols, 1973), a particular species of tRNA (Pongs and Ulbrich, 1976), and smaller effectors (see, for example, Travers, 1976*a,b*), all of which seem to interact with the RNA-P and thereby modify its affinity for promoters of various types. In addition, the RNA-P itself can be modified in several ways (see, for example, Zillig *et al.*, 1976). This leaves us with so many degrees of freedom that no comprehensive model can be suggested.

Even so, an important aspect of the transcription process remains to be considered in relation to protein synthesis *in vivo*. This is the effect on the synthesis of individual proteins of the parameter F_i(max) which expresses the obvious fact that a limit must be imposed on the frequencies of transcription by the physical properties of p_i. In purified *in vitro* systems it has been observed that promoters differ greatly with respect to the time it takes to form an active transcription complex (see, for example, Chamberlin *et al.*, 1976; Seeburg *et al.*, 1977), and experiments indicating that F_i(max) plays an important role *in vivo* are discussed in Section 3.4.4. In molecular terms, the processes that define F_i(max) are not yet well understood, but the very existence of promoters of greatly varying strength poses problems that can be examined without a deeper understanding of the mechanisms involved.

3.4.1 *Actual Frequencies of Initiation*

The many measurements of cgr_p and of the quantities of stable and unstable RNA (see Sections 2.4.1, 2.4.2, and 2.4.3) make it possible to calculate frequencies of initiation of translation as well as transcription at various growth rates.

In the case of translation, the calculation is straightforward: If N *aa* constitute a representative sample of a culture with growth rate μ, the number of ribosomes actively engaged in protein synthesis equals $N\mu \ln 2/cgr_p \times 3600$. Taking the weighted mean molecular weight of the proteins to be about 40,000 at all growth rates (Pedersen *et al.*, 1978*a*), the average translation time comes close to 20 s. Therefore, one active ribosome in 20 must finish the synthesis of a polypeptide each second, and an equal number must initiate a run of translation each second.

A certain small fraction of these ribosomes will attach as number one on a newly initiated mRNA molecule. However, to calculate this fraction it is not enough that we know the mRNA half-life; we must also know the size distribution of the transcriptional units. In principle, this information is contained in the "run-out" experiments introduced by Pato and v.Meyenburg (1970), which tell for how long RNA synthesis continues after initiation of transcription has been stopped by rifampicin. Their experiments clearly show that many mRNA molecules must code for several proteins of average molecular weights, but it is difficult, technically, to derive the actual size distribution of mRNA from curves representing total RNA synthesis.

Thus, frequencies of transcription cannot be calculated as simply and unambiguously as those of translation. First of all, the true frequency with which the RNA-P starts mRNA synthesis at a promoter is not necessarily the frequency we are interested in if, as is usually the case, we measure the synthesis of protein molecules of one kind or another. In such experiments *initiation of transcription is registered only if followed up by translation of complete mRNA cistrons.* Thus, the initiation frequencies we can calculate from the relative yield of, say, β-galactosidase will not include possible abortive initiations, not even those that are followed by production of incomplete polypeptide chains (which in this particular case can be detected by an assay for the "auto-α fragments"; see Section 4.3).

The initiation frequency of mRNA molecules that become productive in protein synthesis can be calculated as follows (see Sections 2.4.2 and 2.4.3). With about 4 and 12% of all RNA in the messenger and tRNA fractions, respectively, and with 80 to 90% of the ribosomes in polysomes, the average spacing of the ribosomes is close to 300 nucleotides along the mRNA. This distance (100 codons) will be covered by a translating ribosome in about 5 s. As an example, consider β-galactosidase with 1021 *aa* per monomer (Fowler and Zabin, 1977) being produced as 1% of total protein: Using the same notation as previously, we obtain $(N\mu \ln 2)/(1021 \times 100 \times 3600)$ for the number of monomers produced per second, and to sustain this rate five times as many productive polysomes are required; half of these must be replaced during one messenger half-life through completion of new transcripts of the *lacZ* gene. In steady states of growth N is approximately 5×10^8 per genome equivalent of DNA, and the half-life of the *lacZ* mRNA is about 60 s; for $\mu = 1$, this gives approximately two initiations per minute (see Section 3.4.4 and Fig. 6).

An upper limit is imposed on such numbers by the parameter F_i (max), but well below this ceiling the frequencies are proportional to μ and to the fractional yield of the protein in question, and inversely proportional to the molecular weight of that protein. When different protein species are compared they may be represented by different numbers of genes (per genome-equivalent of DNA) and appropriate corrections must be made (see Section 2.4.7 and Fig. 3).*

3.4.2 *Coupling between Transcription and Translation*

The polyribosomes were identified by Risebrough *et al.* (1962), Warner *et al.* (1962), and others. Stent (1966) conjectured that transcription *in vivo* required the formation of a complex between the RNA-P itself and the leading ribosome on the

*These calculations are actually fairly precise. At medium to high growth rates the fraction of total RNA residing in polysomes is high; it is quite accurately known, as is the mRNA/total RNA ratio and the cgr_p. The correction for relative gene numbers is only of real significance when high growth rates and long distances between genes are involved. The compound errors are not easily estimated, but I believe the frequency calculation to yield estimates within 10 to 20% of the true value.

mRNA. Thus coupling between transcription and translation was thought to be a functional necessity. Since then it has been shown that the cgr_p can be reduced in various ways without affecting the cgr_{mRNA}. This can be achieved by reducing the activity of elongation factor G with fusidic acid (Bennett and Maaløe, 1975), by *aa* starvation (Engbaek *et al.,* 1973), or by adding α-methylglucoside (αMG) to cultures of relaxed strains (Johnsen *et al.,* 1977) (see Section 4.3). In all these cases we must infer that the distance along the mRNA between the RNA-P and the leading ribosome increases as transcription progresses, and the resulting exposure of a stretch of "naked" mRNA is thought to cause the phenomenon called intracistronic polarity (Pato *et al.,* 1973). Therefore, Stent's model cannot be correct in detail, but I believe that it is essentially correct if the leading ribosome, or some other element necessary to start translation, is considered to be involved in initiating, but not in sustaining, transcription (Maaløe, 1969).

This concept was based on the observation of Forchhammer and Kjeldgaard (1968) that an extract of total RNA, when assayed in an *in vitro* protein-synthesizing system, possessed nearly the same capacity to stimulate protein synthesis per milligram of RNA at all the growth rates tested. Since the assay system did not respond to purified rRNA and tRNA, the stimulating activity could be attributed to mRNA. Thus mRNA seemed to be present in constant proportion to the stable RNA (of which tRNA was a constant and relatively small fraction). This proportionality would be a direct consequence of the suggested coupling between initiation of transcription and translation. Alternatives to a tight coupling of these events are (a) that translation might be initiated some arbitrary time after transcription, a mechanism that would tend to produce intracistronic polarity for which there is no evidence, except under the special conditions mentioned earlier; and (b) that mRNA onto which no ribosome attached would be broken down in the wake of the RNA-P.

A conceivable coupling mechanism can be generated from the observation that in a purified transcription system, only a small percentage of the transcripts that had been initiated actually succeeded in becoming a full-sized message; most transcripts failed to continue beyond a defined and very early termination point (Heyden *et al.,* 1975). The analogy to the phenomenon of *attenuation* seen *in vivo* in the *trp* operon (Yanofsky, 1976) is striking, and the suggestion would be that early termination normally is overcome by the intervention of one or another element involved in initiating translation. Actually, *any* mechanism that makes initiation of the synthesis of a productive mRNA molecule depend on the intervention of some component of the translation system would serve to account for the constancy of the mRNA/total RNA ratio.

Evidence suggesting such a dependency is abundant, but not conclusive: Thus Shin and Moldave (1966) demonstrated that preparations enriched for 30 S particles stimulated *in vitro* transcription from *E. coli* DNA; positive responses have been obtained with factors and combinations of factors known to be involved in initiating translation (see, for example, Revel, 1972; Crépin *et al.,* 1973); and finally, Pongs and Ulbrich (1976) have demonstrated that fMet-$tRNA_f^{Met}$ can interact with the RNA-P to stimulate *in vitro* transcription from λp*lac* DNA. In summary, the complex process by which translation is initiated may, in one way or another, supply a stimulus necessary for messenger transcription to proceed. It would therefore be interesting to know the frequency with which acts of transcription are followed by productive translation. As mentioned earlier, we cannot infer this number from the readily calculated frequency for a particular species of protein, since we do not know the average number of proteins encoded per mRNA molecule. A reasonable guess is that the frequency with which mRNA transcription is initiated is three, four, or five times less than the number one gets by assuming nothing but monocistronic messengers, each coding for a protein of average molecular weight. These oversimplifying assumptions give six to eight initiations of messenger transcription per second at $\mu = 1$; the true number might therefore be as low as one to three (see the end of Section 3.4.1 for an example showing that about two initiations are needed at the *lac* promoter *per minute* to yield 1% β-galactosidase).

3.4.3 Promoter Efficiencies and Constitutive Protein Synthesis

The notion of a constitutive operon as a "probe" for monitoring the general state of repression throughout the genome has been mentioned (Section 3.1). It was first introduced as a possible means by which to test the model for passive control of ribosome synthesis (Maaløe, 1969). The idea was simple enough: If synthesis of the r-proteins, and the rest of the core proteins, were constitutive (that is, did not respond to effectors of any kind), then a constitutive operon, outside the core sector of the genome, might be expected to show the same dependence on μ as do the r-proteins. This argument has led to a great deal of confusion, and I shall explain why I have come to think that it is not generally valid. In doing so, I am *not* trying to prove that passive control alone governs ribosome synthesis; rather, I hope to show that a number of observations that have been interpreted as ruling out this mode of control do not in fact disprove it.

Protein synthesis from a constitutive operon can be described in very simple terms as long as the test system is a large population of cells in a particular state of growth. The measured quantity is usually the fractional yield of the protein in question, or its specific activity if the yield is inferred from an enzyme assay, and we have seen how the requisite frequency of initiation of transcription can be calculated from such data. Note that this calculation gives the *average* frequency of initiation at a particular promoter and at a particular growth rate, F_i, and that there is no simple way of getting at the distribution of the time interval between successive initiations in individual cells. At the end of the next section we shall examine special cases in which this distribution may be revealed.

When μ is introduced as a variable at least two complicating factors must be considered: The first (a) is the individuality or strength of a promoter. This property has already been introduced as the parameter F_i(max), which says that, at a given temperature, the average time interval between successive initiations of transcription at p_i can never be less than $1/F_i$(max). The second factor (b) is the modulating influence of the changes in the background of repressions and derepressions against which an individual operon expresses itself; this property is analogous to the "genetic background" upon which a specific mutational change expresses itself.

(a) It seems that F_i(max) varies enormously among promoters. At the strongest promoters a new act of transcription is initiated almost as soon as the preceding RNA-P molecule has moved far enough along the template to clear the promoter region. For this reason alone the interval between successive starts probably cannot be less than 1 to 2 s. At the other end of the spectrum we find the weakest promoters which seem never to be activated more than once or twice per hour. The existence of promoters with such extremely low F_i(max) values has often been suggested, because it seems meaningless to regulate the synthesis of a control protein such as the *lac* repressor, which is produced in very small amounts, by yet another control protein. A promoter that could only be activated once or twice per hour would solve the problem. Experiments by Steen Pedersen indicate that such promoters in fact exist (personal communication). He selected 14 very weak spots on the two-dimensional gels and calculated the fractional yield of each of these unidentified proteins at μ values ranging from 0.5 to 2.8. In eight of these cases the very small percentage of total protein represented by such a weak spot *decreased* as the growth rate went up. Among all the proteins examined only one in six behaved in this way. Constitutive synthesis from a gene whose promoter has a low and μ-independent F_i (max) would be expected to vary with μ as observed for 8 of these 14 low-yield proteins: Consider once more N *aa* representing the proteins of a steady state system with growth rate μ. To double this amount of protein a certain number of mRNA molecules is required *per doubling time,* more or less independent of μ.* During rapid growth, a promoter with an F_i(max) corresponding to an average of two initiations per hour may not reach the state from which active transcription can proceed even once during a doubling time, whereas in slow-growing cells this will happen several times. What Pedersen's data show is that with increasing doubling time (decreasing μ) the number of productive transcripts from the genes representing the low-yield proteins he measured, also goes up—*not per minute but per doubling time.* This result suggests consitutive synthesis *and* very low F_i(max) values.†

Effects of this kind will be observed if F_i(max) is so low that the promoter p_i "saturates" at some intermediate growth rate. As just argued, a fairly constant number of productive mRNA molecules must be synthesized per doubling time, and this implies that the overall initiation frequency increases approximately as μ (except at very low growth rates; see Section 4.4). Other things being equal, the relative yield of protein p_i, α_i, should therefore remain constant as long as F_i, the frequency of transcription from promoter p_i, can increase in proportion to the overall transcription frequency. As μ increases, F_i(max) may be approached and eventually exceeded, and α_i must therefore gradually decrease. As pointed out earlier, the phase of decrease is all that is observed if p_i is a very "weak" promoter—that is, if F_i(max) is very low. In summary, with increasing growth rate very weak and weak promoters will tend to saturate with the result that an increasing fraction of the mRNA synthesized will originate from operons with strong promoters—that is, promoters that do not saturate even at the highest attainable growth rate of about 2.8 doublings per hour. It can be calculated that at

*As we have seen, only crude estimates can be made of this number, and it *is* subject to variation to the extent that the size distribution of the transcriptional units may change with μ.

†This analysis is deliberately based on F_i(max), since this parameter accounts in a simple way for the relationship between dP_i/dt and μ. Alternatively, the yield from a particular gene could be kept at a low level by strong autorepression, but this mechanism would tend to maintain a constant, μ-independent concentration of its product (Sompayrac and Maaløe, 1973). Some form of translational control could also be involved, such as exceptionally short messenger half-lives, which might be looked for in rifampicin run-out experiments; but again the observed dP_i/dt versus μ relationship would not follow as a direct consequence.

this extreme μ value, transcription of the r-protein genes is initiated every 3 to 4 s; this frequency is not far from the maximum value imposed by the necessity for one RNA-P molecule to clear the promoter region to make room for the next. The reasons for the apparent gross differences between the F_i(max) values are not known; possible mechanisms are discussed in the following section.

(b) When μ is introduced as a variable, the background against which the transcription of a given gene is expressed will also vary—that is, other things will *not* be equal. This aspect of the problem has in fact been discussed in Section 3.1 in terms of the transcriptional cross section. It must be reconsidered now to explain why a constitutive operon outside the core region of the genome may be of limited value as a probe to monitor changes in the sums over the different classes of promoters (SI, SIII, and SIV), discussed in Section 3.1. I want to argue that the limitation is introduced by the parameter F_i(max), and that a particular constitutive gene, or operon, may become totally insensitive as a probe as μ (and therefore the overall initiation frequency) increases. Thus, the protein yield from a constitutive gene whose promoter would saturate at around $\mu = 1.5$ might usefully monitor changes of the transcriptional cross section at $\mu \leqslant 1$. Below this value expansion of the cross section would make the relative yield of that protein go down. On the other hand, at around $\mu = 1.5$, the relative yield would begin to go down for an entirely different reason. Some actual cases are now described.

3.4.4 *Case Studies, in Vitro and in Vivo*

The intricate process by which transcription is initiated continues to be studied in highly purified *in vitro* systems. Much still has to be learned, but strong evidence now exists that F_i(max) differs greatly among promoters. Seeburg *et al.* (1977) used DNA segments of the phage, fd, each carrying a single promoter, to follow the formation of active (rapidly starting) complexes with the RNA-P. Two of their findings are particularly pertinent to our discussion: (a) At concentrations of RNA-P exceeding 10^{-8} M, the reaction is concentration independent with apparent first-order kinetics. (In a cell the size of *E. coli* this corresponds to five to ten free and active RNA-P molecules.) (b) The half-time for formation of the active complexes varied from 20 s to 3 min among the promoters. These half-times are very long compared to diffusion-limited collision frequencies in a bacterial cell, as would be expected if complicating events, such as "opening" of the promoter site, are rate limiting (Chamberlin *et al.*, 1976). To quote Seeburg *et al.* (1977): "it is the relatively slow rate of [active] complex formation, which dictates the frequency of initiation events at each individual [fd] promoter. This conclusion is also supported by determination of the transcriptional activity *in vivo* and *in vitro* of various segments of the fd genome, which indicates that more RNA chains are initiated at the strong promoters [those with relatively short half-times] than at the weak ones."

In this as in many other cases, the comparison between *in vitro* and *in vivo* observations is the great problem and some major differences between the two experimental situations should be emphasized here: In the first place, "accessory" proteins that might take part in the formation of active complexes, by combining either with the template or with the RNA-P, may not be present in the *in vitro* system; and second, supercoiling of the DNA *in vivo* may increase the efficiency of initiation (see, for example, Richardson, 1975). Even with reservations, the *in vitro* observations probably are a reflection, perhaps with some distortions, of the transcription system of the living cell.

On the *in vivo* side we shall first discuss the RNA-P (a), and then the constitutive synthesis of some biosynthetic and catabolic enzymes (b), and finally some experiments suggesting that transcription is initiated at more or less defined intervals (c).

(a) Quantitative assessments of the number of RNA-P molecules per cell (Matzura *et al.*, 1971; Ishihama *et al.*, 1976) long remained uncertain because, on cylindrical SDS-polyacrylamide gels, the β and β' subunits are contaminated by the so-called χ protein (Iwakura *et al.*, 1974). On two-dimensional gels, however, charge differences completely separate the χ, β', and β proteins. The latter subunit has been quantitated by this technique and shown to increase with μ approximately as do the r-proteins (Pedersen *et al.*, 1978*a*). Since the RNA-P core is a $\beta\beta'\alpha\alpha$ complex, and since the α subunits seem to be produced in excess (Engbaek *et al.*, 1976; Blumenthal *et al.*, 1976), the number of β subunits should equal the number of RNA-P core units. Table II shows figures based on the data of Pedersen *et al.* (1977) for total RNA-P and for RNA-P engaged in RNA synthesis. The latter was calculated by dividing the overall rate of RNA synthesis (nucleotides incorporated into mRNA, rRNA, and tRNA per second) by the known cgr_{RNA}. The estimate by difference, of RNA-P not engaged in RNA synthesis, is not very accurate, but it seems clear that nonengaged RNA-P does *not* decrease with μ. Furthermore, the actual numbers for nonengaged RNA-P are very large compared to the number of free and active molecules that, according to the *in vitro* studies of Seeburg *et al.* (1977), would suffice to saturate promoter sites.

The existence of large numbers of nonengaged polymerase molecules is evident from upshift experiments. Thus, in minimal medium, about half the engaged RNA-P molecules may be synthesizing stable RNA and the other half mRNA; after a shift to rich medium, the rate of mRNA synthesis probably remains unchanged for several minutes, as indicated by the fact that dP/dt does not increase at once. In contrast, the rate of rRNA and tRNA synthesis may go up two- or threefold, which means that the total number of engaged RNA-P molecules would have to increase by 50 to 100%. This could not happen within 1 or 2 min unless sufficient polymerase were free in the cell at the time of the shift. It has been suggested by Bremer and Dalbow (1975) and by Dalbow and Bremer (1975) that the overall rate of RNA synthesis might be controlled by the intracellular

TABLE II. RNA-P Core Units per Cell

μ	Total	Engaged	Difference[a]
2.85	2200	1800	400
2.15	2150	1300	850
1.50	1500	800	700
1.16	1350	500	850
0.55	1150	300	850

[a] Differences are accurate to within approximately 100 units.

concentration of free *and* active RNA-P; if so, nearly *all* the nonengaged RNA-P would have to be inactive. Little is known about state and localization of the nonengaged RNA-P molecules; *in vitro* they tend to associate in a loose, nonspecific way with DNA (see, for example, Chamberlin *et al.,* 1976), and the same may be true *in vivo,* as indicated by the very low RNA-P content in DNA-less minicells (Rünzi and Matzura, 1976). However, a loose association is not to be confused with chemical modifications or inactivation of the RNA-P (see, for example, Zillig *et al.,* 1976).

If initiation of transcription and translation is coupled as described in Section 3.4.2, it makes sense to have an excess of RNA-P to prime the promoter sites for the event (initiation of translation) that will set transcription in motion. In spite of many experiments and much speculation nothing definitive is known about the steps that lead to the formation of the active complex between the RNA-P and a promoter, and which precede the actual synthesis of a mRNA molecule. However, the discussion in items (b) and (c) that follow requires that we examine in a general way the features of complex formation that might be responsible for some promoters apparently having low or very low F_i(max) values. The simplest argument would be that very stringent steric specifications had to be met for a collision between an RNA-P molecule and a weak promoter to become effective. Similarly, effective collisions might be rare if the promoter itself only became accessible at intervals by assuming the specific configuration recognized by the RNA-P. Alternatively, and this is the view favored by many (for example, Chamberlin *et al.,* 1976; Seeburg *et al.,* 1977), an initial and rapid binding of RNA-P to a promoter site may be followed by "internal rearrangements" which, sooner or later, may result in the formation of an active complex. It is obviously possible to combine the collision frequency and the rearrangement concepts. These fairly obvious models permit the following reasoning: If a low F_i(max) value is due exclusively to a low collision efficiency, initiation at p_i must be a random-in-time event; the same is not necessarily true if structural rearrangements (before or after binding of the RNA-P) are involved. In the second case, cooperativity, say, in the process of "opening" the promoter, could appreciably narrow down the time distribution of initiations. If so, it would make sense to talk about the "dead time" of a promoter; this parameter would have the dimension of $1/F_i$(max).

The experiments described in item (b) have been analyzed solely on the assumption that F_i(max) varies among promoters. Thus, it makes no difference whether acts of initiation at a promoter are randomly or nonrandomly distributed. In contrast, the data for the *trp* and *lac* operons quoted in item (c) can only be rationalized by introducing more or less well-defined dead times.

(b) Rose and Yanofsky (1972) determined the number of transcripts as well as the enzyme yield from the *trp* operon in an R^- (repressor-deficient) strain at several growth rates. The yield of enzyme per mRNA was constant (that is, there was no indication of translational control); but, after a possible slight increase with μ at the lowest growth rates, the specific activity of the *trp* enzymes dropped two- to threefold as μ increased from about 0.85 in the glucose minimal medium, to about 2.15 in L-broth (compare Fig. 6c). This behavior was ascribed to an unknown mechanism that was called *metabolic regulation.* It should be noted that the study of Rose and Yanofsky (1972) was made before the striking phenomenon of attenuation had been discovered; however, this does not affect the present discussion, since the degree of attenuation is μ independent, as long as tryptophan is not limiting (Yanofsky, 1976).

It is clear that in an R^- strain the *trp* operon does *not* behave the way constitutive genes outside the core sector of the genome had been expected to behave if passive control of synthesis from the core genes played a significant role (see Section 3.4.3). The conclusion must obviously be that passive control is unimportant *or* that the operon studied by Rose and Yanofsky did not constitute a suitable probe. In fact, one may argue that the *trp* genes in the R^- strains are unsuitable simply because they respond to metabolic regulation and hence might not be considered truly constitutive.

However, this argument is very unsatisfactory, implying, as it does, that whenever the yield from genes selected to serve as probes does not show a positive correlation to α_r, the synthesis from those genes is not constitutive. In this way nobody is going to be convinced one way or the other.

It must therefore be recalled that passive control has been identified as part of the logic describing the overall system (Section 3.1) and that the data of Section 3.3 indicate that it contributes positively to the regulation of the synthesis of the core proteins. To this should now be added a second descriptive element: the F_i(max) introduced earlier. This parameter will be considered not only in relation to the data of Rose and Yanofsky but also in relation to the patterns of synthesis of a number of core and noncore proteins, all of which may reflect passive control to a greater or lesser extent.

The four frames of Fig. 6 display the relevant data. The measurements of core proteins in Figs. 6a and 6b are from the two-dimensional gel analysis of Pedersen *et al.* (1978*a*); Fig. 6c illustrates results obtained in our laboratory in a study designed to extend the original observations of Rose and Yanofsky (B. Willumsen, manuscript in preparation); and in Fig. 6d the synthesis of β-lactamase is shown (Engberg and Nordström, 1975; Lindqvist and Nordström, 1970). In each case the relative yields or, if these are not available, the specific enzyme activities are plotted against μ with the α_r curve of Fig. 2 drawn as reference. Information about strains and assays is given in the legend.

Figure 6 calls for one general and several specific comments. First it should be emphasized that the data are not homogeneous. In the Introduction it was emphasized that strains with doubtful genetic relations should not be compared. This rule is violated here mainly in the sense that *E. coli* B is compared with *E. coli* K12 even though the latter grows more slowly than does the former in any medium. In such cases we must rely on trends rather than on quantitative relationships for drawing conclusions. Figures 6a and 6b illustrate core proteins and all the data were obtained with one of the B strains: Figure 6a shows three cases in which core proteins, here the r-protein S1, the elongation factor G, and the arginyl-aminoacyl synthetase, show virtually the same dependence on μ as do the r-proteins as a group. Figure 6b shows two of the synthetases which do not follow the r-proteins all the way. We can conclude that *if* these two synthetases are controlled individually and specifically, they are not under very efficient control. The argument is that if the concentrations of these enzymes are adequate at maximum growth rate, they are unnecessarily high at the lower growth rates.

In Figs. 6c and 6d we look at noncore proteins. As already indicated, this means that the μ scale may not be strictly comparable to that of the α_r reference curve because the data originate from studies with K12 strains. Even so, Fig. 6d shows that the synthesis of β-lactamase, which is constitutive inasmuch as it is not inducible and not affected by addition of cAMP, parallels α_r as far as the comparison between the strains allows. Figure 6c, on the other hand, shows that β-galactosidase (in a strain carrying the promoter mutation, uv5, that appears to abolish all dependency on cAMP), ornithine transcarbamylase, and the *trp* enzymes show more or less the same pattern of synthesis as that observed by Rose and Yanofsky. It should be remembered that in wild type strains the biosynthetic enzymes we look at here are under specific control; in strains *without* a functional repressor it makes little sense to reason about the effects of metabolic conditions in terms of adjusting enzyme synthesis to the needs for their products. In fact, *trpR*$^-$ strains overproduce the *trp* enzymes to such an extent that an adequate flow of tryptophan probably could be maintained even at high growth rates in rich media *without* tryptophan.

To sum up, we have a representative set of core proteins, all of which we may assume to be regulated in the same way, but two of them (Fig. 6b) deviate from the general pattern in that their rates of synthesis do not increase above a μ value somewhere between 1.6 and 1.8. We also have a set of noncore proteins, which all seem to be synthesized constitutively, but nevertheless vary greatly in their relationship to μ. I believe that individual F_i(max) values *could* be assigned to the promoters representing all the proteins of Fig. 6, that would account for the way

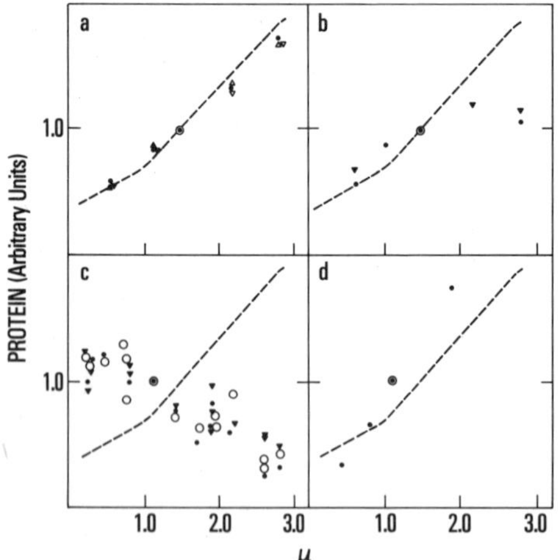

Figure 6. Relative rates of synthesis of individual core and noncore proteins. All the measurements presented are relative amounts of individual proteins or specific activities of enzymes in arbitrary units. Unit values represent data obtained from cultures in glucose minimal medium. In each frame, the α_r/μ relation of Fig. 2 is indicated by a dashed line. (a) Three core proteins: r-protein S1 (●), elongation factor G (▲), and the arginyl-adenosyl synthetase (▼); (From Pedersen *et al.*, 1978*a*). Redrawn and reproduced with permission of the authors and publisher.) (b) Two additional core proteins; the phenyl- (●) and the leucyl-adenosyl synthetase (▼), (From Pedersen *et al.*, 1978a. Redrawn and reproduced with permission of the authors and publisher.) (c) Constitutive synthesis of enzymes: β-galactosidase (▼), ornithine transcarbamylase (○), and the *trp* operon enzymes (●). At each growth rate these three measurements were made simultaneously on a culture of strain CA8050 (K12), which combines the *lacUV5* promoter mutation, an *argR*$^-$ mutation, and a *trpR*$^-$ mutation. The experiments were done by Berthe Willumsen (B. Willumsen and K. v. Meyenburg, manuscript in preparation). Reproduced with the authors' permission. (d) Synthesis of β-lactamase from a chromosomal gene in a K12 strain. (From Engberg and Nordström, 1975. Redrawn and reproduced with permission of the authors and publisher.) According to Anders Byström (personal communication) this strain also carries an "up-promoter" demonstrated to be cis dominant.

some of them deviate from the α_r versus μ relationship.* It is attractive to consider these very different responses to changes of μ as resulting from variation in a common parameter, F_i(max), which *a priori* is expected to vary greatly between promoters. If this view is accepted, the failure of some constitutively synthesized, noncore proteins to respond to μ in the way the r-proteins do, would be understandable.

Finally, we should take a look at the complementary situation: totally, or largely, active control of synthesis of the core proteins. In this case, constitutive synthesis from a noncore gene would be *dis*favored as μ increases, but to a considerably smaller extent than seen in Fig. 6c. Actually, we would have a case of passive, but *negative* control of the constitutive operon selected as probe.

(c) Direct measurements of transcription from the *trp* operon have been made by Imamoto (1968) and by Baker and Yanofsky (1968), and from the *lac* operon by Contesse *et al.* (1970). In these experiments, access to transcription was provided at time zero by relieving repression and by induction, respectively; the rate of mRNA synthesis was followed by hybridizing pulse-labeled RNA to DNA from appropriate transducing phages. The title of the paper by Baker and Yanofsky (1968) reads: "The Periodicity of RNA Polymerase Initiations. . ." and it clearly describes the results obtained in both systems. For technical reasons, the *trp* experiments were made at very high cell densities, which caused transcription to proceed slowly (their data suggest a cgr_{RNA} of about 20, compared to the value of about 60 characteristic of optimal growth; see Section 3.2.1). Although this does not detract from the qualitative significance of the observations, the phenomenon of periodicity is best illustrated for the present discussion by the data of Contesse *et al.* (1970) shown in Fig. 7.

Note first that points on the graph represent the beginning of a 20-s pulse; considering the pool effects, it would be more realistic to move all points over, relative to the ordinate, by some 15 s. With or without this correction, the final rates of synthesis are reached after two or three initial steps, respectively, depending on how much of the *lac* operon is represented on the DNA used for hybridization. The definitive rates must correspond to the steady state of transcription of the *lac* operon, and each step on the curves of Fig. 7 would thus represent initiation by a new RNA-P molecule, *before* the first round of transcription is terminated. The curves suggest that the F_i(max) of the *lac* promoter is such that three RNA-P molecules can be accommodated within the whole operon, and only two within the *lacZ* gene. It is also clear that the *lac* promoters of the many cells in each culture responded more or less synchronously to IPTG; if that had not been the case, initiation at the *lac* promoters of the various cells would have

*On the basis of a precisely defined model, such F_i(max) values could be estimated by curve fitting. If passive control is assumed to govern the synthesis of the core proteins, the frequency of initiation of transcription (followed by translation) must increase with μ not only as does dP/dt but also in relation to the hypothetical changes in the transcriptional cross section. I believe that, for this exercise to be worthwhile, more precise data are needed, especially for the very critical noncore proteins of Figs. 6c and 6d. A word should also be said about elongation factor Tu. The yield of this protein is so large that, at $\mu = 2.8$, the requisite frequency of initiation (assuming one gene and average messenger half-life) would be about five times greater than for the r-proteins—that is, the frequency would exceed the limit imposed by the time needed to clear the promoter region. As is now known, this dilemma is solved (a) by having two, possibly three genes, and (b) by one of these genes producing an mRNA with an unusually long half-life.

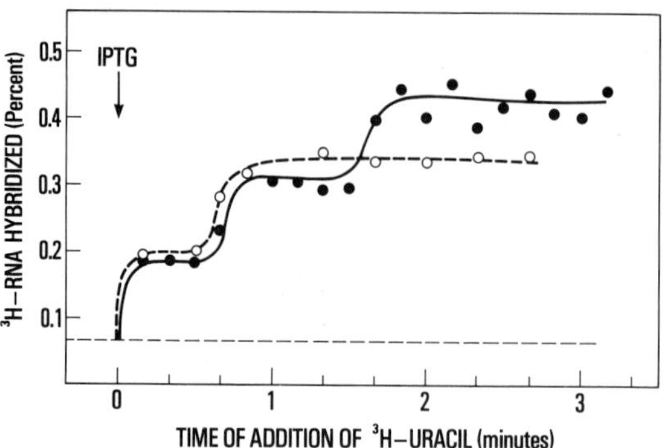

Figure 7. Rate variations of *lac* mRNA synthesis during the early stage of induction. *E. coli* MO was grown at 32°C in medium supplemented with vitamin B_1, charcoal-filtrated vitamin-free casamino acids, and glycerol. When the optical density of the culture had reached 1.0 the culture was induced with 5×10^{-4} M IPTG. From the onset of induction, a series of 5-ml aliquots were pulse-labeled for 20 s with 20 μCi/ml of [^{3}H]uridine (20 Ci/mmol) and poured onto 5 ml of frozen medium containing 20% sucrose and 0.02 M sodium azide. The percent hybridization of each aliquot was determined, after annealing with 1 μg of ϕ_{80}d*lac* DNA (•) and 1 μg of $\phi_{80}EZ_1$ DNA (○) ($\phi_{80}EZ_1$ is deleted for the *y* and *a* genes). The background level, due to non-*lac* mRNA, was determined by a 2-S pulse labeling performed before adding IPTG and is represented by the dashed horizontal line. Each percent value is plotted against the time at which labeling was begun in the corresponding sample. Redrawn from Contesse *et al.* (1970) and reproduced with permission of the authors and publisher.

been more or less evenly distributed, and in a large population of cells the steps would have been obliterated.*

A more serious problem is the discrepancy between the stepwise increase in the synthesis of *lac* mRNA seen in Fig. 7, and the continuous increase in the capacity for enzyme production observed when induction is stopped, either by diluting out the inducer (see, for example, Contesse *et al.*, 1970) or by preventing further initiation of transcription with rifampicin (see, for example, Hirsh and Schleif, 1973). Both the mRNA and the capacity measurements are obtained from experiments of extremely simple design, and I can see no obvious reason why one or the other technique should introduce artifacts.

The notion of rhythm in the initiation of transcription obviously caused great interest when it was introduced some 10 years ago, but doubt about the significance of the data was also expressed. To a certain extent this was because the experiments themselves were considered to be very difficult, but more so, I think, because the assumptions that had to be made to interpret the data were hard to accept. In this connection, it is interesting to see that Imamoto (1968) clearly stated that to generate periodicity transcription would have to be initiated by compounding several events, each of which might be randomly distributed, in order to narrow down the time distribution of the final event. (It is clear from his paper that the drawings showing several RNA-P molecules traveling together should not be taken literally, but were meant to convey the idea of some kind of cooperativity.)

*Hirsh and Schleif (1973) have shown that few, if any, RNA-P molecules in the rifampicin-*resistant* state are attached at the *ara* and *lac* promoters before inducer is added. This, however, does not rule out priming by attachment in a waiting position, in which the RNA-P could remain rifampicin-*sensitive* until the first step or steps in RNA synthesis have been taken. Similarly, the general uncertainty with respect to the details of the process of initiation *in vivo* leaves the elegant EM binding studies by the same authors (Hirsh and Schleif, 1976) outside the frame of the present discussion.

There are several reasons for reopening the discussion of these old experiments. In the first place, Imamoto's idea of a built-in mechanism for reducing the variance of the time distribution no longer strikes one as unreasonable (see, for example, Chamberlin *et al.*, 1976); second, we now have the information necessary to calculate the *average* time interval between successive initiations of transcription in batch cultures, and this can be compared to the frequency indicated by the waves of initiation seen in Fig. 7; and finally, I want to include the very suggestive observation that the number of waves depends on the length of the specific DNA fragment used for hybridization.*

The F_i(max) of the *lac* promoter, estimated from Fig. 7, is about 40 s at 32°C, and probably a little less at 37°C. This agrees well with the average time interval between successive initiations at the *lac* promoter calculated as that required to sustain β-galactosidase synthesis at the differential rate, approximately 0.01, commonly observed in fully induced cultures (see Section 3.4.1).

The figure of 0.01, for which the calculation was made, could be off by a factor of 2 or more either way if it were based on one of the widely diverging estimates that have been reported for the specific activity of purified β-galactosidase. Fortunately, however, the relative quantity of β-galactosidase protein in a fully induced, acetate-grown culture has been measured on two-dimensional gels (S. Pedersen, personal communication). His figure, which is independent of activity measurements, is used here, and the agreement between the directly estimated F_i(max) and the calculated value suggests to me that both approaches offer estimates of the true F_i(max) of the *lac* promoter. Furthermore, the growth rate in the experiment shown in Fig. 7 must have been three to four times that in Pedersen's acetate culture, which indicates that, as expected, F_i(max) does not vary with μ.

3.5 Synopsis

The PSS of a bacterium is a closed system with special properties, three of which have been introduced in this section: (a) control of the development of the PSS is exerted at the level of the substrates for macromolecular synthesis, such as *aa* (Section 3.4.3); (b) the patterns of transcription and of translation, which were compared to the genetic background on which a phenotype is expressed (Section 3.1.1); and (c) the parameter F_i(max), a promoter-specific parameter, which may be akin to the refractory period of an excited neuron (Section 3.4.4).

First, it is argued that the frequency of transcription that is followed up by translation of completed transcripts must be limited by the size of the PSS—that is, by the number of ribosomes that are active in protein synthesis. At medium to

*A comment must be made about this third step. It is well known that the *lacZ* product (β-galactosidase) is greatly overproduced relative to the products of the more distal genes, *lacY* and *lacA*. This might be accounted for by introducing attenuation between the *Z* and *Y* genes—that is, by assuming that only a fraction of the mRNA molecules include transcripts of the distal genes. If so, the third step would be correspondingly diminished and might not be visible at all. It is therefore important to note that Lim and Kennell (1974) have shown that termination of transcription does not occur within the *lac* operon to a measurable extent.

high growth rates this is a large and constant fraction of all the ribosomes in a cell; at low growth rate this fraction decreases, but the mechanism that accounts for this decrease in the overall efficiency of the ribosomes remains unknown.

Second, it is shown that the availability of open promoters for mRNA synthesis (the transcriptional cross section of Fig. 3) increases as μ goes down. It is argued that this increase of the cross section for noncore proteins must contribute to the control of the PSS, whether or not some kind of active control also exists.

Finally, it is pointed out that F_i(max) must vary greatly among promoters. The relative rate of synthesis of a given protein must therefore begin to decrease when growth rates are reached at which the number of effective transcriptions *per doubling time* from the relevant promoter can no longer keep pace with the overall rate of protein synthesis. Beyond this limit, constitutive synthesis of a particular protein is no longer a useful indicator of those changes in the transcriptional cross section that would contribute to passive control of the PSS.

4 *The Role of ppGpp in Regulation of the Protein-Synthesizing System*

The history that gradually focused on the synthesis, breakdown, and effector properties of ppGpp began over 20 years ago (Borek *et al.*, 1955), and it has been reviewed several times, recently by Gallant and Lazzarini (1976). In the present context three facts stand out: first, the early observation by Neidhardt (1963) that the *relA*$^-$ strain which was known to have lost the ability to cut down drastically on RNA synthesis when starved of a required *aa* nevertheless responded "normally" to many changes in the growth conditions; second, the identification by Cashel and Gallant (1969) of the so-called *magic spots* (ppGpp and pppGpp) as implicated in the function of the *relA* gene; and finally, the more recent discovery, made independently by Haseltine and Block (1973) and by Pedersen *et al.* (1973), that rapid synthesis of ppGpp, as well as pppGpp, requires codon specific, uncharged tRNA together with ribosomes, mRNA, and the so-called *stringent factor*. The latter is almost certainly specified by the *relA* gene. It is therefore natural to interpret accumulation of ppGpp above the basal level as a sign of suboptimal charging of one or more species of tRNA—that is, as a sign of substrate limitation in protein synthesis.

In the following discussion I shall not be concerned much with synthesis and breakdown of ppGpp, but rather with its role as effector. Therefore, I shall consider first the ppGpp concentrations during steady state growth and changes in concentration during transients (Section 4.1). Then I shall review some unpublished experiments made as early as 1965–1966 that have acquired considerable relevance in view of recent attempts to implicate ppGpp as a positive effector in some cases (Section 4.2). This will lead to discussions of a novel type of downshift that strengthens the notion that ppGpp is a general inhibitor of RNA synthesis—that is, that it affects synthesis of messenger as well as stable RNA (Section 4.3). Finally, I shall return to the unsolved problems concerning the poor efficiency of the ribosomes at very low growth rates (Section 4.4).

4.1 *ppGpp Concentrations*

4.1.1 *Steady State Concentrations of ppGpp*

From the beginning, ppGpp has been thought to control the synthesis of stable RNA, and particularly of rRNA. Since there is a strong negative correlation between the concentration of ppGpp and rRNA accumulation (Fiil *et al.,* 1972) such a control function requires that the level of ppGpp in the cells decreases as the growth rate goes up. Lazzarini *et al.* (1971) showed that this does occur, but they observed no more than about a twofold reduction in the ppGpp concentration over a fivefold increase in μ. This change in growth rate corresponds roughly to a 20-fold increase in the rate of rRNA synthesis. It was therefore not clear how ppGpp could function as the sole or chief effector.

This aspect of the problem was cleared up by Fiil *et al.* (1972). Using a *relA*$^+$ strain with a temperature-sensitive valyl-tRNA synthetase (NF342), these authors correlated the levels to which ppGpp increased and the rates of accumulation of stable RNA at intermediate temperatures. These rates were established very quickly at the various temperatures and remained constant for at least 60 min; identical results were obtained by radioactive labeling and by chemical measurements of RNA. These basic observations are reproduced in Fig. 8. The quasi steady state rates of RNA accumulation (R) are seen to fall from unity (the value measured in a glucose minimal culture of the parent strain) to less than 0.20, which corresponds to growth with acetate as carbon source. The ppGpp concentration required to reduce R from unity to 0.2 is about 400 units, which is very high compared to the ppGpp levels measured during normal growth even at low μ values; see also Section 4.1.2 and Fig. 9.

It may be added that, at very high growth rates, the rate of RNA accumulation is about 4 to 5 on the ordinate scale of Fig. 8. The linear extrapolation of the plot of $1/R$ against ppGpp, shown as an insert, is not compatible with such high rates of accumulation, since the corresponding concentrations of ppGpp would be negative. Fiil *et al.* (1972) concluded that the rate of accumulation of RNA is not primarily controlled by the ppGpp concentration; instead, ppGpp might, in their

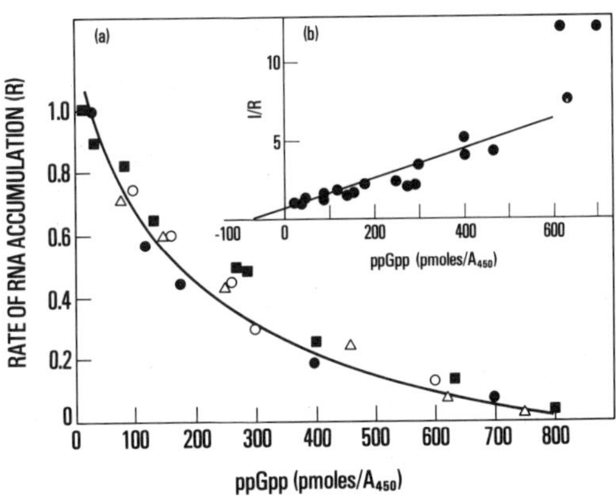

Figure 8. Rates of RNA accumulation relative to the ppGpp levels. The relative rates of RNA accumulation (R), expressed as a fraction of the rate of accumulation at 37°C in a culture of the parental strain (NF321), are plotted against the ppGpp level measured 15 to 20 min after taking the test strain (NF342, *leu*$^-$, *valS*^{ts}) to various intermediate temperatures. Redrawn from Fiil *et al.* (1972) and reproduced with permission of the authors and publisher.

words, "act as a fine control by modulating the stimulating effect of . . . [another] factor."

This factor was introduced in order to balance the production of rRNA against that of the r-proteins. These two major cell constituents are probably synthesized in closely matched quantities (see, for example, Kjeldgaard and Gausing, 1973). A specific mechanism may be required to achieve this because each transcript from the r-protein genes must be paralleled by a total of 20 to 30 transcripts from six or seven genes, each of which specifies a full complement of the three rRNA species. The factor invoked by Fiil and co-workers has not been identified.

The initiation frequencies required to produce enough rRNA during fast growth are extremely high. The need for 20 to 30 sets of rRNA for each r-protein mRNA immediately tells us that every one of the six or seven rRNA genes must, on the average, be transcribed about four times as frequently as are the r-protein genes. At a μ value of about 2.8, transcription must therefore be initiated at an rRNA promoter at intervals of about 1 s;—that is, about as frequently as possible considering that the promoter region *must* be cleared between successive initiations. It has been suggested (Venetianer *et al.*, 1976) that these promoters are special in the sense that they permit four or five RNA-P molecules to be accommodated in "waiting" positions.

For the reasons just mentioned, rRNA can hardly be produced in excess

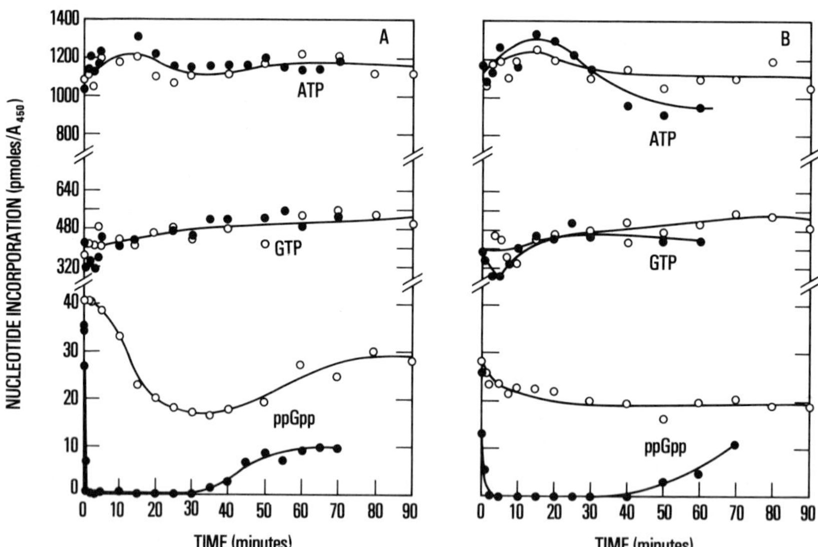

Figure 9. Levels of ppGpp relative to shifts to higher growth rates. Strains NF541 (*rel*$^+$) and NF542 (rel$^-$) were grown exponentially in acetate minimal medium. After three generations of growth with appropriate labeling, part of each culture received glucose to a final concentration of 0.2% and a second part received the components of a rich medium. This is time zero in the graph. Mass increase and accumulation of protein and of RNA were followed by frequent sampling (not shown), and 50-μl samples were withdrawn from the ^{32}P-labeled cultures for pool analyses, as shown on this graph. ○, The acetate-to-glucose shift; ●, the acetate-to-rich medium shift. Redrawn from Friesen *et al.* (1975) and reproduced with permission of the authors and publisher.

when *E. coli* grows at its maximum rate. At lower rates, however, breakdown of excess rRNA occurs (Gausing, 1976; Pedersen, 1976), and under certain conditions of stress, considerable amounts of rRNA are in fact broken down (see Section 4.3).

The extremely high initiation frequencies required at the rRNA promoters during growth at maximum rate might be thought to set an upper limit to μ. However, the number of rRNA genes probably would have been amplified beyond six or seven if that alone had sufficed to push μ_{max} above 2.8. It seems much more likely to me that μ_{max} reflects a state of growth in which α_{ess} has reached a value that cannot be exceeded if the necessary functional balance between core and noncore proteins is to be maintained.

4.1.2 ppGpp Concentrations during Transients

Shifts between media supporting different rates of growth have long been used to study transients (Kjeldgaard *et al.*, 1958). Several experiments of this kind are discussed, and it should therefore be pointed out at the onset that *shifts differ in quality*. In shifts of one kind, there is no need for the cells to adjust to their new condition by *de novo* protein synthesis before the full impact of the shift can be registered. Borrowing Monod's terminology, shifts of this kind might be called *gratuitous;* they include a shift-up produced by adding *aa,* for example, to a glucose minimal culture (no *new* biosynthetic activities are needed to make full use of the supplements and the relevant transport systems seem to be adequate), and a shift-down produced by suddenly restricting the availability of the carbon and energy source (see Section 4.3). Less ideal, sluggish shifts are those in which transfer to the new medium imposes more or less profound changes of the enzymatic equipment of the cells in order that the carbon and energy source(s) in the postshift medium can be used efficiently. The obvious case is a shift-down from rich to minimal medium, in which *de novo* production of many biosynthetic enzymes is a primary necessity. To degrees that are difficult to assess, the same may be true of many shifts between media with different carbon sources, and in all these cases it is difficult or impossible to interpret the *immediate* responses to the shift in simple terms.

With these caveats in mind, consider the shift-up experiments done by Friesen *et al.* (1975), in which the levels of ATP, GTP, and ppGpp were followed all the way from the pre- to the postshift steady state of growth. For reasons that are not clear, it has been difficult to obtain reproducible values for the basal level of ppGpp in different media; thus the day-to-day variations have often exceeded the limits indicated by measurements on independent samples from the same culture (J. D. Friesen, personal communication). Knowing this, it is particularly valuable to have experiments in which the ppGpp level in the same culture has been followed through a shift.

Figure 9 shows data from the paper of Friesen *et al.* (1975). Figures 9A and 9B refer to identical experiments done with a pair of strains isogenic except for the *relA* locus; this has been a standard design in many recent studies. First of all we see that the two strains respond in much the same way to shifts from acetate minimal medium to glucose minimal, or to rich medium. The effects on the ATP and GTP levels are small, if indeed significant, but in shifts to rich medium the ppGpp concentration drops precipitously to values close to zero and stays there for 20 to 30 min; in the shifts to glucose medium the response of the $relA^+$ strain is more pronounced than that of the $relA^-$ strain, but in neither case is it anywhere near so dramatic as in the shifts to rich medium. This may have to do with the latter being of the gratuitous type, whereas the acetate to glucose shift probably is somewhat "sluggish."

The basal levels of ppGpp in acetate, glucose, and rich medium, respectively, that can be read in Fig. 9 are nearly identical in the two strains. Thus, during balanced growth on acetate ($\mu \simeq 0.5$) the level is at 40 to 50 units; it falls to 25 to 30 in glucose and to 10 to 15 in the rich medium. This three- to fourfold decrease, which may or may not be significantly greater than the twofold decrease reported by Lazzarini *et al.* (1971), parallels a five- to sixfold increase in growth rate and about a 20-fold increase in the rate of ribosome synthesis. The concentration range we now consider, 10 to 50 units, should be compared to the figures on the abscissa of Fig. 8, and to the estimate that a ppGpp concentration of some 400 units would be required to lower the rate of rRNA accumulation from its value in a glucose culture to that characteristic of growth on acetate.

Against this background, we shall examine the recent and very striking experiments of Dennis and Nomura (1974). A $relA^+/relA^-$ pair of strains with a temperature-sensitive valyl-tRNA synthetase was used to measure α_r for most of the individual r-proteins at intermediate temperatures—that is, under partial starvation for valine. The results were clear: At 35.5°C all the α_r values were reduced to about two-thirds of their values at the permissive temperature (29°C) in the $relA^+$ strain, and they increased to about 1.5 times these values in the $relA^-$ strain. This is strong evidence for a functional relationship between ppGpp and the rate of synthesis of the r-proteins. But there is more to be said about these experiments.

First, it must be understood that this particular temperature shift in many ways resembles a shift to rich medium, since all the *aa* pools will increase when the flow of *aa* into protein is cut short by reducing the activity of the valyl-tRNA synthetase. The difference, by a factor of 2, between the α_r values at 35.5°C in the $relA^+$ and $relA^-$ strains, should therefore be compared to wild type cultures with μ values of about 2.5 and 1.3, respectively. Second, and more important to our discussion, Dennis and Nomura showed that the shift to 35.5°C elicits a slow reduction in the ppGpp level in the $relA^-$ strain, but a big increase in the $relA^+$ strain. At the time samples were taken for the α_r measurements, the ppGpp level in the $relA^+$ culture was at least ten times above the basal level characteristic of medium to high growth rates. Thus, ppGpp at such high concentrations severely reduces the rate of synthesis of RNA *but also reduces the rate of synthesis of the r-proteins and of a number of other core proteins* (Reeh et al., 1976).

An interesting comparison can be made between the experiments just described and a study of the effects on growth in fusidic acid (Bennett and Maaløe, 1975), a drug known to reduce the translational movement of the ribosomes by interacting with elongation factor G. Fusidic acid thereby reduces the efficiency of the ribosomes and cuts down the consumption of *aa,* causing all the *aa* pools to swell. However, the drug does not affect the charging of any of the tRNA species, and the $relA^+$ strain used in this study was therefore not induced to accumulate ppGpp. The effect on α_r of the "internal" upshift condition produced by fusidic acid was quite similar to that of the temperature shift, in the case of the *relA*⁻ *valS*^{ts} strain of Dennis and Nomura. Thus, two different procedures produced *down*shifts in terms of μ, but were accompanied by internal changes causing α_r to increase as expected for a typical *up*shift.

Transfer to a medium with fusidic acid is an extremely sluggish shift because the drug accumulates slowly in the cells. Nevertheless, transient but extremely high α_r values (0.37–0.38) can be elicited, and they are paralleled by drops in the ppGpp concentrations below the basal levels in rich medium. In summary, we find a simple, inverse relationship between the ppGpp level and μ, suggesting that this unusual tetraphosphate takes part in the overall regulation of growth. However, even the relatively high concentrations of ppGpp measured at low growth rates are themselves very low compared to concentrations that are known to have marked effects on the rates of synthesis of rRNA and r-protein under the usual experimental conditions.

To indicate that ppGpp may play only an indirect role in regulating ribosome synthesis, a second graph from the paper by Friesen *et al.* (1975) is presented as Fig. 10. The two shifts designed to produce identical effects are described in the legend. All that need be said is that the responses in terms of μ as well as RNA and protein accumulation were indeed the same; nonetheless the shift involving addition of five *aa* caused the ppGpp level to drop much faster, and farther, during the transient than it did in the parallel culture. This striking difference was not reflected in the rates of RNA accumulation (α_r was not measured). A similar, transient dissociation of the ppGpp levels has been observed in the downshift experiments of Hansen *et al.* (1975). Again no difference in the rates of accumulation of RNA was seen (see Section 4.3).

4.2 *The Relaxed Syndrome*

We have already seen that experiments in which the responses to shifts of one kind or another of a pair of *relA*$^+$ and *relA*⁻ strains are compared are now

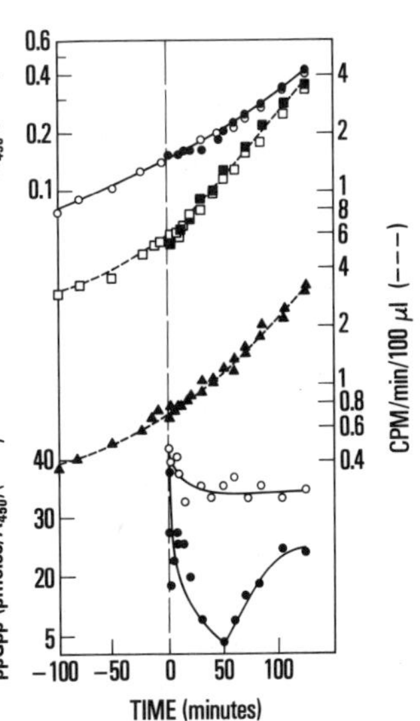

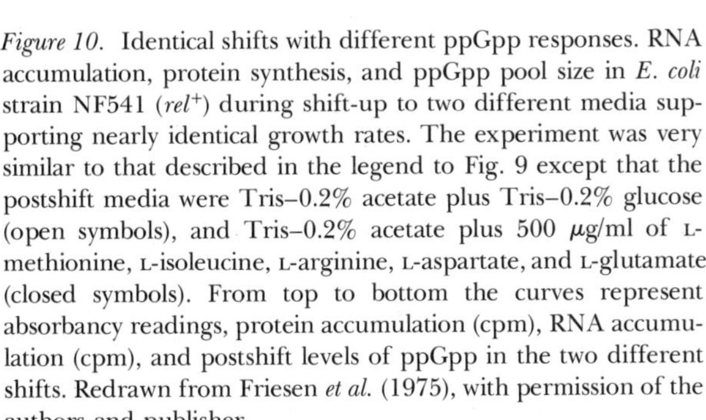

Figure 10. Identical shifts with different ppGpp responses. RNA accumulation, protein synthesis, and ppGpp pool size in *E. coli* strain NF541 (*rel*$^+$) during shift-up to two different media supporting nearly identical growth rates. The experiment was very similar to that described in the legend to Fig. 9 except that the postshift media were Tris–0.2% acetate plus Tris–0.2% glucose (open symbols), and Tris–0.2% acetate plus 500 μg/ml of L-methionine, L-isoleucine, L-arginine, L-aspartate, and L-glutamate (closed symbols). From top to bottom the curves represent absorbancy readings, protein accumulation (cpm), RNA accumulation (cpm), and postshift levels of ppGpp in the two different shifts. Redrawn from Friesen *et al.* (1975), with permission of the authors and publisher.

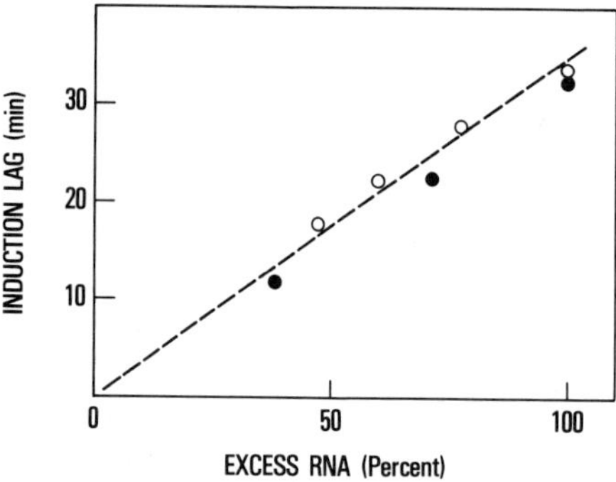

Figure 11. RNA accumulation and the lag in β-galactosidase induction. The "excess RNA" shown on the abscissa accumulates during *aa* starvation as the so-called relaxed particles and reaches a maximum after about 60 min. ●, The induction lag in samples taken during methionine starvation—that is, during accumulation of the excess RNA. ○, The lag in samples taken during incubation in buffer after *aa* starvation—that is, during degradation of the accumulated RNA.

frequently used to test whether a particular function may or may not be under "stringent control." It is therefore pertinent to present some early observations relevant to the interpretation of such experiments.

When a *relA*$^-$ strain is starved for an *aa,* RNA continues to be produced, and it accumulates in large amounts because its turnover is slow. The same happens in a *relA*$^+$ strain when protein synthesis is blocked by chloramphenicol. A common feature in these two cases is that, although protein synthesis virtually ceases, ppGpp does not accumulate; in the first case because the stringent factor is defective, in the second probably because all the tRNA remains highly charged. Another common feature is that the excess RNA appears in the cells as ribonucleoprotein particles (Dagley *et al.,* 1961). It was later shown by Schleif (1968) that the protein moiety of these particles is *not* r-protein.

Yanagisawa (1962) described a different aspect of the syndrome as "the simultaneous accumulation of RNA and of a repressor of β-galactosidase synthesis." By this he referred to the observation that a starved culture of an *relA*$^-$ strain remains uninducible for a considerable time after readdition of the required *aa.* He also demonstrated that this did not apply to constitutive synthesis of the enzyme. The correlation between the two parts of the syndrome was studied in our laboratory by Turnock and Eplov in 1965–1966.* Their key experiment is reproduced here as Fig. 11. It shows that *relA*$^-$ cells which, after a period of histidine starvation, are held in buffer, gradually lose the excess RNA accumulated in the starvation medium and that this loss is closely correlated with the return of the ability to respond to IPTG with the normal, short induction lag. The overall rate of protein synthesis, as measured before the starvation period, was reestablished 2 to 3 min after readdition of the missing *aa.* Moreover, differential labeling showed that the protein profile (from a DEAE column) was the same before and after starvation, except for an excess of synthesis of r-proteins very early after the starvation period, as had been shown by Muto *et al.* (1966). This observation is reminiscent of the effect of partial *aa* starvation on α_r in *relA*$^-$ strains and might have a similar explanation (see Section 4.1.2). Analogous, though less extensive, data exist for induction of tryptophanase synthesis.

The relaxed syndrome is not understood in detail. The salient facts are that

*Geoffrey Turnock, now a lecturer in the University of Leicester, U.K., was a postdoctoral fellow in Copenhagen at that time, and Peter Eplov was a graduate student. They have kindly permitted me to make use of their unpublished data.

the trapping of unidentified, nonribosomal proteins in the "relaxed" or "chloramphenicol" particles is paralleled by the loss of inducibility for at least two enzymes. At the same time bulk protein synthesis remains unaffected. Now that we know about positive control elements, such as the general catabolite activator protein (CAP), it is tempting to speculate that the selective effects on some inducible enzymes might be due to trapping of the CAP protein, or possible analogs. It should soon be possible to identify the "relaxed" particle proteins on two-dimensional gels.

The real purpose in calling attention to the relaxed syndrome, of course, is not to offer an uncertain interpretation. The point is that comparisons between the responses of otherwise identical *relA*$^+$ and *relA*$^-$ strains are giving rise to suggestions about *positive* effects of ppGpp on the synthesis of some proteins (Artz and Broach, 1975; Stephens *et al.*, 1975). This is something quite different from the apparent inhibitory effect of high concentrations of ppGpp on rRNA and r-protein synthesis discussed earlier.

Consider first a *relA*$^-$ strain by itself: During partial *aa* starvation the ppGpp level decreases moderately, whereas inducibility all but vanishes. At this point, it could therefore be argued that ppGpp is essential for β-galactosidase synthesis *in vivo*. Under the right conditions, ppGpp actually stimulates the synthesis of β-galactosidase in a coupled *in vitro* system (see, for example, Aboud and Pastan, 1973, 1975), but this effect depends on the *aa* concentrations in the system, as shown by M. Johnson, N. Fiil, and J. Friesen (personal communication). Nevertheless, β-galactosidase synthesis probably does not depend at all on the ppGpp concentration in growing cells. Thus, addition of IPTG during the time when, after a shift to rich medium, ppGpp can hardly be detected (see Fig. 9) gives normal kinetics and rate of synthesis of β-galactosidase (B. Willumsen, N. Fiil, and J. Friesen, manuscript in preparation).

4.3 *Ribosome and Protein Synthesis during Shift-Down Transients*

An instant shift-down, with the same ideal characteristics as a shift-up from minimal to rich medium, can be effected by suddenly reducing the availability of the carbon and energy source without in any other way changing the preshift minimal medium. It is difficult to use a chemostat setup for this purpose if frequent sampling is desirable, and alternatives have therefore been looked for. The transport mutant isolated and studied by v.Meyenburg (1971) offers one solution. Cultures of this strain require unusually high concentrations of the carbon source (for example, lactose or one of several *aa*) to grow at the same rate as the parent strain. Steady states of growth at low μ values and shiftdowns can therefore be realized by reducing the concentration of the carbon source at the appropriate time. Cell densities from 10^7 to 10^8 ml^{-1} can be used without significantly lowering the concentration of the carbon source, and the genetic stability of this cell line is adequate for such experiments. The potential value of this strain is far from exhausted.

Meanwhile, a different approach has turned out to be extremely useful. The uptake of glucose by *E. coli* cells is competitively inhibited by the analog, α-methylglucoside (αMG). This compound accumulates in the cells as the metabolically inert αMG-6-phosphate (Kessler and Rickenberg, 1963). This sugar phosphate can reach very high concentrations in the cells, but several controls indicate that it is not toxic. Thus, S. Molin (unpublished thesis) found that the growth rates with glycerol or fructose as carbon source were unaffected by αMG, although the αMG-6-phosphate reached high concentrations; also, the sugar phosphate accumulates gradually, whereas the effect of αMG on growth is very rapid (see later).

The relationship between the αMG/glucose ratio and μ on the one hand, and the responses in terms of ppGpp accumulation in *relA*$^+$ and *relA*$^-$ strains on the other, were established by Hansen *et al.* (1975). Extensions of their work have recently been reported by Molin *et al.* (1977) and by Johnsen *et al.* (1977). In these studies, all the shifts were performed by adding αMG in a tenfold excess (by weight) over the glucose in the medium. Within about 30 s this caused the overall rate of protein synthesis (dP/dt) to drop to half the preshift value in *relA*$^+$ as well as *relA*$^-$ cultures. After the shift-down, growth continues for several hours at the rate established at the time of the shift. Thus, a new steady state is approached, and the RNA/protein ratio and α_r correspond to μ exactly as they do in other instances (see Section 3.2.2).

The paper by Molin *et al.* (1977) concerns mainly RNA synthesis and the fate of the newly made ribosomal components. The 5 S RNA, which can be separated from the other RNA species on polyacrylamide gels, was labeled and used as a tag at the promoter-distal end of the rRNA transcript to estimate the cgr_{rRNA} (Molin, 1976) (see Section 2.4.1), and to calculate what fraction of newly synthesized total RNA was represented by rRNA. Simultaneously, the rates of RNA and of protein accumulation were determined by conventional labeling techniques. The α_r was measured according to Gausing (1974), *and* by analyzing total cell lysates on polyacrylamide gels in the presence of 8 M urea. These two methods give estimates of the relative rates of synthesis of r-proteins that wind up in mature ribosomes and of total r-protein synthesis, respectively.

The main results were (a) that the downshift did *not* change the cgr_{rRNA}; (b) that during the transient, both strains synthesize two to three times more rRNA and r-protein than they accumulate in the form of mature ribosomes; and (c) that the *relA*$^-$ strain continues to synthesize mRNA at a rate not significantly different from the preshift rate, whereas in the *relA*$^+$ strain, mRNA synthesis is considerably reduced as early as 2 min after the shift (see later).

First of all, it seems that during the transient a substantial fraction of the rRNA synthesized is broken down. The marked discrepancy between the rate at which the ribosomal components are produced and the rate of maturation of ribosomes is illustrated in Fig. 12. It is obvious that ribosome assembly is restricted during the transient by a mechanism that plays no significant role during balanced growth, when the rates of synthesis of rRNA and of r-protein are well matched and correspond to the rate of maturation. Apparently, the instability of a considerable fraction of the rRNA is *not* due to lack of bulk r-protein, but a deficit of a more subtle kind might be considered. Thus, partial failure to modify, or even to synthesize, some key r-proteins could reduce the rate of assembly relative to the rate of synthesis of the main components of the ribosomes. A second interesting result turned up: When correction is made for the partial instability of the rRNA formed during the transient, then the rate of rRNA synthesis and the levels of ppGpp correlate in much the same way as they do in the experiments of Fiil *et al.* (1972) discussed in Section 4.1.1.

The second new paper on αMG downshifts (Johnsen *et al.*, 1977) concerns protein synthesis. When the flow of energy and carbon is curtailed, protein synthesis must suffer. As early as can be measured, the addition of αMG reduces the cgr_p to about half in cultures of $relA^+$ and $relA^-$ strains. This is remarkable because, as we have seen, the cgr_p is independent of μ during steady states of growth, even in glucose-limited chemostat cultures with μ values as low as 0.1 (compare Section 3.2.1). It is equally remarkable that in the $relA^+$ culture, the cgr_p returns to the preshift value within 2 to 3 min of the shift, whereas it remains at the low value for some 20 min in the $relA^-$ culture. As mentioned, dP/dt stays at half the preshift value throughout the transient and beyond.

Recall that $dP/dt = r_{act} \times cgr_p$ (see Section 3.1), and then consider a point in time, say 5 min after the shift, when the cgr_p has returned to its preshift value in the $relA^+$ culture, while remaining at half that value in the $relA^-$ culture. Since dP/dt is reduced to one-half of the preshift value in both cases, the equation implies that r_{act} must be reduced to half the preshift number in the $relA^+$ culture, while remaining unchanged in the $relA^-$ culture.

The question is now whether a connection can be seen between the potential for ppGpp accumulation and the rapid return to the normal cgr_p in the $relA^+$ culture? The kinetics of accumulation of ppGpp in this strain show that a maximum is reached about 2 min after the shift, and that the concentration then slowly drops (Hansen *et al.*, 1975). This time course suggests that the return of the cgr_p to normal values in the $relA^+$ culture might be a consequence of the buildup of high concentrations of ppGpp. In the $relA^-$ culture, the ppGpp level rises slowly and about 30 min after the shift there is no difference between the strains, either in cgr_p or in ppGpp concentration.

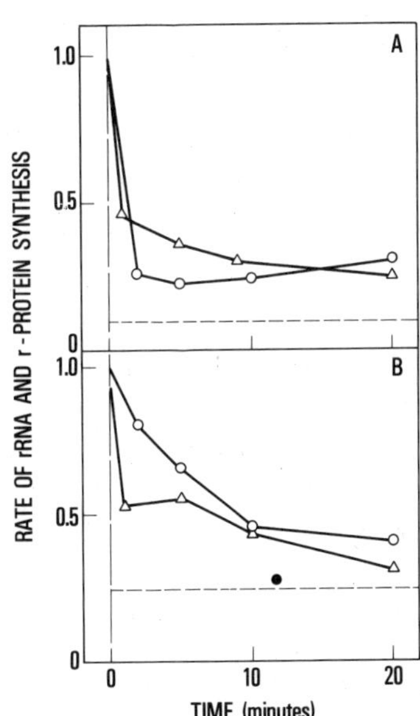

Figure 12. Production and accumulation of rRNA and of r-protein after an αMG downshift. Rates of RNA accumulation (dashed line) and of rRNA synthesis (○). Rates of r-protein synthesis (Δ) were calculated as the products of the relative rates of r-protein and of total protein synthesis, respectively. (A): Strain NF541 (rel^+); (B): strain NF542 (rel^-).

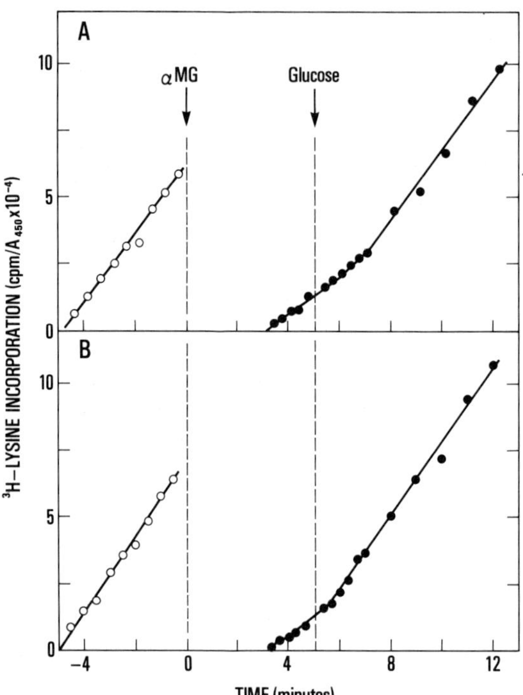

Figure 13. Kinetics of protein synthesis following reversal of an αMG downshift. Exponentially growing cultures were split in two, and one part received αMG at time zero. To both parts [^{3}H]lysine (0.5 μCi/μg, 1 μg/ml) was added at -5 and 3 min, respectively. The downshifted cultures received glucose (to 1%) at 5 min. At the indicated times radioactivity in the acid-precipitable material was determined (cpm/A_{450}). ○, Control; •, downshifted culture. (A): Strain NF1133 (*rel*$^+$); (B): strain NF1134 (*rel*$^-$).

It has been suggested that ppGpp may inhibit the synthesis of mRNA, although to a lesser degree than it inhibits rRNA and tRNA (see, for example, Gallant and Lazzarini, 1976). The observations just described indicate that this may indeed be the case: When the flow of energy suddenly drops, the charging of tRNA probably is reduced and a condition analogous to partial *aa* starvation created.* In cultures of the *relA*$^+$ strain this causes rapid ppGpp accumulation, which may inhibit not only rRNA and tRNA synthesis, but also mRNA synthesis. This crucial argument is supported by two observations: first, that the shift reduces mRNA synthesis in the *relA*$^+$, but not in the *relA*$^-$, culture (see earlier); and second, by the type of experiment illustrated in Fig. 13. Here the downshift was reversed after 5 min; and it was shown that dP/dt returned immediately to the preshift value in the *relA*$^-$ culture, whereas it took several minutes to reestablish the preshift rate in the *relA*$^+$ culture. This was to be expected, *if* the amount of mRNA was reduced in the *relA*$^+$ culture as a consequence of the shift, and if a normal pool of mRNA therefore has to be built up again when the shift is reversed by adding glucose.

This interpretation leads to the following hypothesis: When the flow of energy is lowered, the capacity for protein synthesis is reduced equally in the *relA*$^+$

*The alternative explanation that the shift reduces the synthesis of *aa* is unlikely, since S. Molin (unpublished thesis) has shown that the primary effect of the shift-down on dP/dt is not affected by adding all the *aa* to the medium together with αMG. However, to prove this point conclusively, it would have to be shown that the shift did not inhibit assimilation of *aa* from the medium.

and *relA*$^-$ strains. In the *relA*$^+$ strain this leads to rapid accumulation of ppGpp, which, in turn, inhibits all RNA synthesis, including that of mRNA. Thus, the amount of mRNA decreases as the preshift pool decays. A balance is reached when the mRNA pool is such that only about half as many ribosomes can be accommodated as before the shift. In the *relA*$^-$ strain, on the other hand, ppGpp exerts little, if any, effect on RNA synthesis; the amount of mRNA produced after the shift remains high enough to accommodate all the ribosomes that were active before the shift. In this way, the *common* limitation imposed by the shift on dP/dt is accounted for: in the *relA*$^+$ strain by reducing r_{act}, thus allowing for the cgr_{P} to return to its normal value, and in the *relA*$^-$ strain by keeping r_{act} constant, thus forcing the cgr_{P} to decrease.

The simplest form of this hypothesis assumes that the cgr_{RNA} and the functional half-life of mRNA are both unaffected by the shift. It also suggests that the average distance between neighboring ribosomes on an mRNA cannot be changed much—that is, that a reduction of the cgr_{P} does not lead to closer packing of the ribosomes in a polysome. The first two implications were verified experimentally by Johnsen *et al.* (1977), who showed that the cgr_{RNA} remains unchanged, and that a reduced cgr_{P} increases the lag that always precedes the onset of messenger decay, but does not affect the average functional lifetime of a messenger molecule. The notion of a constant, average spacing of the ribosomes is consistent with the observations, but it can only be rigorously demonstrated by direct measurements of mRNA quantities per ribosome in the polysomes.

As already mentioned, all the stress situations so far examined that reduce the cgr_{P} cause intracistronic polarity (Pato *et al.*, 1973) (see Section 3.1). This was readily demonstrated in the αMG downshift experiment by comparing the relative rates of synthesis of functional and total β-galactosidase on the one hand, and the promoter-proximal auto-α fragment of this protein on the other. Attempts were also made to demonstrate this phenomenon in other proteins with long polypeptide chains, in which it might be expected to be as pronounced as in β-galactosidase; however, with the limited resolution of the method used, no sign of polarity was detected (Johnsen *et al.*, 1977).

The only known difference between the pair of strains used in the αMG downshift studies resides in the *relA* gene, and we can therefore look upon the *relA*$^-$ strain as a kind of "primitive" organism lacking the sophisticated mechanism necessary for rapid accumulation of ppGpp. But there is much more we would like to know about downshifts than that which has been learned so far from the αMG experiments. Thus, it is striking that the ppGpp level reached by both strains some 30 min after the shift is at least twice what is measured in the acetate cultures of the same *E. coli* strain (see Fig. 9), although the growth rate is lower in the acetate cultures than it is in the quasi steady state reached after the downshift. However, this is only an apparent deviation from the rule that the ppGpp level falls as μ increases; 2 to 3 h after the shift, the ribosome number as well as the ppGpp level match the postshift μ value (S. Molin, personal communication).

It is a very puzzling observation, common to all downshifts that elicit a quick response, that the intracellular pools of ATP and other triphosphates do not become depleted. As discussed recently by Gallant (1976), an extremely sensitive "servomechanism" may be required to safeguard against the loss of ATP when the capacity for its synthesis is suddenly reduced. At least one, as yet unidentified, nucleotide (Gallant's *phantom spot*), the level of which changes quite rapidly after various downshifts, has been noted as a possible candidate for a molecule involved in such a servomechanism.

4.4 *The Low Efficiency of Ribosomes at Low Growth Rates*

The problem of the low efficiency of ribosomes at low growth rates was first examined in Section 3.2, but discussion of possible ways of accounting for it was deferred to this section, namely (a) that initiation of translation might become less and less efficient as μ goes down; and (b) that transcription might be reduced so that a decreasing fraction of the ribosomes could be accommodated on the available mRNA.

We have seen that the cgr_p does *not* decrease with μ (Fig. 4). Those ribosomes that are active are therefore fully active, but they represent a progressively smaller fraction of all the ribosomes. It is therefore evident that an unused potential for protein synthesis exists in slow-growing cells. This was demonstrated by Koch and Deppe as early as 1971. Their key experiment is reproduced here as Fig. 14; it shows that within a few minutes of a dramatic shift-up the rate of protein synthesis increased nearly tenfold. This looks like a blowup of the return shift with the *relA*$^+$ strain shown in Fig. 13A. In that case the amount of mRNA had been reduced in the downshift condition and an adequate pool had to be built up again before a full complement of ribosomes could become active. Could not the same interpretation apply to the experiment of Koch and Deppe?

First we must note that the addition of αMG causes the concentration of ppGpp to increase much beyond the levels measured during steady states of growth in acetate cultures ($\mu \simeq 0.5$). It is possible that very high levels of ppGpp are maintained at the extremely low preshift growth rate in Fig. 14 ($\mu \simeq 0.1$); to

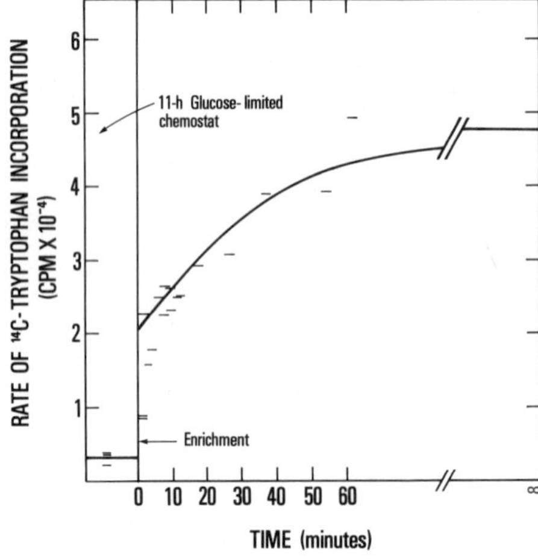

Figure 14. Rates of protein synthesis following a dramatic upshift. Specific protein synthesis rate after a shift-up of an 11-h glucose-limited chemostat. Radioactive tryptophan uptake for 2-min pulses per milligram of dry weight before and after an enrichment that gave a 40-min doubling time (T_2). Also shown are experimental results on a control culture growing with a 30-min doubling time in the enriched medium. Redrawn from Koch and Deppe (1971) with permission of the authors and publisher.

my knowledge this has not been ascertained. However, a stronger, and more direct argument has been introduced by Gausing (1977), when she produced evidence that mRNA constitutes a more or less fixed fraction of the total RNA at μ values all the way down to about 0.23, at which only about one-third of the ribosomes are active.

These observations, together with the constancy of the cgr_p, pose a difficult problem. Since the amount of mRNA is not reduced much, if at all, relative to the number of ribosomes, most of these *could* be accommodated in polysomes. At the same time, the active ribosomes proceed along the mRNA at normal speed and they cannot therefore be encumbered by idling ribosomes. This leaves us with two options: Either these slow-growing cells sense that, say, only one ribosome in three can be allowed to be in polysomes at any given time, or the usual high proportion of the ribosomes (80–90%) enter polysome structures, but two-thirds of them idle along *at the same rate as the active ones.* At the moment there is no way of choosing between these alternatives, but it should be noted that the results of αMG shift-down experiments do *not* conflict with a mechanism in which the frequency of initiation of translation is restricted. Although, as we have seen, only about half the number of ribosomes are active during the transient in cultures of the $relA^+$ strain as compared to the $relA^-$ strain, the reduced cgr_p in the latter strain implies that, *per time unit,* the same number of initiations of translation takes place in both cultures. If a mechanism exists by which the number of ribosomes entering polysomes is restricted at low growth rates, the consequence would be that the average spacing of the active ribosomes would increase since the cgr_p is not reduced in steady state cultures.

Finally, I want to reemphasize that results from *in vivo* experiments, such as those described herein, can form the basis for conclusions only to the extent that we are able to measure the relevant parameters (and in all seriousness, it should be realized that there is no way to know, *a priori,* what parameters *are* relevant).

A deeper understanding of the mechanisms involved requires additional measurements, some of which will have to be done *in vitro,* as in the case of the F_i (max) discussed in Section 3.4.3. Concepts, such as modulation of the promoter specificity of the RNA-P brought about by ppGpp, introduced by Travers (1976*a,b*), and the critical temperature characteristics of individual promoters (Chamberlin *et al.,* 1976), are no doubt important and relevant to many of the problems discussed herein. At the present time, the main difficulty is to bridge the gap between *in vivo* and *in vitro* experiments—that is, to repeat, often under less favorable experimental conditions, a process of the kind that led from the logical construction of the operon model to our present knowledge of the detailed structure of the *lac* operon. To indicate some of the problems we face, consider the fact that an Arrhenius plot of μ versus $1/T$ often is linear from about 20 to 35°C (Ingraham and Maaløe, 1967)—that is, right across some of the promoter "melting temperatures" that have been measured *in vitro.* There are at least two possible ways of reconciling these apparently conflicting results: On the one hand, the cell may contain proteins that interact with promoters to modify their temperature responses; and, on the other hand, the effect of temperature on the readiness of a promoter to open [the F_i(max), in fact] may in many cases be balanced out by changing the degree of intracellular repression exerted on the operon in question. Obviously, we have a long way to go before problems of this kind can be solved.

4.5 Synopsis

From the time of its identification by Cashel and Gallant (1969), ppGpp has been thought to be involved in the control of ribosome synthesis. Consequently, much work has been done to analyze the mechanism of synthesis as well as the function of ppGpp. Most significantly, we know from *in vitro* experiments that the requirements for efficient synthesis of ppGpp in $relA^+$ strains are such that the rate at which this compound is produced reflects the degree of charging of tRNA. Our lack of understanding of the corresponding and much slower process in $relA^-$ strains is almost as conspicuous.

The notion of ppGpp as an effector is supported first by the strong correlation between the high concentrations to which it is raised in "emergency" situations, such as total or partial *aa* starvation, and the degree to which the rate of RNA accumulation is reduced. At the much lower ppGpp concentrations characteristic of steady states of growth, the correlation still prevails but the amplitude of change can hardly account for the differences between rates of RNA accumulation at high and low growth rates, respectively (see Section 4.1.1).

Shift experiments present more serious problems of interpretation. Two cases in point are as follows:

(a) The classical shift-up from a glucose minimal to an *aa*-enriched medium is the first case. Soon after this shift, α_r reaches a peak value, then falls back more than half-way to the preshift value before it again increases to reach the definitive postshift value. This oscillation, for which no reasonable explanation has been offered, lasts almost 20 min—that is, it occurs while the ppGpp concentration is almost zero (see Fig. 9). K. Gausing, who first observed the α_r oscillation, has recently measured the rate of rRNA synthesis during the transient and has shown that drRNA/dt follows α_r (personal communication). In other words, r-protein and rRNA seem to be synthesized coordinately during the period of oscillation, despite the fact that ppGpp is virtually absent.

(b) In a recent paper, Gallant *et al.* (1977) show that a shift from 23°C to 40°C stimulates $relA^+$ cells to increase their ppGpp level as much as does *aa* starvation. In this special situation, rRNA synthesis is *not* affected. This case of total insensitivity of rRNA synthesis to ppGpp is hard to fit into any existing model.

In Section 4.1.1, Fiil *et al.* (1972) were quoted as stating that ppGpp could act by modulating the stimulating activity of some other (still unknown) factor. A possible alternative appears if we consider a common feature of all experiments that bring about accumulation of ppGpp. When protein synthesis is reduced, whether by partial *aa* starvation or by means of an inhibitor, the flow of energy through the system is greatly reduced and the nucleotide balance is affected. In this altered environment, the concentration of ppGpp necessary to reduce the rate of rRNA accumulation by a factor of, say, 2, could be much higher than in normally growing cells.

Most of the experiments discussed in this chapter are simple measurements made on samples from a steady state culture, and it is the remarkable stability and reproducibility of these states that make work of this kind possible. Furthermore, the analyses that have been attempted are based on the general belief that the growth rate and cell composition that characterize a steady state of growth are established and stabilized by the joint effects of a large number of individual controls and couplings.

It is not easy to see how stability is built into a system as complex as a growing cell, and I shall illustrate my own view with two examples. The first is a pure Gedanken experiment in which a cell with the ideal composition corresponding to a particular steady state of growth is allowed to divide. One sister cell is allotted a little more than half the ribosomes and other components of the core system, everything else being evenly distributed. In the cell with an excess of ribosomes the tendency will be to raise the rate of protein synthesis above the level to which the enzyme systems of the cell were previously adjusted. This will cause excessive drain on the *aa* pools, which will diminish, causing internal repression to be relieved to some extent. In terms of the passive control model, the net result is therefore to *reduce* α_r and α_{ess} temporarily, and thus reduce the size of the core system. In the sister cell, similar adjustments will take place, except in the opposite direction. These arguments show that passive control does exhibit the element of feedback acting on the synthesis of the core proteins which is required to confer stability on the overall system. Obviously, an effector molecule, such as ppGpp that monitors the *aa* pool levels via the degree of charging of tRNA, could be imagined to do the same, and it is entirely possible that active as well as passive control operates in growing cells.

The second illustration is again theoretical but it is closer to real life. It is a simulation study of the effects produced in a cell when the core genes, which map relatively close to and on both sides of the origin of replication, are duplicated at some point in the division cycle.

The following parameters were included: the overall rate of protein synthesis, $dP/dt = P\mu \ln 2$, an average amino acid pool size, and the estimated maturation time of the ribosomes. Actual measurements have shown that the content of an amino acid pool typically corresponds to 10 to 15 s of protein synthesis, and that it takes 1 to 2 min to produce a mature ribosome in a glucose minimal culture; at lower growth rates the maturation time increases significantly (Lindahl, 1975). For simplicity it was assumed that duplication of the core genes raised the frequency of transcription from them by a factor of 2.* The simulation program was written by Lauren Sompayrac,† and the results he obtained were more dramatic than we expected. When the growing system is perturbed by duplication of the core genes

*This is approximately true at low growth rates. At maximum growth rate the core proteins make up about half the proteins of the *average* cell and gene duplication cannot possibly double their output (see Fig. 3). If nothing else changed, and if all the core genes were replicated simultaneously, the total transcriptional cross section would increase considerably and the core genes would account for roughly 40 and 60%, respectively, before and after they had been replicated.

†Lauren Sompayrac came to Copenhagen as a theoretical physicist in 1971, but he spent much time in our laboratory. He is now engaged in cancer research in the United States.

it takes about 2 min before *mature* ribosomes begin to accumulate at the new, increased rate, and only then does dP/dt begin to increase faster than before the core genes replicated. From this time on functional ribosomes will continue to accumulate at the new high rate for the full maturation period of 2 min. It turns out that this short period of increase, at twice the previous rate, would suffice to empty the *aa* pools if the rates at which they are fed and drained remained unchanged. This is due to the fact that the rate of flow through an *aa* pool is very high relative to the pool volume. We then introduced a modest amount of feedback inhibition of the activity of the *aa*-synthesizing enzymes (10%), and this turned out to be adequate to buffer the system. This inhibition will be released gradually as the *aa* pool levels go down, and at the same time internal repression will be reduced and the effect of the core gene duplication will be counteracted by expansion of the noncore transcriptional cross section. We now have exactly the same situation as in the Gedanken experiment above, and the same control mechanisms (passive and/or active) can be invoked to damp the effect of the gene duplication. The difference between the two examples is mainly that the second one analyzes a perturbation that, by definition, occurs once in every cell cycle, and suggests that *feedback inhibition may be essential to cope with this perturbation.*

Attempts have been made to register the effect of duplication of the r-protein genes in synchronized cultures (Zaritsky and v.Meyenburg, 1974; Dennis and Young, 1975). No changes were seen comparable to the effect of duplication of, say, the *lacZ* gene. There are two reasons why these experiments have been less informative than might have been anticipated: In the first place, the methods available still leave much to be desired, and data obtained from synchronously dividing cultures are charged with a considerable variance of a technical origin. Second, the theoretical discussion just presented shows that we are far from being able to make quantitative predictions concerning the degree to which duplication of the r-protein genes will be reflected in the relative rate of synthesis, α_r. These two uncertainties reinforce one another, and in my opinion both technical improvements and a better theoretical basis are needed to warrant new attempts to study this problem experimentally.

6 *Afterthoughts*

The subject of this chapter is a diffuse one and the writing has been a long and often difficult process. For the same reason reading may prove to be no easier than the writing. However this may be, work on the manuscript has been both enjoyable and rewarding, in particular, the occasional emergence of new and sometimes simpler ways of looking at or connecting facts, some of which had been known a long time.

There is a direct connection between a more or less theoretical paper I wrote nearly 10 years ago and the present one. Section 2 in the first paper (Maaløe, 1969) described the data then available for analysis, and Section 2.4 of this chapter is an updated list which shows the impressive progress made both qualitatively and quantitatively. Of the many new features, I attach particular importance to the measurements of mRNA quantities and stability, of the chain growth rates, and of the rates of synthesis of many individual proteins. However, it is equally valuable that such critical parameters as α_r and the number of RNA-P molecules have been defined to the point at which their relation to μ can be expressed rather precisely, even at low growth rates.

As in the earlier paper, the analysis of bacterial growth presented here is largely based on parameters that have been studied as functions of μ, whether in steady states of growth or during transients. This somewhat biased attention to *in vivo* measurements made under strictly controlled growth conditions tends to separate the scientists who study the bacterial cell as a unit system from those who analyze subsystems such as the *lac* operon or phage λ. This is understandable because, in either area, the problems and techniques can claim the undivided attention of any scientist. But it is also unfortunate because one of the purposes of driving the analysis of the smaller units through at the molecular level must be eventually to understand how the larger, composite unit works.

What has been done here is to take elements shaped by the classic processes of molecular biology, and define some of their common properties in order to formulate general rules about the behavior of growing bacterial cells. The many parts of this complex unit are almost certainly connected by control loops and by couplings; I believe that the transcriptional cross section and the promoter-specific parameter, F_i(max), are concepts that will prove useful for describing the unit system. Analysis can be based on the common properties of individual controls, couplings, and so on, just as, on a different level, it can be based on the fine structure of an operon.

Throughout the writing of this chapter I have been conscious of a tendency to select for discussion work done in, or in association with, our laboratory. In part this is due to my greater familiarity with our own work than with that of others, and to the extent this is the case I have tried to correct myself. However, another reason for this bias is that so much of our work has been concerned specifically with the problems discussed in the present paper.

Here, at the end, I want to express my gratitude and indebtedness to colleagues at home and abroad. Numerous and very productive discussions have contributed to our present understanding, and several friends have taken the time to read the draft manuscript and to offer their advice; in particular, Robert Schleif's constructive criticism did much to clarify some of the problems discussed in Section 3. During the last few years work in our laboratory received decisive support from the Danish Natural Science Research Council (mostly in the form of fellowships), and from Stiftung Volkswagenwerk (Grant No. 112480).

References

The list of references is not intended to cover all the research touched upon in this chapter. Reference is frequently made to sources that provide more comprehensive coverage, in particular to a recent volume, *Control of Ribosome Synthesis, Alfred Benzon Symposium IX* (N. O. Kjeldgaard and O. Maaløe, eds.), Munksgaard, Copenhagen, 1976. Articles in this book are referred to as *Benzon Symp. IX,* with page numbers.

Aboud, M., and Pastan, I., 1973, Stimulation of *lac* transcription by guanosine 5′-diphosphate, 2′ (or 3′)-diphosphate and transfer nucleic acid, *J. Biol. Chem.* **248**:3356.

Aboud, M., and Pastan, I., 1975, Activation of transcription by guanosine 5′-diphosphate, 3′-diphosphate, transfer ribonucleic acid, and a novel protein from *Escherichia coli, J. Biol. Chem.* **250**:2189.

Andersen, K. B., 1974, Oxygen uptake and energetics of *Escherichia coli,* Lunteren Lectures on Molecular Genetics, 1974 Symposium on the Bacterial Envelope, Lunteren.

Artz, S. W., and Broach, J. R., 1975, Histidine regulation in *Salmonella typhimurium:* An activator–attenuator model of gene regulation, *Proc. Natl. Acad. Sci. U.S.A.* **72**:3453.

Baker, R. F., and Yanofsky, C., 1968, The periodicity of RNA polymerase initiations: A new regulatory feature of transcription, *Proc. Natl. Acad. Sci. U.S.A.* **60**:313.

Bennett, P. M., and Maaløe, O., 1975, The effects of fusidic acid on growth, ribosome synthesis and RNA metabolism in *Escherichia coli, J. Mol. Biol.* **90**:541.

Blumenthal, R. M., Reeh, S. V., and Pedersen, S., 1976, Regulation of the transcription factor and α subunit of RNA polymerase in *Escherichia coli* B/r, *Proc. Natl. Acad. Sci. U.S.A.* **73**:2285.

Borek, E., Ryan, A., and Rockenbach, J., 1955, Studies on a mutant of *Escherichia coli* with unbalanced ribonucleic acid synthesis, *J. Bacteriol.* **71**:318.

Bremer, H., and Dalbow, D. G., 1975, Regulatory state of ribosomal genes and physiological changes in the concentration of free ribonucleic acid polymerase in *Escherichia coli, Biochem J.* **150**:9.

Cashel, M., and Gallant, J., 1969, Two compounds implicated in the function of the RC gene of *Escherichia coli, Nature* **221**:838.

Chamberlin, M., Mangel, W., Rhodes, G., and Stahl, S., 1976, Biochemical studies on the transcription cycle, in: *Benzon Symp. IX,* pp. 22–39.

Chandler, M. G., and Pritchard, R. H., 1975, The effect of gene concentration and relative gene dosage on gene output in *Escherichia coli, Mol. Gen. Genet.* **138**:127.

Coffman, R. L., Norris, T. E., and Koch, A. L., 1971, Chain elongation rate of messenger and polypeptides in slowly growing *Escherichia coli, J. Mol. Biol.* **60**:1.

Contesse, G., Crépin, M., and Gros, F., 1970, Transcription of the lactose operon in *E. coli,* in: *The Lactose Operon* (J. R. Beckwith and D. Zipser, eds.), pp. 111–141, Cold Spring Harbor Lab., Cold Spring Harbor, New York.

Crépin, M., Lelong, J.-C., and Gros, F., 1973, Early steps in the formation of a translation initiation complex on newly transcribed messenger RNA, in: *Protein Synthesis in Reproductive Tissue* (E. Diczfalusy, ed.), pp. 33–51, Karolinska Institutet, Stockholm.

Dagley, S., Turnock, G., and Wild, D. G., 1961, The accumulation of ribonucleic acid by a mutant of *E. coli, Biochem. J.* **88**:555.

Dalbow, D. G., and Bremer, H., 1975, Metabolic regulation of β-galactosidase synthesis in *Escherichia coli.* A test for constitutive ribosome synthesis, *Biochem. J.* **150**:1.

Dalbow, D. G., and Young, R., 1975, Synthesis time of β-galactosidase in *Escherichia coli* B/r as a function of growth rate, *Biochem. J.* **150**:13.

Dennis, P. P., and Nomura, M., 1974, Stringent control of ribosomal protein gene expression in *Escherichia coli, Proc. Natl. Acad. Sci. U.S.A.* **71**:3819.

Dennis, P. P., and Young, R. F., 1975, Regulation of ribosomal protein synthesis in *Escherichia coli* B/r, *J. Bacteriol.* **121**:994.

Deusser, E., 1972, Heterogeneity of ribosomal populations in *Escherichia coli* cells grown in different media, *Mol. Gen. Genet.* **119**:249.

Engbaek, F., Kjelgaard, N. O., and Maaløe, O., 1973, Chain growth rate of β-galactosidase during exponential growth and amino acid starvation, *J. Mol. Biol.* **75**:109.

Engbaek, F., Gross, C., and Burgess, R. R., 1976, Biosynthesis of RNA polymerase, in: *Benzon Symp. IX,* pp. 117–124.

Engberg, B., and Nordström, K., 1975, Replication of R-factor R1 in *Escherichia coli* K-12 at different growth rates, *J. Bacteriol.* **123**:179.

Fiil, N. P., v.Meyenburg, K., and Friesen, J. D., 1972, Accumulation and turnover of guanosine tetraphosphate in *Escherichia coli, J. Mol. Biol.* **71**:769.

Forchhammer, J., and Kjeldgaard, N. O., 1968, Regulation of messenger RNA synthesis in *Escherichia coli, J. Mol. Biol.* **37**:245.

Forchhammer, J., and Lindahl, L., 1971, Growth rate of polypeptide chains as a function of the cell growth rate in a mutant of *Escherichia coli* 15, *J. Mol. Biol.* **55**:563.

Forchhammer, J., Jackson, E. N., and Yanofsky, C., 1972, Different half-lives of messenger RNA corresponding to different segments of the tryptophan operon of *Escherichia coli, J. Mol. Biol.* **71**:687.

Fowler, A. V., and Zabin, I., 1977, The amino acid sequence of β-galactosidase of *Escherichia coli, Proc. Natl. Acad. Sci. U.S.A.* **74**:1507.

Friesen, J. D., Fiil, N. P., and v.Meyenburg, K., 1975, Synthesis and turnover of basal level guanosine tetraphosphate in *Escherichia coli, J. Biol. Chem.* **250**:304.

Gallant, J., 1976, Elements of the down-shift servomechanism, in: *Benzon Symp. IX,* pp. 385–392.

Gallant, J., and Lazzarini, R. A., 1976, The regulation of ribosomal RNA synthesis and degradation in bacteria, in: *Protein Synthesis,* Vol. 2 (E. H. McConkey, ed.), pp. 309–359, Marcel Dekker, New York.

Gallant, J., Palmer, L., and Pao, C. C., 1977, Anomalous synthesis of ppGpp in growing cells, *Cell* **11**:181.

Gausing, K., 1972, Efficiency of protein and messenger RNA synthesis in bacteriophage T4-infected cells of *Escherichia coli, J. Mol. Biol.* **71**:529.

Gausing, K., 1974, Ribosomal protein in *E. coli:* Rate of synthesis and pool size at different growth rates, *Mol. Gen. Genet.* **129**:61.

Gausing, K., 1976, Synthesis of rRNA and r-protein mRNA in *E. coli* at different growth rates, in: *Benzon Symp. IX,* pp. 292–303.

Gausing, K., 1977, Regulation of ribosome production in *Escherichia coli:* Synthesis and stability of ribosomal RNA and of ribosomal protein messenger RNA at different growth rates, *J. Mol. Biol.* **115**: 335.

Gordon, J., 1970, Regulation of the *in vivo* synthesis of the polypeptide chain elongation factors in *Escherichia coli, Biochemistry* **9**:912.

Hansen, M. T., Bennett, P. M., and v.Meyenburg, K., 1973, Intrasonic polarity during dissociation of translation from transcription in *Escherichia coli, J. Mol. Biol.* **77**:589.

Hansen, M. T., Pato, M. L., Molin, S., Fiil, N. P., and v.Meyenburg, K., 1975, Simple downshift and resulting lack of correlation between ppGpp pool size and ribonucleic acid accumulation, *J. Bacteriol.* **122**:585.

Haseltine, W. A., and Block, R., 1973, Synthesis of guanosine tetra- and pentaphosphate requires the presence of a codon-specific, uncharged transfer ribonucleic acid in the acceptor site of ribosomes, *Proc. Natl. Acad. Sci. U.S.A.* **70**:1564.

Henning, U., Dietrich, J., Murray, K. N., and Deppe, G., 1968, Regulation of pyruvate dehydrogenase synthesis: Substrate induction, in: *Molecular Genetics* (H. G. Wittmann and H. Schuster, eds.), pp. 223–236, Springer-Verlag, Berlin and New York.

Heyden, B., Nüsslein, C., and Schaller, H., 1975, Initiation of transcription within an RNA-polymerase binding site, *Eur. J. Biochem.* **55**:147.

Hirsch, J., and Schleif, R., 1973, *In vivo* experiments on the mechanism of action of L-arabinose *C* gene activator and lactose repressor, *J. Mol. Biol.* **80**:433.

Hirsh, J., and Schleif, R., 1976, Electron microscopy of gene regulation: The L-arabinose operon, *Biochemistry* **73**:1518.

Imamoto, F., 1968, On the initiation of transcription of the tryptophan operon in *Escherichia coli, Proc. Natl. Acad. Sci. U.S.A.* **60**:305.

Ingraham, J., and Maaloφe, O., 1967, Cold-sensitive mutants and the minimum temperature of growth of bacteria, in: *Molecular Mechanisms of Temperature Adoptation* (C. L. Prosser, ed.), pp. 297–309, American Association for the Advancement of Science, Washington, D.C.

Ishihama, A., Taketo, M., Saitoh, T., and Fukuda, R., 1976, Control of formation of RNA polymerase in *Escherichia coli,* in: *RNA Polymerase* (R. Losick and M. Chamberlin, eds.), pp. 485–502, Cold Spring Harbor Lab., Cold Spring Harbor, New York.

Iwakura, Y., Koreaki, I., and Ishihama, A., 1974, Biosynthesis of RNA polymerase in *Escherichia coli,* I. Control of RNA polymerase content at various growth rates, *Mol. Gen. Genet.* **133**:1.

Johnsen, K., Molin, S., Karlström, O., and Maalφe, O., 1977, Control of protein synthesis in *Escherichia coli:* Analysis of an energy-source shift-down, *J. Bacteriol.* **131**:18.

Kessler, D. P., and Rickenberg, H. V., 1963, The competitive inhibition of α-methyl glucoside uptake in *Escherichia coli, Biochem. Biophys. Res. Commun.* **10**:482.

Kjeldgaard, N. O., 1967, Regulation of nucleic acid and protein formation in bacteria, in: *Advances in Microbial Physiology,* Vol. 1, (A. H. Rose and I. F. Wilkinson, eds.), pp. 39–95, Academic Press, New York.

Kjelgaard, N. O., and Gausing, K., 1973, Regulation of biosynthesis of ribosomes, in: *Ribosomes* (M. Nomura, A. Tissières, and P. Lengyel, eds.), pp. 369–392, Cold Spring Harbor Lab. Cold Spring Harbor, New York.

Kjeldgaard, N. O., Maaløe, O., and Schaechter, M., 1958, The transition between different physiological states during balanced growth of *Salmonella typhimurium,* J. *Gen. Microbiol.* **19**:607.

Koch, A. L., 1971, The adaptive response of *Escherichia coli* to a feast and famine existence, in: *Advances in Microbial Physiology,* Vol. 6 (A. H. Rose and J. F. Wilkinson, eds.), pp. 147–217, Academic Press, New York.

Koch, A. L., and Deppe, C. S., 1971, *In vivo* assay of protein synthesizing capacity of *Escherichia coli* from slowly growing chemostat cultures, *J. Mol. Biol.* **55**:549.

Kurland, C. G., and Maaløe, O., 1962, Regulation of ribosomal and transfer RNA synthesis, *J. Mol. Biol.* **4**:193.

Lazzarini, R. A., Cashel, M., and Gallant, J., 1971, On the regulation of guanosine tetraphosphate levels in stringent and relaxed strains of *Escherichia coli, J. Biol. Chem.* **246**:4381.

Lehninger, A. L., 1965, *Bioenergetics,* Benjamin, New York.

Lim, L. W., and Kennell, D., 1974, Evidence against transcription termination within the *E. coli lac* operon, *Mol. Gen. Genet.* **133**:367.

Lindahl, L., 1975, Intermediates and time kinetics of the *in vivo* assembly of *Escherichia coli* ribosomes, *J. Mol. Biol.* **92**:15.

Lindahl, L., Jaskunas, S. R., Dennis, P. P., and Nomura, M., 1975, Cluster of genes in *Escherichia coli* for ribosomal proteins, ribosomal RNA, and RNA polymerase subunits, *Proc. Natl. Acad. Sci. U.S.A.* **72**:2743.

Lindqvist, R. C., and Nordström, K., 1970, Resistance of *Escherichia coli* to penicillins VII. Purification and characterization of a penicillinase mediated by the R factor R1, *J. Bacteriol.* **101**:232.

Maaløe, O., 1969, An analysis of bacterial growth, *Dev. Biol. Suppl.* **3**:33.

Maaløe, O., and Kjeldgaard, N. O., 1966, *Control of Macromolecular Synthesis,* Benjamin, New York.

Matzura, H., Molin, S., and Maaløe, O., 1971, Sequential biosynthesis of the β and β' subunits of the DNA-dependent RNA polymerase from *Escherichia coli, J. Mol. Biol.* **59**:17.

v.Meyenburg, K., 1971, Transport-limited growth rates in a mutant of *Escherichia coli, J. Bacteriol.* **107**:878.

Molin, S., 1976, Ribosomal RNA chain elongation rates in *Escherichia coli,* in: *Benzon Symp. IX,* pp. 333–339.

Molin, S., v.Meyenburg, K., Maaløe, O., Hansen, M. T., and Pato, M. L., 1977, Control of ribosome synthesis in *Escherichia coli:* Analysis of an energy source shift-down, *J. Bacteriol.* **131**:7.

Muto, A., Otaka, E., and Osawa, S., 1966, Protein synthesis in a relaxed-control mutant of *Escherichia coli* upon recovery from methionine starvation, *J. Mol. Biol.* **19**:60.

Nath, K., and Koch, A. L., 1970, Protein degradation in *Escherichia coli, J. Biol. Chem.* **245**:2889.

Neidhardt, F. C., 1963, Properties of a bacterial mutant lacking amino acid control of RNA synthesis, *Biochim. Biophys. Acta* **68**:365.

Nomura, M., 1976, Organization of bacterial genes for ribosomal components: Studies using novel approaches, *Cell* **9**:633.

Norris, T. E., and Koch, A. L., 1972, Effect of growth rate on the relative rates of synthesis of messenger, ribosomal and transfer RNA in *Escherichia coli, J. Mol. Biol.* **64**:633.

O'Farrell, P. H., 1975, High resolution two-dimensional electrophoresis of proteins, *J. Biol. Chem.* **250**:4007.

Pato, M. L., and v.Meyenburg, K., 1970, Residual RNA synthesis in *Escherichia coli* after inhibition of transcription by rifampicin, *Cold Spring Harbor Symp. Quant. Biol.* **35**:497.

Pato, M. L., Bennett, P. M., and v.Meyenburg, K., 1973, Messenger ribonucleic acid synthesis and degradation in *Escherichia coli* during inhibition of translation, *J. Bacteriol.* **116**:710.

Pedersen, F. S., Lund, E., and Kjeldgaard, N. O., 1973, Codon specific, tRNA dependent *in vitro* synthesis of ppGpp and pppGpp, *Nature New Biol.* **243**:13.

Pedersen, S., 1976, Stability of nascent ribosomal RNA in *Escherichia coli,* in: *Benzon Symp. IX,* pp. 345–352.

Pedersen, S., Bloch, P. L., Reeh, S., and Neidhardt, F. C., 1978*a,* Patterns of protein synthesis in *Escherichia coli:* A catalog of the amount of 140 individual proteins at different growth rates, *Cell* **14**:179.

Pedersen, S., Reeh, S., and Friesen, J. D., 1978*b,* Functional mRNA half-lives in *E. coli, Mol. Gen. Genet.* (in press).

Pine, M. J., 1970, Steady-state measurement of the turnover of amino acid in the cellular proteins of growing *Escherichia coli:* Existence of two kinetically distinct reactions, *J. Bacteriol.* **103**:207.

Pongs, O., and Ulbrich, N., 1976, Specific binding of formylated initiator-tRNA to *Escherichia coli* RNA polymerase, *Proc. Natl. Acad. Sci. U.S.A.* **73**:3064.

Ramakrishnan, T., and Echols, H., 1973, Purification and properties of M protein: An accessory factor for RNA polymerase, *J. Mol. Biol.* **78**:675.

Reeh, S., Pedersen, S., and Friesen, J. D., 1976, Biosynthetic regulation of individual proteins in *relA*$^+$ and *relA* strains of *Escherichia coli* during amino acid starvation, *Mol. Gen. Genet.* **149**:279.

Revel, M., 1972, Polypeptide chain initiation: The role of ribosomal protein factors and ribosomal subunits, in: *The Mechanism of Protein Syntheses and Its Regulation* (L. Bosch, ed.), pp. 87–131, North-Holland, Amsterdam.

Richardson, J. P., 1975, Initiation of transcription by *Escherichia coli* RNA polymerase from supercoiled and non-supercoiled bacteriophage PM2 DNA, *J. Mol. Biol.* **91**:477.

Risebrough, R. W., Tissières, A., and Watson, J. D., 1962, Messenger RNA attachment to active ribosomes, *Proc. Natl. Acad. Sci. U.S.A.* **48**:430.

Rose, J. K., and Yanofsky, C., 1972, Metabolic regulation of the tryptophan operon of *Escherichia coli:* Repressor-independent regulation of transcription iniation frequency, *J. Mol. Biol.* **69**:103.

Rünzi, W., and Matzura, H., 1976, Distribution of RNA polymerase between cytoplasm and nucleoid in a strain of *Escherichia coli,* in: *Benzon Symp. IX,* pp. 115–116.

Schleif, R., 1967, Control of production of ribosomal protein, *J. Mol. Biol.* **27**:41.

Schleif, R., 1968, Origin of chloramphenicol particle protein, *J. Mol. Biol.* **37**:119.

Seeburg, P. H., Nüsslein, C., and Schaller, H., 1977, Interaction of RNA polymerase with promoters from bacteriophage fd, *Eur. J. Biochem.* **74**:107.

Shin, D. H., and Moldave, K., 1966, Effect of ribosomes on the biosynthesis of ribonucleic acid *in vitro, J. Mol. Biol.* **21**:231.

Sompayrac, L., and Maaløe, O., 1973, Autorepressor model for control of DNA replication, *Nature New Biol.* **241**:133.

Stent, G. S., 1966, Genetic transcription, *Proc. Roy. Soc. Lond. Ser. B* **164**:181.

Stephens, J. C., Artz, S. W., and Ames, B. N., 1975, Guanosine 5′-diphosphate 3′-diphosphate (ppGpp): Positive effector for histidine operon transcription and general signal for amino-acid deficiency, *Proc. Natl. Acad. Sci. U.S.A.* **72**:4389.

Summerton, J. E., 1976, Measurement of the pool size and synthesis rate of the metabolically unstable fraction of RNA in *Escherichia coli* by a method independent of hybridization efficiency and unaffected by precursor compartmentation, *J. Mol. Biol.* **100**:127.

Talkad, V., Schneider, E., and Kennell, D., 1976, Evidence for variable rates of ribosome movement in *Escherichia coli, J. Mol. Biol.* **104**:299.

Travers, A., 1976*a,* Modulation of RNA polymerase specificity by ppGpp, *Mol. Gen. Genet.* **147**:225.

Travers, A., 1976*b,* RNA polymerase specificity and the control of growth, *Nature* **263**:641.

Venetianer, P., Sümegi, J., and Udvardy, A., 1976, Properties of ribosomal RNA promoters, in: *Benzon Symp. IX,* pp. 252–265.

Warner, J. R., Knopf, P. M., and Rich, A., 1962, A multiple ribosomal structure in protein synthesis, *Proc. Natl. Acad. Sci. U.S.A.* **49**:122.

Yamamoto, M., Strycharz, W. A., and Nomura, M., 1976, Identification of genes for elongation factor Ts and ribosomal protein S2 in *E. coli, Cell* **8**:129.

Yanagisawa, K., 1962, The simultaneous accumulation of RNA and of a repressor of β-galactosidase synthesis, *Biochem. Biophys. Res. Commun.* **9**:88.

Yanofsky, C., 1976, Control sites in the tryptophan operon, in: *Benzon Symp. IX,* pp. 149–160.

Yanofsky, C., and Ito, J., 1966, Nonsense codons and polarity in the tryptophan operon, *J. Mol. Biol.* **21**:313.

Yoshikawa, H., and Sueoka, N., 1963, Sequential replication of *Bacillus subtilis* chromosome, *Proc. Natl. Acad. Sci. U.S.A.* **49**:559.

Zaritsky, A., and v.Meyenburg, K., 1974, Synthesis of ribosomal protein during the cell cycle of *Escherichia coli* B/r, *Mol. Gen. Genet.* **129**:217.

Zillig, W., Mailhammer, R., and Rohrer, H., 1976, Structural modifications of DNA-dependent RNA polymerase as means for gross regulation of transcription, in: *Benzon Symp. IX,* pp. 43–54.

Index